RACE AND ETHNIC RELATIONS

American and Global Perspectives

TENTH EDITION

Martin N. Marger

CENGAGE
Learning·

Australia • Brazil • Japan • Korea • Mexico • Singapore • Spain • United Kingdom • United States

CENGAGE
Learning

Race and Ethnic Relations: American and Global Perspectives, Tenth Edition

Martin N. Marger

Product Manager: Seth Dobrin

Content Developer: Liana Sarkisian

Content Coordinator: Rachel McDonald

Product Assistant: Nicole Bator

Media Developer: John Chell

Marketing Manager: Kara Kindstrom

Art and Cover Direction: Carolyn Deacy, MPS Limited

Production Management, and Composition: Naman Mahisauria, MPS Limited

Manufacturing Planner: Judy Inouye

Rights Acquisitions Specialist: Thomas McDonough

Photo Researcher: PMG/Karthik Periyasamy

Text Researcher: PMG/Sharmila Srinivasan

Cover Designer: Ellen Pettengell

Cover Image: © Jeremy Sutton-Hibbert/Getty Images News/Getty Images

For product information and technology assistance, contact us at **Cengage Learning Customer & Sales Support, 1-800-354-9706.**

For permission to use material from this text or product, submit all requests online at **www.cengage.com/permissions**. Further permissions questions can be e-mailed to **permissionrequest@cengage.com**.

Library of Congress Control Number: 2013954985

ISBN-13: 978-1-285-74969-3

ISBN-10: 1-285-74969-3

Cengage Learning
200 First Stamford Place, 4th Floor
Stamford, CT 06902
USA

Cengage Learning is a leading provider of customized learning solutions with office locations around the globe, including Singapore, the United Kingdom, Australia, Mexico, Brazil, and Japan. Locate your local office at **www.cengage.com/global**.

Cengage Learning products are represented in Canada by Nelson Education, Ltd.

To learn more about Cengage Learning Solutions, visit **www.cengage.com**.

Purchase any of our products at your local college store or at our preferred online store **www.cengagebrain.com**.

Printed in the United States of America
1 2 3 4 5 6 7 17 16 15 14 13

BRIEF CONTENTS

CONTENTS

PREFACE

Race and Ethnic Relations: American and Global Perspectives is designed to explore race and ethnic relations in a global context, while covering extensively ethnic groups and issues in American society. The need for such a comparative approach seems especially critical today in light of the increasing ethnic diversity of the United States and most other contemporary societies, as well as the prominence of ethnic conflicts in virtually all world regions. With continuing high levels of immigration, Americans have become increasingly mindful of their society's changing racial and ethnic configuration and its attendant economic, cultural, and political issues. Though usually uninformed about complexities, they have also become at least vaguely aware of comparable issues in other societies. This awareness is episodically heightened by mass media accounts of ethnic conflict in societies as distant and exotic as Rwanda and Kosovo, as well as those closer geographically and culturally, such as Canada and Northern Ireland. With the prevalent forces of globalization and international migration, it has become clear that racial and ethnic issues are no longer confined to specific societies but are linked through social networks, political arrangements, and economic systems.

Curiously, American social scientists have not always kept pace in adapting to the global context of race and ethnic relations. Some continue to focus almost exclusively on the United States, paying only incidental attention to ethnic patterns and events in other societies. In line with this view, texts in the field of race and ethnic relations have ordinarily provided no more than cursory coverage to affairs outside the American sphere—if at all. Students, therefore, often continue to think of racial and ethnic, or minority, issues as uniquely American phenomena.

A growing number of social scientists, however, have come to see the utility and relevance of a more cross-national approach to the study of race and ethnicity. Such an approach distinguished *Race and Ethnic Relations* from other texts in the field when it was first published in 1985. Its objective was to provide readers with a comparative perspective without sacrificing a strong American component. That objective was retained in subsequent editions and remains unchanged in this, the tenth, edition. The book's overriding theme is the global nature of ethnicity and the prevalence of ethnic conflict in the modern world.

At the same time that an international perspective seems more compelling than ever, a close and careful analysis of race and ethnicity in America is imperative. For better or worse, the United States, the most diverse of multiethnic societies, more often than not is a global pacesetter in ethnic relations. More important, most readers of *Race and Ethnic Relations* continue to be American students, who require a solid understanding of their own society, which subsequently can be used as a comparative frame of reference. *Race and Ethnic Relations*, therefore, provides thorough coverage of America's major ethnic groups and issues. My own teaching experience has confirmed that American students commonly acquire a broader and richer comprehension of ethnic relations and issues in the United States when these are presented in a global context and can be viewed from a comparative perspective.

The number of American college and university courses with ethnic content has grown enormously in recent years. This, I believe, is a reflection of the pressing problems and commanding issues of race and ethnicity in the United States and the growing awareness of ethnic divisions and inequalities in an increasingly diverse society. The content of *Race and Ethnic Relations* is comprehensive and thus appropriate for a variety of courses that may be differently titled and structured (for example, "race and ethnicity," "minority relations," "ethnic stratification," "multiculturalism") but that all deal in some fashion with ethnic issues.

The theoretical and conceptual thrust of this edition is unchanged from previous editions: a power-conflict perspective, emphasizing the power dynamics among ethnic groups. Race and ethnic relations are seen as manifestations of stratification and of the competition and conflict that develop over societal rewards—power, wealth, and prestige. In accord with this perspective, I have emphasized the structural, or macro-level, patterns of race and ethnic relations rather than the social-psychological, though the latter are interspersed throughout.

OVERVIEW OF THE BOOK

The intent of the chapters that make up Part I, THE NATURE OF ETHNIC RELATIONS, is to introduce the principal terms, concepts, and theories of the field of race and ethnic relations. These chapters are designed to serve as an analytic framework within which U.S. racial and ethnic groups, as well as those of other multiethnic societies, can be systematically examined.

Part II, ETHNICITY IN THE UNITED STATES, focuses on American society, describing the formation of its ethnic system and its major racial and ethnic populations. Chapter 5 traces the sociohistorical development of the American ethnic configuration and the society's racial/ethnic hierarchy. Chapters 6 through 12 comprise descriptions and analyses of Native Americans, African Americans, Hispanic Americans, Asian Americans, white ethnic Americans, Jewish Americans, and Arab Americans. These are presented within the framework of theories and concepts introduced in Part I. Chapter 13 describes and analyzes what I believe are the most critical ongoing issues of race and ethnic relations in the United States: large-scale immigration and its social, political, and economic effects; the persistent gap between Euro-Americans and racial-ethnic groups; and policies designed to address that gap.

All chapters in Part II reflect the processes and consequences of continued immigration to the United States and the reshaping of the society's ethnic configuration. More specifically, Chapter 5 contains an examination of immigration theories, factors that stimulate international migration, and historical patterns of American immigration. Chapter 7 discusses the increasing diversity of the American black population, which now includes a sizable foreign-born element. Similarly, Chapter 8, Hispanic Americans, includes coverage of more recent Central and South American and Caribbean groups, in addition to the three major Latino groups, Mexicans, Puerto Ricans, and Cubans. Arab Americans, one of the society's increasingly prominent and growing ethnic populations, are examined in Chapter 12, entirely new to this edition. Chapter 13 focuses in large measure on the socioeconomic, political, and cultural impact of ongoing immigration to the United States of non-European peoples, creating, in the process, a rapidly changing ethnic order.

A noteworthy trend in American society is the blurring of racial and ethnic identities as a result of rising levels of intermarriage. The traditional racial/ethnic classification scheme, as a result, has come under more scrutiny and its relevance is increasingly questioned. It now seems apparent that the commonplace racial/ethnic categories employed by various societal institutions are losing analytic significance, though they obviously remain of paramount importance as the building blocks of ethnic stratification. Within the chapters that compose Part II, this trend and its potential consequences are addressed, particularly in Chapters 5, 7, and 13. Also, the notion of "whiteness" and the fluid nature of "race" in America are discussed in Chapter 10.

Part III, ETHNIC RELATIONS IN COMPARATIVE PERSPECTIVE, examines several societies that stand as intriguing and apposite comparisons to the United States. Chapter 14 deals with South Africa, not long ago the most rigidly racist society on the globe. Today, South Africa serves as an important case for students of contemporary ethnic relations, illustrating how oppressive systems of ethnic inequality can undergo fundamental change in a surprisingly brief time with a minimum of violence. Brazil, examined in Chapter 15, is ideally suited to a comparison with the United States, given its past history of slavery and its multiracial composition. Canada, the focus of Chapter 16, is often seen as a northern replica of the United States, but its ethnic ideology and policies are sharply different. In a number of ways, Canada may represent a future model for multiethnic societies. Chapter 17 explores the global nature of contemporary ethnic conflict, focusing first on the increasing ethnic diversity of Western European societies and the ensuing problems of integrating new culturally and racially diverse populations. This is followed by an examination of several relatively recent cases of ethnic conflict: the Rwandan genocide of 1993; the breakup of the former Yugoslavia and the resultant ethnic wars that were waged throughout the 1990s; and the sectarian strife in Northern Ireland, which is most basically an ethnic struggle. Each of these cases demonstrates how in the modern world societies can be consumed with ethnic differences, despite the apparent lack of racial distinctions.

Studying the ethnic composition and dynamics of other societies enables American students not only to explore unfamiliar social terrains but also to reach a more informed understanding of the structure and social forces of their own society and, to some extent, even their own discrete social worlds. Students will surely recognize differences between the United States and other multiethnic societies, but they will also observe patterns that seem intrinsically common to all. All of the cases explored in Part III lend themselves to easy comparison with the United States, and comparative points are drawn throughout each chapter. No prior familiarity with any of these societies is assumed, on the part of either instructors or students.

Each chapter that deals with a specific American ethnic group (Part II) or multiethnic society (Part III) uses the four theory chapters (Part I) as an organizational framework. This enables students to more easily tie together theoretical and descriptive points.

NEW TO THE TENTH EDITION

As in previous editions of *Race and Ethnic Relations,* all statistical materials have been updated, using the latest figures from the U.S. Census Bureau as well as from other data-gathering organizations in the United States and abroad. Many of the statistical data previously presented in tabular form have been converted into more easily read and comprehended graphs, charts, and maps. All chapters contain new or more recent citations, reflecting the continued vast production of empirical and theoretical literature in the field of race and ethnic relations.

In addition to these updates, several new organizational and content features have been incorporated into this edition.

- What I have called "personal/practical application" questions are included at the end of each chapter. These are designed to supplement the "critical thinking" questions, included in previous editions, with hypothetical situations that students may relate to on a more personal level. Both sets of questions are linked to chapter materials.
- To better enable students to comprehend and apply key terms and concepts, these have been highlighted, along with their definitions, as they appear in each chapter. All are also included in the Glossary.
- An entirely new chapter on Arab Americans (Chapter 12) is now included in Part II, Ethnicity in the United States. Although their roots reach back to the nineteenth century, only in the past few decades have Arab Americans become a highly visible part of the American ethnic mix. This chapter, like others in Part II, traces the sociohistorical development of this group, its socioeconomic characteristics, the nature of prejudice and discrimination it has encountered, and its path toward social integration.

ANCILLARIES

Race and Ethnic Relations: American and Global Perspectives, tenth edition, is accompanied by an array of supplements prepared to create the best learning environment inside as well as outside the classroom for both the instructor and the student. All of the continuing supplements for *Race and Ethnic Relations* have been thoroughly revised and updated. I invite you to take full advantage of the teaching and learning tools available to you.

ONLINE INSTRUCTOR'S MANUAL WITH TEST BANK

This supplement offers the instructor brief chapter outlines, key terms and names, new lecture ideas, new questions for discussion, Internet activities, student activities, and additional resources for the instructor. The Test Bank consists of thoroughly updated and revised multiple-choice questions, with answers, and essay/discussion questions for each chapter, all with page references to the text. To access this resource, please log in to your instructor account at http://login.cengage.com.

ONLINE MICROSOFT® POWERPOINT® LECTURE SLIDES

These preassembled **Microsoft PowerPoint Lecture Slides** include graphics from the text, integrated discussion questions, and "quick quiz" questions. This ancillary makes it easy for you to assemble, edit, publish, and present custom lectures for your course. To access this resource, please log in to your instructor account at http://login.cengage.com.

SOCIOLOGY COURSEMATE

The CourseMate for *Race and Ethnic Relations* brings course concepts to life with interactive learning, study, and exam preparation tools that support the printed textbook. Access an integrated eBook, glossary, quizzes, and more in the CourseMate for *Race and Ethnic Relations*. Go to CengageBrain.com to register or purchase access.

ACKNOWLEDGMENTS

As with past editions, I greatly appreciate the support of the editorial and production staff at Cengage. Also to be acknowledged are the reviewers for this edition who contributed helpful insights and suggestions: Kholoud Al-Qubbaj, Southern Utah University; Juliet Bond, Phoenix University; Rohan de Silva, Milwaukee Area Technical College; Monique Diderich, Shawnee State University; Young Kim, SUNY Oswego; and Deidre Tyler, Salt Lake Community College.

ABOUT THE AUTHOR

Martin N. Marger has written widely in the fields of race and ethnic relations, social inequality, and political sociology. He is the author of *Social Inequality: Patterns and Processes*, sixth edition, and *Elites and Masses: An Introduction to Political Sociology*, and coeditor, with the late Marvin Olsen, of *Power in Modern Societies*. His articles have appeared in many sociology and political science journals, including *Social Problems, Polity, Ethnic and Racial Studies*, and the *International Migration Review*. He earned his bachelor's degree at the University of Miami, his master's at Florida State University, and his PhD at Michigan State University. He has taught at Waynesburg University, Northern Kentucky University, and Michigan State University, where he served as associate director of the Canadian Studies Center.

THE NATURE OF ETHNIC RELATIONS

Chapters 1 through 4 introduce some basic concepts, terms, and theories of the field and, in so doing, erect a framework for analyzing ethnic relations in the United States (Part II) and in several other societies (Part III).

Another, implicit, objective of Part I is to explain the sociological approach to race and ethnic relations. This approach is fundamentally different from the manner in which relations among racial and ethnic groups are commonly viewed and interpreted. Sociologists see everyday social occurrences differently than laypeople do, and they describe them differently as well. They go beneath the superficial to uncover the unseen and often unwitting workings of society, frequently exposing the erroneousness of much of what is considered well-established knowledge. The sociologist Peter Berger has put it well: "It can be said that the first wisdom of sociology is this—things are not what they seem" (1963:23). This is particularly so in the study of race and ethnic relations.

For example, most people, if asked, could attempt an explanation of why black-white relations in the United States have been customarily discordant, and they might even venture to explain why conflict is so commonplace among ethnic groups in other parts of the world. They would probably explain that humans are belligerent "by nature" or that there are "inherent" differences among groups, creating unavoidable fear and distrust. Although these explanations are direct and apparently simple to comprehend, they do not necessarily stand up when subjected to sociological analysis. Groups with different cultural origins and physical traits may indeed clash quite commonly, but as we will see, social factors are more significant than innate tendencies when accounting for that discord.

The subject matter of sociology—or any of the social sciences—is not the abstruse world of physics or chemistry but the everyday life of people. Because the objects of their study are so much a part of common human experience, sociologists often seem to make unnecessarily complex what appears to be quite simple. But the application of rigorous theory and methods and the use of precise terminology are the chief distinguishing features of the sociological approach, in contrast to the more unencumbered ways of problem solving that most people employ. In short, sociologists apply a scientific approach to analyzing human relations. And in doing so, they find that much of what is taken for granted as commonsensical is not so simple or common and perhaps not at all "sensical." In studying race and ethnic relations, therefore, it is necessary to establish more precise terms for various racial and ethnic phenomena and to become aware of the major theories and research findings that underlie the sociological approach to this field.

Race and ethnic relations cannot be explained with a single analytic tool or with one general theory. Sociologists are not in agreement on all issues in the field. At times, in fact, they may be sharply opposed. But it is necessary to understand that the very nature of scientific inquiry makes such a lack of consensus an almost foregone conclusion. Scientists, whether physical or social, recognize no absolute explanations or unchanging theories. Like any science, sociology poses more questions than it answers. In the following chapters, therefore, we will find no explanations that have not been questioned, tested, and retested. Though we can obtain no final answers to many puzzling questions using the sociological approach, we can sharpen immeasurably our insight into the whys and wherefores of relations among different racial and ethnic groups. But we must be prepared to accept the frustration that often accompanies the examination of new ideas about what is old and familiar and the use of new methods in observing what has customarily been seen uncritically.

INTRODUCTION

Some Basic Concepts

Every four years, January 20 is an important date in the United States, the day on which a new (or reelected) president is sworn into office. January 20 of 2008, however, carried even greater than usual significance, for it marked the beginning of what seemed like a new era in American history. Barack Obama, a self-identified African American, took the oath of office as the forty-fourth occupant of the White House. He had been elected with what had been, in the contemporary American political context, a solid majority of the vote. No more than a few years before Obama's election, national polls had indicated the improbability of such an occurrence; a black candidate winning the presidency, though perhaps inevitable, seemed many decades in the future. Obama's election, then, evinced a remarkable shift in American race and ethnic relations—the acceptance by a majority of voters of a black man as the nation's most powerful, prestigious, and celebrated figure. That Obama's election in 2008 was not a fluke or the result of unusual political circumstances was confirmed by his reelection—also by a comfortable majority of the vote—in 2012.

The symbolic importance of Obama's election and reelection cannot be overstated. Indeed, many Americans regarded them as milestones, the culmination of generations of struggles to eradicate the social and political inequalities founded on race and ethnicity that had been such a central part of America's heritage. Beneath the veneer of these historic accomplishments and obscured in their celebration, however, were deep and steadfast racial and ethnic divisions, on which the presidential elections of 2008 and 2012—undeniably momentous events—would have no tangible effect.

Despite the fact that an African American today occupies the highest political office in the land, the economic and social discrepancies between blacks and whites remain enormous. Consider that blacks earn, on average, about two-thirds of what whites earn, suffer an unemployment rate double that of whites, are almost three times more likely than whites to live below the poverty line, and are six times more likely to be incarcerated. Similar, if somewhat less severe, discrepancies are evident between whites and other racial and ethnic minorities, particularly Latinos.

Moreover, for the most part, blacks and whites (and to a lesser extent Latinos and whites) continue to live apart, attend segregated schools and churches, and inter-marry at a low rate. Although all of these patterns are slowly changing, leading to a reduction of economic inequalities and a rise in social integration, clearly racial and ethnic divisions continue to create great strains in the social fabric of America.

Along with issues of economic inequality and social segregation that have tradi-tionally plagued American society, new problems of ethnic division and conflict have arisen as the result of a veritable ethnic transformation that has been unfolding since the 1970s. Almost from its very beginnings, the United States was ethnically diverse. But for the past several decades, large-scale immigration has produced a vastly more heterogeneous nation. Literally dozens of new groups, varied in racial and ethnic characteristics, have reconfigured the U.S. population. Virtually no community or re-gion of the country has been unaffected by this newest influx of immigrants. With growing ethnic populations, most of them non-European in origin, has come public debate—often passionate and shrill—over issues of affirmative action, multicultural-ism, and immigration itself.

THE GLOBAL NATURE OF ETHNIC RELATIONS

In recognizing that racial and ethnic issues in the United States are deep-seated and thus not quickly or easily resolved, we should not think that they are uniquely American. Indeed, as we look at the contemporary world, it becomes evident that ethnic conflict and inequality are basic features of almost all modern societies with diverse populations.

RESURGENT ETHNICITY   Social scientists had maintained for many years that industriali-zation and the forces of modernization would diminish the significance of race and ethnicity in heterogeneous societies (Deutsch, 1966). They felt that with the break-down of small, particularistic social units and the emergence of large, impersonal bu-reaucratic institutions, people's loyalty and identity would be directed primarily to the national state rather than to internal racial and ethnic communities. The opposite trend, however, seems to have characterized the contemporary world. Indeed, the past sixty years have witnessed the emergence of ethnic consciousness and division around the globe on a scale unprecedented. The upshot has been an unleashing of ethnic-based conflict, usually contained at a low level, but at times fearsome and deadly.[1]

In developed nations, ethnic groups thought to be well absorbed in the national society have reemphasized their cultural identity, and new groups have demanded political recognition. In Western Europe, ethnically based political movements have emerged in several countries. Throughout Eastern Europe, the late 1980s brought massive economic and political change, rekindling ethnic loyalties that had been suppressed for several decades. And in Canada, the traditional schism dividing English- and French-speaking groups has periodically threatened a breakup of the Canadian nation.

---

[1] The seriousness of ethnic conflict in the modern world is highlighted by the fact that, during the five decades from 1945 to 1994, perhaps as many as 20 million people died as a result of ethnic violence (Williams, 1994). In the past two decades, ethnic conflicts in every world region created additional millions of deaths and displacements.

Moreover, much of Europe, like the United States and Canada, has been affected in recent decades by a greatly increased flow of immigration, made up of people markedly different racially and ethnically from the native population. As these diverse ethnic communities have grown and become permanent parts of the social landscape, social tensions have arisen, sparking outbreaks of aggression and anti-immigrant political movements.

In the developing nations, too, the ethnic factor has emerged with great strength. World War II marked the end of several centuries of imperialist domination of non-Western peoples by European powers, and many new nations were created in Africa and Asia, the political boundaries of which were often carved out of the administrative districts of the old colonial states. Many of these artificial boundaries were drawn with little consideration of the areas' ethnic composition. As a result, the new nation-states often found themselves faced with the problem of integrating diverse cultural groups, speaking different languages and even maintaining different belief systems, into a single national society. The result has been numerous ongoing ethnic conflicts that have erupted periodically in horrendous violence.

In short, racial and ethnic forces have emerged with great power in the modern world. In all societies they are important—in many cases the *most* important—bases of both group solidarity and cleavage. As we look ahead, their impact is not likely to diminish throughout the twenty-first century.

RACE AND ETHNIC RELATIONS

Societies comprising numerous racial, religious, and cultural groups can be described as **multiethnic**. In the contemporary world, multiethnic societies are commonplace, not exceptional. Only a handful of the more than 190 member countries of the United Nations are ethnically homogeneous. "Multiethnicity," notes the sociologist Robin Williams, "is the rule" (1994:50). Moreover, the extent of diversity within many of these societies is very great. As can be seen in Table 1.1, societies that are multiethnic in some degree are found on every continent and in various stages of socioeconomic development.

The study of race and ethnic relations is concerned generally with the ways in which the various groups of a multiethnic society come together and interact over extended periods. As we proceed in our investigation, we will be looking specifically for answers to four key questions.

TABLE 1.1	ETHNIC DIVERSITY OF SELECTED NATION-STATES	
High	Medium	Low
United States	United Kingdom	Finland
Canada	Nigeria	Czech Republic
Australia	Malaysia	Egypt
Israel	Argentina	Japan

BASIC QUESTIONS

1. *What is the nature of intergroup relations in multiethnic societies?* Ethnic relations commonly take the form of conflict and competition. Indeed, we can easily observe this by following the popular media accounts of ethnic relations in the United States and other nations, which are usually descriptions of hostility and violence. However, intergroup relations are never totally conflictual. Ethnic groups do not exist in a perpetual whirlwind of discord and strife; cooperation and accommodation also characterize ethnic relations. Just as we will be concerned with understanding why conflict and competition are so common among diverse groups, we will also investigate harmonious conditions and the social factors that contribute to them.

2. *How are the various ethnic groups ranked, and what are the consequences of that ranking system?* In all multiethnic societies, members of various groups are treated differently and receive unequal amounts of the society's valued resources—wealth, prestige, and power. In short, some get more than others and are treated more favorably. Moreover, this inequality is not random but is well established and persists over many generations. A structure of inequality emerges in which one or a few ethnic groups, called the *dominant group* or groups, are automatically favored by the society's institutions, particularly the state and the economy, whereas other ethnic groups remain in lower positions. These subordinate groups are called ethnic *minorities*. We will be concerned with describing this hierarchy and determining how such systems of ethnic inequality come about.

3. *How does the dominant ethnic group in a multiethnic society maintain its place at the top of the ethnic hierarchy, and what attempts are made by subordinate groups to change their positions?* The dominant group employs a number of direct and indirect methods (various forms of prejudice and discrimination) to protect its power and privilege. This does not mean, however, that subordinate groups make no attempts to change this arrangement from time to time. In fact, organized movements—by African Americans in the United States or Catholics in Northern Ireland, for example—may challenge the ethnic hierarchy. One of our chief objectives, then, will be to examine the ways in which systems of ethnic inequality are maintained and how they change.

4. *What are the long-range outcomes of ethnic interrelations?* When ethnic groups exist side by side in the same society for long periods, either they move toward some form of unification or they maintain or even intensify their differences. These various forms of integration and separation are called *assimilation* and *pluralism*. Numerous outcomes are possible, extending from complete assimilation, involving the cultural and physical integration of the various groups, to extreme pluralism, including even expulsion or annihilation of groups. Usually, less extreme patterns are evident, and groups may display both integration and separation in different spheres of social life. Again, our concern is not only with discerning these outcomes but also with explaining the social forces that favor one or the other.

A COMPARATIVE APPROACH

The study of race and ethnic relations has a long tradition in American sociology, beginning in the 1920s with the research of Robert Park, Everett Hughes, and Louis Wirth. These scholars were among the first to focus attention on the relations among ethnic groups, particularly within the ethnic mélange of large U.S. cities such as New York and Chicago. The sociology of race and ethnic relations has progressed enormously since that time, and it now constitutes one of the chief subareas of the sociological discipline.

With few exceptions, however, American sociologists have continued to concentrate mainly on American groups and relations, often overlooking similarities and differences between the United States and other heterogeneous societies. But if we are to understand the general nature of race and ethnic relations, it is necessary to go beyond the United States—or any particular society—and place our analyses in a comparative, or cross-societal, framework. As we have already noted, ethnic diversity, conflict, and accommodation are worldwide phenomena, not unique to American society. However, because most research in race and ethnic relations has been the product of American sociologists dealing with the American experience, we are often led to assume that patterns evident in the United States are much the same in other societies.

This book adopts a comparative perspective in which the United States is seen as one among many contemporary multiethnic societies. Because readers are likely to be most familiar with American society, however, the center of attention will fall most intently on American groups and relations. Even in those chapters in which other societies are the major focus (Part III), we will draw attention to U.S. comparisons.

A comparative approach not only enables us to learn about race and ethnicity in other societies but also provides us with a sharper insight into race and ethnicity in the United States. It has often been observed that we cannot begin to truly understand our own society without some knowledge of other societies. Moreover, in addition to the differences revealed among societies, similarities may also become apparent. As the sociologists Tamotsu Shibutani and Kian Kwan have pointed out, comparing American ethnic relations with those of other societies "reveals that patterns of human experience, though infinitely varied, repeat themselves over and over in diverse cultural contexts" (1965:21). Discovering such generalizable patterns of the human experience is the ultimate aim of all sociological efforts.

ETHNIC GROUPS

Although they are now familiar and commonly used, the terms *ethnic group* and *ethnicity* are relatively new, not even appearing in standard English dictionaries until the 1960s (Glazer and Moynihan, 1975). Groups generally referred to today as "ethnic" were previously thought of as races or nations, but these terms clearly have different meanings.

CHARACTERISTICS OF ETHNIC GROUPS

UNIQUE CULTURAL TRAITS   Basically, **ethnic groups** are *groups within a larger society that display a unique set of cultural traits*. The sociologist Melvin Tumin more specifically describes an ethnic group as "a social group which, within a larger cultural and social system, claims or is accorded special status in terms of a complex of traits (ethnic traits) which it exhibits or is believed to exhibit" (1964:243). Ethnic groups, then, are subcultures, maintaining certain behavioral characteristics that, in some degree, set them off from other groups, creating cultural divisions within the larger society. Such unique cultural traits are not trivial but are fundamental features of social life such as language and religion.

Unique cultural traits, however, are not sufficient alone to delineate ethnic groups in a modern, complex society. Can we speak of physicians as an ethnic group? Or truck drivers? Or college students? Obviously, we would consider none of these categories "ethnic" even though they are composed of people who exhibit some common characteristics that identify them from others. Clearly, we need further qualifications for distinguishing ethnic groups.

SENSE OF COMMUNITY   In addition to a common set of cultural traits, ethnic groups display a sense of community among members, that is, a consciousness of kind or an awareness of close association. In simple terms, a "we" feeling exists among members. Milton Gordon (1964) suggests that the ethnic group serves above all as a social-psychological referent in creating a "sense of peoplehood." This sense of community, or oneness, derives from an understanding of a shared ancestry, or heritage. Ethnic group members view themselves as having common roots, as it were. When people share what they believe to be common origins and experiences, "they feel an affinity for one another," writes sociologist Bob Blauner, "a 'comfort zone' that leads to congregating together, even when this is not forced by exclusionary barriers" (1992:61).

Such common ancestry, however, need not be real. As long as people regard themselves as alike by virtue of their perceived heritage, and as long as others in the society so regard them, they constitute an ethnic group, whether or not such a common background is genuine. Everett and Helen Hughes have perceptively recognized that "an ethnic group is not one because of the degree of measurable or observable difference from other groups; it is an ethnic group, on the contrary, because the people in it and the people out of it know that it is one; because both the ins and outs talk, feel, and act as if it were a separate group" (1952:156). Ethnic groups, then, are social creations wherein ethnic differences are basically a matter of group perception. Groups may be objectively quite similar but perceive themselves as very different, and the converse is equally true.

Sociologists have debated the relative significance of the cultural element and the sense of community as most critical to the formation of an ethnic group (Dorman, 1980). The argument boils down to a question of whether ethnic groups are objective social units that can be identified by their unique culture or merely collectivities that people themselves define as ethnic groups. Whereas some view the cultural features of the group as its key distinctive element, others argue that stressing its unique culture minimizes the importance of the subjective boundaries of the group that people themselves draw (Barth, 1969). Most simply, the latter maintain that if people define themselves and are defined by others as an ethnic group, they *are* an ethnic group, whether or not they display unique cultural patterns. If this is the case, the cultural stuff of which the ethnic group is composed is unimportant.

Although this point may seem like a relatively minor theoretical one, it is of importance when ethnic groups in a society begin to blend into the dominant cultural system. Sociologists have traditionally assumed that as groups integrate into the mainstream society, the basis of retention of ethnicity diminishes. But whether people continue to practice ethnic ways may matter little as long as they continue to define themselves and are defined by others in ethnic terms. Many Americans continue to think of and identify themselves as ethnics even though they exhibit little or no understanding of or interest in their ethnic culture. Do third- or fourth-generation Irish Americans, for example, really share a common culture with their first-generation ancestors? Wearing a button on St. Patrick's Day proclaiming "I'm proud to be Irish" is hardly a display of the traits of one's Irish American forebears. Yet an ethnic identity may remain intact for such people, and they may continue to recognize their uniqueness (and proudly acknowledge it) within the larger society. Thus, despite the lack of a strong cultural factor, the sense of Irish American identity may be sufficient to sustain an Irish American ethnic group.

We thus have two views of the ethnic group: (1) it is an objective unit that can be identified by a people's distinct cultural traits, or (2) it is merely the product of people's thinking of and proclaiming it as an ethnic group. To avoid the extreme of either of these views, sociologist Pierre van den Berghe defines the ethnic group as both an objective and a subjective unit: "An ethnic group is one that shares a cultural tradition *and* has some degree of consciousness of being different from other such groups" (1976:242). As he points out, it is foolish to think that ethnic groups simply arise when people so will it. Fans of a particular football team may feel a sense of commonality and even community, but they surely do not compose an ethnic group. In short, there must be some common cultural basis and sense of ancestry to which ethnic group members can relate. As van den Berghe notes, "There can be no ethnicity (or race) without some conception and consciousness of a distinction between 'them' and 'us.' But these subjective perceptions do not develop at random; they crystallize around clusters of objective characteristics that become badges of inclusion or exclusion" (1978:xvii). Although ethnic boundaries are very flexible, they are always founded on a cultural basis. At the same time, however, an ethnic group cannot exist in an objective sense independent of what its members think and believe. There must be a sense of commonality, and such a feeling of oneness arises generally through the perception of a unique cultural heritage.

ETHNOCENTRISM The "we" feeling of ethnic groups ordinarily leads naturally to **ethnocentrism,** *the tendency to judge other groups by the standards and values of one's own group.* Inevitably, this produces a view of one's own group as superior to others. The ways of one's own group (*in-group*) become "correct" and "natural," and the ways of other groups (*out-groups*) are seen as "odd," "immoral," or "unnatural." Sociologists and anthropologists have found the inclination to judge other groups by the standards of one's own and to view out-groups as inferior or deficient to be a universal practice.

In multiethnic societies, such feelings of group superiority become a basis for group solidarity. In addition to fostering cohesiveness within one group, however, ethnocentrism also serves as the basis of conflict between different groups. As Bonacich and Modell have explained, "Ethnicity is a communalistic form of social affiliation, depending, first, upon an assumption of a special bond among people of like origins, and, second, upon the obverse, a disdain for people of dissimilar origins" (1980:1). Here we can begin to understand why conflict among ethnic groups is so pervasive and intractable.

ASCRIBED MEMBERSHIP Ethnic group membership is ordinarily ascribed. This means that one's ethnicity is a characteristic acquired at birth and not subject to basic change. Being born a member of an ethnic group, one does not leave it except in unusual circumstances. One might change ethnic affiliation by "passing"—that is, by changing one's name or other outward signs of ethnicity—or by denying group membership. But it is extremely difficult to divest oneself completely of one's ethnic heritage. Through the socialization process, individuals come to learn their group membership early and to understand the differences between themselves and members of other groups. So well internalized is this group identification that one comes to accept it almost as naturally as accepting one's gender. As Hughes and Hughes have suggested, "If it is easy to resign from the group, it is not truly an ethnic group" (1952:156). This understanding of ethnic descent is what creates such sharp and, at times, deadly ethnic divisions (Williams, 1994).

Those who attempt to shed their ethnic identity find that the society will rarely permit this fully. Gordon notes in this regard that in American society a person who attempts to relinquish his or her ethnic identity finds "that the institutional structure of the society and the set of built-in social and psychological categories with which most Americans are equipped to place him—to give him a 'name'—are loaded against him" (1964:29). Once the ethnic categories in a society are set, explains Gordon, placing people into them is almost automatic and is by no means subject entirely to people's volition.

In multiethnic societies where ethnic boundaries are not rigid and where there is much marriage across ethnic lines, like the United States and Canada, the voluntary nature of ethnicity becomes more salient. The ethnic origins of third- or fourth-generation Euro-Americans, for example, may be quite varied. Individuals therefore make decisions about "who they are" ethnically, some in a contrived fashion and others almost unconsciously. People whose family origins may contain Italian, Polish, and Irish elements might emphasize the Italian part and identify themselves as "Italian American," disregarding their other ancestral links. In such cases, the volitional component of ethnic identity is strong. For those whose ethnic identity is based also on physical, or racial, characteristics, however, the capacity to choose becomes more limited. For such people, ascription is paramount.

TERRITORIALITY Ethnic groups often occupy a distinct territory within the larger society. Most of the multiethnic societies of Europe consist of groups that are regionally concentrated. Basques and Catalans in Spain, Welsh and Scots in Britain, and Flemings and Walloons in Belgium are groups that maintain a definable territory within the greater society. Such multiethnic societies are quite different from the United States, Canada, or Australia, where ethnic groups have for the most part immigrated voluntarily and, though sometimes are concentrated in particular areas, are not regionally confined.

When ethnic groups occupy a definable territory, they also maintain or aspire to some degree of political autonomy. They are, in a sense, "nations within nations."

In some societies, the political status of ethnic groups is formally recognized. Each group's cultural integrity is acknowledged, and provision is made for its political representation in central governmental bodies. Such societies are best referred to not only as multiethnic but also as multinational (van den Berghe, 1981).

In other societies, where such multinationality is not formally recognized, certain ethnic groups may seek greater political autonomy or perhaps even full independence from the national state, usually dominated by other ethnic groups. The Basques in Spain represent such a case. With a culture and language distinct from other groups in Spain, the Basques have traditionally seen themselves as a separate nation and have negotiated with the Spanish government at various times to promote their sovereignty. A Basque nationalist movement has been evident since the late nineteenth century, but in recent decades it has taken on a particularly virulent and often violent form (Anderson, 2001; Kurlansky, 1999; Ramirez and Sullivan, 1987). Similar nationalist movements, perhaps not as hostile, have typified many modern multiethnic societies. We will look at one such movement, among French-speaking Canadians in Quebec, in Chapter 16.

Where ethnic groups do not continue to maintain significant aspects of culture (such as language) for many generations and where they are geographically dispersed rather than concentrated, such nationalist movements do not ordinarily arise. This is the case in the United States, where ethnic groups are scattered throughout the society, generally take on dominant cultural ways after a generation or two, and seek greater power within the prevailing political system.

ETHNICITY AS A VARIABLE

GROUP VARIABILITY   Each of these characteristics—unique culture, sense of community, ethnocentrism, ascribed membership, and territoriality—is displayed in varying degree by different ethnic groups. These traits are variables that not only differ from group to group but also change at various historical times within any single ethnic group. Thus we should not expect to find all ethnic groups in a society equally unique in cultural ways, strongly self-conscious and recognized by out-groups, or even ethnocentric. The extent to which ethnic groups are noticeable and maintain a strong consciousness among members depends on both in-group and out-group responses. Some ethnic groups seek rapid assimilation and are accepted into society's mainstream relatively quickly. Others, however, may retain their group identity for many generations because of rejection by the dominant group, their own desire to maintain the ethnic community, or a combination of these two. Jews, for example, have maintained a strong group consciousness in most societies, largely as a result of the historically consistent hostility to which they have been subjected but also because they have consciously sought to preserve their group identity.

INDIVIDUAL VARIABILITY   For people in multiethnic societies, the ethnic group becomes a key source of social-psychological attachment and serves as an important referent of self-identification. Put simply, people feel naturally allied with those who share their ethnicity and identify themselves with their ethnic group. Their behavior is thus influenced by ethnicity in various areas of social life. However, just as ethnicity differs in scope and intensity at the group level, it also plays a varying role for individuals. For some, it is a major determinant of behavior, and most social relations will occur among those who are ethnically similar. For others, ethnicity may be insignificant, and they may remain essentially devoid of ethnic consciousness. For most people in multiethnic societies, however, the ethnic tie is important in shaping primary relations—those that occur within small, intimate social settings such as the family and the peer group. These relations include one's choice of close friends, marital partner, residence, and so on.

There are numerous other groups in modern societies to which people feel a sense of attachment and that provide a source of identification. These include one's social class, gender, age, and occupation. Like ethnic groups, these also become bases of solidarity and societal cleavage. But in multiethnic societies, ethnicity is a primary base of loyalty and consciousness for most people and thus serves as a strong catalyst for competition and conflict. Moreover, ethnicity is usually interrelated and overlaps with these other sources of group identification and attachment.

Most important, ethnicity is a basis of ranking, in which people are treated according to the status of their ethnic group. In no society do people receive an equal share of the society's rewards, and in multiethnic societies, ethnicity serves as an extremely critical determinant of who gets "what there is to get" and in what amounts. In this sense, ethnicity is a dominant force in people's lives whether or not they are strongly conscious of their ethnic identity and regardless of the degree to which ethnicity shapes their interrelations with others.

RACE

Without question, *race* is one of the most misunderstood, misused, and often dangerous concepts of the modern world. It is not applied dispassionately by laypeople or even, to a great extent, by social scientists. Rather, it can arouse emotions such as hate, fear, anger, loyalty, pride, and prejudice. It has also been used to justify some of the most appalling injustices and mistreatments of humans by other humans.

The idea of race has a long history, extending as far back as ancient civilizations. It is in the modern world, however—specifically, the last two centuries—that the notion has taken on real significance and fundamentally affected human relations. Unfortunately, the term has never been applied consistently and has meant different things to different people. In popular usage, it has been used to describe a wide variety of human categories, including people of a particular skin color (the Caucasian "race"), religion (the Jewish "race"), nationality (the British "race"), and even the entire human species (the human "race"). As we will see, none of these applications is accurate and meaningful from a social scientific standpoint. Much of the confusion surrounding the idea of race stems from the fact that it has both biological and social meanings. Although it is impossible to do justice to the controversies surrounding the notion of race in a few pages, several of the more apparent problems attached to this most elusive of ideas can be briefly explored.

RACE AS A BIOLOGICAL NOTION

The essential biological meaning of race is a population of humans classified on the basis of certain hereditary characteristics that differentiate them from other human groups. Races are, in a sense, pigeonholes for categorizing human physical types. Historical efforts at classification, however, have yielded no accord among social and biological scientists. Over the years, the biological understanding of race has created an enormous variation in thought among biologists, geneticists, physical anthropologists, and physiologists concerning the term's meaning and significance.

GENETIC INTERCHANGEABILITY To begin with, the difficulty in trying to place people into racial categories on the basis of physical or genetic qualities stems from the fact that all members of the human species, *Homo sapiens,* operate within a genetically open system. This means that humans, regardless of physical type, can interbreed. If genes of different human groups were not interchangeable, the idea of race as a biological concept might have some useful meaning. But because this is not the case, we see an unbounded variety of physical types among the peoples of the world. As biologist James King has put it, "Nowhere in the world are there two populations of manlike creatures living in close proximity for any length of time with no interbreeding. Wherever and whenever human populations have come together, interbreeding has always taken place" (1981:135).

THE BASES OF RACIAL CLASSIFICATION That a person from one genetic population can interbreed with a person from any other population creates a second difficulty in dealing with the notion of race: answering the question "What are the characteristics that differentiate racial types?" Physical anthropologists distinguish major categories of human traits as either *phenotypes*—visible anatomical features such as skin color, hair texture, and body and facial shape—or *genotypes*—genetic specifications inherited from one's parents. Races have traditionally been classified chiefly on the basis of the most easily observable anatomical traits, like skin color; internal and blood traits have been deemphasized or disregarded. Today, the study of human biological diversity rejects the approach that stresses the classification of individuals into specific categories based on assumed common ancestry and emphasizes, instead, efforts to explain human differences (Kottak and Kozaitis, 1999). This has occurred mainly because there is simply no agreement on which traits should be used in defining races.

Attempts to clearly categorize humans have proved futile because differences among individuals of the same group (or "racial type") are greater than those found between groups (King, 1981; Marks, 1995). As biologist Daniel Blackburn (2000) has explained, all of the popularly used physical features to define races (for example, skin pigmentation, hair type, lip size) show gradients of distribution within population groups within which sharp distinctions cannot be drawn. Despite obvious physical differences between people from different geographic areas, most human genetic variation occurs *within* populations. Even the most inattentive person can see that the skin color of many people who are considered "black" is as light as that of those considered "white," and vice versa. Specific human populations can be distinguished, but, as geneticists Michael Bamshad and Steve Olson have explained, "individuals from different populations are, on average, just slightly more different from one another than are individuals from the same population" (2003:78).[2]

---

[2] The Human Genome Project, which has been mapping DNA, has drawn even sharper attention to the fact that diversity within so-called racial groups is greater than between them.

Physical differences among people obviously exist, and these differences are sta-tistically clear among groups. It is true that, through a high degree of inbreeding over many generations and as adaptations to different physical environments, groups with distinctive gene frequencies and phenotypic traits are produced. There are evi-dent differences, for example, between a "typical" black person and a "typical" white person in the United States. People may be said, therefore, to fall into statistical categories by physical type.

But these statistical categories should not be mistaken for actual human group-ings founded on unmistakable hereditary traits. Racial categories form a continuum of gradual change, not a set of sharply demarcated types. Physical differences be-tween groups are not clear-cut but instead tend to overlap and blend into one an-other at various points. Sociologist William Petersen aptly notes that humans are not unique in this regard: "It follows from the theory of evolution itself that all bio-logical divisions, from phylum through subspecies, are always in the process of change, so there is almost never a sharp and permanent boundary setting one off from the next" (1980:236). Human subspecies, then, are not discrete units but, as King notes, are arbitrarily differentiated parts of one continuous unit, the human species. "No system of classification, no matter how clever, can give them a specific-ity and a separateness that they do not have" (1981:10).

Popular and uncomplicated racial divisions of the human population (for exam-ple, "Caucasoid-Mongoloid-Negroid") are thus overly simplified and essentially dis-cretionary. Moreover, such schemes exclude large populations that do not easily fit into simple arrangements. Where, for example, are East Indians, a people with Cau-casian features but with dark skin, placed? Or groups thoroughly mixed in ancestry, like most Indonesians? Because all human types are capable of interbreeding, there are simply too many marginal cases like these that do not easily conform to any par-ticular racial scheme, regardless of its complexity. In studying human differences, most modern investigators have divided the human species into populations, or sub-populations, as the unit of analysis, not racial groupings (Gould, 1984; Marks, 1995; Tobias, 1995).

The anthropologist Ruth Benedict fittingly remarked that "in all modern science there is no field where authorities differ more than in the classifications of human races" (1959:22). If researchers are in agreement about anything concerning race, it is that racial classification systems are by and large arbitrary and depend on the spe-cific objectives of the classifier. All agree that "pure" races do not exist today, and some question whether they have ever existed (Dunn, 1956; Fried, 1965; Graves, 2004; King, 1981; Tobias, 1995).

The notion of "mixed racial groups," a concept that has become increasingly popular in recent years, does nothing to clarify the issue. To speak of "mixed races" implies, erroneously, that there are "pure" races to begin with. As Blackburn notes, "The very notion of hybrid or mixed races is based on the false assumption that 'African' and 'Caucasian' are pure racial types available for hybridization" (2000:8).

THE SOCIAL CONSTRUCTION OF RACE

Many popular ideas are of dubious scientific validity; race is certainly among these. But as André Béteille has pointed out, "Sociological analysis is concerned not so

much with the scientific accuracy of ideas as with their social and political consequences" (1969:54). Most social and biological scientists today agree that the idea
of race is not meaningful in a biological sense, though this is hardly a settled issue
(Morning, 2005). In fact, with recent advances in genetic research and knowledge
of the human genome, the race concept is receiving renewed attention by many biological and medical scientists (Hochschild et al., 2012; Koenig et al., 2008). The importance of race for the study of intergroup relations, however, clearly lies in its
social meaning. "Races" are socially constructed and that premise cannot be overem-
phasized. How does the process of the social construction of race work?

THE POPULAR BELIEF IN RACE Most simply, people attach significance to the concept of
race and consider it a real and important division of humanity. And, as long as people *believe* that differences in selected physical traits are meaningful, they will act on
those beliefs, thereby affecting their interrelations with others. In a famous maxim,
the sociologist W. I. Thomas keenly asserted that "if men define situations as real
they are real in their consequences" (Thomas and Znaniecki, 1918:79). If, for example, those classified as black are deemed inherently less intelligent than those classified as white, people making this assumption will treat blacks accordingly.
Employers thinking so will hesitate to place blacks in important occupational positions; school administrators thinking so will discourage blacks from pursuing difficult courses of study; white parents thinking so will hesitate to send their children
to schools attended by blacks; and so on.

The creation of such categories and the beliefs attached to them generate what
sociologists have called the **self-fulfilling prophecy** (Merton, 1968). This refers to
*a process in which the false definition of a situation produces behavior that, in
turn, makes real the originally falsely defined situation.* Consider the aforementioned
case. If blacks are considered inherently less intelligent, fewer community resources
will be used to support schools attended primarily by blacks on the assumption that
such support would only be wasted. Poorer-quality schools, then, will inevitably turn
out less capable students, who will score lower on intelligence tests. The poorer performance on these tests will "confirm" the original belief about black inferiority.
Hence, the self-fulfilling prophecy. The notion of black inferiority is reinforced, and
continued discriminatory treatment of this group is rationalized.[3]

The anthropologist Robert Redfield noted that "it is on the level of habit, custom, sentiment, and attitude that race, as a matter of practical significance, is to be
understood. Race is, so to speak, a human invention" (1958:67). The scientific validity of race, then, is of little consequence; rather, it is the belief system of a society that
provides its significance. As Benedict put it, "Any scientist can disprove all its facts
and still leave the *belief* untouched" (1959:99).

THE ARBITRARY BOUNDARIES OF RACE Each heterogeneous society takes whatever are perceived as important physical differences among people and builds a set of racial categories into which those people are placed. But these categories are fully arbitrary.
Different societies will use different criteria with which to assign people racially,

---

[3] Such reasoning and its resultant consequences are precisely what characterized—and for whites,
justified—the system of racially segregated schools in the U.S. southern states before the 1960s.

thereby creating classification systems that may have little or no correspondence from one society to the next. Michael Omi and Howard Winant use the term *racial formation* to describe "the process by which social, economic, and political forces determine the content and importance of racial categories, and by which they are in turn shaped by racial meanings" (1994:61). The social meaning of race, they explain, is constantly subject to change through political struggle.

The arbitrariness of racial categorizing can be seen easily when we compare different societies, each with numerous physical types. The same individual categorized as "black" in the United States, for example, might be categorized as "white" in Brazil. The racial classification systems in these two societies do not coincide. As we will see in Chapter 15, Brazilians do not see or define races in the same way that Americans do, nor do they necessarily use the same physical characteristics as standards with which to categorize people. Obviously, different criteria and different categories of race are operative in each society.

So subject to cultural definition is the idea of race that the selected physical attributes used to classify people need not even be obvious, only the *belief* that they are evident. In Northern Ireland, for example, both Protestants and Catholics sometimes say they are able to identify members of the other group on the basis of physical differences, despite their objective similarity.

Even within societies, the definition of races has never been consistent. Consider the United States, where many European-origin ethnic groups that today are seen as racially the same (Italians, Poles, Jews, and so on) were defined as late as the 1920s as racially distinct. Also, "mulatto," "quadroon," and "octoroon," racial categories neither "black" nor "white," existed in the 1800s but were later discarded. As the historian Matthew Frye Jacobson has written, "[E]ntire races have disappeared from view, from public discussion and from modern memory, though their flesh-and-blood members still walk the earth" (1998:2).

What is perhaps most important regarding the social classification of races is that the perceived physical differences among groups are assumed to correspond to social or behavioral differences. Thus blacks are assumed to behave in certain ways and to achieve at certain levels because they are black; whites are assumed to behave and achieve in other ways because they are white; and so on. "What makes a society multiracial," notes van den Berghe, "is not the presence of physical differences between groups, but the attribution of social significance to such physical differences as may exist" (1970:10). Redfield has drawn an apt analogy: "If people took special notice of red automobiles, and believed that the redness of automobiles was connected inseparably with their mechanical effectiveness, then red automobiles would constitute a real and important category" (1958:67).

It is most critical, then, to look not simply at the racial categories that different societies employ but also at the social beliefs attached to those categories. Such beliefs are the product of racist thinking, which we will consider shortly.

RACE AND ETHNICITY: A SYNTHESIS

As should now be obvious, the term *race* is so charged and misconceived that it is very difficult to employ in a useful analytic manner. To add to the confusion, as noted previously, many groups today defined as ethnic groups were in previous historical periods defined as races.

Because of its confusing usage and its questionable scientific validity, many sociologists and anthropologists have dispensed entirely with the term *race* and instead prefer *ethnic group* to describe those groups commonly defined as racial (Berreman, 1972; Gordon, 1964; Patterson, 1997; Schermerhorn, 1970; Shibutani and Kwan, 1965; Williams, 1979, 2003). In the United States, African Americans, American Indians, Chinese Americans, Japanese Americans, and Mexican Americans have all the earmarks of ethnic groups—unique culture, consciousness of kind, ascriptive membership, and in some cases even territoriality—at the same time that most members of these groups are physically distinct from Americans of European origin. Classifying all these groups as ethnic seems most reasonable because, in addition to their physical traits, consistent and significant cultural traits set them off from other groups.

Sociologist Richard Alba has offered a definition of race as a variant of ethnicity: "A racial group is . . . an ethnic group whose members are believed, by others if not also by themselves, to be physiologically distinctive" (1992:576). In the chapters that follow, we will subscribe to that definition. Thus ethnic groups can include groups identified by national origin, cultural distinctiveness, religious affiliation, or racial characteristics. As we will see, ethnic groups in most modern societies comprise combinations of these national, cultural, religious, and physical traits. For those *groups that are particularly divergent physically from the dominant group*, such as African Americans, Richard Burkey (1978) has suggested the term **racial-ethnic groups**, and that term will be used accordingly.

RACISM

Racist thinking involves principles that lead naturally and inevitably to the differential treatment of members of various ethnic groups. As we will see in Chapter 2, in no society are valued resources distributed equally; in all cases, some get more than others. In multiethnic societies, ethnicity is used as an important basis for determining the nature of that distribution. Ethnic groups are ranked in a hierarchy, and their members are rewarded accordingly, creating a system of ethnic inequality. Groups at the top compound their power and maintain dominance over those lower in the hierarchy. Such systems of ethnic inequality require *a belief system*, or **ideology**, to rationalize and legitimate these patterns of dominance and subordination, and racism has usually served that function.

THE IDEOLOGY OF RACISM

As a belief system, or ideology, racism is structured around three basic ideas:

- Humans are divided naturally into different physical types.
- Such physical traits as people display are intrinsically related to their culture, personality, and intelligence.
- The differences among groups are innate, not subject to change, and on the basis of their genetic inheritance, some groups are innately superior to others (Banton, 1970; Benedict, 1959; Montagu, 1972; Shibutani and Kwan, 1965).

In sum, **racism** is *the belief that humans are subdivided into distinct hereditary groups that are innately different in their social behavior and mental capacities and that can therefore be ranked as superior or inferior.* The presumed superiority of some groups and inferiority of others is subsequently used to legitimate the unequal distribution of the society's resources, specifically, various forms of wealth, prestige, and power.

Racist thinking presumes that differences among groups are innate and not subject to change. Intelligence, temperament, and other primary attitudes, beliefs, and behavioral traits are thus viewed as not significantly affected by the social environment. The failures of groups at the bottom of the social hierarchy are interpreted as a natural outcome of an inferior genetic inheritance rather than of social disadvantages that have accumulated for the group over many generations. In the same manner, the achievements of groups at the top of the social hierarchy are seen as a product of innate superiority, not of favorable social opportunities.

Racist thought is inherently ethnocentric. Those espousing racist ideas invariably view ethnic out-groups as inferior. Moreover, such thought naturally leads to the idea that ethnic groups must be kept socially and physically apart. To encourage social integration is to encourage physical integration, which, it follows, contributes to the degeneration of the superior group.

Ideologies do not necessarily reflect reality; indeed, they are largely mythical. They comprise beliefs that, through constant articulation, become accepted as descriptions of the true state of affairs. We have already discussed some of the scientifically erroneous or dubious principles concerning race. As the anthropologist Manning Nash (1962) has explained, racist ideologies depend on three logical confusions: (1) the identification of racial differences with cultural and social differences; (2) the assumption that cultural achievement is directly, and chiefly, determined by the racial characteristics of a population; and (3) the belief that physical characteristics of a population limit and define the sorts of culture and society they are able to create or participate in.

Racist thought is most prevalent in societies in which physical differences among groups are pronounced, such as differences between blacks and whites in the United States or South Africa. But racism describes any situation in which people's social behavior is imputed to innate, or hereditary, sources. Racist beliefs are therefore not limited to ideas about groups commonly referred to as "races" but can apply to any ethnic group, whether distinguishable by physical characteristics or culture. Jews in Germany during the 1930s, for example, were physically indistinct from other Germans, but this did not prevent the creation by the Nazis of an elaborate racial ideology pertaining to the Jews. Even in the United States, as we will see in Chapter 10, the prevalent view of Irish immigrants to America in the nineteenth century, and later, southern and eastern European immigrants, was that they were racially inferior.

Certain beliefs regarding behavioral and personality differences between men and women are based on the same mode of thinking as racism. Men are assumed to be innately better qualified for certain social roles on the basis of their masculinity, and women are seen as quite naturally occupying other roles suited to their femininity. Although most of the behavioral differences between men and women are attributable more to social learning than to biology, the beliefs regarding the congenital nature of these differences are accepted uncritically by many. The belief in the innate behavioral differences between men and women has been referred to as sexism, but it is obvious that the foundation of this ideology is similar to that of racism.

In recent years, the term *racism* has often been used in a sweeping and imprecise fashion, describing almost any negative thought or action toward members of a racial-ethnic minority or any manifestation of ethnic inequality, not simply as an ideology. However, prejudice and discrimination directed at ethnic group members, as we will see in Chapter 3, take many different forms and need not involve or be based on racist thought.

THE FUNCTIONS OF RACISM

The belief in innate differences among groups is used to justify the unequal distribution of a society's rewards. The place of groups at the top of the social hierarchy and of those at the bottom is explained quite simply as "natural." Racist ideology, then, promotes an ethnic status quo in which one group predominates in the society's economy, polity, and other key institutions and thus receives the greatest share of the society's wealth and power.

Why, for example, do African Americans occupy a disproportionately low number of top positions in all important institutions and own a disproportionately low share of wealth by comparison with Euro-Americans? Racist ideology explains such inequalities as resulting primarily from the inherent inferiority of blacks and superiority of whites. It is asserted that something in the character of black people themselves is at the root of their socially subordinate position, just as something in the character of white people leads very naturally to their social dominance. The same explanation might be used by Protestants in Northern Ireland to account for their historical dominance over Catholics in various areas of social life or by South African whites to rationalize their past dominance of nonwhites.

In the racist mode of thought, the different social and cultural environments of groups are not of major importance in accounting for their differences in social achievement. And, as noted earlier, the perpetuation of beliefs in group superiority and inferiority gives rise to the substantiation of those beliefs through the self-fulfilling prophecy.

The belief in innate group differences leads naturally to actions and policies that are expressions of that belief. The sense of group difference, historian George Fredrickson writes, "provides a motive or rationale for using our power advantage to treat the ethnoracial Other in ways that we would regard as cruel or unjust if applied to members of our own group" (2002:9). If groups are effectively portrayed as inferior, they can not only be denied equal access to various life chances but in some cases enslaved, expelled, or even annihilated with justification. Slave systems of the eighteenth and nineteenth centuries in the Americas were rationalized by racial belief systems in which blacks were seen as incapable of ever attaining the level of civilization of whites. Similarly, the enactment in the 1920s of strict U.S. immigration quotas, favoring northwestern European groups and discriminating against those from southern and eastern Europe, was impelled by an intricate set of racist assumptions. Northwestern Europeans were seen as innately more adaptable to the American social system, and other groups were viewed as naturally deficient in favorable social and moral qualities.

THE DEVELOPMENT OF RACISM

Although beliefs in the superiority and inferiority of different groups have been historically persistent in human societies (recall the universality of ethnocentrism), the belief that such differences are linked to racial types is a relatively new idea, one that did not arise forcefully until the eighteenth century in Europe. Over the following century and a half, a number of political and scientific factors would contribute to the development of the ideology of racism.

EUROPEAN COLONIALISM   During the Age of Discovery, starting in the fifteenth century, lands were conquered by Spain, Portugal, England, France, and Holland in the Americas and Africa, and white Europeans encountered peoples for the first time who were not only culturally alien but physically distinct as well. At first, the justification for subjecting these groups to enslavement or to colonial repression lay not so much in their evident physical differences as in what was seen as their cultural primitiveness, specifically, their non-Christian religions (Benedict, 1959; Fredrickson, 2002; Gossett, 1963). Even the United States engaged, to a lesser extent, in the colonial game, taking possession of the Philippines and several Caribbean islands following its war with Spain at the end of the nineteenth century.

"SCIENTIFIC" RACISM   It was the later development of "scientific" racism, however, that gave impetus to the view that European peoples were superior to nonwhites because of their racial inheritance. Eighteenth-century scholars of various disciplines, including medicine, archaeology, and anthropology, had begun to debate the origin of the human species, specifically, the question of whether the species was one or many. Although inferiority of all non-European cultures was assumed, up to this time most had viewed all human types as subdivisions of a single genus (Rose, 1968; Smedley, 2007). Thought now turned to the much earlier but generally discounted theory of polygenesis, the notion that human groups might be derived from multiple evolutionary origins. No resolution of this debate came until the publication in 1859 of Darwin's theory of evolution, *The Origin of Species*. Darwin was clear in his explanation that differences among humans were superficial and that their more general similarities nullified any idea of originally distinct species or races (Benedict, 1959; Gossett, 1963).

SOCIAL DARWINISM   Scientists of the nineteenth century investigating the idea of race were heavily influenced by Darwin even though he had said little about race per se. Many who studied his ideas, however, drew inferences to human societies from what he had postulated about lower animal species. Darwin's idea of natural selection was now seen as a mechanism for producing superior human societies, classes, and races. Expounded by sociologists such as Herbert Spencer and William Graham Sumner, these ideas became the basis of **social Darwinism**, popularly interpreted as *"survival of the fittest."* Early racist thought, then, was supported by an element of what was considered scientific validity.

"WHITE MAN'S BURDEN" With the scientifically endorsed belief that social achievement was mostly a matter of heredity, the colonial policies of the European powers were now justified. "Race," writes anthropologist Audrey Smedley, "had become a worldview that was extraordinarily comprehensive and compelling in its explanatory powers and its rationalization of social inequality" (2007:267). Native people of color were seen as innately primitive and incapable of reaching the level of civilization attained by Europeans. Economic exploitation was thus neatly rationalized. And, because nonwhites represented a supposedly less developed human evolutionary phase, the notion of a "white man's burden" arose as justification for imposing European cultural ways on these people.

In short, the idea of race appropriately complemented the political and economic designs of the European colonial powers. Sociologist Michael Banton notes that "the idea that the Saxon peoples might be biologically superior to Celts and Slavs, and white races to black, was seized upon, magnified, and publicized, because it was convenient to those who held power in the Europe of that day" (1970:20). Banton adds that the coincidental appearance of these theories with the demise of slavery in the early and mid–nineteenth century provided a new justification to some for subordinating former slaves.

Racism should not be thought of as an opportune invention of European colonialism. As van den Berghe explains, racist thinking "has been independently discovered and rediscovered by various peoples at various times in history" (1978:12). But it was in the colonial era that the idea was received most enthusiastically and firmly established as a social doctrine.

EARLY TWENTIETH-CENTURY RACIST THOUGHT The idea that race was immutably linked to social and psychological traits continued to be a generally accepted theory in Europe and North America during the early years of the twentieth century, aided by the unabashedly racist ideas of writers such as Houston Stewart Chamberlain and Count Arthur de Gobineau in Europe and Lothrop Stoddard and Madison Grant in the United States. Grant's book, *The Passing of the Great Race,* published in 1916, was essentially a diatribe aimed at restricting immigration of southern and eastern European groups, mainly Catholics and Jews, who were at that time most numerous among American immigrants. Grant divided the European population into three racial types—Alpines, Mediterraneans, and Nordics—and he attributed to the latter (from northwestern Europe) the most desirable physical and mental qualities. Consequently, he warned of the degeneration of the American population through the influx of southern and eastern Europeans, whom he classified as inferior Mediterranean and Alpine types.

The impact of Grant's and Stoddard's books was extended by the development of intelligence testing during World War I by American psychologists, whose findings seemed to confirm the notion of inherent racial differences. On these tests, Americans of northwestern European origin outscored, on average, all other ethnic groups. This result was taken by many as empirical evidence of the superiority of the Nordic "race." Intelligence was now seen as primarily hereditary even though little attention was paid to environmental factors in interpreting test results. That northern blacks outscored southern whites, for example, was attributed not to social factors such as better educational opportunities in the North but to the selective migration of more intelligent blacks out of the South. This argument proved false, however, because the same differences were shown between northern and southern whites (Montagu, 1963; North, 1965; Rose, 1968).

BOAS AND THE BEGINNING OF CHANGE   During the 1920s, cultural anthropologists began to question the biological theories of race and to maintain that social and cultural factors were far more critical in accounting for differences in mental ability. Chief among these theorists was Franz Boas, who pointed out the lack of evidence for any of the common racist assertions of the day. Gossett explains Boas's critical role in reversing social scientific thinking on the notion of race: "The racists among the historians and social scientists had always prided themselves on their willingness to accept the 'facts' and had dismissed their opponents as shallow humanitarians who glossed over unpleasant truths. Now there arose a man who asked them to produce their proof. Their answer was a flood of indignant rhetoric, but the turning point had been reached and from now on it would be the racists who were increasingly on the defensive" (1963:430).

RACE AND INTELLIGENCE: RECENT CONTROVERSIES   By the 1950s, social scientists had, for the most part, discarded ideas linking race with intelligence and other social characteristics. But the issue was given new impetus in 1969 when educational psychologist Arthur Jensen (1969) published a report suggesting that heredity was the major determinant in accounting for the collective differences in IQ between blacks and whites. Jensen's theory, methods, and interpretation of findings were all quickly and thoroughly challenged (Alland, 2002; Montagu, 1975; Rose and Rose, 1978).

Twenty-five years after Jensen's report was issued, Richard J. Herrnstein and Charles Murray published a book, *The Bell Curve*, which refueled the debate regarding the impact of race on mental ability and set off much controversy. Herrnstein and Murray presented a kind of social Darwinistic thesis: intelligence, as measured by IQ, is in large part genetic. They premised a strong relationship between IQ and various social pathologies. Following this thinking, those with lower IQs have a greater proclivity toward poverty, crime, illegitimacy, poor educational performance, and other social ills. Because IQ is mostly genetic, they argued, there is no way to change the condition of those with low intelligence through educational reforms or welfare programs. Because lower-intelligence people are reproducing much faster than higher-intelligence people, the society is faced with the possibility of a growing underclass, increasingly dependent on the more intelligent and productive classes.

Perhaps the most controversial aspects of Herrnstein and Murray's book concerned the linkage of IQ and race. The authors pointed out that the average IQ of blacks is fifteen points lower than the average IQ of whites—a differential, they claimed, that held up regardless of social class and even when change in average IQ for groups is taken into account. The implication, then, is clear: blacks are inferior to whites and are thus apt to remain in a state of dependency on the nonpoor and continue to engage in antisocial activities. Herrnstein and Murray therefore questioned the value of welfare payments, remedial educational programs, affirmative action, and other efforts designed to raise the social level of the poor, who, in the United States, are disproportionately black.

As with Jensen earlier, Herrnstein and Murray's thesis, findings, and conclusions were overwhelmingly rejected by mainstream social scientists, who claimed that the authors' methods were flawed and their reasoning specious (Fischer et al., 1996; Fraser, 1995; Jacoby and Glauberman, 1995). Moreover, many viewed the book as much a statement of the political leanings of the authors as a work of social science. Specifically, Herrnstein and Murray were challenged on a number of points. For one, IQ has been shown to measure only certain kinds of intelligence. The developmental psychologist Howard Gardner (1983), for example, has postulated that humans possess at least eight relatively independent areas of intelligence, such as mathematical, musical, interpersonal, and so on. Furthermore, IQ is not fixed but is subject to variation within one's lifetime and, for groups, subject to change over generations. Perhaps most important, the authors did not place sufficient weight on the environmental factors that enable people to express their intelligence, regardless of IQ. Also, the authors reified "race," referring to whites and blacks as if these were clear-cut, distinct genetic groups, ignoring the countless variations within designated racial categories.

Theoretical views like Jensen's and Herrnstein and Murray's have appeared periodically, but the understanding that environment, not racial inheritance, is the key influence in shaping social behavior clearly predominates in social science thinking today (Graves, 2004; Marks, 1995; Nisbett, 2007, 2009). The preeminent view is that genetic differences among human groups are of minimal significance as far as behavior and intelligence are concerned. Those scholars not subscribing to this view have, in van den Berghe's words, "been voices in the wilderness . . . and their views have not been taken seriously" (1978:xxii). Yet as William Newman (1973) points out, ironically, science itself created the myth of race that it is today still attempting to dispel. Moreover, for many decades few sociologists questioned the prevailing ideas about racial inequality. "Both popular and educated beliefs," notes sociologist James McKee, "provided an unqualified confidence in the biological and cultural superiority of white people over all others not white" (1993:27).

CULTURAL RACISM

As scientific thinking on the issue of race has changed, a new, revised version of racist theory has become prevalent. What has been called **cultural racism** rests on the notion that *the discrepancies in social achievement among ethnic groups are the result of cultural differences rather than biogenetic ones* (Banton, 1970; Schuman, 1982). Because of an inability to adapt to the dominant culture, it is argued, economic and social handicaps persist among certain ethnic groups. These inabilities are traced to a people's way of life—its culture—which hinders conformity to the norms and values of the dominant group. Hence, these are seen as "dysfunctional cultures."

The focus on culture rather than physical characteristics has created a type of thinking that, although not racist in the classical sense, takes on some of the same basic features. The ideology of cultural deprivation or dysfunctional cultures is seemingly more benign than "old-fashioned" racism, which focuses on biological superiority and inferiority, but critics have pointed out that the difference between the two is not so sharp (Bobo and Smith, 1998; Ryan, 1975). They argue that cultural deprivation, like traditional racist notions, emphasizes individual and group shortcomings rather than a social system that, through subtle—perhaps even unintentional—discrimination, prevents minority groups from attaining economic and social parity with the dominant group. The socioeconomic discrepancy between whites and blacks, for example, is, in this view, a failure of the latter to conform to the work ethic and to obey institutional authority, not inherent inferiority or racial discrimination (Sears, 2000).

Ali Rattansi points out the artful convergence of cultural racism with classical, biologically based racism in the implication that cultural differences are more or less immutable:

> Thus the supposed avariciousness of Jews, the alleged aggressiveness of Africans and African Americans, the criminality of Afro-Caribbeans or the slyness of "Orientals," become traits that are invariably attached to these groups over extremely long periods of time. The descriptions may then be drawn upon as part of a common-sense vocabulary of stereotypes that blur any strict distinction between culture and biology. (Rattansi, 2007:1045)

Most critical is the notion that such cultural deficiencies are, if not immutable, certainly rigid and not subject to rapid and straightforward change. As Fredrickson has explained, racism need not be limited to biological determinism per se, "but the positing, on whatever basis, of unbridgeable differences between ethnic or descent groups—distinctions that are then used to justify their differential treatment" (2002:137). As we will see in Part III, contemporary ethnic conflicts in various world regions are based primarily on cultural differences, not racial distinctions.

Recognizing cultural racism suggests that there is a wide range of racist thinking that can vary in form, substance, and effectiveness. It also demonstrates how the concepts of race and ethnicity can overlap in shaping the perception of groups and their members' behavior and can, in turn, give rise to racist notions.

In sum, racism is a belief system that has proved tenacious, though modifiable in style and content, in multiethnic societies. Even though popular thinking has drifted away from the old biological racist ideas, more subtle racist notions have taken their place.[4] Racism, then, is a social phenomenon that continues to render an effect where culturally and physically distinct groups meet.

SUMMARY

- The major aspects of the study of race and ethnic relations are (1) the nature of relations among ethnic groups of multiethnic societies; (2) the structure of inequality among ethnic groups; (3) the manner in which dominance and subordination among ethnic groups are maintained; and (4) the long-range outcomes of interethnic relations—that is, either greater integration or greater separation.
- An *ethnic group* is a group within a larger society that displays a common set of cultural traits, a sense of community among its members based on a presumed common heritage, a feeling of ethnocentrism among group members, ascribed group membership, and, in some cases, a distinct territory. Each of these

---

[4] As new genetic information is revealed through studies of DNA, many scientists are concerned that people may begin to disregard environmental influences as they fixate on genetic ones, thereby creating the bases for a reenergized biological racism (Harmon, 2007).

characteristics is a variable, differing from group to group and among members of the same group.

- *Race* is an often misused notion having biological and social meanings. Biologically, a race is a human population displaying certain hereditary features distinguishing it from other populations. The idea is essentially devoid of significance, however, because there are no clear boundaries setting off one so-called race from another. Hence, there can be no agreement on what the distinguishing features of races are, much less how many races exist. Furthermore, the weight of scientific evidence refutes any meaningful relationship of social and mental characteristics and race.

- The sociological importance of race lies in the fact that people have imputed significance to the idea despite its questionable validity. Hence, races are always socially defined groupings and are meaningful only to the extent that people make them so.

- *Racism* is an ideology, or belief system, designed to justify and rationalize racial and ethnic inequality. The members of socially defined racial categories are believed to differ innately not only in physical traits but also in social behavior, personality, and intelligence. Some "races," therefore, are viewed as superior to others.

- The ideology of racism emerged most forcefully in the age of colonialism, when white Europeans confronted nonwhites in situations of conquest and exploitation. The notion that race was linked to social and psychological traits was given pseudoscientific validity in the late nineteenth and early twentieth centuries by American and European observers, a view that was not reversed until the 1930s. Today, scientific thought gives little credence to racial explanations of human behavior, stressing instead environmental causes.

- Even though popular thinking has drifted away from the old biological racist ideas, more subtle racist notions have taken their place. The belief in immutable racial differences may no longer prevail, but the idea that differences in group achievement stem from cultural deficiencies or lack of effort by group members has become a compelling ideology with racist overtones.

CRITICAL THINKING

1. Look at some strongly multiethnic countries and some countries in which there is little ethnic diversity. What historical factors have created heterogeneity in the former and homogeneity in the latter? What accounts for the fact that almost all developed societies of the contemporary world are increasingly multiethnic?

2. Most Americans would not easily accept the idea that "races" are not real, but are social creations. How would you begin to explain this basic sociological postulate to a friend or family member who argues that "you only need to look around you to see that the United States is made up of people of different races"?

3. A few African American political commentators suggested during the 2008 U.S. presidential campaign that Barack Obama was "not black enough." What did they mean by this? Does this comment illustrate differences in the concepts of "race" and "ethnicity?"

4. The U.S. Census has, over the years, changed racial and ethnic categories used in counting the population. How does the creation of racial/ethnic categories by the U.S. Census in effect create "races?" Does the current census provision enabling people to choose more than one race in classifying themselves represent a step in erasing racial and ethnic boundaries, or does it merely create other racial categories and thereby create other boundaries?

PERSONAL/PRACTICAL APPLICATION

1. How conscious are you of your ethnic origins? What role does your ethnic identity play in your life? Does it create any social barriers or, on the contrary, any social privileges?

2. Are your close friends similar to yourself in terms of ethnicity? Explain why this may or may not be the case. Might ethnicity be an unconscious consideration in choosing friendships?

ETHNIC STRATIFICATION | CHAPTER 2

Majority and Minority

A prominent sociologist has suggested that the first questions asked by sociology were these: "Why is there inequality among men? Where do its causes lie? Can it be reduced, or even abolished altogether? Or do we have to accept it as a necessary element in the structure of human society?" (Dahrendorf, 1968:152). Such questions remain fundamental to sociological inquiry and are particularly critical in the study of race and ethnic relations.

Humans are unequal, of course, in many ways. They differ in physical features and in mental capacities, talent, strength, musical aptitude, and so on. All these inequalities are a product of both social learning and genetic inheritance, although the significance of each of these factors is, as we have now seen, not always clear. Perhaps more important, however, people are also unequal in their access to social rewards—that is, various forms of wealth, power, and prestige. These inequalities, all primarily of social origin, are of greatest consequence in accounting for who we are and who we ultimately may be as members of our society.

STRATIFICATION SYSTEMS

In all societies people receive different shares of what is valued and scarce. This unequal distribution of resources creates a system of stratification. A rank order, or hierarchy, emerges in which people are grouped on the basis of how much of society's rewards they receive. Those at the top receive the most of what there is to get, and those at the bottom the least. Societies may comprise any number of strata, but in all cases this system of inequality is structured. That is, stratification is not random, with groups and individuals occupying different positions by chance; rather, social institutions such as government, the economy, education, and religion operate to ensure the position of various groups in the hierarchy. Moreover, the system of stratification in all societies is legitimized by an ideology that justifies the resultant inequality. The pattern of stratification in a society therefore remains stable for many generations.

Modern societies are stratified along several dimensions, the most consequential of which is class stratification, in which groups are ranked on the basis of income, wealth, and occupation. Gender and age are other important dimensions of stratification. Multiethnic societies are also stratified on the basis of ethnicity, and it is on this dimension that we will primarily focus.

POWER AND STRATIFICATION

In a basic sense, power underlies all forms of stratification. Just as differences in wealth, education, occupation, and prestige are mirrors of a society's power arrangement, so too are differences in rank among a society's ethnic groups. Social stratification, then, is a system of unequal distribution of a society's rewards, determined above all by power differentials. In simple terms, those at the top get more of what is valued because they are more powerful; they possess greater power resources in the form of wealth, property, political office, arms, control of communications, and knowledge. The position of others in the stratification system is determined accordingly on the basis of their ability to amass and apply power resources.

STRATIFICATION AND IDEOLOGY

The power of a dominant class or ethnic group is not simply the power of force but also the power to propound and sustain an ideology that legitimizes the system of inequality. Although coercion is always at the root of obedience to authority, and all dominant groups use force when the need arises, coercive techniques are commonly used only in societies where the prevailing system is not accepted by a significant part of the populace. In South Africa under apartheid or in the antebellum U.S. South, for example, whites traditionally enforced their will over nonwhites through blatant forms of repression.

The use of raw force alone, however, cannot be effective in prompting compliance with a system of inequality over long periods. The stability of systems that rely primarily on coercion is always precarious. For government and other supportive institutions of the dominant group to establish and sustain a ruling system that is popularly supported over many generations requires that power be legitimized in less repressive and less direct ways. In protecting their privileges, dominant groups try to engender loyalty and respect in subordinates, not fear (Jackman, 1994). People must come to see the inequalities in power and wealth as just and even socially beneficial. Only then do systems of social inequality attain stability. When this is accomplished, ruling groups need no longer resort to force as the principal means of assuring their power and privilege. Such long-range stability and legitimacy require the development of an effective ideology and its communication through socialization.

Despite their acceptance—usually reflexively—by both ruling groups and masses, the fundamental ideological values tend to accommodate mostly the interests of the society's ruling groups. In the United States, for example, the dominant explanation for social inequality centers on the belief that the society's opportunity structure is open, providing equal chances for all to achieve material success or political power, regardless of their social origins. This presumably being the case, each person controls his or her placement in the social hierarchy. Social success, then, is explained as the result of one's willingness to work hard; failure is the product of lack of ambition or desire to improve oneself. Differences in wealth and power are not denied, but they are seen as a result of individual capabilities and efforts rather than the workings of a class system that ordinarily engenders success for the well-born and failure for the poor. In reality, however, the opportunity structure is hardly equal, and the dominant values of individualism, competition, and achievement favor those who are well off and can easily avail themselves of the opportunities for success. As we will see, just as there are ideologies that explain and rationalize inequalities in social class, so too there are ideologies that explain the differential treatment of groups on the basis of ethnicity. These ideologies generally constitute some form of racism, the basic components of which were discussed in Chapter 1.

To summarize, **social stratification** is *a system of structured inequality in which people receive different amounts of society's valued resources.* This inequality is relatively stable over long periods and gives rise to social classes—groups of people of approximately equal income and wealth. In multiethnic societies, ethnicity becomes an additional—and critical—basis of stratification. Differential power underlies all forms of inequality, and the system is underwritten by an ideology—propounded by the dominant group but generally accepted by others—that justifies differences in social rewards.

ETHNIC STRATIFICATION SYSTEMS

Ethnic stratification, like other forms of stratification, is a system of structured social inequality. In almost all multiethnic societies, a hierarchical arrangement of ethnic groups emerges in which one establishes itself as the dominant group, with maximum power to shape the nature of ethnic relations. Other, subordinate ethnic groups exert less power, corresponding to their place in the hierarchy, extending down to the lowest ranking groups, which may wield little or no power.

Group rank is determined mainly on the basis of distance from the dominant group in culture and physical appearance. Those most like the dominant group are more highly ranked, and those most different are ranked correspondingly low. A system of **ethnic stratification**, then, is *a rank order of groups, each made up of people with presumed common cultural or physical characteristics interacting in patterns of dominance and subordination.* Sociologists ordinarily refer *to ethnic stratification systems* as **majority-minority**, or **dominant-subordinate**, systems. Let us look more closely at the components of these systems, majority and minority groups.

MINORITY GROUPS

DIFFERENTIAL TREATMENT   **Minority groups** are those *groups in a multiethnic society that, on the basis of their physical or cultural traits, receive fewer of the society's rewards.* In a classic definition, Louis Wirth defined a minority group as "a group of people who, because of their physical or cultural characteristics, are singled out from the others in the society in which they live for differential and unequal treatment, and who therefore regard themselves as objects of collective discrimination" (1945:347). Members of minority groups disproportionately occupy poorer jobs, earn less

income, live in less desirable areas, receive an inferior education, exercise less political power, and are subjected to various social indignities. These inequalities are the
result of their social mark—the physical or cultural features that distinguish them.
Moreover, as Wirth pointed out, minority group members are conscious of the fact
that they are differentially treated.

SOCIAL DEFINITION   The physical or cultural traits on which minority status is based are
socially defined. Thus any characteristic may serve as the basis of minority status as
long as it is perceived as significant. Suppose that in society X, hair color is considered a meaningful distinguishing feature. If blond hair is deemed more desirable than
black or brown, those with black or brown hair may be singled out and treated differentially for no other reason than that their hair is not blond. The black-haired and
brown-haired populations of society X are minority groups.

DIFFERENTIAL POWER   Minority groups are afforded unequal treatment because they
lack the power to negate or counteract that treatment. Blond-haired people in our
fictional society can continue to treat nonblonds as less than desirable types and
withhold various forms of social rewards only if they maintain sufficient power to
do so. Minority status, then (like all forms of stratification), is above all a reflection
of differential power. In a dominant-minority system, one group possesses sufficient
power to impose its will on others.

CATEGORICAL TREATMENT   "Minority" denotes a group, not an individual, status. All
those who are classified as part of the group will experience differential treatment,
regardless of their personal characteristics or achievements. Thus people cannot voluntarily remove themselves from their minority position.

SOCIOLOGICAL AND NUMERICAL MEANINGS   The sociological meaning of minority is not the
same as the mathematical definition. Numbers have no necessary relation to a
group's minority status. For example, as we will see in Chapter 14, nonwhites in South
Africa make up more than 85 percent of the population, yet until only a few years ago
they constituted a sociological minority. Nonwhites had almost no access to political
and economic power, were assigned the lower occupational positions, and were afforded grossly inferior opportunities in all areas of social life by comparison with South
African whites, who made up less than 15 percent of the population. Rather than relative size, it is a group's marginal location in the social order that defines it as a minority.

TYPES OF MINORITIES

In its sociological meaning, the term *minority* can be applied to a variety of social
groups. Our chief concern is with **ethnic minorities,** *those groups singled out and
treated unequally on the basis of their cultural or physical differences from the dominant group.* Ordinarily, attention is directed to the most conspicuous ethnic groups,
those with especially marked differences in skin color or those that maintain divergent cultural beliefs and behavior. In the United States today, these are blacks, American Indians, Asians, and Latinos. These are the groups that have experienced the
most blatant and consequential forms of discriminatory treatment.

In addition to ethnic traits, however, other physical or behavioral characteristics are sufficient to set off groups of people from society's mainstream, resulting in differential treatment. To equate minority groups only with highly visible ethnic groups fails to account for the many other types of minorities found in complex societies.

Sex, for example, is a clearly distinguishable physical characteristic that in most societies serves to single out one group—ordinarily, women—for differential treatment. It is in this sense that women may be said to constitute a minority group. Traditionally, women have rarely occupied positions of great political or economic power, have been barred from entrance into many occupations, and have been excluded from numerous areas of social life. Only in recent times have these deeply rooted patterns of sex discrimination changed radically, particularly in developed societies. In many parts of the developing world, however, the subordination of women, often in extreme form, remains starkly evident.

People with physical disabilities—the blind, the deaf, those confined to wheelchairs, and so on—are another evident minority. On the basis of their physical distinctions, they are singled out and given differential treatment in many social contexts. Not until the last few decades had efforts been made to accommodate such people in public buildings, for example, and they had rarely been afforded equal educational and occupational opportunities.

Sexual orientation has traditionally been the basis for extreme forms of prejudice and discrimination in American society. Gay and lesbian people have, in fact, almost classically evinced all the characteristics of the sociological definition of minority. Most obviously they have been given differential treatment. For example, until 2012, openly declared gay and lesbian people were barred from serving in the military. No more than a few years prior to that, many states maintained anti-sodomy statutes that made homosexuality itself a crime. While the civil rights of gay and lesbian people have increasingly been recognized in the United States in the past few years, including same-sex marriage, there remains an image of fundamental difference from the majority, which continues to serve as a social marker and the basis for differential treatment. In many developing societies, the plight of gay and lesbian people is even more severe, and openly declaring oneself homosexual may actually put one's life at risk.

Age constitutes another physical feature that serves to set groups apart for differential and unequal treatment. In all societies, of course, certain limitations are placed on people according to their age. The very young are not expected to fulfill adult roles, and at the opposite end of the life cycle, the aged are not expected to perform as younger people do. But in some cases, this necessary age differentiation exceeds any rational explanation, and people are singled out for differential treatment merely on the basis of their age. Workers are often forced into retirement at sixty-five, even if they remain capable of carrying out their occupational duties and wish to do so. Or elderly people commonly find themselves dealt with in condescending and childlike fashion, having others address them as "dear" or "sweetie" or simply assuming them to be incompetent. Cases such as these constitute age discrimination.

Certain groups are also singled out for differential treatment on the basis of their past behavior. Those acknowledging treatment for mental illness or a prison record, for example, risk social exclusion and discrimination by employers, landlords, and law enforcement agencies.

For each of these groups, a belief system, or ideology, explains and justifies their differential treatment. Women are alleged to be physically or mentally inferior to men or to be unable to perform at the level of men; the elderly are perceived as slow, ineffective, or senile; gays and lesbians are "deviant" or "immoral"; and so on. These generalized beliefs serve as devices to sustain unequal treatment.

DOMINANT GROUPS

In the study of ethnic relations, it has been most common for sociologists to focus primarily on minority groups. Ethnic relations, however, involve not only the problems of those at the bottom of the ethnic hierarchy but also the manner in which those at the top maintain their dominance. Obviously, the existence of minority groups implies a majority group. In this regard, Hughes and Hughes have noted, "It takes more than one ethnic group to make ethnic relations. The relations can be no more understood by studying one or the other of the groups than can a chemical combination by study of one element only, or a boxing bout by observation of only one of the fighters. Yet it is common to study ethnic relations as if one had to know only one party to them" (1952:158).

In the United States, for example, black-white relations for many decades were portrayed as "the black problem in America," and most studies dealt with the social and psychological problems faced by blacks because of their minority status. Only in the 1960s did sociologists begin to show that intergroup relations between blacks and whites were just as much a *white* problem because it was the dominant white group that controlled the character and course of those relations more than did blacks themselves. Our major concern is not the unique character of particular ethnic groups per se as much as it is the social situation within which different groups interact. It is thus necessary to look carefully at the nature and functions of dominant as well as minority ethnic groups.

Most sociologists refer to a society's majority group as the *dominant* group, and that usage is adopted here. This avoids the tendency to think in numerical terms in cases where the dominant group may not be a numerical majority. For consistency, we might also refer to minority groups as *subordinate* groups, implying the existence of a power relationship rather than a numerical one. However, because the term *minority* is so firmly fixed in the sociological literature as well as in popular usage, we will continue to use it.

Most simply, then, we can define the **dominant ethnic group** as that *group at the top of the ethnic hierarchy, which receives a disproportionate share of wealth, exercises predominant political authority, dominates the society's cultural system, and has inordinate influence on the future ethnic makeup of the society.* Let's look more closely at each element of the dominant group's power.

POLITICAL AND ECONOMIC DOMINANCE   Its power advantage in political and economic realms enables the dominant group to acquire a disproportionate share of the society's wealth and to exercise maximum control of key institutions. This does not mean, of course, that all those classified as part of the dominant ethnic group are equally powerful and wealthy. Rather, it means only that members of the dominant ethnic group disproportionately enjoy those privileges. In the United States, for example, white Anglo-Saxon Protestants (WASPs), or Anglo-Americans, historically have been, and to a great extent remain, the dominant ethnic group. But obviously not all members of this category are part of the decision-making elites of government or the economy. They are, however, *disproportionately* holders of the most important positions in those institutions and are able to render decisions that reflect the interests and values of Anglo-Americans generally. Thus, compared with other ethnic groups, Anglo-Americans own a greater share of the society's wealth, earn more income, acquire more and better education, work at higher-ranking and more prestigious occupations, and generally attain more of the society's valued resources.

CULTURAL POWER   In addition to the dominant ethnic group's greater economic and political power, its norms and values prevail in the society as a whole. The cultural characteristics of the dominant group become the society's standards. As the sociologist R. A. Schermerhorn explains, "When we speak of a 'dominant group' we mean that group whose historical language, traditions, customs, and ideology are normative for the society; their preeminence is enforced by the folkways or by law, and in time these elements attain the position of cultural presuppositions" (1949:6).

The cultural supremacy of the dominant ethnic group in a multiethnic society applies to *major* norms and values, not to every element of the society's culture. In any heterogeneous society, certain cultural traits of minority ethnic groups are bound to seep into the mainstream. In the United States, to use a simple example, egg rolls, bagels, tacos, and pizza have become standard "American" fare. But items such as foods represent only minor cultural components. There is no similar acceptance of variety regarding the more basic aspects of culture: language, religious values, political practices, and economic ideology. On these counts, ethnic groups are expected to acculturate to the dominant group's customs and ideals. In the United States, for example, it is assumed that all groups will speak English, will maintain Judeo-Christian ethics, will abide by democratic principles, and will accept capitalist values. Those who do not, remain outside the societal mainstream and may be held in contempt. Although changes even in these areas of culture are usually evident over the long run, such changes are slow, deliberate, and may be accompanied by controversy and open conflict. Bilingual education and bilingual government services in the United States, for example, have often met with strong resistance.

CONTROL OF IMMIGRATION   The dominant ethnic group, given its political, economic, and cultural power, is able to regulate the flow and composition of new members of the society (usually through immigration) and to determine the social treatment of new groups after they have entered the society. Once its preeminence has been established, *the dominant group becomes synonymous with the "host" or "receiving" society*, or what John Porter calls the **charter group**:

> In any society which has to seek members from outside there will be varying judgements about the extensive reservoirs of recruits that exist in the world. In this process of evaluation the first ethnic group to come into previously unpopulated territory, as the effective possessor, has the most say. This group becomes the charter group of the society, and among the many privileges and prerogatives which it retains are decisions about what other groups are to be let in and what they will be permitted to do. (1965:60)

No matter how liberal the immigration policies of a multiethnic society may be, certain restrictions or quotas are always imposed on groups that are highly unlike the dominant group in culture or physical appearance. And in the same manner, those immigrants whose origins are culturally and physically close to those of the dominant group enjoy not only easier entrance but also more rapid and less impeded upward social mobility. Consider, for example, the different social and political responses to Canadian immigrants entering the United States compared to those from Mexico.[1]

RELATIVE DOMINANCE   Dominant group power is never absolute. Although the dominant group is at the top of the ethnic hierarchy, its power and influence are relative, not total. As explained earlier, members of that group are disproportionately in positions of power and influence. Likewise, cultural dominance should not be interpreted as complete. The dominant group has a disproportionate influence on shaping the society's cultural mold, but, as already noted, minority groups obviously make contributions and compel changes. The United States today is hardly a WASP paragon in literature, music, the arts, politics, education, and most other aspects of culture. On the really fundamental aspects of culture, however—such as language, law, and religion—WASP influence remains much stronger and less vulnerable to minority influences.

Also, the ethnic hierarchy is not fixed. Rather, it is always under challenge to some degree; therefore, groups' positions may change and their power may be enhanced or diminished. For example, to speak of the Anglo group as dominant in American society today is not to imply that its position is comparable to what it was in an earlier time when its power was far more thorough. As minority groups over the years have acquired more economic, political, and cultural resources, WASP power has been diluted.

MIDDLEMAN MINORITIES

Certain ethnic groups in multiethnic societies sometimes occupy a middle status between the dominant group at the top of the ethnic hierarchy and subordinate groups in lower positions. These have been referred to as middleman minorities (Blalock, 1967; Bonacich, 1973; Bonacich and Modell, 1980; Turner and Bonacich, 1980; Zenner, 1991).

Middleman minorities often act as mediators between dominant and subordinate ethnic groups. They ordinarily occupy an intermediate niche in the economic system, being neither capitalists (mainly members of the dominant group) at the top nor working masses (mainly those of the subordinate groups) at the bottom. They play such occupational roles as traders, shopkeepers, moneylenders, and independent professionals. Middleman minorities therefore serve a function for both dominant and subordinate groups. They perform economic duties that those at the top

---

[1] Petersen (1980) notes that the dominant group may not always be large and powerful enough to act as a "host" to immigrants. He points to modern Israel as a case in which immigrants have been acculturated not to a host population but to an ideology of Zionism. Similarly, Argentina's development since the late nineteenth century has been marked by the economic and political ascendance of immigrants.

find distasteful or lacking in prestige, and they frequently supply business and professional services to members of ethnic minorities who lack such skills and resources.

Given their intermediate economic position, such groups find themselves particularly vulnerable to out-group hostility, emanating from both dominant and subordinate groups. In times of stress they are natural scapegoats (Blalock, 1967). They are numerically and politically lacking in power and therefore must appeal to the dominant group for protection, which will be provided as long as it is felt that their economic role is necessary. But the role may still be seen as tainted, thus prompting feelings of revulsion and discriminatory actions. Subordinate groups also view middleman minorities with disdain because they often encounter them as providers of necessary business and professional services. Such entrepreneurs therefore come to be seen as exploiters. In the United States, for example, Jews often operated businesses in black ghetto areas of large cities—a role increasingly assumed today by new immigrants, especially Koreans. Conflict between these businesspeople and their neighborhood customers has been common.

Because they stand in a kind of social no-man's-land, part of neither the society's dominant group nor a suppressed minority, middleman minorities tend to develop an unusually strong in-group solidarity and are often seen by other groups as clannish (Bonacich and Modell, 1980). Such in-group solidarity, as well as their business success, creates resentment and antipathy, which in turn sustain a high level of ethnic solidarity.

A number of groups that seem to fit the characteristics of middleman minorities have been evident in almost all parts of the world. Jews in Europe have historically been a classic illustration. As moneylenders in medieval times, they were a generally despised group. They assumed this economic role, however, because for Christians, money lending was regarded as sinful. Such money-lending activity was nonetheless necessary, and as non-Christians, Jews naturally came to fill this occupational niche.

The ethnic Chinese in various Southeast Asian societies, sometimes called the "overseas Chinese," have played a similar role (Chua, 2004; Dobbin, 1996; Freedman, 1955; Zenner, 1991). In the Philippines, for example, the Chinese have traditionally been highly successful in business but have also been the target of much prejudice and discrimination by Filipinos. In Indonesia, too, the overseas Chinese have been prominent as a middleman ethnic group. Although they make up only 3 percent of the population of 240 million, they play a dominant role in the economy, accounting for perhaps 70 percent of all private economic activity (Chua, 2004; Schwarz, 1994). This has bred much resentment on the part of indigenous Indonesians, who view the ethnic Chinese as a cohesive and clannish group—despite the fact that the Chinese have been in Indonesia since the nineteenth century and in recent years have absorbed much of Indonesian culture. Episodes of violence against the ethnic Chinese have been regular occurrences in the past few decades (Chua, 2004; Kristof, 1998; Mydans, 1998).

Obviously, many individuals within groups that have been labeled middleman minorities do not conform to the preeminent characteristics of this type; and even the group as a whole may, at different times, display few of them. But the idea of middleman minorities forces us to consider the many variations among minority groups of a multiethnic society.

THE RELATIVITY OF DOMINANT AND MINORITY STATUS

Minority ethnic status is a relative condition: the treatment of ethnic minorities will vary from group to group and from time to time. Some groups may be consistently singled out and deprived of social rewards, and others may experience only minimal discriminatory treatment. Even for any single group, minority status is not a simple matter to explain. Prejudice and discrimination may be quite strong at certain times and diminished at others depending on various economic, political, and social conditions. Moreover, members of a particular group may experience rejection and denial in some areas of social life but not in others.

Whether a person is part of the dominant group or a minority group depends in all cases on the social context. In some instances people will be part of one, and in other instances they may find themselves part of another. For example, Jews in American society are ordinarily designated a minority group. Yet from the standpoint of blacks, who generally rank lower in economic class and prestige, Jews are part of the dominant white group. Similarly, Appalachian whites who have migrated to northern cities are often discriminated against and derisively referred to as "hillbillies"; they are, then, a minority in that context even though they are usually white, Protestant, and Anglo-Saxon in origin—that is, indisputably part of the society's dominant ethnic group (Killian, 1985; Philliber and McCoy, 1981).

Group membership in a modern, complex society is rarely a simple matter of "either-or" but instead takes the form of combinations that often yield confusing and ambiguous statuses. Different ethnic designations alone can produce equivocal statuses. For example, are Irish American Catholics responded to more generally as whites (that is, part of the dominant group) or as Catholics (that is, part of a minority group)? In different social contexts, either response may be expected. Among blacks, Irish Catholics will be seen and interacted with primarily as white, and their religion or national origin will be of no significance. Among white Protestants, however, their religious identity will be stressed and their racial identity will be unimportant.

ETHNIC STRATA: CLARITY AND MOBILITY

As noted earlier, ethnic stratification, like socioeconomic forms of stratification, is founded on the power of one group over others. The ensuing relations among groups are, therefore, ordinarily relations of conflict (Olsen, 1970). But if the theme of differential power is common to ethnic and socioeconomic forms of stratification, there are also several important differences between the two, namely, the clarity of and mobility between strata.

MOBILITY BETWEEN STRATA In developed societies, socioeconomic classes are relatively open, with porous boundaries between them. Possibilities exist for *individuals to move upward from a lower to a higher class or downward from a higher to a lower class.* Sociologists refer to such movement as **social mobility**. In reality, social mobility from one generation to the next or within one's own lifetime is ordinarily limited. People inherit their class position, and most do not move substantially upward or downward. Nonetheless, equalitarian ideologies proclaim and encourage class mobility. Those born into lower economic classes are, at least theoretically, able to advance.

Ethnic stratification, however, is a system in which the boundaries between strata (ethnic groups) are far more distinct. For most people ethnicity is, as we have seen, an ascribed status; ethnic group membership is assigned at birth and is ordinarily not subject to fundamental change. As a result, ethnic consciousness is more strongly developed among members, and the competition and conflict among ethnic groups are more sharply focused.

In most multiethnic societies, the degree of mobility between ethnic strata is minimal. Particularly where physical differences, such as skin color, are pronounced, extremely wide schisms develop between groups, and the lines of ethnic division remain rigid and relatively impermeable for many generations. This is so even when cultural differences are slight. African Americans, for example, are well assimilated into the dominant culture, yet they remain more rigidly segregated than others. In general, the more visible the differences between ethnic groups, the more clear-cut and inflexible will be the divisions between them. Moreover, physical differences ordinarily relate closely to the degree of inequality between the groups. Wilson notes that "there are no known cases of racial groups in advanced nation states having established equalitarian relationships" (1973:18).

Though it is nowhere common, ethnic mobility differs in degree in various multiethnic societies. In a few, the movement from one ethnic group to another is not unusual. In Mexico and Peru, for example, people who may be Indians can move into the mestizo group, the culturally and politically dominant group, merely by dropping the use of their Indian language and adopting Spanish in its place, by wearing shoes rather than the peasant huaraches, or sandals, and generally by practicing non-Indian ways (van den Berghe, 1978, 1979b). Physical visibility is not a critical factor that prevents such movement. To a lesser degree, the same is true of Brazil. Even in those societies with more rigid boundaries between ethnic groups, mobility may be evident among those close to the dominant group in culture and physical appearance, and such groups may be absorbed into the dominant group over a few generations. But, in general, where groups remain culturally or physically very distinct, mobility between ethnic strata is limited.

CASTE   In some cases, ethnic stratification takes on the characteristics of a **caste system,** *the most rigidly static type of stratification, in which movement from one group to another is highly restricted by custom or law.* The caste system of India is perhaps the most noted, but it is marked not so much by physical distinctions between groups as by people's social descent (Béteille, 1969; Kolenda, 1985). As it is used in non-Indian settings, the idea of caste more generally refers to "a major dichotomous division in a society between pariahs and the rest of the members of the society" (Berreman, 1966:292; see also van den Berghe, 1981). Pariahs are those who are stigmatized on the basis of some physical or cultural feature, and in most multiethnic societies that stigma is skin color or other perceived racial features.

Where ethnic groups are racially defined, relations among them tend toward caste. Endogamy within castes is enforced by custom or law, and interaction between castes in intimate social settings such as peer groups, clubs, and neighborhoods is minimized. Subordinate castes are usually exploited occupationally by the dominant group and experience little or no change in their collective social position.

The United States and South Africa provide the best examples of castelike systems. As we will see in later chapters, in both societies the lines of division between black and white groups are sharply drawn on the basis of perceived physical differences, and endogamy is the norm. Relations between blacks and whites remain chiefly of a secondary nature, that is, in settings that do not call for close personal and intimate contacts.

One interesting case of caste exists in Japan, where a pariah group, the Burakumin, are indistinguishable physically from the rest of the population. Nonetheless, the extreme discrimination imposed on this group has been justified in terms of race. As De Vos and Wagatsuma explain, there is a commonly shared social myth that the Burakumin "are descendants of a less human 'race' than the stock that fathered the Japanese nation as a whole" (1966:xx). In the past, Burakumin were required to wear unique identifying garb, were strictly segregated residentially, and were limited to low-status occupations (Aoyagi and Dore, 1964). Although they are today not as stigmatized as in the past, their social and economic standing remains low, and they still live mostly in isolated communities (Kristof, 1995; Onishi, 2009; Saito and Farkas, 2004). Moreover, many Japanese still take care to avoid marriage with a Buraku individual and may research in great depth the background of a potential mate to assure that there is no trace of Burakumin ancestry.

Northern Ireland, as we will see in Chapter 17, provides another example of a society where its two major religio-ethnic groups, Protestant and Catholic, maintain a castelike relationship. Social interaction other than work usually takes place among coethnics, and contact with members of the other group is limited. Segregated neighborhoods, schools, and recreational activities are the norm, and intermarriage across ethnic lines is severely constrained. As with the Burakumin in Japan, there are no apparent physical distinctions that separate members of the two groups, but through subtle and well-understood social signs, cues, and behaviors, one's ethnic identity is quickly established in all social situations.

The Burakumin in Japan and Protestants and Catholics in Northern Ireland illustrate how presumed group differences—though objectively slight or even nonexistent—can form the basis of strong ethnic divisions. Group separation is maintained through a perception of and belief in profound and unbridgeable differences.

Ordinarily, however, initial placement and subsequent mobility within an ethnic stratification system are affected greatly by visibility. Those who can be stigmatized on the basis of easily perceived differences are more easily discriminated against and kept in a subordinate position than those whose visible differences are negligible. It is for that reason that the Nazis forced German Jews, physically indistinguishable from the rest of the German population, to wear yellow Stars of David on their clothing.

ETHNICITY, CLASS, AND POWER

Although ethnicity and social class are distinct dimensions of stratification, they are closely interrelated. In almost all multiethnic societies, people's ethnic classification becomes an important factor in the distribution of social rewards and, hence, in their economic and political class positions. Where people begin their quest for the society's rewards and what they ultimately achieve depend in some degree on their ethnicity.

The Relationship of Ethnicity to Social Class and Power

That ethnic groups are hierarchically ordered would mean nothing if that rank order were not tied to the distribution of the society's wealth, power, and prestige. The fact that members of one or a few ethnic groups maintain most of the important positions of political and economic power, own an inordinate share of the society's wealth, and enjoy the most social prestige is due neither to chance nor to their greater motivation or innate capabilities. Rather, it is a consequence of the integral link between ethnic stratification and other forms of stratification in the society. In short, the ethnic and class systems are in large measure parallel and interwoven.

CLASS   All aspects of social class—occupation, education, income, and wealth—are closely linked to ethnicity. Those occupying the most powerful and prestigious jobs, those most highly educated, those earning the highest incomes, and those possessing the greatest wealth are statistically more likely to be members of the dominant ethnic group or members of ethnic groups closest to the dominant group in culture and physical appearance.

Of course, success is by no means assured simply due to a favored ethnic background. People's class position at birth, even for those of the dominant ethnic group, is an overarching factor in determining their eventual wealth, power, and prestige. Thus not all Anglo-Americans, for example, are doctors, lawyers, corporation executives, and high-ranking politicians. But for dominant group members, the ethnic factor is removed as an impediment to upward economic mobility; other factors, both individual and structural, will affect their fortunes, but ethnicity will not.

In the same way, we should not think that ethnic minority status automatically relegates one to the bottom rungs of the wealth, occupational, educational, political, and other class hierarchies. The election of Barack Obama as U.S. president is a dramatic illustration of this. But for minorities, the chances of winding up at the bottom are much greater. As we proceed down the ethnic hierarchy, we find increasing political powerlessness, lack of economic opportunity, and social discrimination and exclusion. Members of ethnic groups closer to the bottom of the ethnic hierarchy find the path to social and economic success difficult, regardless of their other, nonethnic, social traits.

In sum, the effect of ethnicity is that minority ethnic group members encounter barriers to the attainment of the various rewards of their society that dominant ethnic group members do not face.

This does not mean, however, that all ethnic minorities are affected to the same extent. For those who are members of low-ranking groups with high ethnic visibility—African Americans, for example—minority status may be the overriding determinant of one's economic and political class position. For other groups more highly ranked and less visible—for example, Polish Americans or Irish Americans—ethnicity may have lost almost all social significance. It is important to remember that there are *degrees* of minority status. The ethnic hierarchy in any society is rarely a simple two-part structure with a dominant group at the top and subordinate groups at the bottom. Rather, minority groups occupy places on a continuum, and the impact of ethnicity varies among them. The question for minority group members, then, is: to what extent does ethnicity become a factor in the allocation of jobs, education, wealth, political power, and other life chances?

SOCIETAL POWER   Because all aspects of stratification are founded on power, whatever changes occur in the distribution of wealth, income, education, and other life chances necessarily depend on the makeup of the society's power elites—those who formulate policies, guide the activities, and decide the significant issues of government, the corporation, education, and other major societal institutions. In analyzing the relationship between decision-making power and ethnicity, the key question to be asked is, how open are the power elites to members of different ethnic groups? To the extent that they are closed, the status of those groups in various dimensions of stratification will remain consistently low.

In most multiethnic societies, there is an apparent relationship between ethnicity and access to important power positions. Some groups are favored over others, and much ethnic conflict centers on the process of filling these elite posts. In the United States, the ideology of equal opportunity does not always conform to the reality of ethnic discrimination in elite recruitment. Today, the power elite is more diverse than at any time in the past, but the fact remains that throughout most of American history, the dominance of white Protestant males has been the general rule at the highest levels of government and the economy.

Similar patterns can be discerned in other multiethnic societies. Israel, for example, is a multiethnic society in which Jews of European origin have generally ranked higher in the society's income, occupational, educational, and political hierarchies than Jews from North African and Middle Eastern societies (Goldscheider, 2002; Simon, 1978; Smooha, 1978). Brazil provides another illustration, where, as we will see in Chapter 15, most positions of institutional power are held by whites rather than by Brazilians of color, though the latter make up fully half the population.

ETHNICITY, CLASS, AND POWER RECONSIDERED

At this point, observations of the social standing of some members of dominant and minority ethnic groups might give rise to wonderment at what ethnic dominance and subordination really mean. Do we not see in the United States, for example, Jews in important power positions or blacks in high-ranking occupations? And, conversely, are not some Anglo-Americans continually at the bottom of the economic class hierarchy? If members of certain minority ethnic groups achieve upper-class standing or significant power, and similarly, if members of the dominant ethnic group remain at the bottom of the economic and political hierarchies, does this not negate the idea of ethnic dominance and subordination? How, in short, can we account for these inconsistent cases?

INDIVIDUAL ACHIEVEMENT   The link between class and ethnicity in any society is not perfect; obviously, some do achieve significant upward mobility despite the handicap of low ethnic rank. But we cannot generalize about groups as collectivities on the basis of the achievement of an outstanding few. Popular interpretations of the relative success or failure of ethnic groups, however, often rely on such faulty generalizations.

For example, a widespread explanation for the inability of certain ethnic groups to rise collectively in economic and political standing is "self-motivation" (Schuman, 1982). It is assumed that individuals themselves, not social forces, are mainly responsible for their social placement. The poor are poor, in this view, because they lack the motivation to improve themselves, and the wealthy are successful because they have greater incentive. That a few from low-ranking minority groups do actually achieve great social success, and that many from the dominant group do not, serve to "confirm" this oversimplistic explanation.

Individual talent, motivation, and ability are, of course, important factors in social success, but only after the competitive field has been severely thinned out by ascribed characteristics such as class origin, gender, and ethnicity. Those with advantages by birth are assured relatively high achievement regardless of their individual capacities. Many in low-ranking groups may have abilities that are necessary to high social achievement, but unless they can gain access to the proper staging of those abilities through quality education, a good first job, and the nurturing of key social connections, they will go unnoticed. One of the basic features of minority groups, it must be remembered, is the social discrimination their members face in important life chances such as these.

Underrepresentation   No matter how many exceptional cases sneak through the ethnic barriers, minority groups remain underrepresented in the society's top wealth, power, and prestige classes. This underrepresentation will vary in degree from group to group, but the lack of equal access to the society's rewards is precisely what defines all such groups as minorities.

Privileges and Handicaps   In accessing the society's rewards, members of the dominant ethnic group—even those at the bottom of the economic and political hierarchies—often retain privileges not enjoyed by members of minority groups, even those of upper-class standing. For example, before the 1960s in most cities of the American South, black entertainers receiving huge salaries still could not stay at the very hotels in which they were performing. A white person, regardless of income or occupation, would not have experienced such exclusion on the basis of skin color. Today, African Americans with high-status occupations—doctors, lawyers, managers—commonly describe the suspicion their presence arouses in suburban, primarily white, neighborhoods or prestigious shopping areas (Cose, 1993, 2011; Feagin and Sikes, 1994).

Ethnic minority group members are burdened in other ways by their racial or ethnic mark. In situations of interaction with dominant group members, they are often viewed as representatives of their "race" or ethnic group. References made in conversation to "your people" or questions of "how do (black people, Jews, Chinese . . .) feel" about a particular issue assume that the minority person, simply by virtue of the fact that he is a minority person, speaks for an entire group. Such forms of address also imply that the minority person is an outsider, not fully part of the social mainstream.

Because they are easily identifiable, minority group members also may suffer the stress of collective remorse when a member of their group is portrayed in a negative light. Korean Americans (and many Asian Americans in general), for example, spoke of the anxiety that they felt after learning of the shooting of thirty-two students and faculty members at Virginia Tech in 2007, collectively holding their breath, hoping that the shooter was not Korean. Many expressed, almost apologetically, their dismay after learning that the shooter was in fact Korean. The killer was constantly referred to by the media as "Asian," and later, more specifically as "Korean," despite the fact that he had grown up in the United States and was fully assimilated. His ethnicity, of course, had no relation to the crime, but for the media, identifying and making repeated reference to it was pertinent. In contrast, consider the media treatment given Timothy McVeigh, who bombed a federal building in Oklahoma City in 1996, killing 168 people. Never was the fact that he was white and born and raised in the United States ever mentioned; it was simply not a relevant detail. Nor did all white people suffer a sense of collective remorse for McVeigh's heinous crime.

CHANGING CLASS POSITIONS   The empirical link between class and ethnicity is complex and subject to frequent change. The class position of particular groups or large segments of those groups may be altered significantly from one generation to another. For example, most members of white ethnic groups in the United States today are part of the middle class or working class, though most of their forebears entered the society at an appreciably lower level. A large proportion of African Americans have also displayed a substantial change in class standing in the past four decades.

Given such changes and the intricacies of the relationship between class and ethnicity, we should not expect to find all members of a particular ethnic group with the same economic and political rank or the same economic and political concerns and attitudes. There is a range of classes *within* each ethnic group even though, for the most part, group members remain clustered at particular levels of the society's general class system.

THE ORIGINS OF ETHNIC STRATIFICATION

How does ethnic stratification arise? And why does it seem inevitable in multiethnic societies? What are the forces that seem to lead almost always to patterns of dominance and subordination when ethnic groups come into contact with one another? Sociologists have suggested several factors that lead to the emergence of ethnic inequality.

FORMS OF CONTACT

To begin with, all systems of ethnic stratification are products of the contact of previously separated groups (Shibutani and Kwan, 1965). Put simply, the composition of multiethnic societies depends on diverse groups coming together in some manner. This entails the movement of people from one area to another but, more fundamentally, from one political unit to another. People may move great distances and remain within the confines of the same political unit, as when a family in the United States moves from New York to California. Ethnic stratification systems, however, are created by the movement of people across national boundaries, usually bringing with them different languages and cultural systems, or by the establishment of new political boundaries. Multiethnic societies are formed through one or a combination of several contact patterns.

CONQUEST   Conquest is a form of contact in which people of one society subdue all or part of another society and take on the role of the dominant group. European colonialism of the eighteenth and nineteenth centuries best exemplifies this pattern. Through greater military technology, the British, French, Spanish, Dutch, Belgians, and Portuguese conquered peoples in a variety of world areas, including Asia, Africa, Australasia, and North and South America. Indigenous groups were brought under colonial rule and became economic appendages of the mother country.

ANNEXATION   A political occurrence in which a part or possibly all of one society is incorporated into another is annexation. As Burkey explains, "If the incorporating society has a dominant group, then the ethnic groups within the incorporated, or annexed, society become subordinate at the point that sovereignty is transferred" (1978:72). Such annexation may occur in a peaceful or a violent manner. The acquisition of the Louisiana Purchase by the United States from France in 1803 illustrates a peaceful annexation. More commonly, however, such acquisitions of territory are made through successful military ventures. For example, the United States acquired most of what is today its Southwest through its war with Mexico in 1846. After the transfer of territory at the conclusion of the war, Mexicans living in those annexed areas became a minority, subject to the dominance of American political institutions.

VOLUNTARY IMMIGRATION   The most common patterns by which ethnic groups come into contact involve immigration. *The migration of peoples from one society to another* may be either voluntary or involuntary. **Voluntary immigration** has been the chief source of ethnic heterogeneity in the United States, Canada, Australia, New Zealand, and Argentina. All white ethnic groups in these societies have been the product of voluntary immigration from European societies during the nineteenth and twentieth centuries. In the United States and Canada, immigrants from Asia, Latin America, and the Caribbean have, in recent decades, contributed even greater diversity to these societies' ethnically varied populations. In the modern world, Brazil and Israel are other societies that have relied heavily on immigration in creating a diverse population. In addition, many previously ethnically homogeneous societies, especially those of Western Europe, have been infused with new racial and ethnic groups as a result of voluntary immigration since the end of World War II.

INVOLUNTARY IMMIGRATION   **Involuntary immigration** *involves the forced transfer of peoples from one society to another.* Such forced movements are best exemplified by the slave trade of the seventeenth, eighteenth, and nineteenth centuries, which brought millions of blacks from Africa to work the cotton and sugar plantations of the United States, Brazil, and the West Indies.

Patterns of immigration, both voluntary and involuntary, to the United States and other societies are discussed in later chapters.

OUTCOMES OF CONTACT

The way in which diverse ethnic groups initially meet has been shown to be a critical factor in explaining the emergence of ethnic inequality and the specific patterns it subsequently takes.

LIEBERSON'S MODEL   Stanley Lieberson (1961) distinguishes two major types of contact situations: those involving subordination of an indigenous population by a migrant group and those involving subordination of a migrant population by an indigenous racial or ethnic group. The first type, **migrant superordination,** is illustrated by *various colonial conquests in which a technologically and organizationally more powerful migrant group subdues the native population.* The second, **indigenous superordination,** is characteristic of most voluntary and involuntary immigrations such as those to North America; in such cases, *the arriving groups are initially made subordinate to a resident dominant group.*

Lieberson maintains that long-term conflict is more likely in societies where the indigenous population at initial contact is subordinate. Native groups less powerful than the arriving colonials are left with few options other than resistance to the new social order imposed on them. This hostility is further strengthened when the conquering group, over time, becomes itself an indigenous group. Where an invading group is successful in subduing the native population, the political and economic systems of the new group are imposed, and warfare and general conflict are likely to result quickly. van den Berghe (1976) also points out that in contact situations between a conquering group and a weaker native group, there is usually a territorial factor, with the defeated group retaining an indigenous area. Indian reservations in North America are an example. Such a territorial base may provide the foundation for a separatist movement, an option not available to immigrant groups who enter as subordinates and typically disperse to different areas.

Situations in which the native group wields greater power and immigrant groups enter as subordinates produce less overt conflict initially. The indigenous group retains control over the size and character of immigration and may encourage quick assimilation, as in the case of most European immigrants to the United States. Moreover, conflict is diminished by the fact that if the immigration is voluntary, dissatisfied immigrants may return to their society of origin.

NOEL'S MODEL   Although the nature of initial group contact is critical, Donald Noel (1968) has pointed out three additional factors that give rise to and shape the eventual system of ethnic stratification: ethnocentrism, competition for scarce societal resources, and an unequal distribution of power.

When culturally or physically dissimilar groups meet, *ethnocentrism* can be expected to typify intergroup attitudes. That is, divergent groups will judge each other in terms of their own culture and their evaluations will usually be negative. The extent of these negative judgments will, however, depend on the degree of difference between the groups: the more dissimilar they are, the more negative the judgment. Studies measuring the degree of acceptance of members of different ethnic groups in the United States, for example, have consistently shown those groups closest in culture and physical appearance to the dominant Anglo group (such as northwestern Europeans) to be ranked more favorably than southern and eastern Europeans such as Italians and Poles and considerably more favorably than African Americans, Mexican Americans, and American Indians (Bogardus, 1959; Owen et al., 1981).

However, ethnocentrism alone, explains Noel, is not sufficient to produce ethnic stratification. Groups may view one another negatively without the necessary emergence of dominant-subordinate relations among them. An additional prerequisite is *competition,* structured along ethnic lines. When groups strive for the same scarce resources, their interrelations take on the characteristics of competition and conflict. Noel posits that the more intense such competition, the greater the likelihood of the emergence of ethnic stratification. Within the competitive arena, those groups with the greatest capacity to adapt to the social and physical environment will end up higher in the ethnic hierarchy. Thus, in the American case, groups emigrating from Europe arrived with different skills, which initially determined their occupational place and subsequently enabled them to climb upward at different rates. In no case, however, was any European group stationed below blacks, because the latter were kept in a basically noncompetitive situation.

Differential power among the various groups is, according to Noel, the final prerequisite for the development of ethnic stratification. Unless one can overpower another, there is no basis for a stable rank order of ethnic groups, even if there is competition and ethnocentrism among them. When a particularly wide power gap exists between competing and ethnocentric groups, the emergent stratification system is likely to be quite durable. Power breeds more power; and once established, the dominant group uses its power to obstruct the competition of other groups and to solidify its dominance. In the end, then, differential power among the various groups is the most critical requirement for the emergence of ethnic stratification (see also Wilson, 1973).

In sum, Noel's theory postulates that competition for scarce resources provides the motivation for stratification, ethnocentrism channels this competition along ethnic lines, and differential power determines whether one group will be able to subordinate others (Barth and Noel, 1972).

MINORITY RESPONSES TO SUBORDINATION

Not all minority groups react similarly to their subordinate status. Some may begrudgingly accept their place and wait for a more just world in the future, and others may struggle relentlessly to reverse their position. A typology of minority response—pluralistic, assimilationist, secessionist, and militant—originally suggested by Wirth (1945), remains an apt and concise model of different strategies of adaptation employed by minority groups in a multiethnic society.

Pluralistic Minorities

Pluralistic minorities *seek to maintain their cultural ways at the same time as they participate in the society's major political and economic institutions.* Some groups that enter multiethnic societies as voluntary immigrants adopt this position for a time after their arrival and appeal to the dominant group to tolerate their differences.

Some groups may carry the pluralistic idea further, opting out almost completely from the larger cultural, economic, and political systems. Certain religious groups in the United States and Canada such as Hutterites, the Amish, and Hasidic Jews have chosen to segregate themselves even though they have not necessarily been rejected by the dominant group. These groups view contact with the larger society as a kind of contamination and a threat to their cultural integrity. Thus they may not use the public schools or participate in mainstream political processes. Even economic matters may be largely self-contained within these groups (Hostetler, 1993; Kephart and

Zellner, 1994; Poll, 1969). Such groups have achieved an accommodation with the dominant group in their society, permitting them to practice cultural patterns that are clearly aberrant in terms of the mainstream culture. Consider, for example, this description of the Amish:

> At best, the Amish seek accommodation, or a state of equilibrium in which working arrangements can be developed whereby they may maintain their unique group life without conflict; in short, they seek a kind of antagonistic cooperation. The Amish group's aim, then, is merely tolerance for their differences, and they are not at all interested in assimilation, "Americanization," or anything that would tend to merge them with the American culture and society. (Smith, 1958:226)

ASSIMILATIONIST MINORITIES

By contrast with pluralistic minorities, **assimilationist minorities** *seek integration into the dominant society*. Wirth explained that such groups crave "the fullest opportunity for participation in the life of the larger society with a view to uncoerced incorporation in that society" (1945:357–58). Thus, whereas the pluralistic minority will usually insist on endogamy and will enforce adherence to the norms and values of the in-group, assimilationist minorities aim for eventual absorption into the larger society.

Most Euro-American groups have maintained assimilation as their long-range objective, though some have been more desirous than others of retaining their ethnic heritage for several generations. Those groups closely similar to the dominant group in culture and physical appearance will ordinarily assimilate more easily and thoroughly than others. Some may even reach the point of complete absorption into the dominant group. For example, ethnic groups of northwestern European origin (Dutch, Scandinavian, and German) have been so completely assimilated over several generations that today they are virtually indistinguishable as ethnic units. However, such cases are exceptional. Rarely are minority ethnic groups completely absorbed into the dominant group, totally relinquishing their culture or physical identity. Moreover, even where assimilation is the group's objective, the prejudice and discrimination the group encounters will often retard such efforts. In Chapter 4, the process of assimilation and the factors that promote it are discussed at greater length.

SECESSIONIST MINORITIES

Secessionist minorities *desire neither assimilation nor cultural autonomy*. Their aim is a more complete political independence from the dominant society. Such groups are usually what were earlier referred to as "nations," not only aspiring to some degree of political autonomy but also maintaining territorial integrity. The separatist movement in the Canadian province of Quebec represents a contemporary secessionist minority. In this case, a substantial element of the Quebec populace seeks not only to retain the French language and French Canadian culture but also to establish an independent Quebec state. A similar movement in the Basque provinces of Spain seeks political autonomy for the Basque people, distinguished as they are from the rest of Spain by language and culture (Douglas et al., 1999; Kurlansky, 1999; Molina, 2010). As we will see in Chapter 17, most nationalist movements are the outgrowths of dissatisfaction on the part of ethnic groups stemming from their minority status and treatment.

In addition to seeking separation from the larger society, a secessionist minority may desire integration with another group or society to which it feels a closer cultural and political similarity. For example, many among Northern Ireland's minority Catholics seek to unite the six counties of Ulster (Northern Ireland) with the Catholic-dominated Irish Republic.

MILITANT MINORITIES

Militant minorities *seek as their ultimate goal not withdrawal, as do secessionist minorities, but rather status as the society's dominant group.* The case of Latvia, a small Baltic country, illustrates this. During the period when Latvia was ruled by the Soviet Union, natives were in some ways reduced to minority status, particularly regarding language and culture, which favored ethnic Russians. With the breakup of the USSR, resulting in Latvian independence, a shift in political power to the native Latvian group occurred. As a result, Latvians not only replaced the Russian political elite but also imposed language and citizenship measures that made non-Latvians, including ethnic Russians, minorities (Bilefsky, 2006; Chinn and Kaiser, 1996; Jubulis, 2001).

It is important to consider that these variable responses of minority groups are not totally or, in some cases, even largely voluntary. Rather, they very much depend on the power of the dominant group to accept or reject minority group aims (Schermerhorn, 1970). It is not only a question of what a minority group initially desires but also what it desires in conjunction with what the dominant group desires for it. A minority group may seek assimilation, for example, but be repelled by the dominant group's aim to keep it isolated. Minority objectives may also change depending on the dominant group's responses. Continual frustration of assimilationist goals, for example, is likely to eventually create a pluralistic or secessionist response.

SUMMARY

- Social stratification is a system of structured social inequality in which groups receive different amounts of the society's wealth, power, and prestige and are hierarchically arranged accordingly.
- In multiethnic societies, ethnicity is a critical basis of stratification. There is, therefore, a hierarchy of ethnic groups in which one, the dominant group, maintains maximum power to determine the nature of interethnic relations. Minority, or subordinate, groups rank in different places below the dominant group, depending on their cultural and physical distance from it.
- Dominant-minority relations are relations of *power*. On the basis of their physical and cultural characteristics, ethnic minorities are distinguished from others and given differential treatment. The dominant group holds disproportionate control of the society's political and economic resources and the ability to shape the society's major norms and values.
- Ethnic rank commonly closely parallels rank in the society's class system. Members of minority ethnic groups are never randomly scattered throughout the economic and political class hierarchies but tend to cluster at specific points.

- Ethnic stratification is the product of contact between previously separate groups. Initial contact may be in the form of conquest, annexation, voluntary immigration, or involuntary immigration. The manner in which ethnic groups meet is a decisive factor in explaining the shape of the system of ethnic inequality that ordinarily ensues. Following contact, groups engage in competition and view one another ethnocentrically; then, ultimately, one imposes its superior power over the others, emerging as the dominant group.
- There are four types of ethnic minorities: *pluralistic* minorities, seeking to maintain some degree of separation from the larger society; *assimilationist* minorities, aiming for full integration into the dominant society; *secessionist* minorities, seeking political autonomy from the dominant society; and *militant* minorities, trying to establish dominance themselves.

CRITICAL THINKING

1. Consider the sociological meaning of **minority**. Then identify several social categories, aside from race and ethnicity, that fit that definition in the United States today. On what bases do they conform to the definition?
2. We saw in this chapter that minority status is flexible, not static. Looking at the development of the American ethnic hierarchy, it is very evident that numerous groups who entered the United States and took their place as minorities are no longer seen or defined that way. What happened to change their status?
3. Social attitudes in the United States toward new immigrants from Europe and Canada are quite different from social attitudes toward immigrants from Latin America and the Caribbean or from Asia or Africa. What accounts for the difference? What factors determine the rank of a new group entering the society?
4. Consider the four major types of ethnic minority discussed in this chapter. Are there cases of all four in the United States today or in past eras? Give some examples of each.

PERSONAL/PRACTICAL APPLICATION

1. Suppose you are part of the society's dominant ethnic group. To what extent are you able to empathize with those who are members of ethnic minorities? Conversely, if you are part of an ethnic minority, can you empathize in any way with those who are members of the dominant group?
2. Stratification, as this chapter explains, is a ranking system and can be based on various social characteristics. Does your socioeconomic rank (your social class) match your ethnic rank?

TOOLS OF DOMINANCE

CHAPTER **3**

Prejudice and Discrimination

We have now seen that in most multiethnic societies, ethnic groups are arranged in a hierarchy in which the dominant group receives a disproportionate share of the society's rewards because of its greater political, economic, and cultural power. It is not enough, however, to simply proclaim that the dominant group is more powerful, though this is, of course, the crux of the matter. We must also look at the techniques by which dominance and subordination are maintained and stabilized in systems of ethnic stratification.

To enforce its power and sustain its privileges, the dominant ethnic group employs certain tools:

- Widely held beliefs and values regarding the character and capacities of particular groups, which take the form of **prejudices**—that is, *negative ideas regarding subordinate ethnic groups and ideas expressing the superiority of the dominant group*. These beliefs often come together in a cohesive ideology of racism or another deterministic notion; but at other times, they are applied to groups in a somewhat disparate, unsystematic fashion.

- *Actions against minority ethnic groups, including avoidance, denial, intimidation, or physical attack*. These actions are termed **discrimination**. Different forms of discrimination may be applied, depending on how threatening the minority group is perceived to be.

In this chapter, we look at some theories and research findings regarding prejudice and discrimination in multiethnic societies. This area of ethnic relations has been the focus of much research, yielding a large amount of empirical data. In looking at prejudice and discrimination, we will necessarily investigate some of the social psychology of ethnic relations, that is, ethnic relations at the individual, or interpersonal, level. Up to this point we have been concerned primarily with ethnic relations at the group, or structural, level, and our approach to prejudice and discrimination will continue to emphasize the group dynamics and consequences of these social phenomena. But here, more than in other aspects of ethnic studies, we must pay closer attention to individual attitudes and motives, recognizing that, as the sociologist C. Wright Mills put it, "Neither the life of an individual nor the history of a society can be understood without understanding both" (1956:3).

PREJUDICE

Most simply, prejudice involves a judgment "based on a fixed mental image of some group or class of people and applied to all individuals of that class without being tested against reality" (Mason, 1970:52). It is, in other words, a generalized belief, usually unfavorable and rigid, applied to all members of a particular group. Although often defined as a prejudgment or preconcept founded on inadequate evidence (Klineberg, 1968), prejudice is, as Berry and Tischler have pointed out, "more emotion, feeling, and bias than it is judgment" (1978:235).

Ethnic prejudices are characterized by several specific features.

- They are *categorical*, or generalized, thoughts. Individuals are judged on the basis of their group membership, not their personal attributes. A prejudicial attitude may be directed at particular individuals, but it is those people's group and its alleged traits that evoke this attitude rather than their individual actions and qualities. Once their group is known, their behavioral traits are inferred. Thus prejudice violates "the norms of rationality" (Pettigrew, 1980: 821).

- They are *inflexible*. As the social psychologist Gordon Allport explains, "Prejudgments become prejudices only if they are not reversible when exposed to new knowledge" (1979 [1958]:9). A prejudice is not simply an error in thought but is an error not subject to correction. Individuals develop emotional attachments to certain beliefs and will not discard them in the light of contrary evidence. People may proclaim that "some of their best friends" are members of a particular ethnic group that they generally view adversely. The implication is that such people do not exhibit the negative qualities ordinarily attributed to members of their ethnic group. But instead of refuting the belief, which logic would dictate, they serve only as "exceptions that prove the rule." Such contrary evidence is recognized but excluded from the generalization; it thus has no correcting effect (Allport, 1979 [1958]).

- They are usually *negative* in content. That is, the specific traits ascribed to targeted groups are considered inferior and socially undesirable. Of course, prejudices may be positive as well as negative. Ethnic group members maintain an overly favorable image of their own group, just as they maintain overly unfavorable images of certain out-groups. Indeed, all ethnic groups express ethnocentric notions regarding their unique character. Moreover, what is often interpreted as prejudice against out-groups may in fact be more a matter of ingroup bias—that is, favoritism toward members of one's own group (Brewer, 1979; Jones, 1996). In studying ethnic relations, however, sociologists and psychologists have been concerned almost exclusively with negative prejudices.

- They are based on erroneous or inadequate group images called *stereotypes*. Because they compose the chief content of ethnic prejudices, let's look more closely at these generalized group images.

STEREOTYPES

Stereotypes were first suggested in 1922 by the political journalist Walter Lippmann, who described them as "pictures in our heads" that we do not acquire through personal experience. In the case of ethnic stereotypes, distinctive behavioral traits of an ethnic group are selected by out-group members who exaggerate them to construct what Shibutani and Kwan call "a shorthand depiction" of the group (1965:86).

Rarely will people claim to dislike people of another ethnic group merely because they are members of that group. Instead, the adverse view will be couched in "rational" terms. One dislikes Jews because they are "shrewd," blacks because they are "lazy," or Italians because they are "loud and uncouth." These mental images of groups thus serve as supports for the negative beliefs that constitute prejudice. Once we learn the stereotypes attached to particular groups, we tend to subsequently perceive individual members according to those generalized images. To stereotype someone, then, "is to attribute to that person some characteristics which are seen to be shared by all or most of his or her fellow group members" (Brown, 1995:82).

RATIONAL STEREOTYPES   Clearly, however, generalizing about groups or objects is a most common pattern of thought. Indeed, it is the very mental technique that facilitates social interaction, particularly in modern, complex societies where we cannot possibly know all the personal characteristics of those whom we encounter daily. Thus, on the basis of some identifying marks such as ethnicity, sex, age, or occupation, we generalize and make judgments about people. In a sense, generalizing on the basis of group membership is a kind of predictive mechanism we use in various social situations.

Consider an example familiar to college students. On the first day of class, students meet Professor X. Professor X is known only as part of the group called professors, about which there are certain general ideas. Professors ordinarily lecture, give examinations, grade students, and so on. Armed with this understanding of the category *professors,* students naturally expect Professor X to behave accordingly. The chances are very great that Professor X will conform to their expectations. In the same way, of course, Professor X may meet his or her students for the first time knowing nothing more about them but that they are students. With this bit of knowledge, however, Professor X can "predict" the actions of the students, using the same technique of categorizing they have used. In this situation, people are expected to perform the roles of professor and student according to the social "script" attached to each.

In most cases, our expectations or predictions about how people will act in different social situations prove correct. As a result, we can interact with people about whom we know nothing more than their social identifications, that is, their group memberships. Individuals occasionally fail to perform in the manner expected, and our predictions are then proved incorrect. But these are not the usual cases. For the most part, we understand correctly what is expected of people in various social roles—professor, student, doctor, father, neighbor, and so on—and what our behavior toward them should be. This is the nature of rational thought and behavior. How, then, do ethnic stereotypes differ from these common forms of generalized or predictive thought?

ETHNIC STEREOTYPES   Ethnic stereotypes differ from rational generalizations in that they are oversimplistic and overexaggerated beliefs about a group, generally acquired secondhand and resistant to change (Harding, 1968; Pettigrew, 1980). Although essential to social interaction, categorical thinking—in this case, thinking in terms of *ethnic* categories—often leads to mental simplification, causing people to overlook individual differences among group members (Yzerbyt and Corneille, 2005). Thus ethnic stereotypes are sustained despite numerous individual cases that clearly refute their validity.

The characteristics attributed to various ethnic groups are established in popular beliefs and become "part of our shared understanding of who 'we' and 'they' are" (Wright and Taylor, 2003:434). That is, they are socially shared representations of groups and they structure and perpetuate intergroup relations. Because stereotypes are *group* depictions, individuals to whom the stereotype is applied are seen not as individuals per se but as representatives of the group of which they are a part.

The content of ethnic stereotypes may change periodically and sometimes radically, depending on different economic, political, and social circumstances. Whatever their content, however, they are conveyed in subtle but effective ways through various socialization agencies, including the family, the school, and the mass media. Thus stereotypes cannot be seen as irrational pictures of ethnic groups held by a numerically insignificant part of the general population. Rather, they are part of the society's heritage, and as Ehrlich (1973) explains, no person can grow up in a society without having learned them.

SELECTIVE PERCEPTION   Stereotypes are reinforced through selective perception. This means that people take note of those cases that confirm their stereotypical pictures and overlook or ignore those that refute them. Those who believe blacks to be lazy or Jews to be deceitful will take special notice of those blacks or Jews who do in fact exhibit such traits but will fail to notice the many more blacks and Jews who do not. Moreover, these stereotypical traits will be inferred, even if they are not evident, so that the observers of an ethnic group will interpret the actions of group members based on their preconceived image.

In addition, stereotypes fail to show how members of the dominant group may share the same negative traits imputed to minority groups, or how the dominant group, through the self-fulfilling prophecy, may contribute to the very creation of these unfavorable traits (Simpson and Yinger, 1985).

UNIFORMITY AND PERSISTENCE OF STEREOTYPES   The pioneer study of ethnic stereotyping in the United States was conducted in 1933 by Katz and Braly, who questioned one hundred Princeton undergraduates on the prevailing stereotypes of various ethnic groups. Their findings indicated a very high degree of uniformity, in some cases as high as 75 percent. Jews, for example, were consistently described as "shrewd," "mercenary," and "ambitious"; blacks as "superstitious," "happy-go-lucky," and "lazy"; and Germans as "scientifically minded," "industrious," and "stolid." Katz and Braly noted that the students had had little or no contact with members of most of the groups they described, an indication that they had acquired the stereotypes by absorbing the dominant culture.

Gilbert replicated this study in 1951 and found a prevalence of many of the same stereotypes. However, he also detected a marked change in attitude. Many students now expressed reluctance to categorize people whom they did not know. A third generation of Princeton students was questioned in 1967, and the researchers in that case also encountered irritation and resentment among those asked to generalize about ethnic groups. Furthermore, they found that the actual content of the stereotypes had now changed considerably, with some groups such as blacks and Jews being assigned more favorable traits than had been the case in 1933 (Karlins et al., 1969). But the uniformity of the application of stereotypes had not changed. In other words, some traditional stereotypes had declined in frequency, but they had been replaced by others, some more favorable than previously. The researchers concluded that although the students in 1967 appeared to be more tolerant and less receptive to stereotyping, the application of ethnic stereotypes was still evident. More recent research revisited the Princeton studies and found that group stereotypes were still quite evident but that the content of those stereotypes had continued to change, with most becoming more favorable (Devine and Elliott, 1995; Madon et al., 2001).

Devine and Elliot (1995), moreover, suggested that the three Princeton studies had all suffered from several methodological shortcomings. First, they did not distinguish between stereotypes and personal beliefs. Knowing an ethnic stereotype does not necessarily mean that one *believes* that these traits actually characterize members of the group. Also, the Princeton studies made no assessment of the respondents' level of prejudice, nor did they employ stereotypical terms that are more prevalent today than they were in 1933, when the study was first conducted. Devine and Elliot sought to correct these shortcomings, although they measured changes in the stereotypes of only blacks, not the full range of groups dealt with in the Princeton studies. They found that there was a clear and consistent negative stereotype of blacks, but that personal beliefs about blacks were equally consistently positive. They concluded that the earlier Princeton studies were actually measuring personal beliefs, not stereotypes, which may or may not be congruent. Although the stereotype is well understood by all, only highly prejudiced people actually endorse it.

An important question posed by Devine and Elliot's study is why stereotypes persist if they don't represent what a great many people believe about a group. Perhaps most important is their durability as part of the society's cultural fabric. Social norms regarding overt expressions of prejudice have changed considerably, yet ethnic stereotypes persist in more subtle and covert, yet effectual, ways.

Stereotype Creation and Change   A noteworthy international study was conducted by Buchanan and Cantril (1953) in the early 1950s that demonstrated how ethnic stereotypes can rise and fall, depending on social and political circumstances. Respondents from eight countries were asked to describe people of other countries by choosing from a list of descriptive adjectives. (This was similar to the technique used by Katz and Braly and others.) The general findings of this study were that people in all eight countries displayed a tendency to use stereotypes in describing other national groups, that their own compatriots were always described in flattering terms, and that the choice of either complimentary or derogatory adjectives depended largely on the current state of relations between the nations.

Based on their evidence, Buchanan and Cantril concluded that stereotypes, rather than preceding people's reaction to a certain group, ordinarily do not exist until objective events demand their creation. Thus they stressed that stereotypes should be thought of not as causative but as symptomatic. As they put it, "Perhaps their important function is the wartime one of providing a rationale within which men are able to kill, deceive, and perform other acts not sanctioned by the usual moral code" (1953:57).

A clear illustration of the creation of stereotypes as a means of rationalizing events is the dramatic change after World War II in the stereotypes of Japanese and Germans held by Americans. During the war, negative images of these two groups— as evil, hostile, and cruel—prevailed; but by the 1960s the groups were seen as clean, efficient, and industrious. The image of Russians was also altered, but in the opposite direction. As World War II allies, the Russians were portrayed in generally positive terms, but this image changed in the late 1940s with the onset of the Cold War and the emergence of the Soviet Union as the United States' chief ideological foe.

THE COMPETITIVE USE OF STEREOTYPES   In the same way that wartime stereotypes are used to rationalize hostility toward enemy nations, stereotypes are used by dominant groups in multiethnic societies to sustain their competitive advantage over challenging or threatening ethnic groups. The negative images of blacks long held by whites in the United States or South Africa have their counterparts in Northern Ireland, where Protestants, the dominant ethnic group, have traditionally held adverse images of Catholics. In these cases, the persistence of negative group images can be explained as a rationale on the part of the dominant group for keeping the minority ethnic group or groups in a subordinate position. As long as groups are perceived as undeserving, their social disadvantages can be justified (Devine and Sherman, 1992).

The competitive use of stereotypes may explain their very content. Simpson and Yinger note, for example, that those groups that have successfully competed with the dominant group cannot be labeled lazy or unintelligent, "so they are pictured as too ambitious, and with a crafty kind of self-interested intelligence" (1985:155). This can be seen clearly in the case of Jews in the United States and other societies where they have exhibited an exceptional ability to achieve economic success. Allport (1979 [1958]) compares the admirable traits of Abraham Lincoln with the disliked traits of Jews and finds them quite the same. Both are generally described as thrifty, hardworking, eager for knowledge, ambitious, devoted to the rights of the average man, and eminently successful in climbing the ladder of opportunity. The key difference, explains Allport, is that the terms used to describe the Jews are often disparaging. Thus thrifty becomes "tight-fisted," hardworking becomes "overambitious," ambitious becomes "pushy," and concerned about human rights becomes "radical" (1958:184). Much the same semantic reversals of positive traits have been used in describing Chinese and East Indians in various Asian and African societies where they constitute economically successful minority groups (Chua, 2004; Hunt and Walker, 1974; Kristof, 1998).

A related notion is "ambivalent stereotypes" (Fiske et al., 1999). The idea is that one can envy and respect high-status groups for their competence but dislike them nonetheless. Asian Americans, as we will see in Chapter 9, are admired as a "model minority" for their economic and intellectual accomplishments; at the same time they are often the victims of subtle discrimination. By the same token, one can disrespect low-status groups for their incompetence but still like them.

THE MASS MEDIA AND STEREOTYPES   With their pervasiveness and enormous impact in modern societies, the mass media—television and motion pictures, in particular—are key conveyors of ethnic stereotypes. For example, until the 1980s, blacks in American films traditionally played subservient characters and rarely were given starring roles except in all-black productions (Brown, 1981; Sterngold, 1998a). Other racial and ethnic minorities have fared little better. American Indians have been portrayed customarily as savages and Hispanics as untrustworthy villains (Coward, 1999; Engelhardt, 1975; Moore and Pachon, 1985). The early years of American television, the 1950s and 1960s, were marked by an almost total exclusion of minorities from the screen except for stereotyped roles. The absence of blacks was due mainly to the fear held by networks and program sponsors of offending white viewers, especially in the South (Sterngold, 1998a; U.S. Commission on Civil Rights, 1977).

In the last four decades, the presence of ethnic minorities in sitcoms, news presentations, and other programming has increased enormously, reflecting proportionately their population segments (Dougherty, 2003; Greenberg and Brand, 1998; Wilson and Gutiérrez, 1995). Nonetheless, the mass media continue to sustain ethnic stereotypes. Media researcher Robert Entman has demonstrated how television news upholds the common stereotype of African Americans, for example, as "poor." The linkage of "black" and "poor" is so strong in the presentation of televised news, explains Entman, that "the white public's perceptions of poverty appear difficult to disentangle from their thinking about African Americans" (1997:33). Moreover, as political scientist Martin Gilens explains, the undeserving character of welfare recipients, in the minds of many white Americans, has come to be associated specifically with African Americans. Historically well-worn stereotypes of blacks—as lazy and irresponsible—were first projected by the mass media in the mid-1960s when many welfare programs were created or expanded. These negative images continue to influence the view of white Americans of the typical beneficiary of welfare (Gilens, 1999). The stereotypical linkage of crime and violence with African Americans is similarly bolstered by television news, especially at the local level.

The mass media, however, may also convey exaggerated positive images. De Roche and de Roche (1991) studied episodes of five police series popular in the 1980s, comparing characterizations of black and white officers. They found that black men in these dramas had been "counterstereotyped." That is, not only were these men not described with negative ghetto stereotypes but they were portrayed in overly favorable terms in comparison with white officers. "In the collection of characters on which we focused, black men are modeled for us as more bourgeois than their white counterparts, more self-directed and affectively well-managed, and more tasteful. . . . They clearly confirm the traditional ideal of a stable, financially responsible husband/father, and this in an era when female-headed, single-parent families are becoming increasingly common in all social sectors" (de Roche and de Roche, 1991:86).[1]

---

[1] The researchers found, however, that blacks in these police dramas tended to remain support characters, not as qualified as whites to head up their organizations.

Although most observers agree that the electronic media are important transmitters of ethnic stereotypes and attitudes, both positive and negative, studies have indicated that their effect may be only to reinforce ideas already acquired before exposure (Sarlin and Tate, 1976; Vidmar and Rokeach, 1974). That is, television may only bolster ethnic stereotypes and attitudes already conveyed in other socialization settings. The electronic media may be unarguably pervasive, but their effectiveness in conveying the same messages to all viewers is by no means certain. Moreover, the extent and nature of television's effect on how children of different racial and ethnic groups interact with each other, as well as how it affects their attitudes toward and knowledge of different groups, are still subject to much debate (Graves, 1999).

SOCIAL DISTANCE

Prejudice involves not simply mental perceptions of ethnic groups (the cognitive dimension of prejudice) but also emotions and a preparedness to act in a certain way toward members of those groups (the affective and conative dimensions). If people believe blacks to be lazy and shiftless, for example, they will also probably feel resentment over welfare payments to blacks, who it is felt obviously do not deserve such assistance. Similar payments to poor whites, however, may go unnoticed or may be seen as merited.

The affective and conative dimensions of prejudice are reflections of **social distance**. Park (1924) first defined social distance as a *degree of intimacy people are prepared to establish in their relations with others*. Feelings of social distance are, according to Williams, "feelings of unwillingness among members of a group to accept or approve a given degree of intimacy in interaction with a member of an outgroup" (1964:29). It is, in a sense, an indication of how acceptable or objectionable are various ethnic groups.

THE BOGARDUS SCALE   In 1925, the sociologist Emory Bogardus constructed a technique to measure social distance between specific American ethnic groups. This social distance scale has continued to be used by sociologists as a general measure of ethnic prejudice. Respondents are asked to indicate whether they would accept a member of an ethnic out-group in varying social contexts, extending from very close encounters to very remote ones. Bogardus asked his respondents to indicate their willingness to interact with members of particular groups in the following situations:

- Close kin by marriage
- Fellow club members
- Neighbors
- Workers in my occupation
- Citizens of my country
- Visitors to my country
- People to be excluded from my country

Each situation represents a lower degree of social intimacy, and assigning an increasing numerical value to each allows a score to be computed for each ethnic group. American social distance scales compiled by Bogardus and others over several generations reveal the relative consistency of the group ranking from year to year. WASPs are at the top of the scale, followed by other northwestern European groups and, in descending order, southern and eastern Europeans, Jews, and various racial-ethnic groups (Bogardus, 1925a, 1925b, 1959; Owen et al., 1981).

Studies of social distance in non-American societies indicate the similar construction of a hierarchy (Bogardus, 1968; Lever, 1968; Pettigrew, 1960), but the basis of social distance may vary from one society to another. Whereas ethnic differences, particularly those with a physical basis, are the most significant criteria of social distance for Americans, in other societies religion, class, or political ideology may be more important factors in separating people.

MEASURING SOCIAL DISTANCE AND PREJUDICIAL ATTITUDES   Feelings of social distance and ethnic prejudicial attitudes in general are difficult to accurately determine, for in real situations individuals do not always think or act as their verbal expressions seem to indicate. Attitudes are elusive and not always subject to clear-cut measurement, no matter how sophisticated the technique. What people tell pollsters, for example, may reflect only what they think are socially acceptable responses, not necessarily their real beliefs and feelings. That is, the pressure to conform to community and other group standards may force individuals to express attitudes that are not genuine. Moreover, the way in which poll questions are worded on racial and ethnic issues may strongly influence the results (Langer, 1989).

A pioneer study by Richard LaPiere (1934) demonstrated the unreliability of people's statements regarding ethnic prejudice. Between 1930 and 1932, LaPiere traveled throughout the United States with a Chinese couple and was refused hotel and dining service only one time. Shortly after his travels, he sent a questionnaire to hotels and restaurants asking whether they would accept Chinese as guests. To his surprise, LaPiere received mainly negative replies from those who responded. A similar finding was obtained in the early 1950s using a black couple as the test case (Kutner et al., 1952).

In sum, prejudice is a complex social-psychological phenomenon that is not always easy to identify and analyze. The negative attitudes people maintain toward various ethnic groups may vary not only in intensity but also in consistency. Moreover, social psychologists have shown that prejudice is multifaceted, encompassing different emotions—anger, fear, disgust, and so on—that may be triggered by the actions of different groups under different circumstances (Cottrell and Neuberg, 2005). And, as we will see, changing situational contexts may force people to alter or disguise their attitudes accordingly.

DISCRIMINATION

Whereas prejudice is the attitudinal element in enforcing ethnic stratification, discrimination is the active, or behavioral, element. **Discrimination**, most basically, is *behavior aimed at denying members of particular ethnic groups equal access to societal rewards.* Thus it goes well beyond merely thinking unfavorably about members of certain groups. Dominant groups carry out actions or enact and enforce legal or customary measures that in some way ensure that minority group members are treated differently and adversely. Recall the most basic aspect of minority status: differential and unequal treatment. Such treatment can involve a wide range of actions and measures in various realms of social life.

Although there are links between the two, prejudice and discrimination must be dealt with as distinct phenomena. There is ordinarily a tendency for prejudicial attitudes to accompany discriminatory behavior; but as we will see, one may occur without the other. Moreover, although the two may be causally related in some instances, in others there need be no cause-effect relationship. In any case, prejudice and discrimination are most frequently mutually reinforcing.

It is important to stress that, like prejudice, discrimination is applied on the basis of group membership, not individual characteristics. Ethnic discrimination, therefore, involves negative actions against individuals that have nothing to do with those individuals' personal status, behavior, or attributes, only their identification with an out-group.

THE RANGE OF DISCRIMINATORY BEHAVIOR

The discriminatory actions that create disadvantages for minority group members vary widely in both form and degree. These variations may be seen on a scale, or continuum (Figure 3.1). The use of derogatory labels in referring to members of ethnic groups ("kike," "wop," "nigger") or phrases with pejorative ethnic references ("to Jew down," "to gyp," "Indian giver") is a relatively minor form of ethnic discrimination; in many cases the language may not even be understood by the user as disparaging. Nonetheless, such terms and phrases contribute to the perpetuation of ethnic stereotypes and render psychological damage of some nature to those who are their targets (Mullen and Leader, 2005).

More serious forms of discrimination with much greater injury to minorities involve the denial of access to various life chances such as jobs, housing, health care, education, justice, and political participation. Minority ethnic groups are placed in disadvantageous positions with regard to these societal rewards and end up receiving less than they would, absent ethnic barriers.

The most severe forms of discrimination involve acts of aggression against ethnic minorities, ranging from isolated incidents of violence to *the attempt to exterminate an entire group*, **genocide**. Examples of the full range abound in the modern world. In the United States, attacks on ethnic minorities have a long tradition, occurring as both selective actions undertaken by individuals or communities (such as lynchings,

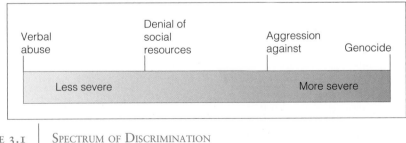

FIGURE 3.1 | SPECTRUM OF DISCRIMINATION

beatings, and bombings) and calculated public policies (such as American Indian removal in the nineteenth century and Japanese American internment in the 1940s).
But the United States holds no monopoly on ethnic aggression. Comparable actions
typify the history, current and past, of most multiethnic societies. Indeed, the nineteenth and twentieth centuries witnessed extremes of ethnic violence and destruction
on a massive scale in a variety of world areas including, among many others, the annihilation of native peoples in Australia, South Africa, and North America; the
slaughter of perhaps a million Armenians by Turks;[2] the systematic murder by the
Nazis of 6 million Jews;[3] and more recently, the massacre of entire populations in
Rwanda and Bosnia.

TYPES AND LEVELS OF DISCRIMINATION

Discrimination is not always overt, nor does it always entail intentional actions of
denial or aggression. Furthermore, there is a vast difference between isolated actions
of individuals and the rational policies of institutions in creating and sustaining patterns of discrimination. The behavior of one café owner in refusing to serve members
of a particular ethnic group is hardly the equivalent of a state policy requiring separate eating facilities for members of that group. The different social contexts in which
discrimination may occur and the often concealed and unintentional forms it may
take require a more precise outline of discrimination.

To simplify matters, we can classify three general types of discrimination: micro-,
macro-, and structural.

MICRO-DISCRIMINATION   *Actions taken by individuals or groups of limited size to injure
or deny members of minority ethnic groups* are perhaps the most easily understood
form of discrimination. This is **micro-, or individual discrimination**. The employment
manager who refuses to hire Asians, the judge who metes out unusually harsh sen-
tences to blacks, and the home owners' group that agrees not to sell houses in the
neighborhood to Jews are examples of discriminators at this level. In these cases, actions are taken by one or a few with the intent to harm or restrict in some way members of a minority group. Notice that the actors are not part of a large-scale
organization or institution but operate within a relatively bounded context.

In cases of **micro-, or individual, discrimination**, the actions taken against minority group members are intentional. Moreover, they appear to be the implementation
of prejudicial attitudes. At first glance we might assume that in the aforementioned
cases the employment manager thinks unfavorably of Asians, the judge of blacks,
and the home owners of Jews. This may in fact be the motivating force behind the
discrimination in all these cases, but we cannot be certain until we understand more
fully the context in which these actions occur. The employment manager, for

---

[2] The Armenian genocide in 1915 is a strongly contested historical event. Turkey denies having carried
out the deliberate killings, maintaining that Armenian deaths—far smaller in number than a million, it is
claimed—were the result of the war that raged in the Ottoman Empire (of which modern Turkey and
Armenia were then a part) at that time.

[3] The literature on the Nazi Holocaust is enormous, but two of the more comprehensive works are
Dawidowicz (1986) and Hilberg (2003). On theories and policies of genocide, see Kuper (1981) and Fein
(1993).

example, may have no ill feeling toward Asians but may feel compelled to carry out what he perceives to be the unwritten yet generally understood company policy of not hiring Asians. The judge may feel that sentencing blacks more harshly will gain her votes among her predominantly white constituency in the next election. And the members of the home owners' group may simply be responding to what they feel are neighborhood pressures to conform. Thus prejudice need not be at the root of even such blatant instances of intentional discrimination. In any of these instances, of course, whether or not the actors' beliefs and attitudes are consistent with their actual behavior does not negate the detrimental effect on those who are the victims of the discriminatory actions: the Asian is still not hired; the black still serves a longer sentence; and the Jew is still denied a home.

MACRO-DISCRIMINATION The preceding cases pertain to the actions of individuals and small groups usually acting in violation of the society's norms. Discrimination, however, may be legal or customary, in which case it is not socially unexpected or disapproved but is legitimized. **Macro-**, or **institutional, discrimination** is *not limited to specific cases of negative actions taken against members of particular groups but is firmly incorporated in the society's normative system.* Social conventions exist in which members of particular groups are legally or customarily denied equal access to various life chances. Discrimination at this level is fundamental to the way a society's institutions work.

In the United States before the 1960s, a well-institutionalized system of discrimination served to effectively block the access of blacks to the same economic, political, and social opportunities afforded whites. In the South, an elaborate system of custom and law legitimized segregated and unequal schools, housing, transportation, and public facilities; kept blacks in low-paying and less desirable jobs; and essentially prevented them from participating in the political process either as voters or as officials. Most of this system was formally established and maintained by specific discriminatory laws, but much of it was also based on the development of customary practices.

South Africa is another case in which a formal and entrenched system of racial segregation and discrimination traditionally served to ensure the power, wealth, and prestige of one group—whites—at the expense of another—nonwhites. As we will see in Chapter 14, this system, called apartheid, has been officially renounced, but many of its essential features are still supported by customs that validate and enforce discriminatory policies and practices.

STRUCTURAL DISCRIMINATION The institutional discrimination that was such a fundamental part of apartheid South Africa or the antebellum U.S. South is today uncommon. In most contemporary multiethnic societies, discrimination is more subtle, less obvious, and more indirect in application. It takes the form of **structural discrimination**, which, unlike micro- and macro-forms, is unintended. It cannot be attributed to the prejudicial beliefs or conforming pressures of individuals, or to the deliberate establishment of a set of rules seeking to withhold privileges or injure members of particular ethnic groups. Rather, *it exists as a product of the normal functioning of the* *society's institutions*. Because of past discrimination of an overt, intentional nature, or because of the spillover effect of direct discrimination in one institutional area into another, certain groups find themselves perpetually at a disadvantage in the society's opportunity and reward structures.

The structural form of discrimination is difficult to observe because ethnicity is not used as the subordinating mechanism; instead, other devices are used that are only indirectly related to ethnicity. This can be illustrated with a few examples.

In recent decades, most new jobs in the United States have been created not in central cities, where they had been concentrated in the past, but in outlying and suburban areas, where shopping centers and offices are built on large expanses of land, where transportation lines, particularly highways, are more accessible, and where taxes are lower. The outlying location of these jobs, however, handicaps African Americans who might qualify for them but who reside disproportionately in central cities. Qualified African Americans are therefore less likely to secure these jobs but for reasons that do not necessarily involve direct discriminatory practices by employers. Businesses choose to locate in the suburbs not because they wish to avoid hiring African Americans but simply because it is in their economic interests to do so. When combined with the more overt discriminatory practices in housing that create the concentration of African Americans in central cities, the outlying location of commerce and industry has the effect of discriminating against African Americans. In other words, the more overt discrimination in one institutional area—housing—has created an indirect discriminatory effect in another—employment.

This indirect form of discrimination can also be seen in the area of education. Because ethnic minority students commonly attend poorer-quality inner-city schools, they are automatically placed at a disadvantage in qualifying for well-paying and promising jobs. Similarly, entrance into top-ranked colleges and universities will be more difficult for them because they will not be adequately prepared to meet the rigid academic requirements. In both cases, there is no necessary intention to discriminate. Decisions by employment managers or college admissions officers may be quite rational, made not on the basis of applicants' ethnicity but on objective employment and academic standards. What is overlooked is that the lower qualifications of minority students are the result of attendance at inner-city public schools that have inadequately prepared them for better jobs and colleges.

Such unintentional, yet effective, ethnic discrimination is repeated in a variety of areas. Bankers, for example, who hesitate to lend to minority entrepreneurs because they fear they are poor credit risks may simply be engaging in practical business tactics. Or the grocery chain that charges higher prices in its inner-city stores than in its suburban stores will explain that such practices are necessary to offset higher operating costs in the inner-city, due to higher rents or insurance rates. Similar patterns emerge in medical services, administration of justice, and other key areas of social life.

Perhaps most significant is that structural discriminatory practices are not only unintentional but largely unconscious. As Harold Baron notes, "The individual only has to conform to the operating norms of the organization and the institution will do the discriminating for him" (1969:143). Discrimination, in other words, does not depend in these cases on the actions of specific individuals or even organizations.

Instead, it is simply a function of the standard working procedures of societal institutions.

Because it is unintentional and largely unwitting, structural discrimination remains difficult to detect; and even when it does become apparent, it is not easy to determine who is ultimately responsible. Its obscurity, therefore, makes it difficult to eradicate. Paradoxically, it is ordinarily carried out by individuals and groups who do not consider themselves to be discriminators. Thus a common impression may be created in which it is minority ethnic groups themselves who are mostly responsible for their subordinate status. The burden of responsibility for social problems, in other words, is placed on the individual or the group, not on patterns of discrimination that may be built into the institutional structure. This is commonly referred to as "blaming the victim" (Ryan, 1975).

THEORIES OF PREJUDICE AND DISCRIMINATION

How do prejudice and discrimination arise, and why are they seemingly so ineluctable in societies where diverse ethnic groups live side by side? Commonsense explanations, and even some scientific thought of earlier times, have accounted for out-group antipathy as a natural, or innate, pattern of human thought and action. The pervasiveness of conflict between different cultural and racial groups and the universality of ethnocentrism among peoples of the world seem to validate the notion that antagonism toward strangers or members of out-groups is simply part of the human psyche. To dislike or fear those who are different from us is, according to this view, both natural and unavoidable.

Today, however, there is general agreement among sociologists and psychologists that, whatever biological preparedness for out-group antipathy humans may possess, prejudice and discrimination are not innate human characteristics. Social factors will dictate if, when, and how those thoughts and actions come into play. A closer look at historical evidence indicates that people of different cultures and physical attributes do not always relate to each other antagonistically and that, when they do, the nature and intensity of their animus are highly variable. Further evidence that ethnic antagonism is not innate lies in social-psychological studies that have traced the various phases of socialization from early childhood through later stages of the life cycle. All indicate that prejudice and discrimination are learned patterns of thought and action.

But if they agree that prejudice and discrimination are not inborn human traits, sociologists and psychologists differ in their explanations of how these thoughts and behaviors arise and are sustained in multiethnic societies. The psychological tradition emphasizes prejudice as the key component in ethnic interrelations; prejudicial thought, therefore, is the focus of most psychological theories. Sociologists have stressed situational factors and power structures as the bases of both negative thought and action toward ethnic groups; prejudice is thus seen not necessarily as the source of hostility but as its outgrowth. Whereas psychologists are likely to see prejudice and, by implication, discrimination as characteristic of certain personalities, sociologists tend to see them as part of the society's normative order, to which individuals are socialized. In the remainder of this chapter, we look at several theories, both psychological and sociological. Because prejudice and discrimination are multifaceted, each theory may contribute to our understanding of these phenomena.

PSYCHOLOGICAL THEORIES

Psychological theories of prejudice have focused on the manner in which antipathy toward out-groups either satisfies certain psychic needs or complements the general personality structure of certain people. In each case, the source of prejudice is traced primarily to individuals rather than to the social forces weighing on them or the groups within which they interact.

FRUSTRATION-AGGRESSION   One of the earliest psychological theories explains prejudice as a means by which people express hostility arising from frustration. This has also been referred to as scapegoating. The essential idea is that people who are *frustrated* in their efforts to achieve a highly desired goal tend to respond with a pattern of *aggression*. Because the real source of frustration is either unknown or too powerful to confront directly, a substitute is found on whom the aggression can be released. The substitute target is a scapegoat, a person or group close at hand and incapable of offering resistance. The aggressive behavior is thus displaced (Dollard et al., 1939). Minority groups in multiethnic societies have served as convenient and safe targets of such displaced aggression. Allport (1979 [1958]) explains that racial, religious, or ethnic groups can be blamed unfairly for a variety of evils because they are permanent and stable and can be easily stereotyped. Therefore, they serve as more of an "all-purpose" scapegoat than those groups or individuals who are blamed for specific frustrations.

The frustration-aggression, or scapegoating, theory of prejudice seems convincing at first glance. It is, as Allport notes, easily understood because of "the commonness of the experience" (1979 [1958]:330). Everyone at times suffers frustrations of their needs and desires, whether simple ones like seeing their favorite football team lose a key game or significant ones like failing to gain a job promotion. When these events occur, people are sometimes inclined to strike out at substitute targets such as their spouse or children. If the frustration is continuous, they may begin to blame more remote groups or institutions like "the government," "bureaucrats," "blacks," "Jews," or "gays."

Despite the common experience of frustration-aggression, however, this theory leaves many questions unanswered. First, under what conditions will frustration not lead to aggression? Psychologists have shown that the frustration-aggression sequence is not necessarily inevitable (Allport, 1979 [1958]; Ehrlich, 1973). Nor is aggression always displaced; sometimes it is directed inward, and at other times it may be thrust on the real source of frustration. This theory does not tell us how and why these responses may be produced.

Another obvious shortcoming of the frustration-aggression theory is its failure to explain the choice of scapegoats or targets of displaced aggression. Why are some groups chosen rather than others? The notion that scapegoats are always "safe goats"—defenseless and easily used—does not hold up when we consider that minority group members may themselves harbor great antipathy for the dominant group or express strong prejudices toward other minorities. A U.S. national survey, for example, indicated that ethnic minorities are more likely than whites to agree to negative stereotypes about other minority groups (National Conference, 1995).

Finally, we must consider the fact that the displacement of aggression on substitute targets can bring at best a very short-lived relief of an individual's anxiety. Unless the actual source is attacked, the feeling of frustration will quickly recur or even intensify. As Allport has put it, concerning frustration, "Nature never created a less adaptive mechanism than displacement" (1979 [1958]:332).

THE AUTHORITARIAN PERSONALITY   The question of why prejudice exerts a strong influence on certain individuals but plays a relatively minor role for others is addressed by the theory of the authoritarian personality. The essence of this theory is that there is a personality type prone to prejudicial thought. In the same way, according to this view, there are basically tolerant personalities.

The authoritarian personality theory was developed after World War II by a group of social scientists determined to trace the psychological foundations of such destructive and regressive movements as Nazism. It subsequently became one of the most widely tested and debated ideas in the social sciences (Kirscht and Dillehay, 1967). In their studies, T. W. Adorno and his associates (1950) found evidence to support the notion that prejudice and political extremism are more generally characteristic of a definite personality type. Such people, they maintained, are highly conformist, disciplinarian, cynical, intolerant, and preoccupied with power. They are particularly authority oriented and are thus attracted to sociopolitical movements that require submission to a powerful leader. Such personality traits extend well beyond people's political beliefs and are reflected in all aspects of their social life. In the family, for instance, authoritarians will subject their children to strong disciplinary action; in their religious beliefs, they will emphasize submission and obedience. In sum, such people strongly support conservative values and resist social change. They are thus more likely to display prejudicial thought and to discriminate when given the opportunity (Altemeyer, 1996).

Other scholars, although not necessarily subscribing to the notion of an authoritarian personality, suggest that prejudice is a general way of thinking for some people. Hartley (1946), for example, found that when purely fictitious groups were presented in a social distance test, people who were prejudiced toward other groups tended to express prejudice toward the fictitious groups as well. Allport also maintains that "the cognitive processes of prejudiced people are in general different from the cognitive processes of tolerant people" (1979 [1958]:170).

In any case, theories suggesting that certain personality types are generally prone to prejudice and discrimination suffer several critical shortcomings. Like the frustration-aggression theory, they fail to tell us how ethnic prejudice and discrimination arise in the first place. The authoritarian personality theory has fallen out of use and is regarded as flawed because it reduces authoritarianism to a personality trait. We see the prejudiced person in action, but not the social conditions that create ethnic rather than other forms of hostility (Duckitt, 1989).

Moreover, the emphasis of this genre of studies has been on patently and intensely prejudiced people, such as members of the Ku Klux Klan or other white supremacist groups. Most prejudice and discrimination, however, is more subtle and less intense and is characteristic of people who cannot be categorized as extremists.

Indeed, in some societies prejudice can be consensual, evident among an entire cross section of individuals with different personality traits (Brown, 1995). For example, studies have shown that the Nazi movement had appeal across a wide social spectrum of German society (Gerth, 1940; Peukert, 1987). Undoubtedly, prejudice and discrimination play a vital part in the thought and actions of many people, but it is necessary to carefully delineate the varying degrees and forms of ethnic antipathy displayed by different individuals.

INDIVIDUALS VERSUS SITUATIONS   The chief criticism of psychological theories in general is that the situations within which people think and act are not given sufficient attention as variables that fundamentally affect the nature of their thought and action. Schermerhorn cautions us to consider that "if research has confirmed anything in this area, it is that prejudice is a product of situations, historical situations, economic situations, political situations; it is not a little demon that emerges in people simply because they are depraved" (1970:6).

However, psychological theories such as frustration-aggression and the authoritarian personality have been popular not only because they are more easily understood than sociological theories emphasizing structural conditions but precisely because they focus the blame for ethnic antagonisms on disturbed individuals—those who are pathological or overtly irrational in behavior and thought. This focus deflects attention from society's normally functioning institutions, which may compel people to think and act negatively toward members of particular ethnic groups. Indeed, as will be seen shortly, prejudice and discrimination are in most cases *conforming* thoughts and actions, not those of a few maladjusted people. As long as ethnic antipathy is thought to be characteristic only of the sick few rather than a proper response to the expectations of the community or society as a whole, it can be seen as eradicable simply by treating those few, not by painfully reexamining established societal institutions.

Psychological theories also tend to focus on negative feelings toward members of a targeted group as the basis of prejudice and discrimination. Researchers have recognized, however, that prejudice and discrimination may take forms based on paternalism, pity, or even affection, in which antipathy toward out-groups is not evident (Eagely and Diekman, 2005; Rudman, 2005). Sexist behavior and thought, for example, does not necessarily imply a dislike of women, and in systems of slavery, slaves may be viewed affectionately by masters. "The central motivator for dominant groups in unequal social relations," notes social psychologist Mary Jackman, "is not hatred, but the desire to control" (2005:89–90). Dominant group members therefore may assure compliance and cooperation of subordinates more effectively by co-opting them rather than by using hostile methods. Virulent forms of prejudice and discrimination may be activated, however, once out-group members begin to challenge their expected social roles and the position of their group in the ethnic hierarchy.

Psychological components of prejudice and discrimination, then, must be seen in conjunction with political and economic structures out of which intergroup relations develop and are sustained. William Julius Wilson asserts that psychological explanations of prejudice and discrimination "prove to have little predictive value when social factors are taken into account" (1973:39). These social factors are the basis of normative theories, to which we now turn.

NORMATIVE THEORIES

Why do we often compliment friends on their new hairdo or clothes when in fact we think they are quite unattractive? Similarly, why do we many times feel obliged to contribute to a class discussion or a business meeting when we really have nothing meaningful to say? Sociologists explain such apparently inconsistent thought and action as a product of situational norms by which we feel compelled to abide. We understand that such actions are expected of us, and in most cases we conform to those expectations even when we have a real desire to ignore or disobey them. Norms are group standards that define how people are expected to act in particular social situations. There are positive sanctions for conforming to and negative sanctions for deviating from them. Because there are norms pertaining to all social situations in which we find ourselves, these social "rules" enable us to predict others' behavior, and in doing so they facilitate interaction. In a real sense, they provide the society with order.

Prejudice and discrimination can be explained within the framework of social norms. Rather than the thoughts and actions of a deviant few, they are conforming responses to social situations in which people find themselves. When negative thoughts about particular ethnic groups and discriminatory behavior toward them are expected, individuals will feel compelled to think and act accordingly. Thus it is to individuals' social environment—the groups they belong to, the cultural and political values and customs operative in their society and community, and the processes of socialization—that prejudice and discrimination can be traced. Obviously, such explanations are considerably different from psychological theories, which focus not on the group contexts of individual thought and action but on the individuals themselves.

In this view, bigots emerge out of the social experiences to which they are exposed. Frank Westie has succinctly explained the essence of normative theory: "Individuals are prejudiced because they are raised in societies which have prejudice as a facet of the normative system of their culture. Prejudice is built into the culture in the form of normative precepts—that is, notions of 'ought to be'—which define the ways in which members of the group ought to behave in relation to the members of selected outgroups" (1964:583–84). Normative theories thus concentrate primarily on the transmission of ethnic prejudices through the socialization process and on the social situations that compel discriminatory behavior (Dean and Rosen, 1955; Williams, 1964).

SOCIALIZATION Prejudice and discrimination can be seen as part of a society's social code, which is passed down from generation to generation. Fear of, dislike for, and antipathy toward one group or another are learned in much the same way that people learn to eat with a knife and fork rather than with their bare hands or to respect others' privacy in personal matters. These standards of behavior are the product of learning processes of which we ordinarily have little cognizance. Socialization is subtle and works in a largely unconscious manner. Prejudice and discrimination, therefore, need not be taught directly and intentionally. If those are the norms and values of the society or community within which the individual interacts, the chances are very great that they will be adopted with little overt instruction. Children at a very early age become aware of these unwritten but clearly understood rules.

Parents are sometimes puzzled by certain expressions and attitudes of their children because they are sure they did not impart these. In their puzzlement, they discount the informal and often undetected ways in which the society's culture is transmitted by the various agents of socialization outside the family. Most learning is accomplished not through direct teaching methods but through observation and imitation. Children—and adults—take cues from their peers and other important reference groups as well as the mass media. Social psychologists have shown that among American children four years of age, ethnic values and attitudes are already beginning to crystallize (Aboud, 1988; Goodman, 1964; Porter, 1971; Ramsey, 1987). By this age, children have been exposed to the society's ethnic hierarchy, particularly the white-over-black element. By age six, children—even those who have had little or no contact with members of other racial and ethnic groups—have a solid conception of racial-ethnic distinctions (Van Ansdale and Feagin, 2001).

Prejudice and discrimination, then, are no more indicators of a defective personality than are one's taste in food or fashion. They are simply products of socialization. If prejudice and discrimination are pervasive in the society or community, the more logical question may not be, "Why do some people display prejudice and discrimination?" but rather, "Why do some people *fail* to display prejudice and discrimination?"

Consider South Africa under apartheid or the pre-1960s American South. Using the normative approach, white prejudice and discrimination against blacks in those settings can be explained not as the product of deviant individuals but as the natural outgrowth of a whole system of racist norms, learned early and thoroughly, that guided people's actions and attitudes. In that system, blacks were not to be thought of or treated in the same way as whites. For whites to avoid almost any contact with blacks beyond the most purely functional (supervising them in a work situation, for example) was correct, expected, and positively sanctioned by societal norms. For whites to deviate from such behavior would have been unusual and responded to with negative sanctions. John Stone (1973), in a study of British immigrants to South Africa in the 1970s, discovered that they frequently changed their attitude toward the segregationist policy of apartheid. Before leaving Britain, the majority was either opposed to or had no opinion about it; after living in South Africa for a time, however, an even larger majority stated that it favored the policy. Stone concluded that the change reflected not the manifestation of latent racist personalities but the need of immigrants to adapt to the ways of their new society: "We are not witnessing the mass attraction of bigoted racialists to a segregationalist's dream, rather we are observing how ordinary people, confronted by a particular social structure, will tend to conform to the attitudes, values, and norms implicit in it" (1973:253).

REFERENCE GROUPS   Even where societal norms dictate fairness toward different groups, prejudice and discrimination may typify the behavior of some people who have been exposed to reference groups that strongly prescribe such behavior.

Reference groups are those that provide individuals with standards by which they shape their own patterns of action and from which they adopt important beliefs and values. In a sense, they serve as models of thought and action. We ordinarily think of the family as a reference group, but there are many others to which we may look for behavioral guidance, even some of which we are not members. In the latter case, we may aspire to membership and thus take on the ways and attitudes of the group. This is most apparent in the early stages of socialization when children begin to identify with particular occupational groups—firefighters, doctors, nurses, and the like—or with sports and entertainment groups they admire. As might be expected, studies have shown that individuals tend to adopt beliefs and values congruent with those of the groups with which they identify.

Applying the concept of the reference group to prejudice and discrimination, such thoughts and actions are normal responses of individuals when called for by their reference groups. No one is immune to the pressures to conform—applied by family, friendship cliques, or other significant groups. The fear of group rejection is constantly present and serves as an effective disciplinary mechanism. Again, this process is subtle, and the individual may not see such conformity as a response to external coercion.

If the person's reference groups change, attitudes and actions can be expected to change accordingly. College students, for example, often face a challenge to their well-formed values when they encounter new ideas from their instructors and classmates. These new ideas often involve social issues like ethnic prejudice. Pearlin's (1954) study at a southern women's college in the 1950s demonstrated the effect of a change in reference groups on students' racial attitudes. Pearlin started with the assumption that white students would find in the college environment sentiments more favorable to blacks than they had found in their precollege experiences. But merely being exposed to new and positive ideas about blacks, he believed, would not in itself reduce prejudicial attitudes. Rather, modification of attitudes was thought to depend more on changes in social relationships. Thus attitudinal changes, Pearlin surmised, were likely to come about only if the students began to identify with new groups holding those favorable attitudes toward blacks. Pearlin's hypotheses were confirmed: those most prejudiced toward blacks remained most strongly affiliated with their precollege membership groups, and those least prejudiced experienced a weakening of such ties and an increasing identification with their new college groups. In short, Pearlin's findings showed that people tend to adopt the attitudes of those groups with which they most strongly identify.

MERTON'S PARADIGM   Given the social, rather than the personality, origins of ethnic prejudice and discrimination, changing social situations can produce fluctuations in individual thought and behavior. Ethnic prejudice and discrimination are thus not constant and unchanging but variable, depending on a number of situational factors: the person's definition of the situation, the compulsion to conform to societal and reference group norms, and the rewards—economic, prestige, political—to be gained by acting and believing in such a manner. That attitudes and actions toward members of particular ethnic groups may fluctuate within different social contexts is demonstrated by Robert Merton (1949) in his well-known paradigm (Figure 3.2).

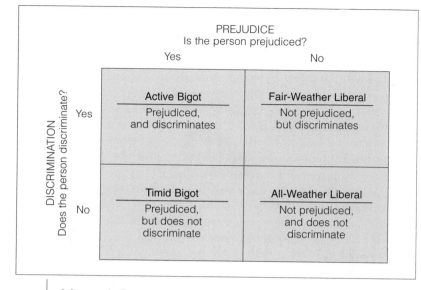

FIGURE 3.2 | MERTON'S PARADIGM

By combining the prejudicial attitudes or lack of such attitudes with the propensity either to engage in discriminatory actions or to refrain from them, Merton suggested four ideal types. First, he denoted unprejudiced nondiscriminators, whom he called "all-weather liberals." These are people who accept the idea of social equality and refrain from discriminating against ethnic minorities. Their behavior and attitudes are thus consistent. A second type, also consistent in behavior and attitude, is prejudiced discriminators, whom Merton labeled "active bigots." Such people do not hesitate to turn their prejudicial beliefs into discriminatory behavior when the opportunity arises. Members of organizations such as the Ku Klux Klan or neo-Nazi parties in the United States or skinhead groups in Europe exemplify such people. Both of these types, consistent as they are in belief and behavior, might indicate by themselves support for the psychological perspective; that is, there are prejudiced people or tolerant people, who may be expected to act accordingly.

Merton's third and fourth types, however, demonstrate the situational context and the effect it may have on people's behavior. In these cases, behavior and attitude are not consistent. Prejudiced nondiscriminators, or "timid bigots," as Merton called them, maintain negative beliefs about ethnic minorities but are precluded from acting out those beliefs by situational norms. If a situation requires fair treatment toward ethnic groups who are viewed negatively by such people, fair treatment will mark their behavior. For example, whites traveling from the American South to northern states before the 1960s would find that the laws and customs of the North required them to interact with blacks in a manner unheard of in their home states. Lewis Killian described how white working-class southerners who had migrated to Chicago responded to blacks in their new environment: "The 'hillbillies' constantly praised the southern pattern of racial segregation and deplored the fact that Negroes were 'taking over Chicago.' In most of their behavior, however, they made a peaceful, if reluctant, accommodation to northern urban patterns" (1953:68). The hotel and restaurant keepers encountered by LaPiere in his previously cited study would also fall into this category.

Unprejudiced discriminators, whom Merton called "fair-weather liberals," also adjust their behavior to meet the demands of particular circumstances. When discrimination is normative in the group or community, such people abide by those patterns of behavior even though they may harbor no prejudicial feelings toward members of the targeted group. To do otherwise would jeopardize their social standing and might even constitute violations of the law. Such a situation was faced before and even after the adoption of civil rights legislation in the 1960s in the American South by many whites who did not share the racial animosity of their neighbors. In a 1987 case, the owner of a pharmacy in Tifton, Georgia, dismissed a black student pharmacist who had been placed in the store as part of her training at the University of Georgia's College of Pharmacy. She was dismissed, the owner explained, because he feared negative customer reaction (*New York Times*, 1987).

It must be remembered that each of Merton's four cases is an ideal type and thus does not reflect perfectly any individual's behavior and attitudes. More realistically, people can be expected to display higher or lower degrees of each. We should also remember that prejudice and discrimination directed at one ethnic group do not necessarily imply the same attitudes and behavior toward others. Those who are antiblack are not necessarily anti-Catholic, and so on. One may be a fair-weather liberal in one instance and a timid bigot in another.

Situational explanations of prejudice and discrimination like Merton's demonstrate that there is no necessary causal relationship between the two. The conventional wisdom has generally assumed that prejudice leads to or causes discrimination. Abundant sociological evidence, however, has shown not only that this sequence need not occur but also that the very opposite is more common (Pettigrew, 1979; Raab and Lipset, 1971). "What we call prejudices," writes the anthropologist Marvin Harris, "are merely the rationalizations which we acquire in order to prove to ourselves that the human beings whom we harm are not worthy of better treatment" (1964:68). Prejudice, then, is used to rationalize discriminatory behavior *after* the fact.

This is an important observation, for it seriously challenges the idea that eliminating discrimination requires a change in attitude, that is, the elimination of prejudice. This was the generally shared opinion of both scholars and policy makers in the United States before the 1960s. Reeducating people was therefore the most frequently proposed remedy for alleviating ethnic hostility. It was thought that if people's faulty ideas about race and ethnicity could be corrected, they would, as a result, be induced to change their behavior. In the past few decades, however, it has become obvious that prejudicial attitudes may have little or no bearing on whether people discriminate against particular groups. Instead, people appear to be motivated to change their behavior toward ethnic groups by laws and other social mechanisms that seriously alter their social situation vis-à-vis those groups. After such situational changes occur, individuals seem to adjust their ideas to fit these new modes of behavior (Ehrlich, 1973; Pettigrew, 1980). Hence, changes in ethnic relations are impelled not by efforts to change attitudes but by changing the structure of those relations.

STATISTICAL DISCRIMINATION   Van den Berghe (1997) has suggested a model of discrimination akin to Merton's that links such behavior to stereotyping. He posits that stereotypes are most often used in a rational way, enabling people to make quick decisions about others in interactive situations where they have minimum information. (Recall this point in our discussion of stereotypes.) *Statistical* discrimination, in which *people are treated negatively on the basis of beliefs about the category of which they are members*, is distinct from *categorical* discrimination, in which *people are treated negatively simply on the basis of being part of a socially assigned category*. The former is rational in that it is responsive to counterevidence, whereas the latter form of discrimination is not. Van den Berghe contends that statistical discrimination is more common than categorical discrimination because it is based on self-interest, whereas the latter is nonrational. Hence, stereotypes are not necessarily to be seen as evidence of prejudice but are simply "guidelines in making statistical discriminations in situations of imperfect information" (1997:5). Where information is in short supply and costly, people will rely on stereotypes.

For example, because blacks appear to commit a disproportionate percentage of violent crimes, whether due to a racist criminal justice system or to class bias, whites stereotype blacks as "criminally inclined" and thus may discriminate against blacks with whom they must interact. In reality, the relationship of crime to such factors as social class and age is stronger than the relationship of crime to race. But most whites, living as they do apart from blacks, are not as attuned to class differences among blacks. It is, then, less costly to discriminate on the basis of race—which is more readily apparent—than class. Banks may make decisions about loans using much the same calculus based on racial criteria, as a result granting fewer loans to blacks or charging them higher interest rates. Again, the decision is a rational one: "Money-lenders are in the business of evaluating risk as cheaply as possible and of adjusting interest rates in direct proportion to assessed risk. . . . [F]rom the perspective of the bank ignoring race would be a costly mistake, unless the bank can develop a more valid and equally cheap discriminator" (van den Berghe, 1997:12). **Racial profiling**, wherein *police select someone for investigation or stronger action on the basis of race or ethnicity*, may be seen in the same way.

Like Merton, van den Berghe demonstrates how discrimination is not necessarily motivated by prejudice. "Modern industrial societies," he notes, "are hot-beds of stereotypy, not because they are populated by bigots, but because people have little else to structure their relationships" (1997:13). Where race and ethnicity create barriers to communication and where members of different groups live in segregated areas, the chances of changing the criteria of statistical discrimination are not great.

POWER-CONFLICT THEORIES

Though the normative theories of prejudice and discrimination appear to go well beyond the earlier psychological theories, they stop short of explaining how or why they arise in the first place. They basically explain the mechanics of prejudice and discrimination, that is, how these social phenomena are transmitted and sustained. To begin to understand their origins we must turn to power-conflict theories.

Most simply, these theories view prejudice and discrimination as emerging from historical instances of intergroup conflict (Bernard, 1951; Newman, 1973). In this view, discrimination serves as a means of injuring or neutralizing out-groups that the dominant group perceives as threatening to its position of power and privilege. Negative beliefs and stereotypes, in turn, become basic components of the dominant group's ideology, which justifies differential treatment of minority ethnic groups. When prejudice and discrimination are combined, they function to protect and enhance dominant group interests. And once established, prejudice and discrimination are used as power resources that can be tapped as new conflict situations demand.

Prejudice and discrimination, in this view, are products of group interests and are used to protect and enhance those interests. To understand negative ethnic beliefs and behavior requires a focus not on individual personalities or even on the constraints and demands of different social situations, but on the economic, political, and social competition among groups in a multiethnic society.

ECONOMIC GAIN   Chief among power-conflict theories are those emphasizing the economic benefits that derive from prejudice and discrimination. Simply put, prejudice and discrimination, in this view, yield profits for those who engage in them. Different groups may be targeted because they present—or are perceived as presenting—a threat to the economic position of the dominant group.

Colonial and slave systems, buttressed by elaborate racist ideologies, are obvious cases in which economic benefits accrued to a dominant group from the exploitation, both physical and mental, of minority groups. We need not look at such historically distant examples, however, to understand the relation between ethnic antagonism and economic gain. In his study of black-white relations in the American South of the 1930s, John Dollard showed that in every sphere of social life—work, health, justice, education—the white middle class realized substantial gains from the subordination of blacks. In exploiting blacks, explained Dollard, southern whites were simply "acting as they have to act in the position within the social labor structure which they hold, that is, competing as hard as they can for maximum returns" (1937:115). Later studies (Glenn, 1963, 1966; Thurow, 1969) concluded that prejudice and discrimination against blacks in the United States continued to benefit at least some segments of the white population.

Resistance in the form of blatant discriminatory efforts can be expected whenever economic advantages appear to be challenged by lower-ranking or more recently arrived ethnic groups. In 1981, for example, white fishermen in the shrimping grounds in Galveston Bay, Texas, encountered competition from immigrant Vietnamese fishermen. About one hundred Vietnamese shrimpers had come to the area during the previous two years, challenging the economic dominance of the whites. Although the situation was eventually resolved, for several months white fishermen, with support from the Ku Klux Klan, engaged in acts of intimidation against the Vietnamese, including physical attacks and arson (Hein, 1995).

MARXIAN THEORY   Some economic-based theories are more specific in suggesting who the beneficiaries of prejudice and discrimination are. Class theorists, in the tradition of Karl Marx, have conventionally held that in capitalist societies, ethnic antagonism serves the interests of the capitalist class—those who own and control the means of economic production—by keeping the working class fragmented and thus easier to control. The basic idea of Marxian theory is "divide and rule." One ethnic element of the working class is pitted against another, and as long as this internal discord can be maintained, the chances that the working class will unite in opposition to the interests of the capitalists are reduced. Capitalists are able to foster ethnic division and ethnic consciousness among the workers, thereby curtailing the development of worker solidarity and class consciousness.

In the United States, for example, the historic conflict between black and white workers has been construed by Marxists as having deflected attention from the common anticapitalist interests of both groups (Allen, 1970; Cox, 1948; Reich, 1978; Szymanski, 1976). Anti-immigrant movements in the past as well as in recent times can be seen in the same way. Ethnic prejudice, therefore, is viewed as a means of sustaining a system of economic exploitation, the benefits of which accrue to the capitalist class. Though capitalists may not consciously conspire to create and maintain racist institutions, they nonetheless reap the benefits of discriminatory practices and therefore do not seek to completely dismantle them.

THE SPLIT LABOR MARKET THEORY   Whereas conventional Marxist thought holds that the profits of ethnic hostilities redound primarily to the owners of capital, others maintain that it is workers of the dominant ethnic group who are the chief beneficiaries of prejudice and discrimination. If ethnic minorities are kept out of desired occupations, the favored workers, rather than the capitalists, are viewed as gaining the most from discriminatory institutions. This is the crux of sociologist Edna Bonacich's split labor market theory (1972, 1976).

According to Bonacich, there are three key groups in a capitalist market: businesspeople (employers), higher-paid labor, and cheap labor. One group of workers controls certain jobs exclusively and gets paid at one scale, and the other group is confined to jobs paid at a lower rate. Given the imperatives of a capitalist system, employers seek to hire workers at the lowest possible wage and therefore turn to the lower-paid sector when possible as a means of maximizing profits. Recent immigrants or ethnic groups migrating from rural areas in search of industrial jobs ordinarily make up this source of cheap labor. These groups can be used by employers as strikebreakers and as an abundant labor supply to keep wages artificially low. Because these groups represent a collective threat to their jobs and wages, workers of the dominant ethnic group become the force behind hostile and exclusionary movements aimed at curtailing the source of cheap labor. Wage differentials, in this view, do not arise through the efforts of capitalists to prevent working-class unity by favoring one group over another, but through the efforts of higher-paid laborers to prevent lower-paid workers—mainly of low-status ethnic groups—from undercutting their wages and jobs. This goal is achieved through various forms of prejudice and discrimination.

The split labor market theory is supported by historical evidence in American society. Successive waves of European immigrants during the nineteenth and early twentieth centuries traditionally served as a source of cheap labor and became the targets of nativist movements, usually backed strongly by labor unions. Following the cessation of European immigration, northward-migrating blacks from the rural South assumed a similar role, touching off periodic racial violence in many cities. Depending on how threatening they were perceived by native workers, various groups at different times were the objects of worker-inspired hostility. For example, efforts in the nineteenth century to restrict the Chinese to particular occupations and to limit their immigration were spurred largely by white workers fearing a deluge of cheaper labor. Lyman notes that after 1850, "[i]t was the leadership of the labor movement that provided the most outrageous rhetoric, vicious accusations, and pejorative demagoguery for the American Sinophobic movement" (1974:70).

In a study of Japanese immigrants in Brazil and Canada, Makabe (1981) found support for the split labor market theory. Japanese immigrants entering Canada, specifically British Columbia, in the pre–World War II years experienced an extremely harsh reception from native Canadians. Makabe explains that among white workers the rejection of the Japanese was unusually cruel. This is accounted for by the fact that the Japanese entered the Canadian economy at the bottom, enabling employers to pay them lower wages and thereby undercutting the more highly paid native workers. Striving for upward mobility, the Japanese found themselves in direct competition with those immediately above them in economic position—white workers. The result was discrimination against the Japanese in the workforce and pressure to halt all Japanese immigration. In Brazil, however, the Japanese experienced a significantly different situation. Rather than entering the labor force in competition with higher-paid workers, they found themselves with skills and financial resources superior to those of most native workers, who themselves were mostly severely disadvantaged former slaves. Little competition and, hence, little conflict arose between them because they did not seek similar occupational positions.

GROUP POSITION Herbert Blumer (1958) posited that prejudice is always a protective device used by the dominant group in a multiethnic society in ensuring its majority position. When that group position is challenged, prejudice is aroused and hostilities are directed at the group perceived to be a threat. As Lawrence Bobo (1999) explains, feelings of superiority among dominant group members toward subordinate group members are not sufficient to produce prejudice and discrimination. What is required in addition is the perception that dominant group privileges and resources are threatened by the subordinate group. In this view, negative stereotypes and discriminatory actions are used by elements of the dominant ethnic group, sometimes directly and other times indirectly, to secure and preserve not only their economic power but their political power and social prestige as well.

Political leaders have long recognized the value of exploiting ethnic divisions for attaining and enhancing their power. For example, until blacks became an electoral factor of some significance in the 1970s, racist politicians in the American South effectively manipulated white fears of blacks to their own ends. After losing the Alabama gubernatorial election in 1958 to a candidate even more avowedly racist than himself, George Wallace declared that he would not be "out-nigguhed again" (Frady, 1968). In the 1988 presidential campaign, George H. W. Bush used the case of Willie Horton, a convicted murderer who raped a woman while on furlough from a Massachusetts prison, to portray his Democratic opponent, Michael Dukakis, as soft on crime. As a black man, Horton's image was intended to elicit white fears of black crime. Similarly, in recent years rightist politicians in France, Germany, and other western European countries have stirred anti-immigrant feelings, particularly against Muslims from North Africa and the Middle East, in appealing to voters.

In addition to economic and political benefits, status privileges may derive from ethnic antagonism. People may enjoy more prestige simply from being a member of the dominant ethnic group, regardless of their social class. In the American South, working- and lower-class whites could take comfort in knowing that they were part of the dominant ethnic group even though economically they were in much the same position as blacks. As Dollard described it, white subordination of blacks consisted of "the fact that a member of the white caste has an automatic right to demand forms of behavior from Negroes which serve to increase his own self esteem" (1937:174). Van den Berghe describes the same well-understood racial etiquette that prevailed in South Africa before the end of apartheid: "Non-Europeans are expected to show subservience and self-deprecation, and to extend to the whites the titles of 'Sir,' 'Madam,' or 'baas.' The Europeans, as a rule, refuse to extend the use of titles and other forms of elementary courtesy to nonwhites, and call the latter by first names (real or fictitious), or by the terms 'boy' and 'girl'" (1967:142).

Wilson (1973) also notes that when the system of ethnic stratification is challenged—that is, when minority groups no longer accept their group position— strong prejudices founded on a racist ideology emerge. Through this ideology, members of the dominant group can, as Wilson explains, "claim that they are in a superior position because they are naturally superior, that subordinate members do not possess qualities enabling them to compete on equal terms" (1973:43). The dominant ideology, incorporating key negative stereotypes of minority ethnic groups, thus reinforces the sense of group position, aids in maintaining patterns of subordination, and serves as a philosophical justification for exploitation.

The resistance of many whites to school busing, residential desegregation, affirmative action, and immigration during the past four decades can be interpreted as the response of those who perceive a threat to their group position. Applying this model, these are negative reactions by members of the dominant group who see their economic, political, and status privileges—their group position—threatened by blacks, Latinos, and other minorities seeking upward social mobility. In recent years, Tea Party and other right-wing activists commonly voiced anti-immigrant sentiments and expressions of resentment toward President Obama and nonwhites generally in response to what they interpreted as threats to their social and economic status (Cose, 2011; Parker and Barreto, 2013; Skocpol and Williamson, 2012).

The theory of group position may also be applied to cases where prejudice and discrimination decline. Alba (2009), for example, postulates that the relatively rapid upward mobility of white ethnic groups—primarily Catholic and Jewish—in the two decades following World War II is attributable to what he calls "nonzero-sum mobility." These groups, previously excluded from various occupations, schools, and neighborhoods, were able to erase the boundary between themselves and the white Protestant majority because the latter enjoyed postwar upward mobility themselves and so did not resist the social and economic advances of the white ethnics. That is, their own privileged group status was not seen as being threatened.

FUNCTIONS FOR MINORITIES   Paradoxically, prejudice and discrimination may serve certain functions for minority groups themselves. Sociologists have recognized that conflict between groups has a unifying effect on the members of each. External threats tend to strengthen group ties and create a sense of solidarity that might not otherwise exist (Coser, 1956). The ability of Jews to survive in various societies in which they were persecuted, for example, has often been attributed to the continued anti-Semitic hostility itself. As constant targets of antagonism, Jews have strengthened their resolve to maintain a group identity and cohesiveness.

Continued prejudice and discrimination directed at a minority group may also contribute to a sense of psychological security for its members. Even though their place is at the bottom, they may take consolation in the certainty and predictability of their social relationships with the dominant group (Levin and Levin, 1982). Moreover, the minority individual's self-esteem may be protected by attributing personal failures to abstract notions like "the system" or "racism" rather than to individual shortcomings.

Prejudice may also serve as a release of frustration for minorities, just as it may for those of the dominant group. Indeed, prejudice should not be seen as characteristic only of dominant groups. Although sociologists have been reluctant to deal with it, prejudice is commonly displayed by minority groups as well, not only toward the dominant group—which seems entirely logical—but also toward other minority groups. Recall the strong antipathy toward each other expressed by American minority ethnic groups, mentioned earlier.[4] If prejudice is normative in the society, minority group members socialized to those norms will be affected in much the same way as members of the dominant group.

The benefits to minority ethnic groups that derive from prejudice and discrimination, however, should not be overdrawn. Clearly, the primary beneficiaries of ethnic antagonism are members of the dominant group.[5]

THEORIES OF PREJUDICE AND DISCRIMINATION: AN ASSESSMENT

As we have now seen, the explanation for prejudice and discrimination in multiethnic societies is complex and by no means agreed on by theorists and researchers. It may very well be that a full investigation of these phenomena requires a multidimensional approach using different aspects of psychological, normative, and power-conflict theories. All may have some validity, depending on which aspect or level of ethnic antagonism is focused on.

In this book, power-conflict theories of prejudice and discrimination are favored because the structural rather than the psychological or small-group dynamics of race and ethnic relations are emphasized. In Parts II and III, therefore, the analysis of prejudice and discrimination in the United States and in other multiethnic societies will view these ethnic attitudes and actions mainly as tools of dominance, developed and used by one group over others in competition for the society's resources.

Although power-conflict theories are stressed, keep in mind that prejudice and discrimination are multifaceted, and therefore other theories cannot be disregarded.

---

[4] Westie (1964) asserts that the prejudice of minority group members has been largely ignored by social scientists. He suggests that this may be a result of the sympathy social scientists usually display for social underdogs. Moreover, interethnic conflict is usually perpetrated by members of the dominant group, and the prejudices of minority group members are seen mainly as responses to these actions. Westie maintains that, however well intended this view may be, it has produced social science literature "which gives the impression that the minority person can 'do no wrong'" (1964:605).

[5] There are also certain negative effects of prejudice and discrimination on the dominant group. See, for example, Bowser and Hunt (1981).

Power-conflict theories will not entirely explain, on the one hand, why some people will not discriminate even when it is profitable to do so or, on the other hand, why some will continue to discriminate when it is no longer beneficial. For such cases, psychological or normative theories may offer additional insight. As Simpson and Yinger point out, not all prejudice and discrimination can be explained by structural variables alone; individuals' responses to group influences are conditioned by their personality and vice versa. Therefore, "[t]he task is to discover how much of the variance in prejudice and discrimination can be explained by attention to personality variables, how much by social structural variables, and how much by their interaction" (1985:29).

SUMMARY

- Prejudice and discrimination are techniques of ethnic dominance. *Prejudice* is the attitudinal dimension of ethnic antagonism. Prejudices are categorical, inflexible, negative attitudes toward ethnic groups, based on simplistic and exaggerated group images called *stereotypes*.
- *Discrimination* is the behavioral dimension and involves actions designed to sustain ethnic inequality. Discrimination takes various forms, ranging from derogation to physical attack and even extermination.
- Discrimination occurs at different levels: micro- (or individual), macro- (or institutional), and structural. *Micro*-discrimination is carried out by a single person or small group, usually in a deliberate manner; *macro*-discrimination is rendered broadly as a result of the norms and structures of organizations and institutions; *structural* discrimination occurs obliquely in an unwitting and unintentional manner and is the indirect result of discrimination in other, more blatant forms.
- There are three major theoretical traditions in explaining the origins and patterns of prejudice and discrimination. *Psychological* theories focus on the ways in which group hostility satisfies certain personality needs; prejudice and discrimination, in this view, are traced to individual factors. *Normative* theories explain that ethnic antagonisms are conforming responses to social situations in which people find themselves. *Power-conflict* theories explain prejudice and discrimination as products of group interests and as tools used to protect and enhance those interests; focus is placed not on individual behavior or on group dynamics but on the political, economic, and social competition among a society's ethnic groups.
- Prejudice and discrimination are multifaceted and therefore require a number of theories to explain their origin and perpetuation in modern societies.

CRITICAL THINKING

1. For the past decade, Muslim Americans, especially those with Arab identities, have been the targets of prejudice and discrimination. What specific types of prejudice and discrimination have they been subjected to? And which theory (or theories) seem most compelling in explaining these attitudes and behaviors?
2. Commercials presented on television today make a conscious effort to use a cast that is a racial/ethnic mixture. Do these commercials accurately reflect the

status—economic, political, social—of the various ethnic groups represented? Do they accurately reflect the true nature of interaction among members of different groups, or are they idealized pictures? Will these depictions, accurate or not, have an impact on the way in-group members think of and relate to out-group members?

3. Consider Merton's paradigm, explained in this chapter. Think of social contexts in which someone might display one or another of these types of thought and action.

4. Structural discrimination, as this chapter points out, is more difficult to detect than blatant, direct forms of discrimination. How would you illustrate structural discrimination to someone who insists that old-fashioned (overt) discrimination is no longer prevalent in American society, and therefore the socioeconomic problems of racial and ethnic minorities must be entirely of their own doing?

PERSONAL/PRACTICAL APPLICATION

1. Think of the concept of social distance, as discussed in this chapter. Describe how far along the social distance scale you are prepared to go with people of other ethnic identities. Why might you accept someone of one particular ethnicity more completely than someone of another?

2. You may have a close friend about whom you lack any knowledge of his or her ethnic background. Once you discover that person's ethnic (or racial) identity, are certain stereotypes almost automatically set off in your mind? Do your perceptions of, and perhaps interaction with, that person change as a result?

PATTERNS OF ETHNIC RELATIONS

CHAPTER 4

Assimilation and Pluralism

The chief question we pursue in this chapter is, "What is the nature of relations among diverse groups in a multiethnic society?" Specifically, when various ethnic groups meet, what is the outcome of that contact? We have already seen that the emergence of a system of ethnic stratification, with its attendant prejudice and discrimination, is most common. But other dimensions and processes of group relations create a variety of interethnic patterns.

Judging from frequent news accounts of violence and general hostilities in multiethnic societies, it would seem that ethnic relations are, as a rule, conflictual. Indeed, almost all theorists have viewed conflict as a fundamental, if not permanent, feature of relations between ethnic groups. Donald Young's observation more than eight decades ago seems no less accurate today: "Group antagonisms seem to be inevitable when two peoples in contact with each other may be distinguished by differentiating characteristics, either inborn or cultural, and are actual or potential competitors" (1932:586).

Though they may not always take overt and extreme form, power and conflict are the primary underlying facets of systems of ethnic inequality. The dominant group may mobilize its power through force, ideology, or both to ensure its dominance, and minority groups will respond with counterforce, accommodation, or submission.

Although ethnic relations are, by definition, relations of conflict, certain qualifications to this maxim must be kept in mind. First, ethnic conflict is not maintained at a constant rate, does not take the same form, and is not based on the same factors from one society to another or even within the same society. It is obviously more intense and sustained in some cases than in others. Ethnic conflict can be reduced, and some societies—though rare—may even achieve a system whereby groups live side by side for long periods in a generally harmonious state. Switzerland provides such an example. Here four languages are spoken, and the society is further subdivided along two main religious lines. Yet, linguistic and religious tolerance have characterized Swiss ethnic relations, and serious conflict has been avoided for many generations.

It should also be understood that conflict is not characteristic only of multiethnic societies, nor is *ethnic* conflict the only or even chief form in all societies, even those that are ethnically heterogeneous. Societal conflict is based on power relations that stem from differences in class, age, gender, and numerous other social factors in addition to ethnicity.

Finally, no matter how discordant or peaceful they may be, intergroup relations in multiethnic societies do not remain fixed. Internal migrations, immigration from other societies, political events, and varying economic conditions continually force revisions in public policies and create the social conditions that lead to change. At particular times, societies may move in the direction of either more harmonious relations among groups or greater separation and more intense conflict.

THEORETICAL MODELS OF ETHNIC RELATIONS

Sociologists have proposed theoretical models that describe patterns of intergroup relations in multiethnic societies (Barth and Noel, 1972; Burkey, 1978; Marden and Meyer, 1978; Schermerhorn, 1970) or cycles of relations through which such societies presumably pass (Bogardus, 1930; Gordon, 1964; Park, 1924). These theories have suggested that ethnic groups follow one of two paths: they either increasingly blend together or remain segregated. Most simply, *groups may become more alike culturally and interact with one another more freely*—this is **assimilation**; or *they may remain culturally distinct and socially segregated*—this is **pluralism**. The latter is of two types, depending on the distribution of political and economic power among groups. In the first type, culturally and structurally distinct groups are relatively balanced and proportionate in political and economic resources; in the second type, they are unequal politically and economically. These differences will be explained later in more detail.

CONFLICT AND ORDER

As models of interethnic relations, assimilation and pluralism are related to the two broad theoretical paradigms in sociology, *order* and *conflict*. These paradigms pertain not simply to specific parts or aspects of societies (like ethnic relations) but to *all* social structures and relations. They are intended to explain generally how societies are sustained and how they change.

Order theorists, whose tradition is most heavily influenced by the late-nineteenth-century French sociologist Émile Durkheim, *see society as a relatively balanced system made up of differently functioning but interrelated parts*. In this view, society is held together and social order maintained through a consensus of values among its groups and through the imperatives of functional interdependence. In contrast, **conflict theorists**, beginning with Karl Marx, *see societies as held together not by broad agreements among groups but by the power of dominant classes and ruling elites to impose their will on others*. Stability and order are maintained through coercion, not consensus. Whereas order theorists stress the manner in which societies maintain cohesion and balance, conflict theorists emphasize the disintegrative aspects of societies and the manner in which they change. As many have pointed out, societies are of course neither wholly ordered nor wholly in conflict, and we must therefore account for both (Dahrendorf, 1959). Nonetheless, sociologists have favored one or the other of these broad theoretical perspectives in their analyses.

Assimilation and pluralism, as models of interethnic relations, correspond closely with these two general sociological paradigms. Order theorists have stressed the assimilation side, emphasizing the ways in which different groups progressively become more unified and indistinct. Conflict theorists, in contrast, emphasize the inequality among ethnic groups and the patterns of dominance and subordination that develop among them. They have thus preferred a pluralistic model of ethnic relations that underscores the persistence of group differences and divisions.

For many years, American sociologists traditionally favored the assimilationist model (Metzger, 1971). The prevailing assumption was that multiethnic societies like the United States tend to gradually but inevitably move toward a fusion of diverse groups. American and world patterns of ethnic conflict in recent decades, however, have seriously undermined this assimilationist bias. Sociologists today increasingly recognize the complexity of interethnic relations and seem to have accepted the inevitability of some degree of pluralism and conflict.

The investigations of American ethnic relations in Part II and of several other multiethnic societies in Part III generally employ a conflict approach; thus the pluralistic aspects of these societies are highlighted. Theoretical frameworks in ethnic relations vary in utility, however, depending on the specific problems to be explained. In general, assimilation, stressing progressive cohesion, applies to groups in multiethnic societies that have entered as voluntary immigrants, whereas some form of pluralism is relevant to those groups that have entered through involuntary immigration, conquest, or expansion. (Recall the discussion in Chapter 2 regarding different forms of initial contact among culturally or physically distinct groups leading to different patterns of ethnic relations.) In the remainder of this chapter, we look more closely at these two general processes and the more specific levels and forms of each.

ASSIMILATION

Most simply, assimilation means increasing similarity or likeness. As Yinger defines it, assimilation is "a process of boundary reduction that can occur when members of two or more societies or of smaller cultural groups meet" (1981:249). Similarly, Harold Abramson defines it as "the processes that lead to greater homogeneity in society" (1980:150). Each of these definitions stresses that assimilation is best seen as a path or trajectory on which ethnic groups may move. It is a process, not a fixed condition or state of relations.

The end point of this homogenizing process is, for an ethnic group, the disappearance of any cultural or racial distinction setting it off from other groups (Alba and Nee, 1999). Or, for the society as a whole, it is "the biological, cultural, social, and psychological fusion of distinct groups to create a new ethnically undifferentiated society" (Barth and Noel, 1972:336). Following this idea to its logical end point, with complete assimilation there are no longer distinct ethnic groups. Rather, there is a homogeneous society in which ethnicity is not a basis of social differentiation and plays no role in the distribution of wealth, power, and prestige. This does not mean, of course, that other forms of social differentiation and stratification such as age, gender, and class do not exist; it means only that the ethnic forms are no longer operative. In essence, a society in which all groups have perfectly assimilated is no longer a multiethnic society.

This complete form of assimilation, however, is rarely achieved. Instead, assimilation takes different forms and is evident in different degrees. In other words, it is a variable that "can range from the smallest beginnings of interaction and cultural exchange to the thorough fusion of the groups" (Yinger, 1981:249). Therefore, in examining the assimilation of ethnic groups, the question that must concern us is not simply "Are groups becoming more alike?" but "To what extent and in what ways are they becoming more alike?"

DIMENSIONS OF ASSIMILATION

Assimilation can be seen in four distinct, though related, dimensions: cultural, structural, biological, and psychological. Our concern is mostly with the first two, but let's briefly examine each of these.

CULTURAL ASSIMILATION   *The cultural dimension of assimilation involves the adoption by one ethnic group of another's cultural traits—language, religion, diet, and so on.* This process is sometimes referred to as **acculturation** (Gordon, 1964; Yinger, 1981). Almost always, weaker (that is, minority) groups take on the cultural traits of the dominant group, though there is ordinarily some exchange in the opposite direction as well. This uneven exchange occurs not only because of the dominant group's superior power but also because of the social advantages for the subordinates in adapting to the dominant group's ways. As van den Berghe explains, "It often pays to learn the ways of the rich, the powerful and the numerous; in the process one becomes more like them and, by that token, often becomes more acceptable to them" (1981:215). The end point of the process of cultural assimilation implies a situation in which the previously distinct ethnic groups are no longer distinguishable on the basis of their behavior and values.

Yinger (1981) has noted that when groups are not highly antagonistic or culturally very disparate, acculturation can be additive rather than substitutive. That is, one group may augment its native culture with select elements of the other's rather than substituting entirely. If the most basic cultural components (like language and religion) are not exchanged, however, the assimilation process can proceed only to a minimal point.

STRUCTURAL ASSIMILATION   Whereas cultural assimilation refers to a blending of behaviors, values, and beliefs, **structural assimilation** refers to an increasing degree of social interaction among different ethnic groups. Specifically, with structural assimilation, *members of minority ethnic groups are dispersed throughout the society's various institutions and increasingly enter into social contacts with members of the dominant group.*

Structural assimilation may occur at two distinct levels of social interaction: the *primary* (or informal) and the *secondary* (or formal). Primary relations are those occurring within relatively small and intimate groups, in particular the family and friendship cliques. Relations among members of these groups are affective, and the group's purposes extend well beyond instrumental goals. There is, most simply, an emotional bond among group members, and relations are therefore close and long lasting. Secondary relations, by contrast, are chiefly within large, impersonal groups such as the school, the workplace, or the polity. These groups are purposeful, designed to fulfill some practical and specific social need; relations among members are thus formal and nonaffective.

Structural assimilation at the primary level implies interaction among members of different ethnic groups within personal networks—entrance into clubs, neighborhoods, friendship circles, and ultimately marriage. In short, people interact in close, personal relations without regard for one another's ethnic identity. Other social traits, such as class, and individual characteristics become more critical than ethnicity. The degree of informal, or primary, structural assimilation of a particular group would be measured using such indicators as rate of intermarriage, club memberships, and residential patterns. The higher the level of interaction with members of the dominant group in these areas of social life, the greater the extent of structural assimilation.

At the secondary level, structural assimilation entails equality of access to power and privilege within the society's major institutions—the economy, government, education, and so on. This means that jobs, housing, schooling, and other key life chances are distributed without regard to people's ethnic affiliation. Essentially, structural assimilation at the secondary level involves, in its ultimate stage, the elimination of minority status. To measure the degree of secondary structural assimilation, we would look at the extent to which a minority ethnic group is approaching parity with the dominant group in the allocation of income and wealth, political power, and education (Hirschman, 1975).

Secondary structural assimilation has often been referred to as **integration** (Burkey, 1978; Davis, 1978; Hunt and Walker, 1974; Simpson, 1968; Vander Zanden, 1983), and we can consider these terms to be synonymous. The essential idea is that *people of diverse ethnic groups are able to participate freely in the various institutions of the larger society, unconstrained by ethnicity.*

This level of structural assimilation involves a legal termination of group discrimination based on ethnicity. Inequality exists, of course, but is founded on bases other than ethnic group membership. With integration, ethnic groups may remain distinct, and members may continue to identify with them. Similarly, out-groups may continue to recognize and respond to them as ethnic groups. What has been achieved, however, is a measure of political and economic equality.[1]

The distinction between primary and secondary levels of structural assimilation is important because it is clear that the entrance of ethnic minorities into formal relations with the dominant group must precede relations within intimate social settings. Groups may achieve a significant degree of secondary structural assimilation without

---

[1] Schermerhorn (1970) has used the term integration in a somewhat different but related way. As he defines it, integration is "a process whereby units or elements of a society are brought into an active and coordinated compliance with the ongoing activities and objectives of the dominant group" (66). The key to integration in this sense is a mutual acceptance of the scope and nature of activities and group objectives by both the dominant group and minority groups. Thus societies may even be integrated around a system in which extreme segregation and inequality are accepted by all parties. Schermerhorn postulates that "when the ethos of the subordinates has values common to those in the ethos of the superordinates, integration (coordination of objectives) will be facilitated; when the values are contrasting or contradictory, integration will be obstructed" (172). See also Kuper (1968).

moving beyond this level into the primary type. Most Euro-American ethnic groups, for example, have reached a point at which they enjoy relatively equal access to jobs, political authority, and other important life chances. They have, in other words, entered into full participation in all institutional areas of American society. Moreover, for most, primary relations are no longer limited largely to the ethnic group. For African Americans and other racial-ethnic groups, however, although they have begun to achieve substantial integration in the economy, polity, and education, that has not been the case in clubs, cliques, neighborhoods, and intermarriage.

PSYCHOLOGICAL ASSIMILATION   Our focus is primarily on how various groups interrelate; our concern, therefore, is chiefly with the cultural and structural dimensions of assimilation. But there is an individual dimension of assimilation, in which attention falls on particular members of an ethnic group rather than on the group as a whole. This is part of the social psychology of ethnic relations and concerns the extent to which individuals have been absorbed into the larger society and identify with it.

With **psychological assimilation**, members of an ethnic group undergo a change in self-identity. To the extent that *individuals feel themselves part of the larger society rather than an ethnic group*, they are psychologically assimilated. As psychological assimilation progresses, people identify themselves less in ethnic terms. Whereas first-generation immigrants and their children will define themselves as Italian American or Irish American, by the third generation ethnicity for most is no longer a major component in response to the question, "Who am I?" This level of assimilation consists not simply of becoming culturally like members of the mainstream society but accepting that society as "home" and thoroughly identifying with it.

Psychological assimilation implies a change not only in self-identification but also in identification by others. Although individual members of an ethnic group may see themselves as simply part of the larger society rather than as ethnics, outsiders may continue to identify them as members of their group, thereby impeding psychological—and structural—assimilation. As Yinger notes, "Prejudice on the part of a dominant group may prevent the granting of full membership in a society to members of minority groups, even though the latter think of themselves only in terms of the larger society" (1981:253). Visibility is, of course, critical here. Those with salient marks of ethnic identity—physical characteristics, in particular—are unable to fully achieve out-group recognition as "nonethnics."

Some individuals may find themselves unable to feel fully part of the larger society *or* the ethnic group. Park (1928) and Stonequist (1937) used the term "marginal men" to describe these people. First-generation immigrants often find themselves in a situation where they are pulled in the direction of the culture of the new society but remain culturally and psychologically tied to the old. Those who are not clearly part of one racial category or another also may be marginal people. In American society in recent years, such individuals have begun to be recognized by themselves and others as constituting a unique "mixed-race" or "multiracial" category.

Although Park, in a classic work (1928), described marginal people as experiencing psychological turmoil, not all of them find difficulty in social adaptation. Some successfully adapt to two cultures, shuttling between them, and others may use their marginality to advantage by serving as middlemen between dominant and minority groups. Moreover, where they constitute a large population, marginal people may form communities of their own, occupying an in-between status in the society's ethnic hierarchy (Shibutani and Kwan, 1965). Recall the description of middleman minorities in Chapter 2.

The social psychology of ethnic identity is a complex matter, the details of which go well beyond our present purposes.[2] Suffice it to say that within any multiethnic society, individuals may vary widely in the extent to which they identify with an ethnic group or choose to disregard ethnicity as part of their self-identity. Much depends not only on visibility but also on the political and social conditions that may affect the individual costs and values of an ethnic identity. For example, whites in American society are permitted much latitude in their choice of personal identity; the same is not the case for racial-ethnic groups.

Biological Assimilation **Biological assimilation,** or **amalgamation,** represents the ultimate stage in the assimilation process. At this point, intermarriage has occurred to such an extent that *there is a biological merging of formerly distinct groups.* They are indistinguishable not only culturally and structurally but physically as well. Some degree of amalgamation is a common by-product of group contact and interaction. But the biological fusion of diverse groups is an outcome of interethnic relations that can occur only over a very long period. Mexico, for one, is a society that seems to have moved far along toward this long-range objective (van den Berghe, 1978); Brazil, as will be seen in Chapter 15, has also progressed in this direction.

Theories of Assimilation

Park's Race Relations Cycle American sociologists in the 1920s, led by Robert Park, were intrigued by the ethnic polyglot that had emerged in cities of the U.S. Northeast and Midwest, and they began to focus their studies on the social forces that brought these groups together or sustained their differences. Park was one of the first to suggest a cycle of race or ethnic relations through which groups would pass in a sequence of stages, leading ultimately to full assimilation.

Park explained that groups first come into *contact* through migration and subsequently engage in *competition,* often characterized by conflict. Out of such competition eventually emerges some form of *accommodation* among the groups, leading finally to *assimilation.* Park maintained that this four-stage cycle pertained to race and ethnic relations everywhere, not simply the United States. Moreover, he saw the sequence as "apparently progressive and irreversible" (1950:150).

A view of ethnic assimilation not unlike Park's race relations cycle has been generally popular in American society. The prevailing thought is that over several generations, group boundaries break down, and the society becomes more homogeneous. However, Park's model has been subject to much sociological criticism over the years. Some have noted the cycle's lack of applicability to many groups (Lyman, 1968b). Although it has seemed to describe fairly accurately the experiences of most European immigrant groups in the United States (as well as in Australia, Canada, Argentina, and other immigrant societies), it does not conform to the patterns

---

[2] The issues of ethnic identity at the individual level are detailed in De Vos and Romanucci-Ross (1975), Horowitz (1975), Romanucci-Ross and De Vos (1995), and Shibutani and Kwan (1965).

displayed by more salient—that is, racial-ethnic—groups and those that have entered involuntarily.

Critics of Park's and other cyclical theories of ethnic relations have also pointed out that such cycles are rarely complete; that is, there are too many truncated instances (Shibutani and Kwan, 1965). Interethnic contact can produce stable outcomes—such as exclusion, pluralism, or continued ethnic stratification—that do not lead inevitably to assimilation. Finally, some have criticized the model's claim of irreversibility (Barth and Noel, 1972; Berry and Tischler, 1978). They note that the cycle may be terminated at any point, and groups may even revert to earlier stages.[3]

GORDON'S STAGES OF ASSIMILATION   Park's model, despite its shortcomings, stands as a precursor to subsequent, more sophisticated theories of the assimilation process. Perhaps the most precise and compelling of these is Milton Gordon's (1964).

Like Park, Gordon explained assimilation as a series of stages, or steps, through which various groups pass. But rather than following a straight line leading from contact to eventual absorption, groups may remain indefinitely at one or another of these stages. The seven stages that Gordon outlined extend from cultural assimilation, the least intense, to civic assimilation, the most complete. At this last stage, there is an absence of prejudice and discrimination against the group, its members have fully adopted the dominant group's values, and they share in the society's power structure without regard to their ethnic origin (Table 4.1). The first two stages, cultural and structural assimilation, however, are the most important.

TABLE 4.1 | GORDON'S STAGES OF ASSIMILATION

Stage	Characteristics
Cultural or behavioral assimilation (acculturation)	Change of cultural patterns to those of host society
Structural assimilation	Large-scale entrance into cliques, clubs, and institutions of host society on primary group level
Marital assimilation (amalgamation)	Large-scale intermarriage
Identificational assimilation	Development of sense of peoplehood based exclusively on host society
Attitude receptional assimilation	Absence of prejudice
Behavior receptional assimilation	Absence of discrimination
Civic assimilation	Absence of value and power conflict

Source: *Assimilation in American Life: The Role of Race, Religion, and National Origins* by Gordon (1964) 67w from p. 71. By permission of Oxford University Press, USA.

---

[3] Geschwender (1978) disputes the interpretation of Park's race relations cycle as absolute, maintaining instead that it is "situationally specific." He sees Park's cycle as an ideal type, not to be taken as inevitable. Rumbaut (1999) also points out that Park recognized that the eventual outcome of relations among racial groups could take several different forms.

As noted earlier, cultural assimilation, or acculturation, denotes the adoption by a minority ethnic group of the dominant group's cultural patterns—language, political beliefs, and so on. But acculturation, explains Gordon, does not assure movement to the next phase. Minority ethnic groups may become very much like the dominant group in behavior and values but still remain structurally segregated.

Structural assimilation, the second stage, is essentially what was previously referred to as primary, or informal, structural assimilation. This stage is the most critical for it is the key to all subsequent stages: "Once structural assimilation has occurred, either simultaneously with or subsequent to acculturation, all of the other types of assimilation will naturally follow" (Gordon, 1964:81). Presumably, as people of minority and dominant ethnic groups interact within close, intimate social settings, the other stages of assimilation necessarily occur, much like falling dominoes: minority group members increasingly intermarry with those of the dominant group, relinquish their ethnic identity, no longer encounter prejudice and discrimination, and fully agree with the dominant group on issues involving values and power conflicts. Indeed, the remaining stages (three through seven) in Gordon's scheme can be subsumed under structural assimilation. The dominant group's acceptance of members of a minority ethnic group into primary relations implies, for example, the absence of prejudice and discrimination and the likelihood of increased intermarriage.

It is with structural assimilation, then, that full assimilation involving all other stages becomes inevitable. But, as Gordon notes, cultural assimilation, the first stage, does not necessarily lead to structural assimilation; it may continue to be the extent of assimilation for many generations (see also van den Berghe, 1981; Wagley and Harris, 1958). Minorities may take on all or most of the cultural ways of the dominant group but still be refused entry into primary relations with its members. African Americans are an evident illustration of this. Though thoroughly assimilated regarding the major elements of the dominant culture (such as language and religion), they remain unassimilated at the structural level (specifically the *primary* structural level). In short, groups may become culturally alike yet remain in relatively segregated subsocieties.

Several criticisms have been made of Gordon's stages of assimilation model. One serious shortcoming lies in its understanding of structural assimilation, stage two, as entailing interaction with the dominant group only at the primary level. Intergroup relations, however, occur at the secondary level as well and indeed are antecedent to primary relations in any significant degree. To what extent do members of a minority ethnic group enter into positions of power in the society's economic, political, and other key institutions? To what extent are they afforded equal opportunities in employment and education? These are significant measures of structural assimilation—at the secondary level—that Gordon does not consider (Marger, 1979). African Americans, for example, appear to have realized substantial integration in recent years in the areas of work and government. They have, in other words, experienced increasing secondary structural assimilation. That they have not accomplished an equivalent level of interaction with whites at the primary level would, in Gordon's view, imply that little structural assimilation had occurred.

Moreover, Gordon seems to suggest that if minorities do not enter into primary relations with the dominant group, it is because the dominant group has held them out. But such social segregation may be in some part voluntary.

Despite these omissions, Gordon's model is valuable to the analysis of interethnic relations primarily because it spells out the intricacies of the assimilation process and its various forms. Assimilation, as Gordon demonstrates, is clearly not a simple, straightforward movement, as earlier theories had seemed to imply.

SEGMENTED ASSIMILATION   Some contemporary immigration theorists have explained that assimilation is more varied than the classical models have presumed. They point out that *assimilation is not a single process and may occur in different domains entailing different reference populations* (Brubaker, 2001, 2004; Fernández Kelly and Schauffler, 1996; Waters, 1999). There are, as a result, diverse outcomes of the adaptation process of immigrants and, especially, their children—hence, the term **segmented assimilation** (Portes and Zhou, 1993, 1994). Ordinarily it has been assumed that movement toward assimilation implies adoption of the dominant group's culture and incorporation into the mainstream social structure. The idea of segmented assimilation suggests, however, that different groups, or elements of groups, may assimilate to an oppositional culture or to another culture clearly outside the mainstream. Rogers Brubaker (2004) explains that the traditional focus of assimilation therefore has shifted from the single question (how much assimilation?) to several questions (assimilation in what respect, over what period of time, and to what reference population?).

Contemporary immigrants in American society illustrate well the notion of segmented assimilation. As will be seen in Chapter 5, most are not European in origin and thus are perceived as racially different; for many, this blocks the path of rapid assimilation into the core culture (that is, "white, middle-class"). Also, the U.S. economy no longer provides opportunities for upward mobility through jobs in manufacturing and other industries as it did for immigrants of earlier periods, thereby preventing the movement of a large segment of the current immigrant population into the mainstream workforce.

Given these conditions, the children of immigrants may still move in the traditional path of assimilation, adopting the dominant culture and integrating into the dominant society. But it is also possible for them to assimilate to a minority culture, thus putting at risk their chances of upward social or economic mobility. Children of West Indian immigrants (from Haiti, Jamaica, and other Caribbean countries), for example, may drift toward African American culture rather than that of the white middle class (Kasinitz et al., 2001). And in a third variant of assimilation, under certain conditions immigrant families may advance economically, yet deliberately seek to preserve their ethnic culture and solidarity. In such cases, by instilling the ethnic culture parents try to protect their children from assimilation into an oppositional culture that may lead to downward mobility (Hirschman et al., 1999). Immigrant Asian families often adopt this path, encouraging their children to excel in education while continuing to adhere closely to traditional ethnic values (Zhou, 2004).

FACTORS AFFECTING ASSIMILATION

Why do some groups in multiethnic societies display a rapid and almost complete assimilation into the larger society, whereas others remain segregated and are the constant targets of prejudice and discrimination? Where assimilation is the prevailing model of intergroup relations, several factors are important in shaping the experience of different groups: how and when a group enters the society, its size and dispersion, its cultural similarity to the dominant group, and its visibility.

MANNER OF ENTRANCE   As we saw in Chapter 2, the way in which a group enters the society is critical in determining its place in the ethnic hierarchy; it is also important in accounting for the nature of the group's long-range societal adaptation. Except for those groups that maintain unfaltering separatist goals ("pluralistic minorities," as they were referred to in Chapter 2), groups that enter voluntarily always make a less conflict-ridden adjustment than those that enter involuntarily or those that are conquered by more powerful invaders. Involuntary immigrants or conquered groups remain in a condition of segregation to one degree or another.

The United States presents a clear illustration of this tendency. European ethnic groups, having entered by choice, were able to assimilate culturally at a pace determined in large part by themselves, no matter how strong the pressures to conform. Furthermore, the option of returning to their society of origin was always present and was, in fact, taken by many. By comparison, African Americans and American Indians were absorbed into the society involuntarily; as a result they could not follow similar paths no matter what their intentions or long-range goals (Blauner, 1972). As noted in Chapter 2, indigenous groups who are made subordinate at the entrance of an invading group (North American Indians) or who enter involuntarily (African Americans) are left with few options other than resistance to the new social order that is imposed on them. Long-term conflict is thus the usual outcome. Subordinate voluntary immigrants (like Europeans to America), in contrast, are more rapidly assimilated (Lieberson, 1961; Schermerhorn, 1970; van den Berghe, 1976; Wilkie, 1977).

TIME OF ENTRANCE   In general, the more recent a group's entry into the society, the more resistance there is to its assimilation (Mack, 1963). Other things being equal, the simple factor of time will ease the fear and suspicion that accompany the entrance of strangers. Groups with alien ways are seen differently after they have lived in the society for several generations. Examples abound in the United States, beginning with the influx of large numbers of Irish immigrants in the early nineteenth century. Given the very substantial assimilation of Irish Americans today, the hostile reception encountered by the first generation is easily forgotten.

DEMOGRAPHIC FACTORS   The degree and rate of assimilation for minority ethnic groups is also affected by their size and the concentration of their population (Blau, 1977; Frisbie and Neidert, 1977). The entrance and assimilation of groups relatively small in number will be resisted less forcefully than that of groups representing a competitive threat. Van den Berghe (1981) suggests that smaller groups are assimilated more easily because they have fewer resources and therefore depend on the larger society and because they necessarily interact more frequently with out-group members.

The concentration or dispersion of ethnic groups also bears on their assimilation. Concentration in particular neighborhoods and geographical areas or in certain occupations tends to retard assimilation because the group is better able to retain its cultural ways and resist intrusions of the dominant group (van den Berghe, 1981).

Dispersal, on the other hand, leads to unavoidable contact and interaction with the dominant group and thus speeds up the assimilation process.

CULTURAL SIMILARITY   No matter how or when the group enters the host society, or what its demographic patterns, assimilation will occur relatively quickly if the group is culturally similar to the dominant group (Berry and Tischler, 1978; van den Berghe, 1981). Those groups in the United States that have followed the assimilation route furthest have, predictably, been culturally closest to Anglo Protestants. In general, the more compatible the culture of the minority group with the dominant group's, the greater will be the depth and speed of assimilation.

VISIBILITY   In almost all multiethnic societies, the most critical factor in determining the degree and rate of assimilation of ethnic groups is visibility. Where physical differences are obvious, manner of entrance, temporal factors, demographic patterns, and even cultural similarity are of much less consequence. For racial-ethnic groups, structural separation remains far more persistent than for groups that are only culturally distinct. Observers of American ethnic relations have long interpreted the retarded structural assimilation of blacks, for example, as a product chiefly of visibility (Park, 1950; Warner and Srole, 1945; Wirth, 1945; Yinger, 1981). The visibility factor is also evident in the case of Asian Americans who may be culturally indistinguishable from the dominant group, but because of their racial distinction are assumed to be outsiders. Second-, third-, and even fourth-generation Asian Americans are commonly faced with having to respond to questions of "Where do you come from?" or to comments like "You speak English so well."

In short, physical differences delay the process of assimilation more than do other factors. Harold Isaacs has poignantly described this dilemma: "An individual can change his name, acquire a new language, ignore or conceal his origins, disregard or rewrite his history, abandon his ancestral religion or convert to another one, adopt a different nationality, embrace new mores, ethics, philosophies, styles of life. But there is not much he can do to change his body" (1989:46). Thus the more visible the group or individual, the longer and more difficult is the process of structural assimilation.

ASSIMILATION AS A SOCIETAL GOAL

Assimilation can be viewed as a goal, or ideal, for which multiethnic societies aim. As such, it is sometimes the basis of public policies designed to reduce the cultural and structural divisions between groups. In the United States, for example, measures intended to lessen segregation in various public spheres (housing, schools, work) and to equalize access to power and privilege (affirmative action programs, voting rights) can be understood as outgrowths of a societal commitment to the eventual achievement of complete assimilation for all ethnic groups. However, the form and ultimate objective of assimilation may vary in the minds not only of policy makers but also of both dominant and minority group members.

Essentially there are two possible forms of complete assimilation for which societies may strive, each involving a different objective and thus a somewhat different path.

- Ethnic groups will assimilate to the dominant ethnic group, adopting its cultural ways and seeking integration into its social institutions.
- Groups will assimilate into an entirely new ethnicity in which all groups surrender their ethnic heritage but in the process create a hybrid society with no dominant group as such.

In the United States and most other multiethnic societies, it is the first form, assimilation to the dominant ethnic group, that has prevailed as the long-range goal. The other, hybrid, option is what has been popularly referred to as "the melting pot." Although that idea has had wide currency, it is not what has prevailed for most of American history. Dominant group assimilation, as we will discover in the chapters of Part II, has been expected of new ethnic groups in the past and continues to be expected of the latest immigrants to America.

As long as one group disproportionately controls power resources in the economic and political realms, that group can dictate the shape and direction of minority group adjustment. It is the dominant group that, in the main, mandates or approves policies of acculturation and integration. In the United States, for example, there has never been any question of whose language, whose religious and political principles, and whose basic institutions will prevail. For most of U.S. history they have been those of the dominant Anglo core group and, despite the challenge in recent decades of a more ethnically diverse population and culture, they continue to be upheld. Ideologically, societies may advocate some kind of ethnic melting pot wherein all groups contribute in proportionate amounts to form a new social system (Brazil, as we will see in Chapter 15, has produced something of a cultural hybrid), but in the modern world such a fusion remains an ideal.

As previously noted, however, we should not think that assimilation is a one-way process whereby minority ethnic groups seek out and become like the dominant group, with no change occurring in the opposite direction. Obviously, the assimilation process will be to some extent mutual, with many aspects of minority cultures becoming part of the dominant culture. But the exchange is far short of equal. On the major elements of culture—language, religion, political beliefs, economic practices, and so on—there is little evidence of mutuality, and at the structural level the exchange is even more lopsided. Those aspects of culture and social structure remain firmly controlled by the dominant group, and minority ethnic groups must adjust to them.

PLURALISM

Like assimilation, pluralism entails several dimensions and forms.[4] In all cases, however, the retention or even strengthening of differences among ethnic groups is presumed. Thus in a general sense pluralism is the opposite of assimilation. Abramson defines pluralism as "conditions that produce sustained ethnic differentiation and continued heterogeneity" (1980:150). In brief, pluralism involves social processes and institutions that encourage group diversity and the maintenance of group boundaries.

---

[4] Pluralism as applied to political systems refers to the relative dispersion of power among various interest groups in a society. This is different from its usage in describing ethnic relations. See Marger (1987).

Just as assimilation occurs in different degrees and at different stages, so too must pluralism be understood as a variable for groups and societies. Ethnic pluralism never entails an absolute separation of groups. Recall the definition of ethnic group as a distinguishable group *within a larger society*. Thus in a pluralistic society there is always some common political or economic system that binds various ethnic groups together. If this were not so, there would not be a multiethnic society but several distinct societies in themselves. Within the broad confines of a common political or economic system, however, groups may differ widely.

CULTURAL AND STRUCTURAL PLURALISM   As with assimilation, we can delineate cultural and structural dimensions of pluralism. **Cultural pluralism** *implies the maintenance of many varied cultural systems within the framework of the larger sociocultural system* (Gordon, 1964). This dimension of pluralism has been referred to in the United States in recent years as **multiculturalism.**

Structural pluralism *connotes not simply differences in culture but also the existence in some degree of segregated ethnic communities within which much of social life occurs for group members.* These ethnic subsocieties, or communities, comprise institutions—schools, businesses, churches, and the like—that duplicate to some extent those of the dominant group. Notice how cultural and structural pluralism are basically opposing counterparts of cultural and structural assimilation.

SYSTEMS OF ETHNIC PLURALISM

Several different types of ethnic relations can be seen among multiethnic societies that are organized around pluralist rather than assimilationist principles.

EQUALITARIAN PLURALISM   Where **equalitarian pluralism** characterizes interethnic relations, *groups retain their cultural and, for the most part, structural integrity while participating freely and equally within common political and economic institutions* (Barth and Noel, 1972; Shibutani and Kwan, 1965). Some have referred to this condition as **accommodation,** wherein *the minority group "desires equality with, but separation from, the dominant group and the dominant group agrees to this arrangement"* (Kurokawa, 1970:131). Technically, of course, if equality with the dominant group is reached, there are no longer dominant-minority relations; there are relations among ethnic groups, but they are not hierarchical and invidious.

Equalitarian pluralism corresponds, like assimilation, to the order model of society, in which balance and cohesion are emphasized. Differences among groups are recognized and even encouraged, but within the framework of a larger set of agreed-on principles. All groups give allegiance to a common political system, participate in a common economic system, and understand a common set of broad ethical values (Williams, 1977). In a sense, ethnic groups become political interest groups that compete for the society's rewards (Glazer and Moynihan, 1970). But these competitive differences do not lead necessarily to serious cleavages and conflict; rather, they are dealt with by a reasonable give-and-take within the context of the consensual rules of the society.

There may be vast cultural differences among ethnic groups, but their essence is not threatened because tolerance of such differences is integral to the social order.

Relations between ethnic groups are thus confined mainly to functional areas such as government and the marketplace, not affective ones such as the family or friendship circles.

CORPORATE PLURALISM   In some societies, equalitarian pluralism is formally declared, and much of the structural and cultural separation is upheld by political authorization. In societies such as Switzerland, Belgium, Malaysia, and to some extent, Canada, *the structural and cultural differences among ethnic groups are protected by the state, and institutional provisions are made to encourage an ethnically proportionate distribution of societal rewards.* Gordon (1975, 1981) calls such cases **corporate pluralism.**

In such pluralistic systems, ethnic units are formally recognized by the government, and political and economic power is allocated on the basis of an ethnic formula. Thus in the political arena legislative seats and other government offices may be apportioned on the basis of ethnicity. Not only is proportionality in the distribution of political benefits assumed, but most important, there is cooperation among leaders of all the significant segments of the plural society.[5] Moreover, on local matters each group maintains a great deal of political autonomy. In the economic realm as well, the objective is an equal distribution of income and jobs among the various groups, proportional to their makeup in the national population (Lijphart, 1977; McRae, 1974; van den Berghe, 1981).

In this type of pluralism, cultural and structural separation are emphasized, not discouraged, and multilingualism is officially sanctioned (Gordon, 1981). Switzerland, with its four official languages—German, French, Italian, and Romansh (the latter spoken by only a very small number)—is perhaps the most obvious and successful multilingual system (McRae, 1983). Canada, with its French- and English-speaking groups, is another familiar multilingual (in this case, bilingual) society, though, as we will see in Chapter 16, multilingualism there has been the source of a severe social divide.

In societies characterized by corporate pluralism, ethnic groups consist mainly of homogeneous, territorially concentrated peoples who have long historic roots in their native area. They have become part of a larger national society either through conquest or by voluntarily relinquishing sovereignty to a central state in order to secure economic and political benefits. Such societies are not at all similar to multi-ethnic societies like the United States, in which ethnic groups have been formed primarily by voluntary immigrants who came from distant societies, severed most of their native roots, and dispersed geographically after arrival.

The cultural divisions among ethnic groups in corporate pluralistic societies are much sharper than those among groups in the United States, where the English language tends to become a great commonizing factor. Structural separation is also stronger, given the territorial concentration of groups. In Switzerland, for example, each ethnic group remains geographically distinct; and in Canada more than 80 percent of French-speaking people reside in one province, Quebec. By contrast, ethnic groups in

---

[5] This type of political arrangement has been described by Lijphart (1977) and others (McRae, 1974; van den Berghe, 1981) as *consociational.*

the United States are scattered, though regional concentrations of particular groups are noticeable (such as Mexican Americans in the Southwest or Swedish Americans in the upper Midwest). Ethnic concentrations in the United States more often take the form of urban pockets or neighborhoods.

With their retention of language and their territorial base, ethnic groups in corporate pluralistic systems are integrated only in their mutual allegiance to a larger national government and the need to participate in a national economic system. Ideally, no single group is dominant, and each is afforded an approximately proportionate share of the society's rewards (Wagley and Harris, 1958). No corporate pluralistic society, of course, has met this ideal, although some have achieved a greater degree of group equality than others.

Equalitarian pluralism seems like an ideal way of managing an ethnically divided society and assuring relative equality among the various groups. But if we can judge from contemporary societies that have attempted this path to ethnic harmony, the results are mixed. The breakup of Yugoslavia is a tragic illustration of an equalitarian pluralist system that failed. Here, several ethnic nationalities, living for the most part in distinct territories, made up a state that provided roughly proportional political power for each. But the system collapsed in 1991, giving rise to one of the most brutal episodes of ethnic warfare that has been witnessed in modern Europe. This case is discussed in more detail in Chapter 17. In the case of Canada, with a number of features of this type of pluralism, the results have been far more benign, though not without persistent low-level conflict, as we will see in Chapter 16. Cases like Belgium and Switzerland, where interethnic relations have been balanced and harmonious for many decades, appear to be exceptional, and even in those societies periodic group conflict is evident. Belgium, for example, experienced a national crisis in 2007 when its two regionally concentrated ethnic communities, French-speaking Walloons and Dutch-speaking Flemings, could not reach an agreement on the country's future direction. So serious was the discord that many feared Belgium as a unified nation-state would be irreparably broken up.

The assertion of Wagley and Harris (1958) that pluralistic aims perpetuate some degree of conflict and the subordination of one group by another is well taken. Indeed, conflict, whether latent or active, seems to be endemic to pluralist systems. The level of conflict, however, is much greater in societies with *in*equalitarian pluralist systems, to which we now turn.

INEQUALITARIAN PLURALISM   Both assimilation and equalitarian pluralism are characteristic of societies that have made the reduction of ethnic inequality a well-established commitment of the state, legitimated by a democratic ideology. With inequalitarian pluralism, however, the outcome of ethnic relations is inequality among the various groups. And, that inequality is sanctioned by the state. Although most characteristic of classic colonial societies and racist regimes like South Africa during its period of apartheid, inequalitarian pluralism can also be seen in some degree in other multiethnic societies where assimilation or equalitarian pluralism is the prevailing type of interethnic relations.

Under **inequalitarian pluralism,** *ethnic groups not only are separated structurally and perhaps culturally but also exist in a state of highly unequal access to power and privilege.* Indeed, the authority and power of the dominant group are key coordinating mechanisms of such systems (Kuper and Smith, 1969; van den Berghe, 1978). In an equalitarian pluralistic system, the various ethnic groups are held together through the consensual allegiance to a common state. In inequalitarian pluralistic societies, however, the state holds the different groups together not through a mutually recognized legitimacy but through coercion. Basically, the state acts to protect the interests of the dominant group. The anthropologist M. G. Smith (1969:33) explains that in a plural society of this type, the state is the agent of the ruling group only. Others have no rights or protection. The majority of people are "subjects, not citizens." Whereas equalitarian pluralism assumes a progressive equalization of political and economic power among groups, the assumption here is quite the opposite. Sustained or increased inequality among groups is a built-in feature of the system, with the dominant group retaining all political authority and the bulk of material wealth.

Social relations between dominant and minority groups are typified by extreme polarization, supported by high levels of prejudice and discrimination. Basically, it is only within the impersonal confines of the economic and political systems that dominant and minority group members interrelate, and those relations are limited to purely functional contacts such as work and government administration. As Kuper and Smith (1969:11) describe it, "Economic symbiosis and mutual avoidance, cultural diversity and social cleavage" characterize the social basis of this type of plural society.

Although the dominant group holds sway, ethnic groups are in a state of economic interdependence. The dominant group needs subordinate ethnic groups to perform physical and menial laboring tasks necessary to economic production; given their relative powerlessness, subordinate groups must meet those demands. Occupational roles are assigned strictly on the basis of ethnicity, with the subordinate group or groups delegated those most onerous and least prestigious (Rex, 1970). The dominant group, as van den Berghe explains, "rationalizes its role in an ideology of benevolent despotism and regards members of the subordinate group as childish, immature, irresponsible, exuberant, improvident, fun-loving, good humored, and happy-go-lucky; in short, as inferior but lovable as long as they stay in their place" (1978:27).

Institutional separatism and duplication characterize inequalitarian pluralistic systems (Smith, 1965; van den Berghe, 1978). This means that each group maintains its own schools, churches, businesses, and so on. In each case, however, there is a great disparity in the quality of dominant and minority group institutions.

PATERNALISTIC AND COMPETITIVE RACE RELATIONS   In its ultimate form, inequalitarian pluralism resembles a caste system in which strict segregation is enforced in all areas of social life (Furnivall, 1948). In these cases, the social distance between groups is maximized in all social situations. What emerges is a system van den Berghe (1978) calls **paternalistic race relations**. *All people understand their social place and, as long as the subordinates do not deviate from their ascribed role, stability is ensured.* In such systems, there is at least some acquiescence of subordinate groups to their inferior status. Actual physical distance need not be enforced in all situations as long as the *social* distance between groups is understood and adhered to. Slavery in the American South, for example, was characterized by a good deal of physical proximity, particularly among slaves who performed household duties (Stampp, 1956).

Inequalitarian pluralistic relations in such an extreme form are realized only in slave or classic colonial systems (Rex, 1970), neither of which is evident in the contemporary world (though South Africa, until the end of its apartheid system, closely approximated them). In those industrialized societies where ethnic divisions are based primarily on race rather than culture, *ethnic stratification bears some resemblance to a caste or colonial system but is less extreme in the segregation of social institutions and relations.* Van den Berghe (1978) has referred to such cases as **competitive race relations.** With the society's economic base changed from agrarianism to industrialism, such competitive relations replace paternalistic relations. Industrialization requires that social roles be assigned more through competition than through ascription. As a result, there are no longer "master-servant" relations as in a caste or paternalistic system; however, competition between the subordinate group or groups and the working-class element of the dominant group emerges. Because the norms of social distance, so well institutionalized in the paternalistic condition, now break down, physical segregation becomes more rigid. Ethnic groups tend to reside and carry out much of their daily activities in homogeneous ghettos. As we will see in Chapter 7, the plight of African Americans following slavery generally conformed to this pattern.

Expulsion and Annihilation   Inequalitarian pluralism may reach the point of **expulsion** or even the **annihilation** of minority ethnic groups. Neither of these outcomes is without precedent in recent Western history. The deportation of Chinese from the United States in the nineteenth century, the detention of Ukrainians in Canada during World War I, and the internment of Japanese in both the United States and Canada during World War II are notable examples of expulsion. As for annihilation, destruction of native groups by white settlers in North America, Australia, and South Africa in the nineteenth century as well as in Latin America earlier are all relevant examples. A more deliberate and methodical case is the genocidal policy of the Nazis, which resulted in the near-destruction of the European Jews. In the 1930s, German Jews were systematically subjected to an almost complete expulsion from every phase of the society's life. Later, Jews were impelled to leave Germany through a terroristic campaign that included physical attacks and the appropriation of their homes, businesses, and wealth. Finally, Nazi policies culminated in the establishment of death camps to which German and subsequently other European Jewish populations were sent to be slaughtered. The campaign of ethnic cleansing by Serbs against Muslims in Bosnia and ethnic Albanians in Kosovo (examined in Chapter 17) are more recent cases of a combination of expulsion and annihilation.

Pluralism in the United States

Although the United States is a multiethnic society that is basically assimilationist in its ethnic ideology and organization, there are pluralistic elements that are evident. Sociologists have pointed out that, despite the forces of assimilation, American ethnic groups still express elements of their ethnic cultures many generations after their immigration. They also note that individuals continue to conduct many of their primary group relations within an ethnic context. The choice of close friends or marital partners, for example, is still strongly influenced by ethnicity. Moreover, many people continue to identify themselves, at least symbolically, in ethnic terms. There is an important difference, however, between continued expressions of ethnic pluralism for Euro-American groups as opposed to racial-ethnic groups in American society.

For Euro-Americans, the persistence of ethnicity is more a matter of individual choice than a collective imperative. Gordon (1975, 1981) has referred to this condition as "liberal pluralism" (as contrasted with corporate pluralism) in which there is no formal recognition of ethnicity in the allocation of government offices or economic rewards and in which individuals are free to express their ethnic identity to whatever extent they choose.

For racial-ethnic groups, pluralism—particularly at the structural level—is mostly involuntary and remains in place largely as a result of the resistance of Euro-Americans. Cultural assimilation for these groups progresses relatively rapidly but structural pluralism remains strongly evident in many areas of social life, including high levels of residential segregation and low rates of intermarriage with whites. This is especially so for African Americans, as we will see in Chapter 7.

THE VARIABILITY OF ETHNIC RELATIONS

The three major patterns of ethnic relations we have examined—assimilation, equalitarian pluralism, and inequalitarian pluralism—are outlined in Figure 4.1. Several points should be kept in mind in looking at these patterns.

- In no case does any one of these models by itself perfectly characterize what occurs in a multiethnic society. Realistically, most societies will exhibit features of two or possibly even all three simultaneously. The United States illustrates this quite well, where assimilation has been the predominant pattern, but a number of features of inequalitarian pluralism pertain to racial-ethnic groups.
- Some degree of conflict is characteristic of all three systems. Although it is obvious in the case of inequalitarian pluralism, conflict is a feature in more subdued fashion of assimilation and equalitarian pluralism as well. It must be remembered that these types of interethnic relations are *movements* in the direction of either homogeneity or ethnic separation, not realized conditions. In either case, some degree of ethnic stratification is evident, and conflict, whether latent or active, is therefore present as well. In societies where assimilation or equalitarian pluralism is more prevalent, the ethnic hierarchy will, of course, be less inflexible and the divisions between groups less acute than in societies where inequalitarian pluralism is the predominant form.
- The outcome of any of these systems, whether assimilation or pluralism, ultimately depends on the aims of both the dominant group and minority groups. As noted in Chapter 2, minority groups have different goals regarding their place in the society and the nature of their relations with the dominant group, but this is only one side of the issue. Perhaps more important are the goals the dominant group wishes for the minorities. When the objectives of dominant and minority groups are congruent, conflict is reduced; when they are in opposition, conflict is unavoidable.

In sum, all these processes and outcomes of ethnic contact are relations of power—power of minority groups and, of course, power of the dominant group.

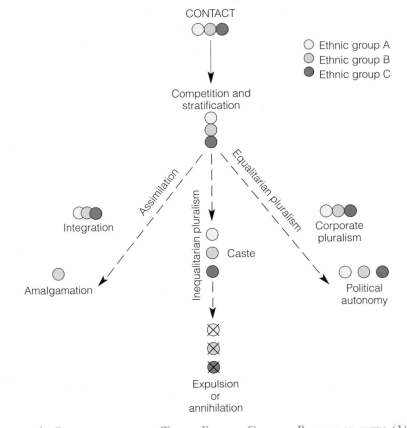

FIGURE 4.1   RELATIONS AMONG THREE ETHNIC GROUPS, BEGINNING WITH (1) INITIAL CONTACT, FOLLOWED BY (2) EMERGENCE OF COMPETITION AND STRATIFICATION, AND (3) MOVEMENT IN ONE OF THREE GENERAL DIRECTIONS

In any situation of ethnic contact, one must first ask, "Who has power and to what ends can they apply it?" On that question hinges the eventual nature of relations among groups.

A TYPOLOGY OF MULTIETHNIC SOCIETIES

In Chapter 2, we explored the manner in which multiethnic societies are stratified, noting that the degree of ethnic stratification can vary from society to society. These different levels of ethnic inequality can now be combined with the processes of assimilation and pluralism to construct a typology of multiethnic societies (Table 4.2). These descriptions are not meant to serve as pigeonholes into which particular societies can be conveniently placed; any society will probably exhibit some features of all. In broad terms, however, a society will be more characteristic of one or the other. The typology will, then, point out major features of multiethnic societies and will serve as a reference that can be consulted in our analyses of specific groups and societies in Parts II and III.

TABLE 4.2 | THREE TYPES OF MULTIETHNIC SOCIETY

Feature	Type of Society		
	Colonial (Segregationist)	Corporate Pluralistic (Multicultural)	Assimilationist
Initial contact between dominant and minority groups	Conquest of indigenous groups by dominant group or involuntary migration of minorities	Annexation or voluntary immigration	Mainly voluntary immigration but involuntary immigration and conquest for salient minorities
Relations between dominant and minority groups	Paternalistic or competitive	Ideally equalitarian, but often competitive	Competitive
Nature of stratification	Caste or castelike; caste and ethnicity overlap closely	Class hierarchy within each ethnic group	Class; class and ethnicity generally overlap, but minority group members are dispersed throughout general class system
Segregation between group	Very rigid; explicitly defined and enforced by tradition and law	Voluntarily rigid; groups often concentrated in distinct territories	Mild and largely voluntary for groups culturally and physically similar to dominant group; rigid and involuntary for salient minorities
Institutional separation among ethnic groups	High except in economy	High except in economy and central government	Low in polity and economy; variable in other areas
Physical and cultural differences between dominant and minority groups	Sharp physical differences; sharp cultural differences, at least initially	Usually slight or no physical differences; key cultural difference usually language or religion	Broad range of physical types; sharp cultural differences initially
Main objectives of ethnic policy	Inequalitarian pluralism	Equalitarian pluralism	Assimilation; some degree of unofficial structural pluralism for salient minorities
Degree of conflict among ethnic groups	High eventually, though usually subdued for long periods	Relatively low except on matters pertaining to cultural and political rights	Mild but at times high between racially distinct groups
Examples	Antebellum U.S. South, colonial India, South Africa under apartheid	Switzerland, Malaysia, Canada (partially)	United States, Brazil, Israel, France

COLONIAL (SEGREGATIONIST) SOCIETIES

Inequalitarian pluralism is the chief feature of societies that can be called **colonial** or **segregationist**. *A dominant group exerts maximum political and economic power and is thereby able to shape the nature of interethnic relations in such a way as to sustain its interests.*

In these societies, the dominant group has ordinarily entered as a conqueror of physically distinct indigenous groups or has brought the minority groups to the society as slaves. Such societies are ordinarily agrarian or preindustrial, with labor-intensive economies calling for a large supply of unskilled workers; minority ethnic groups assume that role. Undergirding this exploitative system is a racist ideology in which the subordinate place of minorities is deemed natural.

Social segregation between dominant and minority groups is maximized, and the rules of interaction are explicitly defined and enforced by both tradition and law. The dominant group and minority groups develop parallel and duplicative institutions (education, religion, recreation), and only within the polity and economy do they interact with frequency. In these contacts, however, the dominant group is clearly in command.

The socioeconomic and ethnic hierarchies closely overlap in these societies so that, for the most part, defining one's ethnicity is tantamount to defining one's class. Indeed, stratification is more castelike, with members of groups automatically afforded privileges (in the case of the dominant group) or inferior status and low access to the society's rewards (in the case of minorities). There is little or no mobility for individuals or groups within this system.

Given the extremely wide power differentials and the well-entrenched racist ideology, conflict usually remains submerged for long periods, and attempts to upset the system are rare and short lived. Eventually, however, the forces of both coercion and ideology break down, and the degree of internal conflict becomes great.

The colonial and slave societies of the seventeenth, eighteenth, and nineteenth centuries are, of course, prime examples of this type. In the modern age, among the world's nations only South Africa under apartheid reflected the major characteristics of a colonial society.

CORPORATE PLURALISTIC (MULTICULTURAL) SOCIETIES

Corporate pluralistic, or **multicultural**, societies are also characterized by cultural and physical separation of ethnic groups, but the key difference between these and colonial societies is the extent to which one group is dominant. In corporate pluralistic societies, *the groups are relatively balanced in political and economic power, and no single one is able to exert its will on all vital societal issues.* Moreover, group segregation, to the extent that it exists, is mainly voluntary.

A basic assumption on which such societies are founded is that ethnicity is not to be discouraged; on the contrary, ethnic groups are expected to retain their identity, uniting only within a common political and economic system. Indeed, the pluralistic basis of the society is usually built into the political framework. Territorial concentration of groups contributes to the retention of ethnic differences and interests.

Given the territorial clustering of groups and the relatively balanced power situation among them, conflict in these societies is ordinarily held to a minimum. Where issues of cultural or territorial integrity arise, however, competition and conflict may become quite intense. In Canada, for example, the dominant English-speaking group is not entirely comfortable with the linguistic rights of French-speaking Quebecers. This has created constant tension between the two groups, at times subdued and at other times heightened.

ASSIMILATIONIST SOCIETIES

Assimilationist societies differ from the first two types in that there is no recognized obligation or objective in protecting the retention of ethnicity. If ethnic groups survive, they do so because of voluntary actions by the groups themselves or because of informal patterns of prejudice and discrimination—not through the designs of political institutions, as in a corporate pluralistic society, or the segregationist dictates of a dominant group, as in a colonial society.

In these societies, *there is a dominant group and a large number of minority ethnic groups, the latter in various stages of cultural and structural assimilation.* The dominant group encourages cultural assimilation—to its ways—and most groups retain only symbolic aspects of their ethnic cultures beyond two or three generations. Structural assimilation, however, is a different matter. Much social interaction, particularly at the primary level, continues to take place within ethnic subsocieties. Highly visible ethnic groups are segregated to some degree in social relations and thus are least structurally assimilated. Those closer to the dominant group in culture and physical appearance are in more advanced stages of structural assimilation, including intermarriage.

Ethnic relations are competitive. Groups compete in the political and economic arenas for power and jobs, as well as in other realms of social life. The nature of socioeconomic stratification is class, not caste; members of various ethnic groups can thus rise (or fall) on their individual merits, at least theoretically. In reality, ethnic groups tend to cluster at certain points in the socioeconomic and political hierarchies so that class and ethnicity overlap to a greater extent than the society's ideology may proclaim. Indeed, for some groups—namely, racial-ethnic groups—castelike features of stratification are often apparent.

In assimilationist societies, a racist ideology explaining the superiority of the dominant group may be openly proclaimed by extremist elements but is more commonly expressed informally and subtly. In modern assimilationist societies, racism is officially and customarily denied, though its undercurrents may manifest themselves in group relations, particularly between the dominant group and racial-ethnic groups.

Because of the more open and competitive stratification system and the lack of an officially sanctioned racist ideology, low-grade conflict among ethnic groups may be more prevalent in assimilationist societies than in either of the other types. In the colonial society, people "know their place" and stick to it; in the corporate pluralistic society, conflict is held in check by a relative balance of political and economic power. But in the assimilationist society, minority ethnic groups compete with the dominant group for social positions and power. In this competition, of course, the dominant group retains substantial power resources and thus great advantages.

SUMMARY

- Although intergroup conflict is intrinsic in multiethnic societies, groups follow one of two general paths: increasing integration or increasing separation. The former is called *assimilation* and the latter *pluralism*. Each is a process through which groups pass as well as an idealized outcome of group relations.
- Assimilation can be viewed at four distinct but related levels: cultural, structural, biological, and psychological. For our purposes, emphasis is placed on the first two. *Cultural* assimilation is the adoption by one ethnic group of another's (usually the dominant group's) cultural traits. *Structural* assimilation is the increasing social interaction between different ethnic groups at both the primary and secondary levels.
- Pluralism occurs in two different forms in multiethnic societies, *equalitarian* and *inequalitarian*. In the former, groups retain their cultural and much of their structural distinctness but participate on an equal basis in a common political and economic system. In the latter, ethnic groups are also structurally separated, but they are grossly unequal in political and economic power.
- *Colonial*, or *segregationist*, societies are those in which groups meet through conquest or involuntary migration of minorities; inequalitarian pluralism prevails, with a castelike stratification system in place.
- In *corporate pluralistic*, or *multicultural*, societies, annexation of territory or voluntary migration brings previously separate groups together. Segregation between them is high, usually in distinct areas, though it is mainly voluntary; equalitarian pluralism is the predominant form of ethnic relations.
- *Assimilationist* societies emerge mainly from voluntary immigration of groups, though other contact situations may also be apparent. The degree of segregation between groups varies on the basis of cultural and physical visibility, as does the level of institutional separation and intergroup conflict. Assimilation of an ethnically varied population is the long-range societal objective.

CRITICAL THINKING

1. Assimilation is a process that is dictated strongly by the dominant ethnic group, but it is not a one-way process; some aspects of minority ethnic cultures become part of the dominant culture. In what ways have contemporary immigrant groups to the United States brought changes to the mainstream, or dominant, cultural system? Consider different elements of culture, like language, food, and music.
2. Some groups that are quite thoroughly assimilated may become targets of discrimination and even extreme oppression. German Jews in the 1930s are such a case. What might bring about such a sharp reversal of ethnic relations?
3. As this chapter explains, expulsion and annihilation are extreme forms of inequalitarian pluralism. An example of expulsion is the internment of Japanese Americans during World War II. Could something similar occur again in the United States? Consider the wave of anti-Muslim sentiment that has periodically erupted since September 11, 2001.

4. Although conflict at some level seems to be an essential characteristic of multi-ethnic societies, there are cases, such as Switzerland or Malaysia, where diverse ethnic groups coexist for long periods in a relatively harmonious and peaceful fashion. What social, political, and economic circumstances must exist to bring about such conditions?

Personal/Practical Application

1. Consider normative theories of prejudice and discrimination (discussed in Chapter 3), specifically socialization theory. If you lived in an inequalitarian pluralistic society and you were part of the dominant group, what might motivate you to voice objection to the system of extreme inequality?
2. As explained in this chapter, the United States has moved increasingly from assimilation as the prevailing ideology of ethnic integration to a form of mild cultural pluralism, referred to as multiculturalism. In your view, are there limits to multiculturalism? For example, some Americans are resentful of the growing status of Spanish as a second language that people can choose in securing government services or in everyday affairs like shopping or watching television.

ETHNICITY IN THE UNITED STATES

In Part II our focus is the United States. Following a description of the development of the American ethnic hierarchy in Chapter 5, several groups, each of them broadly representative of a particular American ethnic experience, are dealt with in depth. Chapter 6 examines the society's indigenous peoples, American Indians, followed by chapters devoted to African Americans, Hispanic Americans, Asian Americans, white ethnics, Jewish Americans, and Arab Americans. These seven (several are pan-ethnic categories that include a number of distinct ethnic groups) represent the full range of the American ethnic mosaic.

Each chapter is organized around four issues, the theoretical basis for which was established in Part I: the establishment of the group as an ethnic minority, its past and present place in the ethnic stratification system, current and past patterns of prejudice and discrimination directed against it, and the degree to which its members have assimilated into the mainstream society.

Chapter 13 concludes this part with an examination of the most salient current ethnic issues of American society.

IMMIGRATION AND THE FOUNDATIONS OF THE AMERICAN ETHNIC HIERARCHY

In the range of its ethnic diversity, the United States is unmatched in the contemporary world. It is a society whose population derives from virtually every region of the world, encompassing people of every imaginable culture, displaying equally varied physical characteristics. Ethnic diversity, however, is not a new American phenomenon. Although its heterogeneity has expanded in recent decades, from the outset of European settlement in the 1600s a continuous flow of newcomers added fresh ingredients to the American potpourri. It is not an exaggeration to describe the United States as the world's greatest experiment in social relations. No society before it and few since have attempted to blend such a varied and steadily changing mix of human elements.

That America is a nation of immigrants is now a well-worn cliché. The historian Oscar Handlin professed that when he began to write a history of immigrant groups in the United States, he discovered that the immigrants *were* American history. With the exception of American Indians and some Mexican Americans, all ethnic groups in the United States trace their origins to other societies. Even those exceptional groups, of course, were at one time migrants to North America. But how the various groups entered American society, the ways they adapted and were responded to, and their initial placement in and subsequent movement along the ethnic hierarchy all differed enormously.

In this chapter, we trace the formation of the United States as a multiethnic society and the development of its ethnic hierarchy. As explained in Chapter 2, groups may make initial contact in several ways, including conquest, annexation, and voluntary and involuntary immigration. That initial form of contact is critical in determining each group's subsequent rate and manner of absorption into the mainstream society. In the case of American Indians, conquest was the nature of contact, and for a few Hispanics, it was annexation. These cases are discussed in greater detail in Chapters 6 and 8. It has been through immigration, however—both voluntary and involuntary—that most groups entered American society and subsequently took their place in the ethnic hierarchy.

We begin with a brief discussion of the dynamics of immigration and some of the themes that sociologists and demographers have put forth in explaining this human phenomenon. We then trace the several major periods of large-scale immigration to the United States, focusing on how, in the process, the society's ethnic configuration has changed. Finally, we describe the contemporary ethnic hierarchy and how the changing makeup of the population is leading to new intergroup relations and trends.

THE SOCIAL, POLITICAL, AND ECONOMIC FACTORS OF IMMIGRATION

As far back as historians and archaeologists are able to trace, humans have been on the move. From prehistoric times forward, human migrations have been impelled by various factors: changing physical conditions, changing economic conditions, political turmoil, trade and commerce, exploration, and war.

THE ONSET OF LARGE-SCALE IMMIGRATION

During the past three centuries, long-distance migration has occurred on a far greater scale and has been more systematic and purposeful than in previous periods of human history. Starting in the late fifteenth century, Spain and Portugal, followed by England, France, and Holland in the early 1600s, established global colonial empires, sending people from the homeland to various parts of Asia, Africa, North and South America, and Australasia. By modern standards, however, these migrations were small scale, involving government administrators, merchants, and soldiers who occupied and secured the colonies. Settlers also arrived, but they were almost always outnumbered by indigenous peoples. At approximately the same time, an intercontinental migration began from Africa, though of an involuntary nature. Over the next three hundred years, around ten million Africans would be transported to North and South America and the Caribbean and another six million elsewhere through the slave trade (Lovejoy, 1996; Manning, 1992).

The nineteenth century marked the onset of international migration as we know it today. Population transfers involving large numbers now began to occur, especially from Europe to the Americas. In the one hundred years after 1820, thirty-three million Europeans would emigrate overseas (Thistlewaite, 1991). Four countries— the United States, Canada, Brazil, and Argentina—became the target destinations of the vast majority of these migrants. Among these four, however, the United States was by far the principal recipient society, absorbing about three-fifths of all European immigration.

THE STRUCTURAL FORCES OF MODERN IMMIGRATION

What stimulated this massive movement, and why did it begin in the nineteenth rather than in previous centuries? Three interrelated factors were pivotal: population growth, industrialization, and technological advances.

POPULATION GROWTH   The emergence of modern technology in the nineteenth century brought about a sharp decline in the mortality rate of European nations. This process is commonly referred to as the **demographic transition.** *Societies, according to this theory, pass through several demographic stages.* In the first, birthrates and death rates are both high, keeping population growth at a relatively stable level. Agrarian societies or those with a very low level of industrial development are characterized by this demographic pattern. In the second stage, birthrates remain at a high level, but with advancing industrialization new technologies give rise to better hygienic conditions, which in turn lead to declining death rates. This sets in motion a spiraling population. Most European societies in the nineteenth century were experiencing this phase of the demographic transition. The European population more than doubled between 1800 and 1900, producing great social strains. Similar conditions are occurring today in many developing societies. The third stage occurs when societies reach a high degree of industrialization and urbanization, forcing birthrates to decline; combined with low death rates, a stable rate of population growth is thereby achieved. Advanced industrial societies—North America, Western Europe, and Japan—reached this stage several decades ago and have entered a fourth stage in which low death rates combined with precipitously low birthrates make immigration the primary means of assuring a replacement population (McFalls, 2003).

INDUSTRIALIZATION   As industrialization expanded in Western Europe in the nineteenth century, the agricultural sector declined, setting off both internal and international migrations. As people were driven off the land, first in Germany and England and later in other countries, they naturally migrated to the cities, where jobs in burgeoning industries were to be found. The overseas migrations of the nineteenth century were thus preceded by internal rural-urban migrations. Later, people in other parts of the continent migrated to those countries of Europe where industrialization was most fully developed, Britain, Germany, and France. The rapidity and revolutionary nature of the changes brought about by the new industrial system outpaced the ability of these countries to accommodate expanding populations, thus encouraging migration abroad. Moreover, the metropolitan societies of Europe supported emigration as a means of coping with social problems at home as well as expanding their interests abroad (Cohen, 1991).

TECHNOLOGICAL INNOVATIONS   Augmenting the demographic stimulants of migration were the transforming technological innovations that now facilitated the relatively rapid and efficient movement of people. With the development of steamships and railroads, people were no longer confined to limited geographic space. Throughout the nineteenth and twentieth centuries, transportation modes continued to become faster and more efficient, creating for most people virtually unlimited movement. "Steamships," writes historian William Van Vugt, "made the world smaller so that emigration was no longer as great a step as it had been" (1999:13). Although immigration had been a constant phenomenon throughout all of human development, its scope now reached epic proportions.

INDIVIDUAL DYNAMICS OF IMMIGRATION

What has been described to this point are the structural, or macrolevel, dynamics of immigration. The decision of people to migrate, however, is not simply the product

of such impersonal, external forces. Not all people in a society in the midst of a population explosion or a wrenching economic transformation choose to migrate. We
must also consider, then, the dynamics of immigration at the individual or microlevel
of analysis. What motivates people to abandon their homeland and move to another
society, often with little or no foreknowledge of what awaits them in their new
home? For individuals and families there can hardly be a more profound and challenging experience than to leave their established surroundings and seek out a new
life in an unfamiliar environment. Writers, historians, sociologists, and journalists
have been intrigued by the immigration phenomenon for decades. This has particularly been apparent in the United States, where immigrants have been not only a
staple focus of social science but a recurrent theme of novelists and folklorists.

Various forces motivate individuals and families to migrate. Political or religious
oppression often encourages people to contemplate leaving. A sense of adventure or
the desire for something new, particularly among young adults, may motivate them
to migrate. But more than anything, it is the promise of economic betterment that
functions as the lure for most immigrants. This is no less the case today than it was
in the nineteenth century at the outset of large-scale immigration to America.

PUSH-PULL FACTORS   Classical theories of immigration stress the rational decisions of
immigrants themselves in shaping migration patterns. The overriding assumption in
this model is that immigration is a deliberate choice made by people once they have
weighed the economic and social costs and benefits of moving. Following along these
lines, those from poorer and relatively closed societies can be expected to move to
richer, more open societies that promise greater economic and social opportunities.

Immigration, in this view, is largely a function of what are commonly described
as **push-pull factors**. *Push factors are those economic, social, and political conditions
in the origin society that exert a stimulus for people to leave.* Economic hardship and
lack of future employment opportunities are the most common incentives for migration, but political oppression, religious suppression, and poor quality of life might
also serve as stimulants. The forces of change emerging in European societies in
the nineteenth century (for example, technological innovations driving farmers off
the land and pressures on the society to accommodate a rapidly expanding population) stimulated out-migration. *Pull factors are those conditions that exert an attraction to a target society.* Jobs, an expanding economy that promises future
opportunities, the availability of land, political and religious freedom, or a higher
quality of life might serve as lures to potential immigrants. The United States,
Canada, Brazil, and Argentina provided this pull for Europeans in the nineteenth
and early twentieth centuries. These countries needed settlers to develop and cultivate their vast empty lands. Later, as industrialization took hold, they required huge
workforces that could build an industrial infrastructure and provide a labor force
for their factories and mines.

This perspective on immigration is based essentially on neoclassical economic
principles. Labor markets operate much as markets for commodities and are shaped
by basic supply ("push") and demand ("pull") dynamics. Thus, where there is a
need for labor or where wages are high, people will move to fill those positions. Similarly, where unemployment is high and wages low, immigration will decline. Potential migrants, it is assumed, engage in cost-benefit calculations, weighing their

circumstances at home against the expected future payoff of migrating. When staying becomes less beneficial than moving, individuals and their families will be inclined to consider migration.

Numerous intervening variables are factored into the migration decision (Lee, 1966). Migration entails acquiring sufficient resources to make the move; families or individuals may want to migrate but lack the financial means to do so. Or, consider information. People may think of migrating, but unless and until they acquire knowledge of a promising receiving society they will go nowhere. This in large measure explains why the abjectly poor, for whom migration would seem a perfectly rational choice, rarely migrate: they lack the necessary financial and informational resources. Psychic costs and benefits may also be considered. Migrating to a new society almost always involves some degree of trauma, giving up familiarity with the home society and adapting to a new one. Not all have the wherewithal to endure such a heavy psychological burden or choose to put their families through the change.

SOCIAL NETWORKS   Although the rational decisions of immigrants themselves may explain much about immigration dynamics, some theorists have pointed out the shortcomings of the neoclassical approach. Immigration decisions, they suggest, are seldom the result of rational calculation alone. Moreover, the market model cannot explain why immigrants choose to go to one society rather than another. In addition, immigration may continue to occur even as traditional push and pull factors diminish. In this view, social and communication networks linking sending and receiving societies are key factors that influence migration decisions. And, these factors serve to perpetuate a migration stream after it has begun.

Once a number of immigrants establish themselves in a new location, they become a link with those left behind. Workers contact relatives and friends in their home villages and towns and inform them of work opportunities in the new location. Today, as we will see later, immigrants learn of employment opportunities and of other attractions of receiving societies not only through contact with migrants who have preceded them but also through advanced means of communication like television and the Internet. Obviously these did not exist in the nineteenth and early twentieth century, making it necessary for immigrants already in America to inform their friends and kin of the work opportunities and for employers to actively recruit workers and encourage them to come.

These social networks create what has been called a **migration chain**, wherein *people tend to immigrate to locations that have already been settled by other family members, friends, or coethnics*. This typified the European immigration streams to America of the nineteenth and early twentieth centuries. Most of the workers in a plant or factory, for example, might have been from one country, but even more specifically from one region or town, having been recruited by employers or by coworkers. These migration streams were strengthened as well by the creation of ethnic institutions—churches, businesses, boardinghouses—that served the needs of expanding ethnic communities.

Demographer Douglas Massey has explained that once these migrant networks are in place, there need be no economic incentive for the flow of migration to continue. Migration becomes self-sustaining in a process he refers to as "cumulative causation": "Every new migrant reduces the costs and risks of subsequent migration for a set of friends and relatives, and some of these people are thereby induced to migrate, thus further expanding the set of people with ties abroad and, in turn, reducing costs for a new set of people, some of whom are now more likely to decide to migrate, and so on" (1999:45).

COMBINING MACRO- AND MICRO-PERSPECTIVES   Theories of immigration are numerous and many emphasize global economic and political structures, the role of government, and transnational dynamics as critical variables in explaining contemporary international migration (Massey et al., 1993; Portes and Rumbaut, 2006). The push-pull and network models described previously focus on the decisions and motivations of individuals and families rather than on immigration as a collective phenomenon. These seem particularly applicable to the waves of voluntary immigrants from Europe to America of the nineteenth and early twentieth centuries. But immigration, like all forms of human behavior, comprises individual actions that occur within and are shaped by social structures.

What Castles and Miller (2009) refer to as the "migration systems approach" has become a more popular way for social scientists of various disciplines to analyze immigration trends, using such a combined perspective. The basic idea is that immigration is the result of both macro- and microlevel structures. Macrostructures—government policies, the global economy, and international political conditions—provide an institutional setting within which individuals and families make decisions about migration. The immigration process, therefore, is always a composite of the dynamics occurring at each level.

SETTLEMENT AND ADAPTATION

The journey of immigrants is, in the end, only a brief span in the total immigration process. Once having entered the new society, immigrants face the complications of adjustment. Immediate challenges—finding suitable housing and earning a living—are daunting by themselves. In addition, however, immigrants must confront longer-term social and cultural obstacles: learning the language of the new society, educating themselves and their children, and becoming accustomed to new norms and values.

THE CONTEXT OF RECEPTION   How immigrants adapt and how well they succeed in a new society is only partially a function of the personal efforts, choices, and skills that they bring to the labor market. The political and social environment they are met with may be receptive or it may be hostile. What government policies are in place to deal with immigrants? What is the attitude toward immigrants among native-born residents? What support systems provided by family, friends, and co-ethnics do immigrants find, and is there a thriving ethnic community to meet their unique cultural needs? A combination of these factors—what Portes and Rumbaut (2006) call the "context of reception"—will determine how and at what speed immigrants move in the direction of either assimilation or pluralism in their adaptive patterns and will affect, in the long term, their ultimate level of success.

SOCIETAL EFFECTS It is important to consider that the impact of immigration falls not only on the newcomers. In fact, mass immigrations have a major impact on all of the receiving society's institutions—business and commerce, work, politics, religion, and education. Also affected are the society's language, cuisine, arts, fashion, and virtually all other aspects of culture.

Migration streams to the United States during the nineteenth and early twentieth centuries had profound economic and social effects, continually adding new elements to the society's original base and, in the process, transforming its character and culture. As we will see later, this transformative process is perhaps even more evident today as the most recent immigrant wave makes its societal mark.

EARLY IMMIGRATION TO AMERICA AND THE ESTABLISHMENT OF ANGLO DOMINANCE

POPULATION COMPONENTS

Despite the human mix that characterized the United States almost from its very founding, before the 1830s the breadth of ethnic diversity was not yet great. When the first European settlers arrived, they encountered an indigenous population that, as we will see in Chapter 6, was both culturally diverse and geographically dispersed. By the third decade of the nineteenth century, the indigenous peoples had been reduced to subordination and no longer posed a serious threat to European dominance.

The only other significant non-European group that was part of the American ethnic picture was composed of black slaves who had been brought from Africa beginning in the late seventeenth century. When blacks first entered the society, the slave system was not yet institutionalized. Most blacks were indentured servants, contracted to serve their masters for a certain period of time, usually several years (Franklin, 1980). This was not essentially different from the status of many white indentured servants who had come from England, Scotland, and Ireland. By the late 1600s, however, black slavery was established in several colonies, and the slave trade was firmly in place. Slavery in the American colonies at first grew slowly but increased rapidly during the eighteenth century. On the eve of the American Revolution, slaves accounted for more than 20 percent of the colonial population (Dinnerstein et al., 2003). Following independence, however, that percentage would begin to decline as Europeans began to enter the society in larger numbers.

Because of their cultural—but especially their physical—distinctness from the European settlers, blacks and American Indians (as well as Mexicans in the Southwest) were, from the first, relegated to the bottom of the emerging ethnic hierarchy; their position would not basically change afterward.

Although their numbers were small, immigrants were not unwelcome. In fact, following the establishment of the American state, a general tolerance for European immigrants was quite evident, and their national origins and even religion were not of great concern. George Washington, speaking to newly arrived Irish immigrants in 1783, expressed this attitude clearly in proclaiming that "[t]he bosom of America is open to receive not only the opulent and respectable stranger, but the oppressed and persecuted of all nations and religions" (quoted in Muller, 1993:19).

It was the English-origin, or Anglo, group—whose defining ethnic features were its northwestern European and Protestant origins—that became the host, or dominant, group. Other, smaller, groups were mainly northwestern Europeans—Scots, Welsh, Scotch-Irish, Dutch, Scandinavians, and German Protestants—all culturally and racially close to this core group and, as a result, relatively quickly absorbed into it. Thus the core group represented a blend of rather similar cultural and racial elements, with the English unmistakably its majority component.

This core group subsequently set the cultural tone of the society and established its major economic, political, and social institutions. All following groups would be required to adapt to an Anglo-Protestant social and cultural framework. It became the standard, notes the historian Arthur Schlesinger Jr., "to which other immigrant nationalities were expected to conform, the matrix into which they would be assimilated" (1998:34). Glazer and Moynihan vividly describe the establishment of a dominant ethnic group in American society with the power to select those who came after it:

> The original Americans became "old" Americans or "old stock," or "white Anglo Saxon Protestants," or some other identification which indicated they were not immigrants or descendants of recent immigrants. These original Americans already had a frame in their minds, which became a frame in reality, that placed and ordered those who came after them. It was important to be white, of British origin, and Protestant. If one was all three, then even if one was an immigrant, one was really not an immigrant, or not for long. (1970:15)[1]

CREATION OF THE DOMINANT ETHNIC CULTURE

Although the Anglo core group has been diluted by successive immigrant waves over many generations and its cultural imprint likewise colored by other groups, its cultural dominance has remained unwavering throughout American history. "The white Anglo-Saxon Protestant," notes sociologist Lewis Killian, "remains the typical American, the model to which other Americans are expected and encouraged to conform" (1975:16). The societal power of the Anglo core group also remains evident, as indicated by its continued overrepresentation among political and economic elites. It is true nonetheless that its power has been diminished in the past few decades by the increasing penetration of non-Anglo groups into these positions (Alba and Moore, 1982; Davidson et al., 1995).

The makeup of the dominant group itself has changed somewhat as other white Protestant groups have melded with it. Today the dominant ethnic group in American society may be said to broadly comprise white Protestants, because those Protestants from other northwestern European societies—namely, the Scandinavian nations, Germany, and the Netherlands—have blended almost imperceptibly with those of British origins. White Protestants, then, have varied national roots, but their common Protestantism and racial character have neutralized any meaningful national differences among them. References to the WASP, Anglo, or Anglo-American group, then, should be understood to mean "white Protestants of various national

---

[1] From Nathan Glazer and Daniel P. Moynihan, *Beyond the Melting Pot*, 2nd ed. (Cambridge, MA: MIT Press, 1970). Copyright © 1970 by MIT Press. Reprinted by permission.

origins." These terms are used interchangeably to denote the dominant American ethnic group and its core culture.

DOMINANT CULTURAL VALUES AND COMPETING IDEOLOGIES   Ideologies and public policies concerning American ethnic relations have historically reflected the cultural prefer-ences and the economic and political interests of the WASP, or Anglo, core group in several ways.

To begin with, the preeminence of Anglo cultural values has consistently under-lain public policies in education, language, law, welfare, and religion. The ascen-dancy of the English language, the English legal system, and, with few exceptions, the Christian faith, was never seriously challenged. From the beginning, the expecta-tion held sway that entering groups—immigrant, conquered, or enslaved—would conform to this core culture.

At certain times—especially in the early years of the twentieth century when eth-nic diversity seemed to reach a high point—the idea of the **melting pot** called into question the inevitability as well as the desirability of Anglo cultural dominance. *The idea was that the cultural differences among the many immigrant groups, so conspicuous at that time, would disappear through a gradual blending process.* Dif-ferent ethnic cultures, it was felt, would eventually fuse into a single "American" cul-ture as a kind of hybrid creation. Although the notion of a melting pot remained symbolically popular for many decades, it never found manifestation in public policy or gained widespread allegiance. In any case, the idea was incomplete because the place of non-Europeans in this ideal social brew was never fully dealt with. The melt-ing pot was belied as well by the fact that the boundaries among the European groups did not disappear, nor were their cultures ever entirely diluted by WASP norms and values.

Another competing ethnic ideology that gained adherents at times is **cultural pluralism**. As explained in Chapter 4, *the objective of cultural pluralism, rather than a fusion of diverse ethnic groups, is the preservation of each.* In the past, cul-tural pluralism in the United States never found more than token acceptance at the level of public policy. In recent times, however, schools, universities, and government agencies have acknowledged the pluralistic nature of American society and have made efforts to affirm multiculturalism as the society's dominant ethnic ideology (Glazer, 1997).

The prevalent ethnic ideology, expressed most forcefully in government policies for most of U.S. history, is assimilation into the dominant group, or what Gordon (1964) has called **Anglo-conformity**. Groups are expected to shed their ethnic uniqueness as quickly and as completely as possible and take on the ways of the core culture. This expectation always guided the prevailing social thought and policy regarding new ethnic groups; it continues to do so even today despite the pluralistic, or multicultural, rhetoric of recent years.

Some have argued that today there is, in reality, no core culture per se. That is, given the rapidly increasing ethnic heterogeneity of American society in recent dec-ades, it has become increasingly difficult to identify a universally acknowledged core culture (Brubaker, 2001). Indeed, as E. Digby Baltzell, a longtime student of the WASP upper class, observed, "[T]he process of cultural pluralism has gone so far that larger and larger segments of the Anglo-Saxon-Protestant majority now see themselves as resentful outsiders" (1991:222). Here it should be remembered that when ethnic groups meet, there is always an exchange, to some degree, between them. There is little debate about the expansion of American cultural diversity and with it, the waning of the traditional dominant culture. But it is equally apparent that an Anglo core cultural base consisting of a few major elements—language, religion, political norms and values—remains in place.

ECONOMIC AND POLITICAL INEQUALITY   The pressures on ethnic groups throughout American history to assimilate culturally have not been complemented by corresponding pressures to assimilate structurally, that is, to enter into full and equal relations with the dominant group in all institutional areas. Although Anglo-conformity has generally been expected of ethnic groups in their behavior and appearance, a pluralistic outcome has more often than not typified the dominant group's expectations of interethnic relations. The more culturally or physically distinct a group, the more its members have been encouraged to remain "among their own" in the intimate areas of social life—marriage, residence, social clubs, and the like (Table 5.1).

Most important, the efforts of minority ethnic groups to attain economic and political equality have traditionally met with resistance, often of an official nature. Through immigration quotas and exclusionary measures, Indian-removal acts, slave laws, institutionalized segregation, antilabor regulations, voting restrictions, and an array of other measures, dominant group interests were protected. Of course, these policies all generated great controversy and conflict, and minority challenges often met with success. But concessions wrested from the dominant group have always been slow, costly, and incremental.

TABLE 5.1 | DEGREE OF ASSIMILATION OF AMERICAN ETHNIC GROUPS

Ethnic Population	Cultural Assimilation	Secondary Structural Assimilation[a]	Primary Structural Assimilation[b]
Anglo-Protestants (core group)	High	High	High
Northwestern Europeans	High	High	High
Irish Catholics	High	High	Moderate
Southern and Eastern European Catholics	High	Moderate	Moderate
Jews	High	High	Moderate
Asians	Moderate	Moderate	Moderate
Latinos	Moderate	Moderate	Moderate
American Indians	Moderate	Low	Moderate
Blacks	High	Moderate	Low

[a]As indicated by entrance into economic, political, and other social institutions at various levels.
[b]As indicated by entrance into primary relations with the dominant group in areas such as residence, club memberships, and intermarriage.

In sum, for most of American history, public policies and publicly proclaimed ideologies have generally corresponded to the dual aims of, on the one hand, cultural assimilation (specifically Anglo-conformity) and, on the other hand, structural pluralism (the latter in a form assuring Anglo dominance). These objectives have reflected the American system of ethnic stratification. Those groups least visibly different from the Anglo core group have been received with less hostility and have been presented with an opportunity structure more open and less limited than have those distinctly nonwhite or non-Protestant.

EUROPEAN IMMIGRATION: THE FIRST AND SECOND WAVES

In the third and fourth decades of the nineteenth century, the relative homogeneity that characterized the United States in its formative years began to change with the onset of large-scale European immigration. The variety of groups that would subsequently enter was nothing short of astonishing and would fundamentally reshape the society from that point forward. Indeed, the start of mass immigration to America marked the beginning of the creation of a multiethnic society whose heterogeneity was unprecedented. In the one hundred years to follow, more than thirty million immigrants would come to the United States, making it, as the historian Maldwyn Jones called it, "the classic country of immigration" (1992:2).

After the colonial period and before the modern era of immigration starting in the late 1960s, the United States experienced two major waves of European immigration. Both contributed great portions of the white population, but the principal groups composing each were markedly different in their national origins and their sociological characteristics. As a result, immigrant groups were not all received by the dominant (or core) group in the same way, and their ensuing absorption into the society also varied.

THE "OLD IMMIGRATION": THE FIRST GREAT WAVE

With immigration before 1820 at a relatively low level, we can think of *the six decades between 1820 and 1880 as the first great wave*, or as historians often refer to it, the **Old Immigration**. This period marks the beginning of large-scale immigration to America, bringing millions who would begin to populate every region of the country. Equally important is that for the first time non-Anglo groups of significant size began to enter the society, although immigration from Britain continued unabated.

As noted earlier, immigration during these years was driven primarily by a combination of labor needs in the expanding American economy and the dislocations occurring in Europe due to the breakdown of the agrarian order and the rise of industrialism. In addition to the declining agricultural sector and emergent industrialization, the Napoleonic Wars created much political and economic instability in Europe, providing an additional catalyst for immigration. The United States held out promise of a new beginning for those who were displaced from their land, who were unable to find a place in the newly emergent economic and social orders, or who were fleeing political revolt. These strong push conditions were accompanied by a growing demand for labor, both agricultural and industrial, in the New World. The demand for workers was made more pressing by the abolition of the African slave trade. Without new settlers, the vast resources of the United States could not be exploited, and as a result, efforts to recruit immigrants were made by state and territorial governments and many emergent industries.

SCOPE AND DIVERSITY   Between 1840 and 1860, more than four million people entered the United States—equal to about 30 percent of the entire free population of the nation in 1840 (Levine, 1992). In proportion to the total population, it was the greatest influx of newcomers in American history. By 1860 the population of most of America's largest cities was nearly half foreign-born. Much of this immigration was driven by the establishment of regular steamship voyages between Europe and America. What had been a dangerous and costly voyage of weeks by sailing ship became a relatively uneventful crossing of days by steamer. Moreover, this new form of transatlantic travel provided mental comfort to immigrants who were no longer faced with what had seemed an irreversible decision. Now immigrants knew they could more easily return if America failed to meet their expectations or if they could not adapt to life removed from the old country.

The immigrants' origins were varied; they included not only those from Britain and other northwest European societies but also, in the West, Chinese; in New England, French Canadians; and in the Southwest, Mexicans. The two most sizable groups arriving during this era, however, were the Germans and the Irish. After the Civil War, most immigrants continued to come from Germany, Ireland, and Britain, but a substantial number came from Scandinavia as well. Altogether, more than two million Swedes, Norwegians, and Danes came to the United States after 1860 (Dinnerstein et al., 2003).

This initial period of large-scale immigration marked the entrance not only of non-British people in large numbers but also, for the first time, significant numbers of non-Protestants. Most of the Irish were Catholics, as were about half the Germans. As a result, nativist—specifically anti-Catholic—actions and rhetoric became widespread. Catholics were seen by many as a seditious element and Catholicism as basically incompatible with American ways (Higham, 1963). By midcentury, however, anti-Catholic nativism had begun to wane as national attention turned to the issue of slavery. Moreover, as the American industrial economy surged, immigration generally was welcomed by industrialists and others of the business community who saw immigrants as a source of cheap labor and a growing consumer market.

Despite early prejudice and discrimination toward the Catholics among them, particularly the Irish, these northwestern European groups were, by comparison with groups that would follow them, advantaged in two ways: they were close enough culturally to the Anglo core group to assimilate within a relatively short historical time, and they were physically indistinct from the Europeans who had preceded them to America, thereby avoiding long-lasting imputations of racial identity. Thus their integration into the mainstream society was not severely constrained, and the resistance of the dominant group to them was, in historical perspective, relatively mild. Here we briefly describe the path followed by the Germans; in Chapter 10 we look more closely at the Irish.

THE GERMANS Germans had settled in America as early as the seventeenth century; they were, in fact, the largest non-British group during the colonial period. But German immigration developed on a massive scale beginning in the 1830s, with the largest numbers coming in two great waves. In just seven years, from 1850 to 1857, almost one million Germans arrived, and between 1865 and 1885, two million more. By the late nineteenth century, Germans were the second-largest ethnic group in the society, exceeded only by the British. Whereas most of the Irish remained in eastern cities, particularly New York and Boston, German immigrants dispersed westward throughout the Northern tier of states from New York to the Mississippi River. Many settled in several large midwestern cities, including Cincinnati, Milwaukee, St. Louis, Detroit, and Chicago. There they established ethnically homogeneous neighborhoods—so-called Kleindeutschlands, or Little Germanys—that would grow with the continued arrival of new immigrants (Dinnerstein et al., 2003; Rippley, 1984).

The class origins of the Germans were more varied than those of the Irish, who were mostly poor and unskilled. Most Germans were tradesmen, industrial workers, small shopkeepers, farmers, and professionals; only a small percentage were unskilled laborers (Keil, 1991). With the growth of manufacturing, German craft workers became the core of the burgeoning U.S. industrial working class (Levine, 1992). Many of the craftsmen and merchants took up residence in small towns, while farmers, attracted by rich lands, settled in large numbers in the midwestern states of Wisconsin, Illinois, and Ohio. In religion, too, the Germans were a diverse immigrant population. Roughly half were Protestant and half Catholic, but there was a small German-Jewish element as well.

In addition to economic circumstances, some German emigrants were driven by political crisis. One small but very influential segment of the German immigration of this period was a group known as "Forty-Eighters." Forced out of Germany because of their participation in an unsuccessful revolution in 1848, they took political refuge in the United States. They subsequently exerted an inordinate influence not only on German American affairs but also on national issues. They were vehemently anti-slavery, as well as anticlerical, and they sowed the seeds of the American labor movement that emerged in the late nineteenth century (Levine, 1992).

THE "NEW IMMIGRATION": THE SECOND GREAT WAVE

Although substantial German and Irish immigration had presented the first real challenge to WASP dominance, the polyglot nature of American society took shape most forcefully as a result of what has been called the **New Immigration**. *During a period extending roughly from the early 1880s until the outbreak of World War I, almost twenty-five million European immigrants came to the United States.* This influx represents the archetypal American immigration, and it is this period that we most often think of when we visualize traditional immigrant life. It is from this second great wave of immigration that most of today's non-Protestant Euro-Americans, or "white ethnics," are descended.

SCOPE AND DIVERSITY These new immigrants not only represented the numerical peak of European immigration to the United States but also radically changed the ethnic character of the society. To begin with, most were from southern and eastern European societies, whose cultures were alien to those of Britain and northwestern Europe. Most important, they were non-Protestant—primarily Catholic and secondarily Jewish. Although most of the earlier Irish were Catholic, in other respects they were not so culturally distant from the Anglo-Americans. Likewise, the Germans may have spoken a different language, but over half were Protestant, and they were not looked on by the dominant Anglo group as cultural inferiors. Finally, the new immigrants were mainly people whose class origins were lower than those of previous groups, except perhaps the Irish. They were chiefly peasants from the poorer states of Europe—southern Italy, Russia, Poland, Greece, and the many parts of the Austro-Hungarian Empire.

The shift in immigrant origins was brought about by a number of factors. First, the growth of industry in Britain and Germany furnished more labor opportunities for their populations. Hence, the pressures to migrate were reduced. German immigration to the United States, for example, dropped from 1.5 million in the 1880s to 0.5 million in the 1890s (Dinnerstein and Reimers, 1999). In addition, economic problems that had been experienced earlier in northern European countries now spread to southern and eastern Europe, creating food shortages and high unemployment and thus an incentive for people to leave.

Pull factors also changed. With the closing of the western frontier, land was no longer the chief lure of immigrants to the United States. The expanding American industrial economy, however, now exerted an even stronger pull. Factories, mines, and mills demanded ever greater numbers of workers, and unskilled immigrants supplied the major part of this need.

An important feature of the New Immigration was that many who went to America expected to return to their origin societies. Consequently, a major portion of the immigration comprised men unaccompanied by their families. Their objective was to earn enough money to enable them to return to their native towns or villages and purchase land or establish themselves in an enterprise. Many did just that. Although records are nonexistent and only rough estimates can be made, perhaps one-third of all immigrants during this period remigrated back to their homelands or to other societies (Thistlethwaite, 1991).

The largest of the new immigrant groups were Italians, mostly from the south of Italy, and eastern European Jews, mostly from Poland and Russia. Both tended to concentrate in cities of the Northeast, especially New York. Another major component of the New Immigration was Slavic groups, including Russians, Ukrainians, Slovaks, Slovenes, Croatians, Serbs, Bulgarians, and especially Poles. Together, these groups made up about four million of the new immigrants. Other central European groups, including Hungarians, Czechs, and Armenians, also were part of the immigration of this era.

Other societies in addition to the United States attracted large numbers of European immigrants during this great age of international migration. These included Australia, Brazil, Canada, and Argentina. In the latter two countries, in fact, immigration rendered an even more profound demographic effect, proportionately, than in the United States (Baily, 1999; Thistlethwaite, 1991). However, in absolute numbers and ethnic variety, American immigration was far greater.

SETTLEMENT PATTERNS   The settlement patterns of the immigrants of this period differed in some respects from the Old Immigration. Whereas many immigrants of the earlier great wave were attracted by the opportunity to own land and therefore settled in rural areas, the new immigrants were mainly industrial workers who settled in cities and towns that supported those industries. As a result, rather than dispersing geographically, the new immigrants tended to concentrate in urban areas, particularly of the Northeast and Midwest. It was here that industrial expansion was occurring and the most alluring work opportunities were being offered. The growth of these industrial centers was rapid and profound. From 1900 to 1920, Milwaukee's population, for example, increased by more than 60 percent, Cleveland's by more than 120 percent, and Detroit's by almost 250 percent (U.S. Census Bureau, 1920). Although it was aided by a steady rural-urban movement of farmers and lumbermen, the primary elements of this population increase were the foreign-born. As seen in Table 5.2, the foreign-born composed more than 30 percent of the population of seven of the fourteen largest cities in 1900.

Once they entered the United States, the different immigrant groups migrated to various cities and regions, determined in large measure by the groups' occupational proclivities. Particular groups seemed to be channeled to specific industries. The Polish immigration illustrates this tendency. Census labor reports show that the overwhelming majority of Polish immigrants were classified as laborers or workers

TABLE 5.2 | FOREIGN-BORN IN THE FOURTEEN LARGEST U.S. CITIES, 1900

City	Total Population	Total Foreign-Born	Percent Foreign-Born
New York	3,437,202	1,270,080	37.0
Chicago	1,698,575	587,112	34.6
Philadelphia	1,293,697	295,340	22.8
St. Louis	575,238	111,356	19.4
Boston	560,892	197,129	35.2
Baltimore	508,957	68,600	13.5
Cleveland	381,768	124,631	32.7
Buffalo	352,387	104,252	29.6
San Francisco	342,782	116,885	34.1
Cincinnati	325,902	57,961	17.8
Pittsburgh	321,616	84,878	26.4
New Orleans	287,104	30,325	10.6
Detroit	285,704	96,503	33.8
Milwaukee	285,315	88,991	31.2

Source: U.S. Census Bureau (1900).

in manufacturing; few were found in agriculture or the professions, in clerical or sales positions, or in building trades (Hutchinson, 1956). In a sense, the Polish immigrant to the United States was a migrant laborer, not so much an immigrant as a "temporary worker," intent on returning home. Historian Caroline Golab (1977) explains that prior to their emigration, Poles already had much experience in the industrial sphere because they had been migrating to various European societies during the late nineteenth century. "It was rare," she writes, "to find a Polish immigrant who had not migrated to some other part of Europe or within his own country before coming to America" (1977:74). Hence, Poles were familiar with town and city life and with machines and factories. Quite logically, then, upon arriving in the United States they migrated to the same industries that had used them in Europe, such as heavy metals and mining. This accounts for the readiness of Poles to settle in cities of the Great Lakes, such as Chicago, Buffalo, Detroit, and Milwaukee, as well as in the anthracite region of western Pennsylvania, where these industries were most prevalent at the turn of the century and in the several decades that followed. As we will see in Chapters 10 and 11, similar factors operated for immigrant Italians and Jews in accounting for their occupational inclinations.

The new immigrants in the cities tended to settle in depressed areas, inhabiting run-down tenements that had been vacated by earlier ethnic groups such as the Irish. The concentration of immigrants in these areas was partially the result of simple economic necessity; such housing was available and affordable. The creation of ethnic enclaves in large cities, however, was also the product of the natural attraction of immigrants to areas inhabited by those who had preceded them to America and came from the same towns and regions. New immigrants sought the security and support of coethnics and thus gravitated to common areas. "Life in the Italian quarter," writes Virginia Yans-McLaughlin, "provided a coherence and familiarity which drew immigrants irresistibly toward it" (1982:118). It was much the same for other groups. As the skeleton of the ethnic community was shaped, others were attracted, creating in turn the need for larger and more complex ethnic institutions.

More than ethnic needs, however, residential clustering and the building of ethnic communities resulted mostly from the attraction of work opportunities in specific locations. In their classic study of Polish immigrants, Thomas and Znaniecki (1918) explained that in large cities the desire to live near work was a stronger consideration than the desire to live near one's fellow ethnics. Thus institutions catering to immigrants—churches, groceries, banks—followed the residential concentration created by employment opportunities. Moreover, one should not exaggerate the ethnic homogeneity of these urban areas, as movies and novels have so often portrayed them. Although there certainly were neighborhood concentrations of immigrants and their children from particular countries, these pockets were never as ethnically pure as has been commonly assumed, nor did they endure much beyond the first two immigrant generations (Hershberg et al., 1979).

RESURGENT NATIVISM AND THE END OF THE SECOND WAVE The sharp cultural and class differences of the new immigrants provoked a strongly antagonistic reaction from the dominant group. Americans questioned how the immigrants would ultimately fit into the society and, abetted by popular notions of social Darwinism, widely assumed their cultural inferiority (Jacobson, 1998).

Prompted by the continued influx, nativist feelings and racist theories reached an apex by the early 1920s, resulting in restrictive quotas enacted by Congress that severely limited large-scale European immigration. The quotas were deliberately designed to favor the sending countries of northwestern Europe and to curtail immigration from southern and eastern European countries. Previously, with no quotas and few stipulations other than health requirements, an open door, in effect, had existed for Europeans. The restrictive legislation represented a symbolic closing of the door. Even prior to the passage of this legislation, immigration from Europe had been reduced by the outbreak of World War I in 1914. Although immigration resumed at high levels for several years after the war, the quotas that were established in 1924 changed the makeup of newcomers back to mainly northwestern Europeans. Moreover, immigration to the United States of all groups dropped precipitously from 700,000 in 1924 to fewer than 300,000 the following year. Not until the early 1970s would the number again approach the level of immigration reached prior to the mid-1920s.

In addition to restrictive quotas, immigration was discouraged by the Great Depression of the 1930s. Unemployment reached record levels, and the United States was therefore no longer seen as an attractive destination. Fewer immigrants arrived during the 1930s than during any decade of the previous hundred years. In fact, more people left the United States during this time than entered as immigrants. As the world again plunged into war in the late 1930s, a further obstacle to immigration was imposed. The historical flow of American immigration can be seen in Figure 5.1.

Figure 5.2 shows the major ancestry groups that currently make up the Euro-American population, which derives primarily from the two main periods, old and new, of European immigration.[2]

THE NEWEST IMMIGRATION: THE THIRD GREAT WAVE

The decade of the 1960s witnessed events that would fundamentally alter the U.S. ethnic order. One, as we will see in Chapter 7, was the black civil rights movement. But something else occurred in the 1960s that, arguably, would prove to have an equally significant impact on American ethnic relations: the revision of immigration laws encapsulated in the Hart-Cellar Act of 1965. This new legislation erased the restrictive country-of-origin quota system that had virtually prohibited certain groups from immigrating to the United States. It also created a system of criteria for admission; priority was now given to those with family ties to U.S. citizens or permanent residents and to those with needed occupational skills.

Several driving forces precipitated this truly radical change in immigration laws. First, the civil rights movement itself provided an initial impetus. As an adjunct to the mounting pressure for an end to discriminatory measures against African Americans, public sentiment now seemed prepared to do the same with immigrants, bringing the reality of immigration laws into line with the rhetoric of equal opportunity. Also contributing to the acceptability of change were the sentiments and efforts of the new presidential administration that entered Washington in 1961. John F. Kennedy,

---

[2] The diversity of the European component of the American ethnic configuration is actually greater than shown in Figure 5.2 because only the largest ancestry groups are included.

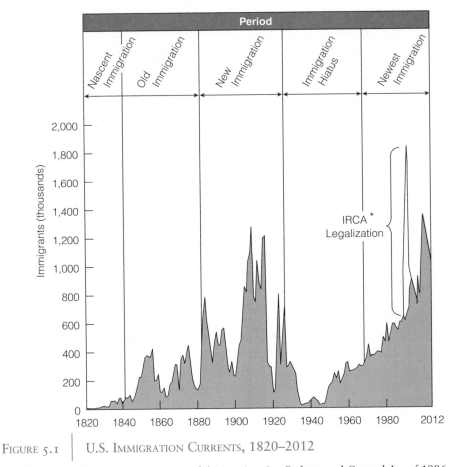

FIGURE 5.1 | U.S. IMMIGRATION CURRENTS, 1820–2012

*IRCA refers to the amnesty provision of the Immigration Reform and Control Act of 1986, under which many unauthorized foreign residents were transferred to legal immigration status.

Source: U.S. Immigration and Naturalization Service (1996); U.S. Department of Homeland Security (2012).

in his 1960 campaign for the presidency, had explicitly pledged to work toward the removal of discriminatory immigration regulations. As with much of the civil rights legislation of the mid-1960s, the revision of immigration laws was enacted in large measure as a tribute to Kennedy after his assassination in 1963.

In 1965, when the revision was passed, few predicted the enormous numbers that would enter the United States as a result, nor could they foresee the profound impact this legislation would subsequently have on the ethnic makeup of the society. The ensuing massive and immensely diverse immigration led to an unprecedented level of heterogeneity of the U.S. population. The two previous periods of substantial immigration, occurring in the nineteenth and early twentieth centuries, had been primarily European in origin. Though resisted by nativist forces, these immigration periods did not essentially threaten the basic cultural and racial makeup of the society. The connotation of *immigrant* in the United States was "of European descent." The

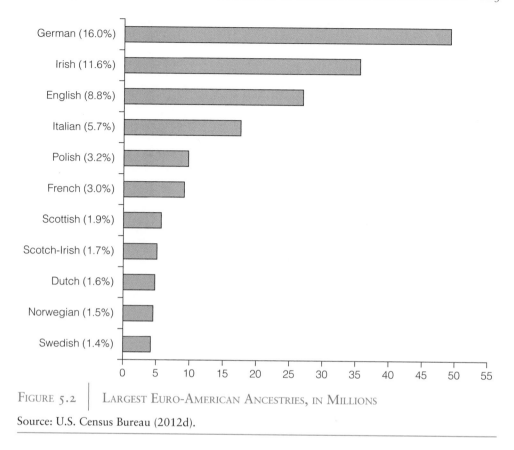

FIGURE 5.2   LARGEST EURO-AMERICAN ANCESTRIES, IN MILLIONS

Source: U.S. Census Bureau (2012d).

modern period of immigration, however, marked the onset of a movement that en-sured that the society's ethnic configuration, and thus its culture, would not be in the future what it had been for most of its history.

CHARACTERISTICS OF THE NEWEST IMMIGRATION

SCOPE   Between 1960 and 2012, thirty-seven million people either entered the United States as legal immigrants or were granted permanent residence. More than nine mil-lion came during the 1990s, and another thirteen million between 2000 and 2012 (Figure 5.1). Today, immigration to the United States averages around one million per year, a number larger than the total immigration to all other developed countries combined.

These figures, of course, do not take into account those who enter unofficially. No one can be certain of their numbers, but government estimates are in the range of eleven million undocumented immigrants currently residing in the United States. Three-quarters are from Latin America, 60 percent from Mexico alone (Hoefer et al., 2012). The number of undocumented immigrants grew rapidly during the 1990s and early 2000s but has leveled off in recent years, mostly as a result of reduced economic opportunities. We will return to the issue of illegal immigration in Chapter 13.

The foreign-born population of the United States has today reached an all-time high. Even during the peak immigration years of the early twentieth century, immigrants numbered only about half of what they do today. However, when looked at in proportionate, not absolute, terms, immigrants today do not have as great a demographic impact as they did in the early 1900s. At that time, immigrants constituted about 15 percent of the total population, compared to around 13 percent today. Even including unauthorized immigrants would not increase that figure significantly. Nonetheless, it is very obvious that immigration in all forms continues to play a large role in U.S. population growth and will play an even greater role in the future (Durand et al., 2006; Passel and Cohn, 2008; Pew Hispanic Center, 2013; Westoff, 2007).

ETHNIC MAKEUP   Numbers alone do not tell the whole story of the current period of American immigration. Perhaps more important, the character of immigration has been radically changed from that of past eras. Whereas the vast majority of immigrants throughout the nineteenth and early twentieth centuries were from European societies, today most are from Latin America and Asia. Figure 5.3 shows how immigration patterns were completely transformed starting in the 1960s.

The ramifications of that change have had a ripple effect throughout virtually all economic, political, and social institutions.

Figure 5.4 shows the leading countries of origin of immigrants to the United States. Clearly countries of Latin America and Asia predominate. But the sheer diversity of non-European immigrants is quite astonishing. Among the Asians, for example, are Filipinos, Koreans, Indians, Pakistanis, Chinese, Vietnamese, Cambodians, Laotians, and Hmong. From the Middle East have come Palestinians, Iraqis, Iranians, Lebanese, Syrians, Egyptians, and Israelis. Those from Africa include Somalis, Ethiopians, Ghanaians, and Nigerians. And from Latin America, the national origins of the immigrants cover the entire Central and South American regions, in addition to the Caribbean.

MOTIVATING FORCES   What has prompted this newest large-scale immigration to the United States? Obviously the changes in immigration law in 1965 provided the favorable legal conditions and thus the necessary vehicle. But essentially the same push-pull factors that motivated past immigrants to America have been functioning for the newest wave. Although many have come as political refugees (mostly from Southeast Asia, Cuba, and Central America), most immigrants have responded to the population and economic pressures of their societies of origin. Most, in other words, seek economic betterment and improved social conditions. Other factors, of course, may play a role in motivating immigration, including political forces, cultural preferences, and physical security. But most commonly, job opportunities will be the primary incentive (Gozdziak and Martin, 2005; Hirschman and Massey, 2008; Saad, 1995).

Immigration to the United States, as well as to other industrialized societies of North America, Europe, and Australia, accelerated in the past five decades, mostly because of the widening gap between rich and poor nations. The economies of the latter, many of them newly independent states, were generally incapable of supporting rapidly growing populations, thereby creating a migration push. Most of the world's population growth has occurred since the end of World War II, and the

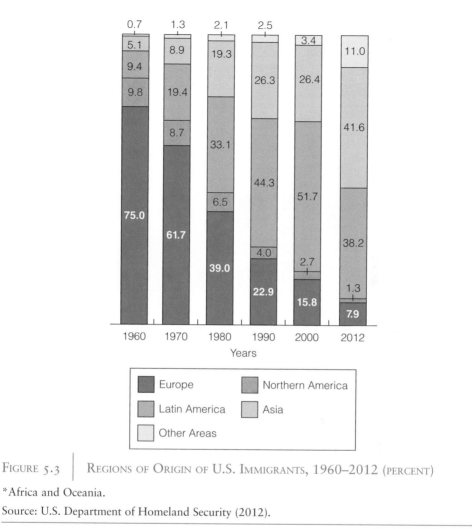

FIGURE 5.3 | REGIONS OF ORIGIN OF U.S. IMMIGRANTS, 1960–2012 (PERCENT)

*Africa and Oceania.

Source: U.S. Department of Homeland Security (2012).

overwhelming majority of that growth has taken place in the less developed countries. Moreover, the gains in economic development made by many of these countries during the 1960s and 1970s slowed or reversed in the 1980s due to economic recession, growing debt, and internal political conflicts. Those conditions induced further pressures to migrate to the wealthier countries, like the United States, whose advanced economies promised better employment opportunities and a superior quality of life.

If poverty and overpopulation, as in the past, have provided the major push factors for the contemporary flow of immigration to the United States and other industrialized countries, pull factors are also similar to those that stimulated past immigration waves: the changing structure of the U.S. economy, especially the expansion of low-wage jobs, but also high-status technical jobs.

SETTLEMENT PATTERNS Although the Newest Immigration has affected virtually all U.S. regions and states, it has had the most radical demographic effect on a few states,

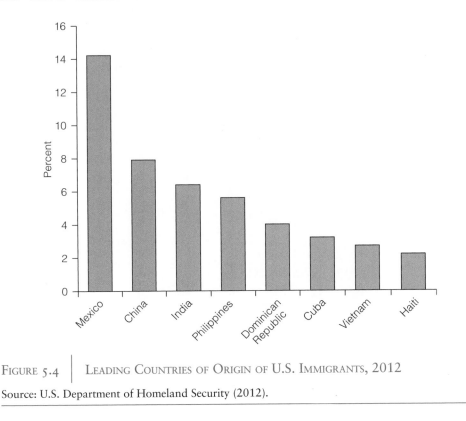

FIGURE 5.4 | LEADING COUNTRIES OF ORIGIN OF U.S. IMMIGRANTS, 2012

Source: U.S. Department of Homeland Security (2012).

namely California, New York, Illinois, Texas, and Florida. So great has the change been in California and Texas that they have been transformed into states where non-Hispanic whites are no longer the numerical majority. Similarly, though immigrant groups have dispersed widely, specific groups remain concentrated in particular regions. Asians, for example, have settled mostly in California; Latinos in the Southwest and Florida; and almost all new groups have large communities in New York City. Table 5.3 shows the states with the largest immigrant populations.

Like so many American immigrants of past generations, many today come with few expectations of remaining permanently in the United States. A great number do leave, and many (particularly from Mexico) engage in a continual back-and-forth movement. But also like their predecessors, many who at first see their immigration as temporary become firmly ensconced in their new society. The result has been, as in earlier eras of immigration, the flourishing of ethnic enclaves in cities that have attracted large numbers of new immigrants. Some, namely New York and Chicago, remain meccas for new immigrants, much as they were in the past. But others—Sunbelt cities like Los Angeles, Miami, and Houston—are now focal points of the Newest Immigration (Iceland, 2009).

Although most of the newest immigrants have settled in large urban areas, virtually no U.S. city or town has been unaffected. Many immigrants have leapfrogged over the central cities and settled in suburban communities, a pattern not common in

TABLE 5.3 | STATES WITH LARGEST IMMIGRANT POPULATIONS

State	Number of Immigrants (Millions)	Immigrants as Percent of Total Population
California	10.1	27.0
New York	4.3	22.2
Texas	4.1	16.4
Florida	3.7	19.4
New Jersey	1.8	21.5
Illinois	1.8	14.0
Massachusetts	1.0	14.9

Source: U.S. Census Bureau (2012d).

the past (Singer, 2010). As an example, consider Bridgewater, New Jersey, an affluent township forty miles from New York City. Seemingly immune to ethnic changes occurring in the city and its surrounding communities, Bridgewater had been a solidly white, middle-class enclave. Its Asian population had been almost invisible in 1980, but twenty years later Bridgewater was 10 percent Asian, and its schools were teaching students speaking forty-one native languages (Chen, 1999). Numerous small towns not in metropolitan areas are even more unexpected destinations for new immigrants. In Maine, for example, Lewiston's population of thirty-six thousand was augmented in 2001 by more than two thousand Somali immigrants (Belluck, 2002). Garden City, Kansas, would hardly be imagined as an immigrant destination, but, drawn by jobs in its meatpacking plants, immigrants—most from Mexico—changed the town's ethnic composition from 80 percent non-Hispanic white in 1980 to less than 50 percent by 2000 (Stull and Broadway, 2008). Immigrant communities such as these have become commonplace in all parts of the country.

CLASS RANGE Another evident feature of the Newest Immigration is its class range. Unlike past great waves of immigration, in which most newcomers were either poor or of modest means, the latest influx is made up of individuals and families who span the entire class spectrum, from low-level, unskilled workers to highly educated professionals. Those from Latin America tend to have less education and fewer skills, whereas those from Asia generally have more education and higher occupational skills (Mather, 2007). The result of this socioeconomic variety has been equally diverse patterns of settlement and adaptation among immigrants.

GENDER During previous large-scale immigration periods, male immigrants outnumbered females among most groups. The Newest Immigration is more gender balanced. In fact, the majority of immigrants from some countries (the Philippines, Korea, Jamaica, and the Dominican Republic, for example) are women (Rumbaut,

1996). Immigrant women have disproportionately filled jobs in the low-wage service sector—such as hotel and restaurant workers—and in certain labor-intensive industries. In San Francisco, for example, the garment industry is the largest manufacturing sector, employing more than twenty-five thousand workers; 90 percent are women, most of them Asian immigrants (Louie, 2000).

CONTEMPORARY IMMIGRATION AS A GLOBAL PHENOMENON

As vast and diverse as immigration to the United States has been during the past four decades, it has hardly been a uniquely American phenomenon. Indeed, looking at the world's developed nations reveals that almost all have been affected in a similar way. The numbers may be smaller than in the United States, but the relative impact has been no less deeply felt and, in some cases, has produced even greater social strain. In addition to the traditional immigrant-receiving societies of Canada, Australia, and New Zealand, many western European nations now have large immigrant populations and, as result, have emerged as multiethnic societies. For countries such as Britain, France, and Germany, this has represented a profound transformation and a wrenching societal experience, the consequences of which are still unfolding. Even relatively homogeneous societies like Italy, Spain, Sweden, and Japan have begun to evince greater ethnic diversity due to immigration. And, in all of these societies, the changing ethnic composition has brought forth heated public debate, just as in the United States: Are immigrants becoming too numerous? Is the society's cultural integrity threatened? Are immigrants taking jobs from native workers? We will revisit these issues as they have affected the United States in Chapter 13 and western European nations in Chapter 17.

To recognize that we are in an age of immigration should not cause us to forget that immigration, as noted at the outset of this chapter, has been ongoing since the very emergence of human societies. Nonetheless, it would hardly be exaggerating to declare that today we are in a period in which the dimensions of immigration are greater, and its consequences loom larger, than in any previous era in human history. And there appears no likelihood that the tide of international migration will recede in the foreseeable future. As long as there is a global stratification system, well-off nations with strong economies will attract people from nations with weak economies and severe population pressures. It seems apparent, then, that the issues and social consequences of large-scale immigration and increasing global diversity will be with us for many decades.

THE AMERICAN ETHNIC CONFIGURATION

It is now obvious that the newest immigration has thrust the United States into a period in which ethnic heterogeneity is historically unprecedented. Never before has America been more diverse, and that diversity will in all likelihood continue to expand. The traditional ethnic dominance of Euro-Americans, therefore, is no longer a given. As shown in Figure 5.5, if current demographic trends continue, Euro-Americans will make up less than half the American population by the middle of this century. This eventuality is assured in light of recent census data indicating that deaths now exceed births among non-Hispanic white Americans.

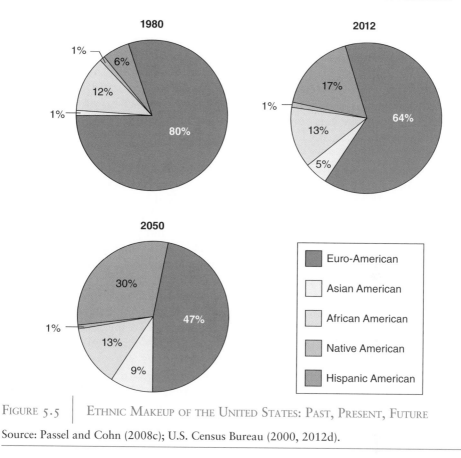

FIGURE 5.5 | ETHNIC MAKEUP OF THE UNITED STATES: PAST, PRESENT, FUTURE

Source: Passel and Cohn (2008c); U.S. Census Bureau (2000, 2012d).

"OFFICIAL" RACIAL AND ETHNIC CLASSIFICATIONS

The demographic projections in Figure 5.5 must be studied with some caution. To begin with, these are merely estimates demographers have constructed on the basis of current and, presumably, future demographic trends. But demographers have no way of predicting what kind of fertility and mortality rates might be unleashed by economic and social forces. Few were able to foresee the post–World War II baby boom, for example. Nor can they be sure of future immigration rates since these are determined mostly by federal legislation, making them a key political issue subject to all kinds of political influences.

CHANGING RACIAL AND ETHNIC CATEGORIES Another, perhaps more significant, factor makes those projections less compelling than they initially seem. Even if demographers were capable of making accurate predictions, their estimates do not take into consideration that the categories currently used to classify the population by race and ethnicity are likely to change in the future.

In 1977, the U.S. federal government, as a means of monitoring and enforcing civil rights legislation and affirmative action programs, adopted a set of ethnic categories by which people could be classified and counted (Wright, 1994). These were subsequently adopted by virtually all government agencies, including the U.S. Census Bureau, as well as by educational institutions, corporations, and the mass media. Five racial-ethnic categories were put in place, and they have been changed only in minor ways since then: white, black, Hispanic, Asian/Pacific Islander, and American Indian/Alaskan Native. These have become familiar categories, appearing on various educational and employment applications, census forms, and surveys. Unfortunately, rather than clarifying the conceptual morass of race and ethnicity in American society, these categories have further confounded the issue and contributed to even greater confusion.

For example, consider that Hispanic—an ethnic designation—is a separate category, alongside "racial" categories—black, white, Asian, American Indian. Or consider that extremely broad ranges of cultures and physical types are lumped together into each of these categories. For example, to speak of "blacks" in the United States in the aggregate is to assume a homogeneity that does not exist. Overlooked is the ethnic variety among U.S. blacks. Although blacks whose ancestry is traceable to American slavery predominate, today Haitians, Jamaicans, other Caribbeans, many Latinos, and even a growing population of African immigrants are also part of the American "black" population. Culturally these groups have little in common; they may even evince physical differences. Likewise, the term *white* has no significance beyond a reference to white ethnic groups in the aggregate. To speak of Polish Americans, Jewish Americans, and Irish Americans as part of a common group is to falsely meld groups whose cultural traditions are quite distinct.

What the current U.S. classification system represents, then, is an extremely crude and bewildering attempt to count the population by ethnicity. It also demonstrates the arbitrariness of racial or ethnic classification. Race and ethnicity, it should be remembered, are social constructs; they are not biologically fixed divisions of the human population or of the population of any particular society. The ways that societies choose to subdivide and classify their populations can and do change. In the United States, racial and ethnic classifications have been subject to repeated modifications, redefinitions, and replacements (Lee and Bean, 2010). Moreover, it should be remembered that census questions ask people to identify *themselves* in terms of race and ethnicity. Given the complexities of trying to determine anyone's actual racial and ethnic history, census data must be seen as measures of *identity*, not *ancestry* (Perez and Hirschman, 2009).

FADING ETHNIC BOUNDARIES   An even more important caveat to be considered in looking at these projections is that—as we will see in the next few chapters—much intermarriage is occurring across racial and ethnic lines, the effect of which is to further blur these already fuzzy and politically influenced categories. Consider golfer Tiger Woods as an example. Woods has been touted as a great black athlete. But is he "black"? His father was African American, his mother is Asian American, and his ancestry also includes Euro-American and Native American. As Woods himself explains, "Growing up, I came up with this name: I'm a 'Cablinasian,'" a blend of Caucasian, black, Indian, and Asian (Fletcher, 1997). President Obama, of course, is another obvious case, his mother a white woman from Kansas, his father a black man from Kenya.

As the population becomes more diverse, cases such as that of Tiger Woods and President Obama are becoming commonplace. Indeed, they may be exemplars of

America's ethnic future. Continued racial and ethnic mixing are reshaping racial and ethnic boundaries and identities and will continue to do so in the coming decades. The result is that the familiar racial-ethnic categories—white, black, Hispanic, Asian—no longer make much sense (if they ever did) in classifying people and simply no longer work in accurately describing American society.

Although the current classification system may be far less than satisfactory, we remain thoroughly reliant on racial and ethnic data gathered by the U.S. Census Bureau, as well as by other public and private agencies and organizations that employ the official racial-ethnic scheme. Until sociologists and other social scientists can collaborate with policy makers to construct a more rational and meaningful way of conceptualizing ethnic boundaries, we are constrained to use the categories currently in place.

"MIXED RACE" PEOPLE The failure of the commonly used racial-ethnic classification scheme to accurately reflect the increasingly mixed blend of the U.S. population has led in recent years to calls for the creation of a "mixed race" category. Presumably this would account for those individuals who fall through the cracks of the current system, that is, those whose parentage reflects more than one racial-ethnic category. The 2000 U.S. Census stopped short of creating such a category, though it did, for the first time, give people the choice of identifying themselves with more than one racial category—an option chosen by 2.4 percent of the population (Lee, 2001). A similar option was provided in the 2010 Census and was chosen by 2.9 percent of the population (U.S. Census Bureau, 2012f).

The creation of a mixed-race category would more comfortably fit a growing segment of the population, but it would do nothing to make the racial-ethnic classification scheme any less arbitrary; it would only create another category that carries no more real meaning than any of the others. Moreover, creating a mixed-race category, as explained in Chapter 1, raises the implication that there are "pure" races to begin with. As one researcher has put it, "[I]nsofar as there is no biological foundation for the three or four acknowledged races, neither is there a biological foundation for mixed-race groups" (Zack, 2001:52).

In reality, of course, there has been a mixed-race population in American society from the very beginning. In earlier times, when the essential racial division was black/white, the so-called **one-drop rule** arbitrarily crystallized this racial dichotomy. In effect, one had to be either black or white; there were no in-between categories.[3] *Blacks were defined as those having black ancestors—that is, "one drop of black blood."* But the percentage of black ancestry (what fraction constituted one drop?) varied from state to state. In some it was one-fourth; in others, one-eighth, one-sixteenth, or even one-thirty-second (Davis, 2001). At the turn of the last century, the one-drop rule was essentially the law of the land; even today it affects racial classification (Hollinger, 2005). Most of the U.S. black population, in fact, has some white ancestry, and a small portion of the white population has some black ancestry (Blackburn, 2000; Smedley, 2007).

---

[3] In fact, the U.S. Census during the mid-nineteenth and early twentieth centuries did provide an "intermediate" racial category, "mulatto," which was eliminated by 1930 when the one-drop rule defined as black "anyone with any black ancestry" (Davis, 2001; Lee and Bean, 2010).

CONTINUED ETHNIC DIVERSITY   One unmistakable fact does emerge from the confusion accompanying attempts to classify and quantify the American population by race and ethnicity: the United States continues to evolve into an ever-more heterogeneous society. In the twenty-first century, America's ethnic character will be more culturally and physically diverse than it was in the twentieth century. And despite the efforts of some to preserve groups' social boundaries, the trend toward a fading of those racial and ethnic lines seems irreversible.

Despite the virtually unparalleled heterogeneity of the United States today, it is interesting—if bemusing—to consider how misinformed most Americans are about the ethnic makeup of their society. Most cannot describe the actual size of various ethnic groups with anything even approaching accuracy. This has been demonstrated by national surveys over the past decade, in which Americans grossly exaggerate the size of racial and ethnic minorities. The typical American estimates the size of African American, Hispanic American, and Asian American populations to be more than twice as large as they actually are (Carroll, 2001). This seriously distorted picture of the American ethnic configuration is bound to influence people's attitudes and actions regarding ethnic-related issues such as immigration, affirmative action, and multiculturalism.

THE AMERICAN ETHNIC HIERARCHY TODAY

The analysis of America's ethnic groups that follows in the next seven chapters suggests that the American ethnic hierarchy today can be viewed as divided into three comprehensive tiers:

1. The top tier comprises white Protestants of various national origins, for whom ethnicity has no real significance except to distinguish them from the remainder of the ethnic hierarchy.
2. The intermediate tier comprises white ethnics of various national origins (mostly Catholic and Jewish) and many Asians, for whom ethnicity continues to play a role in the distribution of the society's rewards and continues to influence social life—but in both instances, decreasingly so. In fact, for some white ethnic groups in this tier, ethnicity has lost almost all social relevance, as they blend into the top tier.
3. The bottom tier comprises racial-ethnic groups—blacks, Latinos, American Indians, and some Asians—for whom ethnicity today has significant consequences and for whom it continues to shape many of the basic aspects of social and economic life.

This ethnic hierarchy has remained relatively stable for the past century and a half. Although the distance between many of the groups has been reduced, their rank order has, with few exceptions, not been basically altered.

Arranging ethnic groups in such a rank order, of course, masks the more specific class and cultural differences among them, as well as the internal differences within each. But in a very general sense, each of the country's ethnic groups can be placed in one of these tiers, based on a combination of the factors we will look at in upcoming chapters: the collective place of the group in the society's economic and political hierarchies; the extent to which prejudice and discrimination remain significant for the

group; and the extent to which group members have entered into full social participation with members of the dominant group.[4]

Perhaps the most important aspect of this three-part American ethnic hierarchy
is that the gap between the bottom tier of groups and the other two is much greater
than the gap between the top and intermediate tiers. Thus, except for Asians and
some Hispanics, the American ethnic system today seems to be increasingly dichoto-
mized—white and nonwhite. The overriding issues of American ethnic relations,
therefore, remain focused on the economic, political, and social disparities between
whites and nonwhites and the policies intended to reduce them. The crux of the matter is relatively simple: will the gap between whites and nonwhites continue in its
present form, will it be reduced substantially, or will it perhaps widen? The issues
of American race and ethnic relations are largely subsumed by this pivotal question.

SUMMARY

- Following its conquest of the indigenous peoples and enslavement of blacks, the
  Anglo core group established its social and political dominance in America and
  exercised the power of selection over those who came afterward.
- American ideologies and public policies regarding ethnic groups and issues have
  consistently reflected Anglo cultural preferences and Anglo economic and political interests. Based on those thoughts and actions, an ethnic hierarchy has been
  shaped, and the rank order of groups has not essentially changed.
- The ethnic diversity of the American population blossomed forcefully during the
  approximately one hundred years between the 1820s and 1920s. Two massive
  waves of European immigration took place. The *Old Immigration,* during the
  first fifty years of that period, brought mainly groups from northern and western
  Europe, the largest among them the Irish and Germans. During the second fifty
  years, the *New Immigration* consisted primarily of groups from southern and
  eastern Europe. The cultural distinctiveness of the latter provoked a nativist
  movement that ultimately resulted in the passage of restrictive immigration laws,
  ending, in effect, further large-scale European immigration.
- The contemporary period of immigration, referred to as the *Newest Immigration,* was launched by the passage of immigration reforms in 1965, which ended
  country-of-origin restrictions. Whereas previous periods of heavy immigration
  had consisted mostly of European groups, the Newest Immigration is made up in
  largest part of groups from Asia and Latin America.
- The scope of the Newest Immigration has exceeded even that of the early 1900s,
  previously the heaviest period of immigration to the United States. As a result,
  an ethnic reconfiguration of the U.S. population is occurring. Asian and Latin
  American groups are growing, and Euro-American groups are declining in size.
- The American ethnic hierarchy today is made up essentially of three comprehensive tiers: white Protestants of various national origins at the top; white

---

[4] Using a combination of these factors means that some will offset others. For example, although Jews and
Asians rank higher in their aggregate economic class position than northwestern Europeans, they still do
not fully interact at the primary level with those groups, and the extent of prejudice and discrimination
directed against them is much greater. Hence, when all these factors are considered, Jews and Asians are
part of the intermediate tier of groups rather than the top.

ethnics of various national origins, along with most Asian Americans, in the middle; and Hispanic Americans, African Americans, and American Indians at the bottom.

CRITICAL THINKING

1. In the early formation of the American nation, English influences predominated in shaping basic institutions, like language, law, politics, and economy. Why were American Indians or other ethnic groups unable to exert similar influence?
2. Examine the ethnic makeup of the top leadership of America's major political, economic, educational, and cultural institutions. Based on your findings, are Anglo-Americans still the dominant ethnic group? In what ways has their political, economic, and cultural power been diluted?
3. Does "Anglo-conformity," discussed in this chapter, still characterize the expected path of assimilation for new immigrants to America? Is the "melting pot" model more applicable to U.S. society today than in past eras?
4. Do an analysis of your city or town in terms of its ethnic makeup. Has it changed during the past several decades and, if so, in what ways? Which ethnic groups are present today that were absent in past times? What factors contributed to those groups' migration to your community?

PERSONAL/PRACTICAL APPLICATION

1. As explained in this chapter, the United States is in the midst of a demographic transformation in terms of race and ethnicity. How will the changing U.S. racial/ethnic landscape affect you?
2. Consider the process of voluntary immigration, as described in this chapter, and try to put yourself in the place of immigrants. What might motivate you to consider leaving familiar surroundings and move to another city or state or even another country, not knowing what might await you once you arrived there?

Native Americans

If the United States is a country of immigrants, what can be said of American Indians (or Native Americans)? As the original inhabitants of the continent, surely they cannot be considered immigrants in the conventional sense, like Europeans of the nineteenth and early twentieth centuries or Asians and Hispanics of today. But in a sense, even these native peoples can be seen as an immigrant group when considered in a long-range historical context. From twelve thousand to fifty thousand years ago, humans reached the Western Hemisphere from Northeast Asia with the help of glaciers that, from time to time, covered most of what is today Canada and Alaska, forming a land bridge between the two hemispheres. Some may have come by water as well as land. These migrations probably occurred over thousands of years, and it is unlikely that the migrants came from similar racial origins (Goodman and Heffington, 2000). How humans first came to inhabit what is today North and South America remains an unresolved question among archaeologists. But what is no longer much debated is that American Indians originated in another geographic locale.

Realistically, of course, it would seem absurd to refer to American Indians as an "immigrant" group, despite their ancient migration from another part of the globe. Their cultures and societies developed here, not on another continent. Clearly, American Indians are North America's indigenous peoples, and in this they maintain a unique status among the ethnic populations of the United States and Canada. As Ronald Wright has put it, "Native Americans . . . are American in a way that no others can be. Even if we suppose that their ancestors arrived 'only' 15,000 years ago . . . they have been here thirty times longer than anyone else. If we call that time a month, Columbus came yesterday" (1992:x).

CONQUEST AND ADAPTATION

Because American Indians inhabited the continent prior to the explorations of Europeans, they occupy a singular place in the shaping and development of the American ethnic hierarchy. Unlike any other group, their placement into the society's ethnic scheme and their subsequent adaptation were shaped by conquest.

EUROPEAN CONTACT

When Europeans crossed the Atlantic beginning in the late fifteenth century, they found two continents inhabited by millions of people.[1] Native peoples extended over the entire Western Hemisphere, from the Arctic to the tip of South America. These "Indians," as the Europeans called them, were by no means homogeneous but were organized into hundreds of tribes or nations, some quite advanced in social organization and technology. They practiced a wide variety of subsistence patterns ranging from hunting and gathering to intensive agriculture (Crawford, 2001). They were different from the Europeans, however, and the white invaders ethnocentrically interpreted these physical and cultural differences as proof of inferiority and lack of civilization.

Estimates of the indigenous population north of the Rio Grande at the time of Columbus's arrival range from one million to eighteen million (Crawford, 2001; Thornton, 1995; Ubelaker, 1988). Like the more numerous natives of what is today Latin America, these peoples spoke many different languages, conformed to disparate cultural systems, and were organized in a variety of economic and social forms. There were, then, no "American Indians" per se but rather a great number of diverse ethnic groups, viewing themselves as different from one another as from the European invaders (Lurie, 1991).

EMERGENCE OF AN ETHNIC HIERARCHY The establishment of European supremacy over the North American Indians conforms closely to the theories of the emergence of dominance and subordination explained in Chapter 2. As a conquered population, the subordination of Indians occurred through a highly conflictual process. An indigenous group invaded by a more powerful settler group can only capitulate or, more commonly, resist. This ordinarily makes for a high level of conflict and rejection of assimilation into the dominant group.

The four major conditions for the emergence of ethnic stratification—contact, ethnocentrism, competition, and differential power—are clearly seen in the case of Indian-white relations at the outset of European settlement and afterward.

The Native Americans' physical and cultural distance from European standards inevitably produced an ethnocentric white view. Of particular importance to the North American settlers was that the Indians were "heathen." Moreover, they were preliterate and at a technological level far below that of the Europeans. A combination of these traits contributed to the image of Indians as savage and inferior and, consequently, to the rationale for their exploitation and annihilation. As Alvin Josephy has written, "In the Western Hemisphere, Indians, to the whites, were all the same, and the newcomers disagreed among themselves only over the extent to which the native populations differed from, or seemed to be inferior to, Europeans" (1969:279).

Contact and even ethnocentrism among alien ethnic groups will not in themselves create dominant-subordinate relations, however. Competition over valued

---

[1] The number of indigenous people inhabiting North and South America at the time of European discovery is disputed among anthropologists and historians. Estimates range from less than ten million to more than a hundred million (Dobyns, 1966; Kroeber, 1939; Utter, 1993). As one researcher has declared, "Prior to the late nineteenth century, all numbers are uncertain, and a uniquely correct number of natives never will be had" (Jaffe, 1992:2).

resources is a third factor, and in the case of Indian-white relations, land provided the necessary ingredient. The settlers' objective was to secure land for agricultural production. From the outset, divergent cultural perceptions of "property" led to misunderstandings and hostilities between Indian and settler. Indian lands were held communally, not as private property. Whites consequently perceived them as basically unoccupied. Furthermore, most Indians did not engage in European-style farming, which led to the view that these lands were underutilized and thus justifiably occupied by those who would cultivate them in a "proper" manner.

In the end, however, it was the differential power between Indians and settlers that proved the crucial element in establishing dominance and subordination. In particular, the deciding factor in the European conquest of indigenous peoples in all parts of the Western Hemisphere was the Europeans' technological skills, which gave them a tremendous advantage in war. When the vastly superior arms of the colonialists were combined with the inability of Indians to resist the diseases introduced by the Europeans, it was inevitable that whites would prevail in the struggle for resources and that Native populations would decline.

DISPLACEMENT AND DEPOPULATION

From the beginning of European settlement to the end of the eighteenth century, Indian-white relations in North America centered primarily on the fur trade. During that time, three major colonial powers vied for the economic resources of the Western Hemisphere: Britain, France, and Spain. Each had different colonial objectives. Spain, chiefly concerned with exploiting the mineral wealth of its possessions, concentrated its activities primarily in the southern part of the hemisphere. In North America, the Spanish influence was eventually confined to the present-day American Southwest. Indigenous peoples in what are today Mexico, Central and South America, and parts of the Caribbean were brutally exploited as a labor force, first through enslavement and later through debt peonage (Cumberland, 1968).

France's economic activities, by contrast, focused mostly on supplying the European market with furs. Its activities extended throughout most of the northeastern and central parts of the North American continent, including the St. Lawrence waterway, eastern Canada, the Great Lakes, the Ohio Valley, and the Mississippi River Valley. As a result, the French necessarily nurtured close relations with North American Indian nations who aided them in their economic goals. Through the fur trade, Indians were integrated into the emergent North American economy.

THE BRITISH TRIUMPH   The British, too, were engaged in the fur trade and, like the French, entered into alliances with Indian nations. The British-French rivalry in the fur trade led to deadly conflict among the Indian nations attached to each country. Moreover, as a result of their economic and political relations with the Europeans, native Indians' cultural institutions and forms of social organization were radically transformed in numerous ways.

More than the Spanish and the French, the British valued land as a resource, and their colonial activities eventually became focused on that goal. Britain was interested in having colonists permanently settle the land and farm it rather than simply exploit it for minerals and furs. With the triumph of the British in North America in the late eighteenth century, the crux of Indian-white relations shifted from the fur trade to issues of land. The British colonial government as well as its successor, the newly independent United States, sought to control the acquisition of lands by American settlers and to protect the rights of Indians, but both governments failed. The quest for land by British colonists proved insatiable, leading ultimately to the displacement of the indigenous population.

At the start of the nineteenth century, Indian-white relations evolved into a confrontation in which the rapacious quest of white settlers for land led to ever-greater efforts to dispossess Indians of lands they were occupying. Two paths to this objective became apparent: Indians would be separated from their lands either through assimilation (by encouraging them to abandon their communal patterns of landholding in favor of private property) or through removal (by relocating them through negotiated purchase of land or through conquest) (Cornell, 1988). These two paths were not mutually exclusive; rather, they were complementary, not only in a practical sense but also as a way of satisfying the objectives of settlers, who wanted land, and reformers, who sought to "civilize" the Indians with American culture. This dual objective, notes historian Robert Berkhofer, was a constant pattern throughout most of American history: "From the founding of the nation until recent times, and some would include today as well, United States policy makers placed two considerations above all others in the nation's relation with Native Americans as Indians: the extinction of native title in favor of White exploitation of native lands and resources and the transformation of native lifestyles into copies of approved White models" (1979:135).

By the late nineteenth century, most of the remaining Native population had been forcibly resettled in reservations west of the Mississippi. It was not coincidental that reservations were located on lands that whites deemed virtually worthless for farming or grazing.

DEPOPULATION   Indian societies were physically reduced through a combination of exposure to European diseases, armed conflict, starvation, and the breakup of cultural systems that had traditionally provided for social and material needs. Although Indians did make many cultural adaptations to the white presence, doing so did not prevent the eventual destruction of their social forms and the succession of white dominance (Wax, 1971). By the time Americans had gone beyond the Appalachian Mountains, Indians had already adopted several new technologies from the Europeans. Firearms now aided in hunting and, on the plains, the horse, introduced by the Spanish in the Southwest, had thoroughly transformed the Indians' way of life. Despite these adaptations, however, Indians remained politically fractured into many scattered tribes. This made them vulnerable to American settlers' intrusions and, ultimately, divestiture.

As for diseases, perhaps nothing more severely affected the Indians' physical condition than the epidemics brought by Europeans, which decimated many tribes from the very outset of their contact with settlers and explorers. These epidemics continued, repeatedly, throughout the nineteenth century (Thornton, 1995; Washburn, 1975). Native populations lacked immunity to smallpox, measles, cholera, typhoid, malaria, and other diseases; their effects were devastating.

GOVERNMENT POLICIES AND AMERICAN INDIANS

Stephen Cornell (1988) has suggested that the "Indian problem" for Euro-Americans from the time of settlement to the present has had three aspects: economic—how best to secure Indian resources, especially land;[2] cultural—how best to assimilate Indians into the dominant, non-Indian, culture; and political—how best to control Indians to bring about solutions to the first two problems. The counterpart to the Indian problem was a "Euro-American problem" for Indians. "In its essence," notes Cornell, "this problem seems to have been tribal survival: the maintenance of particular sets of social relations, more or less distinct cultural orders, and some measure of political autonomy in the face of invasion, conquest, and loss of power" (1988:7). It is around these two conflicting problems that Indian-white relations have revolved for the past three centuries.

The solution to the Euro-Americans' economic, cultural, and political "Indian problem" was displacement and depopulation of Indian tribes, as well as cultural decimation through forced assimilation. All of these strategies were underwritten by government edict. Several phases of the history of relations between American Indians and the federal government can be delineated. Each corresponds to shifting objectives and specific measures, always undertaken with two implicit goals: (1) eliminating the Native American population as an impediment to western settlement and the needs of an expanding American economy, and (2) eradicating Native American cultures and political forms.

SEPARATION The earliest U.S. policy toward Indians grew out of the struggle for land that had characterized relations between Indians and the English settlers. Various agreements between Indians and colonists essentially held that the Indian nations were independent political entities and were to be dealt with on that basis. This policy continued after U.S. independence. Specific bounded territories were promulgated, separating Indian nations from Americans. Thus the paradoxical status of Indians was, as Edward Spicer describes it, "a people within the territory of the new nation and yet outside the processes of its political life" (1982:179).

The legal status of Indian lands, however, quickly became hazy as settlers coveted these lands and began to infringe on them. Indian lands were ceded irreversibly by treaties, usually entered into by Indians under duress or in ignorance of their meaning, through fraudulent schemes perpetrated by whites, or by sheer force. Moreover, treaties themselves provided no protection of those lands left to the Native Americans because renegotiations and violations became standard practice whenever disputes arose. Where there was desirable land, whites eventually took it. Clashes occurred repeatedly; the solution, from the American perspective, was to engage the Indians in negotiation and land exchanges for their removal to lands farther west, where it was assumed these border issues would no longer arise.

REMOVAL The Indian Removal Act of 1830 called for the relocation of all tribes living in the eastern United States to lands set aside for them west of the Mississippi

---

[2] It should be remembered that at the outset of contact between Indians and Europeans, it was not land but other resources—in North America, furs—that Europeans sought. From the late eighteenth century on, however, land was the major objective.

River. This policy was pursued vigorously by President Andrew Jackson, who represented the sentiments of settlers who saw Indian occupation of desired lands as an obstacle to the development of white dominance. Removal was rationalized by the view that Indians were not using land as "God had intended"—that is, to farm in the way of the white man. As Governor William Henry Harrison asked, "Is one of the fairest portions of the globe to remain in a state of nature, the haunt of a few wretched savages, when it seems destined by the Creator to give support to a large population and to be the seat of civilization?" (quoted in Hagan, 1961:69).

This ten-year movement marked a time of great hardship and suffering for Indians and resulted in the virtual destruction of many tribes, particularly those that resisted removal. Perhaps the most infamous episode involved the forced migration of the Cherokees in 1838, an affair referred to as the "Trail of Tears." In the summer of that year, the U.S. Army rounded up all sixteen thousand Cherokees and held them for months in disease-infested camps. The march west to Oklahoma extended over the autumn and winter, and by its end, four thousand had died (Wright, 1992). As one historian has described it, the removal of Indians at best "resembled closely the pioneering experience of thousands of their white contemporaries" and at worst "approached the horrors created by the Nazi handling of subject peoples" (Hagan, 1961:77). The Indian Removal Act was perhaps the most devastating single action taken by the federal government in destroying Indian cultures and societies.

THE WESTERN INDIANS   After the American Civil War, the U.S. government took steps to put an end to Indian resistance that stood in the way of western development. Thus the locus of Indian affairs shifted to the tribes of the plains, the Southwest, and the Far West. The result was the further dispossession of Indian lands through military campaigns as well as the signing of treaties that were quickly broken by the government when their terms impeded settlement.

A key part of this effort was the suppression of the Ghost Dance religious movement. The Ghost Dance religion had been founded by a prophet, Wovoka, of the Paiute tribe. Believers were convinced that they could drive whites from Indian lands and restore their traditional way of life by following certain rituals, particularly a ceremonial dance. In the late 1880s, the movement spread across the plains. The U.S. government viewed the Ghost Dance as a serious threat; in 1890, following the killing of Sitting Bull, one of the movement's leaders, the government ordered troops to move on the Ghost Dancers. This led to the slaughter of hundreds of Sioux Indians at Wounded Knee near the Pine Ridge Reservation in South Dakota (Brown, 1972). This incident represented the culmination of military force against Indians, and from that point forward, Indian resistance ceased. Wounded Knee became a symbol of the federal government's destructive anti-Indian policies.

In addition to military force, the Plains Indians were decimated by disease and by the extermination of the buffalo (or bison), which had been the very heart of their way of life. They depended on it not only for food but also for clothing and shelter. The herds were drastically reduced when white hunters shot them, harvesting meat and hides. In a matter of only a few years, as many as fifteen million buffalo were virtually extinguished. The passing of the buffalo further diminished the power of Western Indians to resist white settlement. The choice they now faced was between depending on government handouts and adopting the Euro-American way of life (Wexler, 1995).

ALLOTMENT AND COERCIVE ASSIMILATION   By the latter part of the nineteenth century, armed resistance to federal policy had ended. Government officials and reformers now advocated a policy that they felt would once and for all solve the Indian problem: Reservation lands would be broken up and allotted to individual tribal members. Unallotted lands would then be sold to non-Indians. This policy was formalized in 1887 with congressional passage of the Dawes Act (also called the General Allotment Act). The policy of allotment further diminished those lands remaining in Indian possession, leaving them with even fewer productive resources. Furthermore, the administrative effects of the Dawes Act led to an even greater intrusion of the BIA (the Bureau of Indian Affairs, the government's major agency for dealing with Indians) into Indian affairs, further diminishing native cultures and weakening tribal political institutions.

As a means of coercing Indians into an acceptance of their now virtual powerlessness in the face of the federal government, a policy of forced assimilation was put into place. The policy of breaking up Indian lands and assigning title to individuals was, by itself, intended to make over Indian families, using white settlers as the model: "hard working and economically motivated" (Spicer, 1982:183). But part of the policy also consisted of specific attempts to assimilate Indians into the mainstream society. In cooperation with religious groups, efforts were made to eradicate native cultures and impose the American way of life. This meant destroying native religions and putting Christianity in their place, denying Indians native languages and substituting English, and generally reducing education to an effort to "civilize" Indians. Native American children were placed in government-run boarding schools, which indoctrinated them with the belief that Euro-American culture was vastly superior to "primitive" tribal cultures. Separated from their families for years at a time, the children were taught to speak English, wear Western clothing, and pray as Christians; failure to behave accordingly merited stern punishment from school authorities (Adams, 1995). The BIA commissioner in 1889 bluntly stated his bureau's intent: "The Indians must conform to 'the white man's ways,' peaceably if they will, forcibly if they must" (quoted in Cornell, 1988:56).

At the start of the twentieth century, then, Indians were impoverished and virtually at the mercy of the federal government, whose paternalistic policies continued to reflect white ethnocentrism. By that time, the Indian population had been reduced to fewer than 250,000. Indians were now officially wards of the federal government and were thus no longer seen or dealt with as separate nations.

A NEW APPROACH TO INDIAN AFFAIRS   In the 1930s, an effort was made to reverse the decades of neglect and exploitation that had characterized government policies vis-à-vis American Indians. It was also an attempt to create a more pluralistic system of Indian-white relations. John Collier, enlightened and sympathetic toward Native causes, was appointed to head the BIA during President Franklin D. Roosevelt's first term. Collier proposed major changes in Indian policy that included tribal self-rule and efforts to encourage Native cultural preservation. Most important was his proposal to end the allotment system created by the Dawes Act, which had resulted in the loss of much Indian land. Congress passed the Indian Reorganization Act (IRA) in 1934, which included a number of Collier's recommendations. It was not met with enthusiasm by all Indians, however. Some were committed to assimilation, and others were not prepared to accept at face value any measure offered by the federal government. Still others saw the act as simply one more form of paternalism. Most tribes, however, approved it. The IRA, though failing to fundamentally alter the status and condition of Native Americans, did result in halting some of the more egregious aspects of past Indian policy. It prohibited further allotments of tribal lands and encouraged Indians to establish self-governing systems within their tribes. Moreover, Collier reduced the heavy-handed control of Indian life by the BIA and reduced efforts aimed at forced cultural assimilation.

TERMINATION AND ITS REVERSAL   In the 1950s, another turnabout in the government's approach to Indian affairs emerged with a new policy known as "termination." In essence the objective was, once again, assimilation. Termination essentially meant the end of the federal government's responsibility to provide various social, educational, and economic services to Indians; it also ended government protection of Indian lands and property held in trust for the tribes (Senese, 1991; Wilkinson, 2005). In a sense, Indians would now be entirely on their own. Moreover, the tribal organization of Indian life was seen as a relic of the past, and it was felt that Indians should be dealt with on an individualistic, not a communal, basis. The idea was that they should be treated no differently from other citizens. To accomplish this, it was proposed that the reservation system be dismantled and the federal government's unique role toward Indians be severed. No longer would treaty obligations be recognized, and the sovereign status of tribes would be eliminated.

Most Indians strongly resisted termination, realizing that the ultimate result would be the loss of tribal lands as well as Native cultures. The failure of termination led to a return to a policy more in line with the IRA during the next three decades. Reversing its position again, federal policy toward American Indians in the 1970s moved in the direction of greater autonomy for tribal governments.

RED POWER   A major influence on the government's decision to reverse the policy of termination was the Red Power movement. During the late 1960s and early 1970s, Native Americans put pressure on the federal government to address Indian issues and needs and to reassert Indian rights. Moreover, as with other ethnic groups, American Indians during that time began to reaffirm their ethnic identity. These ideas and actions, in total, came to be known as the **Red Power** movement. Specifically, *the aim of Red Power was to end the federal policy of termination, return Indians to traditional cultural ways, and review and revitalize Indian communities*. As it took shape, Red Power in many ways emulated similar movements occurring at that time among African Americans and Latinos. It brought the grievances and general plight of Native Americans to public awareness through numerous incidents—marches, protests, sit-ins, and demonstrations.

Perhaps the most publicized and attention-grabbing event of the Red Power movement was the nineteen-month occupation of Alcatraz Island, an abandoned federal prison in San Francisco Bay, beginning in 1969. Many Indian activists saw this as the spark that ignited the Red Power movement. Following Alcatraz, several other occupations took place in various locations around the country. The siege at Wounded Knee in the Pine Ridge Reservation in South Dakota in 1973, however, overshadowed the others. The conflict actually began as an internal power struggle among the Oglala Sioux tribe. Eventually it involved not only the two sides of the tribe but also federal law enforcement officials, the BIA, local citizens, nationally prominent entertainers who lent their celebrity endorsement, national religious and legal organizations, and the national news media (Josephy et al., 1999). One of the major groups involved was AIM, the American Indian Movement. AIM had been founded in 1968 by a group of young Indian men, including Dennis Banks, and had been one of the major organizing units of the earlier Alcatraz occupation.

Wounded Knee culminated in a standoff between AIM and federal officials that continued for more than two months, with sporadic violence resulting in several deaths and many injuries. Banks and Russell Means, another activist leader, were subsequently arrested and charged with inciting to riot and conspiracy. The choice of Wounded Knee was hardly random. This location held great symbolic significance for American Indians because it had been the site where the U.S. cavalry had massacred hundreds of Sioux men, women, and children in 1890. After Wounded Knee, the militancy of Red Power subsided relatively quickly, however, and was replaced with a more legalistic approach to preserving and enhancing Indian cultural and economic interests.

Alcatraz and Wounded Knee took on great symbolic meaning for Native Americans and helped to stimulate ethnic reidentification among Indians who had drifted into an urbanized assimilation path. A renewal of ethnic pride and consciousness became evident, and a cultural renaissance emerged on many Indian reservations (Kelly and Nagel, 2002).

SELF-DETERMINATION In 1975, Congress passed the Indian Self-Determination and Education Assistance Act, permitting tribes to take control of numerous federal programs on reservations. This marked a revolutionary break with the past, releasing Indian tribes from the strict control and supervision of the BIA (Josephy et al., 1999; Wilkinson, 2005). The effect was a renewal of tribal sovereignty and, in effect, a reversal of the policy of termination. Many of the reservations' social and economic functions were now assumed by tribal governments.

In the 1980s, the Reagan administration made severe cuts in federal assistance to American Indians. As with all other federal agencies, the Reagan approach was to remove the federal government from activities and turn these over to the private sector. In 1983 alone, aid was cut by more than one-third, affecting programs on all reservations (Worsnop, 1992). These policies reflected a philosophical rejection of self-determination that had seemed to hold sway in the 1970s, but they were short-lived. By the late 1980s and throughout the 1990s, self-determination once again dominated government policy as Congress and the courts made decisions that strengthened the sovereignty of Indian nations (Josephy et al., 1999; Wilkinson, 2005). In 1990, for example, Congress enacted the Native American Languages Act, intended to promote the use and development of Indian languages.

Recent politics of American Indians have focused primarily on political self-determination and on the protection of remaining Indian lands. Indians have, in the past two decades, generally experienced greater empowerment than at any time since the nineteenth century. Much of this power has been the result of the leverage that tribes have been able to exert in their control of valuable minerals and other resources that are now part of Indian-controlled areas. As much as 30 percent of U.S. energy sources, such as oil, gas, and coal, may be located on tribal lands (Harvard Project, 2008). Moreover, many tribes, as we will see, have become adept at business affairs and have developed numerous enterprises on reservations.

A consistent theme of power imbalance seems to run through the history of U.S. government policy vis-à-vis Native Americans. At all times, whatever the issues or whichever administration formulated policy, Indians themselves lacked the political power to defend their interests. They were always reactors to, never initiators of, policies affecting their welfare. At certain times the object of government measures was to assimilate Indians to the cultural ways of whites, whereas at other times, Indians were seen as unassimilable and thus to be left alone. In either case—with few exceptions—gross neglect typified the attitudes of policy makers, and each succeeding generation of Native Americans has felt the consequences of that neglect.

AMERICAN INDIAN DEMOGRAPHICS

It is difficult today to accurately measure the American Indian population because definitions of who is "Indian" are not clear and unequivocal. As with other ethnic groups, the U.S. Census counts anyone as Indian who declares him- or herself so. Within the Native American population itself, Indian identity is left to each tribe to define. Thus a variety of criteria may be used in establishing a person's tribal membership. Often blood quantum is a qualification; but in some cases, one need only trace his or her lineage to earlier tribal members (Snipp, 2002; Thornton, 1987).

In some cases, the indeterminate nature of Indian identity has led to internal tribal disputes centering on who should qualify for membership and thus be entitled to tribal benefits. In the Cherokee Nation, for example, there are descendants of black slaves owned by Cherokees, freed blacks who were married to Cherokees, and the children of mixed-race families known as black Cherokees, all of whom were part of the migration to Oklahoma in 1838 (Nieves, 2007a). In 2007, in a highly controversial move, the Cherokee Nation revoked the tribal citizenship of about three thousand of the descendants of slaves owned by the tribe. Tribal leaders later restored partial rights to some of these black members, though not full citizenship.

POPULATION GROWTH Despite the problems in accurately determining Indian identity, there is little doubt that Indians are a relatively young and growing ethnic population. The "official" Indian population today (including Alaska natives) is 2.9 million, which is less than 1 percent of the total U.S. population. Nonetheless, this represents a significant growth over the 1980 recorded Indian population of 1.4 million and a fivefold increase from 550,000 in 1960. As noted in Chapter 5, beginning in 2000 the U.S. Census enabled people to declare more than one racial identity. If those who declared some Indian ancestry are included with those who identified themselves as American Indian only, the population of this category jumps to about 5.2 million, or 1.7 percent of the U.S. population (U.S. Census Bureau, 2012f).

Much of the increase in the Native American population is the product of a high birthrate, but a larger part is accounted for by the growing tendency for people to declare their Indian affiliation (Sandefur et al., 2001). By the late 1960s, "Indianness" had risen in popularity. As a result of the Red Power movement, Indian activism inspired new pride among Native Americans, as did the more favorable portrayal of Indians by the mass media (Nagel, 1995). In addition to the new positive public perception of this group, many now claimed Indian ancestry in order to be included in lucrative judgments being awarded to tribes in land-claims cases. "Every announcement of a large judgment," notes historian William Hagan, "seemed to trigger the memories of some Americans that their family trees included an Indian, usually a chief's daughter, a princess" (1992:284). Many discovered that declaring oneself "Indian" also resulted in eligibility for various federal assistance programs, including health and education. Indian identity, then, remains problematic, with no systematic way to verify the validity of genuine Indianness. This is hardly a recent problem, however; it extends well back in American history (Hagan, 1992).

Today the federal government officially recognizes 564 Indian tribes in the United States; almost 300 other groups are seeking official recognition as tribes (Bureau of Indian Affairs, 2009; Peterson, 2004). Not all American Indians, however, are actually enrolled members of these tribes. Figure 6.1 lists the ten largest American Indian tribes.

RESERVATION INDIANS The American Indian population is split between those living on and those living off the approximately 300 federal Indian reservations. The

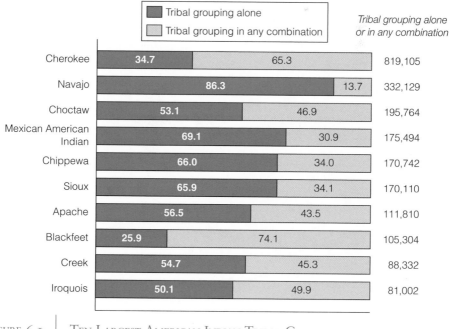

FIGURE 6.1 | TEN LARGEST AMERICAN INDIAN TRIBAL GROUPINGS

Source: U.S. Census Bureau (2012f).

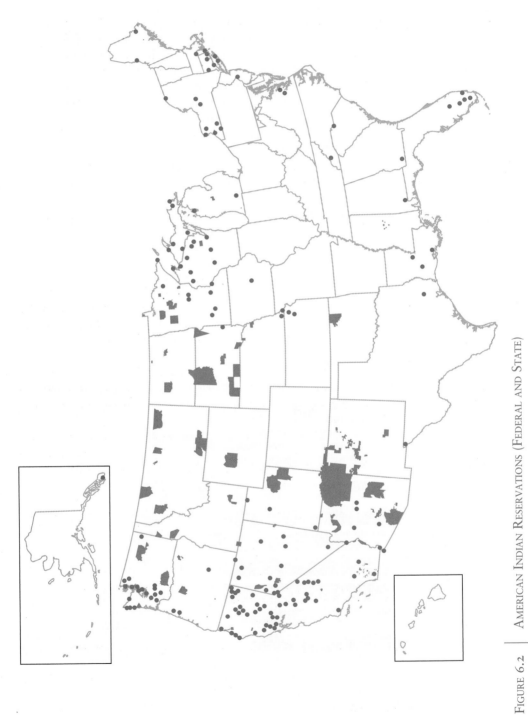

FIGURE 6.2 | AMERICAN INDIAN RESERVATIONS (FEDERAL AND STATE)

Source: U.S. Census Bureau (2012f).

TABLE 6.1	METROPOLITAN AREAS WITH THE LARGEST AMERICAN INDIAN POPULATIONS*	
	American Indian Population	Percent of Total Metro Population
Phoenix	99,000	2.4
New York	93,000	0.5
Los Angeles	91,000	0.7
Tulsa, OK	77,000	8.2
Albuquerque, NM	52,000	5.9
Oklahoma City	51,000	4.1
Riverside–San Bernardino–Ontario, CA	46,000	1.1
Dallas–Fort Worth	43,000	0.7

*Alone or in combination with other racial groupings.

Source: U.S. Census Bureau (2012b).

largest of these reservations is the Navajo, covering 16 million acres of land in Arizona, New Mexico, and Utah, with a population of around 175,000. Most other reservations, however, are much smaller. The second largest is the Pine Ridge Reservation in South Dakota and Nebraska, with a population of 19,000. Figure 6.2 shows the geographic size and location of American Indian reservations.

URBAN INDIANS Almost three-quarters of all Native Americans live off reservations, most of whom reside in urban areas. Indeed, the movement of American Indians to cities has been a continuous process. Federally sponsored urban relocation programs begun during the 1950s impelled much of this movement.

Table 6.1 and Figure 6.3 show those states and metropolitan areas where Indians are most numerous. As is evident, Indians are concentrated in a few states, mostly in the West. The largest populations are in California, Oklahoma, and Arizona.

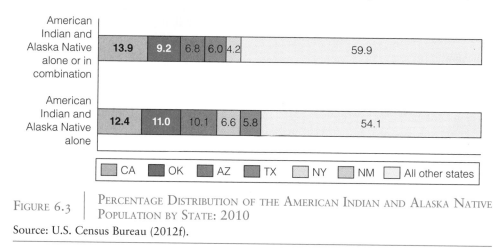

| FIGURE 6.3 | PERCENTAGE DISTRIBUTION OF THE AMERICAN INDIAN AND ALASKA NATIVE POPULATION BY STATE: 2010 |

Source: U.S. Census Bureau (2012f).

AMERICAN INDIANS IN THE CLASS SYSTEM

SOCIOECONOMIC STATUS

Wherever American Indians reside, on reservations or in urban areas, they are below the national average on most socioeconomic measures—income, education, occupation, employment, health care, mortality, and housing.

ECONOMIC MEASURES   Table 6.2 indicates some of the economic discrepancies between American Indians and the general population. As can be seen, median household income is well below that of the total population, and the percentage of people in poverty is twice as high. The social and economic conditions on reservations are especially dismal and remain resistant to significant change. Two of the poorest counties in the United States are Sioux reservations in South Dakota.

Income and poverty rates vary somewhat from tribe to tribe, but in no case do they even come close to the rates for whites or even the national average. Median household income for American Indians is about two-thirds of the median household income for non-Hispanic whites. The comparative poverty gap is even wider, with the American Indian rate more than double the rate for whites. Of particular consequence is the serious Native American unemployment rate. For reservation Indians, this has been the result of a combination of poor education and geographical isolation. But the rate of unemployment for American Indians, whether on or off reservations, is well above the national average and remains high in good times and bad (Austin, 2009). Although urban Indians are less impoverished than reservation Indians, they remain concentrated near the bottom of the class system, experiencing persistent joblessness and difficulty in adjusting to urban economies that no longer can absorb workers with low skills and education. The urban migration of American Indians is not likely to decline radically and thus will continue to render profound economic—as well as social and cultural—effects on Indian life.

TABLE 6.2 | SOCIOECONOMIC STATUS OF AMERICAN INDIANS, 2011

	American Indians*	Total Population	Non-Hispanic Whites
Education			
High school graduates	78.9%	85.4%	88.0%
College graduates	13.3%	28.2%	29.9%
Median household income	$35,192	$50,502	$53,444
Professional and managerial occupations	26.2%	35.7%	37.7%
In poverty	29.5%	14.3%	13.0%
Unemployed	10.0%	8.7%	5.6%

*Alone, not in combination with other racial groupings.

Source: U.S. Census Bureau (2012d).

EDUCATION Regarding education, Native Americans again find themselves at or close to the bottom of the social hierarchy. For example, American Indian students have the highest high school dropout rate among all ethnic groups. Achievement levels are among the lowest as well. More than any other factor, the low level of human capital among Native Americans, primarily in the form of education, accounts for the socioeconomic discrepancy between Indian and white populations (Waters and Eschbach, 1995).

From very early on the federal government has had some role in Indian education. Missionary schools had been established by the early nineteenth century, and the government assisted these schools in their efforts. The BIA operated its own school system with schools on or near reservations. After the Civil War, many boarding schools were established for Indians. As noted earlier, mental and physical abuse were common in these schools, and suppression of Indian cultural ways was a fundamental objective. The failure of the boarding school system was eventually recognized, and Indian students were increasingly channeled into Indian day schools and the public school system (Thornton, 2001).

By the mid-twentieth century most observers agreed that the system of educating Indian children was pathetically ineffective and in many ways perpetuated the low social and economic status of Native Americans (Josephy et al., 1999). A U.S. Senate report in 1969, summarizing a lengthy investigation of the Indian education system, concluded that "our national policies for educating American Indians are a failure of major proportions. They have not offered Indian children—either in years past or today—an educational opportunity anywhere near equal to that offered the great bulk of American children" (cited in Josephy et al., 1999:186). Federal control has undergone slow change over the past several decades. By the late 1960s, several states had assumed responsibility for educating Indian children. Today American Indians have gained a greater measure of control over the education of their children, and this trend is likely to continue (Thornton, 2001). Nonetheless, the educational gap between Indians and the larger society remains a problem of serious proportions. A Department of Education task force studying Indian education concluded that "[i]t is evident that the existing educational systems, whether they be public or federal, have not effectively met the educational, cultural, economic, and social needs of Native communities" (U.S. Department of Education, 1991:12).

Despite this gap, positive trends are evident on most measures of education for American Indians. High school completion in 1980 was 56 percent; it had risen to 66 percent in 1990 and today is 79 percent. Similarly, the numbers of Native American students attending and graduating from college and pursuing graduate and other professional degrees have steadily risen. In 2011, 13 percent of adult Native Americans were college graduates, compared to 11 percent in 1996 (U.S. Census Bureau, 1997; 2012d).

HEALTH Economic insecurity and low educational achievement translate into other measures of deprivation. Although improved, the health of American Indians is still worse than that of the general population by almost every indicator. For example, Indian deaths from tuberculosis are more than 500 percent higher than for other Americans (Indian Health Service, 2010). Doctors, nurses, and dentists are in short supply on reservations, and for urban Indians, access to health care providers is typically far more difficult and inadequate than for the general population (U.S. Commission on Civil Rights, 2004; Worsnop, 1992). Poor diet also contributes greatly to Indian health problems, particularly on reservations. The rate of diabetes, for example, has reached epidemic proportions. Furthermore, suicide and alcoholism rates are considerably higher among Native Americans than among the general population. American Indians fifteen to twenty-four years old commit suicide at three times the national average for that age group (Nieves, 2007b). And the mortality rate from alcoholism among American Indians is over six times the rate among the rest of the U.S. population (Indian Health Service, 2010).

In sum, although new policies and favorable court decisions have clearly improved some of their economic and social conditions, Native Americans, both reservation and urban, continue to lag behind the general population on most measures of socioeconomic status. Despite progress, the proportionate gap between American Indians and the general population has not closed appreciably. Life expectancy of Indians, for example, has risen nearly ten years (from 65.1 to 74.5 years) since 1976. Yet the difference in life expectancy relative to other Americans has changed very little (Indian Health Service, 2006; U.S. Commission on Civil Rights, 2004). Similar wide disparities remain between Indians and the general population in education, occupation, income, poverty, and unemployment, as we have seen.

Continued Native American economic and social deprivation is in large measure explained as a result of the lack of attention and priority given to Indian needs, especially by the federal government. For example, the Indian Health Service, which is the primary health care provider for Native Americans, spends less than $2,000 per patient per year; this is about half of what the government spends on prisoners and less than 40 percent of what it spends on the general population (U.S. Commission on Civil Rights, 2004). That issues pertaining to American Indians often seem like an afterthought to policy makers may be in large part a result of the fact that their population base is simply not large enough to translate into effective political power. Hence, other minority ethnic groups may be better able to bring their needs to the attention of both public and private agencies and therefore garner proportionately greater benefits from assistance programs.

RECENT ECONOMIC PROJECTS

Indian activism of the past three decades has been met with measures designed mostly to quell militancy rather than to address the unique socioeconomic problems of Native Americans. Moreover, as vital resources such as oil, natural gas, coal, and uranium have been discovered in recent years on federal Indian reservations, Native Americans have experienced new pressures threatening their control over their remaining lands. Indian lands are estimated to contain nearly 30 percent of the nation's coal reserves west of the Mississippi, as much as half of potential uranium reserves, and up to 20 percent of known natural gas and oil reserves (Grogan et al., 2011). In some cases, those resources have served as a valuable political tool in pursuing Indian political and economic goals. Some tribes, however, object to mining on their lands, regardless of its value.

In recent years, many Indian tribes have become more aggressive in pursuing projects designed to foster greater economic development and independence. Because

American Indian nations are considered sovereign, the U.S. government does not tax tribes or their wealth. This has provided opportunities to establish businesses of various kinds, some of which have been extremely successful and have resulted in significant economic benefits to Indian reservations. The Lac du Flambeau Chippewa tribe in Wisconsin, for example, distributes Ojibwe brand pizza in seven states, and the St. Regis Mohawk tribe of New York produces widely sold fishing products. In Arizona and New Mexico, the Navajo reservation grows produce that it sells in twenty-one states as well as Mexico and Canada, generating millions of dollars of revenue (Grasser, 2013; Trueheart and McAuliffe, 1995).

A number of tribes have begun to generate stable income streams from off-reservation investments, which have subsequently been used to create employment opportunities on the reservation. And, several Indian nations have become significant economic forces in relatively poor and rural settings. Business success among tribes, however, has not been universal, and most Indian enterprises begin from a low base. Moreover, the future of Native American economic development is heavily dependent on the ability of tribes to maintain sovereignty over their resources and investments, which have been vulnerable to legal and political challenges by state and federal governments (Harvard Project, 2008).

INDIAN CASINOS   The largest and most significant of contemporary Indian enterprises is legalized gambling ("gaming"). The economic revenues accumulated and the jobs created by gaming have far exceeded those of other Native American projects. In many cases, gaming has turned what were economically dormant Indian nations into communities with a solid economic base from which other aspects of socioeconomic improvement have been launched. In 1988, the federal government enacted the Indian Gaming Regulatory Act (IGRA), which provided for the operation of gaming by Indian tribes, with the objective of promoting tribal economic development and self-sufficiency. Since then, 460 Indian casinos and bingo parlors run by 240 tribes in thirty states have been established. In 2011, the industry produced over $27 billion in revenues, almost double revenues in 2000 (National Indian Gaming Commission, 2012).

Many Indian nations have used their profits from gambling enterprises as a springboard to other forms of tribal economic development. Also, the emergence of a thriving gaming industry on many reservations has led to the return migration of many Native Americans who had been living in urban areas (Johnson, 1999).

The boom in casino gambling has resulted in an unanticipated twist of good fortune for some Indian tribes, which are shifting suddenly from near destitution to substantial wealth. Funds have been used to construct schools, health clinics, and housing, and to capitalize other projects. The Pequots, for example, are buying back their land and are investing in nongaming businesses. The Seminoles of Florida, arguably the most enterprising tribe in the gaming industry, are Florida's largest operator of casinos, from which they realized $800 million in revenue in 2005. Although most of the tribe's business is its casinos, it also runs citrus, cattle, and tobacco operations. In the Seminoles' most ambitious venture, they purchased Hard Rock International, the famous music-themed chain of hotels, restaurants, and casinos, in 2006 for $964 million (de la Merced, 2006). They now own more than 130

Hard Rock Cafes worldwide, including two in Las Vegas, and more than a dozen hotels in four states and seven countries (Stutz, 2012).

Ironically, the success of some of these operations has redounded to non-Indians by providing jobs as well as tax revenues to states. The Mashantucket Pequots of Connecticut are a community of only a few hundred, but they operate a casino—the world's largest—that employs thirteen thousand people and is exceeded only by the federal government as a contributor to the state's treasury. Its annual revenue is over $1 billion (Barlett and Steele, 2002; Carstensen et al., 2000; Harvard Project, 2008).

Some tribes have sought to take casino gambling one step further and establish operations in areas near cities that are not on official Indian lands. This has been referred to as "off-reservation gambling." One such large-scale off-reservation project has been run by the Sault Ste. Marie Tribe of Chippewas, who generated $351 million in revenue in 2006 from their casino in downtown Detroit (Rivlin, 2007).

Despite the real and potential financial rewards of casino gambling, however, these have not yet proved to be the economic salvation of American Indians. The fabulous success of the Seminoles and a few other tribes can be deceptive since the popular perception of newfound wealth does not characterize the vast majority of the Indian population. Clearly, the benefits have not been felt for all Indians, even of those tribes that operate flourishing casinos. In fact, among all the tribes with casino gambling, only a few near major population centers have thrived. Only twenty-three tribes account for nearly half of all revenue generated by tribal casinos (Rivlin, 2007). For the remainder, gambling revenues have not been great enough to appreciably upgrade the quality of life. Furthermore, huge financial payoffs have commonly accrued not to tribal members but to outside—non-Indian—investors, who in some cases have pocketed more than 40 percent of casino profits (Barlett and Steele, 2002; Pace, 2000). Nor has the unemployment rate on most reservations been significantly affected because so many of the casino jobs are occupied by non-Indians. Above all, it must be remembered that two-thirds of all American Indians belong to tribes that do not own Las Vegas–style casinos. There is also fear among some Indian leaders that the attraction of tourists and the outside professional and managerial personnel who help operate the casinos will result in a further deterioration of tribal traditions and values (Pulley, 1999).

In sum, the introduction of gaming has been successful in providing economic resources to many Indian communities, enabling them to invest in projects designed to improve their standard of living (Cornell, 2008; Harvard Project, 2008). But whether casino gambling or other enterprises prove to be long-term sources of tribal revenue is not at all certain. The immense success of a few tribes has obscured the continuing deprived condition of those that have not benefited from profitable enterprises. The fact remains obvious that as a part of the American ethnic hierarchy, Native Americans are still in a very low position.

PREJUDICE AND DISCRIMINATION

The unique history of Indians as they encountered Europeans, compared to that of other groups who entered mainly as immigrants, shaped the nature of prejudice and discrimination to which American Indians have been subjected. The idea of Indians as "savages" corresponded to the threat that Indians posed to early settlers.

Furthermore, in many cases it served as justification for divesting them of lands. But once their defeat was accomplished and Indians posed no economic or political threat to Euro-Americans, an elaborate racist ideology to justify the continuation of their subordinate status was no longer needed. Moreover, following the Indians' defeat, the bulk of their population was separated from the mainstream society on isolated reservations. However, the unique political status of Indian tribes and their placement on distant reservations led to the perpetuation of negative stereotypes, attitudes, and actions, though somewhat less vehement and impassioned than in earlier times.

THE WHITE VIEW OF NATIVE AMERICANS

From the start of contact between Europeans and Natives in North America, ethnocentric views prevailed on both sides. For Europeans, the distinctness of specific tribes was blurred by a perception of uniform inferiority. "That almost no account in the sixteenth century portrays systematically or completely the customs and beliefs of any one tribe," notes Berkhofer, "probably results from the newness of the encounter and the feeling that all Indians possessed the same basic qualities" (1979:26). This tendency to see Indians collectively, rather than as separate and culturally varied units, carried forward to the modern period and has been challenged only in recent times.

The view of American Indians as inferior served as a justification for expropriating their lands as settlers pushed westward. As noted earlier, in the perspective of Euro-Americans, land was to be settled and developed. Indians, therefore, stood in the way of American expansion. Those who fought Indians became folk heroes; those whom they fought were subhuman savages. "Over the course of the nineteenth century," notes John Coward, "the story of western expansion became part of a celebrated national myth where heroes rose to their civilizing task and Indians were worthy—and exotic—impediments to progress and empire" (1999:6).

One unusual aspect of the white view of Indians is that the "savage" characterization has, throughout American history, often been accompanied by admiration, yielding the "noble savage." Thus, on the one hand, Indians could be brave, innocent, and industrious, and on the other, cruel, treacherous, and lazy. But as Berkhofer has explained, "Beneath both the good and bad images used by explorer, settler, missionary, and policy maker alike lay the idea of Indian deficiency that assumed— even demanded—that Whites do something to or for Indians to raise them to European standards, whether for crass or idealistic motives" (1979:119).

However positive some characterizations of American Indians were, the dominant white view was always one of inferiority. Indians lacked those practices and institutions that, from the Euro-American standpoint, constituted civilization: a work ethic, effective government, a defensible culture, and almost all kinds of technology (Coward, 1999). Thus, in the confrontation between European and Native, the ultimate outcome in the European view was the triumph of "civilization" over "savagery." As Berkhofer notes, from the middle of the eighteenth century to very recent times, for whites, "the only good Indian was indeed a dead Indian—whether through warfare or through assimilation" (1979:30).

NATIVE AMERICAN STEREOTYPES AND THE MASS MEDIA   The use of stereotypes in describing Native Americans proved critical in the production of a rationale for stripping them of their lands and for efforts at dissolving their cultures. From the early nineteenth century, the mass media played the major role in creating and communicating those stereotypes. As depicted in newspapers, which were aligned with both government and business interests, Indians were obstacles to economic growth and national expansion. Newspapers were also eager to offer their readers "colorful and exciting stories that confirmed the correctness of American values and goals" (Coward, 1999:39). Journalists, in their stories, conformed to a standardized, simplified view of Indians and their place in American society that could be easily understood by readers. Newspapers sustained the dual image of Native Americans as both romantic and savage, but in all cases as fundamentally different from Euro-Americans and thus not in accord with Western civilization (Coward, 1999; Weston, 1996).

In more recent times, negative stereotypes have endured largely through the portrayal of American Indians in motion pictures. Most Americans—indeed, most people of the world—have acquired their beliefs about North American Indians through films. Though created as entertainment, they nonetheless have shaped a perception of Indian culture that has been stubbornly resistant to change. Even American Indians themselves have drawn from these films in constructing their own views of their cultural heritage (Price, 1973).

The savage, hostile Indian, in confrontation with white settlers pushing westward, has been a common movie theme since the introduction of films in the early part of the twentieth century. From the beginning, films have consistently depicted Indians with a set of degrading stereotypes. "The prototype of the Hollywood Indian," notes Rita Keshena, "was treacherous, vicious, cruel, lazy, stupid, dirty, speaking in ughs and grunts, and often quite drunk" (1980:107). In the 1960s, films began to depict a more sympathetic view of American Indians, but one no less founded on inaccurate and simplistic stereotypes.

Not only have American films typically portrayed Indians as brutal and primitive but they have mostly continued to blur the cultural distinctions among them. Historically, films have presented Indians not as individuals or even as members of different tribes but as faceless parts of the collectivity labeled "Indians." Moreover, the Hollywood image of Native Americans has been derived mainly from those Indian societies with a historical reputation for violence, particularly the horsemen of the plains and Southwest. Tribes with more passive and nonviolent reputations have been ignored. As Price (1973) notes, the emphasis on Plains Indians such as Apaches and Cheyennes is ironic because they did not evolve until after whites introduced the horse and were not similar to most other American Indians. North American Indians did not ride horses before Columbus, and most still did not hunt from horseback in the mid-1800s. Furthermore, most were not hunters at all but agricultural people.

In recent years, the use of Indian names by sports teams has raised additional issues with Indian stereotypes. Team names such as "Braves," "Redskins," and "Chiefs" are seen by many as demeaning, contributing to the perpetuation of the Indians' image as savage and warlike. A so-called Indian war chant—used by fans of the Florida State University athletic teams (called "Seminoles") as well as the Atlanta Braves baseball team—for example, evokes images of belligerence and brutality.

The sports mascot issue seemed to resonate with the mainstream American media in the 1990s, when several prominent newspapers editorialized sympathetically with those protesting the use of these names (Weston, 1996). One newspaper, the *Portland Oregonian,* announced a policy of no longer using nicknames of sports teams that were deemed racially offensive. Some college teams voluntarily changed their names and symbols. Stanford University changed its athletic teams' name from "Indians" to "Cardinal," St. John's University teams went from "Redmen" to "The Red Storm," and the Miami of Ohio "Redskins" became the "Red Hawks." Others, like the University of North Dakota "Fighting Sioux," resisted criticism (Davey, 2009; Lapointe, 2004). In 2005, the NCAA (National Collegiate Athletic Association), the governing body of U.S. college sports, brought this issue to further light when it banned Indian mascots, logos, and nicknames from postseason tournaments. This fell short of abolishing Indian references entirely, but put more pressure for change on universities still using Indian themes. A few universities (including Florida State) won appeals of this decision after contending that they had received permission from Indian tribes to use their names (Lapointe, 2006). The University of North Dakota, which had continued to challenge the change edict and had engaged in a four-year legal battle, finally agreed in 2010 to drop its Indian nickname and logo, a decision reaffirmed in 2012 following a statewide referendum on the issue. Efforts at changing the Indian names of professional sports teams have not met with similar success. In 2013, for example, a movement was begun to pressure the Washington Redskins of the National Football League to change their name and logo; the team's owners indicated that they had no intention of doing so (Maske, 2013).

Violence Against American Indians

Violence against American Indians has a long and bloody history. As described earlier, no other group in the American ethnic system ever experienced the kind of genocidal actions that ultimately led to the decimation of the Native population.

Although systematic violence no longer characterizes relations of American Indians with the dominant group, individual forms of violence remain a serious problem. Government reports have shown that the rate of racially motivated hate crimes is higher against American Indians than against any other ethnic minority, including African Americans (Harlow, 2005). Rape against Indian women occurs at an alarming rate, and the vast majority of such cases involve non-Indian perpetrators (Duthu, 2008). Violent attacks are particularly evident in majority-white cities that abut Indian reservations. In these settings, Indian and white cultures collide, often calling up racist notions that subsequently lead to incidents (Buchanan, 2007).

ASSIMILATION AND AMERICAN INDIANS

Cultural Assimilation

Throughout the nineteenth century and the first half of the twentieth, U.S. policy toward American Indians was designed to destroy native cultures and political entities in hopes of creating assimilation into the dominant Euro-American culture and society. A major effort in this regard was the General Allotment Act of 1887. As noted earlier, the thrust of this legislation was to break up Indian tribes into individual farming units, hoping to accelerate assimilation. The driving force of the policy was the belief that once Indians were fully assimilated, the "Indian problem" would disappear.

From 1887 to 1934, the efforts of the BIA, supplemented by those of Christian churches, which were active on most reservations, were directed energetically toward assimilation. Although the annihilation of Indian cultures and of Indian self-governing institutions was not fully realized, the effects on Native American culture and social life were nonetheless very great. Moreover, during this period the policy of forced assimilation engendered deep distrust of the federal government and extreme hostility toward whites (Spicer, 1980).

The assimilation objective changed in 1934 with the appointment of John Collier to head the BIA. Collier, as mentioned earlier, championed the Indian Reorganization Act (IRA), which proved to be a watershed, bringing with it an entirely new approach and direction to Indian-white relations. Collier, it should be remembered, encouraged the preservation of Native cultures. But as Cornell explains, despite Collier's efforts and what became known as "the Indian New Deal," the basic assumption of Indian policy—eventual assimilation—was not essentially altered. Rather, this new approach was merely "a means of easing the process of change and giving Native Americans some measure of control over it" (1988:93). By the early 1950s, blatant assimilation sentiments again prevailed with enactment of the policy of termination.

CURRENT PATTERNS   For most of American history, then, forced cultural assimilation—that is, the disintegration of Native cultures—has been the primary objective of dominant political and social institutions. That objective has never been fully accomplished, but the continual assault on Native cultures has resulted generally in a significant dissolution of Indian cultural influences on individuals, especially regarding language. Dozens of tribal languages continue to exist and function, and there are efforts to revive their use, at least as a way of affirming Indian identity (Harvard Project, 2008; Hitt, 2005). But their practice is common mostly on reservations, and the proportion of Indians not also speaking English is low. Quite clearly, English today is the prevailing language among Native Americans (U.S. Census Bureau, 2007d, 2010b). Many of those who are younger and have migrated to urban areas have been cut off from other tribal cultural influences in addition to language, such as religion; hence, in some ways, their ethnic distinctness becomes increasingly difficult to sharply define.

Despite the steady dilution of Native American culture, the emergence of a strong pan-Indian movement in recent years has emphasized common Indian values and practices, which may serve as a shield against more complete cultural assimilation. In the past four decades, American Indian art and other manifestations of Indian culture have seemed to proliferate. Many new tribal museums have been established, and numerous tribal and urban Indian powwows and arts festivals are held annually. Also evident is an explosion of Native American literature, music, and film, as well as the founding of the National Museum of the American Indian in Washington, D.C. (Josephy et al., 1999).

There is, then, a somewhat contradictory pattern of cultural assimilation that has been occurring among American Indians. As Native Americans have increasingly moved off the reservations into urban areas and have intermarried with non-Indians, they have shed many tribal traditions. However, as Josephy and his colleagues (1999) have explained, a countervailing pattern has emerged, bringing about a kind of "retraditionalization." The same urban Indians who enter the mainstream society return to the reservation periodically to take part in traditional practices as a means of reconnecting with past roots and Indian identity. These oppositional trends have led to internal debates among Indians, raising questions of "Who is an 'authentic' Indian?" "Who is entitled to tribal membership?" "Who should represent Indians before the federal government?" and "Who should be eligible for affirmative action benefits?" (Glaberson, 2001; Tobar, 2001). Those are questions that will continue to plague efforts at building a more cohesive American Indian ethnic population.

STRUCTURAL ASSIMILATION

In examining the structural assimilation of Native Americans, it is important to keep in mind that this is an ethnic population divided into two demographic parts: those living on and those living outside the reservations. In either case, as already seen, Native Americans today remain at or near the bottom of the society's occupational, educational, and income hierarchies. This collective position has stifled rapid movement into mainstream economic and political institutions.

SECONDARY STRUCTURAL ASSIMILATION   Regarding various measures of secondary structural assimilation, the distinction between reservation and nonreservation Indians is an important consideration. This is particularly apparent in looking at the participation of Indians in the political system. Those on reservations are subject to laws made by the tribal government; as a result, politics there are intratribal. Those off reservations do not represent a large enough group in any area to be able to carry much political weight. Thus American Indian representation in government at all levels has been minimal. The economic success of the gambling industry for some American Indians, however, has led to more concerted and systematic lobbying to protect and enhance Indian needs (Dao, 1998). This has necessarily pulled Native Americans into some mainstream political issues.

Although substantial secondary structural assimilation has not occurred, one aspect of Indian life that seems to augur increasing assimilation in the future is the expansion of education. As noted earlier, a growing number of young Native Americans attend college and, as this occurs, more will inevitably take their place in the business and political worlds.

In considering secondary structural assimilation, it must be kept in mind that American Indians remain in a somewhat peculiar legal limbo. In some regards they have regained aspects of political sovereignty, but in others they are still beholden to the federal government (Kehoe, 1999; Wilkinson, 2005).

PRIMARY STRUCTURAL ASSIMILATION   At the primary level of assimilation, those Native Americans who have migrated to the cities appear to have begun to integrate into the larger society. Perhaps the most telling indication of this is the high, and increasing, rate of intermarriage among American Indians and whites (Eschbach, 1995; Qian and Lichter, 2007; Stevens and Tyler, 2002). Well over half of married Indians are married to non-Indians, a percentage considerably higher than for any other nonwhite racial-ethnic group (Lee and Edmonston, 2005; Snipp, 2002). More than 75 percent of the members of the Cherokee Nation, for example, have less than one-quarter Cherokee blood; most are of European ancestry (Nieves, 2007a). The extremely high rate of intermarriage among American Indians is particularly strong among those in urban areas.

What this high and, apparently, unflagging rate of intermarriage means for the future of American Indians can only be speculated upon. In combination with the assimilating forces of urban life, it may indicate a decline of Indian ethnicity. Relatively undeterred by racial identity, Indians can more easily move into the mainstream—unlike African Americans, Asian Americans, and some Latinos. Intermarriage may also reduce the number of persons who, in the future, will identify themselves ethnically as Indians. However, the existence of tribes and of relatively cohesive Indian communities on reservations, where intermarriage rates remain lower, would seem to preclude the complete disappearance of American Indians as part of the American ethnic system (Eschbach, 1995; Qian and Lichter, 2007). At the same time, it is unlikely that steady movement in the direction of assimilation will be seriously impeded as Native American ethnicity is increasingly diluted.

INDIANIZATION   The idea of an American Indian ethnic group has, from the outset of European-Native contact, been a construct of whites. The extreme cultural and social diversity of the Native population was dissolved into a collective notion that created a single category. Among Indians, supratribal consciousness, though emergent in the early part of the twentieth century, did not become a well-developed movement until the 1950s and 1960s. This may have been largely a result of Indian migration to the cities (Cornell, 1988, 1990; Fixico, 2000). In the urban environment, Indians discovered that tribal identities and interests no longer held much usefulness as they found themselves in political and economic competition with other ethnic groups. This Indian ethnic consciousness, developed in the urban communities, reached back to the reservations through returning migrants.

In recent years, then, American Indians have begun to coalesce into a coherent ethnic group, in a kind of "cross-tribalism" or "pan-Indianism." Tribal identities remain, but increasingly Indians have begun to develop an Indian consciousness that transcends tribal divisions. In a sense, what is emerging is an "American Indian" ethnic group (Nagel, 1995). This unity may result in greater political effectiveness and a stronger public awareness of Indian issues and problems.

SUMMARY

- The conquest by European settlers of Native Americans took the form of contact, ethnocentrism, competition over land, and the imposition of white dominance through superior force. Indians were exploited and largely annihilated in the nineteenth century. By the start of the twentieth century, American Indians were essentially wards of the state.

- Although the Indian population was decimated in the nineteenth century through war, disease, and the breakup of tribal cultures, in recent decades it has been on the rise, due largely to the tendency of more people to claim Indian ancestry.
- Almost three-quarters of all Indians live outside of reservations, and most of these are in urban areas.
- On various measures of socioeconomic status—income, education, occupation, health—the gap between American Indians and Euro-Americans is extremely wide. Urban Indians are somewhat better off economically than those living on reservations, but all Indians rank near the bottom of the stratification system.
- Some tribes have benefited substantially from the establishment of gambling casinos, but the long-term benefits of these economic projects are not yet evident.
- Native Americans have been viewed by whites since the early nineteenth century as both savage and noble. Indian stereotypes, which were promulgated in the twentieth century primarily through films, have been strongly resistant to change.
- Attempts have been made repeatedly throughout various historical periods to force assimilation of Native Americans into the dominant culture. Although these attempts were never completely successful, Indian cultures have been diluted over many decades.
- Regarding structural assimilation, American Indians have intermarried with Euro-Americans at a very high rate. This, as well as the assimilating forces of urban life, may lead to a further decline of Indian cultures and institutions.
- In the past two decades, a pan-Indian movement has emerged, which has led to a renewed sense of Indian ethnic identity and efforts to revitalize Indian cultures.

CRITICAL THINKING

1. One of the paradoxes of Indian-white relations is the fact that whites have seen Indians as both menacing and admirable. Are there other American ethnic groups that have been viewed in such seemingly contradictory ways, either in the past or currently?
2. Prejudice and discrimination against Native Americans has declined markedly in recent decades. In fact, other Americans today often express compassion for the continued plight of American Indians. Why has this change in thought and behavior occurred, unlike the far more obdurate prejudice and discrimination toward African Americans?
3. Most universities and many high school athletic teams that had traditionally used Indian nicknames have changed them. A few, however, retained those names after receiving approval from Indian tribal leaders. Why would some Indian tribes not object to the use of their name by athletic teams? Why have professional sports teams that use Indian names and mascots been immune to pressures brought against schools and universities?
4. Compare the history of Indian-white relations in the United States to that of other societies in which settlers encountered indigenous peoples, such as Canada, Australia, New Zealand, and most of Latin America. Were relations more or less benign? Were the long-term outcomes similar? Frame your analysis in terms of the variable patterns discussed in Chapter 4.

PERSONAL/PRACTICAL APPLICATION

1. If you were a Native American, how would you respond to the issue of sports teams calling themselves "Indians" or "Redskins" or "Chiefs" (all actual team names)? Would you respond similarly if a team used *your* ethnic identity as a nickname?

2. Speak to your friends or family about American Indians in comparison with other racial or ethnic minorities. Are their views more sympathetic to the minority status of American Indians? If yes, what is the explanation for this?

AFRICAN AMERICANS

The experience of African Americans[1] is unique among American ethnic minorities. No other group entered the society as involuntary immigrants, and no other group was subsequently victimized by two centuries of slavery. The vestiges of these social facts account for the uninterrupted inequalities between whites and blacks throughout American history and the agonizing nature of the adjustment of blacks to a predominantly white society.

Throughout most of its history, the American racial-ethnic system was, essentially, a binary structure, in which persons were classified as either black or white. Asians and other groups distinctly non-European or non-African in origin were simply too small in population to significantly complicate this system. Obviously that is no longer the case. With the society's growing ethnic diversity in the past several decades, race and ethnic relations have taken on a much broader scope and become more complex. Yet African Americans remain the most visible ethnic group, a fact that must be combined with the heritage of slavery in accounting for the extraordinary character of the black experience. Hence, despite the emergence of a more heterogeneous ethnic configuration, it is the place of blacks in the racial-ethnic system that continues to present the major challenge to American ethnic relations.

DEVELOPMENT OF THE BLACK MINORITY

The history of blacks in American society is complex and subject to a variety of interpretations. There is no disagreement, however, that it is a history marked above all by the continued stress of accommodation to the culture and institutions of a white-dominated society. Stress between whites and blacks has varied in intensity at different times, but it has never been far from eruption into overt and oftentimes

---

[1] Whether *African American* rather than *black* is a more appropriate and acceptable term for the American black population is a debated point. Surveys consistently indicate that most people do not have a strong preference for either (Gallup Organization, 2001c; Jones, 2013b; Sigelman and Tuch, 2005). In this book, the terms are used interchangeably. Technically, however, whereas *black* stresses the racial distinction that sets off this group from others, *African American* emphasizes the group's ethnic identity—that is, its culture and historical origins (Patterson, 1997).

violent hostility. With the understanding that conflict has been a constant underlying theme of black-white relations, we can subdivide the history of those relations into three broad epochs, each characterized by a somewhat different pattern: slavery, the Jim Crow period, and the modern era.[2] During these three periods, relations were shaped most fundamentally by social and economic trends and structures of the larger society. As those trends and structures changed, black-white relations, in turn, took on new form and meaning.

Slavery: Paternalistic Domination

Blacks first entered American society when they landed in Virginia in 1619. Their legal status, however, was undetermined for at least forty years. Until the 1660s, the status of blacks as servants was not essentially different from that of many others, including some whites.[3] Various forms of bondage had been prevalent in all the colonies almost from the founding of the society (Handlin, 1957). Perhaps the most common form was voluntary servitude, or indenture, in which a person was bound by contract to serve a master for a certain length of time, usually four to seven years (Jordan, 1969). In exchange, the servant's passage to the colonies was paid. A great number of people, particularly from Scotland and Ireland, entered the country in this manner.

Although no specific date marks the establishment of chattel slavery, slave codes and statutes evolved piecemeal, so that by the 1660s most southern states had enacted laws defining blacks as slaves rather than as indentured servants. Most of the features of American slavery were now settled and, at least in the slave-holding states, the ambiguity of black status was resolved.

THE CHOICE OF BLACKS   Other groups in the colonies seemingly could have served as slaves; only blacks, however, were inevitably to qualify. Other groups were treated harshly, and even the enslavement of some was seriously considered. The Irish were dealt with severely, and attempts were made to enslave Indians. Why, then, in the end, were only blacks chosen for indeterminate bondage (that is, slave status)? The historian Oscar Handlin gives us some clues. In the mid-seventeenth century, the colonies were in dire need of labor; immigration from Europe therefore had to be stimulated. As a result, the colonies adopted legislation that promised better conditions for servants and expanded the opportunities for release from servitude. To encourage immigration, the master-servant relationship gradually evolved into a contractual one. But blacks did not benefit from the liberalization of the bondage system. Because they were involuntary immigrants, the application of these policies could have no effect on their numbers. In short, there was nothing to be gained by placing blacks under this legislation. "To raise the status of Europeans by shortening their terms would ultimately increase the available hands by inducing their compatriots to emigrate; to reduce the Negro's term would produce an immediate loss and no ultimate gain" (Handlin, 1957:13).

---

[2] This broad historical subdivision of the black experience conforms generally to that delineated by William Julius Wilson (1980).

[3] Jordan (1969), however, maintains that some blacks were enslaved even before the 1660s.

The choice of blacks was dictated as well by white ethnocentrism. Most simply, their physical and cultural traits were more obviously distinct than those of other groups. Especially significant to the English settlers was what they interpreted as the heathenism of blacks. This served as a vital part of the rationalization for enslavement. Indeed, as historian Winthrop Jordan (1969) points out, the religious difference between whites and blacks was of greater importance, at least initially, than the physical distinction. As Christians, white indentured servants could not be dealt with as nonhumans. Yet the religious factor is not in itself a sufficient explanation for black enslavement, because the colonists made no distinction between black nonbelievers and those who had been converted to Christianity; both groups qualified as slaves. Jordan concludes that it was an aggregate of qualities, the total of which set blacks sufficiently apart from others, that made them the most likely candidates for slavery. Their lack of Christianity and unique physical appearance were vital, but other traits that the English interpreted as savage and bestial "were major components in that sense of difference which provided the mental margin absolutely requisite for placing the European on the deck of the slave ship and the Negro in the hold" (1969:97). The prevalent white images of blacks, explains Jordan, can be traced to the first contacts between the English and Africans in the sixteenth century. From the first, "Englishmen found blackness in human beings a peculiar and important point of difference" (1969:20). Black skin became a key visible mark for identifying people who were believed to be defective in religion, savage in their behavior, and sexually wanton.

But what of Indians? Were not they, like blacks, nonbelievers and even physically distinct? On these bases, they were as subject to white ethnocentrism as were blacks and thus, logically, could have been eligible for slavery. But Indians maintained certain advantages. First, they were unused to settled agriculture and therefore not easily suited to plantation labor. Moreover, unlike blacks, Indians were familiar with the terrain and could more easily escape to their own people (Stampp, 1956). The most important difference between blacks and Indians in regard to their ultimate status, however, was the power resources each could command. Put simply, Indians could offer greater resistance to enslavement. As William Wilson (1973) notes, blacks were forced to live in a foreign land, lacked organization, and were scattered about the countryside. In no way did Indians face a similar situation. Indians remained culturally and politically organized in nations, whereas blacks, following enslavement in Africa and the brutal middle passage, were dispersed and purged of their native cultures. As a result, they remained more dependent on white settlers in the way of culture and sheer survival and posed little collective threat (Jordan, 1969).

In sum, the cultural and physical differences of blacks were important rationales for their enslavement, but in the end the critical factor was differential power (Geschwender, 1978; Noel, 1968). Although black skin and "heathenism" gave rise to white ethnocentrism, this did not predetermine the choice of blacks. Rather, blacks simply lacked a viable community from which they might have mustered sufficient counterforce to resist slavery. Color became the visible symbol of slave status, and gradually a racist ideology was developed to buttress the system.

The American form of slavery that evolved was unique, having had no precedent in seventeenth-century England (Elkins, 1976; Handlin, 1957). Indeed, it was a system different from any earlier forms of slavery. The Portuguese and Spanish had already enslaved blacks in their colonies and had maintained slaves at home since the early fifteenth century. But the status of slaves in these societies differed somewhat from that of their American counterparts. As we will see in Chapter 15, in Brazil and other colonies of Portugal and Spain, slaves maintained certain property and family rights and were often freed. In the American form, however, the slave was essentially an object not to be afforded common human privileges.

In the final analysis, the evolution of slavery in the United States was primarily a consequence of economic rationality, prompted above all by the demand for cheap labor in the underpopulated colonies. "The use of slaves in southern agriculture," explains historian Kenneth Stampp, "was a deliberate choice (among several alternatives) made by men who sought greater returns than they could obtain from their own labor alone, and who found other types of labor more expensive" (1956:5). Although it had existed in all the colonies at one time or another, slavery by the middle of the eighteenth century was confined to the South, where plantation agriculture had become the foundation of the economy. In the North, political and moral factors had been influential in the decision to abandon slavery, but more important, the system was seen there as a hindrance to continued industrialization.

THE DEVELOPMENT OF A RACIST IDEOLOGY   As explained in Chapter 3, negative belief systems regarding ethnic groups are more likely to follow than precede the development of systems of discrimination. This is clearly seen in the case of American slavery. The pseudoscientific form of racism—the belief that blacks are innately and permanently inferior to whites—did not arise forcefully until well into the nineteenth century. Racial prejudice was evident before that time, of course, but black inferiority was seen as a product of slavery itself, not of inherent deficiencies (Fredrickson, 1971). Before the nineteenth century, beliefs in cultural, not biological, inferiority were in effect (Wilson, 1973). Blacks were seen as heathens and as savages but not necessarily innately so.

During the 1830s, slavery came under serious attack. Drawing on the environmentalist thought of the eighteenth-century Enlightenment, abolitionists argued that blacks should be freed to develop their capacities to the fullest. During the three decades between 1830 and 1860, proslavery theorists, in response to abolitionist thought, developed a racist belief system replete with ideas pertaining to irreversible physiological differences, such as cranial size and shape, which allegedly explained mental and physical inferiority. Also stressed in these new racist theories were the failure of blacks to develop what Westerners considered a "civilized" life in Africa and the dangers of **miscegenation** (or **intermarriage**)—that is, *racial interbreeding*, in this case between blacks and whites; many feared this would lead to racial degeneracy (Fredrickson, 1971, 2002).

Racist ideology was not limited to the South, however. Most northerners were likewise unprepared to accept blacks as equals and subscribed to basically similar ideas. Fredrickson notes that "a common Northern dream depicted an all-white America where the full promise of equality could be realized because there was no black population to be relegated to a special and anomalous status" (1971:323). Even antislavery forces included at least one school of thought (whose supporters included Abraham Lincoln) holding that only by establishing a colony for blacks in the Caribbean or in Africa would the issue of slavery and the racist fears of white Americans be solved.

After the 1830s, almost all whites of whatever persuasion agreed that blacks were inferior to whites in certain fundamental qualities, especially intelligence and initiative, and that those differences were essentially unchangeable. Because of these differences, an integrated society, it was maintained, was not feasible. Barring their elimination, therefore, continued subordination of blacks was natural and necessary (Fredrickson, 1971). These ideas did not change fundamentally until well into the twentieth century.

The development of a racist ideology, then, is best seen as a response of southern slaveholders to the threat of abolition of the slave system, but a response that gained national acceptance. Perhaps most significant in contributing to the general acceptance of biological racism were the ideas of social Darwinism, which gained widespread recognition in the late nineteenth and early twentieth centuries. Indeed, the fullest flowering of the racist ideology did not occur until after slavery had been abolished (Woodward, 1974).

MASTER-SLAVE RELATIONS The American brand of slavery was a system of paternalistic domination, creating, in some degree, a father-child relationship between master and slave (Elkins, 1976). The total control of the slave's existence by the plantation master was explicitly legitimized by the slave codes of the various slaveholding states. Slaves lacked virtually all legal rights. They could not own or inherit property, testify in court, hire themselves out, or make contracts. Slave laws clearly provided that marriage between slaves held none of the rights of marriage between free people. As a result, families could be broken up in trade with no consideration given to keeping husband, wife, and children intact as a single unit. Finally, laws forbade teaching slaves to read and write.

Enforcement and interpretation of the slave laws were left almost entirely to the master, not the courts (Fogel and Engerman, 1974). With such wide discretion, utter cruelty was not uncommon; but it was felt that brutal treatment would be controlled through public opinion, by the master's sense of decency, and above all, by economic interest (Elkins, 1976). Slaves, representing a significant financial investment, would not be physically or emotionally abused to the point of incapacity. Historians Robert Fogel and Stanley Engerman (1974) explain, for example, that although slave marriages were not legally recognized, masters nonetheless encouraged their slaves to marry and establish families to ensure an atmosphere of stability. Thus slave families were not separated if this was avoidable (see also Gutman, 1976).

As noted in Chapter 4, systems of paternalistic ethnic relations do not require the maintenance of great physical distance between groups. Rather, it is the *social* distance that is of greatest importance. In the pre–Civil War South, a rigid stratification system was enforced through a well-understood racial etiquette, in which physical contact was not uncommon but the social places of whites and blacks were not violated. Social intimacy might often be seen between master and slave in the form of concubinage between white men and black women or the raising of white children by black "mammies." It was out of this apparent physical closeness that emerged the myth of racial harmony in the Old South. But such intimate personal contacts in a paternalistic system are always relationships between unequals. Moreover, the degree of physical intimacy between the two groups should not be overstated. On large plantations, where most black slaves were held, personal relations with masters were limited to household servants. The majority of slaves were field hands who rarely came into physical contact with the white master or his family.

Although most blacks in the South remained in rural slavery, a few hundred thousand free or quasi-free blacks, mostly concentrated in cities, lived within the slavery regime. But these free blacks were scarcely better protected by law than slaves, and the degree of discrimination mounted against them was not significantly less. They could make contracts and own property, but in most other respects their civil rights were not much greater than slaves' (Stampp, 1956; Woodward, 1974). Furthermore, blacks living in the North, though not subjected to slavery, were by no means accepted by whites into the mainstream society. As Wilson describes it, "Whites rejected slavery as an acceptable institution in the North but were unwilling to endorse the view that blacks should receive social, economic, and political equality" (1973:94).

SLAVERY AND THE WHITE POPULATION   The immediate effects of slavery on blacks were obvious, but the impact of the system on whites must be considered more carefully. Most whites in the South did not accrue direct benefits from the slave system. In fact, most whites were small farmers who owned no slaves. In 1860, there were 385,000 owners of slaves among 1,516,000 free families in the South. Moreover, the typical slave owner held only a few slaves. In 1860, 88 percent held fewer than twenty, 72 percent held fewer than ten, and 50 percent held fewer than five (Stampp, 1956). Obviously, large slave owners—the planter class—represented only a tiny fraction of the southern white population. Most of the slaves, however, were concentrated in those few large plantations.

For the elite planter class, slavery provided substantial economic and social rewards; for smaller slaveholders, it offered correspondingly more modest benefits. But what did slavery provide for nonslaveholders, and why did they so universally accept it? Clearly, it could provide them no direct economic payoff, but it could serve as a means of controlling potential economic competition from blacks. In addition, it furnished to poor white farmers a kind of prestige that could be derived from the thought of belonging to a superior "race" (Stampp, 1956). In short, both fear of competition from blacks and social-psychological rewards account for the acceptability of slavery by nonslaveholders. Moreover, the power of the ruling planter class to impose its racist ideology on all whites, regardless of social class, was substantial. As explained in Chapter 2, nonruling groups ordinarily accept for the most part the prevailing social system in the belief that it works in their interests and is justifiable, even though it may work primarily in the interests of the powerful few.

In sum, the two hundred years of slavery were a period in which the subservience of blacks to whites was legitimized in the South and informally recognized in the North. In addition, a racist ideology emerged out of the slave system that continued to influence the perception of blacks by whites well into the twentieth century.

RACIAL CASTE: THE JIM CROW ERA

For a brief historical moment following the Civil War, it appeared that blacks might overcome the crippling social effects of slavery. With the abolition of slavery and with the protection of the federal government, freed blacks began to acquire some of the power resources necessary to enter into the mainstream society with whites. They now exercised the vote and were even elected to high offices in state and federal governments. An educational system was also established that promised still greater potential opportunities. The decade of Reconstruction from 1865 to 1877 marked the end of paternalistic relations between blacks and whites and the beginning of competitive relations. But it was this competition that eventually led to the system of segregation as a means of controlling the black threat to white economic and social advantage. And to reinforce that control, a racist ideology was developed into a more sophisticated and generally acceptable belief system.

THE END OF RECONSTRUCTION   The process of denying blacks competitive power resources had begun well before 1877, but that year serves as a historical benchmark ending Reconstruction. The results of the presidential election of 1876 were disputed, and as a means of breaking the impasse, the Democrats and Republicans agreed to a compromise: federal occupying troops, guaranteeing protection to blacks, would be removed from the South in exchange for the electoral votes of southern Democrats for the Republican candidate, Rutherford B. Hayes. This was not simply a symbolic action; it enabled whites to establish further control over the black population without fear of interference by the federal government. By that time, the planter class had already reasserted its traditional place of power, and the economic dependence of blacks on their former masters was revived with the advent of the sharecropping system. With the removal of the federal government's oversight, the drive to reinstitute white dominance was given greater impetus.

Perhaps the most significant mechanism in this process was the disenfranchisement of blacks. When it became apparent that they might hold the balance of political power between competing white factions, efforts to deny voting rights to blacks began in earnest. This put an end to the incipient alliance that had begun to form within the Populist movement between poor blacks and poor whites. Through various constitutional provisions, blacks were effectively disenfranchised throughout the South by the first decade of the twentieth century. In South Carolina, for example, a clause was adopted that called for two years' residence, a poll tax of $1, the ability to read and write any section of the U.S. Constitution or to understand it when read aloud, the owning of property worth $300, and the disqualification of convicts (Franklin, 1980). Such measures, of course, automatically disqualified most blacks from the registration rolls.

THE GROWTH OF SEGREGATION   Following disenfranchisement, the path was cleared for the enactment of the entire array of **Jim Crow** *measures designed to separate blacks and whites in almost all areas of social life: housing, work, education, health care, transportation, leisure, and religion.* Trains, waiting rooms, drinking fountains, parks, theaters, hospitals, and other public facilities were now segregated. Access to hotels, restaurants, and barbershops was limited to those serving either blacks or whites, and by 1885 most southern states had enacted laws requiring racially separate schools. With the 1896 Supreme Court decision in the case of *Plessy v. Ferguson,* the "separate but equal" doctrine was upheld, and the system of segregation in the South was securely in place.

Segregation as it now appeared was not the same as what had been practiced during slavery. As van den Berghe (1978) explains, roles and institutions that had previously been racially complementary now became racially duplicate. Rather than whites and blacks maintaining different statuses and roles within the biracial plantation, separate economic and social institutions now served each racial group in a castelike system. Although state laws requiring separate facilities for the two groups stipulated that these would be equal, the Supreme Court never defined "equality" while repeatedly upholding the "separate but equal" principle. As a result, the most consistent fact of this system was that separate facilities were inherently unequal; those reserved for blacks were grossly inferior or inadequate, usually by design.

In addition to physical separation, there was the need for a system of racial etiquette, for blacks and whites obviously could not be totally isolated from each other. The etiquette was intended to make clear the caste positions of dominant and subordinate groups. Whites could not shake hands with blacks, for example, nor would blacks be addressed as "mister" but rather as "boy" or by first name. And most important, racial endogamy was rigidly enforced. Indeed, prohibition of interracial marriage was one of the first formal measures enacted by southern states in the formation of the Jim Crow system.

Racial segregation was maintained by both force and ideology. As to the former, blacks in the South could be dealt with in an almost totally unrestrained fashion. Where legal techniques were found inadequate, extralegal measures were used. With the rise of vigilante groups such as the Ku Klux Klan, violence and physical intimidation became the order of the day. Lynching, the most extreme technique employed to maintain black subservience, was resorted to with increasing frequency. From 1884 to the outbreak of World War I, more than thirty-six hundred lynchings occurred, most in the South and with most of the victims black (Franklin, 1980). Fear and humiliation were integral parts of everyday life for blacks in the southern states.[4]

Conditions in the North, though not supported by legal mechanisms as in the South, were not significantly different. Although black civil rights were acknowledged, segregation in virtually all areas of social life typified black-white relations. Indeed, as the historian C. Vann Woodward (1974) points out, segregation actually emerged first in the North before the abolition of slavery. Despite the free status of blacks in the northern states, their inferiority and inability to assimilate had not been questioned in any quarter. Segregation was a means of assuring that blacks understood and kept "their place," and it was well established in all the free states by 1860. Moreover, blacks in most of these states had been effectively disenfranchised.

The system of segregation in both the South and North was abetted by the fullest development of a racist ideology, replete with ideas of racial superiority and inferiority, supported by scientific thought. Woodward (1974) notes that a further contributing factor to the vehement racism of this era was American imperialistic ventures at the turn of the century, involving nonwhite peoples in the Philippines and the Caribbean. The negative stereotypes of blacks that emerged from slavery—lazy,

---

[4] A poignant description of the constant terror and degradation to which blacks were subjected is provided in the works of Richard Wright, one of the great black writers of the twentieth century. See especially *Black Boy* (1945) and *Uncle Tom's Children* (1938).

childish, irresponsible, uncouth—became even more potent. In short, ideas of white—specifically, Anglo-Saxon—supremacy were at a zenith at the turn of the century and for two decades afterward.

INTEGRATION OR SEPARATISM?   Early during the period of caste, black leadership strongly debated whether to adapt to or challenge the system of segregation and denial of black rights. One school of thought, reflected most clearly by Booker T. Washington, contended that the most realistic course was to accept a form of separatism based on the idea of self-improvement. In his famous "Atlanta Compromise" address in 1895, Washington deemphasized the effects of racial discrimination and called for blacks to learn industrial skills and trades that would help strengthen the black community internally. Opposed to this strategy of accommodation and withdrawal were those led by W. E. B. Du Bois, who argued for the complete acceptance of blacks into all areas of the society and advocated militant resistance to white racism. Du Bois contended that it was not solely the responsibility of blacks, nor was it in their capacity, to alter their collective place in American society, but that it was primarily the responsibility of whites, who held the power to effect such change. His efforts were instrumental in the founding in 1909 of the National Association for the Advancement of Colored People (NAACP), which eventually became the chief organization working for black civil rights.[5]

The alternative strategies of integration or separation were debated by succeeding generations of black leaders. The theme of separatism reemerged strongly in the 1920s when Marcus Garvey, a Jamaican immigrant, led a movement advocating the return of American blacks to Africa. Integrationist thought prevailed among most blacks, however, although the separatist issue, as we will see, surfaced again in the 1960s.

NORTHWARD MIGRATION AND URBANIZATION Relations between blacks and whites took on a particularly confrontational and violent tone during the early 1900s. This was brought about by demographic and economic changes in the society that placed the two groups in direct competition.

Industrial expansion in the North created an increased demand for labor, and along with immigrants from Europe, blacks from the South began to fill these places. If jobs were the key "pull" factor impelling black out-migration from the South, the declining cotton economy and the continued enforcement of Jim Crow were the chief "push" factors. The demographic changes involving the black populace were profound. In 1910, almost 90 percent of blacks were living in the South, but eighty years later, little more than half remained there. Streams of blacks from the South to the North and West reached epic proportions from 1940 to 1970, with almost 1.5 million leaving in each of these three decades. Moreover, this migration was almost wholly to the cities, making blacks an increasingly urbanized population.

With their movement to the cities, black workers began to pose a labor threat to white workers in the North, and the violence and intimidation that had characterized southern race relations became a national phenomenon. Adding to the tension created by job competition was the discontent expressed by the many black World War I

---

[5] Washington's and Du Bois's major ideas concerning the place of blacks in American society and their approach to change are contained in Rudwick and Meier (1969).

veterans who had been exposed to more liberal social conditions in Europe. As well, racist ideology, premised on the belief that blacks were inferior to whites, was no less accepted by most people in the North as in the South. As Fredrickson notes, "The migration of blacks . . . from the directly oppressive conditions in southern rural areas to the somewhat freer atmosphere of the urban North did not alter the conviction of most whites that they were lower-caste people, born to serve" (2002:93). The upshot of the rising competition for work and housing, the growing impatience of blacks, and the pervasiveness of racism was an increase in the frequency and severity of black-white hostilities. Serious race riots occurred in numerous cities throughout the first two decades of the century (Rudwick, 1964; Rudwick and Meier, 1969; Tuttle, 1970).

In sum, black-white relations from the end of Reconstruction in 1877 to the 1940s involved a system of restrictive competition (Wilson, 1973). The forces of industrialization created the potential for significant changes in the structure of race relations, and blacks were now theoretically capable of competing with whites, in contrast to their previous slave status. But, in fact, strict segregation in most areas of social life and the denial of political participation severely limited their accumulation of competitive resources—jobs, education, and housing. Wilson has noted that when groups lack the necessary political resources, whatever opportunities are created by industrialization are offset by new controls imposed by the dominant group. "Race relations become competitive, but only in the narrowest sense" (1973:64). Blacks could offer little resistance because of the strength and resolve of whites of all classes to maintain the discriminatory system through both legal and extralegal mechanisms. Furthermore, the system was buttressed by a scientifically endorsed racist ideology, which posited the innate inferiority of black people.

FLUID COMPETITION: THE MODERN ERA

Whereas the Jim Crow period was one of restrictive competition between blacks and whites, World War II marked the emergence of more fluid competitive relations. In such a system, according to Wilson (1973), the minority group's power resources are increased to the point at which it can begin to challenge the dominant group's authority and invalidate racist stereotypes. Members of the minority begin to move into positions formerly occupied only by the dominant group; in the process, institutional racism is exposed and challenged, and individual racism is seriously undermined. As these changes occur, some members of the dominant group begin to question the legitimacy of racial discrimination and either refrain from practicing racist norms or openly attack the racist system. As Wilson explains, "The greatest threat to any form of racism, then, is the significant entry of minority members into upper-status positions within the larger society" (1973:59).

The mid-twentieth century was clearly that period in which the acquisition of power resources by blacks impelled the most forceful challenge to the American racist system. This challenge involved the efforts of organized pressure groups as well as unorganized forces. Together they made up what became the most compelling U.S. social movement of the twentieth century.

THE NASCENT CIVIL RIGHTS MOVEMENT Several factors account for the emergence of the movement for black rights after World War II. Much of the impetus had been established in the 1930s, when the Roosevelt administration began to appoint blacks to non-policy-making government advisory positions (Franklin, 1980; Myrdal, 1944). Moreover, blacks benefited from the administration's relief programs even though these continued to be administered in a discriminatory fashion (Katznelson, 2005; Roediger, 2005). Another significant inroad made by blacks in the 1930s was their entrance into the labor movement, specifically the Congress of Industrial Organizations (CIO). Unlike the American Federation of Labor (AFL), the CIO was relatively nonexclusionary in its organizing activities. Representing industries in which most black unskilled laborers were employed (steel, automobiles, mining, clothing), the CIO now afforded blacks labor union participation in significant numbers for the first time.

Wartime conditions in the 1940s opened new areas of work to blacks and provided an even greater push to the northward migration that had been occurring since the early part of the century. Jobs in the defense plants of cities like Detroit, Chicago, Philadelphia, Cleveland, and New York attracted a new surge of migrants from the South, whites as well as blacks. Blacks, however, continued to suffer the effects of discrimination in the distribution of these jobs. As a result, black labor leader A. Philip Randolph organized the March on Washington movement and threatened to mobilize millions of blacks in protest. Hoping to squelch mass unrest, President Franklin D. Roosevelt in 1941 issued an executive order prohibiting racial discrimination in federal jobs and created the Fair Employment Practices Committee to monitor the order. Randolph's movement was an early sign of the effectiveness of black protest, which would reach much greater heights after the war.

In addition to the economic opportunities World War II provided, it presented white political leaders with a dilemma: how could the United States support a racist system at home while it was at war abroad with the most flagrantly racist regime, Nazi Germany? This inconsistency was difficult to reconcile without a greater commitment to black rights. Moreover, following the war, racist policies proved an international liability in dealing with the emerging nations of nonwhite Africa and Asia. As a result, the federal government now seemed more amenable to efforts at ending patterns of discrimination. One of the first institutions to be desegregated was the armed forces.

THE *BROWN* DECISION AND THE END OF JIM CROW Perhaps the most momentous single government action during the 1950s that stimulated the push for black advancement was the Supreme Court decision of 1954 in the case of *Brown v. Board of Education of Topeka*. In that case, the Court ruled invalid the separate but equal doctrine, which had been upheld in the famous *Plessy v. Ferguson* decision of 1896. The 1954 decision basically removed the legal foundation of segregation and thus marked the beginning of the end of the entire Jim Crow system. Influenced by the testimony of social scientists, the Court made clear in its ruling that separate schools were inherently unequal and imposed an inferior status on black children, causing irreparable psychological change. School segregation laws, the Court declared, therefore violated the equal protection clause of the Fourteenth Amendment of the Constitution. This clause stipulates that no state shall deprive people of life, liberty, or property without due process of law, or deny them the equal protection of the law. The *Brown* ruling was subsequently used as the legal basis for court decisions that struck down other segregated institutions.

The significance of the *Brown* decision cannot be overstated, for it provided the spark to ignite a surge of hope among blacks that had begun to build in the 1940s. Lewis Killian has described its psychological impact on blacks: "The authority of the law of the land was now cast on the side of change, not on the side of the status quo. Any Negro with even the vaguest idea of what the Supreme Court had said now had good reason to hope for an improvement in his life and, even more so, in the life of his children" (1975:43). The drive for integration and an end to societal discrimination may be said to have begun with *Brown*. One year after this decision was handed down, the civil rights movement was symbolically launched with the boycott by blacks of the segregated bus system of Montgomery, Alabama.

NONVIOLENT PROTEST From the late 1950s to about 1964, nonviolent protest and civil disobedience were the chief tactics of the civil rights movement. The basic idea was that individuals had a moral duty to disobey the law when it was clearly unjust. This was a strategy that had been applied many times historically in a variety of social settings, its most notable success occurring in the 1940s in India, where Mahatma Gandhi led the movement for independence from Britain (Sibley, 1963). The technique was to actively oppose the law but in a peaceful fashion. The proponents of nonviolent protest, Martin Luther King Jr. in particular, emphasized that it was necessary to win over the opposition through friendship and understanding, not through defeat or humiliation. Those participating in sit-ins at segregated lunch counters, freedom marches, or freedom rides were therefore prepared to accept suffering without retaliating in kind (King, 1964).

Hostile white retaliation against demonstrators and civil rights workers, much of it in the South carried out by civil authorities, included shootings, beatings, firebombings, and even killings. A pivotal year was 1963, when several key events occurred. Police brutally suppressed black demonstrators in Birmingham, Alabama; a Birmingham church was firebombed, killing four black children; and National Guardsmen were used to carry out the court-ordered desegregation of the University of Alabama. These events were given prominent media attention and galvanized public and governmental support for black civil rights. This supportive mood climaxed in the summer with a march on Washington of more than two hundred thousand people. One year later, the most comprehensive civil rights measure since the Civil War was enacted, prohibiting discrimination in voting, public facilities, schools, courts, and employment. And in 1965 the Voting Rights Act was passed, ending the systematic disenfranchisement of southern blacks.

SHIFTING FOCUS FROM THE SOUTH In essence, the civil rights acts of 1964 and 1965 signified the demise of official segregation and discrimination in the United States. But the end of de facto racial separation and inequality was not yet in sight. Moreover, black leaders became increasingly aware of the discrepancy between what had been accomplished in a legal sense and what had not been accomplished in the actual improvement of life conditions for most blacks. This was sharply evident in northern cities, where blacks were concentrated in depressed ghettos. The black movement, accordingly, shifted its focus from the South to the urban areas of the Northeast and Midwest. This shift signaled a move not only in the geographic locus of protest but also in the goals and tactics of the movement itself.

Securing legal rights was not the primary aim of black protest in the North, as it clearly had been in the southern states. Technically, those rights had been in place for many decades. Instead, emphasis was placed on economic issues—jobs, housing, education, and all other life chances—which, through discriminatory actions, were being denied or poorly provided to most blacks. These issues, however, did not lend themselves to the kind of direct amelioration that had been the case with previously state-approved segregation and denial of civil rights. Legal maneuvering and sit-ins were inapplicable to socioeconomic problems.

As a result, the black movement passed temporarily through a more militant phase, in which many black leaders now proclaimed that ending racism required systemic changes in economic and political institutions, changes that purely nonviolent protest had failed to address. The shift away from defensive and nonviolent tactics produced not only more militant rhetoric but also serious urban violence throughout the late 1960s (Bergesen, 1980; Jacoby, 1998; National Advisory Commission on Civil Disorders, 1968).

BLACK POWER   Complementing the substantially changed strategy of black protest during this period was a reformulation of the movement's ideological goals. Until the mid-1960s, the chief aims of black protest were the abolition of segregated public facilities, the achievement of legal equality, and the encouragement of eventual racial integration at all societal levels. It was the third of these objectives that now came under attack from black leaders like Malcolm X and Stokely Carmichael, who pronounced *the need for independence of black action and the establishment of black control of black community institutions*. This was the essence of the idea of **black power**. Integration was rejected in favor of a separatist position, coalitions with white liberal organizations were severed, and whites were generally purged from the movement. Racial prejudice and discrimination were seen not simply as the attitudes and actions of individuals but as structural features of the American social order. The notion of institutional racism now became popularized (Killian, 1975). The ideology of black power (or black nationalism) was espoused most vehemently by the Student Nonviolent Coordinating Committee (SNCC) and, later, by the Congress of Racial Equality (CORE). Less militant black protest organizations, such as the National Association for the Advancement of Colored People (NAACP) and King's Southern Christian Leadership Conference (SCLC), however, remained steadfastly committed to integration.

Black power was widely interpreted as the advocacy of militant tactics against the white power structure, and the stridency of its proponents represented what seemed to many whites a kind of racism in reverse. Black power, however, comprised a twofold campaign: to assert control over the development of the black community with minimal interference by whites and to instill self-pride among blacks, who, in the view of black power advocates, had for generations internalized the white stereotyped image of black people.

The black community in the United States was seen essentially as a colonial people dominated by white institutions—hence the rejection of integrationist objectives. Stokely Carmichael and Charles V. Hamilton stated this position: "The fact is that integration, as traditionally articulated, would abolish the black community. The fact is that what must be abolished is not the black community, but the dependent colonial status that has been inflicted upon it" (1967:55). Blacks were urged to build genuinely black communities by taking control of school boards, businesses, and political bodies in their areas of the central cities. Only by closing ranks and asserting power within their own communities, it was felt, could blacks emerge as a viable political force, able to compete within a pluralistic system.

The development of a black communal identity and pride that had long been stifled by white oppression prompted an acceptance of African cultural roots that had previously been consciously shrouded. As Isaacs contends, the slogan for which the black radicalism of this period is likely to be remembered is "Black is beautiful." "In both its literal and its symbolic meanings," he notes, "it became the password to a measure of self-acceptance by black Americans that generations of earlier leaders and tribunes of the people had sought in vain to achieve" (1989:85–86).

The militant phase of black advocacy was short-lived. In the late 1960s, public concern with black civil rights in general was increasingly displaced by preoccupation with the Vietnam War. Most important, with the advent of the Nixon administration, it became clear that the federal government's response to black nationalist goals and tactics would be severely repressive (Button, 1978). The militant stage of the movement came to an end, according to Killian, not because the racial crisis had passed but because the white power structure effectively demonstrated the dangers of such measures by imprisoning, assassinating, or forcing into exile the most "dramatically defiant" black leaders (1975:155).

Though ephemeral and seemingly unsuccessful in achieving its long-range goals, the black power phase of the civil rights movement may, in retrospect, be judged more positively. First, black nationalism in the 1960s created a degree of self-pride among blacks that had not been achieved by previous nationalist movements such as Garvey's in the 1920s. Second, short-run profits in the form of increased government assistance followed the urban turbulence. Although it is commonly believed that violence is counterproductive to groups employing such techniques in political protest, historical evidence seems to reveal the contrary. When ruling elites make concessions to challenging groups—or conversely, when challenging movements are effectively stifled—violence or the threat of violence on the part of the successful group looms large in the picture (Gamson, 1975). Francis Fox Piven and Richard Cloward (1971) maintain that protest movements like the one for black civil rights do not gain establishment concessions because of the successes of their organization and leadership but because of the social disturbances they create. Relief programs such as welfare and unemployment payments are set up not so much because of moral incentives as because of the need to contain the social disorder created by mass protests. This conclusion seems to be borne out by events of the 1960s. Following the riots of that decade, the federal government extended numerous programs to the cities designed to address those social conditions that were believed to be at the core of black discontent. As with the aid provided to the unemployed of the Great Depression, write Piven and Cloward, "[t]he expansion of the welfare rolls was a political response to political disorder" (1971:198).

Of course, when concessions are granted under these circumstances, they may be only superficial, momentary, or symbolic; that is, their long-range effect may not be great. As Piven and Cloward (1971) explain, government relief to the poor and unemployed is usually withdrawn once social order has been restored. Such was clearly the case with aid to the urban ghettos. In his study of government responses to the riots of the 1960s, James W. Button (1978) shows that once the level of hostilities subsided and the areas were relatively pacified, the programs were either discontinued or emasculated by underfunding.

ASSESSING THE CIVIL RIGHTS ERA By the early 1970s the "Second Reconstruction" had ended, and blacks began to take stock of their advances and failures of the previous two decades. Most important, the legal infrastructure of the system of segregation in the South had been dismantled. No longer could state and local ordinances prevent blacks from entering public facilities previously reserved for whites. Unlike the first Reconstruction following the Civil War, white vigilante groups like the Ku Klux Klan were limited in their influence and were not supported either materially or morally by most whites. Most important, the federal government accepted the commitment to black civil rights and effectively countered state and local governments that sought to sustain segregation in some form. This was in sharp contrast to the abandonment of southern blacks by the federal government during the post–Civil War period.

A second major achievement of the civil rights movement of the post–World War II era was the expansion of black economic opportunities and the consequent emergence of a sizable black middle class. But it is also in the economic realm that the movement's most glaring failures were to be found. As we will see in the following section, blacks accumulated much greater competitive resources, such as white-collar jobs, middle-level income, and higher education, than at any previous time. These were not distributed equitably among the black population, however. Thus, although the black middle class expanded considerably, the conditions of a significant black underclass, submerged mainly in urban ghettos, were not seriously affected.

Finally, if the decline of biological racist ideas was clearly on the horizon as early as the 1930s, the postwar period of black-white relations, despite its volatility, seemed to further diminish the persuasiveness of this ideology. The notion of innate black inferiority—biological racism—was no longer given credence by the vast majority of white Americans.

BLACK DEMOGRAPHICS

POPULATION Representing about 13 percent of the population—forty-one million people—African Americans were until recently the largest ethnic minority in the United States, a distinction now held by Hispanic Americans. This figure includes all those who identify themselves as either entirely or partially black. The latter, however, make up only a tiny portion (about 7 percent) of the total (U.S. Census Bureau, 2012a). Much of the population growth among African Americans in recent years stems from the entrance of newer ethnic elements—black immigrants.

BLACK ETHNIC DIVERSITY As we will see, the contemporary African American population is internally divided between middle- and working-class families, on the one hand, and poor families, on the other. Indeed, so wide has the gap emerged between these two elements that almost four in ten African Americans today no longer see blacks as a single racial-ethnic group (Pew Research Center, 2007a).

It is becoming increasingly difficult to describe the American black population not only in broad class terms but in ethnic terms as well. Though the overwhelming majority of American blacks trace their heritage to slavery, an increasing minority are voluntary immigrants or their descendants. Indeed, more Africans have entered the United States since 1990 as voluntary immigrants than entered as slaves before slave trafficking was outlawed in the early nineteenth century (Roberts, 2005). Almost all foreign-born blacks come from Caribbean and African nations. Specifically, the most sizable element of contemporary black immigration to the United States has comprised people from the West Indies, in particular, Haiti, Jamaica, and Trinidad and Tobago.

The foreign-born black population rose by almost seven times between 1960 and 1980 and more than tripled between 1980 and 2005. Today more than 8 percent of the U.S. black population is foreign-born, and that figure is estimated to rise to almost 10 percent by 2020. Between 2000 and 2006, immigration contributed at least 20 percent of the growth of the U.S. black population (Kent, 2007). In three states, New York, Massachusetts, and Minnesota, more than one quarter of the black population is foreign-born. In New York City alone, over one-third of its large black population is foreign-born (Bernstein, 2005). Although New York and Miami have been the major destinations for black immigrants, Jamaicans, Haitians, Somalis, Ethiopians, and Nigerians can be found increasingly in a variety of U.S. cities and towns (Kent, 2007; Logan, 2007).

As a result of this new immigration, the American black population is evolving into a somewhat broad, multicultural ethnic category similar to that of Asian Americans and Hispanic Americans, though certainly to a lesser degree. Despite racial similarities, black immigrants culturally have little in common with African Americans. As the American black population becomes more culturally diverse, one can no longer make assumptions about group characteristics simply on the basis of racial designation. Moreover, black ethnic diversity has created dilemmas of group and individual identity. The question is "Who is African American?" Traditionally this term has been applied to blacks whose ancestry is traced to American slavery. But does it now also include those blacks born in the Caribbean or Africa, many of whom prefer to describe themselves as "Jamaican American" or "Nigerian American" or "African"? As one Ethiopian-born activist explains, "If I walk down the streets, white people see me as an African American. Yet African Americans are saying, 'You are not one of us.' So I ask myself, in this country, how do I define myself?" (Swarms, 2004:14). Or, as a Haitian immigrant in Miami describes the reaction of his fellow high school students: "I was just another brother until I opened my mouth" (Fears, 2002:31).

Although the reality of racial categorization in American society often brings foreign-born blacks and African Americans together, sharp cultural differences have at times made for tense relations between these two elements of the black population. Some in the traditional African American population see foreign-born blacks as a new source of competition for jobs, education, and other social resources and are reluctant to view them as genuine coethnics. In fact, immigrant blacks generally rank higher than native-born blacks on various measures of socioeconomic status including education, income, and employment (Logan and Deane, 2003; Model, 2008).

Black economic and cultural differences have sometimes translated into intragroup animosities. For example, psychologist Marvin Dunn, who has studied black diversity in Miami, notes the common stereotypical views of African Americans among Caribbean- or African-born blacks: "They are violent, they don't respect their elders, they have no sense of family, they don't want to work. They depend on welfare." By the same token, African Americans often say of black immigrants, "They're here to take our jobs. They'll work for nothing. They're cliquish. They smell. They eat dogs. They think they're better than us" (Fears, 2002:31).

GEOGRAPHIC PATTERNS As explained earlier, in the early and middle decades of the twentieth century, blacks moved in large numbers from the rural South to the urban North. In the past two decades, however, this pattern has begun to reverse itself as more blacks remigrate to southern states (Frey, 2010). As seen in Figure 7.1, in six states, all of them located in the South, blacks represent one-quarter or more of the total population.

African Americans, whether in the North or the South, reside mostly in urban areas. Table 7.1 shows the urban areas with the largest African American

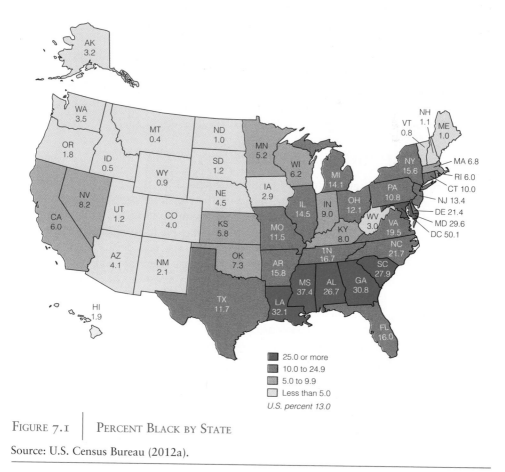

FIGURE 7.1 | PERCENT BLACK BY STATE

Source: U.S. Census Bureau (2012a).

TABLE 7.1	METROPOLITAN AREAS WITH LARGEST AFRICAN AMERICAN POPULATIONS (1,000s)	
	Number	Percent of Total Population
New York	3,363	17.8
Atlanta	2,773	33.6
Chicago	1,722	18.2
Washington, D.C.	1,438	25.8
Philadelphia	1,242	20.8
Miami	1,169	21.0
Detroit	980	22.8
Dallas–Fort Worth	962	15.1
Houston	935	15.9
Los Angeles	908	7.1

Source: U.S. Census Bureau (2012a).

communities. Within metropolitan areas, blacks traditionally had been heavily concentrated in central cities rather than suburbs. That trend changed during the 2000s, and today more than 50 percent of blacks live in suburbs (Frey, 2010). Whites, however, are still far more likely to be suburban residents than blacks or other minorities.

AFRICAN AMERICANS IN THE CLASS SYSTEM

In the post–civil rights era, attention has been focused on the speed and extent of African Americans' movement toward parity with whites in various dimensions of social, economic, and political life. Where do African Americans stand in the society's class and power hierarchies, and how has that position changed during the group's sociohistorical development, particularly the last few decades? Answers to these questions must be approached cautiously because there are various measures of stratification, and data can be marshaled to support a variety of often contradictory views. As explained in Chapter 2, minority status does not necessarily indicate the position of a group at the time its members enter the society in large numbers but rather the patterns that develop to keep them in a disadvantaged position generation after generation. Such patterns are clearer for African Americans than for any other ethnic group in American society.

THE ECONOMIC STATUS OF AFRICAN AMERICANS

The fact of slavery weighs so heavily on the black experience that its lingering effects continue to make African Americans a special group in the American ethnic hierarchy. Obviously, blacks' collective status under slavery was at the lowest rung of the economic class system, even for those who were free in the North or South. Following the abolition of the slave system, this status changed little, as we have seen. Indeed, the economic standing of blacks was even more tenuous following slavery because they were denied the opportunity to compete on an equal basis with whites. Clearly, they did not fare well in resisting a system of blatant discrimination and repression that kept them in a subservient position.

Even after large-scale migration to northern cities starting in the second decade of the twentieth century, a system of direct and intentional discrimination effectively thwarted serious competition with the white population. For the most part, blacks were shut out of higher-skilled, more prestigious, and higher-paying jobs. Recall the split labor market theory of discrimination, discussed in Chapter 3. Not until the 1950s and 1960s did blacks begin to move out of the lowest-level jobs in significant numbers. In effect, then, meaningful comparisons with other ethnic groups can be made only for the post–World War II period.

African American economic progress in the last five decades is debatable; social scientists as well as policy makers, both black and white, have reached no consensus on either its extent or substance. There is, however, relatively broad agreement regarding several general patterns: (1) the 1960s was a decade of substantial improvement in the economic status of blacks collectively, as measured specifically by income, occupation, and education; (2) by comparison, in the 1970s and 1980s the rate of progress was curtailed except for a few specific subsets of the African American population; (3) in the 1990s and 2000s, the economic status of blacks rose but, despite black upgrading, the economic gap between blacks and whites collectively remained wide; (4) the economic progress of the modern period created a substantial black middle class but had little impact on blacks at the bottom of the economic scale; and (5) the deep economic recession of the late 2000s severely impacted black families of all social classes.

Let's look at some specific variables of economic class, comparing the position of blacks and whites.

INCOME   Depending on how the data are interpreted, there are grounds for both positive and negative views regarding black income. The real income of black families increased substantially after World War II, especially during the 1960s (Farley and Hermalin, 1972; Levitan et al., 1975). And the enormous expansion of the black middle class is undeniable. This is the positive picture. But the negative view arises when we look at the gap between median black income and median white income. As shown in Figure 7.2, this gap has changed only slightly in the past several decades. In 2011, median black household income was about 60 percent of median white household income—not significantly different from what it was in the 1960s.

During the past five decades, the income gap between blacks and whites vacillated: during some periods black median income rose relative to whites but at other times blacks seemed to fall further behind whites (Holzer, 2001; Mishel et al., 2009; Smith, 2001). What has remained constant throughout, however, is the income inequality between blacks and whites at all points of the class hierarchy. Thus poor blacks are generally poorer than poor whites, and affluent blacks are less well-off than affluent whites.

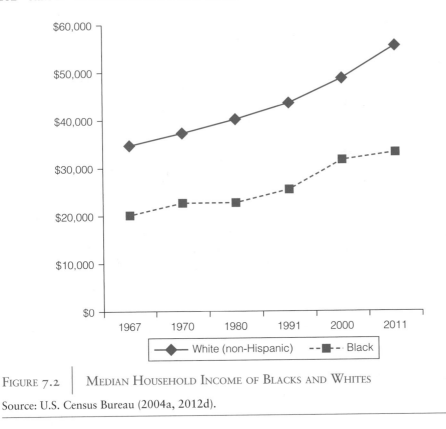

FIGURE 7.2 | MEDIAN HOUSEHOLD INCOME OF BLACKS AND WHITES

Source: U.S. Census Bureau (2004a, 2012d).

However, this general picture should not obscure the fact that comparisons be-tween black and white income are subject to a number of variables, including region and urban location. More important, the black population is polarized between rela-tively stable middle- and working-class people, on the one hand, and those in pov-erty, on the other.

POVERTY   Another mixed indicator of the current economic status of African Ameri-cans is the poverty rate. The discrepancy between white and black poverty has de-clined, but blacks are still almost three times as likely to be in poverty as whites (Figure 7.3).

An important factor in accounting for these statistics, both positive and negative, is changes in the African American family structure. Those families that have experi-enced income increases have been dual-breadwinner families. Among such families there are only slight differences between blacks and whites. Similarly, much of the halt in the relative rise of black income from the late 1970s through the early 1990s is attributable to the increasing number of female-headed, single-parent families, which earn considerably less than husband-wife families (Mather, 2010; Farley, 1993). Today more than one-half of black families are single-parent, most of which are female-headed. Median income of female-headed black families is less than half of median income of married-couple black families (U.S. Census Bureau, 2012d).

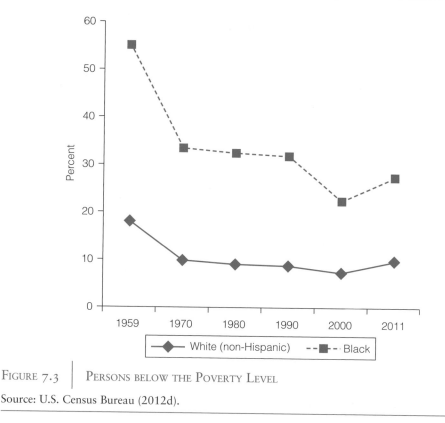

FIGURE 7.3 | PERSONS BELOW THE POVERTY LEVEL

Source: U.S. Census Bureau (2012d).

WEALTH Wealth, or property, refers to the economic assets that families possess—homes, bank accounts, stocks and bonds, real estate, and so on. Wealth is extremely important as a determinant of class position because it can be invested and thus earn income, or it can be accumulated and passed down to other family members. It also reduces families' reliance on wages and salaries, thus enabling them to withstand financial crises like the loss of a job or a serious medical emergency. Indeed, some contend that wealth is the single most significant factor in accounting for the persistent difference between black and white socioeconomic status. "Understanding the racial wealth gap," writes Thomas Shapiro, "is the key to understanding how racial inequality is passed along from generation to generation" (2004:6).

Regarding wealth, differences between blacks and whites are much greater than differences in income. Blacks are less likely than whites to own any type of wealth, and the holdings of those blacks who do own wealth are far smaller than for white wealth owners. For example, less than 7 percent of black households own stocks compared to 27 percent of white households (Pew Research Center, 2011; Shapiro, 2004). In fact, the typical white household has assets more than twenty times those of the typical black household (Figure 7.4).

Recent studies indicate that the wealth gap between blacks and whites not only failed to narrow over the past two decades but actually widened significantly (Pew Research Center, 2011; McKernan et al., 2013; Pendall et al., 2012; Shapiro

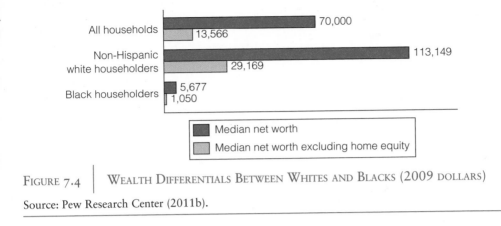

FIGURE 7.4 | WEALTH DIFFERENTIALS BETWEEN WHITES AND BLACKS (2009 DOLLARS)

Source: Pew Research Center (2011b).

et al., 2010). In 2009, the gap was almost twice as wide as it was in 2005. Moreover, the studies showed that middle-income whites own substantially more wealth than high-income blacks.

Home ownership—the most significant form of wealth for most Americans—differs substantially among white and black families. Although the gap in home ownership rates had begun to close in the 1990s, with the housing bust of the late 2000s, rates fell steeply for African Americans (and Latinos), far more than for whites. In 2010, fewer than half of black and Hispanic families owned homes, compared to three-quarters of white families (McKernan et al., 2013; Pendall et al., 2012). Since most black wealth is in the form of home equity, the housing crisis proved disastrous to many families who had assumed costly and risky subprime loans—targeted at the African American community. As a result, many who had borrowed against their home equity lost the little wealth they had accumulated or lost their homes when they could no longer make mortgage payments (Kochhar et al., 2009; Oliver and Shapiro, 2008; Wright, 2009).

Most black income, then, even within the middle class, does not originate from property or self-employment. And, for single-parent families in particular, more common among African Americans, the presence of only one breadwinner makes it extremely difficult to accumulate wealth from savings (Campbell and Kaufman, 2006; Mather, 2010). A recent study, in fact, showed that single black (and Hispanic) mothers with children of any age have less than 1 percent of the wealth of single white mothers with children of any age. Single-parent, female-headed families with children younger than age eighteen, the study showed, have median wealth of *zero* (Chang, 2010).

Lacking significant financial assets with which to cushion the impact of a pay cut or a job loss, most black families are apt to struggle during hard economic times, such as the deep recession of the late 2000s. And, lacking wealth, they are less able to give their children the financial assistance that provides advantages and opportunities (such as buying a home or helping with school tuition) that translate into upward mobility.

OCCUPATION If income and wealth are the chief determinants of life chances, work is, for most people, the medium through which income and wealth are acquired.

Occupationally, African Americans as a group have improved their place over the past several decades, progressively moving into better-paying and higher-status jobs. In 1890, most black males were in agricultural jobs, and females were in domestic and personal service occupations. By 1930, the structure of black employment had changed substantially, with an enormous drop in agriculture and an increase in manufacturing. This was due, of course, mainly to the out-migration of rural southern blacks to industrial cities of the North. But the most significant occupational changes occurred between 1940 and 1970. Whereas in 1940 about one-third of all employed blacks were farmworkers, by 1960 only 8 percent were, and by 1970, 3 percent. By contrast, in 1940, only 6 percent of all employed blacks were in white-collar occupations. By 1970, 24 percent were white-collar workers, and by 2000, well over half were.

Although it is clear that blacks have been increasingly integrated into the mainstream workforce, black-white disparities remain evident. Blacks are underrepresented in jobs at the top of the occupational hierarchy and overrepresented at the bottom. A far higher percentage of whites are in managerial and professional occupations, whereas blacks predominate in service jobs such as cleaners, cooks, security guards, and hospital attendants. Moreover, the apparent occupational advancement of blacks need not be seen only as a product of the decline of discrimination in employment. Much of it is due to changes in the occupational structure itself as the economy shifted from an agricultural to an industrial base and subsequently to a service base.

Furthermore, the discrepancy between blacks and whites in rate of unemployment is wide (Figure 7.5). Since the end of World War II, black unemployment has

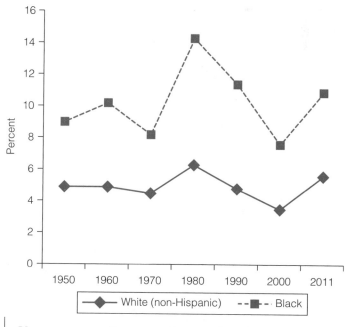

FIGURE 7.5 | UNEMPLOYMENT RATES FOR PERSONS AGE 16 AND OLDER

Source: U.S. Census Bureau (1979, 2004b, 2012a).

been, except for a few years, at least double that for whites. When one considers that government unemployment statistics do not account for those discouraged people who have given up the search for a job, the unofficial number and percentage of unemployed is much higher.

EDUCATION As with income and occupation, there is room for interpreting educational data for African Americans both positively and negatively. Measured by years of schooling, blacks have increased their educational achievement both absolutely and relative to whites during the last five decades. Especially noticeable gains were achieved in the number of high school graduates. Among the adult population, 82.5 percent of blacks in 2011 had completed high school, compared with 88 percent of whites (Figure 7.6). Also, today there is only a slight difference in the high school dropout rate for blacks and whites.

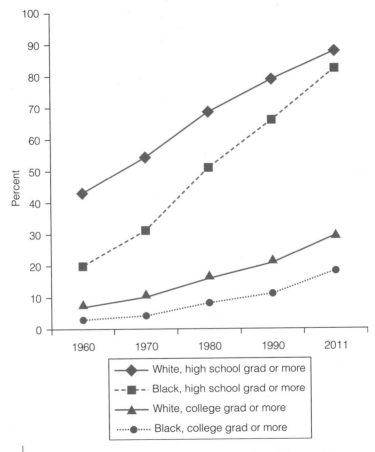

FIGURE 7.6 | EDUCATIONAL ATTAINMENT OF PERSONS AGE 25 AND OLDER

Source: U.S. Census Bureau (1979, 1997, 2012a).

Substantial gains have also been evident in black college attendance and graduation. In 1967, fewer than 300,000 black eighteen- to twenty-four-year-olds (13 percent) were attending college; by 2008 more than 1.3 million (32 percent) were (Fry, 2009). As shown in Figure 7.6, whereas in 1980, 8 percent of adult blacks had earned at least a bachelor's degree, in 2011, over 18 percent had done so. Black college attendance, however, has vacillated in the past two decades, rising in some years and declining in others. A similar trend is apparent in the number of blacks earning graduate degrees.

Despite the narrowing differential in years of schooling, black achievement continues to lag behind white achievement at all school levels—elementary, secondary, and college (Dillon, 2009; Kane, 2004; Massey et al., 2003; Neal, 2006; Schmidt, 2007). Moreover, this gap seems to hold across class lines. Lower educational achievements are, of course, translated into fewer opportunities for occupational advancement.

Explanations vary for racial differentials in educational achievement. At one extreme, some argue that much of the difference between blacks and whites is due to genetic factors (Herrnstein and Murray, 1994; Jensen, 1969, 1973). This view is overwhelmingly rejected by the vast majority of social scientists and educators. At another extreme, many maintain that the schools themselves are inherently racist in both content and technique and thus guarantee the failure of blacks (Bowles and Gintis, 1976). A third position emphasizes a variety of social and environmental factors. The essential argument is that family background and individual and peer-group attitudes are more important determinants of academic performance than either genetic inheritance or the characteristics of the schools (Coleman et al., 1966; Fordham and Ogbu, 1986; Jencks et al., 1972; McWhorter, 2000).

Another factor that negates much of the increase in access to higher education for African Americans is the far lower percentage of high school graduates who attend elite or highly competitive colleges and universities. Most attend less competitive four-year institutions or two-year colleges, whereas whites predominate at the most selective schools. This pattern is evident even among equally qualified white and African American high school students (Carnevale and Strohl, 2013). The result is a perpetuation of inequality in occupations and earning power, with advantages going to those who attend more selective schools.

EXPLAINING THE BLACK-WHITE ECONOMIC GAP

It is clear that on all economic measures, blacks as a group—despite their advances—continue to lag behind whites. The questions that such statistics prompt are whether the persistent gap between blacks and whites is attributable to lingering racism, whether deeper structural forces impede the attainment of economic equality, or whether something unique about the black experience contributes to the group's continued plight. The latter question is covered later in this chapter when comparing African Americans with other American ethnic groups. The first two questions have stirred a lively debate among sociologists for several years. The issue is essentially this: Does racial discrimination continue to affect the economic status of blacks more than their socioeconomic background and current circumstances? That is, do blacks succeed or fail mainly on the basis of their skills and education, or does the racial factor still play a major role in their economic destiny?

In considering this issue, remember that the black population is not homogeneous in terms of economic status. Indeed, some have suggested that there are really "two black Americas"—one that has established itself solidly within the middle and working classes and one that is isolated and not fully part of the mainstream workforce. The latter is what has been referred to as an "underclass" and constitutes perhaps one-third of the total black population (Hochschild, 1995; Smith, 2001; Wilson, 1980). What is really in question, then, is the persistence of a much larger low-income proportion of the African American population, compared to the white population.

CONTINUED DISCRIMINATION Some see the economic gap between whites and blacks as the result of continued discrimination in labor, housing, and credit markets (Hacker, 1995; Pager and Karafin, 2009; Steinberg, 1995). Andrew Hacker, for example, takes this position: "To be black in America is to know that you remain last in line for so basic a requisite as the means of supporting yourself and your family. More than that, you have much less choice among jobs than workers who are white" (1995:110). The idea is that whites still harbor negative stereotypes of blacks, which prevents full acceptance of blacks into all levels of the workforce and into stable and secure neighborhoods. Employers, for example, often see blacks as less dependable and less trustworthy than members of other ethnic groups and, even for low-skill jobs, prefer to hire immigrants (Bertrand and Mullainathan, 2004; Giuliano, 2008; Wilson, 1996).

In a related argument, others point to continued residential segregation as preventing the movement of blacks out of inner-city areas, where decent jobs are scarce or where there is no accessible transportation, to areas where jobs do exist. Residential segregation is sustained, they argue, by the continued refusal of whites to accept more than a token representation of blacks in primarily white suburban neighborhoods. As a result, poor blacks remain trapped in neighborhoods that are resistant to racial change, perpetuating a cycle of poverty (Massey and Denton, 1993; Sampson, 2009).

A family's residential environment affects not only poor families but those that are near or part of the middle class as well. A recent study shows that only a small percentage of white children live in high-poverty neighborhoods throughout childhood, but a majority of black children do. This is found to be a critical factor in accounting for the lower rate of upward economic mobility of blacks and their considerably greater chance of experiencing downward mobility. Living in a socially disadvantaged environment makes it difficult for black families that have achieved middle-class status to retain and pass on to their children the advantages of their class position (Sharkey, 2009).

DYSFUNCTIONAL CULTURE Another explanation for the continued high proportion of African Americans in poverty is found in the notion of "dysfunctional culture." The essential idea is that blacks living in a ghetto subculture find it difficult to thrust themselves out of that environment and to take advantage of available opportunities. The environment to which observers refer is one in which there are mostly families without fathers, neighborhoods infested with drugs and crime, and unusually high rates of unemployment. Perhaps most important, the work ethic has disappeared.

The underclass—not entirely, but primarily, black—therefore lacks the attitudes that lead to meaningful employment in the mainstream workforce (Kaus, 1995). Others explain that a black ghetto culture is an adaptive mechanism to life in low-income neighborhoods but becomes a handicap when manifested in a mainstream social environment (Anderson, 1999; McWhorter, 2000).

STRUCTURAL FORCES Among a majority of sociologists, the most compelling explanation for the black-white economic gap is structural rather than the result of either white racism or cultural failures. William Julius Wilson (1980, 1987, 1996, 2009) is a major proponent of this position. His contention is that although discrimination based on race accounted for the denial of necessary economic resources to African Americans in earlier times, today those impediments are of less importance. Most critical in determining the contemporary economic fate of African Americans are class factors—education, occupational skills, and pathways into the job market.

The American economy, it is explained, has been restructured in the past several decades, favoring those with higher education and technical skills. Individuals who lack those occupational requisites find themselves shut out of an increasingly high-tech workforce and tossed aside or forced into low-paying, dead-end jobs. African Americans have made up a disproportionate part of that unskilled, low-income part of the workforce. Moreover, industry and commerce have moved to areas on the urban fringe so that any available entry-level jobs are outside the range of the central city, where most low-income African Americans live. The result is a semipermanent state of unemployment (Cawthorne, 2009; Sharkey, 2009; Stoll, 2006; Wilson, 1999).

The lack of job opportunities creates a vicious cycle in which blacks are forced to remain in inner-city ghettos, attending inferior schools and thereby reinforcing their disadvantaged position in the labor market (Haskins, 2009). Moreover, high rates of joblessness, Wilson explains, "trigger other neighborhood problems that undermine social organization, ranging from crime, gang violence, and drug trafficking to family breakups and problems in the organization of family life" (1996:21). Several studies support this view (Harkins, 2009; Holzer, 2009).

BLACK MIDDLE-CLASS FRAGILITY In comparing black and white economic status, the question arises as to whether middle-class blacks have attained the equivalent status of middle-class whites, or whether their middle-class status is marginal, making them "second-class" Americans despite their personal achievements (Conley, 1999; Lacy, 2007; Landry, 1987; Pattillo-McCoy, 1999).

It is clear that the path to middle-class status for blacks has been much more tortuous than for whites, and their middle-class position remains far more tenuous. Some have even suggested that the black middle class itself is subdivided into a "fragile" lower middle class that continues to lag behind the white middle class on most measures of socioeconomic status, and a "stable" middle class, virtually the same as its white counterpart in occupation, income, education, and housing (Lacy, 2007; Patillo, 1999).

The severe economic recession of the late 2000s revealed the precariousness of the black middle class. African American families at all class levels were strongly impacted by this downturn, but middle-class families in particular felt its effects, losing much of the modest gains they had experienced in earlier years. As previously noted, the differences in assets (wealth) held by middle-class blacks and whites are striking, with most black wealth concentrated in home ownership, which proved devastating in the recession.

No matter how it is measured, however—income, home ownership, unemployment—the decline in economic security for African Americans, regardless of their social class, during the recession of the late 2000s was much sharper than for whites (Weller and Logan, 2009). In the second decade of the century, there is no indication that the economic disparities between whites and blacks will be reduced in the near future.

African Americans and Societal Power

As noted in Chapter 2, whatever changes occur in the distribution of the society's resources—education, income, jobs—necessarily depend on access to important decision-making positions in the polity and economy, those two institutions in which power lies most heavily. To what extent have African Americans begun to move into such positions?

Political Power   The gains made by African Americans in the realm of politics in the past four decades have been largely a product of increased electoral participation. Although an increasing number of blacks had begun to register and vote in the South during the 1940s, legal impediments as well as threats of violence still discouraged most from trying to exercise the ballot. As Gunnar Myrdal explained, "It is no test of the franchise that some Negroes are permitted to vote in a given community, for what is permitted to a few would never be permitted to the many" (1944:489). With the advent of the civil rights movement in the 1960s, however, enormous gains were achieved in black voter registration in the South, aided by the civil rights legislation of 1964 and 1965. In only eight years, between 1964 and 1972, more than two million blacks were newly registered in the southern states.

In the North, the expansion of black electoral power was not a product of the civil rights movement but rather of the demographic changes that had occurred during the three decades from 1940 to 1970. The transformed racial composition of cities of the Northeast and Midwest accounted more for the rise in black electoral power than any increase in black voting or concessions of power by whites. As the central cities became increasingly black and the suburbs increasingly white, blacks in the central cities found themselves with the numerical power to influence the outcome of local elections.

The upshot of increased political participation in the South and altered urban demographics in the North was a sharp rise in the number of black elected officials, particularly at the local and state government levels. Whereas in 1970 some fifteen hundred blacks had been elected to political office, that number had swelled to over nine thousand by the early 2000s (U.S. Census Bureau, 2004b). Today, the election of black officials at state and local levels of government is no longer exceptional. And, increasingly, black officials are elected in primarily white constituencies. In 2007, about 30 percent of the nation's 622 black state legislators represented predominantly white districts (Swarns, 2008).

Moreover, the black presence in national politics has become very apparent. The black vote is an important swing factor in many elections as well as a significant electoral force in its own right. And, black political power has risen as black political participation has increased. Participation has continued to rise so that today there are only marginal differences in black and white voting rates. In fact, black voters proportionately outnumbered white voters in the 2012 presidential election (File, 2013).

The rising black presence in national politics has continued in the early years of the new century. Although the presence of African Americans in high-ranking cabinet and advisory positions of presidential administrations had been evident in Clinton and Bush administrations, the election of Barack Obama to the presidency marked a culminating point in national politics for African Americans. The 2008 and 2012 presidential elections are indicative of a profoundly changed American electorate, seemingly no longer strongly driven by racial/ethnic identity. Consider that only fifty years earlier, fewer than 40 percent of Americans had said that they were willing to vote for a black person as president. In 1999, 97 percent of whites said they were willing to do so, but until Obama's election, those sentiments had never really been tested (Newport et al., 1999).

ECONOMIC POWER The increasing presence of African Americans in positions of political power has not been duplicated in the corporate world. Relatively few African Americans serve on the board of directors or in top-level executive posts of the largest corporations, and these few are often token rather than bona fide decision-making appointments.

Corporate management is an area of authority in U.S. society that is not amenable to change through the electoral process; African Americans and other ethnic minorities, therefore, are dependent on the inclinations of the corporate elite to grant them a share of power. Although they are still disproportionately underrepresented, an encouraging pattern is emerging in which blacks continue to increase their presence in executive and managerial positions (Stodghill, 2007). In 2012, thirteen Fortune 500 corporations were headed by an African American CEO, compared to only two in 1999 (Domhoff and Zweigenhaft, 2012).

PREJUDICE AND DISCRIMINATION

The modern period of black-white relations has seen profound changes in patterns of racial prejudice and discrimination. Whereas some form of biological racism was a solidly entrenched ideology in all parts of the nation well into the 1930s, by the 1960s a majority of Americans had rejected the notion of innate black inferiority. This shift in attitude accompanied the demolition of the legitimated system of racial segregation in which overt and direct forms of discrimination, both individual and institutional, were approved and defended by the state. But as these traditional forms of discrimination were abolished, they were replaced by covert and indirect forms, which are less amenable to alleviation.

Before the civil rights movement of the 1960s, discrimination against African Americans in much of the United States was de jure, that is, legitimized and enforced by local, state, and even the federal government. Where it was not enforced by law, it was upheld by well-understood customs. Today, however, prescribed discrimination has given way to de facto forms, which are the result not of direct antiblack measures but of institutional arrangements, demographic trends, and housing patterns.

RESIDENTIAL DISCRIMINATION

The most evident, persistent, and consequential forms of racial discrimination in American society are found in the area of housing. Moreover, the patterns of discrimination here are direct and deliberate to a far greater extent than in the areas of work and education, which are largely structural. The effects of lingering racial discrimination in employment, for example, are debatable. Employers may continue to discriminate against blacks, but the extent and impact of those actions as well as their intent are difficult to precisely ascertain. Legal measures, in any case, have made them uncommon. In education, too, although urban school systems are as segregated today as they were before the 1954 Supreme Court decision barring segregated schools, these patterns are maintained more by residential segregation and the past drawing of racially distinct school districts than by the conscious design of whites. Discrimination in housing, by contrast, is clearer, and its intent is not easily disguised. It is well recognized that less progress in breaking down racial barriers has been made in this sphere of social life than in any other (Brown et al., 2003; Massey and Denton, 1993).

Because the patterns are so clear-cut, focusing on housing is essential in looking at current forms of discrimination leveled at African Americans. However, there are additional important reasons for concentrating more intently on residential discrimination. Its ripple effects touch every other area of social life (Farley et al., 1979; Holzer, 2001; Massey, 2004; Massey and Denton, 1993; Sigelman et al., 1996; Steger, 1973). Where people live determines in largest part the schools they attend (and thereby the quality of their education); the jobs they have access to (and thus their occupational destiny); the benefits they receive from public institutions like hospitals, recreational facilities, and transportation systems (and, as a result, their health and well-being); and the commercial establishments they have access to (and thus their style of life). Moreover, the society's residential arrangements determine the essential character of interethnic relations. With continued segregation in housing, whites and blacks are prevented from interacting at a personal level, thus reinforcing racial attitudes and perpetuating structural pluralism. Though it is difficult to select one element of the system of American ethnic relations and call it the "key" to the maintenance of conflict and misunderstanding, residential segregation is unquestionably among the most fundamental. It would hardly be an exaggeration to proclaim that the present pattern of housing in the United States is the chief obstacle to progress in all other aspects of race and ethnic relations.

THE FORMATION OF BLACK GHETTOS   Wide-scale residential segregation has been a continuous fact of urban life in the United States since blacks began to move to cities in large numbers in the early twentieth century. As with European immigrants before them, they were concentrated in urban ghettos, but unlike those of the European groups, black ghettos did not progressively break up with each succeeding generation (Hershberg et al., 1979; Kantrowitz, 1973, 1979; Massey and Denton, 1987, 1993; Roof, 1978, 1979; Spear, 1967; Taeuber, 1975).

Black-white residential segregation patterns changed little throughout most of the remainder of the century despite greater government activity in promoting desegregation, the development of more liberal social attitudes, and the emergence of

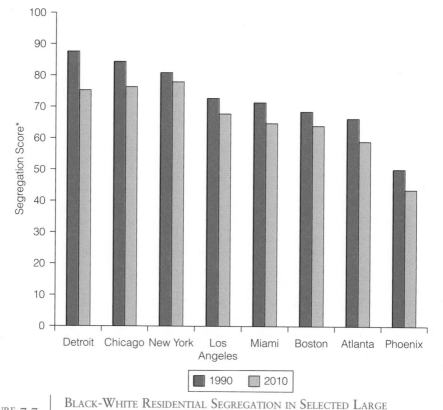

FIGURE 7.7 | BLACK-WHITE RESIDENTIAL SEGREGATION IN SELECTED LARGE METROPOLITAN AREAS, 1990 AND 2010

*The segregation score or index of dissimilarity is a measure of the extent of residential segregation between two groups. It indicates the percentage of one group that would have to move to achieve total integration; that is, a population mix that corresponds to each group's proportion of the population of a designated area. For example, if a metro area were made up of 80 percent whites and 20 percent blacks, the two groups would be represented in each neighborhood in that same proportion. The index can range from 100, indicating total segregation, to 0, indicating total integration. Thus, the higher the index, the greater the degree of segregation.
Source: Frey, 2010b.

more middle-class blacks with sufficient incomes to afford better housing (Farley, 1985; Levitan et al., 1975; Massey and Denton, 1987, 1993; Orfield, 1985; Roof, 1979; Van Valey et al., 1977; White, 1987). Figure 7.7 shows the level of black-white residential segregation in eight of the largest U.S. metropolitan areas. As can be seen, segregation did decline somewhat during the two decades between 1990 and 2010. But the simple fact remains that most African Americans, regardless of their social class or the geographic area in which they live, continue to reside in predominantly black neighborhoods. And this "residential color line," notes Thomas Shapiro, "is the key feature distinguishing African Americans from all other groups in the United States" (2004:141).

Migration to the suburbs, a movement followed heavily by whites since the end of World War II, has been replicated to some degree by blacks in recent years. In fact, as noted earlier, a slight majority of African Americans now live in suburbs rather than central cities (Frey, 2010). However, when blacks do move to suburban areas, often they again find themselves in racially segregated communities (Adelman, 2005; Charles, 2003; DeVita, 1996; Massey and Denton, 1987, 1993; Pattillo-McCoy, 1999).

RACE OR CLASS? Might these patterns be more a result of class than of race? That is, might the persistence of black residential segregation merely reflect the lower economic status of blacks generally? If this were so, we could expect to find blacks and whites of similar social class residing in racially mixed neighborhoods. But studies have shown that blacks of every economic and educational rank are more highly segregated from whites than are Hispanics and Asians (Iceland and Wilkes, 2006; Logan, 2011; Massey and Denton, 1993).

Class factors, however, do play a role in determining patterns of residential segregation. For one, middle-class blacks are more likely to live in less segregated neighborhoods (Adelman, 2005). Even in the most highly segregated urban areas, middle-class blacks are less segregated from whites than are poorer blacks. Also, the resistance of whites to black incursions is reduced when blacks are perceived as higher in social class. As Alba and his colleagues explain, "Whites are reluctant to accept blacks as neighbors, but that reluctance is eased when the socioeconomic characteristics of African Americans are superior to those of whites themselves" (2000:557). Thus, there is evidence to support the notion that whites avoid black neighbors for reasons related more to class than to race (Alba et al., 2000; Harris, 1999; Iceland and Wilkes, 2006). However, stereotyping still plays a strong role in the makeup of neighborhoods and limits black choices. As Alba and his colleagues (2000) point out, because blacks are stereotyped as poor, and because whites are more willing to accept black neighbors who are perceived as superior to themselves in social class, blacks live in less affluent communities than do whites of the same socioeconomic status.

Might a part of the continuation of residential segregation be the preference of blacks themselves to remain in racially homogeneous areas? There is evidence that continued racial segregation in housing is in some measure voluntary (Patterson, 2006; Thernstrom and Thernstrom, 1997). But this observation must be carefully qualified. For more than two decades, studies have shown that most blacks favor residential integration and would prefer racially mixed neighborhoods (Krysan and Farley, 2002; Massey, 2001; Pattillo, 2005; Rawlings et al., 2004; Sigelman et al., 1996). Surveys show that whites, on the other hand, are willing to accept residential integration only if the number of blacks is minimal (Schuman et al., 1997). The preference of blacks to live among coethnics, then, is motivated largely by a perception of racial prejudice on the part of whites and thus the tendency to avoid majority-white neighborhoods (Charles, 2003; Krysan and Farley, 2002).[6]

---

[6] Although there is agreement among analysts that blacks seem to prefer mostly black neighborhoods, the debate centers on the reasons for that preference: a strong desire to live among coethnics, apprehensions about white hostility, uneasiness about white cultural influences, or the view that integration yields few benefits.

THE MAINTENANCE OF RESIDENTIAL SEGREGATION Despite the enactment of many state and federal antidiscrimination statutes since the 1940s, residential segregation is maintained by a set of discriminatory practices, both individual and institutional. Past government policies created much of the framework for its current patterns. Beginning in the 1930s, the federal government encouraged home ownership among middle- and working-class people by establishing the Federal Housing Administration (FHA) as well as other housing-related agencies. The idea was to furnish cheap financing by providing government-backed mortgage insurance. From the outset, the FHA officially discouraged integrated residential areas by refusing to guarantee loans for homes that were not in racially homogeneous areas, a policy that was not changed until 1962 (Abrams, 1966; Katznelson, 2005; Massey and Denton, 1993; Roediger, 2005).

The real estate system has been a key element in the perpetuation of segregated housing. Until the 1950s, real estate boards, almost entirely white in membership, followed a strict code that restrained the renting or selling of property in white areas to blacks. Using various techniques, white real estate agents would steer blacks to black areas and whites to white areas to enforce this code. Banks and other lending institutions cooperated in the system by "redlining," that is, designating certain areas within which real estate loans would not be made. Another device used by real estate brokers was "block-busting." By spreading word through a white neighborhood of an impending black influx, agents would frighten whites into selling their homes cheaply. These homes were subsequently sold to blacks at inflated prices. In the process, all-white areas were transformed quickly into all-black areas.

Where such tactics were unsuccessful, zoning regulations were established, specifying certain types of housing that could be erected in particular neighborhoods; these were designed to exclude low-income, mainly black, units. Restrictive covenants, agreements made by home owners not to sell to members of particular groups, were also used to bar blacks (and in some cases other minorities) from white residential areas (Roediger, 2005). Until the Civil Rights Act of 1968 prohibited discrimination in housing, such covenants had been widely applied and were even supported by law until 1948. When all else failed, violence and intimidation were frequently used to deter blacks from buying or renting in all-white neighborhoods.

Although these practices have been curtailed by antidiscriminatory legislation, some are still evident (steering, redlining) whereas others have been replaced by more subtle measures that support segregation (Massey, 2001; Massey and Denton, 1993; U.S. Department of Housing and Urban Development, 2013; Yinger, 1996). The discrimination African Americans face in securing home mortgages, for example, is an important factor in sustaining segregated neighborhoods. Regardless of income, lending institutions—banks and mortgage companies—are less inclined to approve blacks than whites, especially when homes are located in suburban areas (Heath et al., 2002; Shapiro, 2004). Despite antidiscriminatory statutes, such institutional forms of residential discrimination are difficult to detect and to effectively police. Ironically, black borrowers were disproportionately the victims of the housing crisis of the late 2000s. As banks were pressured to lend to people who ordinarily would not have qualified for a home mortgage, unsuspecting black and other minority families became easy prey for banks and other lenders. By selling these loans (so-called subprime mortgages) they reaped enormous profits from fees, knowing that eventually many of these families would be unable to meet the loan obligations and their homes would be foreclosed.

In sum, housing patterns between blacks and whites are, as has been repeatedly demonstrated, qualitatively different from those between other groups. Massey and Denton (1993) have described this uniquely disadvantaged black urban environment as "hypersegregation"—that is, segregation across several dimensions. No other ethnic group, they show, experiences such multidimensional segregation. Although there has been a reduction in recent decades of the intensity of black-white residential segregation (Farley, 2011), a majority of the African American population continues to live in densely populated, segregated neighborhoods of America's largest urban areas (Iceland, 2009; Logan et al., 2004; Massey, 2001; Morin, 2013; Wilkes and Iceland, 2004). This environment shuts out most contact with whites except through participation in the labor force. Thus African Americans living within the largest urban black communities remain highly segregated and spatially isolated. Although racial discrimination in other institutional realms of American society has broken down to a great extent, it remains tenacious in residential patterns and continues to render a severely negative impact on African Americans.[7]

CHANGES IN RACIAL ATTITUDES

As the nature of racial discrimination has changed in the past several decades, so have the attitudes of whites toward blacks. Although there is much room for optimism regarding these attitudinal shifts, there are also indications that some racist beliefs are still firmly rooted in American thought, albeit in modified form. Moreover, there is ample evidence that blacks and whites do not see the same reality regarding the current state of race relations or the nature and causes of economic and social problems faced by African Americans.

Attitudinal changes during the past forty years have been measured by extensive and frequent surveys among blacks and whites. At the outset, it should be noted that survey data regarding attitudes must be accepted cautiously as representative of people's real beliefs and feelings. At best they are rough descriptions, not precise readings. Because they are designed to draw collective or general pictures, they will not necessarily reflect the nuanced views of individuals. Obviously, blacks and whites as collectivities are internally variegated, and views among individuals will differ on the basis of age, gender, education, and numerous other social variables. Moreover, as explained in Chapter 3, surveys, particularly those dealing with socially sensitive issues like race and ethnic relations, may reflect mainly what people believe interviewers want to hear or what they feel to be the most socially acceptable response, not necessarily their true feelings. Finally, attitudes may appear inconsistent from item to item or from survey to survey. For example, public opinion studies over the years have shown a strong acceptance of integrated schools but an equally strong rejection of school busing (Gilbert, 1988; Greeley and Sheatsley, 1974). Interpretations of such inconsistencies are often problematic. In this case, the antibusing sentiment

---

[7] Marston and Van Valey (1979) argue that there are some positive effects of residential segregation for blacks, namely, that it creates and sustains an institutional structure that caters to blacks, induces ethnic group consciousness, and in some cities produces local political clout.

may be an indication of covert racism, or it may indicate other, nonracist attitudes such as commitment to the neighborhood school, fear of the dangers of busing, or resentment over the inconvenience caused by busing (Greeley and Sheatsley, 1974). It may also reflect a particular view of what government should or should not do to accelerate racial equality (Sniderman and Piazza, 1993). With these caveats in mind, let's look at some specific patterns.

STEREOTYPES   First, it is clear that negative stereotypes of blacks have diminished and that less support is given to blatant forms of prejudice (Brink and Harris, 1963, 1967; Campbell, 1971; Greeley and Sheatsley, 1974; Krysan, 2008; Pettigrew, 1975, 1981; Schuman et al., 1997; Schwartz, 1967; Sheatsley, 1966; Williams, 1977). Table 7.2 shows the decline in recent decades of two of the most historically unyielding black stereotypes.

It is true nonetheless that traditional negative stereotypes of blacks—poor, violent, irresponsible—are still potent, if diminished (Feagin and Sikes, 1994; Saulny and Brown, 2009; Waters, 1999). Although Table 7.2 shows their decline, it also shows their persistence among a very substantial percentage of whites. Few African Americans do not at times become aware of the continued virulence of these stereotypes and of the reality of racial profiling. Even middle-class blacks commonly relate incidents of being the targets of watchful eyes of security guards in upscale stores or of police in predominantly white neighborhoods. The journalist Brent Staples explains that although few Americans today subscribe to crude racist beliefs, black stereotypes have been embedded in the society's culture for centuries. "This makes it easy for people to see the world," he writes, "through a profoundly bigoted lens without being aware they are doing so" (2012:SR10).

SOCIAL DISTANCE   White attitudes toward racial integration and various levels of social distance have also changed in the past five decades. Most whites now seem committed (at least verbally) to integration in most social areas entailing secondary relations. Only a tiny minority of whites nationwide object to contact with blacks on the job, in schools, or in public facilities.

But far less attitudinal change is noticeable when intimate contacts are involved. Few blacks today, regardless of social class, are likely to socialize with their white fellow employees outside the formal work setting (Sigelman et al., 1996). And, as we have seen, most whites are opposed to living in areas with a high percentage of blacks (Carroll, 2007b; Farley, 2011; Krysan and Faison, 2011; Pew Research Center, 2007a; Schuman et al., 1997).

TABLE 7.2 | WHITE AMERICANS' STEREOTYPES OF BLACKS (PERCENTAGE)

Stereotype	1990	1998	2008
Whites work harder than blacks (yes)	—	42	38
Whites are more intelligent than blacks (yes)	57	28	26

Source: Krysan and Faison (2011).

On the other hand, an overwhelming majority of whites no longer object to interracial dating and marriage. This represents a significant change in the past several decades. In 1987, less than half of whites approved of black-white dating; in 2007, 83 percent approved. Although there is a marked discrepancy among different age cohorts—younger Americans are more tolerant than their parents and grandparents—even among those over sixty years of age there is a clear shift toward acceptance (Pew Research Center, 2007c).

Similarly, regarding marriages between blacks and whites, American attitudes have changed dramatically. In 1958, almost all whites (and blacks as well) disapproved of interracial marriages; approval gradually rose over the next few decades, but well over half of whites still disapproved in 1983. By 2013, only 16 percent disapproved (Figure 7.8).

ATTITUDES AND ACTIONS   Although white attitudes concerning discrimination and integration are apparently increasingly liberal, ambivalence arises when translation of those attitudes into concrete actions is suggested. This is evident in various ambits of social life, including politics, jobs, housing, and education. Whites agree that blacks should have equal rights and privileges in all these areas but are less committed to approving measures necessary to guarantee them (Pettigrew, 1979; Pew Research Center, 2007c; Schuman et al., 1997; Tuch and Hughes, 2011).

An obvious inconsistency between attitude and action lies in the realm of housing. Although most whites favor racially integrated neighborhoods, white preferences for the *degree* of integration are quite different from those of blacks. Whereas blacks seem to prefer a relatively even racial mix, whites prefer a mix in which whites are in the majority (Farley, 2011; Gallup Organization, 2003; Schuman et al., 1997; Sigelman et al., 1996). In a sense, there is a circular movement in which prejudice

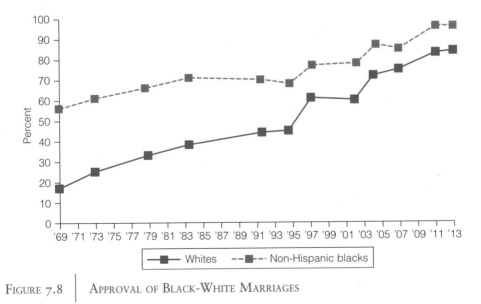

FIGURE 7.8 | APPROVAL OF BLACK-WHITE MARRIAGES

Source: Newport (2013b). Copyright © 2013 Gallup, Inc. All rights reserved. The content is used with permission; however, Gallup retains all rights of republication.

feeds on segregation, and vice versa. Increased contact seems to moderate racial prejudice, especially if the groups are of the same social class. As mentioned earlier, a key reason for whites' fears of black penetration of their neighborhoods is the lower social class of blacks, not necessarily negative racial attitudes alone (Alba et al., 2000; Farley et al., 1979; Iceland and Wilkes, 2006). Whites are concerned that their neighbors will be poorer and will have lifestyles very much at odds with their own. Although blacks moving into white neighborhoods are generally similar to their white neighbors in socioeconomic status, fixed negative images of blacks (as "poor," "lazy," "squalid") often prevent whites from perceiving the class distinctions among them—hence the continued resistance to integrated residential areas. Without social contact, prejudicial attitudes and images persist, in turn leading to intransigence on matters of neighborhood desegregation.

Some go further than merely pointing out the reluctance of whites to link attitudes with actions toward blacks or the inability to differentiate class and race, maintaining that white attitudes themselves have not really changed as radically as they appear. In the view of Picca and Feagin (2007), for example, although expressions of racism are no longer as commonly overt as in the past, they have only "gone backstage," that is, confined to private settings where whites interact with other whites, especially friends and relatives. Whites, they suggest, have learned different racial performances, depending on the social setting.

DOMINATIVE AND AVERSIVE RACISM   Pertinent to this issue is the distinction between **dominative** and **aversive** racism. *With dominative racism, actions are taken to oppress racial minorities and keep them subservient. Aversive racism, by contrast, is characterized by inaction wherein racial minorities are simply ignored and avoided when possible*; relations at best become "polite, correct and cold in whatever dealings are necessary between the races" (Kovel, 1970:54). Continued residential segregation provides the backdrop for such aversive discrimination in various areas of social life. Although obvious and blatant forms of discrimination are no longer prevalent, more benign black-white relations are in large measure founded on intergroup avoidance (Pettigrew, 1980).

Aversive racism is more difficult to detect and to deal with since it is not overt but subtle in expression and often unconscious and unintentional (Bonilla-Silva, 2003; Dovidio, 2001; Dovidio and Gaertner, 1998). In such cases, people will deny their prejudice and may express politically correct racial views, but when called upon to support substantive change (racially mixed housing or some form of affirmative action, for example), they will balk.

Others have suggested similar forms of contemporary racism, subtle and not sharply focused as the blatant racial discrimination of the past. Lawrence Bobo (2004) refers to "laissez faire racism" as a combination of several factors: a continuation of negative stereotyping of blacks, blaming blacks themselves for the persistent black-white socioeconomic gap, and resistance to policy changes that address the gap. Similarly, some refer to "color-blind racism" as the tendency for white Americans to maintain that the United States, in the post–civil rights era, is now a meritocratic society in which color no longer matters (Bonilla-Silva and Dietrich, 2011; Forman, 2004). Hence, whatever shortcomings remain for blacks are the result of individual effort, obviating the need for compensatory measures like affirmative action. These subtle forms of racial discrimination lead to what Tyrone Forman calls racial apathy on the part of whites, that is, a lack of concern over racial inequalities. The election of a black president, moreover, has solidified the belief among whites that issues of racial inequality are no longer pertinent.

In his classic work dealing with black-white relations in the United States, the Swedish social scientist Gunnar Myrdal (1944) described the disjuncture between white attitudes and actions vis-à-vis blacks as "an American dilemma." The dilemma, according to Myrdal, is the conflict between, on the one hand, adherence to the "American Creed," entailing a belief in equality and Christian precepts, and on the other, group prejudice and discrimination based on material and status interests. The dilemma remains evident today in many ways.

CONTRASTING VIEWS OF PROBLEMS AND SOLUTIONS

Whites and blacks do not share a common view of the problems facing African Americans. Whites today tend to view blacks as better off economically and socially than blacks view themselves; they also see the extent and pace of racial change in the United States as far greater and more rapid than do blacks. Most whites believe African Americans enjoy equal opportunities in education, jobs, housing, and access to credit; African Americans have very different views. Moreover, whites see the current state of black-white relations more positively than blacks do. Nonetheless, African Americans today have an increasingly optimistic view of future conditions and relations, a trend that may have been spurred in large measure by the election of a black president (Carroll, 2006, 2007b; Gallup Organization, 2001b, 2003, 2010a; Jones, 2004, 2008; Pew Research Center, 2010a, 2010b; Saad, 2007a, 2007b; Smith, 2000).

BLACK AND WHITE VIEWS OF RACIAL PROGRESS   Surveys indicate that whereas whites see problems of racial inequality as more of a "mopping up," with basic changes having been made, blacks perceive deeply rooted discrimination continuing to affect their chances in work and education (Bobo, 2004; Dohorty, 2013; Gallup Organization, 2001b; Jones, 2006; Krysan, 2008; Pew Research Center, 2007c, 2010b; Roper Center, 2001; Smith, 2000; *Washington Post*–ABC News, 2012). Whites define racial progress mostly in political terms. If laws have been changed that previously denied blacks access to critical institutions, then it is assumed that an equal opportunity structure has been created and that blacks are equipped to take advantage of those opportunities. Most blacks, in contrast, do not see simply a few remnants of the past that need attention but a well-entrenched system that continues to block their socioeconomic advancement. Table 7.3 shows the wide gap in black and white views of African American discrimination. Note also the absence of change in these views in more than a decade, despite Barack Obama's election and reelection.

Also, blacks see a justice system that treats blacks differently than whites. The gap between whites and blacks in their perception of how blacks are dealt with by the police in their communities remains particularly wide (Pew Research Center, 2010b; *Washington Post*–ABC News, 2012). A recent national survey revealed that whereas 68 percent of blacks say the American justice system is biased against blacks, only 25 percent of whites agree (Newport, 2013). And the hugely

TABLE 7.3	PERCENTAGE SAYING "THERE IS MUCH DISCRIMINATION AGAINST AFRICAN AMERICANS"	
	2001	2013
Whites	15	16
Blacks	48	46

Source: Doherty (2013); Pew Research Center (2010b).

disproportionate percentage of blacks who are incarcerated in America's prisons is seen by African Americans as stark evidence of a skewed criminal justice system.

DISCRIMINATION VERSUS INDIVIDUAL EFFORT   Although whites generally recognize the disadvantaged status of a large proportion of the black population, there are striking differences in black and white explanations for that depressed condition. Most whites continue to view the problems as the result of blacks' own shortcomings, not social conditions or institutional forms of discrimination (Lipset, 1987; Schuman et al., 1997; Sniderman and Piazza, 1993). The majority, however, no longer attribute these deficiencies to biological inferiority (Schuman, 1982; Schwartz, 1967; Sniderman and Piazza, 1993). Most often, whites see black disadvantages as a product of "lack of ambition," "laziness," "irresponsibility," or "failure to take advantage of opportunities." The prevalent attitude is, "If we did it, so can they." Indeed, some hold that the image of blacks as lacking ambition and being socially irresponsible is the most common negative black stereotype (Sniderman and Piazza, 1993). Blacks, by contrast, see their plight as deriving primarily from discrimination or a lack of the same opportunities as whites (Gallup Organization, 2003).

Structural forces that strongly affect people's place in the stratification system (such as the society's changed occupational order or its basically altered demographic patterns) are not easily perceived. Thus social success or failure is apt to be interpreted as primarily the result of individual effort. Moreover, the belief that each member of society is personally responsible for his or her social lot is a basic component of the prevailing American ideology. The society's opportunity structure is pictured as open, providing equal chances for all to achieve material success or political power regardless of their class or ethnic origin. As a result, it seems only natural that whites would be more likely to attribute the lower social and economic achievement of blacks to individual and group characteristics than to the workings of a class and ethnic system that automatically favors the wellborn, especially those who are also white, and disadvantages the poor, especially those who are black. In addition, the more blatant forms of discrimination, which long held blacks in subservience—slavery, the Jim Crow system—are easier to perceive and understand by whites than are the indirect and largely covert forms of structural discrimination that continue to render debilitating effects. Moreover, even at the individual level, discrimination in less easily measured forms—poor service, racial slurs, fearful or defensive behavior, lack of respect, racial profiling—is a commonplace experience for most blacks (Feagin and Sikes, 1994; Morin and Cottman, 2001; Picca and Feagin, 2007) and renders a further injurious impact.

AFRICAN AMERICANS: ASSIMILATION OR PLURALISM?

Chapter 4 pointed out that assimilation and pluralism are processes and outcomes of interethnic relations that are not necessarily mutually exclusive for any particular group in a multiethnic society. Rather, a group may display aspects of both simultaneously. In the case of African Americans, certain patterns of assimilation seem apparent even though inequalitarian pluralism has generally been more characteristic of their historical and even contemporary experience.

CULTURAL ASSIMILATION

Because the African cultural heritage was largely destroyed under slavery, blacks had little choice about the adoption of the dominant group's major cultural traits, especially religion and language.[8] Some hold that in light of this historical fact, cultural assimilation has been as fully accomplished for blacks as for any American ethnic group (Gordon, 1964; Pinkney, 2000). However, although it is true that blacks have adopted the dominant group's major cultural forms, it is equally evident that an African American cultural variation has evolved in the United States, one that clearly sets blacks apart from Anglo-Americans and other ethnic groups (Hannerz, 1969; R. Taylor, 1979). African Americans have developed a linguistic style, have molded a unique version of Protestantism, and have created black art forms, especially in music.

REVERSE ASSIMILATION   As explained in Chapter 4, cultural assimilation is commonly a reciprocal process in which minority ethnic groups not only absorb the dominant culture but also transform it to some degree with their own contributions. This has clearly been the case for African Americans in recent decades. Indeed, the sociologist Orlando Patterson maintains that the black influence on the prevailing American culture is today not simply evident but pervasive (1997). Blacks, he suggests, dominate American popular culture—including music, dance, language, sports, and youths' fashion. "So powerful and unavoidable is the black popular influence," he writes, "that it is now not uncommon to find persons who, while remaining racists in personal relations and attitudes, nonetheless have surrendered their tastes, and their viewing and listening habits, to black entertainers, talk-show hosts and sitcom stars" (1995b:24). As a reflection of this cultural drift, until her retirement in 2012, Oprah Winfrey had been chosen as America's favorite TV personality for many years (Corso, 2010). "Blackness (or nonwhiteness)," notes journalist Leon Wynter, "now suffers less and less of a discount in the marketplace, while whiteness commands less and less of a premium" (2002:136).

ASSIMILATION OF THE FOREIGN-BORN   As explained earlier, the American black population is increasingly diverse culturally, and assimilation of the growing foreign-born element must be analyzed apart from African Americans. Many in the immigrant

---

[8] A long-standing debate among sociologists and historians concerns the degree to which blacks retained elements of their native African cultures under American slavery. Some, like Frazier (1949), assert that blacks were essentially stripped of African cultural ways; others, like Herskovitz (1941), maintain that remnants remained intact.

generation from West Africa and the Caribbean seek to maintain their ethnic identity and remain apart from the African American community. In some part this is in recognition of the American ethnic hierarchy, in which African Americans are stigmatized and in terms of status remain in the lower tier. Separation may also be based on the desire to maintain one's ethnic identity and culture. The reality of racial categorization in the United States, however, acts as a counterforce, drawing African- and Caribbean-born immigrants into the larger black population. Increasing intermarriage with African Americans also has an integrating effect. Some foreign-born blacks eventually even come to identify themselves as "African American" (Halter, 2007). All of these responses can be seen as aspects of segmented assimilation.

STRUCTURAL ASSIMILATION

Structural assimilation at the secondary level has begun in earnest for African Americans as they increasingly acquire better jobs, higher incomes, and more education, and enter into power positions. For the black underclass, however, secondary structural assimilation has been stymied, and the prognosis for future movement in this direction is bleak. Even for the black middle class, of course, it is still only in the incipient stages.

Structural assimilation at the primary level—as measured by residential patterns, club memberships, friendship cliques, and intermarriage—remains low. Whites today do have more personal contact with blacks than they did in the past, but that contact continues to be minimal and casual. Residential integration is a very basic prerequisite to other aspects of primary and even secondary assimilation, but in this area, as we have seen, African Americans have not followed the usual path of American ethnic groups. Though residential dispersion accompanied upward class mobility for other groups, this has not occurred for African Americans, who continue to experience high levels of segregation despite advances in income, occupation, and education.

Interracial marriage—the most important indicator of primary structural assimilation—also remains at a relatively low—albeit growing—level. Intermarriage of whites with other ethnic minorities is rapidly increasing but this trend is not as strong in relation to African Americans (Lee and Bean, 2010; Lee and Edmonston, 2005; Qian and Lichter, 2007; Wang, 2012).

Blacks and whites may increasingly work, shop, vote, and attend school together, but the formal interrelations in these areas of social life do not extend very far into more personal relations as neighbors, church members, close friends, or marital partners. Clearly, the social boundaries that continue to thwart change are stronger for African Americans than for any other American ethnic group.

ASSIMILATION: AFRICAN AMERICANS AND OTHER ETHNIC GROUPS

Is the experience of African Americans so exceptional that they cannot be compared using the same analytical models and criteria applied to other ethnic groups? This is an issue of some debate in sociology. The question is essentially this: is the assimilation, or "ethnic" model—so clearly applicable to white ethnic groups, examined in

Chapters 10 and 11—relevant to African Americans, or are they best viewed in terms of some form of inequalitarian pluralism? Let's briefly examine the two sides of this issue.

THE ASSIMILATION MODEL   In the 1960s and 1970s, many analysts applied the assimilation model to African Americans, seeing their social evolution as generally comparable to that of white ethnic groups. In this view, African Americans until the 1960s were denied entry into the competitive mainstream but afterward had to contend with other ethnic groups for jobs, education, housing, and political power. Changed societal conditions and group cultural differences were acknowledged, but the most important difference between African Americans and other ethnic groups, in this view, was simply the time that African Americans had entered the competition: they were latecomers (Glazer, 1971; Glazer and Moynihan, 1970; Handlin, 1962; Kristol, 1970; Sowell, 1981). The process, therefore, might be longer and more painstaking, but eventually, African Americans, like other ethnic groups, would be assimilated into the mainstream of American society.

THE INEQUALITARIAN PLURALISTIC MODEL   Some aspects of the assimilation model continue to be applied to African Americans. But a more prevalent view today regards the black experience as unique among American ethnic groups. Even exponents of the assimilation model have conceded that the path of societal integration followed by African Americans is exceptional vis-à-vis not only past immigrants but even new immigrant groups (Glazer, 1997).

In explaining the noticeably different patterns of assimilation of African Americans compared with other ethnic groups, emphasis is placed on the fundamentally divergent African American ethnic experience; no other group arrived so nearly completely as involuntary immigrants, none was subjected to two centuries of slavery, none experienced the depth and persistence of racial prejudice and discrimination, none was so totally divested of its native culture, and none was so indelibly marked.

Some social scientists in the 1960s and 1970s went further in distinguishing the African American experience, contending that it was best portrayed as "internal colonialism"—that is, characterized by features similar to those of nineteenth century European colonialism: blacks' entry into the society was forced, their indigenous culture was destroyed, they were oppressed by the dominant group, and they were the target of a racist ideology (Blauner, 1969, 1972; Franklin and Resnik, 1973). Other ethnic groups did not face similar circumstances.

Those who see African Americans as singular among American ethnic groups argue that the vestiges of slavery have immeasurably handicapped them in their efforts to compete with whites. The abolition of slavery did not change established relations of dominance and subordination. Blacks continued to be relegated to the lowest occupational levels and were denied fundamental political rights well into the 1960s. Furthermore, the economic and political structures of American society, they argue, have changed so basically from what they were for entering European immigrant groups that comparisons with them are rendered vacuous. African Americans simply cannot utilize economic and political institutions as European immigrants earlier did to secure better jobs, housing, and political clout and, thus, to effectuate assimilation.

ASSIMILATION AND VISIBILITY As noted in Chapter 4, visibility is usually a decisive factor in the rate and degree of structural assimilation for ethnic groups in multiethnic societies. The greater the visibility, the slower and less intense is the assimilation process. This would seem to be of overriding importance in the case of African Americans.

Despite its abatement in the past two decades, the racial stigma of African Americans remains an ever-present fact in the United States. As Hughes and Hughes put it more than six decades ago, "Few are the situations which both Negro and white Americans enter, side by side, unlabelled and undifferentiated. Even a monk or a hero must be named by his color unless he is white" (1952:28). Isaacs has poignantly expressed the significance of visibility as it affects the lives of black people in America:

> For black Americans, more clearly perhaps than for any other group, the element of color and physical characteristics lies at the very center of the cluster that makes up their basic group identity. It is the one element around which everything else in their lives has been made to revolve, the heart of the identity crisis that is with them every hour of every day and which they need more than anything else to resolve. (1989:66)

The factor of racial visibility may decline in significance in the future, however, as the dividing lines of traditional racial and ethnic categories continue to fade. Even the perceived racial division between blacks and whites is no longer seen by many people as clearly as it had been in the past. When asked about President Obama's racial identity, only a bare majority of both black and white Americans agreed among themselves. Most blacks said Obama is "black," but more than one-third identified him as "mixed race." Among whites the pattern was reversed, with about half answering that he is "mixed race," whereas just a quarter identified him as "black" (Pew Research Center, 2010b).

In sum, although the importance of ethnicity has diminished for most whites in American society, for blacks it continues to be a crucial social characteristic, influencing almost all aspects of social life. Though the relations between blacks and whites are no longer the almost exclusive focus of ethnic analysts and policy makers—as they had been for most of the twentieth century—they will likely remain the major issue of interethnic relations in the United States for many generations.

SUMMARY

- The black experience is unique among American ethnic groups. No other group entered the society so completely as involuntary immigrants, and no other group was subjected to such fully institutionalized degradation. The after effects of slavery continue to influence the patterns of black-white relations and the place of blacks in the social hierarchy.

- Black-white relations can be divided into three major periods. *Slavery,* lasting from the 1600s until 1865, was characterized by paternalistic domination in which blacks were totally subservient, by law in the South and by informal processes in the North. Following the brief interlude of Reconstruction after the Civil War, black-white relations entered the *Jim Crow* era. Although no longer slaves, blacks were forced back into a subservient economic role and, through legal and extralegal measures, were denied the vote as well as other basic civil liberties. Moreover, a fully legitimized system of segregation was developed in the South, abetted by the application of a racist ideology. An era of *fluid competition* between blacks and whites commenced in the 1930s, and in the 1960s the Jim Crow system was demolished in the wake of the black civil rights movement.

- As a collectivity, blacks in the past five decades have made significant strides in income, occupation, and education. However, they continue to lag behind whites on all measures of socioeconomic status. Furthermore, the gains of the past fifty years have not been experienced in a balanced fashion within the African American community. One segment has taken its economic place in the working and middle classes, but the other segment is an underclass—chronically unemployed, undereducated, and acutely indigent.

- The economic problems of poor blacks have been explained as, in some part, a continuation of racial discrimination, but primarily as a function of the lack of relevant skills and education in a restructured American workforce.

- The levels of prejudice and discrimination against blacks have historically been more severe and persistent than those against any other ethnic group in American society. White attitudes today are more tolerant than in the past, and discrimination is confined mostly to the structural variety. Housing, however, is an area that has not been as strongly affected by changes in black-white dynamics.

- Although African Americans have reached a high level of cultural assimilation and are in the incipient stages of secondary structural assimilation, primary structural assimilation remains at a low level. Little black-white interaction occurs outside formal social settings such as work and education.

- Although some in the past maintained that African Americans were simply latecomers to the urban environment and would eventually display assimilation patterns much like those of white ethnic groups, most today see African Americans as a group whose historical experience and current conditions are so distinctive that comparisons with white ethnic groups are baseless.

CRITICAL THINKING

1. In what ways might the election and reelection of a self-identified African American as U.S. president affect the social and economic future of African Americans? How might having an African American in the White House affect white perceptions of and attitudes toward African Americans? How would you respond to someone who states that "since an African American is the U.S. President, African Americans can no longer make claims of racial discrimination?"

2. Explain how the relative lack of wealth of African Americans impedes their upward mobility, even among the black middle class, whose incomes might not be much different from whites'.

3. Middle-class white neighborhoods usually remain overwhelmingly white because homeowners fear that if blacks move in they will bring poverty and its attendant problems. For black families to purchase homes in such neighborhoods, of course, would require that they would also be middle class. But white fears

rest on the stereotype of blacks as "poor." Explain this as a case of *statistical discrimination*, discussed in Chapter 3.

4. As this chapter notes, the black population of the United States is becoming more varied in ethnicity. What social advantages (or disadvantages) might blacks from African and Latin American/Caribbean countries have over African Americans? Consider black immigrants in terms of the theory of *segmented assimilation*, discussed in Chapter 4.

Personal/Practical Application

1. This chapter explains that blacks and whites may have different ideas about the ideal racial makeup of their neighborhood. If you marry and have children, what factors would you consider in choosing a neighborhood in which to live and raise a family? Would race or ethnicity be one of those factors?

2. Despite a radically changed racial environment in the United States since the 1960s, many argue that race consciousness and social distance among blacks and whites remain strong. How evident is racial segregation in your university (dorms, eating areas, fraternities and sororities, and so on)?

HISPANIC AMERICANS

There are more people of Mexican origin in Los Angeles than in all but one city in Mexico; in New York City, there are more Puerto Ricans than in San Juan; and only in Havana are there more Cubans than in Miami. Together, Hispanic Americans number fifty-two million, almost 17 percent of the total U.S. population. And the Hispanic American ethnic category is growing rapidly and substantially. So rapid has the increase been, in fact, that today it has surpassed African Americans as the country's largest ethnic minority.

Hispanic Americans constitute several distinct ethnic groups, linked by a shared language and a cultural heritage derived, in the main, from Spanish colonialism. By far the largest group among them is Mexican Americans; the rest are people from the Caribbean and Central and South America, as well as a small group whose origins are traced to Spain (Figure 8.1). Because Mexican Americans, Puerto Ricans, and Cubans are the oldest and, for Mexicans and Puerto Ricans, the largest elements of the Hispanic American population, in this chapter we focus mainly on those groups.

It is important to underscore the cultural variety among the various Hispanic American groups. These distinctions are often obscured by the common tendency of the media, government agencies, and the general public to lump together all those with Latin American origins, using the terms *Hispanic* or *Latino*.[1] South and Central America and the Caribbean are geographically vast, and the specific countries that make up these regions are each distinct in their historical development, their racial-ethnic composition, their cultural practices, and even their language. As to the latter, although these countries may be linked by a common linguistic heritage, anyone who

---

[1] Although the term *Hispanic* has been most frequently used in the past few decades, *Latino* is today also commonly used to describe those of various Latin American origins. Although some studies have found a preference for *Hispanic* over *Latino* (Gallup Organization, 2003; Tucker and Kojetin, 1996), others indicate that a majority of Hispanic Americans prefer neither term, choosing instead to identify themselves with their country of origin—"Mexican," "Cuban," and so on (de la Garza et al., 1992; Pew Hispanic Center, 2002, 2009; Taylor et al., 2012). The U.S. Census itself has, at different times, used a variety of terms to describe the Hispanic American population. In this chapter, *Hispanic* and *Latino* are used interchangeably.

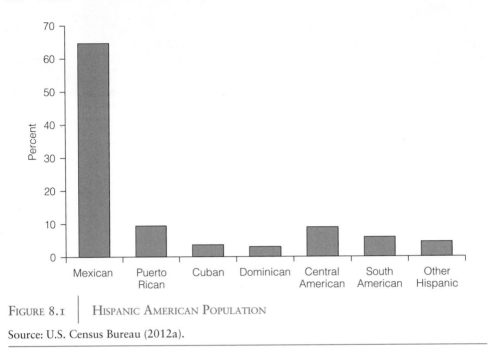

FIGURE 8.1 | HISPANIC AMERICAN POPULATION

Source: U.S. Census Bureau (2012a).

has studied the Spanish language even minimally knows that the Spanish spoken by Mexicans is not precisely the same as the Spanish spoken by, for instance, Colombians or Argentinians, or, much less, Spaniards.

THE "IN-BETWEEN" STATUS OF HISPANICS

If a common theme runs through the unique histories and experiences of the several Hispanic groups in the United States, it is their intermediate ethnic status between Euro-American groups, on the one hand, and African Americans, on the other. In several important respects, Hispanics are an ethnic minority "in between."

MANNER OF ENTRY

Hispanic Americans have entered the society in a variety of ways. Although Mexicans have come to the United States in the last hundred years as voluntary immigrants, they retain a heritage of having been absorbed into American society originally as a conquered group. The area that today is the American Southwest was part of Mexico before being annexed by the United States in 1848 following the Mexican War. Mexicans living on those lands were subsequently incorporated into the United States, much in the fashion of classic European colonialism.

Puerto Ricans are a group not clearly part of either voluntary immigration or conquest. Puerto Rico became a territory of the United States in 1898 following the Spanish-American War, and in 1917 the inhabitants of the island were given the status of American citizenship. This enabled them to immigrate freely to the mainland with no restrictions. Puerto Ricans therefore are not technically immigrants, even though they come to the mainland from a distinctly foreign culture.

The movement of Cubans to the United States in the past five decades has been a voluntary immigration. Yet it has been different from those of most past groups in that it was initially impelled by political rather than economic motives. Moreover, the class characteristics of immigrant Cubans and the reception given them by the host society were unusual. As the sociologist Alejandro Portes has put it, "Seldom has a foreign group come to the United States so well prepared educationally and occupationally and seldom has this country received one so well" (1969:508).

The national origins of other groups that make up the contemporary Hispanic American population are varied, but the majority are first-generation immigrants who have come to the United States in the past four decades. Some, particularly those from Central American countries, have entered as refugees, escaping political violence, but most are voluntary immigrants who, like those before them, have come in pursuit of improved economic opportunities.

HISPANIC AMERICANS AND RACE

Perhaps the most obvious and consequential in-between feature of Latinos is their racial status. Members of all Hispanic American groups include a range of physical types.

Mexico is racially amalgamated to a greater degree than perhaps any modern society (van den Berghe, 1978). Over the past two centuries, its population has been transformed into one dominated by mestizos, a physical type combining European and Indian traits. Although many are either basically European, or at the other extreme, purely Indian, the vast majority of Mexicans today are racial hybrids. Most Mexican immigrants to the United States are mestizo, making them physically distinct from the Anglo majority but still "the least divergent racially of all non-Caucasian groups" (Murguía, 1975:51). So ambiguous is the racial status of Mexican Americans that the U.S. Census Bureau has in the past classified them both as a separate "race" and as part of the white population. In any case, most Mexican Americans are physically distinct enough to be perceived by many Anglos in racial terms. (The term *Anglo* is used in the Southwest and in other areas with a large Hispanic population to distinguish any non-Hispanic white.)

Puerto Ricans' in-between racial status is equally complicated. Whereas Mexicans are mostly mestizo, Puerto Ricans inherit a racial background that is a combination primarily of European and African but also some Indian elements. The large African population originally brought to the island as slaves intermarried with whites over several generations, and today, Puerto Ricans cover the entire color spectrum. On the island, racial distinctions are not so acute, and one's color is seen along a scale or continuum (Duany, 1998; Fitzpatrick, 1987; Padilla, 1958; Rodríguez, 1996). But problems arise in the United States, where there is a racial dichotomy, allowing for no intermediate categories. Many who are "white" on the island may therefore find themselves classified as "black" when they come to the mainland. The differential treatment they subsequently receive on the basis of their darker skin (by American standards) is frustrating and problematic because interracial marriages are commonplace in Puerto Rico, and social treatment of darker groups is more equitable than on the mainland. Similar problems are encountered by Dominicans who immigrate to the United States. They may consider themselves white, but in the American context they are often perceived as black (Duany, 1998).

In both Mexico and Puerto Rico, racial distinctions are often a function of class and cultural differences rather than of physical type alone. Those who rise in socioeconomic status or who adopt Western cultural ways may, accordingly, be "whitened." As we will see in Chapter 15, this is also the case in Brazil. In the United States, however, racial distinctions are based on skin color or other physical characteristics, in addition to ancestry, and are not modified by changes in class or culture.

Among the Hispanic groups in the United States, Cubans are the least racially heterogeneous, most falling into the American category of "white" (Diaz, 1980). Cuba itself is a racially variegated society, however, suggesting that the Cuban immigration has been racially selective (Portes et al., 1977). Most early Cuban immigrants were of higher social status in Cuba and hence whiter in color, whereas many of those who followed have been of lower class origins and of darker color (Pedraza-Bailey, 1985).

Racially and culturally, those from Central America constitute a very diverse collectivity. In addition to six different national origins, the Central American population includes indigenous, Afro-Caribbean, European, and mestizo peoples. Latinos from Central American countries like Guatemala, El Salvador, and Nicaragua generally evince mestizo origins, not unlike most Mexicans, but many are comparable to those from Caribbean countries, combining African and European origins.

Those from South American countries exhibit the entire color range. Most Peruvians, for example, resemble Mexicans, but Argentinians are indistinguishable from Euro-Americans. Many from coastal areas of Colombia closely resemble those from Puerto Rico or the Dominican Republic, but other Colombians are racially European.

The fact that Latinos manifest such a broad range of physical types calls into question the very notion of race as it applies to these groups. In recent years, "Hispanic" or "Latino" has sometimes been used as a racial designation. Police and news agencies, for example, commonly describe crime suspects in that manner (for example, "a 20-year-old Hispanic male . . ."). But the concept of Hispanic racial features is rendered meaningless in light of the fact that individuals within the various Hispanic groups exhibit every conceivable physical (that is, "racial") characteristic. And, Latinos themselves, regardless of national origins, see themselves differently in terms of race and ethnicity and the way they are perceived by others. "Divisions are evident," notes Rubén Rumbaut, "between regions and groups, within groups, and even within families" (2011:7). Although "Hispanic" or "Latino" is often interpreted as a racial designation, it remains basically a cultural—that is, ethnic—description. Like other racial and ethnic designations, however, it is arbitrary and subject to change, as group identities and census categories are modified. The confusion and complexities of race and ethnicity as they pertain to Latinos is a perfect illustration, as discussed in Chapter 1, of how races and, to some extent, ethnic groups, are essentially social creations.

THE DEVELOPMENT OF THE LATINO MINORITY

Despite broad similarities, Mexican Americans, Puerto Ricans, Cuban Americans, and the more recent Latino groups each display a unique history of immigration as well as unique patterns of settlement and adaptation.

MEXICAN AMERICANS

Relations between Mexicans and whites (Anglos) have, historically, been characterized consistently by conflict, albeit of varying intensity. Early in the contact between these groups, a pattern of Anglo domination and Mexican subordination emerged. And, as with American Indians, the vehicle through which domination was initially established was physical conquest.

THE MEXICAN WAR Mexico gained its independence from Spain in 1810, and at that time its territory encompassed an area as far north as what is today Colorado. Most of this northern Mexican region, far removed from the administrative capital of Mexico City, was inhabited by Indians and a smaller number of Mexicans. As a means of populating the area, the Mexican government in 1821 granted permission for foreigners—mainly Americans—to settle in the area that today is Texas. By 1830 about twenty thousand Anglos were living in the Texas region, constituting a numerical majority. Culturally and physically distinct from the indigenous Mexicans and Indians, the Anglo settlers from the outset translated their differences into contempt and hostility toward these groups.

The Mexican government could exercise only limited control over the Texas colony, and pressures from Anglos for independence grew steadily. In 1835–36, the colony revolted and established itself as an independent republic. Mexico never recognized the independence of Texas; when it became clear that the United States favored annexation of the territory, much friction was created between the American and Mexican governments, leading eventually to war in 1848.

The war proved disastrous for Mexico, which lost more than half its territory. The United States acquired what are today the states of California, Colorado, New Mexico, Nevada, Utah, and most of Arizona, in addition to Texas, which had been annexed previously. Under the Treaty of Guadalupe Hidalgo, ending the conflict, Mexicans living in what was now American territory were guaranteed the rights of American citizenship if they chose to remain. Most did, and it was from this nucleus that the Mexican American ethnic group evolved.

The significance of the Texas rebellion and the Mexican War cannot be overstated as factors in shaping subsequent relations between Mexican Americans and Anglos in the Southwest. The borderlands remained an area of overt conflict long after the conclusion of the war, and this protracted hostility intensified the mutually negative perceptions of Anglos and Mexican Americans (Grebler et al., 1970). Anglo-Mexican conflict did not characterize all of the Southwest to the same degree. In New Mexico, relations between the two groups were relatively peaceful by comparison with Texas, where the strife reached its most violent and bloody form. But in all cases, the war left a residue of antagonistic feelings and group images among Anglos and Mexicans that continued to color interethnic relations for many generations.

THE EMERGENCE OF ETHNIC STRATIFICATION   Although their property rights and political rights were guaranteed by the Treaty of Guadalupe Hidalgo, Mexicans in the newly acquired American territory found themselves increasingly displaced by Anglos. Their property was taken through official and unofficial force and fraud, and they were gradually transformed into a colonized workforce serving the area's labor needs. Mexicans were also dispossessed politically, as Anglos firmly established themselves in power. The scenario of heavy immigration of Anglos followed by the divesting of Mexicans' economic and political power was played out somewhat differently from area to area. In New Mexico, for example, the indigenous Mexican population, called Hispanos, did retain some political and economic power. These people were descendants of the original Spanish settlers dating back to the 1600s, and they did not identify themselves with the later-immigrating Mexicans (DeSipio, 2006; Nostrand, 1973). In general, however, Anglos perceived Mexicans as a conquered—and therefore inferior—people who quite naturally were relegated to subordinate ethnic status (Alvarez, 1973).

The development of ethnic stratification in the Southwest seems clearly in accord with Noel's (1968) theory, discussed in Chapter 2. All three factors necessary to the emergence of an ethnic hierarchy—ethnocentrism, competition, and above all, differential power—came into play after 1821 in the confrontation of Anglos and Mexicans (McLemore, 1973). Anglos settling the region were most intent on securing land held by Mexicans—hence the ensuing competition. With them, they also brought their racial attitudes, which led them to see Mexicans as an inferior people whose exploitation could therefore be justified. Finally, with their numerical power—and following annexation by the United States, their political power—Anglos became the dominant group, displacing Mexicans both legally and by force.

MEXICAN IMMIGRATION   As a result of the Texas and Mexican wars, the first Mexicans were incorporated into American society through conquest. But subsequent generations of Mexicans entered the United States as voluntary immigrants. Indeed, it must be remembered that the original Mexican Americans, those living on annexed lands, were a relatively small group, perhaps no more than seventy-five thousand (McWilliams, 1968). More than twice that number of Indians were living in the same territory. The overwhelming majority of Mexican Americans, then, entered as voluntary immigrants, motivated by the same push-pull factors as other immigrant groups.

The flow of Mexican immigration during the nineteenth and twentieth centuries was dictated primarily by the labor needs of the American Southwest (Camarillo, 1979; Tyler, 1975). Most simply, movement across the border was heaviest during periods when the demand for cheap labor was high. Although immigration was evident during the remainder of the nineteenth century following the Mexican War, it reached enormous proportions during the twentieth century with the expansion of the American Southwest and the corresponding inadequacies of the Mexican economy. The pressures of a continually growing population and the political destabilization brought about by the Mexican Revolution in 1910 further impelled migration northward.

Between 1910 and 1930, the development of railroads, mining, and agriculture, all labor-intensive industries, spurred the first great wave of Mexican immigration into the United States (McWilliams, 1968). As the chief source of unskilled labor in the Southwest, Mexicans became a subordinate labor force. They were commonly kept in debt by their employers, paid less than others for the same work, concentrated in the least desirable occupations, often used as strikebreakers, and laid off most easily in distressful times (Barrera, 1979).

When they represented an excess labor supply, Mexicans were driven back across the border. During the 1930s, for example, not only were their jobs abolished but they served as convenient scapegoats for the area's unemployment and growing welfare rolls. Deportations were launched in which undocumented immigrants were rounded up and sent back to Mexico. Repatriations also took place in which Mexicans, many of whom had been born in the United States, were induced to leave the country.

During World War II, a severe labor shortage once again prompted the importation of Mexican workers. But by the early 1950s the deportations and repatriations of the 1930s were grimly repeated in a policy called Operation Wetback, designed to stem the flow of undocumented Mexican immigrants. In five years, 3.8 million people were sent back to Mexico (Moore and Pachon, 1976). Relatively few were actually deported in formal proceedings; most were simply rounded up and expelled. In the process, Mexicans were once again reminded of their tenuous status in the United States. People who simply "looked Mexican" were stopped and required to present evidence of their legal status. Although there is some question about the official number actually apprehended and deported (Garcia, 1980), Operation Wetback in any case created fear and suspicion among the Mexican American populace.

Today Mexicans are the country's largest immigrant group, and their numbers are swelled by a continual flow of undocumented entrants. During the past three decades, Mexicans have constituted about a quarter of all legal immigrants to the United States and more than half of all immigrants who enter without papers (Alba, 2013; Hoefer et al., 2010). As in the past, economic conditions are the chief motive of this migration.

THE CHICANO MOVEMENT AND THE DEVELOPMENT OF ETHNIC CONSCIOUSNESS As with other ethnic groups, in the 1960s and 1970s Mexican Americans engaged in significant political activity as part of a period of growing ethnic consciousness. Perhaps the two most dramatic episodes of Mexican American political activism during this period were the movements led by César Chávez and Reies López Tijerina. Chávez's objective was to organize migrant farmworkers, and Tijerina focused on regaining communal lands that he charged had been illegally taken from Hispanos in New Mexico.

Chávez's well-publicized movement of this period consisted of efforts to force growers in California and the Southwest to recognize his United Farm Workers union (UFW) as the representative of thousands of migrant laborers, most of them Mexican Americans. Past efforts had been made to organize farmworkers in the Southwest, some even dating back to the early years of the twentieth century, but Chávez was the first to succeed. Through his organizing activities, Chávez became something of a folk hero to Mexican Americans, and his charisma thrust him into the forefront of the larger **Chicano movement,** *aimed at securing greater social,* *political, and economic rights for Mexican Americans* (Solis-Garza, 1972; Steiner, 1970; Stoddard, 1973).

The movement led by Tijerina and his organization, the Alianza de las Mercedes, was more openly militant, and its objectives were less pragmatic than Chávez's. Tijerina argued that several million acres of land in New Mexico had been fraudulently taken from Hispanos following annexation of the territory at the conclusion of the Mexican War, and his aim was to recover them. To draw attention to his cause, he provoked several confrontations with federal and state officials, leading to his arrest and subsequent imprisonment (Knowlton, 1972). Although the Alianza helped in creating a national awareness of the Chicano movement, its goals were too visionary to yield concrete benefits.

Chávez's and Tijerina's movements were the two most prominent elements of the larger Chicano movement of the 1960s and early 1970s, which extended well beyond specific personalities and organizations. *Chicanismo* was very similar to the black nationalist movement and was spawned in the same general atmosphere of minority activism in the 1960s. Its major goal was to galvanize Mexican American political action in order to assert group rights and expand economic and social opportunities. Many of the same tactics employed by the black movement, including legal cases, protests, boycotts, and direct encounters with authorities, were adopted.

Like its black nationalist counterpart, the Chicano movement also represented an assertion of militant cultural pluralism. Rather than assimilation, the idea was to affirm the unique Mexican American culture and to instill in Mexican Americans a sense of ethnic pride and awareness (Macías, 1972; Steiner, 1970). Younger activists began to refer to themselves as Chicanos rather than Mexican Americans and to proclaim a more militant and antiestablishment position. The term *Chicano* had in the past referred to lower-class Mexicans, but it now began to represent an ethnic viewpoint (Macías, 1972; Moore and Pachon, 1976). Just as *black* had been substituted for *Negro*, *Chicano* came to symbolize an ethnic identity that repudiated the negative group image that had been imposed for generations by Anglos.

Just as black militancy subsided in the 1970s, so too did the Chicano movement. But as with black power, the long-range effects of *Chicanismo* were more substantial than they first appeared. Dramatic socioeconomic gains were not forthcoming, as governmental programs aimed at depressed Mexican American communities were short-lived. Intangible benefits, however, were more meaningful and lasting. A positive group identity emerged among Mexican Americans, and ethnic unity was instilled through the idea of *La Raza*.[2] Most important, national attention was drawn for the first time to Mexican Americans as a sizable minority group.

Puerto Ricans

The formation of the Puerto Rican minority in the United States differed considerably from the development of the Mexican American community. To begin with, Puerto Ricans are a relatively recent immigrant group. Even though they had

---

[2] *La Raza* (literally, "the race") is not similar in meaning to the North American idea of race but is a somewhat vague concept denoting a common heritage or peoplehood.

maintained a presence on the mainland for most of the twentieth century, their numbers were quite small until after World War II. The greatest influx occurred during the 1950s, when nearly 20 percent of the island's population moved to the mainland. Whereas in 1940 only one thousand Puerto Ricans migrated, in 1953 almost seventy-five thousand did so (U.S. Commission on Civil Rights, 1976). In the 1970s, as the U.S. economy stagnated and low-skill jobs disappeared, the migration to the mainland ceased and, in fact, a net return migration was sustained throughout the decade. The 1980s, however, saw a second major flow of Puerto Ricans to the mainland, creating an increased population (Boswell and Jones, 2006).

The push-pull factors impelling Puerto Ricans to migrate have been similar to those for other voluntary immigrant groups: a surplus population in an economically depressed society seeks economic betterment in a society that promises improved conditions. Puerto Rican migration has therefore fluctuated, depending on economic conditions both on the island and on the mainland. An unusually high birthrate coupled with a seriously underdeveloped economy created great pressures for out-migration from the island after World War II. Two factors increased the appeal of the mainland United States. First, with citizenship rights, Puerto Ricans could enter the country with no restrictions, quotas, or other legal steps. Second, with the establishment of airline service, transportation between the mainland and the island became relatively cheap and rapid. The migration process for Puerto Ricans, then, was considerably less complicated than that for other groups.

Although the absolute size of the Puerto Rican migration to the mainland is not among the largest compared with past groups, it is numerically significant in relation to the island's population. By 1960 there were one-third as many Puerto Ricans on the mainland as in Puerto Rico, and by 1970 one-half as many (Bahr et al., 1979). In 2009, there were more Puerto Ricans on the mainland than on the island (Lopez and Velasco, 2011). Moreover, migration statistics do not accurately indicate the frequent back-and-forth movement between island and mainland, so it is quite reasonable to assume that a much larger number of islanders have, at one time or another, migrated. Indeed, during the three decades between 1940 and 1970, it is unlikely that a single Puerto Rican family was unaffected by this movement (Fitzpatrick, 1987; U.S. Commission on Civil Rights, 1976).

To a limited extent, Puerto Ricans participated in the ethnic movements of the 1960s and early 1970s. Several groups emerged at that time, advocating control by Puerto Ricans of their communities and emphasizing ethnic identity and pride. The Young Lords, for example, a militant group in Chicago and New York, imitated the activities of the Black Panthers in organizing Puerto Rican ghettos and calling for a redress of squalid conditions (Lopez, 1980). As with black and Chicano political activism, Puerto Rican militancy resulted in drawing the serious attention of national authorities to the Puerto Rican community for the first time. But as with blacks and Chicanos, programs of federal and community agencies directed at Puerto Ricans dwindled with the passing of militancy in the 1970s.

Today the Puerto Rican population is increasingly a second- and third-generation ethnic group, with two-thirds having been born on the mainland. This is especially evident in New York, where the largest number of Puerto Ricans continues to reside. In 1950, more than 81 percent of all Puerto Ricans on the mainland lived there, and in 1970, 62 percent did so. During the 1970s and 1980s, Puerto Ricans began to disperse, with large communities developing in California and Florida (Torrecilha et al., 1999; Whalen, 2005). By 2010, less than 20 percent of Puerto Ricans lived in the New York metropolitan area (Logan and Turner, 2013). Other significant Puerto Rican communities are found in Chicago, Philadelphia, and northern New Jersey, especially Newark and Jersey City. The remainder are primarily in cities of the Northeast. In addition, many—even some born on the mainland—have returned to Puerto Rico (Dockterman, 2011).

CUBAN AMERICANS

The conditions of immigration for Cubans and their subsequent settlement patterns have been unlike those experienced by either Mexicans or Puerto Ricans.

Given its geographical proximity to Cuba, the United States often served as a haven for Cuban exiles during times of political turmoil before and after the Cuban movement for independence from Spain in the late nineteenth and early twentieth centuries. In fact, one of the key figures of Cuban political history, José Martí, sometimes called Cuba's George Washington, spent much of his exile during the 1880s in the United States. Small colonies of Cubans had for many years been present in Key West, Tampa, Miami, and New York.

THE CUBAN REVOLUTION AND LARGE-SCALE IMMIGRATION   In the 1960s, following the revolution led by Fidel Castro, Cubans began to come to the United States in significant numbers. Whereas pull factors of migration were stronger for Mexicans and Puerto Ricans, push factors seemed more crucial for early Cuban immigrants. Their chief motivation was political refuge rather than economic betterment. Nonetheless, the underlying economic objectives of the émigrés should not be overlooked. Because the revolution imposed a socialist system in Cuba, the power and economic opportunities for most of the early immigrants—who were mainly middle-, upper-middle-, or upper-class people—had been severely constricted. As Portes has explained, "The wealthy, the successful, the powerful, the educated saw their status challenged, their influence radically curtailed, and their economic position continuously menaced" (1969:506). Departure from Cuba, then, was not simply politically inspired but was calculated in the hope of reestablishing a relatively high socioeconomic status in the United States and returning to Cuba when political conditions were favorable. For later Cuban immigrants, economic factors in the decision to migrate were even more clearly apparent (Portes et al., 1977; Wong, 1974).

Cuban immigration from 1960 onward was aided immeasurably by American policies that in effect permitted the entrance of Cuban refugees without restriction. The political objectives of this policy are not difficult to understand. The chilled relations between the Cuban and American governments soon after the Cuban Revolution were followed with strongly belligerent actions by the United States, including an American-sponsored invasion of Cuba by Cuban exiles at the Bay of Pigs in 1961 and imposing a trade embargo against Cuba in the same year. The Cuban missile crisis of 1962 firmly established the Castro regime as a basic element in the Cold War, and from that point forward Cuba became a bête noire for successive American administrations. Cuba was portrayed in the United States as a "communist stronghold" and the Cuban people as "politically enslaved." The emigration of large numbers of Cubans, therefore, was seen as confirmation of this perspective, and the refugees were welcomed enthusiastically.[3]

Large-scale Cuban immigration ceased in 1973 but resumed for a brief period in 1980, when more than 125,000 people came to the United States. The conditions under which this second immigrant wave left Cuba were graphically described by the American media, and the manner in which they arrived—most in small boats— further dramatized their entrance. They were dubbed *Marielitos,* in reference to the port of Mariel, from which most of them departed Cuba. These immigrants were generally lower in social class than the original Cuban refugees, and many were black or mulatto. As a result, they did not find adjustment to American society as smooth or their reception as enthusiastic. They experienced higher rates of unemployment, low-paid work, and dependence on charity and public welfare. Moreover, the presence of a criminal element among them served to taint the entire group (Corral and Viglucci, 2005; Portes et al., 1985).

Except for part of the second wave, most Cuban immigrants entered the United States with a strong family and kin support system that aided their adjustment considerably (Portes et al., 1977). Indeed, following the initial immigration in the early 1960s, succeeding immigrants were encouraged to come to the United States by relatives or friends who could offer them physical and economic assistance (Fagan et al., 1968). Although the presence of friends and relatives has also characterized the Mexican and Puerto Rican migrations (recall the network theory of immigration, discussed in Chapter 5), the support system of Cubans has been stronger and more complete. Moreover, Cubans have experienced substantial economic success and have therefore been able to offer greater assistance to those who have followed them. Mexicans, Puerto Ricans, and most other Latin American groups, by contrast, have remained collectively in a far less secure economic position.

As we will see, the Cuban American population is markedly distinct from most other Latino groups in demographic and socioeconomic features. It is older—the median age is forty-one compared to twenty-seven for the rest of the Latino population— and the rate at which Cubans acquire U.S. citizenship (60 percent) is more than double the rate for other Hispanics (Pew Hispanic Center, 2006a). In terms of income, occupation, and education, it more closely resembles the non-Hispanic white population, and it has not been subject to the prejudice and discrimination that Mexicans and Puerto Ricans have historically experienced.

DOMINICANS, CENTRAL AMERICANS, AND SOUTH AMERICANS

In the past two decades, the Hispanic American population has been infused with newer groups, from Central and South America and from the Caribbean, whose numbers in the United States previously were very tiny.

DOMINICANS   Those from the Dominican Republic constitute the largest of the newer Latino groups, about 3 percent of all Hispanic Americans. Dominicans began

---

[3] For thirty-five years, the U.S. federal government maintained an open-door policy for Cubans who fled to the United States. Under the policy, they were given immediate permanent resident status—terms afforded no other immigrant or refugee group. The policy was effectively terminated in 1995 with the elimination of Cubans' automatic acceptance as political refugees (Nackerud et al., 1999).

arriving in large numbers in the 1970s, and their percentage of the Latino population has continued to grow. Between 1990 and 2000 the number of Dominican immigrants nearly doubled (Grieco, 2004). More than half of all Dominicans have settled in the New York metropolitan area, where they are one of its largest immigrant groups. The remainder are in just a few states, mostly in the Northeast and Florida (Boswell and Jones, 2006; Logan and Turner, 2013).

When looked at in comparison with other Hispanic groups, Dominicans have fared more poorly, resembling in many ways the Puerto Rican patterns of adaptation and socioeconomic status (Hernández, 2007; Kasinitz et al., 2006; Reimers, 2006). Most Dominicans work in low-paying service or factory jobs and are handicapped by low levels of education. Roughly half of Dominican families are headed by women, and they suffer a poverty rate considerably higher than the overall poverty rate in New York, where most reside (Bernstein, 2005). The racial factor has also affected this group because most identify themselves as "mulatto" or "black" (Pessar, 1995).

CENTRAL AMERICANS   Immigrants from Central American countries entered the United States in large numbers starting in the 1980s, mostly because of political unrest in the region. Many arrived as refugees and many were undocumented. Immigration from Central America continued throughout the 1990s and 2000s, though no longer as the result of political unrest. Today Central American immigrants come to the United States primarily as a result of the push-pull factors of employment and living conditions. Over half have come to the United States since 1990 (Logan and Turner, 2013; Stoney and Batalova, 2013; U.S. Census Bureau, 2007c).

Together, those from Central American countries constitute about 9 percent of the Hispanic population in the United States. The largest specific groups within this category are Salvadorans (more than 40 percent), Guatemalans, Hondurans, and Nicaraguans. As noted earlier, racially and culturally, Central Americans make up a diverse category.

On socioeconomic measures, compared to other Latino groups, Central Americans most closely resemble Mexicans: their level of education is low, their household income is low, and they occupy mostly unskilled occupational positions. There are, however, noticeable differences among the various Central American groups. Nicaraguans have higher incomes, higher levels of education, and are more apt to work in managerial and professional jobs than Guatemalans, Hondurans, or Salvadorans (Chinchilla and Hamilton, 2007; Logan and Turner, 2013; Stoney and Batalova, 2013).

SOUTH AMERICANS   Those whose origins are South American countries make up almost 6 percent of Hispanic Americans. The largest specific components of this subpopulation are from Colombia (by far, the largest, representing about one-third of the South American category), Ecuador, and Peru. There is also a sizable Brazilian population in the United States, though this group is technically not part of the "Latino" world. It is difficult to estimate the size of these groups because so many have come to the United States as undocumented immigrants, but it is certain that their numbers are substantially larger than census figures would indicate (Margolis, 1998). Like others who preceded them, these immigrants come in pursuit of economic betterment, and many harbor expectations of returning to their origin societies after working for a period in the United States.

A large number of immigrants from South America in recent years have been drawn to the United States as a result of economic uncertainties in their home countries. In the early 2000s, economic conditions created hardships especially for middle-class families in countries like Argentina and Venezuela who were well educated and had skilled occupations but who found themselves increasingly impoverished or even unemployed. Thus the push factors motivating migration to the United States and other countries with strong economies intensified (Forero, 2001; Marrow, 2007; Price, 2006).

In their social and economic characteristics, South Americans in the United States exhibit stronger similarities to Cubans than to other Latino groups. On average they are better educated, are concentrated in higher occupational categories, and earn higher incomes. Still, it is important to keep in mind that the category "South Americans" comprises not only a culturally diverse population but one with noticeable socioeconomic differences among its various groups.

HISPANIC AMERICAN DEMOGRAPHICS

POPULATION TRENDS   Hispanic Americans constitute a rapidly growing ethnic minority. Between 1990 and 2010, the Latino population increased by 130 percent, far greater than the U.S. population in total. This is accounted for by a combination of continued heavy immigration and a comparatively high birthrate.

In addition to their rapid growth rate, Hispanics in the United States constitute a relatively young population. The median age of the Hispanic American population is twenty-seven, compared to forty for the non-Hispanic white population and thirty-seven for the total U.S. population. About one in three Hispanics is under eighteen years of age compared with one in five non-Hispanic whites. At the other end of the age spectrum, whereas about 5 percent of Hispanics are sixty-five or older, about 15 percent of non-Hispanic whites are in this age bracket. These demographic facts indicate that Hispanics are likely to account for almost half of U.S. population growth in the next fifty years (Durand et al., 2006).

GEOGRAPHIC TRENDS   In recent decades, the Hispanic population has been rapidly dispersing widely throughout all regions of the United States. Indeed, few states and regions do not now contain a Latino community of some size. This represents a sharp break with past patterns of settlement, in which Hispanics entered and mostly remained in the Southwest (Brown and Lopez, 2013; Parisi and Lichter, 2007; Parrado and Kandel, 2008; Smith and Furuseth, 2006; Zuñiga and Hernández-León, 2005).

This dispersion has been particularly evident among Mexican Americans. Mexican communities in the industrial cities of the Midwest had been in evidence for several decades (Aponte and Siles, 1994), but the movement of Mexicans into other regions, particularly the Southeast, has now become a substantial flow (Brown and Lopez, 2013; Durand et al., 2006; Frey, 2010; Smith and Furuseth, 2006). Consider that in Atlanta and Memphis, Hispanics (mostly Mexicans), never in the past more than a minor element of the population in those cities, now dominate the construction and landscaping trades (Schmitt, 2001b). Mexicans, along with some Central Americans, have become the main source of labor for meatpacking plants in Omaha, Nebraska, and numerous other smaller midwestern cities. And they make up the bulk of the workforce in poultry

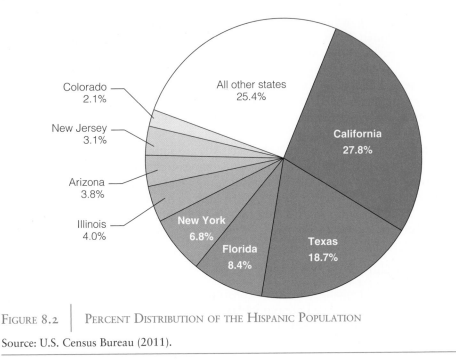

FIGURE 8.2 | Percent Distribution of the Hispanic Population

Source: U.S. Census Bureau (2011).

plants in Delaware, North Carolina, and Arkansas and seafood-processing plants along the East Coast (Durand et al., 2006; Viglucci, 2000). Mexicans have also become a significant part of New York's large Latino population (Bernstein, 2005).

Despite the growth of Latino communities throughout the United States, the vast majority of Hispanic Americans are still concentrated within just a few states. In fact, almost two-thirds of all Hispanics live in just four states—California, Texas, Florida, and New York (Figure 8.2). Like African Americans, Latinos are primarily urban dwellers. Table 8.1 shows the metropolitan areas with more than 1 million Latinos.

Specific Latino groups are similarly concentrated in just a few states and major metropolitan areas. Although Mexican Americans can now be found in virtually every state and region, fully one-half still live in two states alone, California and Texas. Cubans are the most concentrated of the three major Hispanic groups, with about half residing in the metropolitan Miami area. Other sizable Cuban communities are found in New York, northern New Jersey, Tampa, and Los Angeles, but in no way do they resemble the Miami community. Dominicans, like Cubans, are concentrated in a single locale, New York City, where they now constitute the city's largest immigrant group.

Among the Central American groups, the major concentrations of Salvadorans and Guatemalans are in Los Angeles, where about one-fifth of all Central American immigrants reside. The largest cluster of Nicaraguans is in Miami, where they make up one of the city's largest Latino groups. In the past, Miami was primarily a Cuban American city, but in recent years it has become a magnet for all Hispanic groups, including tens of thousands from virtually every country in Latin America and the

TABLE 8.1	METROPOLITAN AREAS WITH MORE THAN ONE MILLION LATINOS (IN MILLIONS)	

	Number	Percent of Total Population
Los Angeles–Long Beach, CA	5.7	45
New York–Northeastern, NJ	4.2	24
Miami–Hialeah, FL	1.6	66
Houston–Brazoria, TX	2.0	36
Chicago, IL	1.9	21
Riverside–San Bernardino, CA	2.0	47
Dallas–Fort Worth, TX	1.7	28
Phoenix, AZ	1.1	30
San Antonio, TX	1.1	55
San Francisco–Oakland–Vallejo, CA	1.1	22
San Diego, CA	1.0	32

Source: Motel and Patten (2012).

Caribbean (Alberts, 2006; Illa, 2010; Logan and Turner, 2013). In addition to Los Angeles and Miami, other large Central American populations are found in New York, Houston, and Washington (Logan and Turner, 2013; Miyares, 2006).

South Americans are the most dispersed among Latinos. Nonetheless, half of all South Americans in the United States are found in just three metropolitan regions: New York–Newark, Miami–Fort Lauderdale, and Los Angeles (Logan and Turner, 2013). In several New York City boroughs, there are distinguishable Colombian, Ecuadorian, Peruvian, and Argentine concentrations. Colombians have been settling in New York since the 1960s and have made a significant imprint on the residential and commercial life of certain neighborhoods. But even this group is composed mostly of those who have come to the United States only in the past two decades (Guarnizo and Espitia, 2007).

LATINOS IN THE CLASS SYSTEM

SOCIOECONOMIC STATUS

As in other societal dimensions, Hispanic Americans seem to occupy an intermediate position—below Euro-Americans, but in several ways above African Americans—in the degree of economic success they have achieved and in the paths to upward mobility they are following. However, keep in mind that "Hispanic American" is an umbrella category encompassing several distinct groups. Thus important differences will be evident among them.

TABLE 8.2 | HISPANIC AMERICAN SOCIOECONOMIC STATUS, 2011

	Median Household Income	Percent of People in Poverty	Percent Unemployed
Non-Hispanic white	$53,444	13.0	5.6
Hispanic	$39,589	25.8	8.4
Mexican	$38,884	27.5	8.2
Puerto Rican	$36,460	27.4	9.8
Cuban	$38,808	19.3	8.0
Dominican	$32,815	28.5	9.9
Central American	$40,307	24.9	8.7
South American	$50,578	14.0	6.7

Source: U.S. Census Bureau (2012a).

INCOME, WEALTH, AND POVERTY   As shown in Table 8.2, despite moderate gains during the past two decades, Hispanic household income remains only about three-quarters of non-Hispanic white income. Though income for all Hispanic groups is substantially lower than for whites, there is a range among the specific groups. Cuban families and those from South America on average earn more than Mexicans, who earn slightly more than Puerto Ricans. Dominicans are the lowest income-earning group among all Hispanics, and a larger percentage is below the poverty line.

Like African Americans, Latinos lag far behind whites in wealth (Figure 8.3). The median net worth of Latino families is about 5 percent of whites', and less than half

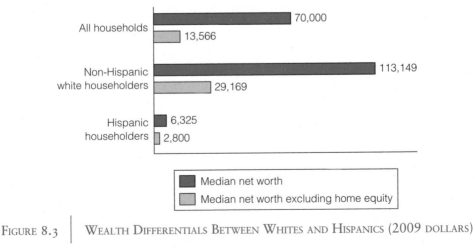

FIGURE 8.3 | WEALTH DIFFERENTIALS BETWEEN WHITES AND HISPANICS (2009 DOLLARS)

Source: Pew Research Center (2011b).

of Latino families own their homes, compared to three-quarters of whites (Pendall et al., 2012). Possessing little wealth, Latino families were among those most severely affected by the deep economic recession of the late 2000s (Kochhar et al., 2009; Mather and Jacobsen, 2010). Between 2007 and 2010, Hispanic family wealth dropped by over 40 percent, compared to 11 percent for whites (McKernan, 2013).

With income and wealth well below average, Latinos—especially Mexicans and Puerto Ricans—are, predictably, disproportionately among America's poor. One-quarter of all Latinos are below the poverty line, a rate almost twice that of non-Hispanic whites (Table 8.2). As with income, however, there is a poverty range within the Hispanic American category: Mexicans, Puerto Ricans, Dominicans, and Central Americans below the poverty line are more numerous than Cubans and South Americans. Indeed, among a significant part of the Latino population, economic conditions are much akin to those of the black underclass (Reimers, 2006).

Occupation The low incomes and high poverty rates of Latinos, particularly Mexicans, Puerto Ricans, and Dominicans, reflect their generally lower occupational levels. Except for Cubans and most South Americans, Latinos are underrepresented in the higher-status occupational categories and overrepresented in the lower ones. For example, whereas almost 40 percent of non-Hispanic whites are in managerial or professional occupations, less than 20 percent of Hispanics are. At the other end of the occupational hierarchy, 16 percent of non-Hispanic whites but 23 percent of Hispanics are engaged in service jobs (U.S. Census Bureau, 2012a). Hispanics also exhibit high unemployment rates compared to non-Hispanic whites (Table 8.2). And, as with African Americans, official statistics understate the extent of this problem.

Again, it is important to keep in mind the social class differences among Latino groups. There are similarities among Mexicans, Puerto Ricans, Dominicans, and Central Americans, but Cubans and, to some extent, South Americans exhibit noticeably higher statuses on various measures of income, occupation, and education. For example, whereas only 16 percent of Mexicans are in managerial and professional occupations, the comparable figure among Cubans is 31 percent, not radically different from the percentage of non-Hispanic whites.

What accounts for the relatively low levels of income and occupation of most Hispanic American groups? Although employment discrimination against Latinos has historically been a factor for particular groups, especially Mexicans in the Southwest and Puerto Ricans in New York, today low levels of education and their relatively recent entry into the labor force seem most critical (Camarillo and Bonilla, 2001). Large proportions of all Latino groups have immigrated to the United States since 1965. As latecomers to the American postindustrial economy and with generally lower educational levels, they have necessarily occupied the least skilled and lowest-paying jobs. Like African Americans, they have become disproportionately part of the most expendable element of the workforce, generally lacking the requirements for positions above the semiskilled or unskilled levels. As an unusually young population, Mexican Americans, Puerto Ricans, Dominicans, and most Central Americans are most likely to lack marketable skills and work experience. They also lag in educational attainment. Hence, the economic gap between those groups and non-Hispanic whites remains wide.

TABLE 8.3	HISPANIC AMERICAN EDUCATIONAL LEVEL, PEOPLE AGE TWENTY-FIVE AND OLDER, 2011			
	Less Than High School (%)	High School Graduate Only (%)	Some College (%)	Bachelor's Degree or Higher (%)
Non-Hispanic white	12.0	28.8	29.3	29.9
Hispanic	36.8	27.1	22.9	13.2
Mexican	42.2	27.0	21.2	9.6
Puerto Rican	24.9	29.8	29.1	16.1
Cuban	23.1	28.9	23.8	24.2
Dominican	33.7	25.9	25.1	15.3
Central American	46.5	24.5	18.8	10.2
South American	15.6	26.0	27.5	30.8

Source: U.S. Census Bureau (2012a).

EDUCATION Perhaps the most serious factor in accounting for the lower occupational status and earnings of Hispanic Americans in general is their lower educational level. As shown in Table 8.3, only 63 percent of Latinos twenty-five years of age and over have completed high school, compared with almost 90 percent of non-Hispanic whites. The dropout rate among Latinos is also significantly higher compared to that of both whites and blacks. Advances in recent years have begun to close this educational gap, however. The proportion of younger Latinos who have completed high school is substantially higher compared to older Latinos, and the proportion attending college has also increased markedly (Cárdenas and Kerby, 2012; Fry and Taylor, 2013; Pew Hispanic Center, 2009).

Although Latinos are narrowing the educational attainment gap with other racial and ethnic groups, they still lag behind in college completion rates and remain the least educated. Moreover, as with African Americans, Hispanic American students are severely underrepresented in strongly competitive colleges and universities. Those high school graduates who go on to higher education do so primarily at less selective four-year schools or two-year community colleges (Carnevale and Strohl, 2013). However, as with income, it is important to consider the differences in educational attainment within the Latino category. As is seen in Table 8.3, for example, the discrepancy between Mexicans and Central Americans, on the one hand, and Cubans and South Americans, on the other, is very great.

Reasons for the lower educational attainment of Latinos, particularly Mexican Americans and Puerto Ricans, have been debated. Some contend that inappropriate cultural values fail to encourage successful school performance; others maintain that schools themselves, Anglo dominated as they are, do not address the unique needs of Latino students. One point about which there is general agreement is that language difficulties intensify the academic problems of these students (Celis, 1992; Moll and Ruiz, 2002; Orfield, 2002). Many Mexican American students in the

Southwest are from families with non-English-speaking backgrounds, making their school experience trying, at best. Language problems create or exacerbate much of the high dropout rate and poor academic performance among Puerto Ricans and Dominicans as well.

In sum, Mexican Americans, Puerto Ricans, and Dominicans are underrepresented at the higher levels of the mainstream economy and overrepresented among the poor. And, because they face the same increasingly technical and high-skilled job market as African Americans with a correspondingly low skill level, they will continue to make up a large segment of the poor in the foreseeable future. In looking at the socioeconomic levels of these groups, it should be noted that these are aggregate figures reflecting the groups' general position in the American stratification system. There will obviously be an entire range of occupational and income positions within each group despite their disproportionate concentration at the lower levels. Moreover, evidence indicates that when adjustments are made for differences in human capital, especially education and English-language proficiency, there is little difference between Latinos and whites in occupational status and earnings (Duncan et al., 2006).

Cubans: A Special Case

Cubans are unlike most other Latino groups in their adaptation to the American economic system. Indeed, few groups in American ethnic history have matched the phenomenal socioeconomic rise of the contemporary Cuban American community in so short a time. On all measures of socioeconomic status, Cubans are clearly well in advance of Mexicans and Puerto Ricans despite their more recent arrival. Their income and educational levels are higher, their unemployment rates are lower, and they are well represented across the occupational spectrum. What accounts for these striking differences?

CLASS BACKGROUND Cuban immigrants to the United States, particularly the first wave, brought with them a class background markedly different from that of most other Latinos. A disproportionate number derived from the middle and upper strata of prerevolutionary Cuban society. In their study of early Cuban refugees, Fagan, Brody, and O'Leary (1968) found that professional and semiprofessional occupations were heavily overrepresented among this group, whereas fishing and agricultural jobs—those predominating in Cuba—were hardly represented at all. The occupational skills Cubans brought with them enabled them to work themselves more easily into the mainstream economy and climb rapidly even though most of them experienced downward mobility when they first entered the United States. Indeed, research has shown that among younger Cubans in the United States, the rate of upward mobility has been similar to that of immigrants from English-speaking countries (Massey, 1981a). In addition to their high occupational status in Cuba, the refugees were far better educated than most Cubans, and most were urban residents. Succeeding immigrant cohorts displayed a somewhat lower class background but still represented a privileged sector of Cuban society (Portes et al., 1977).

Cubans in the United States, then, differ radically from most other Hispanics in their social origins. Unlike Mexicans and Puerto Ricans, early Cuban immigrants were likely to have been occupationally successful and well educated before emigrating. These higher skill and educational levels translated into rapid upward mobility in the United States.

RACE   The more rapid ascent of Cubans in the American ethnic system is explained in some part by the fact that most immigrant Cubans have been white and therefore have not been exposed to the added handicap of racial discrimination in the labor market, with which many Mexicans, Puerto Ricans, and Dominicans have had to deal.

THE MIAMI ETHNIC ENCLAVE   Another advantage Cubans have had over other Hispanic groups has been the creation of an institutionally complete community in Miami, where most Cuban Americans reside. An ethnic enclave economy was established by early Cuban immigrants, composed of numerous small businesses serving the ethnic community. In addition, a small manufacturing sector emerged that subsequently employed other Cuban immigrants who arrived with fewer skills and less financial capital. The presence of an ethnic subeconomy not only created jobs for these immigrants but also enabled them to avoid the secondary labor market of the mainstream economy, where jobs ordinarily are temporary and do not provide for occupational advancement. Many of the Cuban immigrant workers who were employed by Cuban business owners eventually acquired the skills and capital needed to open their own establishments, which created an expanding Cuban subeconomy (Portes and Bach, 1985; Stepick and Stepick, 2002; Wilson and Portes, 1980).

Neither Mexicans nor Puerto Ricans have been as demographically concentrated as Cubans, thus making the development of similar economic arrangements impracticable. Moreover, neither group contained a substantial entrepreneurial class, possessing both business capital and skills, that could re-create in the United States the position it held in its country of origin and thereby serve as the nucleus of an ethnic subeconomy.

U.S. GOVERNMENT POLICY   Cuban economic success in the United States also has been very much a product of the fact that Cubans were received as political migrants—that is, as refugees—rather than as economic migrants. As a result, the U.S. government provided aid to early-arriving Cuban immigrants that helped them in reestablishing their class position (Pedraza-Bailey, 1985; Stepick and Stepick, 2002). Mexicans and other Latinos, by contrast, have traditionally been received as economic migrants and have never been given similar government assistance.

CUBANS AND OTHER LATINOS   Despite the extraordinary achievement of Cubans relative to other Hispanic Americans, their economic success in the United States should not be overstated. Although Cubans are certainly in a very different category from Mexicans, Puerto Ricans, and Dominicans, their median income is still below Anglos', and their poverty and unemployment rates are higher. Moreover, even though they are well represented in the higher-level white-collar occupations, most are still in clerical, semiskilled, or service jobs. Moreover, the economic gap between Cubans and other Hispanics is closing (Betancur et al., 1993; Pérez-Stable and

Uriarte, 1993; U.S. Census Bureau, 2007c). Many newly arriving Latinos from South American countries, for example, bring the same economic resources with them that early-arriving Cubans brought: high educational and occupational skills. And, like Cubans, they do not face a racial disadvantage.

The economic success of Cuban Americans has been lauded so often that an image has arisen of an ethnic group made up entirely of families that have prospered in the United States. But this popular image belies an increasingly evident internal socioeconomic discrepancy. The first waves of Cuban immigrants who came in the 1960s and 1970s have fared very well. More recent waves, however, arrived with fewer economic resources and have not experienced the very favorable reception given the earliest immigrants. The result is an increasingly evident income disparity within the group (Eckstein, 2009; Pérez, 2007).

LATINOS AND POLITICAL POWER

Latino political power in the United States is still in its infancy. However, Hispanic Americans represent a growing and increasingly powerful political force. This was evidenced most clearly in the presidential election of 2012. Most political observers attributed the Obama victory in large measure to the resoundingly strong Latino vote, about three-quarters of which went to the Democratic candidate. Although Latinos have not yet mobilized as effectively as have African Americans in recent years, the rapidly growing Latino population assures that Hispanic American political influence will continue to blossom.

At the national level, Hispanic Americans are increasingly evident on the political scene, including 38 members of Congress and several cabinet appointments. In state and local politics, they have begun to flex their political muscle where they have been traditionally a large, or majority, element of the population. As early as 1981 San Antonio had elected a Mexican American mayor, and in 2005, Antonio Villaraigosa, the son of Mexican immigrants, was elected mayor of Los Angeles. This marked the first time a Mexican American had served in that office in modern times. Cubans have emerged as the dominant political force in Miami with its large Hispanic population, holding most high-ranking local political positions (Pérez, 2007; Stepick et al., 2003). In California and Texas, where the largest part of the Hispanic American population resides, Latinos are increasingly winning elections not only at the local level but to state offices as well. Although the Hispanic electorate in any state does not yet correspond to the Hispanic percentage of the population, its influence is becoming more apparent. By 2011, almost six thousand Latinos held publicly elected offices at various government levels—local, state, and federal (National Association of Latino Elected and Appointed Officials, 2011).

Hispanic voting rates have lagged behind those of whites and blacks. This is mostly a product of the large percentage of immigrants in the Latino population, as well as the fact that the population is younger, which makes for fewer eligible voters. In 2012, Hispanics made up 17 percent of the general population, but comprised less than 9 percent of all voters (Lopez and Gonzalez-Barrera, 2013). However, an expanding population base and continued assimilation will probably stimulate increased political activism. In the 2008 national election, for example, although the Latino voting rate was significantly lower than for non-Hispanic whites and blacks, almost ten million Latinos voted, a 40 percent increase from 2000 (Campos-Flores and Fineman, 2005; U.S. Census Bureau, 2010f). In a number of states, Hispanics already constitute a critical swing vote (DeSipio, 2006). The significance of the Latino vote in the 2012 presidential election, in which a record number of Latinos participated, has already been noted. The Latino vote is likely to double in size within a generation in large measure a result of the fact that the Latino population is younger than any other ethnic group. Thus, those eligible to vote will increase markedly in the next few decades (Taylor et al., 2012).

That Latinos are a growing element of the population and a potential electoral force has not been ignored by the two major political parties at both local and national levels. Presidential candidates, for example, now routinely make appeals to the Latino vote. Moreover, Hispanic Americans give indications of greater political independence, rather than voting strictly along party lines (Johnson, 2004). This has enhanced their appeal to both parties. In 2007, a debate was held in Miami among the Democratic presidential candidates, who addressed mostly issues of concern to Hispanics; what made it unique is that it was conducted in Spanish and televised on the Spanish-language network, Univision. Following the debate, one Latino political figure proclaimed his vision of the Latino political future: "We're just getting started. Pretty soon the rest of the country will start looking like Miami. And just imagine what will happen in 2050, when 6 of the 10 largest US cities start with 'Los' or 'San'? Like us or not, here we come" (Lovato, 2007).

LATINOS IN THE CORPORATE WORLD

As with African Americans, the role of Hispanic Americans in leadership positions in the American economy is limited. Very few of the largest U.S. corporations have Latinos on their boards of directors or in executive positions. Among CEOs of Fortune 500 companies, only six—little more than 1 percent—are Hispanic Americans (DiversityInc, 2013). Nonetheless, there is evidence that an increasing number of Latinos are moving through the corporate elite pipeline (Zweigenhaft and Domhoff, 2006). Also, an increasing number of Hispanic-owned firms are growing in economic significance, though most enterprises owned by Latinos are small, family-operated businesses (U.S. Census Bureau, 2006).

A disproportionate number of highly successful Latino entrepreneurs and corporate executives come from Cuban American backgrounds (Whitefield, 2000; Zweigenhaft and Domhoff, 2006). As noted, the greater economic success of Cuban Americans by comparison with other Latinos is the result primarily of their generally higher class background and the entrepreneurial skills and resources many brought with them from Cuba. In Miami, Cuban Americans are the economically dominant ethnic group, in both large and small enterprises. Cubans are president of the city's largest bank, owner of the largest real estate developer, managing partner of the largest law firm, and publisher of the largest newspaper (Cardwell, 1998; Navarro, 1999).

Other successful Hispanic corporate executives exhibit social backgrounds similar to the Cubans': they usually are products of at least middle-class families and are well educated. Moreover, as Zweigenhaft and Domhoff describe them, some have had an elite education "that gave them the social connections and educational credentials to move quickly into responsible positions in the corporate community"

(2006:148). Few appear to have come from humble origins or did not use family backing or elite educations to rise to their positions. All, furthermore, are racially and culturally compatible with their Euro-American counterparts. What distinguishes them more than any other characteristic, however, is their international orientation. Having grown up in various Latin American countries, they have a worldly background that reflects the kind of globalism within which contemporary corporate executives must function, and they are usually fluent in a language other than English.

PREJUDICE AND DISCRIMINATION

Because of their in-between minority status, Latinos have not been subjected to the dogged prejudice and discrimination aimed at African Americans, but neither have they been dealt with as European immigrant groups were. John R. Howard refers to such intermediate groups as "partial minorities" and calls attention to their paradoxical position: "By being somewhat below the threshold of public consciousness with regard to 'minority problems' they escape some of the pervasive indignities visited upon blacks, but, precisely because of this it becomes difficult to arouse public indignation at the depredations they do suffer" (1970:8–9).

More precisely, an entire range of attitudes and actions has characterized the relations between Anglos and the several Hispanic groups. Variations have turned on each group's historical context, its geographical location, and its racial makeup. Those close to the dominant Anglo group, like most Cubans, have experienced little prejudice and discrimination; at the other extreme, dark-skinned Puerto Ricans and Dominicans have often encountered ethnic antagonism equal in form and degree to that suffered by African Americans.

HISPANIC STEREOTYPES

For as long as they have been part of the American ethnic mosaic, Mexicans and Puerto Ricans have labored under strongly negative stereotypes. These have arisen in response to the manner in which both groups became part of the society (more so in the case of Mexicans than Puerto Ricans) and as manifestations of American racist ideology.

MEXICAN AMERICANS Feelings of racial superiority on the part of Anglos and policies reflecting those beliefs have been obvious since the American conquest of the Southwest in the nineteenth century. More than anything, the historical circumstances that originally brought Mexican and Anglo into confrontation fashioned a set of group images that have remained stubbornly persistent.

As noted in Chapter 3, negative group images are ordinarily formed as a result of some form of interethnic competition, usually economic. In the Southwest, land and labor were the critical components in the development of mutual antagonism between Anglos and Mexicans. "To the early American settlers," writes Carey McWilliams, "the Mexicans were lazy, shiftless, jealous, cowardly, bigoted, superstitious, backward, and immoral" (1968:99). These images subsequently served to rationalize the dispossession of Mexicans' lands and the exploitation of their labor.

The negative stereotyping of Mexicans was abetted by the fact that they were a conquered people. To the Anglos this only confirmed even more positively their inferior nature. Correspondingly negative stereotypes of Anglos naturally arose in this situation. To the Mexicans, Anglos were "arrogant, overbearing, aggressive, conniving, rude, unreliable, and dishonest" (McWilliams, 1968:99).

Also critical in the development of negative stereotypes of Mexicans were the racial attitudes Anglo settlers brought with them to Texas. From the beginning, Anglos viewed Mexicans as innately inferior on the basis of their racial characteristics (Moore and Pachon, 1976). Mexicans were different in language, culture, and religion, but they were also physically distinct from the Anglos. As a mestizo people, they were not blacks, but neither were they whites. This in-between racial status led to the classification of Mexicans as a separate racial group, one perhaps a notch above blacks but clearly inferior to the Anglos.

The poverty and illiteracy of those Mexicans with whom Anglo settlers had the most dealings further strengthened these negative views. McWilliams notes that most of the Anglos who settled in the Southwest borderlands in the mid-nineteenth century were middle-class people who failed to find a Mexican middle-class counterpart. "If a larger middle-class element had existed," he explains, "the adjustment between the two cultures might have been facilitated and the amount of intermarriage might have been greater" (1968:75). Although upper-class Hispanos were initially seen in a different light, they too were eventually absorbed into the adverse image.

The negative stereotypes of Mexicans that developed during and after the period of conquest were maintained and even strengthened not only in the Southwest, but also in the society generally. Mexicans were included among the "undesirable" groups denoted by the Dillingham Commission in the early 1900s, along with southern and eastern Europeans, falling somewhere between blacks, on one hand, and "white" Americans, on the other. It is instructive to note that despite their undesirable status, Mexicans were so vital as a cheap labor supply in the Southwest that they were exempted from the restrictive immigration laws drafted at the time, largely on the basis of the commission's report.

The stereotypes developed in the nineteenth century continued to influence the views of Anglos in the Southwest and elsewhere throughout most of the twentieth century. In his study of a south Texas community in the early 1970s, Simmons (1971) found that Anglos believed Mexicans to be unclean, prone to drunkenness and criminality, deceitful in dealings with Anglos, hostile, and unpredictable. Among favorable images of Mexicans were their "romantic" rather than "realistic" attitude and their fun-loving nature. But even the positive traits implied a childlike and irresponsible character, making the Mexicans appear deserving of their subordinate status.

Television and movies played a large role in perpetuating negative stereotypes of Mexican Americans, much as they did for African Americans and American Indians. In Hollywood films, Mexicans traditionally were portrayed as villains in confrontation with Anglos; even when they were allied with the Anglo heroes, their roles were ordinarily subservient and often ludicrous (Berg, 2002; Petit, 1980; Wilson and Gutiérrez, 1995). Television commercials at times also perpetrated the villainous image of Mexicans (Martinez, 1972; Wagner and Haug, 1971). For example, in one of their ads, the Frito-Lay Company featured a cartoon character, the "Frito Bandito," who cunningly stole Fritos corn chips from the kitchens of unsuspecting housewives, prompting the suggestion that shoppers buy two bags. As TV producers and mass-media advertisers have become more attuned to a growing Latino market, such stereotypical images are no longer common. Ten years after it was forced to remove its Frito Bandito ad, for example, the Frito-Lay Company advertised Tostitos, another corn chip product, this time with a character who spoke with a lilting Spanish accent and presented a romantic image of a distinguished, cultured Latino (Wilson and Gutiérrez, 1995).

PUERTO RICANS   Although they have not experienced a history of confrontation and conflict with Anglos, as Mexicans have in the Southwest, Puerto Ricans have nonetheless commonly been stereotyped negatively. This is mainly a product of their racial distinction and their generally lower-class position. Because much of the Puerto Rican population in the United States is dark-skinned, they often experience the same forms of prejudice as non–Puerto Rican blacks.

The dilemma of Puerto Ricans regarding their racial status in the United States is expressed dramatically by Piri Thomas, a dark-skinned Puerto Rican who has written of his New York experience. Thomas tells of an early incident that thrust on him the American race ethic. After applying for a door-to-door sales job, he is told by the employment manager that he will be called when "the new territory is opened up." His white friend applies immediately afterward and is told to start "on Monday." Realizing what has happened, Thomas relates the story to his black buddy, who tells him that "a Negro [sic] faces that all the time." "I know that," Thomas replies, "but I wasn't a Negro then. I was still only a Puerto Rican" (1967:104).

CUBANS   Concentrated in a single city and having arrived more recently, Cubans have not had the kind of national exposure that Mexican Americans or even Puerto Ricans have had. Moreover, as a group that is for the most part physically indistinguishable from the Euro-American population, Cubans automatically enjoy a status that precludes racially based stereotypes. Their generally higher class position has also provided some insulation from negative images.

Although the attitudes of Anglos toward Cubans in Miami were initially benign, subsequent events created an increasing bifurcation between the two groups. One area that has fueled Anglo resentment is language. Increasingly, Miami has become a bilingual city, and this has met with Anglo resistance. In 1980, a referendum was held to bar Dade County (Miami) from using public funds to conduct official business in any language but English and from promoting any but "American" culture. The measure, which narrowly passed, polarized the community along ethnic lines (Boswell and Jones, 2006; Burkholz, 1980; Castro, 1992). The county commission in 1993 voted unanimously to repeal the 1980 antibilingualism ordinance.

The referendum on bilingualism disclosed the depth of the anti-Cuban sentiment among Anglos. A survey conducted immediately following the vote revealed that nearly half of those who had voted for the ordinance did so as an opportunity to express their protest, not because they saw it as a good idea. Furthermore, more than half the Anglos who supported the measure said they would be pleased if it "would make Miami a less attractive place for Cubans and other Spanish-speaking people" (Tasker, 1980). In recent years, Miami has become a predominantly Hispanic city, but language continues to draw the strongest divide between Hispanics and Anglos. As one Anglo expressed it, "Sometimes, I feel like I'm a visitor in my own country" (Morgan, 1998a).

An additional source of animosity has been Cubans' rabid and sometimes virulent politics, specifically as they relate to Cuba. Nothing better illustrated this than the Elián González affair in 2000. It was perhaps the most polarizing single event separating the Cuban and Anglo populations in Miami, and even more broadly, the United States as a whole. Following his rescue at sea after his mother, along with several others, had drowned in an attempt to sail to Miami from Cuba, six-year-old Elián was temporarily placed in the custody of his great-uncle's family. The family subsequently refused to release the child for return to his father's custody in Cuba, which set off a five-month standoff with the U.S. Justice Department. The incident created a national drama—most Cuban Americans felt the child should not be returned to his father in Cuba, but most Anglos in Miami, as well as in the rest of the country, felt otherwise (Viglucci and Marrero, 2000). Many held that Elián was being used by the Miami Cuban community as a political pawn in their longstanding ideological war with Fidel Castro. Elián was eventually reunited with his father, but the incident left Miami bitterly divided.

DISCRIMINATORY ACTIONS

The in-between status of Hispanic Americans—between Euro-American ethnic groups, on the one hand, and African Americans, on the other—becomes more evident in looking at patterns of discrimination. Hispanic Americans have lacked the heritage of a legal system that, in the case of African Americans, made discrimination legitimate. As a result, such actions have ordinarily been extralegal rather than formally endorsed by the state. Moreover, although discrimination against Latinos has been widespread and often intense, it has never been as profound and intransigent as that experienced by African Americans.

This is illustrated by Pinkney's study (1963) of one community with sizable Mexican American and black populations. General patterns of discrimination for both groups, he found, were similar but were much stronger in degree for blacks. For example, greater acceptance of Mexican Americans was evident in public accommodations such as hotels and restaurants as well as in residential areas and social clubs. Anglos also expressed greater approval for granting equal rights in employment and residence to Mexican Americans. Pinkney attributed these differences in part to differences in skin color: Mexican Americans, closer to the dominant Anglo group, were more acceptable. A more recent study revealed much the same pattern. In a series of experiments, John Dovidio and his colleagues found that in various social environments, Latinos do experience differential treatment, though not necessarily the most blatant forms that have traditionally characterized anti-black discrimination. Also the social distance of Latinos from Anglos was found to be markedly lower than for blacks from whites (Dovidio et al., 2010).

That it has not been as rigid, intense, and formalized as that aimed at African Americans should not disguise the severity of discrimination to which Latinos in some circumstances have been exposed, particularly Mexican Americans in the Southwest. Many of the actions against them in areas such as South Texas have been similar in degree and scope to those experienced by blacks, including segregated restaurants, churches, schools, and other public facilities (Burma, 1954; Grebler et al., 1970). Roberto Suro describes this social environment:

> In South Texas in the 1950s, discrimination against Latinos was different because it was not based on the single immutable characteristic of skin color and because it was not primarily a matter of law. Culture and convenience dictated that poor, dark-skinned Mexicans should not mix with whites, that they should go to inferior schools, and that they would always work at the bottom of the labor force. And while segregation applied to all blacks, the discriminatory culture of South Texas distinguished among Latinos, treating some like Mexicans and others not. (Suro, 1998:81)

Physical violence against Mexicans in the Southwest also has a long history. McWilliams notes that in the mining camps of the nineteenth century, any crime committed was immediately attributed to Mexicans, for whom lynching was the accepted penalty. "Throughout the 1860's," he writes, "the lynching of Mexicans was such a common occurrence in Los Angeles that the newspapers scarcely bothered to report the details" (1968:130).

POLICE RELATIONS An often turbulent relationship has traditionally existed between Latinos—Mexican Americans in particular—and various agencies of the criminal justice system (Moore and Pachon, 1976; Morales, 1972; U.S. Commission on Civil Rights, 1970). Although the level of discrimination has historically been high in Texas, several of the most serious large-scale incidents of the modern era have taken place in Los Angeles. These include the notorious pachuco, or zoot suit, riots of 1943 and the East Los Angeles riot of 1970. In both cases, the police played major roles in either precipitating or sustaining the hostilities.

The events surrounding the zoot suit riots have been well documented (Garcia and de la Garza, 1977; McWilliams, 1968). During the 1940s, the Los Angeles press waged a virulent anti-Mexican campaign, creating a rancorous mood among Anglos. Sensational accounts were presented of Mexican American youth gangs whose members took to wearing bizarre outfits called zoot suits. Following an incident involving Mexican American youths and sailors stationed in Los Angeles, two hundred servicemen began a four-day rampage, in which they drove through the Mexican area in taxis, randomly attacking those on the streets. The Los Angeles police did virtually nothing to halt the violence and in fact arrested many of the victims. Serious rioting continued for almost a week, ending finally with the intervention of the military police. This incident permanently damaged Mexican-Anglo relations in Los Angeles for decades afterward, and it confirmed the ever-precarious status of the Mexican American community in an Anglo-dominated environment. Although other such incidents have rarely garnered as much public attention, police brutality (or indifference, as the case may be) in dealing with Mexican Americans has been historically commonplace.

The lack of trust in the police and the courts among Hispanic Americans is not something of the distant past. In fact, a national survey indicated that Latinos have a view of the criminal justice system not unlike that of African Americans. Fewer than half of Latinos express confidence that police officers will treat them fairly or will not use excessive force on suspects, and fewer than half believe that the courts will deal with them fairly (Lopez and Livingston, 2009).

In recent years, tension between Latinos and law enforcement agents has centered mostly on efforts by state and local government to identify those who are undocumented. In 2010, about 5 percent of Latinos (8 percent of men, but only 2 percent of women) said that they had been stopped by police or other authorities and asked about their legal status in the United States (Lopez et al., 2010). Although this represented a decline from two years earlier, Latinos in general, Mexicans in particular, remain concerned about the use of these measures and their impact on friends and family (Lopez et al., 2010).

RESIDENTIAL SEGREGATION   For Latinos, discrimination in housing has been generally less severe than for African Americans (Alba, 2009; Camarillo and Bonilla, 2001; Grebler et al., 1970; Iceland, 2009; Massey, 1981a, 1981b, 2001; U.S. Commission on Civil Rights, 1976). Studies have shown that Hispanic-Anglo segregation is lower than black-white (Anglo) segregation (Farley, 2011; Iceland, 2009; Iceland and Wilkes, 2006; Logan et al., 2004; Logan and Zhang, 2009; Massey, 2001; Mumford Center, 2001a).

Although residential segregation of Latinos may not be as severe as for African Americans, a recent analysis has shown little change in the index of dissimilarity for Latinos over the past two decades, leading the researchers to conclude that Latino segregation in general is persistent "at a moderate to high level" (Logan and Turner, 2013). Closer analysis, however, reveals that while overall Hispanic segregation is unchanging, there are clear differences among specific groups. In 1990, segregation of Puerto Ricans, Cubans, and Dominicans from whites was quite strong, stronger than for Mexicans. But in the two decades after 1990, there were large declines for these three groups while Mexican segregation remained very evident.

CURRENT ISSUES AND TRENDS   Today, anti-Latino discrimination seems focused mostly on large immigrant populations in specific locations, particularly on those who have entered the United States without papers. At times, however, this has had a spillover effect on all Hispanic Americans. As noted earlier, this has been especially evident in recent years as harsh measures intended to identify unauthorized immigrants have been enacted by various state and local governments. Many worry about the effects of these measures on their employment and housing opportunities and also about deportation of family members.

In sum, discrimination against Latinos has been ongoing throughout American history, but it has differed in scope and intensity for different Latino groups and differed as well by region. In any case, its severity has not been the same as that experienced by African Americans, today or in the past. Nonetheless, Latinos in general are conscious of ethnic discrimination aimed at them and express growing concerns that this is a continuing problem in American society. Latino discrimination may also be fueled by growing anti-immigrant attitudes as well as the awareness on the part of non-Hispanics of the increasing demographic power of Latinos. A national survey of Latinos in 2010 revealed that about one-third said that they or a family member had experienced discrimination based on race or ethnicity in the past five years. Almost two-thirds felt that discrimination was a major problem that impeded their success in America, an increase from 54 percent in 2007 (Lopez et al., 2010). It is of note that 70 percent of foreign-born Latinos were aware of discrimination and felt that it was a major problem, compared to about half of native-born Latinos.

Moreover, Americans generally have come to see Hispanics as the racial/ethnic group most often subjected to discrimination. Almost one-quarter of Americans say that Hispanics are discriminated against "a lot," a share higher than observed for any other group (Pew Research Center, 2010).

HISPANIC AMERICAN ASSIMILATION

In Chapters 10 and 11, we will see that Italian Americans, Jewish Americans, and other ethnic groups of European origin are best analyzed within the framework of the assimilation model of ethnic relations, wherein ethnicity becomes less intense with each succeeding generation. African Americans, by comparison, were seen in Chapter 7 as a case that more closely fits the inequalitarian pluralistic model; rather than decreasing in significance, ethnicity continues to shape intergroup relations and to affect in a basic manner most life activities of group members. Where do Hispanic Americans fit on the continuum of ethnic relations, and which model is best suited to their experience in American society? Similar to African Americans, there is some disagreement over whether they are basically like previous immigrant groups who have gone through stages of assimilation or whether their American experience evokes some form of pluralism.

As with other aspects of their American experience, it is important to keep in mind the diversity of the Hispanic population because patterns of assimilation and pluralism have not been entirely the same for the various groups and subpopulations within them. There is evidence to support the view that Latinos in general have experienced a relatively low but increasing level of both cultural and structural assimilation. More specifically, however, the various Hispanic American groups exhibit a wide range of assimilation, extending from Hispanos in New Mexico—who have strongly assimilated both culturally and structurally into the dominant group—to black Puerto Ricans and Dominicans, who have displayed a low level of structural assimilation.

CULTURAL ASSIMILATION

Unlike African Americans, Hispanic Americans were not stripped of their native cultures on entering the society and have retained many ethnic traits, especially the Spanish language. Nonetheless, Latinos are moving toward cultural assimilation, though not necessarily in the same manner as Euro-American groups of the past.

LANGUAGE Latino cultural assimilation is strongly evident in language usage. Although Spanish is the most widely used non-English language in the United States, only a minority of Hispanic Americans speak only Spanish, and most of these are first-generation immigrants. As Latinos move to the second and third generations, English becomes their primary language (Alba, 2004; Pew Hispanic Center, 2002, 2009; Portes and Rumbaut, 2006; Portes and Schauffler, 1996; Rumbaut et al., 2006). Moreover, even among those who continue to use Spanish in the home, the majority use English in other social contexts, like work and friendship groups. De la

Garza and colleagues (1992) found that not only did most U.S. citizens from each of the three major Latino groups use English or a mixture of English and Spanish at home but they preferred English-language media to either bilingual or Spanish-language media. Furthermore, over 90 percent of Latinos supported the proposition that citizens and residents of the United States should learn English. Later studies have confirmed even more strongly the view among an overwhelming majority of Latinos of the importance of learning and speaking English (Pew Hispanic Center, 2006c). Clearly, then, contrary to a common public perception about the maintenance of Spanish, most Latinos prefer English and support its usage.

A national survey of Latinos indicated that language is strongly correlated to acceptance of mainstream American values. That is, on key social issues, the views of Latinos whose primary language is English (mostly of the third generation) are more in line with those of the society as a whole than of Latinos who either speak mainly Spanish or are bilingual (Pew Hispanic Center, 2004b). For example, 47 percent of Spanish-dominant Latinos find divorce acceptable, whereas 67 percent of English-dominant Latinos do, close to the same percentage (72 percent) as non-Latinos. This pattern is evident on other social issues such as abortion, homosexuality, and having children outside marriage.

PROXIMITY TO ORIGIN SOCIETY   One factor contributing to slower cultural assimilation among Latinos is the proximity of their society of origin. This is especially evident for Mexican Americans. More than four decades ago, McWilliams observed that "[i]n migrating to the borderlands, Mexicans have not found immigrant colonies so much as they have 'moved in with their relatives'" (1968:38). Today that observation seems even more compelling. In a sense, the Mexican immigrant in the Southwest never really leaves Mexico. The frequency and ease of travel between the United States and Latin America have created a similar phenomenon for other Latino groups (Garcia-Passalacqua, 1994). Immigrant networks have made back-and-forth movement so commonplace that family units are best characterized as transnational, residing simultaneously in two societies.

Cubans present something of a unique case in this regard. Because early immigrants came not expecting to stay and take on the life of the new society, they clung to the old culture more strongly than most other ethnic groups. This pattern is not uncommon in the case of political exiles, whose commitment to their adopted country is never complete. Resistance to cultural assimilation is particularly evident among older, first-generation Cuban Americans, but this is changing with the second and third generations, for whom there are no thoughts of return (Goodnough and Gonzalez, 2006). In their analysis of Cuban Miami, Stepick and Stepick concluded that "[i]n many respects, the children of Cuban immigrants have been assimilated in characteristic American fashion. They have abandoned the self-identity label of American, but neither do they identify as Cuban. Instead, they strongly prefer the hyphenated label Cuban-American" (2002:84).

RESIDENTIAL CONCENTRATION   Another important factor that has contributed to retention of the ethnic culture among some Latinos is their residential concentration. This is especially apparent for Cubans. Unlike Mexicans and Puerto Ricans, Cubans in the United States, especially those in Miami, have established a well-knit institutional structure that caters exclusively to the unique social and cultural needs of the group. Raymond Breton (1964) has suggested the idea of "institutional completeness" to describe the extent to which the needs of an ethnic group are met by institutions within the ethnic community—businesses, churches, newspapers, schools, and so on. Ethnic communities may range from those that are institutionally complete, in which members need to make no use of the host society's institutions, to those that are institutionally incomplete, in which the network of interpersonal relations is almost totally within the context of the host society. Cubans in Miami appear to be at the extreme of institutional completeness. There is not even a need to learn English, because Miami is very much a bilingual city. Television stations broadcast and newspapers publish in Spanish; bilingual public schools are common; and food stores, restaurants, and businesses of all kinds cater to the unique ethnic needs of the Cuban and other large Latino communities.

To a lesser degree this characterizes other Latino communities. New York City's huge Dominican population, for example, is heavily concentrated in one neighborhood, Washington Heights, within which almost all its cultural needs are met. More than other Latino groups in the city, Dominicans are residentially segregated and have the least exposure to non-Hispanic whites (Levitt, 2007).

Hispanic residential segregation may be related to high levels of immigration. That is, immigrants tend to cluster in enclaves among coethnics and, as the immigrant generation gives way to their children, increased residential dispersion usually occurs. After examining ethnic residential patterns, John Iceland (2009) concluded that, since Latino immigration levels are high and many of the immigrants are of low socioeconomic status, increasing levels of Latino residential segregation might be expected in the short and medium term. Over the long term, however, as succeeding generations experience upward mobility, segregation is likely to decline.

RACE AND ASSIMILATION For most Puerto Ricans and an element of the Dominican and Cuban populations, a racial factor complicates the assimilation process. For those darker in skin color, the incentive to retain the Spanish language and culture is a product of their perceived need to separate themselves from the African American population. It is more advantageous, they feel, to be identified with the Latino populace than to be lumped with non-Hispanic blacks. For those who are neither clearly white nor black by American racial standards, the problem is even more acute, for they often find themselves marginal to both categories. For example, among the Cuban American population, a relatively substantial portion who came as part of the 1980 refugee wave are, in the American sense, racially defined as black, as are many who were later arrivals. But they find themselves in a marginal position, peripheral to the larger Cuban community and unable to assimilate into the African American community (Hay, 2009; Nordheimer, 1987; Ojito, 2000).

FUTURE PATTERNS Despite the problems of language and proximity to their origin societies, Latinos do appear to be moving along a trajectory of cultural assimilation. The Latino National Political Survey revealed much evidence of this movement (de la Garza et al., 1992). Regarding social networks and social distance, for example, although members of each specific national-origin group felt closer to and interacted more frequently with coethnics (that is, Mexicans with Mexicans, Cubans with Cubans, and so on), most felt closer to Anglos than to any of the other Hispanic national-origin groups.

The survey found that most Latinos do not identify themselves as members of a panethnic group—Hispanic or Latino—but as members of their specific national group or simply as Americans. Although members of the various groups expressed common views on some issues and attitudes, on many they also expressed marked differences. Hence, "to the extent that the Hispanic political community exists, there is scant evidence that it is rooted in alleged distinctive cultural traditions such as Spanish-language maintenance, religiosity, or shared identity" (de la Garza et al., 1992:13–14). In sum, the survey found that Latinos were well within the mainstream of American political thought and action. "There is no evidence here," the study concluded, "of values, demands, or behaviors that threaten the nation's cultural or political identity" (1992:16). Similar findings were reported in a national survey of Latinos in 2002 (Pew Hispanic Center, 2002).

A more recent study revealed a relatively steady movement along an assimilation trajectory even for current Latino immigrants (Waldinger, 2007). The vast majority were shown to maintain only a moderate level of engagement with the societies from which they come. Although there are variations by country of origin, most see their future in the United States; only among the most recent arrivals are continued close ties to their home countries evident. In general, the longer the immigrants reside in the United States, the more strongly they identify with it.

Moreover, as with African Americans, cultural assimilation is not entirely a one-way process. Rather, there is a growing interchange of Latino and American core cultures. Latino cultural influences are increasingly evident in American music, language, fashion, and cuisine: Salsa and Latin jazz are now standard musical genres; pop artists like Jennifer Lopez, Ricky Martin, and Marc Anthony appeal to a broad range of Americans; Spanish is today the most popular foreign language, taught nationwide in primary and secondary schools and in every U.S. college; and Americans prefer salsas and picantes as often as they do that most American condiment, ketchup (Fabricant, 1993; Garcia, 1999; La Ferla, 2001). In a sense, what we are witnessing is continued assimilation alongside ethnic cultural affirmation—a phenomenon similar to that experienced by earlier generations of immigrants.

STRUCTURAL ASSIMILATION

Despite shortcomings in income, occupation, education, and political power, each succeeding Hispanic American generation has improved its status over that of the previous one, and it appears that Latinos are moving into greater participation in the society's mainstream economic and political systems. The pace and extent of secondary structural assimilation, however, have been uneven for the three major groups. Puerto Ricans have proceeded more slowly and less effectively than Mexicans, who in turn have not progressed as rapidly as Cubans.

SECONDARY STRUCTURAL ASSIMILATION Because Latinos entered the United States on a large scale only since the 1970s, they have not been able to use the economic and political systems as leverage in their pursuit of upward mobility and political power as effectively as earlier immigrants had done. By the time Latinos had begun to enter American society in significant numbers, the unskilled factory jobs that prior groups had used as economic springboards were already dwindling, and the local political machines, which had doled out many benefits to immigrants, no longer had much impact. The economic and political conditions faced by these groups, then, are closely parallel to those encountered by African Americans. Integration into mainstream economic, political, and social institutions, therefore, has been impeded simply by these structural circumstances.

The factor of race in the case of Puerto Ricans, Dominicans, and, less so, of Mexican Americans may also account for some of the lag in secondary structural assimilation (Rodríguez, 1996). As noted earlier, prejudice and discrimination, though not as well rooted and intense as those experienced by African Americans, are continuing facts for Latinos, particularly in areas where they constitute a numerically significant part of the population. One possible scenario for the Latino population is for those who are racially defined in the American context as "black" to move toward the black side of the U.S. "color line," and those lighter-skinned to move toward the white side, in what was described in Chapter 4 as segmented assimilation (Bonilla-Silva, 2004; Frank et al., 2010; Suárez-Orozco and Páez, 2002).

Cubans in the United States have clearly exceeded other Latino groups in both their pace and degree of economic success. However, their upward economic mobility in the past four decades has been not so much within the mainstream economic system as within a parallel Cuban economy, particularly in the Miami area. Nonetheless, given their class advantage—and, for most, their racial indistinctness—assimilation at the secondary level should be a less painstaking and lengthy process than it has been for either Mexican Americans or Puerto Ricans.

Primary Structural Assimilation   At the primary level of structural assimilation, Latinos have clearly advanced further than African Americans. We have already seen that on one important measure, residential integration, Latinos have begun to show strong signs of dispersion. Over half now live in suburbs, for example (Frey, 2010).

As explained, the rate at which members of an ethnic minority group marry those of the dominant group is perhaps the best evidence of how far the assimilation process has advanced. Although marriage across ethnic lines has occurred increasingly among Euro-American ethnic groups, it has not occurred significantly among African Americans. As in other aspects of comparison, Latinos seem to fall somewhere between these two extremes.

Past studies indicated a trend toward increasing marital assimilation for Hispanics (Fitzpatrick, 1987; Grebler et al., 1970; Jaffe et al., 1980; Mittelbach and Moore, 1968; Murguía and Frisbie, 1977; U.S. Commission on Civil Rights, 1976; Valdez, 1983). Later data indicate a continuation of that trend. In fact, evidence suggests that Latinos are following the same generational patterns of intermarriage as European-descent groups did in the first part of the twentieth century (Lee and Bean, 2010; Qian and Lichter, 2007; Stevens et al., 2006; Wang, 2012). In 2010, more than one-quarter of Hispanic newlyweds married someone whose race or ethnicity was different from their own (Wang, 2012).

ASSIMILATION: LOOKING AHEAD

In accounting for the patterns of adjustment of Hispanic Americans and their relations with Anglos, again we return to the theme of intermediacy, or in-betweenness. Unlike Euro-Americans, who obviously fit the traditional assimilation model, or African Americans, who in large measure do not, Latino groups fall somewhere between these two extremes.

At present, only for Cubans and some from South American countries among the Hispanic American groups does the traditional ethnic, or assimilation, model seem largely relevant (Arias, 2001). For Mexicans, Puerto Ricans, Dominicans, and most Central American groups, the nature of their entry into the society, their racial characteristics, and their class handicap make the assimilation model less clear-cut and impose at least some aspects of inequalitarian pluralism (Frank et al., 2010; Massey, 2008).

Some analysts maintain that Latinos, for the most part, may be following the general path of assimilation followed earlier by Euro-American groups, though their slower and more tortuous movement in that direction is occurring with little notice. Linda Chavez (1991), for example, argues that their ongoing assimilation is concealed by the continued entrance of new immigrants in great numbers who remain more attuned to the ethnic culture and language. She also contends that Hispanic American leaders, strongly committed to multiculturalism, are responsible for generating heightened ethnic consciousness in Latino communities in their attempts to retard assimilation.

In considering the future of Hispanic Americans, it is well to reiterate a point emphasized at the outset of this chapter: There are extremely diverse class, ethnic, and racial origins among those who make up this panethnic category, as well as great differences among its generations. As with Euro-Americans and, as we will see in the following chapter on Asian Americans, it is misleading to describe the broad composite of groups that make up this category in holistic terms. To speak of "Hispanic Americans" (or even "Latinos") as if this were a term with intrinsic meaning rather than a convenient Census Bureau category is overly simplistic and inaccurate. Hence, the kind of adaptation Hispanic Americans make in their absorption into American society—as well as the imprint they make on the society and its culture—will undoubtedly be disparate, though uniformly strong.

SUMMARY

- As part of the American ethnic system, Hispanic Americans, or Latinos, occupy an intermediate status between Euro-Americans and African Americans.
- Three major groups make up the Hispanic American minority—Mexicans, Puerto Ricans, and Cubans—but groups from other Latin American and Caribbean societies have, in the past three decades, become a large and significant component of this panethnic category. Each group is concentrated regionally, and each is primarily urban.
- Mexicans, the largest Latino group, have entered American society mostly as voluntary immigrants, but their roots in the United States reach back to Spanish colonial times before the American conquest of the Southwest. Puerto Ricans

are technically American citizens, yet they have arrived from a foreign culture, much as other immigrant groups have. Cubans are more recent arrivals and differ from Mexicans and Puerto Ricans in class origins and racial features. Newer groups from Central and South America and the Caribbean have entered as voluntary immigrants, though some, particularly from Central American countries, arrived as political refugees.

- Latinos generally are below average on all measures of socioeconomic status and are a large element of the American poor. There are differences among the major groups, however, with Cubans and South Americans in a relatively higher position than Mexicans, who are somewhat higher than Puerto Ricans, Dominicans, and most Central Americans.

- Latinos are in a less developed stage of political participation than are African Americans and, like the latter, are underrepresented in top positions in all institutional areas. Their potential power, however, is considerable.

- The level of prejudice and discrimination directed at Hispanic Americans has not been as intense as that experienced by African Americans but has been more severe than that suffered by Euro-American ethnic groups. Mexican Americans have commonly been the chief targets of ethnic antagonism in the Southwest, where traditionally most have lived.

- Compared with both African American and Euro-American ethnic groups, Latinos have displayed a slower pace of cultural assimilation, owing largely to their relatively recent entrance into American society. However, movement in that direction is clear, evidenced especially by the regularity of language assimilation.

- The speed and extent of secondary structural assimilation for Latinos have generally resembled African American patterns; but at the primary level, Latinos have outpaced African Americans, demonstrated by increasing residential and marital assimilation.

CRITICAL THINKING

1. The category "Hispanic American" or "Latino" is, in fact, internally diverse, and most members of the specific groups prefer to be identified as members of those groups. Is it possible that a common, panethnic, identity might eventually emerge among Latinos of diverse origins if schools, media, government, and businesses continue to refer to them as part of a collective category?

2. As this chapter noted, newspaper and television news stories commonly refer to a crime suspect's racial identity. Among the descriptive terms used is *Hispanic*. How would you explain that descriptions like this are based on racial and ethnic stereotypes that not only provide little in the way of accuracy, but may actually be misleading and thus counterproductive?

3. Spanish is used by large numbers of people in many areas of the United States and is now the country's second most spoken language. Is it possible that the United States will become a de facto bilingual country in the future? Or does the pattern of language assimilation among Latinos make this development unlikely?

4. Think of ways in which Latino culture has been infused into the mainstream American culture. How does this illustrate the bidirectional nature of ethnic assimilation?

PERSONAL/PRACTICAL APPLICATION

1. How do you think you would respond if a police officer stopped you for no apparent reason and asked for your proof of citizenship? Would you be offended, or would you feel that the officer was performing a rightful duty?

2. As explained in this chapter, Hispanic Americans exhibit a wide range of physical characteristics, making "racial" identification virtually meaningless. Nonetheless, Hispanic Americans are very commonly portrayed in racial terms by the media, by law enforcement agencies, by government officials, by university administrations, and other societal institutions. How might this be accounted for? Does this demonstrate how races are socially created?

Asian Americans |

CHAPTER 9

Asians have been in the United States since the early part of the nineteenth century. Only in recent decades, however, have they emerged as a sizable element of the American ethnic configuration. Like other ethnic populations, Asians have displayed variant patterns of settlement and adaptation and have been received by the dominant group in different ways, depending on social, political, and economic circumstances. As it has been with other groups, our objective is to trace the historical development of Asians in the United States and examine their place in the stratification system, the forms of prejudice and discrimination they have encountered, and their patterns of assimilation into the larger society.

At the outset, it is important to stress that Asian Americans constitute an extremely diverse panethnic category. In some ways it is even more disparate than the Hispanic American category, most of whose groups share a common language and religious tradition. To lump together those from numerous Asian societies as "Asian Americans" is therefore a serious oversimplification. In fact, Asian Americans consist of at least a dozen distinct groups, with extremely diverse cultural and even physical features. Currently the largest groups are Chinese, Filipino, and Asian Indian. But substantial populations of Japanese, Koreans, and Vietnamese, as well as small numbers from other Asian societies, are included. The term *Asian American*, therefore, must be understood as nothing more than a convenient category that enables us to look at the general characteristics of the various Asian groups together, in comparison with other ethnic populations in the United States.

IMMIGRATION AND SETTLEMENT

The Asian experience in the United States can be divided into two distinct and divergent parts. The first consists of the old Asian immigration, occurring roughly from the middle of the nineteenth to the early years of the twentieth centuries. The Chinese were the first to arrive, followed by the Japanese and, in much smaller numbers, Koreans and Filipinos. These first Asian immigrants were mostly unskilled laborers, recruited for construction or agricultural work. Following restrictive measures, very few additional Asians entered the society until the revision of immigration laws in 1965. That marked the onset of the second period of immigration to the United States from Asia, which continues to the present.

The second, and current, immigration has been markedly different from the earlier movement in that most of the new immigrants have been noticeably higher in class origin, and many have been well educated and occupationally skilled. Moreover, the new Asian immigration has been considerably more diverse in national origin, made up of people from almost every contemporary Asian society. Today, except for the Japanese, Asians in the United States are predominantly first-generation immigrants.

The First Wave

Chinese Americans   The Asian American experience may be said to begin in the mid-nineteenth century. Unfavorable social and economic conditions and political unrest at that time in China, combined with the lure of labor opportunities in Spanish, Portuguese, Dutch, and British colonies, created a swell of Chinese emigration, mostly from the country's southeastern coastal provinces.[1] The United States was also among the destinations of these emigrants, mostly single males, some of whom were lured by the discovery of gold in California in the 1840s (Dinnerstein and Reimers, 1999; Lyman, 1970). In the 1860s, the construction of the transcontinental railroads drew many additional Chinese laborers to the United States. Like other immigrants before them, most intended to return after bettering themselves economically.

In a very short time, the influx of Chinese produced one of the most hostile movements in American ethnic history. The cast of an anti-Chinese movement had been set even before the Chinese landed in the United States, notes Stanford Lyman, having been "preceded by a richly embellished but almost entirely negative stereotype" (1974:56). Perceived by white workers as a labor threat, the Chinese were forced out of their jobs and commonly driven from one town to another. In addition to the antagonism they encountered from hostile white workers, the Chinese were severely restricted by legal measures used to limit their occupational and residential movement. These culminated in the Chinese Exclusion Act, passed by the U.S. Congress in 1882 and designed to preclude Chinese immigration for a ten-year period. The act was extended for another ten years in 1892, however, and the ban against Chinese was made permanent in 1907. These measures marked a watershed in American ethnic history: for the first time, a specific group was formally barred from entrance. In fact, the Chinese Exclusion Act was the precursor of numerous anti-immigration laws that would be enacted in the following decades.

Although the exclusionary laws never fully suspended Chinese immigration, they effectively checked any significant population growth of the Chinese in America. When the initial act was passed in 1882, there were about 125,000 Chinese in the United States. For the next six decades, the Chinese population not only failed to grow but actually declined. By 1910 it had dropped to around seventy thousand, a figure that changed only slightly until the 1970s. This decline was the result not only

---

[1] Chinese immigration over many generations has constituted one of the great diasporas. Today Chinese communities can be found in almost every country of the world (Poston et al., 1994). Ethnic Chinese minorities have played an especially important economic role in most Southeast Asian societies (Chua, 2004; Minority Rights Group, 1992).

of the restrictive legislation itself but also of the overwhelmingly male composition of the Chinese population that had made up the mid-nineteenth-century immigration.

Because the Chinese immigrants were almost always male, there existed no possibility of a natural population increase and thus no creation of a stable and thriving ethnic community. The severely unbalanced sex ratio, in addition to the relentless discrimination they experienced, forced the Chinese at the turn of the century into urban ghettos, the familiar "Chinatowns," from which they did not begin to disperse until well after World War II.

By the early 1960s, the Chinese population had begun to grow somewhat, due to the repeal of the Chinese Exclusion Act in 1943 as well as to the entry of war brides, refugees, and some scientific personnel after World War II. This growth also helped correct the unbalanced sex ratio. As with other non-European groups, however, it was the revised immigration legislation of 1965 that created a large-scale Chinese population increase in the United States. Not only were more immigrants admitted, especially from Hong Kong and Taiwan, but the Chinese population was augmented as well by a large number of Vietnamese refugees who, though having lived in Vietnam, were ethnically Chinese. We look more closely at these new immigrants, along with others from Asia, in later sections of this chapter.

JAPANESE AMERICANS The formal exclusion of the Chinese provided for the entrance of Japanese immigrants beginning late in the nineteenth century. In their early immigration and settlement patterns, the Japanese in some ways resembled the Chinese who preceded them. Most of the first Japanese immigrants were males who were confined to lower-status occupational positions. And, like the Chinese, they soon came to be seen as a labor threat in California, where most had settled. A reaction of severe antagonism culminated in formal legislation that prohibited their further entrance. In 1907–1908, the United States and Japan entered into a Gentlemen's Agreement, which effectively restricted further Japanese immigration to the United States. Under the terms of the agreement, only nonlaborers and relatives of resident Japanese would be permitted to enter. The Oriental Exclusion Act of 1924 carried the restriction a step further, effectively barring *all* subsequent Japanese immigration.

Despite some similarities, in a number of important ways the early Japanese immigrant experience differed from that of the Chinese. To begin with, although they initially worked in many of the same jobs as the Chinese, a large percentage of the early Japanese immigrants gradually turned to farming, where they carved out a niche for themselves in the California economy. More important, the sex ratio of the Japanese immigrant population, though heavily male in the early years, eventually was balanced with the introduction of a substantial female immigration. The Gentlemen's Agreement of 1907–1908 had permitted wives and families of men already in the United States to join them. As a result, the social difficulties created by the demographic imbalance in the case of the Chinese did not materialize. With families intact, the base was provided for a second generation and thus a natural increase in the Japanese American population. By 1940 nearly two-thirds of the Japanese in the United States were native-born (Kitano and Daniels, 1995). Today, unlike other Asian American groups, the Japanese have not increased their numbers through large-scale immigration. As a result, the Japanese American population has stabilized and is rapidly being surpassed by other, more recent, Asian groups.

The demographic and historical characteristics of the Japanese American group today set it apart in some ways from others that make up the Asian American population. Perhaps most important, because most are native-born, Japanese Americans have a longer history in the United States and are generationally divergent from others.[2] As a result, Japanese American assimilation is more advanced than for other Asian groups.

THE SECOND WAVE

The Chinese and Japanese were the first and largest of the Asian groups to enter the United States. Not until the liberalized immigration law of 1965 did a new stream of Asian immigrants augment these two communities. In addition to the Chinese, the number and variety of Asian societies represented by the post-1965 stream were considerably greater than any previous immigration, and it is these groups that today make up the major components of the Asian population in the United States. Although small numbers of all these groups had been in the society since the late nineteenth century, none could be said to represent a sizable community comparable to those of the Chinese and Japanese. Other than the Japanese, no more than half the total population of any Asian American group is native-born. The history of these new Asian American groups, therefore, is essentially of the last four decades.

KOREANS   In the early years of the twentieth century, a few thousand Koreans were recruited to work the sugar plantations in Hawaii. Ironically, they were needed to replace Chinese workers who had been barred by the 1882 Chinese Exclusion Act. Further Korean migration to the United States in any significant numbers did not occur again until the late 1950s, following the Korean War, when many came either as refugees or war brides. Prior to this time, the Korean population on the U.S. mainland never exceeded two thousand (Light and Bonacich, 1988).

In the post-1965 period, a formidable growth of the Korean ethnic group in the United States occurred. Between 1970 and 1980, the Korean American population increased five times—from 70,000 to 355,000. By 2000, Korean Americans numbered over one million. Like other Asian groups, Koreans have concentrated in California, where almost one-third reside, and in the New York metropolitan area, but in recent years there has been an increasing dispersion to other regions (Reimer, 2006). Contemporary Korean immigrants have tended to be from the urban middle class of Korean society, including many college-trained professionals. What's more, most have come as family units (Kitano and Daniels, 1995; Min, 1991).

Koreans in the United States have been especially prominent as owners of small businesses, usually operated as family enterprises (Kim, 1981; Light and Bonacich, 1988; Min, 1988, 1995, 2007). Indeed, Koreans have one of the highest rates of small-business ownership among all U.S. ethnic groups (Hoffman and Marger, 1991; Zhou and Kim, 2007). Korean-owned and -operated convenience stores, greengroceries, dry cleaners, manicure shops, and liquor stores have become common fixtures in many large American cities.

---

[2] Although the Chinese have an even longer history in the United States, most of the current Chinese American population is foreign-born.

Even though Christians are a minority in Korea, about three-quarters of Korean immigrants have affiliated with Christian churches in the United States (Min, 2007). Korean immigrants' Christian faith is a significant group characteristic. Indeed, as one study describes it, "[t]he Korean church is perhaps the single most important ethnic institution anchoring this community" (Zhou and Kim, 2007). In addition to religious needs, it provides educational and economic assistance and has been vital in maintaining group solidarity.

The flow of immigration between 1970 and 1990, which had created such a substantial increase in the Korean American group, declined considerably during the 1990s, and some who had immigrated earlier chose to return to Korea. Economic opportunities in Korea were seen as more promising, and the political instability that had characterized Korea in the past had been reduced. These factors diminished the appeal of immigration to the United States. They also created an incentive to return to Korea. For those who had been successful in the United States, having earned a university degree or having run a profitable business, return represented an opportunity to enjoy even greater success in their country of origin, where they may have felt more culturally at ease. Return was also appealing for those who had experienced increasing difficulties in adapting to American culture or who found that operating a small business, so common among Korean Americans, was simply too risky in terms of physical safety as well as economic security (Belluck, 1995; Min, 1995). The image of Korean shops being burned and looted during the Los Angeles riots of 1992 and the frequent harassment of Korean shop owners in other cities made Korean Americans more conscious of ethnic conflict, a condition not encountered in a homogeneous Korea.

Despite the drop-off in new immigrants and the return of many in the past two decades, Korean Americans remain a substantial part of the Asian American population. In fact, the overall growth of this group since 1970 has been spectacular, rising to more than 1.4 million today.

FILIPINOS Few people think of Filipinos[3] when they consider the makeup of America's Asian population. Yet Filipinos are one of the largest Asian American groups. As with others, the liberalization of immigration laws in 1965 accounts for their rapid and substantial increase. Until recently, only Mexicans had traditionally outnumbered Filipinos as immigrants to the United States.

The Philippines was a colony of Spain for more than three centuries before becoming an American possession in 1899. As a result, most Filipinos are Roman Catholic and have Spanish surnames. After the Spanish-American War, the United States took possession of the Philippines, and the American cultural influence became dominant. Following independence in 1946, political ties between the two countries remained close, and American cultural and economic influences continued to be strong. English is one of the country's major languages and is spoken by most educated people. Historically and culturally the Philippines has had little in common with other Asian countries. Geographically, however, it is part of the Pacific Rim, and thus Filipinos are considered Asian people.

---

[3] There is no "F" sound in Tagalog—the language, other than English, that is most widely spoken in the Philippines. Hence, rather than Filipino, Pilipino is sometimes used as a preferred term of reference.

No group better illustrates the link between immigration and labor needs than do the Filipinos. Like the early Koreans, a few thousand Filipinos came to Hawaii in the early 1900s, recruited as agricultural workers to replace the excluded Chinese. Many more immigrated to the mainland in the 1920s as California farm producers faced a labor shortage created by newly restrictive quotas limiting cheap labor from Mexico. By 1930, almost fifty thousand Filipinos were living on the mainland of the United States, most in California (Mangiafico, 1988). Because the Philippines at that time was a U.S. territory, Filipino immigrants were actually considered American nationals though not U.S. citizens. Because of this, they were not subject to the same kind of quota restrictions as other Asians until 1935, when the Philippines was granted deferred independence.

Like the Chinese before them, early Filipino immigrants were almost entirely single males. And with the general anti-Asian social and political climate on the West Coast, where most settled, Filipinos found themselves subjected to the same kinds of discriminatory actions as the Chinese and Japanese. Following the outbreak of World War II, the Philippines was a critical ally in the Pacific war against Japan, and as a result, the social climate in the United States reflected a more tolerant view of Filipinos.

Along with discrimination, the low level of education of Filipino immigrants produced a restricted range of occupational opportunities. Most worked as seasonal agricultural laborers and in the service sector in low-status positions such as restaurant workers, hospital attendants, and hotel workers.

Between 1960 and 1970, the number of Filipinos in the United States nearly doubled, and equally spectacular growth occurred in the 1970s and 1980s. Today's Filipino population is over 2.6 million, and those from the Philippines continue to make up a very significant percentage of America's new immigrants. Cultural compatibility—especially language—and U.S.-Philippines political ties are important factors that account for the continued attraction of the United States for Filipinos.

The social characteristics of the new immigrants are considerably different from those of the earlier period. The strong U.S. military presence in the Philippines following World War II had enabled many Filipino women to immigrate as wives of U.S. servicemen. But most Filipino immigrants are now families with children, seeking the promise of better economic opportunities. Recent arrivals have also been better educated than those of earlier decades, and many come with professional credentials, enabling them to move into highly skilled occupational fields like medicine and education (Cariño, 1987). Nursing is the most common occupational niche of Filipino immigrant health care professionals, who make up a majority of foreign-born nurses in the United States (Choy, 2007). Filipinos' generally higher occupational positions, along with English proficiency, have dictated against the formation of cohesive and distinguishable ethnic enclaves comparable to Chinatowns (Agbayani-Siewert and Revilla, 1995).

VIETNAMESE The entrance of Vietnamese into the United States has been unlike that of any other Asian group. Most have come as political refugees since the end of the Vietnam War. Prior to that time, very few Vietnamese had been resident in the United States.

Between 1966 and 1975, some twenty thousand Vietnamese arrived in the United States, but the Vietnamese immigration reached epic proportions following the fall of the South Vietnamese government in 1975. In the nine years between 1975 and 1984, more than seven hundred thousand Southeast Asian refugees came to the United States, most of them Vietnamese (Gardner et al., 1985). Perhaps the most dramatic entrance of these refugees occurred in 1979 and 1980 when thousands fled Vietnam, Laos, and Cambodia in crude boats. The United States absorbed more than two hundred thousand of these so-called boat people in a period of fifteen months.

Most of the refugees coming in the second wave were unlike the earlier Vietnamese in that they were relatively unskilled, less educated, and spoke little English (Wong, 1986). For example, almost 80 percent of the Vietnamese coming to the United States between 1965 and 1969 were college graduates, as opposed to less than 16 percent of those who arrived between 1975 and 1980 (U.S. Commission on Civil Rights, 1988). Their adjustment to American society, as a result, was far more difficult, and today they remain the most economically depressed element of the Asian American population. The Vietnamese, then, display a bipolar class breakdown—those who have successfully adapted to American society and are at a comfortable economic level, and those at the opposite end of the class hierarchy, disproportionately among the unemployed, poor, and socially alienated (Weiss, 1994).[4]

Though the heaviest concentration of the almost 1.5 million Vietnamese Americans is in California, the group has dispersed widely throughout the United States. Substantial Vietnamese populations can be found today in Houston, New Orleans, and Arlington, Virginia. The largest communities, however, are in Orange County, California. In the adjoining cities of Westminster and Garden Grove, thousands of Vietnamese have created a Little Saigon, with a full range of institutions serving the ethnic enclave.

OTHER SOUTHEAST ASIANS   During the 1980s, the Asian American population expanded not only in size but in diversity as well. In addition to the Vietnamese, several hundred thousand refugees from other Southeast Asian countries arrived in the United States following the end of the Vietnam War in 1975. The largest among these were Laotians (mostly Hmong tribespeople from the mountains of Laos) and Cambodians.

These groups, like many Vietnamese, have not easily adapted to American society. The intention of U.S. policy when they entered as refugees was to resettle them in various regions and cities. The expectation was that this would accelerate their adaptation and would avoid any negative impact on the labor market in communities that might absorb a large concentration of refugees.

The resettlement of the Hmong, in particular, has been fraught with serious problems (Beck, 1994; Hein, 2006). Culturally, the group had not been exposed to Western ways before emigrating from a relatively isolated region of Laos, and the class and educational characteristics of most of the refugees were significantly lower than those of other Asian groups, further contributing to their problems of adaptation.

---

[4] As a result of subsequent immigration waves, the Vietnamese in America today constitute not only a class-diverse group but also one that reflects a cross section of Vietnamese society in terms of language, religion, and region of origin.

ASIAN INDIANS[5]   Like the Chinese, Indians have a long heritage of immigration to societies in all parts of the world. Until recently, however, the United States had not been one of their principal destinations. In the nineteenth century, most had gone as indentured servants to various parts of the British colonial empire, including South Africa and a number of Caribbean societies. In the early years of the twentieth century, a few thousand workers from India were recruited as agricultural laborers in California and Washington, but because of immigration restrictions, the Indian population did not expand. Like other Asian immigrants at this time, they experienced extremely virulent forms of discrimination. In this racially hostile atmosphere, the relatively small Asian Indian population remained isolated in agricultural communities, mainly in California.

As with other Asian groups, the liberalized immigration measure of 1965 represented the point at which Indians began to enter the United States in significant numbers. And like the others, the new immigrants were sociologically quite distinct from their earlier counterparts. Those who have come in recent decades represent, for the most part, a highly educated and occupationally very skilled population. Indeed, their educational level, compared with that of other ethnic groups, is quite remarkable. More than 70 percent of the adult Indian American population are college graduates, and almost 40 percent have earned graduate or professional degrees. Clearly, Indian Americans are one of the most highly educated segments of the American population. What's more, their income and occupational levels are far above average.

Most Indians coming to the United States represent a select subset of the Indian population—more educated, more skilled, and more generally privileged. They are part of the worldwide migration of those with educational and occupational skills leaving developing societies in favor of the industrialized societies of the West, where economic and professional opportunities are more plentiful. In fact, many of the Asian Indian immigrants of the recent period have been students at American universities who, after completing their studies, find employment and remain in the United States. Asian Indian software engineers, scientists, and other highly skilled technical workers constitute a huge segment of the workforce of the American computer industry (Boxall, 2001). The relatively high occupational status and high level of education of Indian immigrants in the past two decades have enhanced the speed and extent of their economic and social adaptation to American society. Indians have also benefited from the fact that on arrival most are fluent in English and have already been exposed to Western values.

In the past decade, the Asian Indian community in the United States has become more diverse, with a larger commercial class augmenting the predominantly professional and technical character of the first post-1965 immigrants.[6] Hence, in addition to their place in skilled occupational areas, Asian Indians have played a significant role as entrepreneurs in the small-business sector; for example, more than half of all motels in the United States today are owned by Asian Indians (Dhingra, 2012;

---

[5] Increasingly, Asian Indians are referred to as "Indian Americans," though the U.S. Census continues to use "Asian Indian."

[6] The post–World War II Indian immigration to Britain, by contrast, comprised not only a much larger but also a more heterogeneous population, with many unskilled and poorly educated people among them (see Watson, 1977).

Varadarajan, 1999). This entrepreneurship is comparable to the economic path followed by the recent generation of Korean immigrants. In California's Silicon Valley, where their presence in the computer industry is so evident, immigrants who are not high-tech workers—taxi drivers, restaurant workers, truckers, and so on—have begun to create a more class-diverse Asian Indian community (Boxall, 2001).

It is important to understand that those classified as "Asian Indian" represent a group that is internally quite diverse culturally. India itself is an ethnic potpourri, with virtually dozens of languages and cultural communities. The religious division among Hindus, Muslims, and Sikhs is only the most apparent and profound variance. Moreover, the national and geographical origins of Asian Indians in the United States are varied. Most have come from India, but sizable numbers have come from Pakistan, Bangladesh, and Sri Lanka. The term *South Asian* has been used to refer to this culturally and nationally diverse collection of groups. Ethnic Indians have come as well from countries where Indians had settled earlier in the nineteenth century as part of the Indian diaspora. Among these are Indo-Caribbeans, from Caribbean countries including Trinidad and Tobago and Guyana (Sengupta and Toy, 1998), and those from East African countries who were forced to flee oppressive governments in the 1970s.

ASIAN AMERICAN DEMOGRAPHICS

Several features stand out in the current demographic patterns of Asian Americans.

GROWTH AND DIVERSITY   Today, Asian Americans remain a relatively small part of the total American population—not quite 5 percent—but one that is rapidly growing not only in numbers but also, as we will see, in economic and cultural significance. Between 1970 and 1980 the Asian American category grew by 128 percent; between 1980 and 1990, by 108 percent; and in the 1990s, by 69 percent. In the decade between 2000 and 2010, its growth was over 45 percent, four times faster than the total U.S. population and faster than any other racial/ethnic category (U.S. Census Bureau, 2012c). It is projected that the current Asian American population of more than 14 million will almost double by 2020; by 2050, Asian Americans may make up as much as 13 percent of the total population (Martin and Midgley, 2010). This rapid and sizable population growth is attributable mainly to the fact that Asians now account for more than a third of all immigration to the United States.

Figure 9.1 shows the breakdown of the Asian American population today. In 1970, the Chinese and Japanese formed the bulk of the population, but today these two groups together are less than one-third. With 23 percent, the Chinese category remains the largest component. The Indian American population has grown more rapidly than any of the other Asian ethnic groups during the past decade and is now the second largest, with 20 percent, slightly larger than the Filipino population. The Vietnamese and Korean populations have also continued to grow steadily, each with about 10 percent. In contrast, the Japanese American population is growing much more slowly because of very limited immigration from Japan and a low fertility rate.

REGIONAL CONCENTRATION   Although they have begun to settle in all regions of the United States, Asian Americans are most heavily concentrated in just a few states:

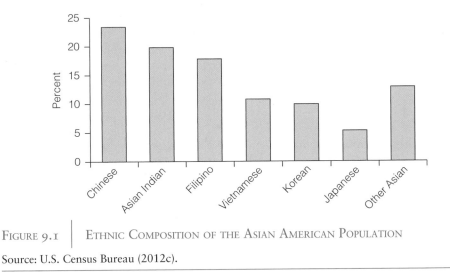

FIGURE 9.1 | ETHNIC COMPOSITION OF THE ASIAN AMERICAN POPULATION

Source: U.S. Census Bureau (2012c).

California, Hawaii, New York, New Jersey, and Texas (Figure 9.2). Except for Hawaii, California by far has a higher percentage of Asians as part of its population than any other state, 13 percent. This is over one-third of the entire Asian American population. Among the specific groups, only Indian Americans have not clustered in California, having dispersed more widely.

Like other ethnic populations that have arrived primarily in the post-1965 era, Asian Americans dwell mostly in large urban areas. Indeed, urban residence characterizes virtually the entire Asian American population, with one-third found in just three metropolitan areas—Los Angeles, New York, and San Francisco. As shown in Table 9.1, Asians make up nearly 15 percent of the total metropolitan population of Los Angeles, and in San Francisco they are more than 23 percent.

SUBURBANIZATION One of the more intriguing demographic patterns exhibited by Asians is their rapid rate of suburbanization. Like blacks and Hispanics, Asians, especially those coming since 1970, have been leaving the central cities in ever greater numbers or, as immigrants, have gone directly to the suburbs, bypassing more traditional central-city destinations (Alba et al., 1999; Frey, 2010). Indeed, a majority or near majority of every Asian group lives in suburbia (Logan, 2001). In the Los Angeles metropolitan area, many of the Asian enclaves are now found in suburbs rather

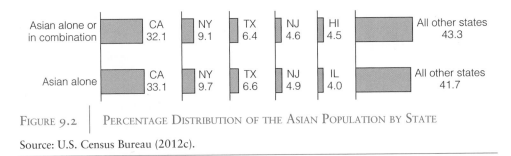

FIGURE 9.2 | PERCENTAGE DISTRIBUTION OF THE ASIAN POPULATION BY STATE

Source: U.S. Census Bureau (2012c).

TABLE 9.1 | METROPOLITAN AREAS WITH THE LARGEST ASIAN POPULATIONS

	Asian Alone Population (in Thousands)	Percent of Total Metro Area
Los Angeles–Long Beach–Santa Ana, CA	1,885	14.7
New York–Northern New Jersey–Long Island, NY–NJ–PA	1,878	9.9
San Francisco–Oakland–Fremont, CA	1,006	23.2
San Jose–Sunnyvale–Santa Clara, CA	572	31.1
Chicago–Naperville–Joliet, IL–IN–WI	533	5.6
Washington–Arlington–Alexandria, DC–VA–MD–WV	517	9.3
Honolulu, HI	418	43.9
Seattle–Tacoma–Bellevue, WA	393	11.4
Houston-Sugar Land-Baytown, TX	389	6.5
San Diego–Carlsbad–San Marcos, CA	336	10.9

Source: U.S. Census Bureau (2012e).

than in the central city. Monterey Park is an interesting example. In the early 1970s, it was a Los Angeles suburban town of mainly Anglos and Hispanics. Now the population is over 60 percent Asian, mainly middle class, and most of its businesses are Asian-owned (Chowkwanyun and Segall, 2012; Fong, 1994; Horton, 1995).

ASIAN AMERICANS IN THE CLASS SYSTEM

SOCIOECONOMIC STATUS

What is perhaps most striking about the Asian American population today is its unusually high ranking on most measures of socioeconomic status: income, occupational prestige, and level of education. Indeed, as a recent wide-ranging analysis of Asian Americans has put it simply, today they constitute "the highest-income, best-educated" racial group in the United States (Pew Research Center, 2013:1).

The socioeconomic success of Asians in the United States, however, must be looked at with some caution. There are differences among specific groups as well as among members of the same group. For example, although 67 percent of employed Asian Indians and over half of Chinese are professionals or managers, less than 30 percent of Vietnamese are in such occupations (Table 9.2). The income and educational levels of the latter are also considerably lower than those for other Asian American groups. Another internal difference is the tendency toward economic polarization. That is, members of specific Asian American groups tend to cluster at the two extremes of the economic hierarchy, either at or near the top (such as professionals and business owners) or at or near the bottom (service workers and other low-status job classifications).

TABLE 9.2	ASIAN AMERICANS IN PROFESSIONAL AND MANAGERIAL OCCUPATIONS (%)
Total U.S. population	35.7
White, non-Hispanic	37.7
Asian	48.5
Asian Indian	66.5
Chinese	54.4
Filipino	42.4
Japanese	53.6
Korean	46.1
Vietnamese	29.7

Source: U.S. Census Bureau (2012a).

INCOME   The household income of Asian Americans far exceeds that of most other minority ethnic categories and, with the exception of the Southeast Asians, surpasses most Euro-American groups as well. In 2011, median household income for Asian Americans was $68,000, which was $15,000 more than for non-Hispanic whites (U.S. Census Bureau, 2012a).

Figure 9.3 shows significant internal differences within the Asian category, however. As is seen, the median income for Koreans is little more than half of what it is for Asian Indians. Moreover, although most Asian groups have higher family incomes than whites, they also have more family members in the labor force. And, despite their high socioeconomic status relative to whites, Asian Americans do not control a commensurate amount of wealth (Campbell and Kaufman, 2006).

The same internal discrepancy is evident regarding poverty. The Asian American poverty rate is lower than for the total U.S. population but higher than for whites.

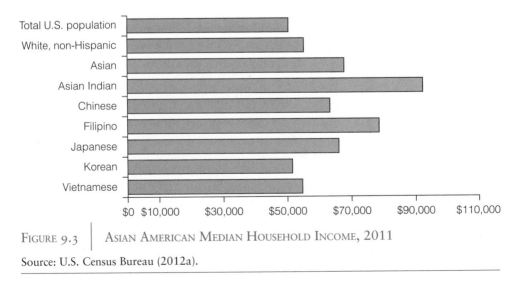

FIGURE 9.3   ASIAN AMERICAN MEDIAN HOUSEHOLD INCOME, 2011

Source: U.S. Census Bureau (2012a).

However, it is extraordinarily high for two relatively small Southeast Asian groups, Cambodians and Hmong.

OCCUPATION The rate of labor force participation for Asian Americans is higher than that of any other ethnic category, and the range of occupations held by Asian Americans is broad, extending from working-class jobs to professions such as medicine and engineering.

Here the split between those of previous generations and the more recent immigrants is very evident. In striking contrast to the earlier Asian immigrants, most of whom were unskilled laborers, Asians today are disproportionately among the most highly skilled. A key feature of the immigration reform of 1965 was the preferential treatment given to family members and to those with valuable occupational training. Because Asians had been virtually excluded for several decades, most of the early post-1965 immigrants entered not as family members but as skilled workers. As a result, many of the recent arrivals are found in the professions or in highly skilled jobs. Since the 1970s, almost half of working Asian Americans have reported professional and managerial occupations (U.S. Census Bureau, 2007b; U.S. Commission on Civil Rights, 1988). For the past twenty years, Asian immigrants to the United States have been helping to fill shortages in a number of professional and technical areas, especially computer engineering and medicine.

High occupational rank is characteristic of native-born and foreign-born Asian Americans alike. Except for the Southeast Asian groups, Asian American men are more heavily represented in white-collar occupations and less represented in blue-collar occupations than are non-Hispanic white men. Nearly half of Asian Americans are in managerial and professional occupations, a rate considerably higher than for other ethnic minorities and even for non-Hispanic whites (Table 9.2). At the other end of the occupational hierarchy, the percentage of Asians in service jobs is about the same as for non-Hispanic whites and well below the percentage for blacks and Hispanics (U.S. Census Bureau, 2012a). However, the sharp stratification within the Asian American population is, again, very evident in the occupational world. At the same time that Asian Indians, Filipinos, Japanese, Koreans, and Chinese are disproportionately in high-ranking professional and technical fields, a high percentage of Vietnamese and other Southeast Asians are in low-paying, unskilled occupations.

Despite their comparatively large numbers in professional and technical positions, Asian Americans are underrepresented in managerial positions. And at the very top of the corporate ladder, their numbers are as small as for African Americans and Hispanic Americans. In 1990, 0.3 percent of senior corporate executives were Asian Americans (U.S. Commission on Civil Rights, 1992) and were still estimated to be less than 1 percent in 2004 (Zweigenhaft and Domhoff, 2006). Asian Americans have often spoken of a glass ceiling, referring to the limits to which they are able to rise in the corporate world. For example, in California's Silicon Valley, the major center of the American computer industry, Asians make up as much as one-third of the engineering workforce. But they are not well represented in upper management at the larger companies. Some, as a result, choose to start up high-technology enterprises of their own (Pollack, 1992). This was a route taken earlier, in the 1950s, by the late An Wang, Chinese American founder of Wang Laboratories, for many years one of the country's largest computer companies. As Zweigenhaft and Domhoff conclude, it is still "the most frequent pathway for Asians to take to become directors of *Fortune*-level companies" (2006:189).

Several explanations have been offered to account for the underrepresentation of Asian Americans in executive posts. Some see them at a disadvantage in vying for administrative jobs because of a lack of communication skills. Also, Asian Americans are often stereotyped as docile and thus not prepared for leadership positions (Min, 2006; Zweigenhaft and Domhoff, 2006). Despite their scant numbers in the executive suites of the American corporate economy, however, the extraordinarily high level of education of most Asian Americans will likely accelerate their movement into top management positions at a faster rate than that of other minorities. Furthermore, as the number of Asian Americans in the general population grows, corporations will find it in their interests to seek out and place more Asian Americans in higher-ranking positions, much as they have done with African Americans and Hispanic Americans.

ASIANS IN THE SMALL-BUSINESS SECTOR   A salient feature of the economic role played by Asian Americans today is the unusually large number who choose to own small businesses. Indeed, Asian American business ownership exceeds by far that of any other ethnic population as well as for all Americans (U.S. Census Bureau, 2010e; Zhou, 2009).

Several factors seem to account for the high rate of participation of Asians in the small-business sector. Most of their firms are family-operated businesses, in which several or even all family members participate. As a result, labor costs are minimal, and relatively small amounts of capital are needed. Many immigrants thus find small enterprises attractive as opposed to the corporate sector, where there are limited opportunities for nonwhites, especially those who do not speak English fluently. Owning a small business gives immigrants some independence and enables them to avoid having to deal with what are seen as unreceptive dominant economic institutions.

EDUCATION   The differences between Asian Americans and other minority ethnic groups regarding education are striking. All Asian American groups are either close to or exceed the average level of schooling of the white population. In the percentage of college graduates, however, the difference between Asian Americans and all other ethnic groups is extraordinary. Asian Americans—both men and women—are more likely to have earned a college degree than non-Hispanic whites. As seen in Figure 9.4, half of all adult Asian Americans hold at least a bachelor's degree, compared to 30 percent of non-Hispanic whites. Moreover, 21 percent of adult Asians hold advanced degrees, almost twice the percentage of whites. Many Asian immigrants, of course, are highly educated before they enter the United States, making them distinctly different from most contemporary Hispanic immigrants as well as Asian immigrants of past eras. Notice in Figure 9.4 that despite the extraordinarily high educational levels for Asian Americans generally, the Vietnamese are below average, and this holds for other small Southeast Asian groups—Cambodian, Hmong, Laotian—as well.

In academic achievement, Asian American students are well above the norm on almost every measure. For example, Asian American high school students score higher than students from any other ethnic group in mathematics scholastic aptitude tests and are second only to whites on verbal aptitude tests. In New York, Asian enrollment in the city's most elite specialized high schools has soared; of the fourteen

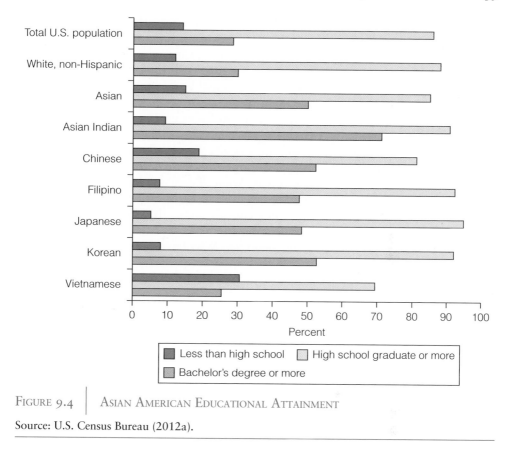

FIGURE 9.4 | ASIAN AMERICAN EDUCATIONAL ATTAINMENT

Source: U.S. Census Bureau (2012a).

thousand students enrolled in these schools, which require a grueling entrance exam, 60 percent are Asian (Spencer, 2012). Predictably, Asian students enter top universities far out of proportion to their numbers. At Harvard and Yale, for example, Asian Americans constitute 15 to 20 percent of U.S. students; at MIT, 27 percent; and at elite University of California campuses they outnumber all other groups, including whites (Gordon, 2013). Moreover, Asian Americans have the highest college graduation rate among all racial and ethnic groups (Min, 2006). And, Asian Americans earn professional degrees far out of proportion to their population (Ryu, 2009).

Science, medicine, and other highly technical fields are academic areas in which the excellence of Asian American students is especially apparent. But they are entering the humanities and the arts in increasingly greater numbers and are excelling in those fields as well. For example, Asians and Asian Americans make up a large and growing percentage of students at the most prestigious music schools such as Juilliard, Eastman, and the New England Conservatory (Wakin, 2007).

ETHNIC AND CLASS FACTORS IN UPWARD MOBILITY

The economic and educational success of Asian Americans is, as we have now seen, extraordinary. In looking at the status of the Asian American population, however, several caveats are in order. First, we need to remember the diversity of this ethnic category. There are differences among specific groups. For example, as previously noted, more than half of all employed Asian Indians, Chinese, and Japanese are professionals or managers, in sharp contrast to the occupations of most Vietnamese and other Southeast Asians. The income and educational levels of the latter are also considerably lower than those for other Asian American groups. Second, not only is the Asian American population made up of many different groups but each group itself is internally varied, stretching over an entire range of social class. Third, the Asian American population consists of two historical components—those who came as early as the mid-nineteenth century and those who have come since 1965. As we have now seen, many of the more recent Asian Americans have come with class and educational backgrounds very different from those of earlier ones.

Taking these cautionary points into consideration, the economic and educational record of most Asian American groups remains remarkable when compared to other parts of the American ethnic hierarchy. Why have Asian Americans moved up the economic ladder more quickly than other ethnic groups, and why do they seem to outdistance others in the realm of education? Is it because of a unique cultural heritage, or are class factors more important? Might Asian Americans even be innately different from other racial and ethnic groups? Attempts to explain the unusually high rates of Asian American mobility, educational achievement, and social adaptation have employed several different lines of thought.

INNATE SUPERIORITY?   Some contend that Asians are innately superior. A few highly controversial studies have suggested, for example, higher average IQs of Chinese and Japanese compared to others (Herrnstein and Murray, 1994; Lynn and Vanhanen, 2002). As explained in Chapter 1, most social scientists do not recognize the validity of such genetic arguments, whether applied to seemingly high-achieving or low-achieving groups. Indeed, such explanations border on racist notions. As noted in Chapter 1, social scientists have emphasized either structural opportunities or cultural factors in explaining why some ethnic groups seem to display greater collective achievement than others. The intelligence factor is one that nonetheless confounds observers because on certain measures of intelligence, Asian Americans seem to consistently outperform other ethnic groups.

CULTURAL FACTORS   Many sociologists and psychologists point to cultural factors in accounting for high achievement among Asian Americans. Stressed in this explanation are values that emphasize the importance of education and the cohesiveness of the family. Almost all studies that have addressed this issue seem to converge on one point: Asian Americans are prepared to work arduously and, in turn, are able to instill in their children a great motivation to work hard. Hence, Asian American business owners, it is often observed, work longer hours than their competitors, and time spent on homework or other school activities by Asian American students considerably exceeds that spent by non-Asian students (Butterfield, 1986; Kim, 1981; Light and Bonacich, 1988). In his study of Koreans in New York, Illsoo Kim traces their business success largely to cultural factors. "An emphasis on economic motivation and mobility," he writes, "has contributed to acculturation in the new land because it was already a deep part of contemporary Korean culture and values" (1981:295).

He claims that values such as industriousness and the control of impulses and emotions, which correspond closely to dominant American values, were carried by Korean immigrants to the United States. A similar argument has been advanced regarding the economic success of Japanese Americans, who it is claimed brought values from Japan that were highly compatible with white, Protestant, middle-class American values (Caudill and De Vos, 1956).

That cultural factors are of key importance in explaining Asian American educational and economic success is suggested by a study, conducted by University of Michigan researchers, investigating the adaptation of Vietnamese refugees. At the time of their arrival in the United States, most of the subjects of this study (part of the boat people of the late 1970s) had completed little formal education, few had any marketable skills, and only one in a hundred spoke fluent English. The researchers found that despite these handicaps, many had achieved economic independence after only a few years. Furthermore, whatever success they had realized was almost entirely a product of their own efforts; the assistance of outside agencies had played a very limited role (Caplan et al., 1989).

The academic achievement of school-age children from these families was more remarkable still. Despite their severe language deficiencies and family poverty, after an average of only three years they were experiencing success in American schools. On national standardized tests of academic achievement, 27 percent of the refugee children scored in the ninetieth percentile on math achievement—almost three times better than the national average. And despite their somewhat lower scores in English-language proficiency, they earned higher grade point averages than their peers (Caplan et al., 1991, 1992). The researchers found that this rapid educational success was closely linked to a cluster of core values deriving from the Confucian and Buddhist traditions of East and Southeast Asia: education and achievement, a cohesive family, and hard work. These, they found, "are, if not identical, congruent with what have been viewed as mainstream middle-class American beliefs about the role of education in getting ahead" (1991:172).[7]

Other researchers have also emphasized the importance of Confucian values in explaining the rigid work ethic that Asian Americans and their children seem to exhibit. Kim (1981), for example, views Confucianism as a significant factor in accounting for Korean business success. Zhou and Bankston (1998) also find this to be an important part of the explanation for the educational success of Vietnamese children. In addition to the sanctified place of education and the family, Confucian thought stresses that people can always be improved with effort and instruction, values well suited to the competitive worlds of American business and schooling.

In assessing the achievement of Asians and Asian Americans compared to other Americans (and Westerners in general), psychologist Richard Nisbett (2009) attributes most of the difference simply to greater effort on the part of the former. Notions of higher innate intelligence are deemed essentially meaningless. "Asian intellectual accomplishment," he writes, "is due more to sweat than to exceptional gray matter." The tendency for Asian students to work harder and to invest more time in school

---

[7] The researchers did find a critical difference between the two value systems, however, in their orientation to achievement: "American mores encourage independence and individual achievement, whereas Indochinese values foster interdependence and a family-based orientation to achievement" (Caplan et al., 1992:41).

work is closely related to the cultural factors discussed previously, in particular fulfilling the expectations of the family and upholding its honor. Nisbett explains:

> Asian families are more successful in getting their children to achieve academically because Asian families are more powerful agents of influence than are American families—and because what they choose to emphasize is academic achievement (159).

Accomplishment on behalf of family, then, is a greater incentive to success than achievement for one's self, the more common American motivational force.

STRUCTURAL FACTORS Although most observers today recognize the importance of unique cultural factors in explaining Asian American success, many also see a fortuitous fit between the opportunity structure of contemporary American society and the class background of these immigrants. We have already noted the disproportionate number of highly skilled and educated people who make up the Asian immigrant population. Hence, the existence of many doctors, engineers, and other highly trained professionals among them is largely a function of their preimmigration class position. In some ways, today's Asian immigrants represent a select subset of the population of their origin societies. What has been referred to as a "brain drain" consists of the out-migration of highly skilled and educated people from developing societies. Such people recognize the more abundant opportunities to employ their skills and training in the United States as well as the possibility of earning substantially greater income. Hence, many are not poor, underprivileged immigrants but are already imbued with achievement-oriented values along with the class qualifications to implement those values.

The importance of a propitious opportunity structure, therefore, cannot be overstated in explaining the rapid and substantial upward mobility of Asian Americans, especially the more recent arrivals. Many are people who bring their skills and training with them, enabling them to adapt more easily than others with fewer skills and lower educational backgrounds. Structural opportunities, then, in addition to a cultural heritage compatible with American achievement values, are critical in accounting for the relatively rapid success of so many Asian Americans.

In this view, the importance of human capital, especially education, also accounts for the relative success of groups with longer histories in the United States, namely Japanese and Chinese Americans. Sakamoto and colleagues (1998) suggest that William J. Wilson's thesis (discussed in Chapter 7) of the declining significance of race applies to Asian Americans in the last half century. In occupation and other measures of socioeconomic status, the disadvantages of racial distinctness, they claim, have been substantially reduced—if not eliminated entirely—for these groups. Their success is explained primarily as the result of high educational achievement.

Zhou and Kim (2007) suggest that it is a combination of both cultural values and structural conditions that account for the exceptional educational success of Asian Americans. That is, cultural values that stress education need the additional support of ethnic social structures to fully flourish. In their study of Chinese and Koreans in Los Angeles, they found that beyond the public schools, community institutions—Chinese- and Korean-language schools, ethnic media, ethnic churches, and ethnic businesses—played an important role in transmitting and reinforcing Confucian educational values, thereby helping ensure the success of children in high school and ultimately having them gain entrance to prestigious universities.

ASIAN AMERICANS AND POLITICAL POWER

Even though their accomplishments in the economic and educational spheres of American society are increasingly evident, Asians have not exhibited a similar prominence in the society's political institutions. In the recent past, Asian American representation in the political world was limited mainly to Hawaii's congressional and senatorial seats and a few congressional seats from California. There is much evidence to indicate that Asian Americans are on the brink of much fuller participation in the political world, both as an electorate and occupying positions of leadership (Nakanishi, 2001; Pew Research Center, 2013).

PARTICIPATION   Participation of Asian Americans in political life has traditionally been limited. This has been attributed to a number of factors. The nationally diverse and factionalized nature of the Asian American population has made it difficult to build a power base that can be mobilized for political action. Moreover, because such a large proportion of the Asian American population is foreign-born, many are not U.S. citizens and thus not eligible to vote. Also, among older Asian Americans, especially those who have been victimized by past discrimination, the tendency to remain as invisible as possible continues to militate against strong political activism (Gross, 1989). Finally, many of the newest Asian immigrants come from societies where citizen participation in politics is not institutionalized and where there exists a profound distrust of government generally.

As their population base widens and as the newer immigrants experience progressive assimilation, Asian Americans are beginning to play a more active political role, and not only in those areas where they constitute a sizable proportion of the population or where they are a significant minority. The number of Asians elected to political office at state and local levels continues to grow, and today thirteen members of Congress identify themselves as Asian American (Manning, 2013).

Asian American elected officials are unique compared to African American and Latino elected officials in that they tend to emerge from political districts where, aside from Hawaii, their group does not account for a substantial portion of voters (Lai et al., 2003). One of the most visible indications of an increasing Asian American presence on the American political scene was the 1996 election of Gary Locke as governor of Washington. Locke is a second-generation Chinese American whose election marked the first time an Asian American had been elected to a governorship outside Hawaii. He subsequently served as U.S. secretary of commerce in the Obama administration.

Asian Indians offer another case of entering the American political scene without relying on a large coethnic population base. Despite their relatively recent arrival in American society in significant numbers, they have utilized a few social advantages—namely a high level of education and most being English speakers—to propel them into state and local politics. States and towns with a miniscule Asian Indian population have elected Asian Indians to state legislatures and to mayorships in small but growing numbers (Vitello, 2007). Most prominently, second-generation Indian Americans were elected governor of Louisiana in 2007 and South Carolina in 2010. Asians are a tiny percentage of both states.

POTENTIAL POWER   Although Asian Americans still represent only a small percentage of the national population, their increasing visibility, numbers (they are the fastest-growing electorate in the United States), and financial power will necessarily force politicians to take notice of them, particularly in states like California, where they are strongly concentrated. Already, for example, Asian Americans have established themselves as significant campaign donors (Kirschten, 1992; Vitello, 2007). Perhaps most important, future issues may arise that will have a mobilizing effect on the Asian American population. Widespread occurrences of discrimination, for example, real or perceived, are apt to solidify the disparate Asian American ethnic groups and to create a more potent political base. The same is true of immigration issues, which are of key importance to all elements of the U.S. Asian population, in much the same way as they are to Latinos (Constable and Lazo, 2013).

Interethnic competition and hostilities, such as the boycotts against Korean shops by blacks or the attacks on Korean businesses during the Los Angeles riots of 1992, may also stimulate Asian American political participation. A government investigation of the contributions of Asian nationals to the Democratic Party in 1996, causing investigators to look primarily at contributors with Asian-sounding names, as well as Republican proposals to curtail immigrant family reunification and deny welfare benefits to legal immigrants prompted many Asians to become citizens and thus eligible to vote (Rodriguez, 1998). Moreover, the growing importance of U.S. economic and political relations with Asian nations may create issues that will redound to the Asian American population, pushing them increasingly into the domestic political arena.

PREJUDICE AND DISCRIMINATION

Among voluntary immigrants to American society, none suffered more severe forms of prejudice and discrimination in their settlement experiences than Asians. Although they were not enslaved nor were genocidal measures taken against them, they were constant targets of all forms of social and physical abuse. Indeed, if there is a common thread running through the early history of Asian American groups, it is the experience of rampant prejudice and discrimination of the most vehement and often violent nature.

The Anti-Asian Heritage

Anti-Asian sentiments have lengthy historical roots, reaching back, as we have seen, to the entrance of the Chinese into the American West in the mid-nineteenth century. The intensity of early anti-Asian actions traditionally seemed to hinge on the perceived threat of Asian immigrants to the jobs of native workers. The anti-Chinese movement, which began in the mines of California, spread to other West Coast areas and eastward to other parts of the country (Lyman, 1974). Everywhere, Chinese were met with derision and hostility. They enjoyed hardly any legal rights and were common targets of assault and even murder. "The bewildered Chinese," note tenBroek, Barnhart, and Matson, "speaking little or no English and ignorant of Western customs, became the victims of every variety of fraud and chicanery, abuses encouraged by the absence of an active public opinion which might have alerted the police and the courts, and by the law prohibiting the testimony of Chinese in cases involving whites" (1968:15).

Sinophobia   The standard panoply of negative ethnic stereotypes—dirty, immoral, unassimilable—was applied to the Chinese. In addition, they were seen as sly, untrustworthy, and inscrutable. Much of the negative stereotyping of the Chinese, and later the Japanese, stemmed from their labor role and their alleged impact on the job market. The term *coolie labor,* for example, denoting undignified work, derives from this period. That the Chinese were inferior to whites was beyond question in most people's minds. Perhaps the most devastating aspect of Chinese stereotyping, however, involved the notion of a yellow peril. Chinese immigration, in this view, was an invasion of people who were loyal only to their country of origin and who, if not stopped, would eventually take over the United States (tenBroek et al., 1968).

This sinophobia, fueled especially by labor leaders, culminated in the previously mentioned Chinese Exclusion Act of 1882. By 1910, the anti-Chinese movement had achieved its primary objectives: Chinese workers had been all but eliminated from the labor market, and restrictive legislation was in place that effectively barred further Chinese immigration. Moreover, the ban on future immigration, particularly of women, seemed to ensure that the Chinese American population would, in the long run, disappear.

THE ANTI-JAPANESE MOVEMENT   Many of these same attitudes and actions were cast on the later Japanese immigration. At the time of their first substantial entrance in the 1880s, the Japanese were received enthusiastically by large farm owners and manufacturers, who saw them as a natural substitute for the Chinese laborers who had been excluded by the restrictive measure of 1882. So long as they occupied only menial or unskilled jobs, for which there was a shortage of workers, they were seen as filling a need. But like the Chinese before them, they soon found themselves the targets of labor groups who viewed them as potential competitors. Anti-Chinese agitation easily shifted to the Japanese, with many of the same stereotypes now applied to them, including the yellow peril (tenBroek et al., 1968). Despite the Gentlemen's Agreement between the United States and Japan in 1907–1908 limiting Japanese immigration, as well as a California measure restricting Japanese ownership of land, anti-Japanese ferment never fully subsided.

In many ways, the anti-Japanese movement of the 1910s and 1920s was merely a continuation of the earlier anti-Chinese movement. Despite their modest group size (in California, where most were concentrated, they numbered only seventy thousand in 1919, little more than 2 percent of the state's population), the Japanese were seen by many as a threat to white dominance. The Japanese in California had become successful farmers, and although they controlled only about 1 percent of California's farmland, their highly efficient agricultural practices produced more than 10 percent of the value of the state's crops (Kitano and Daniels, 1995). Many whites were alarmed at the growing Japanese economic influence and began to agitate for restrictions not only on further immigration but also on the right of the Japanese to own land in the state. California newspaper publishers William Randolph Hearst and V. S. McClatchy emerged as two of the leading figures in the anti-Japanese movement, outlining in their newspapers the themes of the Japanese "threat." Their propagandistic attacks claimed that the Japanese refused to assimilate, that their birthrate was so great they would eventually outnumber whites, and that their "low standard of living" presented an unfair advantage in economic competition with whites (McClatchy, 1978).

All these charges, of course, had been applied to southern and eastern European immigrants as well, but the anti-Japanese campaign brandished a unique fear of an immigrant "invasion." Following World War I, a full-blown exclusionist movement developed in California, designed to isolate those Japanese already in the United States and, as one California politician described it, "to influence Congress and the administration at Washington to enact such legislation, even if the amendment of the Constitution be necessary, as will protect the white race against the economic menace of the unassimilable Japanese" (quoted in Daniels, 1977:84).

The vehemence of the anti-Japanese movement in California was unmatched against any other American ethnic group outside the South. Much of this hostility was the result of the racial distinctiveness of the Japanese. Southern and eastern European groups, though the targets of virulent racist attacks, were not so indelibly marked. Perhaps of greater importance, however, the Japanese were economically successful in both agriculture and business; therefore, they were resented and easily targeted as a competitive threat to the majority population. The Japanese fear was further fueled by the growing disrepute of Japan as an aggressive, militaristic international power during the early decades of the century. The Japanese military threat, covered sensationally by the press, served as cause for suspicion of Japanese Americans. Finally, that the Japanese were concentrated in California provided an opportune setting for a movement aimed at an Asian group. As one historian describes it, "California, by virtue of its anti-Chinese tradition and frontier psychology, was already conditioned to anti-Orientalism before the Japanese arrived" (Daniels, 1977:106).

JAPANESE INTERNMENT   Fueled by the shock of the attack on Pearl Harbor and reinforced by the stereotype of Japanese treacherousness, the anti-Japanese movement culminated after the start of World War II with the internment of the Japanese population living on the West Coast. No other series of events in American ethnic history, other than the enslavement of African Americans and the genocidal measures employed against Native Americans, was comparable to this action. Two months after the Japanese attack on Pearl Harbor in 1941, President Franklin D. Roosevelt, acting on the recommendation of Secretary of War Henry L. Stimson, issued Executive Order 9066, authorizing the U.S. Army to remove any group viewed as a security risk. Almost all those of Japanese ancestry living in California, Oregon, Washington, and Arizona—120,000—were affected. They were subsequently rounded up and sent to internment camps (called "relocation centers") in several Rocky Mountain states and in Arkansas. Having no time to prepare, most were forced to quickly liquidate their businesses; many abandoned their homes and possessions, losing their life savings. They were permitted to carry with them only a single suitcase of personal belongings.

The Japanese Americans were held in the detention camps almost until the end of the war. In an ironic twist, thousands of incarcerated Japanese men were permitted to enter the U.S. Army as a special unit; in the European theater, where they fought the Nazis, they became one of the most highly decorated battle units in American military history.

Several aspects of the Japanese incarceration stand out. To begin with, most of the detained had been born in the United States and were American citizens. Hence, people were denied their fundamental civil rights solely on the basis of their ethnicity. In his final report, the commanding U.S. Army general responsible for carrying out the Japanese removal put the matter bluntly: "The Japanese race is an enemy race and while many second and third generation Japanese born on United States soil, possessed of United States citizenship, have become 'Americanized,' the racial strains are undiluted" (quoted in tenBroek et al., 1968:263). Second, that the removal of the Japanese was racially motivated was demonstrated by the fact that Americans of neither German nor Italian ancestry were similarly treated despite the war against Germany and Italy. Moreover, after the war it was acknowledged that the Japanese on the West Coast had never presented a military threat, and in fact their incarceration may actually have retarded the war effort. The emptiness of the security issue is reflected in the fact that the larger Japanese population in Hawaii was never removed despite its more strategic location. In the end, the exclusionist movement that had begun in California in the earlier years of the century had accomplished its objective—removal of the Japanese. The fear, suspicion, misperception, and envy bred by decades of negative stereotypes were released in a grievous deed, responsibility for which was shared by nativist groups, farmers, entrepreneurs, the military, the courts, and state and federal politicians.

Not until 1983 did a congressional committee formally acknowledge the political injustice that had been dealt the Japanese Americans, recommending a formal apology and compensation of $20,000 in damages to each survivor of the incarceration. Legislation to accomplish this was enacted in 1988.[8] The long-term effects of the internment camps, of course, could not be erased so easily, particularly for first- and second-generation Japanese Americans. For them, American democracy had failed to provide its promise of protection for all citizens. Even for the third and fourth generations, the World War II internment remains the paramount event of the history of the Japanese in America.

Although the Japanese internment is the most infamous case of direct institutional discrimination, all Asian groups have been subjected to various legal forms of discrimination at one time or another. Indeed, during the early years of the century, anti-Chinese and anti-Japanese measures were usually designed to apply to all Asians. In California, intermarriage of whites and Asians, for example, was barred, and residential segregation was enforced by restrictive covenants aimed specifically at Asians. Even federal laws were designed to handicap Asians. Perhaps most revealing, it was not until 1952 that federal legislation was passed making all Asian immigrants eligible for U.S. citizenship (U.S. Commission on Civil Rights, 1988).

PREJUDICE AND DISCRIMINATION IN THE MODERN ERA

Expressions of prejudice and acts of discrimination against Asian Americans in recent years are in no way comparable to the widespread and vehement forms these took in the nineteenth and first half of the twentieth centuries. Nonetheless, negative Asian stereotypes as well as occasional violent incidents have been an integral part of the modern Asian American experience.

---

[8] A similar internment was carried out by the Canadian government at the start of World War II against the twenty-three thousand people of Japanese origin living in the West Coast province of British Columbia, 75 percent of whom were Canadian citizens (see Adachi, 1976; Daniels, 1981; Sunahara, 1980). In much the same fashion as the U.S. Congress, the Canadian parliament in 1988 moved to formally apologize to Japanese Canadians and to make a reparation payment of $19,325 to each survivor (Burns, 1988).

PREJUDICE A 2001 national survey sponsored by the Committee of 100, a national organization of leaders of the Chinese American community, revealed a mixed view of Americans, both positive and negative, toward Asian Americans generally and Chinese Americans specifically (Yankelovich Partners, 2001). The great majority of Americans believed Chinese Americans to have strong family values, to be honest in business, to be patriotic Americans, and to place a higher value on education than most other U.S. groups. At the same time, however, about one-third of Americans felt that Chinese Americans were more loyal to China than to the United States and had too much influence in the U.S. high-tech sector. Moreover, nearly half believed that Chinese Americans might be passing secret information to China. As with other studies of ethnic prejudice, it was concluded that prejudice toward Chinese Americans was most closely correlated with social class—those at lower educational and income levels were more commonly prejudiced.

Regarding attitudes toward all Asian Americans, the study indicated that almost one-quarter of Americans are uncomfortable with the idea of an Asian American as president of the United States; only 15 percent expressed a similar feeling about an African American president, and only 11 percent about a Jewish American president. One-quarter of Americans indicated that they would disapprove to some degree if a member of their family married an Asian American, as opposed to 34 percent expressing disapproval of marriage to an African American and 24 percent to a Hispanic American. One in six (17 percent) also expressed at least some displeasure with a substantial number of Asian Americans moving into their neighborhood.

DISCRIMINATION In the 1980s and early 1990s, anti-Asian sentiment was fueled by the success of Asian Americans relative to other ethnic groups, and also by the economic prowess of Asian societies, especially Japan, and their ability to compete with the United States in the American as well as the global market. Business leaders, labor officials, and politicians accused Japan of unfair trade practices, which they claimed had created many of the ills of American industry. Advertisements encouraging people to buy American-made products were tinged with anti-Asian sentiments, and many politicians used economic nationalism as a primary component of their campaigns. These actions redounded to Asian Americans (Mydans, 1992a).

Perhaps the most compelling and dramatic illustration of the impact of economic nationalism on Asian Americans was the killing in Detroit of Vincent Chin in 1982. While celebrating his impending wedding with friends at a bar, Chin, a young Chinese American engineer, was verbally accosted by two white autoworkers, one of them unemployed at the time. Assuming that Chin was Japanese, they told him that he and one of his friends, also an Asian American, were "the reason we're all laid off." A fight ensued, after which the two chased Chin and clubbed him to death on the street. The attackers were fined $3,780 and given three years' probation after one pleaded guilty and the other pleaded no contest to charges of manslaughter. The leniency of the sentence outraged the Asian American community, and the U.S. Justice Department subsequently filed civil rights charges against one of the attackers. He was convicted by a federal jury of violating Chin's civil rights and sentenced to twenty-five years in prison. That conviction was later overturned on appeal, however, and he was acquitted of all charges, ending any further criminal prosecution

(Moore, 1987). The Chin killing and its aftermath still resonate strongly among many Asian Americans several decades after the event (Clemetson, 2002).

In addition to the Chin case, several other incidents in the 1980s provoked strong reactions. Each involved violent attacks on Asian Americans that resulted in only nominal punishment by the courts (U.S. Commission on Civil Rights, 1986). Other less dramatic but no less disturbing anti-Asian actions, many motivated by unadulterated racist sentiments, were reported in the 1980s and early 1990s (Noble, 1995; U.S. Commission on Civil Rights, 1992).

Two events of the late 1990s seemed to galvanize Asian Americans around issues of perceived discrimination. In one case, some contributions given to the Democratic National Committee (DNC) during the 1996 presidential election raised by a Chinese American former official of the DNC were later found to be illegal, having been channeled by overseas Asian political influence seekers. Subsequent investigation of illegal contributions cast suspicion on Asian American contributors who had done nothing wrong (Tajitsu Nash and Wu, 1997; Wang, 2003). In another incident, Wen Ho Lee, a Taiwanese-born U.S. citizen and former government-employed nuclear scientist, was accused of espionage in 1999. He was indicted on several dozen felony counts for downloading classified nuclear design and testing information at the Los Alamos National Laboratory in New Mexico. Subsequently all but one charge was dropped in what was seen as a major embarrassment for federal prosecutors (Farhi, 2006; Wu, 2003). Both of these cases caused much chagrin among Asian Americans, many of whom saw them as blatant instances of anti-Asian stereotyping and discrimination.

ASIANS AND OTHER ETHNIC MINORITIES   In addition to harassment and hostilities issuing from the white majority, Asian Americans have sometimes been targets of violence by inner-city blacks and Latinos. As previously noted, many Asian immigrants turn to small business, and they often operate groceries, restaurants, liquor stores, and laundries in minority areas of large cities. In this, they are playing a middleman minority role. And as we saw in Chapter 2, such middleman business owners frequently become the targets of hostility for their customers. Tensions between Asian business owners and blacks in inner-city areas have been especially high in recent years. Animosities have arisen from the perception of blacks that Asian shopkeepers often deal with them disrespectfully and have no interest in the black communities in which they operate, other than earning a profit. Hence, they see Asians as exploiters. Shopkeepers, in turn, complain of the frequency of harassment, theft, and vandalism of their businesses. Spike Lee's 1989 film, *Do the Right Thing,* is a graphic depiction of the tensions that often accompany the establishment of Asian (mostly Korean) small businesses in African American neighborhoods.

In a widespread pattern, black boycotts of Asian-owned businesses occurred in several cities in the 1980s and 1990s including New York, Philadelphia, Washington, and Los Angeles. During the Los Angeles riots of May 1992, shops owned by Koreans in the South Central district, where most of the disturbances occurred, seemed to be particular targets of looters and arsonists. Before the riots, relations between Korean shopkeepers and their black customers had been volatile for years and seemed to reach a high point when a Korean-born grocer shot to death a fifteen-year-old black girl whom the merchant had accused of stealing a bottle of orange juice. Black resentment grew even stronger when the woman grocer, who was convicted of voluntary manslaughter, was given only a probationary sentence (Mydans, 1992b; Ong et al., 1992). Discordant relations between Koreans and blacks were most dramatic in Los Angeles as a result of the 1992 riots, but a similar undercurrent was evident in the inner cities of other urban areas where Koreans operated small businesses serving minority residents (Chesley and Gilchrist, 1993; Gold, 2010; Goldberg, 1995; Wilkerson, 1993).

Friction between Koreans and blacks in business relationships should not be seen as unique, however. Similar animosities have arisen among blacks and merchants of other ethnic groups operating their enterprises in inner-city areas. In Detroit, for example, most inner-city neighborhood groceries and liquor stores are owned by Arab Americans or Chaldean Americans, who are seen in much the same way as Koreans in other cities (Dawsey, 1992; Gerdes, 1993). Moreover, conflict between neighborhood coethnics and shop owners of an outsider ethnic group has a long tradition in many countries and has been evident in diverse parts of the world in recent times (Chua, 2004; White, 1993).

MODERATING LEVELS OF PREJUDICE AND DISCRIMINATION   Today, there seems little chance of a return to the extreme levels of prejudice and discrimination that characterized the Asian American experience of earlier decades. Labor market discrimination against Asian Americans, for example, is no longer a serious issue (Sakamoto and Xie, 2006). One of the more apparent differences today is that negative stereotypes of the past have given way to a generally positive collective image. Indeed, Asian Americans are increasingly seen as role models for other ethnic groups. As sociologist Peter Rose has put it, "The pariahs have become paragons, lauded for their ingenuity and industry and for embodying the truest fulfillment of the 'American Dream'" (1985:182). Moreover, despite isolated cases of blatant and direct discrimination, such as those that occurred in the 1980s and 1990s, most instances of discrimination today are more subtle and indirect.

Occasional episodes of violence against Asian Americans have been motivated in recent years primarily by identification (or misidentification) as Muslims. Some Asian Americans—namely, South Asian Muslims, mostly from Pakistan and Bangladesh—found themselves victimized by the anti-Muslim hysteria that characterized the American response to the terrorist attacks of 9/11. Sikhs from India, although not Muslim, were also harassed through the mistaken identity of their traditional turbans (Bakalian and Bozorgmehr, 2009). Such incidents occurred with some frequency in the immediate aftermath of the September 11 attacks, but since then they have been isolated cases. For example, in 2012 in another case of mistaken identity, a gunman attacked a Sikh temple outside Milwaukee, killing six persons.

Perhaps the most compelling evidence of the decline of anti-Asian discrimination is the view of Asian Americans themselves. A 2013 comprehensive study (Pew Research Center, 2013) concluded that, in general, "Asian Americans register fairly low levels of concern about group discrimination" (97). Only about one-in-eight respondents in the study stated that discrimination is a major problem; about half said that it is a minor problem, and about a third viewed it as not at all problematic (Figure 9.5). Moreover, a majority said that their Asian identity was not a hindrance to getting into college, finding a job, or getting a promotion. Another national survey of Asian Americans (Ramakrishnan and Lee, 2012) indicated that only a tiny percentage view race and racism as important problems for the Asian American

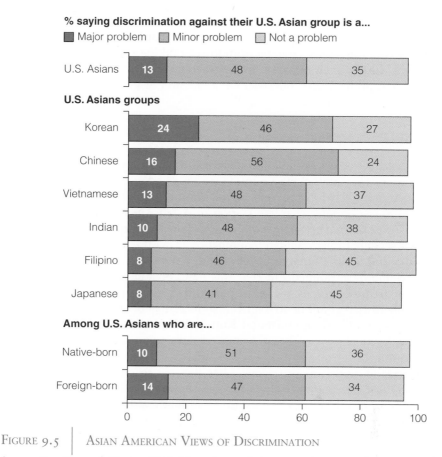

% saying discrimination against their U.S. Asian group is a...

FIGURE 9.5 | ASIAN AMERICAN VIEWS OF DISCRIMINATION

population. These views contrast very strongly with those of African Americans and Hispanic Americans, who see discrimination in various social areas as a serious on-going problem for their groups.

One key indication of a lower rate of discrimination against Asian Americans than against other nonwhite groups today is their more rapid rate of residential integration, exceeding that for both blacks and Latinos. Even though their presence in American society is more recent than either of those two groups, the level of Asian-white residential segregation is comparable to that of Hispanics from whites and much lower than the segregation of blacks from whites (Bayer et al., 2004; Logan, 2011; Logan et al., 2004; Massey, 2001; Mumford Center, 2001a; Pew Research Center, 2013). Ironically, the success of Asian Americans in education has created a potential backlash from majority whites and other ethnic minorities who see themselves being surpassed in competition for the society's rewards. For example, because Asian Americans qualify for admission in such large numbers, some of the most prestigious universities have limited their enrollment through the use of nonacademic criteria. Despite significantly higher GPAs and scores on entrance examinations, Asian applicants are often rejected in favor of lower-ranking students of other ethnic origins (Golden, 2007). Such measures are markedly similar to the quotas used by Ivy League schools in the 1920s to restrict the admission of Jews, who also qualified in disproportionate numbers.

ASIAN AMERICANS AS A "MODEL MINORITY"

The clear movement toward assimilation, demonstrated most powerfully by the extraordinary success they have achieved in the realms of education and the economy and by their comparatively less problematic adaptation to American society, has led in recent years to the popular characterization of Asian Americans as a **model minority**—that is, *an example of success that other minority ethnic groups might emulate*. Although the label is seemingly complimentary, many Asian Americans today see it as condescending in a number of ways and, paradoxically, a case of harmful stereotyping. Some have pointed out that this new, positive image tends to gloss over the shabby treatment of Asians in the past and disguises the serious social and economic problems that continue to confront some elements of the Asian American population (Chao et al., 2010; Hurh and Kim, 1989; Wu, 2002). Obscured is the fact that some, particularly Southeast Asians, remain poor, uneducated, and relatively unsuccessful.

In addition, the model minority image has become so well entrenched that it places tremendous social and psychological pressures on Asian Americans, especially young people. Students, for example, must do exceedingly well in their studies and in other pursuits to uphold this image and may even feel that they must outperform their non-Asian peers. This becomes especially problematic for recent Asian immigrants whose command of English may be minimal and who therefore suffer added burdens in an educational environment that is not attuned to the unique cultural needs of Asian students (Bernstein, 1988).

The model minority image has also seemed to create a new set of stereotypes, which though generally positive, nonetheless create the same kinds of group effects as other, negative, stereotypes: they induce the perception of commonality of traits among all individual ethnic group members. A Chinese American schoolteacher expresses the frustration produced by the new stereotypes: "Years ago they used to think you were Fu Manchu or Charlie Chan. Then they thought you must own a laundry or restaurant. Now they think all we know how to do is sit in front of a computer" (quoted in *Time*, 1987). Moreover, as explained in Chapter 3, positive traits attributed to ethnic groups may easily be manipulated so that they become negative in content and thus serve as the rationales of prejudicial thought and discriminatory behavior. If Asian Americans today are seen as "brainier" and "more ambitious" than others, such seemingly positive traits may easily be transformed into negative ones like "shrewd" and "pushy" should Asians come to be seen as outcompeting the dominant group.

The characterization of Asian Americans as a model minority may also work to deflect attention from the structural problems of other minority ethnic groups in the society. If Asian Americans can attain success despite their racial visibility, it is argued, even race cannot be viewed as an overwhelming handicap in American society. As Asian Americans are seen as hardworking, persevering, and highly motivated, the failure of other minority racial and ethnic groups to reach comparable levels of achievement is attributed to a lack of determination and hard work rather than to structural problems created by changing economic and social institutions or to different group histories (D'Souza, 1995).

ASIAN AMERICAN ASSIMILATION

As with economic stratification, in a number of ways assimilation for Asians in the United States exhibits a dual pattern. On the one hand, within all Asian American groups those who are highly educated and who hold prestigious jobs are increasingly moving toward both cultural and structural assimilation. On the other hand, those less affluent and educated retain much of their traditional ways and have had more difficulty moving beyond the ethnic enclave.

In examining assimilation among Asian Americans, it is important to consider that much of the Asian American population today is foreign-born. It is thus too early to marshal solid evidence regarding the rate and extent of assimilation for those who have not gone beyond the first generation. Most evidence must necessarily come from Chinese and Japanese Americans, both groups well into the third and fourth generations. With these limitations in mind, let's look at the patterns and prospects of Asian American assimilation.

Chinese and Japanese Patterns

Chinese American Assimilation   As in other societies to which they have migrated, the Chinese in America traditionally exhibited a resistance to assimilation. The characteristic features of Chinese ethnic communities worldwide seemed to be social and cultural exclusiveness and a low level of absorption into the larger society (Lyman, 1968a, 1974; Purcell, 1980). Much of this self-imposed isolation was the product of Chinese community organizations, which in the new society reflected extended family patterns, known as lineages. Traditionally, much of Chinese social life in the United States was organized around various kin, clan, and secret societies.

Additional reasons have been suggested for the historical lag in assimilation among Chinese Americans. For one, the early demographics of this group—almost all males without wives and families—prohibited a second generation from establishing stable families and moving along an assimilation path comparable to that of other immigrant groups (Kitano and Daniels, 1995; Lyman, 1974). Moreover, the Chinese were prohibited from acquiring citizenship and thus remained legally marginal. Finally, the levels of prejudice and discrimination were unusually harsh, barring Chinese from closer relationships with members of the dominant group or from entering into mainstream institutions. For their first one hundred years, the Chinese were a relatively isolated ethnic group in American society.

Despite past patterns of insularity, it appears that Chinese Americans today are moving rapidly in the direction of assimilation, both cultural and structural. Regarding cultural assimilation, a study of Chinese Americans in six Southern California counties found that although 79 percent spoke mostly Chinese at home, 68 percent spoke English when conducting personal business; among those between the ages of eighteen and twenty-nine, 88 percent spoke English outside the home. Only 16 percent of the respondents' children spoke only Chinese (*Los Angeles Times*, 1997).

A more recent study of Chinese Americans in New York found a similar proclivity among the second generation to discard the parental language (Kasinitz et al., 2008).

JAPANESE AMERICAN ASSIMILATION   Among Asian American groups, the Japanese have experienced a more profound assimilation than others. This is partly because they have resided in the United States for several generations and their numbers have not been swelled in recent years by new immigrants. Patterns of assimilation among Japanese Americans, however, differ on the basis of generation. Japanese Americans refer to each generation with a particular name. The first, or immigrant, generation is called *Issei;* the children of the Issei, or second generation, are referred to as *Nisei;* and their children, the third generation, are called *Sansei.* Japanese Americans are now well into the fourth generation, referred to as *Yonsei.*

The Issei represent a generation that did not assimilate. In this, they did not differ essentially from other immigrant groups. However, the much more severe prejudice and discrimination directed against them produced a greater disincentive to venture out of the ethnic milieu and attempt to interact with the larger society.

By contrast, Nisei, and especially Sansei and Yonsei, are highly assimilated culturally, as indicated by the very thorough adoption of dominant cultural traits (Levine and Rhodes, 1981; Onishi, 1995). This is not surprising in light of the high level of education among Japanese Americans of the second generation and beyond, and the great strides each generation has made in socioeconomic advancement.

Regarding secondary structural assimilation, we have already seen the marked success of Japanese Americans in economic and educational realms of the society and their increasing movement into various institutional areas. Residential patterns of Japanese Americans also suggest a relatively fluid movement into the mainstream society (Kitano, 1976; Levine and Rhodes, 1981).

At the primary level of structural assimilation, there is strong evidence that Japanese Americans have begun to leave the ethnic community and forge close associations with non-Japanese. This is particularly apparent among those beyond the second generation (Levine and Rhodes, 1981; Onishi, 1995; Qian et al., 2001). The most telling indicator of primary structural assimilation is intermarriage, and in this, several studies have found significant increases, especially for the Sansei (Kikumura and Kitano, 1973; Kitano and Daniels, 1995; Lee and Fernandez, 1998; Levine and Rhodes, 1981; Montero, 1980; Tinker, 1973). Among his sample of Japanese Americans, Montero (1980) found that 40 percent of the Sansei had married a non-Japanese. This compared to only 10 percent of the Nisei and 1 percent of the Issei. Kikumura and Kitano (1973) found a similar pattern and concluded that Japanese Americans chose marital partners as frequently from outside the ethnic group as from within. Data on out-marriage of Asian Americans in Los Angeles between 1975 and 1989 showed a rate of more than 50 percent for Japanese Americans (Kitano and Daniels, 1995). More recent studies confirm the strong trend toward exogamy among Japanese Americans, indicating that, among Asian American ethnic groups, they are most likely to intermarry with whites (Qian et al., 2001; Qian and Lichter, 2007). The comprehensive 2013 Pew study of Asian Americans indicated that more than half of new marriages of Japanese Americans are with non-Asians, a rate higher than for any other Asian American group.

In general, these studies indicate a progressive trend toward primary structural assimilation and an increasing dilution of the Japanese American ethnic community with each succeeding generation. In light of their rapid and extensive assimilation, some have even begun to question the ability of Japanese Americans to maintain a viable ethnic community into the next (Yonsei) generation (Montero, 1980; Onishi, 1995).[9] Others, however, maintain that despite very strong assimilation tendencies, Japanese Americans do not appear to be losing all cultural distinctiveness and identification (Kitano, 1976; Woodrum, 1981). One study, in fact, concluded that Japanese Americans appear to have been able to successfully combine high levels of structural assimilation with a strong retention of ethnic identity and group cohesion (Fugita and O'Brien, 1991).

THE FUTURE PATH OF ASIAN AMERICAN ASSIMILATION

In examining issues of stratification, the internal cultural, racial, and class variability of the Asian American population was emphasized. In looking ahead to future assimilation patterns, these differences once again must be kept in mind. As with stratification, the rate of assimilation among Asian Americans will differ from group to group. Vietnamese and other Southeast Asian groups, most recent and least prosperous, currently appear to be that part of the Asian American population experiencing the most tortuous path toward cultural and structural assimilation. At the other extreme, Japanese Americans, one of the oldest and most prosperous among Asian American groups, continue to move most directly and rapidly in that direction.

Asian American structural assimilation is evident at the secondary level in occupational and residential patterns and at the primary level in the substantial rise in intermarriage. As to residential patterns, Asian Americans are the least segregated among all American racial/ethnic categories. More than others, they are likely to live in racially mixed neighborhoods. Only 11 percent of Asian Americans currently live in a census tract in which Asian Americans are a majority (Pew Research Center, 2013).

Japanese American intermarriage was noted earlier, but for Asian Americans generally, intermarriage is noticeably higher than for African Americans and Hispanic Americans (Lee and Bean, 2010; Lee and Edmonston, 2005; Passell, 2010; Pew Research Center, 2013; Qian and Lichter, 2007; Wang, 2012). Between 2008 and 2010, almost 30 percent of recent Asian newlyweds married a non-Asian (Pew Research Center, 2013). Asian groups with the lowest level of intermarriage are those with the newest immigration cohorts, such as Asian Indians, Vietnamese, and recent Chinese. This may be seen as solid evidence of fading social boundaries between Asian Americans and the larger society with each successive generation. Indeed, as with Latinos, Asian Americans give evidence of following generational patterns of intermarriage that are similar to those of Euro-Americans during the early twentieth century (Lee and Bean, 2010; Stevens et al., 2006; Wang, 2012).

Often overlooked in considering assimilation among Asian Americans generally is the fact that almost half of this population identify themselves as Christian (Kosmin

---

[9] Montero (1980) notes that, ironically, the remarkable upward economic mobility of the Japanese in America may lead to a diminution of the traditional ethnic values that helped to stimulate that advancement in the first place; hence, the loss of those values may create a leveling off of the achievement of future generations.

and Keysar, 2006; Pew Forum on Religion and Public Life, 2012a). Among Koreans, a majority do. This religious commonality with the society's dominant group has no doubt served for many as a valuable support in advancing cultural assimilation.

For the new immigrants, the occupational and educational achievements of many will enhance the prospects for both cultural and structural assimilation. Already it is evident, for example, that Asian Americans are dispersing throughout the urban areas in which they are concentrated. More than 60 percent of Asians lived in suburbs in 2008, a higher percentage than for any other ethnic minority, despite their relatively recent arrival in large numbers. And, that suburban trend continues unabated (Frey, 2010).

Another strong indicator of continuing Asian American assimilation is individuals' social networks. Among all Asian Americans, less than half say that all or most of their friends are of the same ethnic origin. Most telling, however, is the difference between native-born Asian Americans and immigrants. Whereas half of the foreign-born say that their friendship network is mainly with coethnics, only 17 percent of native-born Asian Americans say the same (Pew Research Center, 2013).

IMPEDIMENTS TO ASSIMILATION In Chapter 11, we will see that the assimilation of Jews in the United States has advanced as far as in any society where Jews have lived in large numbers. Yet the religious divergence between Jews and non-Jews remains a factor that continues to impede full assimilation of Jewish Americans. The same can be said for Asians, whose physical distinctiveness makes them even more indelibly conspicuous within the larger society. Because of this visibility, non-Asian Americans continue to see Asian minorities as somehow not fully American, as "outsiders"—even though their ancestors may have been in the United States for several generations, they may speak only English, and they have no more ties to Asian societies than Swedish Americans have to Sweden or Italian Americans to Italy. A twenty-four-year-old U.S.-born Detroit resident expresses the common frustration of Asian Americans who are always questioned about their origin. "Where are you from?," she's constantly asked, followed by "Where are you *really* from?" In the eyes of whites, as she puts it, "[w]e're the perpetual foreigner" (Ding, 2008). Because of racial visibility, then, ethnic identity and consciousness among Asian Americans—regardless of the extent of their assimilation—is not likely to fully erode.

As will be discussed in Chapter 13, the preeminence of multicultural ideology in schools, universities, and the media has made it easier for ethnic minorities to preserve or perhaps reaffirm aspects of the ethnic culture. This is no less the case for Asian Americans. Multiculturalism is also helping to create a consciousness of what may have been a quite inchoate Asian American panethnicity. Tamar Jacoby (2000) maintains that young, college-educated Asian Americans appear to be experiencing a kind of ethnic reaffirmation in large measure as a product of the oppositional multiculturalism to which they are exposed in their university experience. Moreover, the creation of institutionally complete ethnic enclaves in some areas where Asian Americans constitute a significant portion of the population provides a further defense against rapid assimilation.

Despite these impediments to assimilation, Asian Americans, like Hispanic Americans, appear to be reconciling the competing demands of ethnic retention and incorporation into the societal mainstream. Asian Americans seem determined to be both

Asian and American, seeing no inherent incompatibility between the two and no conflict in balancing them. Jacoby quotes a young Stanford graduate, who asserts that "I'm determined to hold on to my Chinese heritage and my Chinese culture. But there's absolutely nothing about America that I reject because of my ethnicity." And a Chinese business owner echoes this view, claiming that assimilation and ethnic retention are "two parallel tracks, and there's no reason to choose between them" (Jacoby, 2000:25). A poignant illustration of this came in 2013 when a second-generation Indian American was named Miss America. For the talent portion of this quintessentially American competitive event, she performed classic and contemporary Indian dances.[10]

ASIAN AMERICAN PANETHNICITY   As explained, the term *Asian American* is a misnomer given the cultural and physical variety of the population to which it refers.[11] Yet a social and political coming together of the various elements of this broad and extremely diverse ethnic category may be becoming a reality. The continued use of the collective *Asian American* by government, academic, and media institutions already reflects the perception of a more coherent ethnic population than actually exists. This is not unlike the use of the term *Hispanic American* or *Latino* to collectively describe the various groups of Latin American origins. And it is not unlike the process of ethnic group formation that characterized the American experience of many European groups of the earlier immigration periods. Only in the American context were Italians from various provinces in Italy, for example, designated and dealt with collectively as "Italians." Through this common treatment by Americans, an Italian American group consciousness was engendered, and eventually Italians in America identified themselves primarily as "Italian" rather than as "Calabrian," "Sicilian," or "Neapolitan."

Weighing against the further development of Asian panethnicity, of course, are the radical cultural and racial differences among the various groups that make up the Asian American population. Moreover, immigration experiences and patterns of adaptation to American society have not been the same for these groups. It would be hardly meaningful to equate the class and cultural interests of Asian Indians on the one hand and Cambodians on the other, for example. Nonetheless, Asian Americans of diverse ethnic origins may increasingly recognize this collective designation as a useful political strategy, just as Latinos are beginning to do (Espiritu, 1992; Shinagawa and Pang, 1996). As separate and dispersed groups, none can exert the kind of political leverage at the national or even local level that a more consolidated position may produce.

Asian Americans, then, will continue to wrestle with issues of ethnic identity and assimilation. A majority of Asian Americans today identify themselves not as "Asian Americans" but as members of specific ethnic subgroups: "Japanese Americans," "Vietnamese Americans," and so forth (Pew Research Center, 2013). This is similar to the ethnic self-identities of Hispanic Americans, as we saw in Chapter 8. Nonetheless, American racial categorization may drive members of disparate Asian groups toward some collective Asian American identity, while they remain cognizant of their particular ethnic heritage. In her study of second-generation Chinese and Korean

---

[10] In response to numerous bigoted comments made on social media following her selection, she declared that "I have to rise above that. I always viewed myself as first and foremost American." It is also noteworthy that the first runner-up in the competition was Chinese American.

[11] Jacoby points out that "the very term 'Asian American' traces back to the 1960s, when it was coined, in the spirit of 'Black Power' and 'Chicano Studies,' as a kind of battle cry" (2000:22).

Americans, Kibria (2002) found elements of such a dual identity pattern. Beyond ethnic identities at any level, however, there are undeniable forces of assimilation thrusting Asian Americans into the societal mainstream.

SUMMARY

- Asian Americans are a diverse collection of distinct ethnic groups, the largest among them Chinese, Filipino, and Asian Indian.
- Asian immigration to the United States consists of two periods. The first began in the mid-nineteenth century with the immigration of Chinese manual laborers. Exclusionary measures beginning in 1882 halted further Chinese immigration and provided an opening for Japanese workers to replace them in the early part of the twentieth century. Discriminatory laws brought further Asian immigration to a virtual halt in the 1920s.
- The second period of Asian American immigration extends from 1965, when immigration laws were liberalized, to the present. About half of Asian Americans today are first-generation immigrants who stem from this second wave.
- Many of those of the second immigrant wave have come to the United States with much higher occupational skills and more education than earlier immigrants.
- Asian Americans are heavily concentrated in a few states and urban areas. The largest populations are in California and the New York metropolitan area.
- Asian Americans, as a whole, are at or near the top of the stratification system in terms of income, education, and occupation. There are important differences between and within the various Asian groups, however.
- Whereas high levels of prejudice and discrimination marked the Asian American experience during the first wave of immigration, Anti-Asian beliefs and actions today are comparatively mild and indirect. In recent years, Asian Americans have been portrayed as a model minority because of their exceptional economic and educational attainments.
- High levels of cultural assimilation characterize those Asian American groups who have entered the third and fourth generations—specifically, the Japanese and Chinese—and there is also evidence of increasing rates of both primary and secondary structural assimilation among them. There are also strong indications of cultural and structural assimilation among Asian Americans of more recent immigration.

CRITICAL THINKING

1. Analyze the factors that account for Asian American economic and educational success, in comparison with other ethnic populations that have entered the United States in large numbers in the past fifty years.
2. Japanese Americans before and during World War II were the targets of some of the most severe prejudice and discrimination ever endured by any American ethnic group. Although most were U.S. citizens, they were commonly seen as "the enemy." Are there parallels to that historical case with the present-day plight of Muslim and Arab Americans?

3. There is strong evidence that Asian Americans are assimilating structurally more quickly than blacks or Latinos. What accounts for this? Consider the various dimensions of structural assimilation, discussed in Chapter 4.

4. Asians in recent decades have immigrated in large numbers to several societies in addition to the United States, including Canada, Australia, and the United Kingdom. Has their experience in those societies paralleled that of Asian Americans? What are the evident similarities and differences, and how can these be explained?

Personal/Practical Application

1. Consider the idea of the "model minority," discussed in this chapter. Can you relate to the social pressures that Asian Americans often describe in meeting social expectations?

2. As explained in this chapter, intermarriage among Asian Americans is higher than for any other racial minority group. What might explain this?

WHITE ETHNIC AMERICANS

In this chapter, we look closely at white (or Euro-American) ethnic groups, specifically those described in Chapter 5 as part of the second tier of the American ethnic hierarchy, whose origins are chiefly part of the classic period of immigration that occurred during the nineteenth and early twentieth centuries.

WHY STUDY WHITE ETHNIC GROUPS?

It is important to look carefully at white ethnic groups for several reasons.

- They are a substantial element of ethnic America. Like other panethnic designations, of course, the very term *white ethnic* is only a categorical convenience, linking together numerous distinct groups. Nonetheless, there are many apparent similarities in the original immigration and settlement patterns these disparate groups followed and in their movement toward assimilation into mainstream American society.
- Because ethnicity has sharply declined in importance for white ethnic groups, it is easy to assume that it is no longer a factor in the social, political, and economic life of those who identify with them. But, as we will see, this assumption seems exaggerated and premature.
- None of these groups has reached parity with the dominant Anglo group in economic status and political power, none has entered fully into primary relations with the Anglo group, and none has completely lost its cultural form through assimilation. Hence, although white ethnics are becoming less clearly defined as their cultural and structural assimilation continue unabated, they remain discernible ethnic populations in American society.
- The social and economic disparities between Euro-American and racial-ethnic groups are a troubling issue of American society and have generated much debate among social scientists, policy makers, and the general public. Some see the gap as a product of internal group differences and others as the result of

inequitable social and economic opportunities. To shed light on this issue, it is necessary to understand how the characteristics and social experiences of white ethnic groups have been similar to and different from those of the country's racial-ethnic groups. What cultural systems and social and economic skills did each bring with them? What was the nature of the opportunity structure open to each? What obstacles to social integration for each were dictated by the dominant group? To what extent did each seek assimilation or strive to preserve ethnic uniqueness? What was the nature and level of prejudice and discrimination directed at each? As we will see, although there are similarities between white ethnic groups and racial-ethnic groups on all these questions, there are also enormously consequential differences.

In much of this chapter, we focus primarily on two groups, Irish Americans and Italian Americans. The Irish are the largest group that came to America as part of the Old Immigration, that period roughly from the 1830s to the 1880s, whereas the Italians constituted the largest element of the New Immigration, from the 1880s to the 1920s. However, it is not only their size that makes it logical to study these two white ethnic groups in some detail. The Irish, as we will see, entered at the very bottom of the ethnic hierarchy but have advanced to the point where today they are, arguably, the most assimilated white ethnic group. In fact, in the estimation of some, they are now part of the dominant group. As for Italian Americans, in a number of ways they constitute the prototypical white ethnic group, illustrating well much of the general experience of Catholic ethnic groups of the New Immigration.

Religion is the most critical factor distinguishing white ethnics in American society. As Catholics and secondarily Jews (who will be examined in the next chapter), they were immediately set apart from the Protestant majority at the time of their entrance and given a strongly negative reception. Also, the fact that they were noticeably distant from the Anglo core group in other cultural ways and that they were generally poor at the time of arrival helped to set them on a more tortuous path of adaptation than had been experienced by earlier immigrants.

IRISH AMERICANS

The immigration of the Irish represents in several ways a watershed point in the history of immigration to the United States.

- Irish immigration constitutes the first large-scale movement to America of non-Protestants. Roughly half of the German immigrants of the Old Immigration were also Catholic, but in numbers they were far overshadowed by the Irish.
- The immigration from Ireland beginning in the mid-nineteenth century represents the first great influx of low-skilled workers destined for the mills and factories of growing American cities, rather than the farms and small towns that had been the more usual destinations of previous immigrants.
- Irish immigration gave rise to a virulent nativist movement aimed at limiting the entrance to America of non-Protestants. In the decades following the Irish influx, efforts to restrict immigration would continue, culminating in the 1920s in the cessation of large-scale immigration from Europe.

- Irish immigration prompted a redefinition of "race" in American society. Irish Catholics at first were not considered fully part of either the white or black population, and this indeterminate racial status was not resolved for several decades. Other European groups that would follow the Irish, those of the New Immigration, would present a similar dilemma of racial classification.

IMMIGRATION AND SETTLEMENT

We can better understand the Irish American experience by first examining the status of the Irish in their origin society before the great migration began.

ENGLAND IN IRELAND   Although a pattern of English domination reaches back to the twelfth century, it was in the early part of the seventeenth century that England embarked on its thorough colonization of Ireland. The most comprehensive colonizing effort was made in Ulster, the six counties that eventually would be set off from the rest of Ireland as Northern Ireland. There, Irish lands were confiscated and given to Scottish Presbyterians and English Anglicans to settle on. The native Catholics were either driven out of Ulster or subdued into a totally subordinate position. In the remainder of Ireland, the population was reduced to a slavelike status.

In the English view, the native Irish were uncivilized people against whom the most brutal tactics could be employed. Indeed, historian George Fredrickson (1981) suggests that the experience of the English in Ireland foreshadowed their colonizing practices and policies toward indigenous peoples in North America. If the natives were uncooperative in ceding their lands, the "savage" image could be invoked, and they could then be either exterminated or confined to reservations. Unlike the settlers' experience with the indigenous peoples of North America, however, the Scottish and English settlers could not so easily dispose of the Catholics. For one, there were simply too few settlers to impose total control. More important, English domination was not accepted easily by the subordinate Irish, and their resistance to colonial rule proved continually problematic. As Marjorie Fallows describes it, Ireland was "too close to England to be ignored" but "too rebellious to be tamed" (1979:16).

In 1641, Irish Catholics revolted in a massacre of Protestants, an action that was avenged a few years later with a countermassacre by the armies of the English dictator Oliver Cromwell. Afterward, confiscation of remaining lands held by Catholics began, and by the early 1700s, Catholics possessed only a tiny portion of all land in Ireland.

THE SCOTCH-IRISH   At the end of the seventeenth century, James II, the last Catholic English monarch, was defeated by William of Orange. What followed William's victory was an even more sustained suppression of the Catholic populace. The Catholic clergy was also restricted, as was the Irish educational system. A series of statutes called Penal Laws were enacted, intended to divest Catholics of all political and economic power. Though brought to bear most severely against the Catholic population, many of these oppressive laws were also applied against the Scottish Presbyterians, who had settled mostly in Ulster. This gave rise to the mass emigration of thousands of Ulster Presbyterians to the United States. These settlers were to be known in America as the Scotch-Irish and would make up a major element of the U.S. colonial population in the eighteenth century.

The immigration of Irish Protestants must be seen as distinct and fundamentally different from the immigration of Irish Catholics. In America, Irish Protestants assimilated early on and afterward were in no essential way distinguishable from the Anglo dominant group. Indeed, it was Irish Protestants themselves who appropriated the term *Scotch-Irish* as a way of differentiating themselves from the subsequent waves of Irish Catholics (Kenny, 2000).

THE FAMINE IRISH   Although they were the earliest immigrants from Ireland, the Scotch-Irish did not form the mold of Irish ethnicity in America. Rather it was the much larger and later-emigrating Catholic population of the south of Ireland who became the standard bearers of Irish ethnicity in America and who remain the group of reference when speaking of Irish Americans.

The mass immigration of Irish Catholics to America beginning in the 1830s was marked by almost classic push-pull factors. Potatoes had been the Irish staple food and was, literally, the means of subsistence for the peasant population. Although crop failures had been endured in the past, the blight that struck in the late 1840s was nothing short of catastrophic. Famine and disease were rampant, resulting in perhaps a million deaths (Miller, 1985). For most, the choice was simple: emigrate or perish. As historian Lawrence McCaffery has written, most Irish Catholics arriving in the United States during this period "were running away from destitution and oppression rather than rushing toward freedom and opportunity" (1997:7). When combined with the push of a deteriorating rural economy, the pull of North America with its seemingly unlimited labor opportunities was irresistible. The response was overwhelming. Between 1845 and 1855, about one-fourth of Ireland's entire population emigrated overseas, almost 1.5 million sailing to the United States and another 340,000 to Canada. Many of those who initially landed in Canada later found their way to the United States.

The journey across the Atlantic was, for this generation of Irish immigrants, almost as harrowing as the passage of African slaves being transported to America. Disease and starvation ensured that a large percentage would never make it to their destination. In 1847, 30 percent of those bound for Canada and 9 percent of those destined for the United States perished while on board ship or shortly after landing (Kenny, 2000). So many died making the crossing—which could take as long as six weeks—that the vessels carrying Irish immigrants became known as "coffin ships."

In the decades following the Great Famine, push and pull factors stimulating Irish emigration—enduring rural poverty in Ireland and the promise of jobs elsewhere—did not essentially change. As a result, Irish immigration to North America continued; more than 2.5 million Irish left for America between 1855 and 1900. With the continued flow of immigration, large Irish American communities took shape, which in turn financed the passage of future immigration from the island. Hence, many leaving Ireland had their passage paid for by relatives or friends already settled in North America (Meagher, 2001).

SETTLEMENT AND WORK   The Irish were in many ways unprepared for life in an industrial, urban society. Most were poor peasants who lacked the skills and education that could be converted into rapid occupational mobility and economic success in the new society. As a result, most entered the labor force in unskilled jobs, where they and their children remained. The Irish were, as Edward Levine describes them, America's "first truly proletarian labor force" (1966:59). They found work in railroad and canal construction, as stevedores on docks of major ports, and in the various building projects of the expanding cities. Wherever they settled, the Irish were disproportionately concentrated in the lowest-paid, least-skilled, and most dangerous and insecure jobs (Glazer and Moynihan, 1970; Wittke, 1967).

The majority of Irish immigrants settled in cities of the Northeast, especially New York and Boston, where work was plentiful in manual laboring occupations that were compatible with their low skill levels. This was a settlement pattern unlike that of other groups of the Old Immigration, like Germans and Irish Protestants (the Scotch-Irish), most of whom scattered to other regions. The overwhelming settlement of Irish immigrants in urban areas seems a paradox in light of the fact that most were rural or small-town peasants in origin. Their poverty, however, almost automatically ruled out establishing themselves as farmers. Their immediate need was to secure work as unskilled laborers, which they found in the cities. Moreover, their rural experience had been mostly as workers familiar only with potato cultivation, not as owners or managers (Jones, 1992). Irish tenant farmers in Ireland also had not acquired the skills needed for the style of farming prevalent in America. Most were field laborers who knew only how to work small bits of land using simple tools, an agricultural skill set not suited to large independent farms requiring knowledge of planting and harvesting (Levine, 1966; McCaffrey, 1997).

The squalor of living conditions endured by early Irish immigrants to America was comparable to that found in European cities of the nascent industrial age. These teeming urban slums produced Irish death rates that were actually higher than they were in Ireland (Wittke, 1956). With the added burdens of blatant discrimination by Anglo Protestants in almost every aspect of social and economic life, many turned to drink and to crime. High crime and alcoholism rates, in turn, further strengthened the popular image of the Irish as incorrigible, corrupt, violent, drunken, and incapable of ever becoming acculturated to American—that is, Protestant—ways. And, they gave added stimulus to what had already been a virulent anti-Catholic nativism.

WOMEN IN THE WORKFORCE A unique aspect of the Irish immigration to America was its large component of single women. In fact, as many women came as men, all seeking work. After the Great Famine, a system of land inheritance was institutionalized in Ireland in which only one son could acquire the family farm. Because marriage was often dependent on having land as a means of supporting a family, men commonly married late and many did not marry at all. This obviously reduced the chances of marriage for Irish women. The lure of America for young Irish women, therefore, was twofold: the greater likelihood of finding a husband and readily available employment.

Irish immigrant women found their occupational niche as domestic servants. In fact, almost all Irish women who worked did so as domestics. Stephen Steinberg (1989) has addressed the curious fact that working women of other immigrant groups did not engage in this work, typically choosing instead the garment trades or textile factories. Since so many Irish immigrant women, unlike those of other immigrant groups, were unmarried and had no families in America, Steinberg explains, working as live-in domestic servants was a perfectly logical option: "[I]t provided them with a roof over their head and a degree of personal security until they were able to forge a more desirable set of circumstances" (1989:160). There was no cultural proclivity for Irish women to opt for this occupation any more than there was a culturally based distaste for it among those of other ethnic groups. The concentration of Irish women in this occupation was simply a rational choice, necessitated by economic need. Second-generation Irish women, more economically secure, were no more inclined to choose domestic service than women of any other ethnic group.

POWER AND SOCIAL CLASS

The Irish in America utilized two institutions to establish their place in the society: the Catholic Church and the Democratic Party. Ultimately, they came to dominate both.

URBAN POLITICS   For Irish Americans nothing stands out more clearly in their adaptive patterns than their great success in the political realm, particularly at the local level. It was in urban politics that the Irish made their initial and longest-lasting mark on American society. No ethnic group ever more successfully exploited urban politics as a means of establishing their place in the society and as a stepping-stone to upward mobility.

In New York and other cities, the Irish took command of the Democratic Party and created an effective political machine. The Irish-led urban machines of the nineteenth century are often depicted as corrupt institutions that ruled cities through voter fraud and by rewarding friends and coethnics with political spoils. This is not an inaccurate picture. But party bosses and their organizations that relied on patronage and graft had been prevalent in many large American cities long before the Irish came upon the scene. Most important for Irish immigrants, the machines served as key supports for individuals and families adapting to city life, seeking to find their place in a trying and often hostile social environment. In return for loyalty at the polls, the machines helped Irish workers overcome the employment discrimination they often faced by providing government jobs and by negotiating contracts with Irish-owned enterprises. Police and fire departments in New York, Chicago, and other cities, for example, became heavily Irish in makeup. The machines also provided money, housing, and food to destitute immigrants who had no other means of support.

The urban political machines, then, were crafted to serve Irish interests and, later, the interests of Poles, Italians, Jews, and other groups that followed them. They were undeniably corrupt, but the means by which they operated must be placed into historical context. This was a time when governments did not provide aid to needy or sick families. A welfare state did not exist to give succor to those who lacked the skills and financial resources to survive in a capitalist system.

Whether the means used by political bosses were corrupt, therefore, must be weighed against their effectiveness in providing what struggling immigrants required. "The politician," notes Marjorie Fallows, "was a pivotal figure in the complex pattern of loyalties that marked the Irish neighborhoods—a realist who was less concerned with abstract notions of public service and good government than with the practical concerns of keeping himself in office by building up a patronage system that would bind politician and constituent together in a web of mutual obligation and loyalty" (1979:113).

How is it that largely uneducated peasants became so adept at capturing the urban political system and utilizing it to their advantage? Some have attributed Irish political success to their experience with English governmental institutions in Ireland, making it logical that they would adroitly deal with the American system based essentially on a similar legal structure (Levine, 1966). Moreover, the fact that the Irish spoke the language of the dominant group upon arrival proved a huge asset that other groups could not quickly utilize.[1]

Others explain the ability of the Irish to take command of urban machine politics as a natural outgrowth of a culture brought from Ireland. Glazer and Moynihan (1970) point out that under the Irish Penal Laws, Catholics had almost no legal rights. For redress, therefore, they had to look outside the legal system, that is, to more personalized agents. This dependency was transferred to the urban setting in America. One could not rely on the legitimate government but needed rather to turn to informal authorities within the Irish American community. This is essentially how city political machines, like the ill-famed Tammany Hall in New York, functioned. Also, familiarity with the Roman Catholic Church provided the Irish with a disciplined hierarchical order that could be used as an organizational model (Dinnerstein et al., 2003).

The ability to relate to potential voters at a personal level served as another valuable asset for Irish American political leaders. Historian Carl Wittke has written that "Irish wit and adaptability, a gift for oratory, a certain vivacity, and a warm, human quality that made them the best of good fellows at all times—especially in election campaigns—enabled the Irish to rise rapidly from ward heelers to city bosses, and to municipal, state and federal officials of high distinction" (1967:157). McCaffrey (1997) suggests that, in addition to such interpersonal skills and knowledge of English, the Irish succeeded in American urban politics because this was an institutional area not monopolized by the Anglo-Protestant elite, who were more inclined to pursue business and the professions.

By the 1960s, the power of most urban political machines led by Irish politicians had faded. One notable exception was Chicago's Democratic machine, led by Richard J. Daley. Until his death in 1976, Daley, born and raised in a working-class Irish American neighborhood, governed the city as mayor in the style of an old-time political boss, dispensing jobs, housing, and various urban projects in return for loyalty at the polls (Royko, 1971). As a consummate power broker, he cobbled together a diverse coalition of various ethnic groups and labor unions that enabled him to hold the reins of political power for twenty-one years. His son, Richard M. Daley, was elected Chicago's mayor in 1989, a position he would retain, like his father, for more than two decades.

NATIONAL POLITICS   The prominence of Irish Catholics in American politics during the nineteenth and early twentieth centuries was confined mainly to cities with large white ethnic populations. At the national level the Irish did not play an equally conspicuous role. This began to change in the 1930s when Franklin Roosevelt appointed

---

[1] This point is disputed by Greeley (1981), who contends that many early Irish immigrants did not speak English or didn't speak it well. Moreover, Anglo Protestants also spoke English but obviously this did not lead to their dominance of urban politics in the nineteenth century.

Irish Catholics to Cabinet and other high-level administrative positions. But the greatest achievement of national political power occurred in 1960 with the election of John F. Kennedy as U.S. president. Kennedy's election marked a milestone not only for the Irish but for American Catholics generally. No Catholic had been elected president, and only one Catholic before Kennedy had even been nominated as a presidential candidate by either party. That occurred in 1928 when Al Smith, a plain-speaking, self-made Irish American Catholic, was chosen by the Democratic Party to run against Republican Herbert Hoover. The election proved to be one that would expose the enduring strength of anti-Catholicism in the United States in the early decades of the twentieth century.

Smith had grown up on the sidewalks of New York City and with the help of Tammany Hall rose through the ranks of the Democratic Party, eventually being elected governor of New York. In the presidential campaign, a number of issues separated Smith and Hoover, in particular Prohibition (which Hoover favored and Smith did not), but the issue underlying all else was religion. Despite Smith's strong assurances that his Catholicism would have no influence on his administration, vicious innuendos and rumors ceaselessly plagued his candidacy: "The Pope will move to Washington as soon as Smith wins"; "A tunnel connecting the White House to the Vatican will be built"; "Protestant marriages will be annulled." Protestant ministers openly preached sermons against Smith, filled with virulent anti-Catholic rhetoric. Smith's association with Tammany and the fact that he was an Irish American who favored ending Prohibition only magnified the fears of Protestant America, particularly in rural areas and small towns.

Although Smith carried the vote of most of the country's large cities—whose populations were heavily Catholic working class—he was rejected overwhelmingly in the rest of the country. Even if religion had not been a major issue, Smith likely would have lost since the United States at the time was relatively well off economically. "General Prosperity," said Hoover afterward, "was on my side" (quoted in Boller, 2004:227). But Smith and many others saw the outcome as an expression of the country's still powerful anti-Catholicism. No political observer denied that religion had played a significant role in his defeat.

More than three decades after Smith's failed bid for the presidency, once again the American electorate demonstrated its reluctance to place a Catholic in the highest political office. The 1960 campaign of the Democratic nominee, John F. Kennedy, showed that anti-Catholic prejudice remained sufficiently widespread and deep-seated to strongly influence a national election.

Kennedy was part of a wealthy Irish American family that had played a prominent political role for several generations in Boston. Reminiscent of the Smith candidacy in 1928, Kennedy stated unequivocally his belief in the absolute separation of church and state and offered his assurance to skeptical voters that on public matters he did not speak for the Catholic Church, nor did the Church speak for him. Addressing a meeting of Protestant ministers in Houston, Kennedy declared, "I am not the Catholic candidate for President, I am the Democratic Party's candidate for President who happens also to be a Catholic" (quoted in Sorensen, 1965). But the issue persisted throughout the campaign. As in 1928, bigots and extremists on the right continued to issue warnings that the Pope would be governing America if Kennedy were elected. Anti-Catholic tracts were distributed to millions of homes, along with radio and TV attacks and countless mailings. Kennedy won by the barest of margins, despite the fact that long-term Democrats who were Protestant abandoned the party in droves.[2]

Kennedy's presidency seemed to put an end once and for all to any obstacles to Catholics in their quest for the nation's highest elective offices. Irish Catholics in prominent political positions on the national political scene are now accepted as commonplace, and their ethnicity has virtually no bearing on their public images. In recent years, Irish Americans have served as vice president, as Speaker of the House of Representatives, as Senate Majority Leader, in numerous Cabinet positions, and as justices of the Supreme Court.

THE CHURCH   Aside from urban politics, the realm within which Irish Americans would make their strongest impression was the Catholic Church. As the first sizable Catholic immigrant group and one that possessed that priceless asset, the English language, it was perhaps only logical that the Irish would come to dominate the Church in America. By the time other Catholic immigrant groups—Italians, Poles, Slavs—began to arrive in America as part of the New Immigration, the Irish had already taken control of the Church hierarchy. This created discontent and resentment on the part of other Catholic ethnics, who often felt alienated from parishes and priests who did not speak their language and had little understanding or sympathy for their unique religious needs (Levine, 1966).

A vital component of Irish domination of the Catholic Church in America was the establishment of a system of parochial schools at all educational levels. The desire to educate Irish children in parochial schools was in large measure motivated by the anti-Catholic discrimination that typified American public schools in the mid-nineteenth century. For succeeding generations, the Catholic school system served not only as a bulwark against what were seen as dangerous Protestant influences but as the key institution in building and sustaining an Irish American community amid the surrounding Anglo majority.

As late as the 1970s, sociologist Andrew Greeley (1972) reported that 35 percent of the clergy and 50 percent of the hierarchy were still Irish despite the fact that Irish Americans were only 17 percent of all Catholics. In recent years the influence of the Irish in the American Catholic Church has begun to wane as Hispanics have become its population base, and parishes have adapted to radically changed demographic conditions. Nonetheless, the residue of Irish dominance still permeates the Church's educational and administrative structures.

SOCIAL CLASS   As already noted, most of the Irish who came during the great surge of immigration in the mid-nineteenth century were poor and unskilled. Indeed, severe impoverishment upon arrival in America is now one of the most frequently noted characteristics of this group's immigrant legacy. The entrance of Irish Americans solidly into the middle class, therefore, is a story of relative success. The movement from deep poverty to middle-class status, however, took place over a period of several generations.

---

[2] In one of the closest presidential elections in American history, it was the efforts of Daley's political machine in Chicago that enabled Kennedy to win Illinois, thereby assuring him the necessary electoral votes to win the presidency.

In the latter decades of the nineteenth century, Irish Americans began to move up the occupational ladder as they took advantage of the opportunities created by expanding industrialization. Their social ascent was aided as well by the entrance of new immigrant groups whose members began to replace them on the bottom rungs of the class hierarchy. Still, Irish progress was largely incremental, moving within the working class. If the immigrant generation worked in mainly unskilled jobs in manufacturing and construction, their sons (the second generation) moved into occupations as artisans and craftsmen (Hershberg et al., 1979). The rise of labor unions in many new industries, often working in collaboration with urban political machines, was a major factor in facilitating Irish working-class upward mobility.

The Irish, along with other white ethnics, did not experience substantial movement into the middle class until the post–World War II era with its booming economy. Like other working-class white ethnics, hundreds of thousands of Irish Americans took advantage of the GI Bill, which enabled them to attend colleges and universities largely at government expense and to purchase homes with government-guaranteed low-interest mortgages.

By the end of the twentieth century, Irish Americans as a collectivity ranked somewhat lower than other white ethnic groups in median household income, though the absolute differences among the groups were not great (Hacker, 1997). Today, Irish Americans are a solid element of the American middle classes. As can be seen in Table 10.1, on most dimensions of social class the Irish are at or above the national average. It is of note that white ethnics generally have been upwardly mobile in recent decades as indicated by their increasing net worth. Lisa Keister's research (2007) has shown that non-Hispanic whites raised in Catholic families have accumulated relatively high wealth in their lifetimes, despite being raised in comparatively disadvantaged families.

Prejudice and Discrimination

As the first immigrants to enter American society on a massive scale whose religious faith separated them fundamentally from the Anglo-Protestant majority, the Irish early on became the objects of rampant prejudice and discrimination. Much of the hostile reception given them was, by present-day standards, based on racist beliefs and attitudes. To understand the virulence of these beliefs and behaviors, we again need to consider the image of the Irish under their English masters in Ireland.

Irish Stereotypes   In Chapter 4, inequalitarian pluralism was described as a form of ethnic relations in which dominant group members require those of the subordinate, or minority, group to perform the onerous physical laboring tasks of society that they seek to avoid. In rationalizing this system, subordinate group members are regarded as innately inferior, not fit to perform work beyond the most simple and menial, and as childlike, requiring that their immature and irresponsible behavior be held in check by the dominant group. This describes the essential nature of ethnic relations in colonized Ireland. Among the English, a set of negative stereotypes portrayed the Irish as an almost subhuman people. Irishmen were seen as brutish and animalistic, ignorant, pugilistic, drunken, and thoroughly incapable of ever becoming a civilized participant in the mainstream society.

TABLE 10.1 | SOCIOECONOMIC STANDING OF EURO-AMERICAN ETHNIC GROUPS, 2011*

	Median Household Income	Percent College Graduates	Percent Professional or Managerial Occupations
Total U.S. population	$50,502	28.5	36.0
Russian[†]	$70,310	57.5	55.4
Greek	$62,873	41.0	46.4
Scottish	$62,670	43.0	47.8
Italian	$62,286	35.1	42.2
Polish	$61,846	37.6	42.9
Czech	$61,758	40.4	46.8
Danish	$61,669	40.8	47.3
Welsh	$61,188	42.5	49.2
Swedish	$60,703	40.5	45.5
Hungarian	$60,537	40.6	46.7
Norwegian	$60,535	36.9	42.7
German	$58,556	33.8	41.0
English	$57,355	37.7	45.4
Irish	$57,319	33.3	40.9
Portuguese	$57,127	23.7	34.0
Scotch-Irish	$56,242	41.6	47.3
French	$54,688	30.8	40.2
Dutch	$52,791	29.3	39.3

*European-origin groups with more than one million who claim sole ethnic ancestry.

[†]A large percentage of those who claim Russian ethnic origin are Jewish.

Source: U.S. Census Bureau (2012a).

These harshly negative images of the Irish carried over into North America. Unlike the earlier-arriving Irish Protestants (the Scotch-Irish), who largely escaped rejection, they were poor peasants and, more importantly, Catholics, a combination of social characteristics that would set them apart sharply and fuel the nativism of this period. "Nativists in Britain and the United States," writes McCaffrey, "despised Irish Catholics as cultural and political subversives, treacherous agents of authoritarianism, ignorance, and superstition, the key ingredients of popery" (1997:2). With anti-Catholicism as the core of nativists' antagonism, the growing waves of Irish immigrants were portrayed as the carriers to America of a potentially polluting influence. As a result they were subjected to bitter and often militant prejudice and discrimination, expressed by "No Irish Need Apply" notices commonly displayed by landlords and employers.

The caricature of the Irish as ignorant, drunken, and quarrelsome was a staple of early and mid-nineteenth-century American society. Newspapers would be filled with descriptions of the brawling and rioting of Irish workers. Popular illustrated magazines regularly featured cartoons that portrayed the Irish as "apelike creatures with large mouths, thick lips, protruding lower jaws, jagged teeth, short noses with gaping nostrils, sloping foreheads, receding chins, and dangling arms" (Ross, 2003:133). The totality of these images conveyed above all that the Irish were inferior, not only culturally but perhaps even biologically. The fact that most were poor, uneducated, and living in squalid urban conditions reinforced those beliefs.

Greeley points out that much of the same negative portrayal and imagery of blacks in American society was, in the nineteenth century, aimed at the Irish:

> Their family life was inferior, they had no ambition, they did not keep up their homes, they drank too much, they were not responsible, they had no morals, it was not safe to walk through their neighborhoods at night, they voted the way crooked politicians told them to vote, they were not willing to pull themselves up by their bootstraps, they were not capable of education, they could not think for themselves, and they would always remain social problems for the rest of the country. (1972:119–20)

In the twentieth century, Irish stereotypes were modified to convey a less threatening image. The Irish were seen as fun-loving, comic, and affable, but still as excessive drinkers and brawlers. The characterization of the hard-drinking, fighting Irish has survived even today as evidenced by media portrayals and in popular culture.

THE IRISH AND WHITENESS   It is important to consider how the Irish, the first among many American immigrant groups who did not closely conform to the dominant group, were viewed in terms of race and ethnicity. Their case demonstrates the fluid nature of "race" and how the meaning of this concept can be altered. The racial status of the Irish was indeterminate at the time of their entrance into American society and for many decades afterward. In the labor hierarchy as well as in all other realms of social and economic life they were not considered black, but neither were they qualified to be members of the dominant white (that is, Anglo-Protestant) group. Historian Noel Ignatiev writes that "[s]trong tendencies existed in antebellum America to consign the Irish, if not to the black race, then to an intermediate race located socially between black and white" (1995:76).

David Roediger (2005) similarly describes the Irish of the mid-nineteenth century as the first of those groups considered racially "in-between," who only later would be assigned to the "white race." The Catholic and Jewish immigrants that would follow them, as part of the New (second wave) Immigration were characterized as "races" and would encounter the same question of classification: where did they fit in the American racial configuration? The indeterminacy of the racial status of members of those groups would remain an issue for several decades before they acquired the full status of "whiteness."

Roediger chronicles how the racial status of the Irish and other immigrant groups was eventually decided as a result of government policies, Census designations, and employer preferences that ensured the distinction of white ethnics from African Americans. But the Irish themselves—and later, other white ethnics—shortly after entering the United States came to understand the need to distinguish themselves from blacks.

The new immigrants, Roediger explains, quickly became attuned to the American racial system, and were made aware of white supremacy, characteristically demonstrated by the Irish. They understood that as lowly as they might be viewed and dealt with, they still might become naturalized citizens and thus, at least symbolically, enter the white world. "Even at a low point of their racialization," Roediger writes, "the new immigrants remained in-between, or 'conditionally white'" (2005:144).

Above all, it was in the world of work that the racial distinction between the Irish and free blacks was ultimately determined. Ignatiev has written that in all U.S. historical periods "the 'white race' has included only groups that did 'white man's work'" (Ignatiev, 1995:112). Thus, in order to authenticate their "white" status, Irish workers refused to work with blacks and monopolized occupations that had previously been filled by free black laborers.

THE IRISH AND AFRICAN AMERICANS The historical relations between Irish Americans and African Americans have commonly been characterized as harsh and confrontational. This relationship stems from the fact that in the early decades of immigration, blacks were frequently the chief competitors of the Irish for unskilled jobs and were often brought in by employers to depress the wage scale or used as strikebreakers against striking Irish workers. More basically, only blacks stood below them in the racial hierarchy. Labor violence, often pitting Irish against black workers, was not uncommon.

In 1863, the tension between Irish and black workers erupted in one of the most horrific riots in American history. The Draft Act of that year, intended to expand the Union Army in the ongoing Civil War, permitted men to evade military service by buying the services of substitutes to take their place. The impoverished Irish saw this as imposing an unfair burden on them. More important, they did not support one of the key rationales of the war, an end to slavery, since this meant the introduction of freed blacks into the labor market and, therefore, additional competition for unskilled jobs. Anticonscription riots broke out in many northern cities, but the most violent and deadly occurred in New York, where for five days, mobs roamed the streets, killing blacks and burning and pillaging. Although many Irish fought for the Union in the Civil War, they did so, as Ignatiev (1995) explains, primarily to establish their claim to citizenship and secure their place in the racial hierarchy as "white."

In recent decades, opposition to school busing for the purposes of racial integration has sometimes reflected an Irish American tinge. In Boston in the mid-1970s, for example, vivid pictures were transmitted throughout the country of white, largely Irish, rioters in South Boston screaming racial epithets at school buses carrying black children into what they saw as their territory. Opposition to affirmative action has also been seen by some political observers as a major factor in explaining the movement of a large segment of the Irish American electorate in the past two decades to the Republican Party, abandoning what for more than one hundred years had been its traditional loyalty to the Democrats (McCaffrey, 1997).

Although the historical animosity of the Irish toward African Americans is well documented, the comparison of Irish Americans with other white ethnic groups on racial issues in recent years reveals little difference. In fact, Irish American attitudes toward neighborhood and school integration in the early 1970s were more favorable than among any other white ethnic group, with the exception of Jewish Americans (Greeley, 1977).

ASSIMILATION

"If one is convinced that the most appropriate goal for American culture is the homogenization of the various components of the total population into one relatively WASP-like group, then the story of the American Irish is a great success." That is how Greeley (1972:256) has described the status of Irish Americans in the U.S. ethnic system. More than any white ethnic group, the Irish appear to have been most thoroughly absorbed into the mainstream of American society.

CULTURAL ASSIMILATION   Looking at contemporary Irish Americans in terms of Gordon's assimilation model (Chapter 4), it appears that they have reached the point of full cultural assimilation. Many, in fact, would argue that, to all intents and purposes, Irish-American ethnicity exists today only in symbolic form. The Irish, as Greeley puts it, "are the only European immigrant group to have over-acculturated" (1972:263). Or, as McCaffrey has described them, Irish Catholics have become "America's favorite ethnics" (1997:192).

   Though Irish Americans can be said to have come close to complete cultural assimilation, one group characteristic remains distinct and strong enough to continue to serve as an identity marker: religion. Thus, despite the almost total lack of a discernable Irish American ethnic culture—St. Patrick's Day has become no less an American holiday than St. Valentine's Day—Catholicism remains an evident cultural feature that distinguishes Irish Americans from the dominant Anglo group.

STRUCTURAL ASSIMILATION   By the 1950s, anti-Catholic prejudice and discrimination had waned considerably and any remaining suspicion of American Catholicism—still overwhelmingly dominated by Irish Americans—was effectively put to rest by the election to the presidency of John F. Kennedy in 1960. Today there appear to be no impediments to the movement of Irish Americans into positions of power and influence in government and in the corporate world. All indications are that secondary structural assimilation is a process that will continue unabated.

   In realms of power and influence aside from politics, such as the corporate world and academia, the Irish presence is no less evident than that of other white ethnic groups (Greeley, 1972). Alba and Moore (1983) found that by the 1970s, Irish Catholics were well represented in key institutions of American society (see also Greeley, 1981). The names of Irish Americans are commonly found among the rosters of top corporate executives, in literature and the arts, and on university faculties.

   At the level of primary structural assimilation, only religion continues to serve as an obstacle. Here, too, however, there is evidence to indicate a steady movement in the direction of complete assimilation. Intermarriage for Irish Americans is higher than for any other white ethnic group (Alba, 1976; Fallows, 1979). The resulting ethnic mixture, Fallows writes, "has diluted the meaning of Irish American identity" (1979:149). This trend of increasing interethnic marriage is characteristic of all white ethnic groups, including, as we will see in the next section, those of southern and eastern European origin (Alba, 1990; Sherkat, 2004).

   What should not be overlooked in the success of the Irish is the fact that they made their initial entrance into American society early on, before the second great wave of immigration, and that from the outset their English-language skills and familiarity with an Anglo ruling class gave them an automatic head start, notwithstanding the virulent prejudice and discrimination they encountered. As Greeley has pointed out, "The American Irish had the shortest way to come of any of the immigrant groups—for all the impoverishment of the Famine, the Atlantic crossing, and the early years in the slums" (1972:258).

Another factor that accelerated the structural assimilation of Irish Catholics—and of white ethnics in general—was the termination in the 1920s of large-scale immigration from traditionally Catholic countries in Europe. This gave white ethnic groups time to absorb American ways without the need to attend to a constant flow of new immigrants clinging to the old culture. In addition, the cessation of large-scale European immigration squelched the concerns of Protestant Americans who feared being overwhelmed by non-Anglo immigrants.

The important question regarding Irish assimilation is whether other white ethnic groups, specifically those of the second great immigration wave of the late nineteenth and early twentieth centuries, are following in their path. As we will see in the next section, all indications are that they are, though not necessarily in the same form or at the same pace.

ITALIAN AMERICANS AND OTHER WHITE ETHNICS

Although Irish Americans were the first of many non-Anglo, non-Protestant groups that would make the United States an ethnic polyglot, the term *white ethnics* is often used in a more focused sense to refer to groups of the New Immigration. As noted in Chapter 5, these were immigrants who came in the late nineteenth and early twentieth centuries primarily from southern and eastern European societies, especially Italy, Poland, Russia, Greece, and the various states that made up the Austro-Hungarian Empire.

Small numbers of southern and eastern Europeans were part of American society almost from its founding, and a relatively small-scale immigration proceeded throughout the first three-quarters of the nineteenth century. Before 1880, however, about 85 percent of all immigrants came from Britain, Germany, Canada, and Scandinavia. By 1896, immigrants from Italy and the Russian and Austro-Hungarian empires accounted for more than half of total immigration (Ward, 1971). As part of the New Immigration, Italy contributed by far the largest number. Because of its size and because it represents in many ways the archetypal white ethnic experience, we will focus much of the following discussion on the Italian American population.

IMMIGRATION AND SETTLEMENT

For Italians, the large-scale movement to the United States after 1880 represented the second great emigration. Earlier in the nineteenth century, Italians had immigrated mainly to Argentina and Brazil and only secondarily to the United States. Many of these early immigrants, to the United States as well as to South America, were from northern Italy (Baily, 1999; Foerster, 1919).

IMMIGRATION Beginning in the 1880s, both the origins and destinations of Italian emigrants changed. With the depletion of coffee plantations in Brazil and political turmoil in Argentina, most Italians now found the United States more attractive. And whereas the northern Italian provinces had previously supplied most of the immigrants, southern Italy now became the major region of out-migration. This change in regional origin fundamentally altered the class and cultural character of the Italian population in the United States. By comparison with the more prosperous and literate immigrants from the North, those from the South were poor and uneducated. Moreover, whereas the northerners had been mostly artisans, shopkeepers, and professionals, those from the South were mainly agricultural peasants.

Southern Italy had historically been an impoverished, backward area, subjected to constant economic and political exploitation by the region's ruling groups; but unusually severe conditions were created in the latter nineteenth century. Depressed wages, little or no opportunity to improve a low agricultural output, little industrialization, poor health conditions, neglect by the central Italian government, and heavy indirect taxation all served to create a strong impetus for emigration. These push factors, combined with the pull of a burgeoning American industrialism calling for unskilled labor, resulted in a veritable exodus of southern Italian peasants. Between 1880 and 1915, four million Italians went to the United States, half of them in the single decade from 1900 to 1910. Between 80 and 90 percent were from the area of southern Italy known to Italians as the *Mezzogiorno*, and about one-quarter of these were from Sicily (Foerster, 1919; Gambino, 1975; Lopreato, 1970; U.S. Census Bureau, 1975).

One extremely important characteristic of Italian immigrants of this time was that, unlike earlier Irish immigrants, most who went to the United States did not expect to stay permanently. These were primarily men unaccompanied by their families, intent on earning money to purchase land after returning to their native villages. Many did go back. In fact, no other sizable immigrant group displayed such a high rate of return migration. Between 1908 and 1914, almost one-third of those who arrived from Italy eventually returned or migrated to another society. Between 1915 and 1920, more than half returned, and between 1925 and 1937, 40 percent did so (Learsi, 1954). Like the Italians, most Greek immigrants during this period also came with intentions of returning after acquiring sufficient savings to establish themselves in a comfortable life in their home villages (Lopata, 1976).

Settlement   Most Italians, like the majority of all immigrants of the New Immigration, settled in cities. By 1920, three-quarters of America's foreign-born population were urban dwellers, including almost 90 percent of Russians, 85 percent of Italians and Poles, and 80 percent of Hungarians (Ward, 1971). The attraction of the cities was twofold. Most important, it was here that the great masses of low-paid unskilled laborers were needed. Secondly, most of the immigrants entered in a chain migration process, joining friends and relatives who had preceded them to these areas.

Italian immigrants in the cities tended to settle in depressed areas, inhabiting run-down tenements that had been vacated by earlier ethnic groups such as the Irish. The cluster of Italians in these areas was partially the result of simple economic necessity; such housing was available and affordable. However, as with other groups of the New Immigration, the creation of ethnic concentrations in New York and other cities was also the product of the natural attraction of immigrants to areas inhabited by those from their regions in Italy who had preceded them to the United States. Italian immigration was network driven, in which those from one village gravitated to locations where previous emigrants from the village had settled. There was little identity with the Italian nation per se. Rather, immigrants identified with the province from which they came or simply their village. As a result, Italian sections of American cities were internally subdivided into "Little Calabrias," "Little Sicilys," and the like, and neighborhoods were often settled by those from a common village or town in the province (Vecoli, 1964). Settlement patterns were similar even outside the large cities (LaGumina, 1988).

URBAN ADJUSTMENT   Because cities like New York were at this time establishing their industrial infrastructures—roads, rail lines, subways, factories, housing—opportunities abounded for unskilled labor. It was there that these immigrants, like the Irish before them, found their occupational niches. Italians made up three-quarters of the building laborers in New York City at the turn of the century and almost the entire labor force constructing the city's subway system (Glazer and Moynihan, 1970).

THE ITALIAN AMERICAN POPULATION TODAY

With almost 18 million people claiming either full or partial Italian ancestry, Italian Americans are one of the largest among America's white ethnic groups (see Figure 5.2). Although it is far more dispersed than in earlier times, the Italian American population continues to be heavily urbanized and concentrated in the states and large cities of the Northeast (Table 10.2). Italian Americans in the New York–New Jersey metropolitan region, for example, number more than 2.5 million, one of the largest elements of its ethnically diverse population.

WHITE ETHNICS AND SOCIAL CLASS

The southern and eastern Europeans who arrived in America as part of the New Immigration are well into the third generation; many are even into the fourth generation. Given the length of time since these groups' large-scale entry, some evaluation can be made of their progress on various socioeconomic measures, both in absolute terms and in relation to other American ethnic groups. How far have they come, what means have they employed in seeking upward mobility, and how might we account for their current status in the overall stratification system? Again, we use Italian Americans as that group that best illustrates the white ethnic experience.

TABLE 10.2 | STATES WITH HIGHEST PERCENTAGE OF ITALIAN AMERICANS

State	Italian American Population	Percent of Total Population
Rhode Island	199,580	19.0
Connecticut	683,411	19.0
New Jersey	1,480,876	16.8
Massachusetts	937,916	14.2
New York	2,674,040	13.7
Pennsylvania	1,559,649	12.2

Source: U.S. Census Bureau (2012a).

INCOME AND OCCUPATION   In terms of income, occupation, and education, Italian Americans have "made it" in American society to a greater degree than many ethnic groups and to a far greater degree than African Americans, Latinos, and American Indians. Although upward mobility was minimal for the first two generations, the third and fourth generations are now exhibiting solid evidence of middle-class status. As seen in Table 10.1, the median family income of Italian Americans is well above the national average. As with Irish Americans, especially significant improvement occurred during the 1950s, a period in which the American middle class expanded greatly.

Little occupational mobility was evident for Italian Americans and other white ethnics between the first and second generations (Greeley, 1977; Yans-McLaughlin, 1982). That is, most of the children of immigrant parents remained in working-class jobs. In the case of Italians, because children were expected to contribute to the family's support, they entered the labor force early, precluding the more extensive education needed for higher-status jobs. Most Italian children thus followed their fathers into manual laboring occupations. But those of the third and fourth generations, with greater education, have moved in significant numbers into all occupational areas—including the top professions of medicine and law, where they had been severely underrepresented (Alba, 1985; Glazer and Moynihan, 1970).

As with other indicators of socioeconomic status, in educational attainment Euro-American ethnic groups have in recent years converged, with marginal differences among them. Alba (1995) has shown that for men of southern and eastern European ancestry born between 1916 and 1925, there remain some apparent disadvantages when compared to the educational levels of those of northwestern European ancestry. But when those born between 1956 and 1965 are looked at (these would be third- or fourth-generation ethnics), southern and eastern European origin men are at or above the average of northwestern European origin men.

Italian Americans remain more solidly working class than Jews, Irish, and several other white ethnic groups. Moreover, some have argued that Italian American mobility is a reflection not so much of the group's progress per se as of a change in the society's occupational structure. As more white-collar and professional jobs have been created, more Italian Americans (and other white ethnics as well) have filled them (Alba, 1985; Gallo, 1974). But the extent of their mobility is in any case quite extraordinary considering how relatively ill equipped they were at the time of their arrival to succeed in the American economic system.

In sum, the class position of white ethnics is on most measures at or above the national average. Moreover, for the third and later generations, parity with the society's core ethnic group, northwestern Europeans, has nearly been reached (Alba, 1985, 1995, 2009). Overall, then, the socioeconomic differentials between various Euro-American groups seem to be shrinking.

POLITICAL AND ECONOMIC POWER

Having looked at the place of Italian Americans and other white ethnics in the economic realm, let's look briefly at the power hierarchy. To what extent do members of these groups occupy positions of authority in the society's two key power institutions, government and the corporation?

NATIONAL POLITICS   Patterns of participation at the highest levels of government parallel patterns in the economic realm: the first two generations made little progress, but in recent years, white ethnics have begun to achieve important positions in significant numbers.

Catholics in general—and southern and eastern European Catholics in particular—were traditionally underrepresented in the highest offices of the federal government (Alba and Moore, 1982; Mills, 1956; Schmidhauser, 1960; Stanley et al., 1967). Figure 10.1 shows how radically this historical pattern has changed over the past several decades. Catholics, most of whom are white ethnics, today are actually overrepresented in the U.S. Senate and House in proportion to their share of the total population. Moreover, the choice of Michael Dukakis, the son of Greek American immigrants, as the Democratic nominee for president in 1988 seemed to mark a culmination in the national political fortunes of southern and eastern European ethnic groups, indicating their fullest entrance into American politics.

What is perhaps even more indicative of how radically and relatively quickly the role of religion has changed as a determinant of membership in the American political elite is the present-day makeup of the U.S. Supreme Court. No more than a few decades ago, the Court was described as having a "Catholic seat" or a "Jewish seat," implying the need for at least token membership by non-Protestants. In 2013, not a single Protestant served on the Court, with six of its judges Catholic and three Jewish.

Looking at Italian Americans specifically, prior to the 1960s, their presence at the highest levels of the federal government was limited. Not until 1950 was an Italian American elected to the Senate, and not until John F. Kennedy's administration in 1962 was an Italian American named to a Cabinet post. Since then, Italian American representation in the Congress and at the higher levels of the executive branch has steadily increased. By 1980, at least twenty-five Italian Americans were serving in the Congress, compared with four in the 1930s and eight in the 1940s

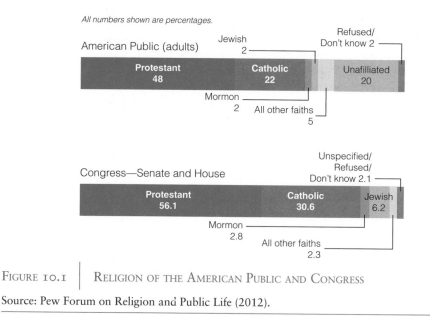

FIGURE 10.1 | RELIGION OF THE AMERICAN PUBLIC AND CONGRESS

Source: Pew Forum on Religion and Public Life (2012).

(Rolle, 1980). In 2012, six U.S. senators and twenty-five members of the U.S. House of Representatives were of Italian ancestry (NIAF, 2012). Moreover, two Italian Americans currently serve on the Supreme Court, a sharp change from the past in light of the fact that no southern or eastern European Catholics had served on previous Courts (Lieberson, 1980; Schmidhauser, 1960).

STATE AND LOCAL POLITICS   Having arrived after the Irish had secured their place in the urban power structure, groups of the New Immigration began their movement into state and local politics only after World War II. Much of their success depended on the size of the white ethnic population in any particular city or state. Italian Americans, for example, attained substantial power in states and communities of the Northeast, where they constituted a large proportion of the electorate (Dinnerstein et al., 2003; LaGumina, 1988). Similarly, in cities and states of the Midwest, with large eastern European communities, white ethnics by the 1950s and 1960s commonly occupied high political offices (Lopata, 1976; Marger, 1974).

THE CORPORATE WORLD   In the past, the number of Italian Americans in the executive suites of America's largest corporations had been well below their proportion of the general population. A study of Chicago's largest corporations, for instance, revealed that although Italians in the 1970s made up almost 5 percent of the city's population, they made up less than 2 percent of these corporations' directors and 3 percent of their officers (Barta, 1979). In New Haven, Connecticut, Wolfinger (1966) found that although Italians in 1959 made up 31 percent of the registered voters and held 34 percent of major political offices, they made up only 4 percent of the city's economic elite.

Similar patterns have been evident for other white ethnics. In the past, few entered into the top ranks of economic power. For example, in the previously cited study of Chicago's top corporations, Poles, that city's largest white ethnic group, were shown to be even more severely underrepresented than Italians. Although they made up almost 7 percent of the city's population, they made up only 0.3 percent of these corporations' directors and 0.7 percent of their top officers. Numerous studies have shown the historically consistent domination of the American business elite by those of northwestern European and Protestant origins (Baltzell, 1991; Davidson et al., 1995; Marger, 1974; Mills, 1956; Newcomer, 1955). Sociologist Suzanne Keller (1953), for example, found that 89 percent of top business leaders in 1900 and 85 percent in 1950 were Protestants, most of British descent; 7 percent in each year were Catholics; and 3 and 5 percent, respectively, were Jews. By the 1970s, little had changed—Catholics made up 14 percent and Jews 6 percent of the corporate elite (Burck, 1976). Alba and Moore in 1982 found a continuation of Anglo dominance of economic power positions. However, Anglo overrepresentation in power elites (including those of the corporate world) now seems in decline, and Catholics are steadily moving toward parity (Christopher, 1989; Davidson et al., 1995; Zweigenhaft and Domhoff, 2006).

PREJUDICE AND DISCRIMINATION

In the United States today it is hard to visualize the fervor of prejudice and discrimination directed at white ethnic groups in the nineteenth and early twentieth centuries. Anti-Semitism (prejudice and discrimination aimed at Jews) has a long history in

American society, as we will see in Chapter 11, but, as was noted earlier, anti-Catholicism also has a lengthy heritage. It is important to emphasize, however, that the prejudice and discrimination directed at white ethnics never was as intense, widespread, or institutionalized as that aimed at racial-ethnic groups—African Americans, Latinos, American Indians, and Asian Americans. Though at times intense and even violent, discriminatory actions were ordinarily limited to derogation and denial of access to certain jobs, schools, and neighborhoods. Never were white ethnics denied legal rights, as were blacks and, to a lesser degree, Asians; nor were they ever so thoroughly excluded from the larger society.

ITALIAN AMERICANS: IMAGES AND ATTITUDES   Italians suffered the usual antagonisms directed at Catholics generally, but anti-Italian feelings ran especially high as part of the newly inspired nativism and jingoism that arose in response to the new immigrants of the late nineteenth century. Indeed, with the arrival of hundreds of thousands of southern Italian peasants, Italians now replaced the Irish as the chief target group of the country's well-worn anti-Catholicism.

Much of the prejudice and discrimination leveled at white ethnics was founded on racist ideologies. The early years of the twentieth century were a time when ideas of social Darwinism prevailed among scholars and policy makers alike, and new racial theories were widely espoused and endorsed. Those groups composing the New Immigration—overwhelmingly Catholics and Jews—were therefore described in racial terms, that is, as morally and even physiologically inferior to Americans of northwestern European origin. The new immigrants were perceived not simply as different from those who had preceded them but as threatening to the very social and biological fabric of the society (Higham, 1963). Italians were seen as the most degraded of all the entering groups.

The poor and illiterate peasants from southern Italy were received quite differently than the earlier northern Italian immigrants. Whereas the latter had been seen as honest, thrifty, law abiding, and generally "molto simpatici to the American character," the southern Italians were viewed as degenerate, ignorant, lazy, dirty, destitute, violent, superstitious, and criminally oriented (Train, 1912, cited in Moquin and Van Doren, 1974). Northern Italians, too, seemed to share the prevalent American view of their southern counterparts as a generally repugnant people (Gambino, 1975). One journalist of the time put the matter quite bluntly: "What shall we do with the 'dago'?" Note his description of the Italian as almost animalistic in nature:

> This "dago," it seems, not only herds, but fights. The knife with which he cuts his bread he also uses to lop off another "dago's" finger or ear, or to slash another's cheek. He quarrels over his meals; and his game, whatever it is, which he plays with pennies after his meal is over, is carried on knife at hand. More even than this, he sleeps in herds; and if a "dago" in his sleep rolls up against another "dago," the two whip out their knives and settle it there and then; and except a grunt at being disturbed, perhaps, no notice is taken by the twenty or fifty other "dagoes" in the apartment. He is quite as familiar with the sight of human blood as with the sight of the food he eats. (Train, 1912, cited in Moquin and Van Doren, 1974:261)

The most serious elements in the negative image of Italians in America concerned their "criminal nature" and the existence of a widespread Italian criminal community, known as the Mafia, or the Black Hand. Few accounts of Italian immigrants during the late nineteenth century neglected the presumed criminal orga-nizations among them. By the 1890s, the image of the criminally prone Italians and their association with the Mafia was well established (LaRuffa, 1982).

The strong anti-Catholic sentiments of a large segment of the American populace as well as government leaders aroused fear of Italian immigrants and not infrequently incited hostile actions against them. Anti-Italian antagonism was also partially a re-sponse of established workers, who saw the immigrants as a cheap labor force and thereby a threat to their jobs. Killings and lynchings of Italians, mostly in the South but in other regions as well, were numerous between 1890 and 1915. The most seri-ous of these incidents occurred in New Orleans in 1891. Following the murder of the city's police superintendent, it was assumed that Sicilian immigrants were responsible, and hundreds of Italians were arrested. Nine were subsequently brought to trial, but none was found guilty. Anti-Italian sentiment was so feverish, however, that eleven of the suspects were taken from their jail cells by a mob and murdered; officials made no attempt to intervene. Afterward, local newspapers and business leaders expressed approval of the action (Gambino, 1977; Higham, 1963; Nelli, 1970).

Another especially notorious incident occurred in 1927. This was the infamous trial of Sacco and Vanzetti. Two Italian immigrants, Nicola Sacco and Bartolomeo Vanzetti, were implicated in an armed robbery in Braintree, Massachusetts. Numer-ous witnesses attested to their lack of involvement in the crime, but despite the dubi-ous evidence against them, they were found guilty. After numerous appeals, they were executed in 1927. The case created a storm of protest among civil rights groups throughout the country and remained a cause célèbre for many decades afterward. It emerged not only as an ethnic controversy but also as an issue of symbolic class war-fare. Those defending the two men's innocence saw their trial as a conflict between the oppressed working classes and the wealthy WASP elite (Baltzell, 1964).

The guilt or innocence of Sacco and Vanzetti is still debated, but historians gen-erally agree that their trial was considerably biased and that the judicial process was seriously abused. Indeed, the perversion of justice in the case was finally acknowl-edged in 1977 when the state of Massachusetts officially proclaimed the trial unfair. In addition to the strong anti-immigrant sentiments of the time, the fact that the two defendants were avowed anarchists undoubtedly contributed heavily to their convic-tion. This was the era of the Red Scare, when government officials fanatically de-nounced and harassed union and socialist activities, in which Italians often took part. To many, Sacco and Vanzetti represented an affirmation of the perceptions held by much of the American public that Italian Americans were radical, conspira-torial, and criminal. These perceptions seriously colored the two men's arrest, trial, conviction, and, finally, execution (LaRuffa, 1982).

The Sacco and Vanzetti case had an enormous and lasting effect on Italian Americans, and for many years it influenced their view of the state and the judicial process. As Gambino has put it, "The outcome of the affair was simply another con-firmation of the ancient belief of the Italian immigrants that justice, a very important part of their value system, had little to do with the laws and institutions of the state" (1975:122).

ENDURING ITALIAN AMERICAN STEREOTYPES   Although they are rarely expressed today as vir-ulently and hostilely as they were during the period of immigration, many of the early

Italian American images remain dynamic elements of American ethnic mythology (Alba and Abdel-Hady, 2005). Few would be hard put to describe the "typical" Italian American. Gambino observed the consistency of responses of non-Italians to his query about the Italian American image: "the Mafia; pizza and other food; hard hats; blue collar; emotional, jealous people; dusky, sexy girls; overweight mammas; frightening, rough, tough men; pop singers; law and order; pastel-colored houses; racists; nice, quiet people" (1975:352). These stereotypes are reinforced by media depictions, especially television and films, and are frequently used in marketing products, despite the fact that they do not reflect the reality of contemporary Italian American life (Gambino, 1998; Lichter and Lichter, 1982; Mitrano and Mitrano, 1996).

In 2009–2010, the reality series *Jersey Shore*, shown on MTV, seemed to illustrate the obdurate nature of these stereotypes. All of the hackneyed features of an Italian American, mostly working-class, ethnic subculture, created in large measure in 1977 by the film *Saturday Night Fever,* were on display: tough-talking, sexually obsessed young men and women wearing tight jeans and ostentatious jewelry, speaking in heavy Brooklyn dialect. One television critic described it as "less a reality show than a cartoonlike comedy that magnifies the vulgarity and bravado of a particularly colorful ethnic stereotype" (Stanley, 2011). The show's theme created controversy in the Italian American community, with some deriding it as the prototypical negative depiction of Italian Americans. Others, mostly younger Italian Americans, saw the show and its reference to "Guido culture" as less disapproving, simply reflective of an ethnic style. As one of the show's protagonists put it, "I was born and raised a Guido. It's just a lifestyle, it's being Italian, it's representing family, friends, tanning, gel, everything" (Cohen, 2010).

The most enduring negative image Italian Americans have had to deal with concerns the connection to organized crime. Because the Mafia is synonymous with organized crime and Italian Americans are identified with the Mafia, the logic, to many, becomes simple: "Italians = organized crime" (Iorizzo and Mondello, 1980). Although few deny the considerable involvement of some Italian Americans with organized crime, Italians have held no monopoly on such activities. As many have explained, crime has been a traditional avenue of upward mobility for members of most ethnic groups in American society (Albini, 1971; Bell, 1962; Ianni, 1974). Italian domination of organized crime did not emerge until the late 1920s, when Prohibition made trafficking in liquor profitable. Before that time, much of organized crime was controlled by the Irish. As illicit drugs replaced liquor as the major underworld-supplied item, Italian organized crime turned to more legitimate business areas, leaving the drug traffic to newer urban ethnic groups (Cook, 1971; Ianni, 1974; Kleinknecht, 1996).

Even though Italians did not introduce organized crime to the United States, and never were more than a tiny percentage of Italian Americans ever engaged in such activities, the connection of Italian Americans with crime families remains steadfast in the public imagery. Movies and television have been important purveyors of this stereotype (Carilli, 1998; Cortés, 1994). Dating from the 1930s, the association of Italians and organized crime has remained a common Hollywood theme. Movies such as *The Godfather, Married to the Mob*, and *GoodFellas*, and television programs such as *The Sopranos* are reflective of this genre. A survey in 2000 questioned teenagers of different ethnic backgrounds on their perception of movie and television stereotyping. When asked specifically to identify the role a character of Italian background would be most likely to play in a movie or on television, most (including Italian Americans) responded "crime boss" (NIAF, 2001). Recall the relationship of media depictions of ethnic groups with the creation and persistence of ethnic stereotypes, discussed in Chapter 3.

PREJUDICE AND DISCRIMINATION TODAY   Surveys have regularly recorded the periodic changes in prejudice toward blacks, Jews, Latinos, and Asians. By contrast, there has been little study of shifting public attitudes toward Italian Americans and other white ethnics. Beyond the persistence of certain negative stereotypes, what little evidence exists indicates a significant decline in prejudice and discrimination against these groups in recent decades. Cases of rejection in occupation and neighborhoods on the basis of ethnicity, however, were still evident well into the 1950s. In 1960, for instance, it was discovered that home owners in the exclusive Detroit suburb of Grosse Pointe had maintained a rigid point system for screening potential home buyers, based on personal traits such as accent, name, swarthiness, and especially ethnicity (*Detroit News*, 1990; *Time*, 1960). Out of one hundred points, Jews were required to score a minimum of eighty-five, Italians seventy-five, Greeks sixty-five, and Poles fifty-five. Blacks and Asians did not count at all!

Antidiscriminatory legislation has made such cases today uncommon, but prejudice and discrimination in subtle forms fueled by unrelenting stereotypes is not entirely something of the distant past (Crispino, 1980; LaGumina, 1996). For example, one study (Alba and Abdel-Hady, 2005) looked at the degree to which Italian Americans have been afforded entry into the American intellectual elite, as indicated by their election to the American Academy of Arts and Sciences, composed of the most honored people in the academic, artistic, and scientific worlds. It found that Italian Americans were noticeably underrepresented and tentatively concluded that a process of ethnic exclusion, driven by traditional demeaning Italian American stereotypes, might account for this. And, in 2010, charges were leveled at the City University of New York (which recognized Italian Americans as an official affirmative action category for employment), claiming that Italian Americans were underrepresented in university jobs (Foderaro, 2010).

ASSIMILATION

Whereas the applicability of either assimilationist or pluralist models of ethnic relations in the cases of African Americans and Hispanic Americans has been widely debated, for Italian Americans, as well as for other white ethnic groups, there is little argument that assimilation has been the preeminent pattern of their adaptation to American society and their relations with the dominant group.

Although Italian Americans illustrate quite well the assimilation model of ethnic relations, their collective rate and extent of assimilation have differed in the cultural and structural dimensions. They have substantially adopted the dominant cultural ways but have not entered into equal relations with the dominant group to a corresponding degree at either the primary or secondary level.

CULTURAL ASSIMILATION   Italian Americans have, with each succeeding generation, shed Italian cultural ways increasingly in favor of those of the core culture. Indications of cultural assimilation are apparent in such areas as family patterns, language usage, and diet. In his study of Italian Americans in Boston's West End, Gans (1982) found that even by the second generation most no longer remained attached to the major features of the southern Italian culture. This generation had retained only traditional cooking and foods as well as some use of the Italian language. At the conclusion of his study in 1962, Gans predicted that with each following generation the traditional Italian culture would be further diluted; subsequent research has confirmed his prediction. In a study of Italian Americans in Bridgeport, Connecticut, Crispino (1980) found that even in such areas as food and cooking, acculturation had taken place to a very great extent in later generations. Whereas 44 percent of first-generation respondents continued the tradition of making their own pasta (which, Crispino asserts, is "a more discriminating test of Italianness than cooking habits in general"), only 15 percent of the fourth generation did so.

Even intensely strong family ties and loyalties—the sine qua non of the Italian American cultural system—are weakened by the third generation or have begun to be replaced by more common American patterns (Alba, 1985; Crispino, 1980; Gans, 1982). For example, whereas divorce is either taboo or limited in the first two generations, it becomes more accepted in the third.

Perhaps the most telling evidence of the cultural assimilation of white ethnic groups in American society is the loss of the mother tongue. This is especially apparent in the case of Italian Americans. In Crispino's study (1980), 72 percent of second-generation respondents (both parents born in Italy) spoke Italian, but only 14 percent of the third generation (both parents born in the United States) and 10 percent of the fourth generation (those most thoroughly assimilated culturally) did so. This pattern of language conversion is typical of all white ethnic groups. Whereas many older members of these groups still speak a European mother tongue in the home, among younger people, usually by the third generation, hardly any do (Alba, 1995). Today, Italian remains one of the most common non-English languages spoken in U.S. homes (around 750,000), but this represents a tiny fraction of the total U.S. Italian American population and is likely to decline with each passing generation.

Observers of the Italian American ethnic group have uniformly noted the relationship between social class and cultural assimilation (Crispino, 1980; Gambino, 1975; Gans, 1982; Lopreato, 1970; Roche, 1982). As class rises, the level of cultural assimilation rises correspondingly. Today the Italian American ethnic culture—to the extent that it is evident—remains essentially a working-class phenomenon. As they continue to leave the working class, moving into middle-class occupations and neighborhoods, Italian Americans become indistinguishable from other middle-class Americans. This is especially so among those of the third generation and beyond. These assimilation patterns are evident among other white ethnic groups as well.

STRUCTURAL ASSIMILATION   At the secondary, or formal, level of structural assimilation, we have seen that Italian Americans have begun a large-scale movement into the American mainstream. Their occupational, income, and educational levels are at or above the average for the society as a whole (Table 10.1). Most important, they have begun to enter positions of power within most institutional realms, though still in only small numbers at the very top echelons.

Within the context of personal social settings (that is, at the primary group level), the pace and degree of structural assimilation for Italian Americans is related to social class and generation. With upward mobility, close friendships are founded on common class interests and lifestyles, not ethnic or kinship ties (Alba, 1994; Crispino, 1980; Gans, 1982; Roche, 1982).

A key indicator of primary structural assimilation is residential integration. Historians and sociologists have often noted the long-lasting cohesiveness and stability of Italian American neighborhoods (Glazer and Moynihan, 1970; LaRuffa, 1988). But this too appears to be mostly a class-related phenomenon, rather than something uniquely Italian American. Working-class people are generally more stable residentially primarily because their opportunities for spatial mobility are more limited than those of middle-class people. Therefore, because most first- and second-generation Italian Americans were working class, they were inclined to remain in their original central-city neighborhoods. With mobility into the middle class, however, there has been as much propensity among Italian Americans as among other ethnic groups to leave the ethnic neighborhood for the more heterogeneous suburbs (Alba, 1995, 2009; Crispino, 1980; Gans, 1982; Martinelli, 1989).

Moreover, the residential solidarity of Italian Americans has often been exaggerated. Studies have shown that in Chicago and Philadelphia, for example, Italians from the outset moved frequently from their original areas of settlement, regardless of their social class. The "Little Italys" of various cities were in any case apparently never as homogeneous as has commonly been assumed (Hershberg et al., 1979; Lopreato, 1970; Nelli, 1970). At an early point of the New Immigration, some groups, including Italians, were highly segregated in northern cities, to be sure. But this pattern changed quickly and radically in the early years of the twentieth century (Lieberson, 1980). Today, even where Italian American ethnic communities are evident, their homogeneity is evaporating. In the greater New York region, Alba found that "[t]he Italians, some of whose ethnic communities are still conspicuous, reside mostly in non-ethnic areas, and their continuing suburbanization is eroding the most ethnic Italian neighborhoods" (1995:12).

INTERMARRIAGE   The most critical indicator of primary structural assimilation is the extent to which members of an ethnic group have intermarried with those of the dominant group or outside their own ethnic group. For white ethnics, there appears to be an increasing tendency toward intermarriage with each successive generation; in fact, most marriages today involve some degree of ethnic intermixing (Alba, 1995; Lee and Bean, 2010; Lieberson and Waters, 1988; Sherkat, 2004). As Waters has noted, "For white ethnics, the longer a group has been in the United States and the greater the percentage of its members in later generations, the lower the in-marriage ratios" (1990:104). By the 1990s, almost three-quarters of younger Italian Americans had spouses without Italian ancestry, and for Polish Americans the out-marriage figure was even higher (Alba, 1995).

THE FUTURE OF ETHNICITY FOR WHITE ETHNICS

THE DISAPPEARANCE OF WHITE ETHNICITY?

It is clear that the strength of ethnicity for Italian Americans and other white ethnic groups diminishes as they move increasingly from the working class to the middle class. With upward mobility, social class rather than ethnicity becomes the most significant determinant of job, school, neighborhood, lifestyle, friendship circle, and even marital partner. For these groups there is also strong evidence to support what has been called **straight-line theory**, which posits that *movement toward more complete assimilation proceeds irreversibly with each successive generation* (Gans, 1979; Sandberg, 1974).[3]

Does this spell the eventual disappearance of the Italian American ethnic community and others like it? That development may lie in the future, but it is unlikely in the lifetimes of contemporary white ethnics or their children. Whereas certain social forces are propelling assimilation relentlessly, others serve to retard the process.

SYMBOLIC ETHNICITY   In any case, the form of ethnicity for current and future generations of Italian Americans and other white ethnic groups in the United States will not be the same as it was for their parents or grandparents. Although they will probably continue to assert an ethnic identity and retain a few major elements of the ethnic culture, their acknowledgment of ethnicity will be largely expressive rather than instrumental. Gans (1979) has suggested the idea of **symbolic ethnicity** to describe the continued and at times even accentuated expression of ethnicity among those of the third and fourth generations. Ethnicity, he explains, becomes *a matter of voluntary personal identity, but it is largely devoid of vitality as a determinant of social behavior*. It is, as Richard Lambert (1981) has put it, vestigial, something of the past that people can identify with but that has a very limited effect on their daily lives.[4]

Distinctive and substantial white ethnic communities that once made up much of the urban landscape of cities like New York, Chicago, and Boston are fast disappearing. For most white ethnics, upward mobility continues to impel residential dispersion to heterogeneous suburbs, which, in turn, seems to lead irreversibly to a decline of ethnic awareness and identity. For these groups, ethnicity appears destined to function largely in a symbolic fashion.

EURO-AMERICANS: A NEW ETHNIC GROUP?

As ethnicity for Euro-Americans continues to decline as a consequential factor in various areas of social life, the boundaries of white ethnic groups become less and less distinct. Ethnic consciousness has, as a result, progressively weakened. In fact, Lieberson and Waters (1988) have found that a substantial and growing portion of the white population is apparently unaware of its ethnic origins, does not identify with any of the commonly understood ethnic groups, or simply refuses to identify its ethnic ancestry. They refer to this grouping as "unhyphenated whites."

One of the forces contributing to this blurring of ethnic lines and the tendency for Euro-Americans to deemphasize ethnicity, at least in an instrumental fashion, is the enormous intermixing of ethnic groups that has been occurring continuously for the past 150 years. Today, it is increasingly difficult for people to affirm their ethnic

---

[3] Gans, who first suggested the straight-line theory, has suggested that it be renamed "bumpy-line theory" to account for the unexpected movements that the assimilation process may take, sometimes even reverting to a more expressive ethnic identity. The direction of the movement—toward more thorough assimilation—does not change, however (Gans, 1992).

[4] For a rejoinder to Gans's thesis, particularly as it applies to Italian Americans, see Vecoli (1996).

ancestry without declaring two and often several additional categories. For example, they may describe themselves as "Irish" but "Italian" as well. And, as earlier noted, ethnic intermarriage continues unabated. The result is that ethnic identity for Euro-Americans has become largely a matter of choice. And as ethnicity becomes more voluntary, resulting in fading ethnic boundaries, "in a sense," writes Richard Alba, "a new ethnic group is forming—one based on ancestry from anywhere on the European continent" (1990:3).

Sociologist Mary Waters (1990) studied these fading ethnic divisions and found that, increasingly, white ethnics make conscious choices about their ethnic identity. Not only can they choose their ethnic affiliation but they can choose whether to emphasize ethnicity in their lives at all. "Ethnicity," she writes, "is increasingly a matter of personal preference" (1990:89). Given the strong rate of intermarriage among white ethnics, each generation's ethnic background becomes more complex and difficult to clearly define. Individuals—if they think of ethnicity at all—therefore begin to identify themselves in ethnic terms that reflect only a part of what may be their very elaborate ethnic heritage. People tend to stress only one element of their complex ethnic ancestry for any number of reasons. They may know more about one parent's family history than another, or their surname or physical characteristics may conform to stereotypical images of the chosen group, or they may perceive certain of their ancestral lines as more socially desirable than others (Waters, 1990). In any case, for white ethnics, ethnicity is increasingly voluntary.

Yet the fact remains that most Euro-Americans do choose and continue to affiliate with some ethnic group. Thus ethnic identity, which may be in large part uncertain or even fictive, continues to play a symbolic function for many whites however little bearing it has on their social behavior or economic standing. They may carefully pick and choose their ethnic affiliation, but, as Waters explains, "[i]t is clear that for most of them ethnicity is not a very big part of their lives" (1990:89). Ethnicity becomes mostly something that provides a feeling of community as well as an identification that may set one off as unique and interesting. It has no impact on vital life decisions like where to live, whom to associate with, or even whom to marry.

The tendency for individuals to make choices about their ethnic identity demonstrates that ethnicity is not fixed; it is fluid and can change in significance within one's lifetime or from one generation to the next.

In considering these "ethnic options," as Waters refers to them, it is important to understand that the kind of freedom to select the content and strength of one's ethnic identity is not the same for racial-ethnic groups. Whereas for Euro-Americans the question, "What are you?" can be answered as the individual chooses, for members of racial-ethnic groups the answer has already been given by out-group members. African Americans, Asian Americans, Native Americans, and many Latinos do not, therefore, enjoy the same kind of options in the realm of ethnicity. "The reality," notes Waters, "is that white ethnics have a lot more choice and room for maneuver than they themselves think they do. The situation is very different for members of racial minorities, whose lives are strongly influenced by their race or national origin regardless of how much they may choose not to identify themselves in ethnic or racial terms" (1990:157). Failure on the part of Euro-Americans to acknowledge that difference often leads to their belief that ethnicity need play no greater role for African Americans or Hispanic Americans or Asian Americans if they so choose.

This creates misconceptions about the continuing strength of ethnicity and its impact on the lives of members of racial-ethnic groups.

SUMMARY

- Starting in the 1830s, Irish Americans were the first non-Protestant immigrants to enter American society in large numbers. They were almost all poor and unskilled at the time of their entrance and experienced severe prejudice and discrimination, primarily as a result of their Catholic faith. Nativism of the early and mid-nineteenth century was mostly an anti-Catholic movement, and the Irish, as a result, became its primary targets.

- The Irish used the urban political machine and the Catholic Church as the two institutions to establish their place in America. No other group, before or after, so effectively used the local political system to advance their cause as the Irish. The Catholic Church in America was basically shaped and led overwhelmingly by Irish Americans.

- Irish Americans have assimilated more completely than any other white ethnic group. Cultural assimilation occurred relatively quickly and structural assimilation has occurred over several generations. Today, aside from religion, the Irish are hardly distinguishable from the dominant Anglo group.

- Italian Americans are representative generally of Euro-American ethnic groups that derive from the classic (or New) immigration period from the late nineteenth to the early twentieth centuries.

- Italian Americans, having entered the society at the bottom of the class hierarchy, had displayed significant upward mobility by the third generation. Today, they are at or above the national average in income, occupation, and education, and they have begun to enter positions of political and economic power. This holds true for other white ethnic groups as well.

- Beginning with large-scale immigration in the late nineteenth century, Italians in the United States were the objects of prejudice and discrimination, sometimes of a particularly virulent nature. Perhaps the most persistent negative stereotype that Italian Americans have had to deal with involves their identification with organized crime.

- Italian Americans today display a relatively high level of cultural assimilation, and although it is less thorough and rapid, structural assimilation is also occurring steadily.

- As the assimilation process continues, for Italian Americans as for other white ethnic groups, ethnicity is increasingly reduced to a symbolic function and becomes less a determinant of social behavior. As ethnic boundaries weaken, a new, Euro-American ethnic group may be emerging.

CRITICAL THINKING

1. Nativism in the nineteenth and early twentieth centuries focused primarily on Catholics and Jews. Compare nativism then with its more current forms, directed primarily at Hispanic immigrants and Muslims. Are the claims,

arguments, and political rhetoric of present-day nativists similar to those used earlier against Catholics and Jews?

2. If, as this chapter explains, the Irish were redefined as white ("became white") after several decades, will this occur for Asian ethnic groups and most Latinos, now considered "nonwhite?" What social, political, and economic forces would bring about that reclassification?

3. What is the likelihood that contemporary immigrant groups to the United States will assimilate in a manner similar to that of white ethnic groups of the nineteenth and early twentieth centuries? Consider the social and economic circumstances faced by Irish and Italian immigrants, and the various factors that affect assimilation, discussed in Chapter 4.

4. Consider the prevalence and strength of anti-Catholicism in America prior to the early twentieth century. Today, anti-Catholicism seems little more than a historical footnote. What occurred to bring about such a radical shift in social thought and behavior?

Personal/Practical Application

1. Most Euro-Americans today can claim to be ethnic hybrids, that is, children of multiple ethnic origins. As this chapter explains, they can literally choose their ethnic identity. If you are non-Hispanic white and are questioned about your ethnicity, how do you identify yourself? How did you decide on your ethnic self-identity?

2. Given the growing prevalence of intermarriage among all racial and ethnic groups, increasingly those who are not Euro-American can also begin to "choose" their racial-ethnic identity. However, is their freedom of choice the same as for Euro-Americans? Explain this.

Jewish Americans

Jews in America are a relatively tiny ethnic group—little more than 2 percent of the general population. But for their distinctive religious identification, they might logically be analyzed within the context of white ethnic Americans (Chapter 10), with whom they share many of the same sociohistorical features. Yet, they warrant close study independent of other American ethnic groups for several reasons.

In many ways, Jewish Americans are a unique case among American ethnic groups. As people who for centuries had suffered persecution and dislocation in almost all European societies, Jews found the United States a haven that provided for the first time a social atmosphere in which Jewish identity could be retained without fear of official repression and from which there was no need to contemplate flight. Jews in America also found their economic circumstances relatively unconstrained, in sharp contrast to the situation they had faced in Europe. So well nurtured were the opportunities offered them that Jews today represent the prototypical American ethnic success story. In only two generations, Jewish Americans collectively accomplished a truly phenomenal upward mobility that makes theirs a compulsory case for the study of ethnic relations. Indeed, they present a compelling comparative case, against which other groups' similarities and differences can be scrutinized and assessed. Why have Jewish Americans, as a collectivity, succeeded so evidently, and can their story be relevant to the experiences of other American ethnic groups, both historical and contemporary?

Unlike most American ethnic groups, Jews do not identify with a common homeland, having come from many nations. The broad geographical native region of the majority of Jewish Americans, however, is the same—eastern Europe—and a common language, Yiddish (a combination of mostly German and Hebrew elements), supersedes most national differences. Jews in the United States are best seen as an ethnic group: they clearly display a unique cultural heritage, a consciousness of community, and a group identity. Regardless of their degree of commitment to Jewish religious practices and beliefs (many are nonbelievers), and regardless of their national origins, Jews recognize themselves as a people with common roots and with common cultural features, that is, as an ethnic group.[1] They are seen in that fashion, as well, by the larger society.

That Jewish Americans lack a common homeland does not in itself constitute a significant difference from many white ethnic groups in the United States. In most cases, immigrants from southern and eastern Europe arrived with little national consciousness; regional or village boundaries generally defined their sense of peoplehood. Only in the United States, as a means of social adjustment and in response to out-group recognition, did they adopt a national identity.

IMMIGRATION AND SETTLEMENT

In some ways, Jewish patterns of immigration and settlement in the United States paralleled those of Italians. Both ethnic groups were primarily part of the New Immigration, and following large-scale entrance, both found themselves initially at the bottom of the class hierarchy. Like the Italians, Jews initially settled mostly in the large cities of the Northeast. Many of the occupations of the two groups during the major period of immigration also overlapped, as did their residential areas. As non-Protestants who had emigrated from relatively backward societies, both groups were made-to-order targets of the nativist and racist ideas of the turn of the century. Furthermore, because both groups remained concentrated primarily in eastern cities, they did not blend easily into the larger society. The prejudice and discrimination they faced, therefore, proved more pernicious and dogged than those aimed at other white ethnic groups. Despite the similarities in their mode of entrance and settlement patterns, however, important differences between these two groups were also apparent. These differences led, as we will see, to different rates of mobility in the second and third generations and ultimately to different positions in the class system.

IMMIGRATION PATTERNS

Jewish immigration to America has a lengthy history. In fact, a few Jews supposedly accompanied Columbus on his first voyage. Three fairly distinct periods of Jewish immigration can be identified.

SEPHARDIC JEWS   The earliest immigrants to the United States included Sephardic Jews, whose ancestors were Spaniards and Portuguese. When the Jews were expelled from Spain and Portugal in the late fifteenth century, many migrated to England, Holland, Brazil, and the Caribbean islands. The first American Jews came from all these locations and even from the Iberian peninsula (Glazer, 1957). The Sephardim are to be distinguished from the Ashkenazic Jews of central and eastern Europe, who constituted almost all Jewish immigration into the United States after the colonial period.

---

[1] A national survey in 1990 found that most Jewish Americans defined their Jewish identity in ethnic or cultural, not religious, terms (Kosmin et al., 1991). This was confirmed again in a 2013 comprehensive survey of Jewish Americans. Almost two-thirds said that being Jewish is mainly a matter of ancestry and culture, not religion (Pew Research Center, 2013c). Thus most Jewish Americans consider themselves Jews primarily as members of an ethnic group.

GERMAN JEWS   A second and more sizable Jewish immigration to the United States was made up of German Jews who came with the great wave of other German immigrants during the 1840s and 1850s. These Jews were socially and economically several notches below the earlier Sephardic Jews, who by that time were relatively prosperous and respected. Although most were merchants, their trade was at a level considerably less significant than that of their Sephardic predecessors. Many, in fact, were peddlers who moved westward with the country's expansion. Indeed, at the middle of the nineteenth century, most of America's twenty thousand itinerant traders were German Jews (Lipset and Raab, 1995). Dispersing widely to all parts of the country, German Jews settled in both large cities and small towns. Where they settled they commonly established clothing and dry goods or general stores, the vestiges of which are seen today throughout the United States (Brockman, 2001; Pressley, 1999). A few developed into large national chains. Many familiar department store names, including Macy's, Bloomingdale's, Saks Fifth Avenue, and Neiman-Marcus, stem from German-Jewish founding families.

Collectively, the German Jews displayed rapid and substantial upward mobility; by the 1880s, most were middle-class entrepreneurs. Among the most successful of them emerged a small, aristocratic upper class which intermingled socially with the Gentile upper class and even intermarried among them (Baltzell, 1958, 1964). By the turn of the century, German Jews were so thoroughly assimilated that they had lost much sense of ethnic identity. Only their religious affiliation distinguished them from other Americans (Herberg, 1960).

EAST EUROPEAN JEWS   By 1880, Jews were still only a tiny fraction of the American population, numbering no more than a quarter million. In the next three and a half decades, however, two million Jews would come to the United States. It is primarily from this wave of immigration that the contemporary Jewish American population derives.

Jews coming to the United States after 1880 differed markedly from those who had settled earlier. To begin with, most were not from Germany but from eastern Europe, particularly Russia and Russian-occupied Poland. One of the key motivating factors behind the mass emigration of Jews from Russia was the severe oppression they were subjected to under tsarist policies. Following the assassination of Tsar Alexander II in 1881, bloody pogroms were sanctioned, and onerous restrictions were placed on Jews in all areas of social and economic life. America, therefore, was seen not only as a destination of economic opportunities but also as a political haven, a society that promised liberation and acceptance (Biale, 1998). Unlike the Italian emigration, which at first comprised mainly single men expecting to return, the Jewish emigration from Russia was a movement of family units who harbored few thoughts of returning to their homeland.

The East European Jews also differed from earlier immigrant Jews in their class origins: they were poorer and less educated. Jewish life in eastern Europe had been confined to all-Jewish villages and towns; as a result, these immigrants were extremely provincial in contrast to the relatively urbane and worldly German Jews. Striking differences between the two groups also lay in their attitudes toward religion. The East Europeans came with a version of Judaism far more traditional than the modernized and Americanized Reform Judaism of the Germans. To the German

Jews, Judaism was simply a faith that was not to interfere with assimilation into the core American culture and society. To the East Europeans, coming from the shtetl, or Jewish village, where religion governed all aspects of life, Judaism was an entire social world. As Glazer and Moynihan explain, "In practice, tone, and theology, the Reform Judaism of the German Jews diverged from the Orthodoxy of the immigrants as much as the beliefs and practices of Southern Baptists differ from those of New England Unitarians" (1970:130).

In addition to differences in class and religious practice, East European Jews did not disperse geographically as had the earlier German Jewish immigrants. Instead, they remained in a few large cities, especially New York, where they were concentrated in lower-class ghettos. In this environment, the East European Jews, like the Italian immigrants of that time, experienced severely harsh conditions of work and residence. Many labored in garment factories, commonly called sweat-shops, where they worked long hours for scant salaries. In 1911, one-third of Jewish heads of household earned less than $400 a year; and in 1914, the average hourly wage of male clothing workers was about 35 cents. Even by 1930, garment workers were still earning an average of only $24.51 a week—less than the pay of workers in stockyards and slaughterhouses (Howe, 1976). As a result, most families became work units in which women and children as well as men labored fifteen or eighteen hours a day, often taking piecework home with them.

Like the Italians who streamed into New York and other large eastern cities at the same time, the East European Jews accepted housing conditions that were frequently brutal and demeaning, crowding into run-down tenements abandoned by previous ethnic groups. On the famous Lower East Side, 350,000 people per square mile jammed into squalid rooms and apartments in which sanitation facilities were meager at best. These degraded conditions of life only further confirmed the racist views of those who saw immigrant Jews of that time, along with other southern and eastern Europeans, as innately degenerate.

In sum, the urban environment into which immigrant Jews entered the United States in the late nineteenth and early twentieth centuries required struggle and stamina, which could be sustained only by the expectation that life conditions would improve—possibly for them but surely for their children. "The generation that entered the immigrant ghetto," writes Ben Halpern, "was confronted by one overwhelming task: to get out or enable the next generation to get out" (1958:36). This goal proved attainable for most. One important factor in accounting for the determination of immigrant Jews to succeed in American society was their understanding that there was no return to the homeland; for most Jews, the commitment to the United States was final.

Given their striking differences in class, religion, and culture, it is little wonder that the German and East European Jews were initially antagonistic. The Germans viewed the new immigrants as illiterate, uncouth, and provincial greenhorns, who could only cause embarrassment to the American Jewish community and produce a backlash of anti-Semitism. By the 1890s, however, the divisions between the two groups were evaporating. Inspired partially by humanitarian motives and partially by the concern that they might be lumped with these poor and religiously traditional Jews in the eyes of their Christian neighbors, German Jews now accepted responsibility for uplifting and assisting their East European cohorts (Wirth, 1956). By the end of large-scale immigration in the early 1920s, the subdivision of the American Jewish community into German and East European groups was no longer apparent (Glazer and Moynihan, 1970; Yaffee, 1968).

JEWISH AMERICAN DEMOGRAPHICS

Since religion is not an item in the U.S. Census, the Jewish American population can only be estimated. All surveys indicate that the number falls between 5 and 6.5 million, roughly 2 percent of the U.S. population (Pew Research Center, 2013c; Sheskin and Dashefsky, 2011; United Jewish Committee, 2003). This is a declining figure in both absolute and relative terms. Demographers have projected a continuing decrease in numbers, due to an unusually low birthrate in combination with an increasing rate of intermarriage (Della Pergola et al., 2000). Even though they are only a small part of the American population, and a declining one as well, Jews in the United States constitute more than 40 percent of the world's Jewish population. Augmenting the Jewish American population in the last three decades have been immigrant Jews from the former Soviet Union and from Israel. Together, however, these groups constitute only a small portion of Jews in the United States.

From the time of their large-scale arrival in the United States in the late nineteenth and early twentieth centuries, Jews had traditionally been concentrated in northeastern states. Although they still reside predominantly in that region, they have dispersed in recent decades to the South, especially Florida, and to California in the West. Table 11.1 shows the states with the highest percentage of Jews.

Jews, like Italians, are among the most urbanized of American ethnic groups. In fact, half of the entire Jewish population in the United States is found in just seven large urban areas: New York, Los Angeles, Miami, Chicago, Boston, Washington,

TABLE 11.1 | STATES WITH LARGEST JEWISH POPULATIONS

State	Estimated Jewish Population	Estimated Jewish Percent of Total
New York	1,635,020	8.4
California	1,219,740	3.3
Florida	638,635	3.4
New Jersey	504,450	5.7
Illinois	297,935	2.3
Pennsylvania	294,925	2.3
Massachusetts	277,980	4.2
Maryland	238,000	4.1

Source: Sheskin and Dashefsky (2011).

and Philadelphia.[2] More than 20 percent are in the New York metropolitan area alone (Sheskin and Dashefsky, 2011).

JEWISH AMERICANS IN THE CLASS SYSTEM

SOCIOECONOMIC STATUS

The fact that most Jews of the New Immigration entered the society in relative poverty made their rapid upward mobility quite remarkable. As a group, Jews rose virtually from the bottom to the top of the stratification system in two generations. On all measures of social class—income, occupation, and education—Jews collectively rank higher than almost all other American ethnic groups.

The high socioeconomic position of Jewish Americans is well evident in family income and wealth. The median household income of Jews has far exceeded the national average for several decades (Chiswick, 1993; Cohen, 1987; Massarik and Chenkin, 1973; Smith, 2005; Wilder and Walters, 1998). By 1990, the median Jewish American household income was 34 percent higher than for the society as a whole (Kosmin et al., 1991; United Jewish Committee, 2003). Table 11.2 shows the significantly greater percentage of Jewish families that fall into the higher income brackets. Almost 70 percent of Jewish households are above the national median of $50,000. One-quarter of Jewish Americans have household income above $150,000, compared to only 8 percent of all U.S. households. Similarly, Jewish Americans far exceed national averages in family wealth (Keister, 2003).

Commensurate with their relatively high income, Jews display a disproportionate concentration in higher-status occupations, namely, the professions, business ownership, and managerial positions. Although the German Jews had, for the most part, established themselves as part of the middle class by the late nineteenth century,

TABLE 11.2 | SOCIOECONOMIC STATUS OF JEWISH AMERICANS

	Jewish Americans (%)	U.S. Population (%)
Income		
Households with income below $30,000	20	36
Households with income above $100,000	42	18
Education		
High school or less	17	42
College graduates	58	29
Graduate degrees	28	10

Source: Pew Research Center (2013c).

---

[2] Studies of Jews in small towns and rural areas include Rogoff (2001), Rose (1977), and Weissbach (2005).

the rise of the East European Jews since the cessation of large-scale immigration in the 1920s has been even more striking (Chiswick, 1984). Only in New York City, containing the largest concentration of Jews in the United States, did there remain a substantial Jewish working class until the 1970s (Glazer and Moynihan, 1970; Goldstein, 1980; Pritchett, 2002; Sklare, 1971). Quite simply, as Calvin Goldscheider has stated, "poverty and unskilled occupations are largely uncharacteristic of contemporary American Jews" (1986:4). A recent survey (Ukeles et al., 2013), however, showed a surprising number of Jewish households in New York City (one in five) living in poverty. These were mainly Hasidic (ultra-Orthodox) families, Russian immigrants, and seniors.

Finally, the extremely high level of educational attainment among American Jews is well recognized, having been noted repeatedly in sociological studies (Goldscheider, 1986; Lehrer, 1999; Smith, 2005). Indeed, the high income and occupational levels of Jews can be accounted for in large measure by their similarly high educational level. As shown in Table 11.2, 58 percent of Jews, compared to 29 percent of adult Americans in general, are college graduates; more than one-quarter of Jews have earned a graduate degree, compared to 10 percent of the general population. Only Asian Indians among other ethnic groups exhibit a higher percentage of college graduates. If education is the key to upward mobility in American society, Jews are the classic case of an ethnic group having converted educational opportunities into economic betterment in a relatively short time.

ETHNIC AND CLASS FACTORS IN UPWARD MOBILITY

The saga of Jewish upward mobility is often proclaimed as the great success story of American ethnic history. Indeed, no other ethnic group has seemed to rise so far so quickly. As a collective rags-to-riches case, Jewish Americans are often held up as an example for other groups to emulate. What has provided for this extensive collective mobility in only two generations? Here we might compare the Jews with other white ethnic groups, such as the Italians, that entered the United States on a large scale at precisely the same time yet have not exhibited a similar degree of upward mobility. This comparison may give us strong clues to why different ethnic groups experience varying rates of success in the United States and in other multiethnic societies as well.

Herbert Gans has explained that in looking at the mobility patterns of ethnic groups, it is essential to try to understand whether what happens to them is "ultimately more or less a function of their characteristics and culture than of the economic and political opportunities which are open to them when they arrived and subsequently" (1967b:8). That is, patterns of mobility among Jews, Italians, or any ethnic group can be seen as the result of either the group's cultural traits or the opportunities to which the group is exposed, or perhaps a combination of those factors. Some of the same explanations of the economic and educational success of contemporary Asian Americans, discussed in Chapter 9, have been offered to account for Jewish success.

INNATE ADVANTAGES? Jewish intellectual and occupational achievement in the United States, and in other societies as well, has been so consistently high that some have concluded that Jews, collectively, are innately advantaged. That is, there is a genetic basis for Jewish accomplishment in science, the arts, and commerce. High intelligence among Jews, specifically Ashkenazi Jews, has been confirmed by IQ tests. "Jews have the highest average IQ of any ethnic group for which there are reliable data," notes psychologist Richard Nisbett (2009). The result is a tremendous overrepresentation among those who are eminent in virtually all academic fields, including medicine, science, mathematics, law, and the arts.

But is high IQ among Jews genetically based? A few have suggested that there is in fact reason to believe that over many centuries through a process of natural selection, Jews have acquired an elevated intelligence (Cochran et al., 2005; Murray, 2007). After reviewing some of the most frequently cited genetic explanations, however, Nisbett (2009) concludes that there is very little convincing scientific evidence to substantiate claims of innate advantage. Recall similar biological arguments put forth to explain high Asian achievement (Chapter 9), none of which are given much credence. More fruitful explanations focus on varying ethnic cultures and the opportunities that ethnic groups are afforded.

GROUP CULTURE   Similar to a prevalent explanation of the comparatively high level of intellectual and occupational success among Asian Americans, Jewish culture is seen by some as the key to explaining the rapid and profound success of Jews in America.

The substance of this view is contained in what Bernard Rosen (1959) termed the "achievement syndrome." Groups differ, he explained, in their orientation toward achievement. In comparing Jewish immigrants with Italian immigrants, the former are portrayed as adhering to a culture that encouraged independence and stressed achievement values and high educational aspirations in children, whereas the latter did not. Italian Americans traditionally stressed accomplishment only as it benefited the family in some concrete fashion. Individual achievement, however, was deemphasized because it destabilized the structure of the family unit and threatened its authority. Glazer and Moynihan write that "that form of individuality and ambition which is identified with Protestant and Anglo-Saxon culture, and for which the criteria of success are abstract and impersonal, is rare among American Italians" (1970:194–95). Among Jews, however, it was preeminent. Thus, instead of plying their father's trade, as did most second-generation Italians, Jews aspired to higher occupational positions. As Philip Gleason explains, "Unlike Catholic peasant immigrants, the Jewish immigrants did not regard intensive education and professional careers as something beyond their experience or capacity; they saw them rather as their natural lines of aspiration" (1964:168).

Perhaps the most frequently cited cultural value in explaining Jewish upward mobility, particularly as it contrasts with the more limited mobility of other ethnic groups of the New Immigration, is the group's attitude toward education. The value of learning has a rich tradition in Jewish culture extending as far back as biblical times, and it permeates all aspects of Jewish life (Fauman, 1958; Gleason, 1964; Strodtbeck, 1958). Because of this key cultural trait, it is held, Jews have quite naturally achieved proportionately greater success in business, academia, science, and the arts than have other ethnic groups.

The Jewish penchant for education is contrasted sharply with the traditional southern Italian apprehension about this institution. Rather than a means to social and economic advancement, education was viewed by immigrant Italians as having

little practical value and in some ways being antithetical to family and group interests (Covello, 1967). Hence, they did not avail themselves of the educational opportunities in the United States as quickly and exhaustively as did the Jews. Some have
observed that even later generations of Italian American parents, including those of
the middle class, have not overly pressured their children for achievement. "By their
norms," concludes one study, "motivation as well as ability should come from the
child, not the parent" (Johnson, 1985:189).

Intense family loyalty as part of the southern Italian culture was especially consequential for immigrants and their children. With an allegiance to family above all
else, individual Italian Americans found it difficult to pursue occupations that did
not concretely improve the family's welfare. The most important work objective
was to find a steady, secure job that would contribute to the family's immediate
well-being and that would not take the individual outside the realm of the family's
influence. Such values restrained upward job mobility (Gambino, 1975; Gans,
1982).

Obviously, the choice of fulfilling short-range work goals instead of seeking the
long-range advantages of continuing education was not unique to Italian immigrants
and their children. But the norm was perhaps more firmly rooted among them. There
is little doubt, then, that their skeptical view of formal education contributed heavily
to their slower and more limited upward mobility compared to Jewish Americans.

GROUP OPPORTUNITIES   Disputing the contention that ethnic groups differ in achievement primarily because of internal group values, some maintain that differences depend mostly on the opportunity structure that groups encounter on entering the
society and the skills they bring to that structure (Gans, 1982; Steinberg, 1989).
The essence of this view is that class factors in general explain more about mobility
than do ethnic cultures.

Although not denying differences in group culture, sociologist Stephen Steinberg
(1989) argues that Jewish immigrants came to the United States with an advantageous occupational background that enabled them to cultivate the American economic system better than other ethnic groups could. Russian Jews entering the
United States were not peasants but were urban people who had been engaged in
manufacturing and commerce. Consequently, although they were certainly poor,
they were equipped with skills that would serve them well in the expanding industrial economy. Italians and most other groups of the New Immigration were, by
comparison, mainly agricultural peasants. Moreover, Jewish occupational skills complemented almost perfectly the needs of the rapidly growing garment industry, where
so many Jewish immigrants found employment. "Jews," notes economist Eli Ginsberg,
"had the good fortune to be in the right place at the right time" (1978:115). Moreover, as the economy later shifted from agriculture and manufacturing to services,
Jews benefited from their background in trade and commerce.

Nathan Glazer (1958) also points out that East European Jews who came to the
United States had a long heritage of middle-class occupations, including the professions and buying and selling, by contrast with other immigrant groups. Although
tsarist oppression had driven Jews from their traditional trades, their link to them
was not broken. Thus they came to the United States with little capital but not with
the limited horizons of the working class. In effect, the East European Jews were part

of the working class for only one generation. They were neither the sons nor the fathers of workers (Herberg, 1960).

Steinberg also questions the attribution of Jewish occupational success to the high value attached to learning. Although he does not deny that the higher rate of literacy among immigrant Jews compared, for example, to that of southern Italians was instrumental in their more rapid mobility, he maintains that substantial educational advancement followed economic mobility rather than the converse. After looking at historical data, Steinberg concludes that for Jews, "economic success was a precondition, rather than a consequence, of extensive schooling" (1989:136).

In a similar fashion, the presumably lower educational and occupational aspirations of immigrant Italians can be accounted for by class as much as by ethnic factors. That is, their attitudes toward education and work, particularly among the first and second generations, may have reflected their primarily working-class status, not simply their southern Italian heritage. Gans (1982) explains that many of the behaviors and values he observed among second-generation Italian Americans in Boston's West End were not so much ethnic traits as working-class traits. Thus the limited mobility aspirations among working-class Italians might be expected as well among working-class families of other ethnic groups, even working-class Jews. As Gans has observed, "I continue to be impressed by how similarly members of different ethnic groups think and act when they are of the same socioeconomic level, are alike in age and other characteristics, and must deal with the same conditions" (1982:277–78).

Looking at Italian American achievement and attitudes toward education as the products primarily of a unique ethnic trait also fails to take into account the opportunity structure in the old country, which dictated limited aspirations. "To whatever extent Southern Italians exhibited negative attitudes toward education," writes Steinberg, "in the final analysis these attitudes only reflected economic and social realities, including a dearth of educational opportunities" (1989:142).

Historian John Briggs (1978) also discounts the idea of a predisposed negative view of education among southern Italians. He attributes their rejection of schooling in the United States, instead, to their expectation of an imminent return to Italy. It was not a lack of interest in education but the lack of relevance of American schooling that led to its widespread repudiation, he contends. Schooling in the new country could be of little value if one's stay was only temporary.

To put the matter simply, those who interpret different patterns of mobility among immigrant Jews and Italians in class rather than ethnic terms stress that, whatever the extent of a group's occupational and educational background or tradition, only when it is combined with a favorable opportunity structure can it enhance upward mobility. For Jews, there was a fortunate compatibility between the skills and experience they brought with them and the needs and opportunities of the American economic structure. And as Steinberg suggests, "It is this remarkable convergence of factors that resulted in an unusual record of success" (1989:103). Similarly, Samuel Baily (1999), in his study of Italian immigrants in Buenos Aires and New York City, concludes that the slower upward mobility of Italians must be seen as a product of the skill level most of them possessed, in combination with the job opportunities afforded by the host society. Valid comparisons among ethnic groups can be made, then, only when consideration is taken of the society's constantly changing opportunity structure.

GROUP CULTURE AND GROUP OPPORTUNITIES: A SYNTHESIS Evidence can be marshaled to support both the group culture and opportunity structure arguments regarding ethnic mobility; but as with most issues in sociology, the truth more likely lies somewhere between the two.

On the one hand, had Jews not possessed at least some propitious cultural traits, including among them a reverence for learning, it is not likely that such rapid mobility would have been possible even within the context of a favorable opportunity structure. Educational opportunities in the United States existed for other European ethnic groups, but it seems evident that immigrant Jews availed themselves of these opportunities more thoroughly and avidly than did others. Moreover, Jews in other immigrant societies such as Canada, Argentina, and South Africa have displayed, as a group, economic and educational success comparable to that attained in the United States and have pursued similar occupations.

On the other hand, there is little question that immigrant Jews brought with them a wider range of occupational skills that automatically placed them a notch ahead of other ethnic groups in the labor market. As we saw in Chapter 8, the rapid economic success of Cuban immigrants in the United States in the past four decades, compared to that of other contemporary Hispanic groups, poses a quite similar case, as does the upward mobility of many Asian groups.

POLITICAL AND ECONOMIC POWER

THE POLITICAL REALM Jewish participation at the higher levels of government has been relatively greater than that of Italians and most other southern and eastern European groups. In fact, Jewish representation in the federal government is far out of proportion to their share of the general U.S. population (Alba and Moore, 1982; Pew Forum on Religion and Public Life, 2008; Zweigenhaft and Domhoff, 2006). Jewish entrance into the realm of American political power is a relatively recent phenomenon, however.

A few Jews were elected to Congress as early as the nineteenth century, but only in the past few decades have they been elected in meaningful numbers. Before 1945, only fifty-nine Jews had ever served in either house. The number of Jews in the U.S. Congress increased significantly after 1974 (Goldberg, 1996; Zweigenhaft and Domhoff, 1998, 2006). Today, in light of their minor proportion of the general populace, Jews are overrepresented in the federal legislature, making up 11 percent of the Senate and 5 percent of the House (see Figure 10.1). They are also strongly represented as federal political appointees and civil servants and have occupied a number of Cabinet positions, particularly in Democratic administrations (Zweigenhaft and Domhoff, 1998, 2006). Moreover, Jews play an important role in political campaign financing at the national level (Goldberg, 1996).

Jews today experience few impediments to elective office at any level of government. In fact, a 1987 Gallup poll revealed that 89 percent of Americans would vote for a well-qualified Jewish presidential candidate; only 46 percent felt that way fifty years earlier (Gallup, 1982, 1988; Quinley and Glock, 1979). The acceptance of Jews in higher politics is demonstrated as well by the fact that Jewish candidates are today commonly elected to office by overwhelmingly non-Jewish constituencies (Wyman, 2000). In recent years, both U.S. senators from Wisconsin, for example, were Jewish, despite the state's Jewish population of only one-half of 1 percent. The selection of Joseph Lieberman, an Orthodox Jew, as the vice presidential candidate on the Democratic ticket in 2000 symbolically shattered any remaining barriers to Jewish American penetration of the very highest levels of government. Jewish Americans have consistently shown a greater propensity than other ethnic groups for political participation, and this may account in some measure for their increasing prominence in elective offices.

THE ECONOMIC REALM With regard to economic power, Jews in the past were underrepresented in the American corporate elite (Alba and Moore, 1982; Korman, 1988; Zweigenhaft and Domhoff, 1982, 2006). The strong Jewish participation in the business world traditionally was in small or medium-size firms (most of them self-owned), not the larger, more powerful corporations that dominated the American economy (Institute for Social Research, 1964; Kiester, 1968; Marger, 1974; Ward, 1965). Furthermore, where they had been successful in the corporate world, Jews more commonly entered the elite through the growth of their own companies rather than by climbing the organizational ladder of well-established concerns.

The past underrepresentation of Jewish Americans in the top echelons of the corporate world can be explained partly as discrimination and partly as a propensity for Jews to prefer independence in business and the professions. The two are related. Faced with discriminatory hiring practices, which did not decline significantly until after World War II, Jews naturally turned to self-owned businesses or independent professions such as law and medicine.

The historical dearth of Jews in the corporate elite is ironic in light of the commonly held assumption among most Americans that Jews wield inordinate economic power. Studies beginning in the 1930s consistently showed this to be a misconception (Burck, 1976; *Fortune*, 1936; Keller, 1953; Newcomer, 1955; Warner and Abegglen, 1963). To the extent that the public illusion of excessive Jewish economic power persisted, it may be explained to some degree by the conspicuous nature of certain industries in which Jews did exert a dominant influence, most notably movies and television, retail sales, and real estate (Glazer and Moynihan, 1970; McWilliams, 1948; Selznick and Steinberg, 1969).

The traditional pattern of Jewish absence from the corporate elite has changed markedly, with Jews increasingly entering the top executive positions of major corporations (Christopher, 1989; Klausner, 1988; Silberman, 1985; Zweigenhaft, 1984, 1987; Zweigenhaft and Domhoff, 2006). In their study of the American power elite, Zweigenhaft and Domhoff (2006) conclude that, despite underrepresentation in some business sectors, Jews are actually overrepresented in the corporate elite. Whatever discrimination in large corporations and law firms continues to exist may be simply the lingering effect of past occupational barriers against Jews that, for the most part, have been formally taken down. Indeed, a study of the late 1980s concluded plainly that by then the disadvantage of Jews in the executive suites of major corporations had "all but disappeared" (Klausner, 1988:33).

PREJUDICE AND DISCRIMINATION

For centuries Jews have commonly been victims of ethnic bigotry in Western societies. As non-Christians they were particularly vulnerable to scapegoating during times of social adversity and were generally relegated to the role of economic and political outcast. Jews were made especially ready targets for oppression after the thirteenth century, when they were forced into separate communities in almost every European society in which they resided. The term *ghetto* was first used to describe the area of Venice in which Jews were compelled by custom and law to live. It was subsequently applied to the Jewish quarter of all European cities.[3] Jews were forbidden to own land or join craft guilds and maintained an ever-tenuous civil status. The capricious nature of persecution and the lack of political security forced them to remain constantly vigilant and prepared to flee. Because of the need to remain mobile, they turned to occupations such as tradesman, moneylender, or professional, which provided for easy movement of capital and skills (Fauman, 1958; Glazer, 1958). These occupational roles often carried social stigmas, which further provoked resentment and hostility from those who were required to deal with them.

THE ELEMENTS OF ANTI-SEMITISM

A set of distinct and consistent negative stereotypes, some of which can be traced as far back as the Middle Ages in Europe, has been applied to Jews. Among the most common of these are that Jews are monied, dishonest, and unethical; clannish, prideful, and conceited; and power hungry, pushy, and intrusive (Quinley and Glock, 1979). It is the connection of Jews with money, however, that appears to be the sine qua non of **anti-Semitism**—that is, *prejudice and hostility toward Jews*. As Glock and Stark put it, "Perhaps the most constant theme in anti-Semitism from medieval times down to the present is of the Jew as a cheap, miserly manipulator of money, forever preoccupied with materialism, and consequently possessing virtually unlimited economic power" (1966:109). As late as 1995, more than a quarter of non-Jewish whites supported the assertion that "[w]hen it comes to choosing between people and money, Jews will choose money"; even larger percentages of African Americans, Latinos, and Asian Americans agreed (National Conference, 1995).

That Jews have more money than most people is, as Selznick and Steinberg (1969) explain, accepted as fact among a majority of non-Jews, though this in itself does not necessarily imply anti-Semitism. Nonetheless, even unprejudiced non-Jews exaggerate the extent of Jewish wealth, and it is, of course, only a short step to interpreting this belief in a negative manner. Although Jews in American society are, as we have seen, collectively higher in socioeconomic status than most other ethnic groups, this does not mean that all Jews are therefore wealthy. Like members of all ethnic groups, Jews will be found at every class level, even the very bottom.[4]

Closely akin to the beliefs regarding the connection of Jews with money are those concerning Jews and business. "Of all the crimes and misdemeanors charged to the Jews," notes Stephen Whitfield, "the worst has been an uncanny or even supernatural capacity to make money, perverted into avarice" (1995:90). The shrewd and unscrupulous businessperson has been a tenacious negative

---

[3] On the development of European ghettos, see Wirth (1956).

[4] For a description of the Jewish poor, most of whom are elderly, see United Jewish Committee (2003).

stereotype of Jews for ages. Derivative of this image is the pejorative American phrase, to "Jew down," which means, roughly, to bargain craftily or to buy something more cheaply than the seller initially offers it. The beliefs regarding Jews and business deal not only with ethics, however, but also with power and manipulation. Jews have been commonly seen as maintaining excessive influence in the business world and as controlling large segments of the economy of the United States and of other societies. The fallaciousness of this image has been confirmed repeatedly; and as noted earlier, in the past Jews have been underrepresented in the higher executive positions of the most powerful American and multinational corporations.

The attributions of Jewish power are often blatantly contradictory. One of the staples of crude anti-Semitism concerns the notion of world conspiracy. Curiously, Jews have been portrayed on the one hand as leaders of a world capitalist conspiracy and on the other as leaders of a world communist conspiracy. This inconsistency has never deterred anti-Semites from expressing both beliefs when the situation calls for it. Indeed, as McWilliams so aptly suggested, there is a tone of "demagogic genius" in this contradictory charge, "for it permits an appeal to the dispossessed and a threat to the rich to be voiced in a single sentence" (1948:92).

ANTI-SEMITISM IN THE UNITED STATES

Historically, three general periods of anti-Semitism in the United States can be defined, each characterized by a different level of intensity and content.

THE EARLY PERIOD   During the early decades of Jewish settlement, anti-Jewish prejudice and discrimination were minimal. Although there were anti-Semitic groups in colonial America, they were not excessively vocal or influential (Blau and Baron, 1963). During the era of the Old Immigration, between 1820 and 1880, German Jews were not essentially distinguished from their Christian counterparts except in religion; and as noted earlier, they mostly assimilated into the mainstream society. Like the early northern Italians, German Jews (and the earlier Sephardic Jews) were too insignificant numerically to evoke intense hostility. Furthermore, they were geographically dispersed and established themselves in economically secure positions, thereby avoiding the attention of nativist groups, which targeted the much larger, city-concentrated, and destitute Irish Catholic population.

This relatively concordant relationship with the Anglo-Protestant majority began to change, however, with the entrance of the East European Jews in the 1880s. A harbinger of the changed attitude toward all Jews occurred in 1877 when Joseph Seligman, a prominent financier and political adviser, was denied entrance to the Grand Union Hotel in Saratoga, New York, a noted upper-class resort (Baltzell, 1964; McWilliams, 1948). From that point on, Jews were commonly barred from certain resorts, clubs, and college fraternities.

THE DEVELOPMENT OF POPULAR ANTI-SEMITISM   Like the southern Italians, immigrant East European Jews became prime subjects of the racist ideologies of the late nineteenth and early twentieth centuries. Just as the criminal tendency attributed to Italians was simply part of "their nature," the association of Jews with money was seen as deeply rooted in this group's heritage and perhaps even as genetic. Consider this brief passage from Jacob Riis's influential essay of 1890, *How the Other Half Lives:*

> As scholars, the children of the most ignorant Polish Jew keep fairly abreast of their more favored playmates, until it comes to mental arithmetic, when they leave them behind with a bound. It is surprising to see how strong the instinct of dollars and cents is in them. They can count, and correctly, almost before they can talk. ([1890] 1957:84)

This description, it should be noted, was not necessarily written in a derogatory vein. Indeed, Riis's essay was essentially sympathetic to the plight of both immigrant Italians and Jews. Yet its racist assumptions are obvious. The racist element of anti-Semitism was, of course, carried to its ultimate by Hitler and the Nazis during the 1930s and 1940s, but its genesis can be traced to much earlier ideas and their exponents in the United States as well as Europe.

During the 1920s, American anti-Semitism was strongly abetted by the activities of the Ku Klux Klan and of Henry Ford, the noted industrialist. The Klan's anti-Semitic and anti-Catholic propaganda and actions during this time were as widespread and severe as its antiblack campaigns. As for Ford, his anti-Semitic program lasted from 1920 to 1927, during which time his newspaper, the *Dearborn Independent*, issued virulent attacks on Jews.[5] The "world Jewish conspiracy" idea was a regular component of the *Independent*, which used the notorious *Protocols of the Elders of Zion* as its basis. This was a well-known anti-Semitic tract circulated originally in the early 1900s by the Russian secret police. Its thesis asserted a Jewish conspiracy to achieve world domination through control of finance and banking, and it was used by the Russian tsars as a rationale for their oppressive policies and actions against the Jews. Though its contentions were repeatedly disproved, this manifesto was subsequently circulated widely, and its theme became a standard item of the litany of anti-Semitism. Ford later apologized to the American Jewish community for his attacks, saying that he had been misled, but many of the ideas espoused by the *Independent* remained in wide circulation during the 1930s.

Although the anti-Semitism of this era was extremely vituperative, it rarely generated violence. Most of the discrimination against Jews involved their expulsion or exclusion from various areas of social life, including clubs and resorts, neighborhoods, schools and colleges, and certain occupations (McWilliams, 1948). As Jews began to occupy a disproportionate number of seats in elite colleges and professional schools in the early 1920s, restrictive policies were enacted, limiting their entrance. Strict Jewish quotas were established at Harvard and other Ivy League institutions as well as at medical and law schools throughout the country (Baltzell, 1964; Bloomgarden, 1957; Korman, 1988). Jews were also blatantly discriminated against in the professions, especially medicine, law, and academia. Jewish doctors, for example, were often refused appointments to hospital staffs and were denied clinical privileges. Law firms and university faculties imposed similar restrictions.

---

[5] McWilliams (1948) writes that the rise of the Ku Klux Klan and Ford's anti-Semitic campaign were not unrelated. The Klan began to attract a mass following in 1920 when Ford launched his attack.

The rising level of anti-Jewish prejudice and discrimination in the 1920s and 1930s can be seen as part of the popular anti-immigrant and isolationist sentiments that characterized this period. But it can also be seen as a response to Jewish upward mobility. Before the early years of the twentieth century, Jews were fairly well removed from the general occupational structure, concentrating mainly in a few industries, such as the garment trade, or operating their own small businesses. By the second decade of the century, however, second-generation Jews were beginning to break out of the insular Jewish community and challenge—most often, successfully—members of the dominant group, particularly in the more prestigious occupations and educational settings. Artificial barriers, in the form of quotas and other discriminatory policies, were therefore thrown up. The anti-Jewish attitudes that had emerged earlier in the 1870s among the well-to-do have been interpreted in much the same manner: as social barriers that arose in response to competition for privilege (Higham, 1963; McWilliams, 1948).

DECLINING ANTI-SEMITISM   In the 1950s and, particularly, the 1960s, anti-Semitism in the United States declined precipitously. The anti-Jewish prejudice and discrimination that persisted took on more camouflaged and restrained forms. This downtrend has been unremitting.

The decline in recent decades of negative Jewish stereotypes and of anti-Semitism generally among the American public has been documented by numerous studies and opinion polls (Anti-Defamation League, 1998, 2009a; Chanes, 1994, 2006; Gallup, 1982; Glock and Stark, 1966; Quinley and Glock, 1979; Selznick and Steinberg, 1969; Sheskin and Dashefsky, 2011; Stember, 1966; Williams, 1977). The unscrupulous businessperson, for example, is no longer a foremost Jewish image among most non-Jews. Nearly half the respondents in polls taken in 1938 and 1939 believed that Jewish businesspeople were less honest than others. A 1952 poll reported a similar percentage. By 1962, however, that stereotype had diminished considerably: 70 percent felt that Jewish businesspeople were no different from others, although the notion that they were shrewd in business remained strong (Stember, 1966). By the early 1980s, the percentage of those adhering to the stereotype had further declined (Raab, 1989), and in 2009 only 12 percent of Americans held to the view that Jews are less honest in business than others (Anti-Defamation League, 2009a).

A significant change has also been registered in the perception of Jewish economic and political power. Public belief in the extensiveness of Jewish influence seemed to peak in the 1930s and 1940s. Forty-one percent of Americans in 1938 thought that Jews had too much power, and 58 percent thought so in 1945 (Quinley and Glock, 1979). In 1952, 35 percent of Protestants and 33 percent of Catholics still answered yes to the question, "Do you think the Jews are trying to get too much power in the United States or not?" But in 1979, the percentages of Protestants and Catholics answering yes were only 12 and 13, respectively (Gallup, 1982). Opinion polls throughout the 1980s indicated similarly that only a small percentage viewed Jews as having inordinate power and influence (Smith, 1991). In 2009, only 13 percent of Americans still embraced the stereotype of excessive Jewish power (Anti-Defamation League, 2009a). Stereotypes regarding "Jewish power," however, remain steadfast among those who are most strongly anti-Semitic. Also, one-third of

Americans continue to believe that Jewish Americans are more loyal to Israel than to the United States (Anti-Defamation League, 2009a).

In the realm of personal relations, negative attitudes toward Jews have also declined markedly. In a 1948 survey, 22 percent expressed the view that they would prefer not to have Jewish neighbors. By 1964, only 8 percent answered similarly, a figure that has never risen. Even intermarriage with Jews, perhaps the most sensitive area of interethnic relations, was viewed negatively by only 34 percent of non-Jews by 1981 (*Public Opinion*, 1987; Smith, 1991).

It is important to note that even during those times when it was intense and wide ranging, anti-Semitism in the United States differed from European forms in that it was never institutionalized, and no official anti-Jewish measures were ever enforced (Konvitz, 1978). Had this not been so, it is unlikely that Jewish Americans as a group could have attained their high position in the society's stratification system as relatively quickly as they did.

In sum, it appears that prejudice and discrimination against Jewish Americans remain persistent among a small segment of the U.S. populace, but they are mainly random in occurrence, not organized and widespread as in the 1920s and 1930s. In 1991, for example, the exclusion of Jews by the Kansas City Country Club prompted Tom Watson, the professional golfer, to resign from the club in protest. Such restrictive policies of certain fraternal groups, recreational clubs, and similar social organizations is a carryover from previous eras and is anomalous to the general tendency toward more ethnically open social organizations (Braverman and Kaplan, 1967; Selznick and Steinberg, 1969; Slawson and Bloomgarden, 1965).[6]

Blatantly racist anti-Jewish rhetoric is rare except among those who are part of the right-wing lunatic fringe of American sociopolitical life, and the same is true of occasional acts of violence against Jews. In the 1980s and 1990s, a rise in black anti-Semitism was evident (Anti-Defamation League, 1998; Golub, 1990), and some contend that it is among African Americans that anti-Semitism remains strongest (King and Weiner, 2007). But even this may represent what Robin Williams calls "situational antagonisms rather than deep-seated traditionalized hostilities" (1977:6) and appears to be declining (Anti-Defamation League, 2009a).

Furthermore, despite the persistence of certain traditional negative Jewish stereotypes, positive images of this group are also common in American society (Anti-Defamation League, 2009a; Martire and Clark, 1982; Quinley and Glock, 1979). Jews are ordinarily seen as intelligent, religious, and family oriented and are often admired for their accomplishments in science and the arts. A 2006 national poll showed that 77 percent of Americans had a favorable opinion of Jews whereas only 7 percent saw them unfavorably (16 percent voiced no opinion). This rating was slightly higher than that given to Catholics and substantially higher than the rating of Evangelical Christians (Pew Forum, 2006). Of course, as explained in Chapter 7, surveys such as this do not always reveal people's true feelings, given their reluctance to voice what they feel are socially or politically incorrect opinions of minority groups. So, the level of Jewish acceptance may not be as high as indicated. In any case, it seems evident that positive views strongly outweigh negative ones.

---

[6] A study of Jewish life in Los Angeles (Sandberg, 1986) showed that Jews themselves perceived more anti-Jewish discrimination in private clubs than in any other area of social life.

ACCOUNTING FOR THE DECLINE What can account for this noticeable decline in anti-Semitic thought and action during the past few decades? Though such social trends are never attributable to a single factor, the general rise in educational level among the American populace does stand out as the single most important agent. Research and public opinion polls indicate consistently that anti-Semitism is related to educational level. To put the matter simply, the lower the educational attainment, the greater the degree of anti-Semitism. For example, surveys of attitudes toward Jews have repeatedly found college-educated people more tolerant than those with less formal education (Anti-Defamation League, 1998, 2009a; Chanes, 2004; Gallup, 1982; Gilbert, 1988). Thus, as the level of education has increased for Americans generally, anti-Jewish attitudes and actions have waned correspondingly (Glock and Stark, 1966; Lipset, 1987; Quinley and Glock, 1979; Selznick and Steinberg, 1969).

THE TENACIOUSNESS OF ANTI-SEMITISM Although anti-Semitism in the United States today is probably as mild and subdued as it has been at any time or place in modern Western history, this does not mean that more malignant and intense anti-Jewish thought and action might not be evoked under certain circumstances. Prejudice and discrimination, it should be remembered, are tools applied at varying times and degrees to protect the social, economic, and political advantages of some groups over others. In recent decades, for example, the overrepresentation of Jews in leadership positions of the mass media, particularly television and the movies, has produced frequent charges with anti-Semitic intimations of Jewish "control" of the media. Also of concern have been the propaganda and violent actions of rabidly anti-Semitic right-wing extremist groups, which, though small in numbers, have used the Internet as a medium for spreading their messages.

Events in the Middle East in the 2000s seemed to give rise to a resurgence of anti-Semitism in Western Europe, the effects of which, many feared, would eventually reach the United States. Most of the serious actions against Jews were instigated by militant Muslims in response to Israeli actions against Palestinians in the occupied territories (Eberstadt, 2004).[7] A parallel situation—most Western European countries have much larger Muslim populations—simply did not exist in the United States. Nonetheless, even in the United States questions were raised as to whether anti-Israeli or anti-Zionist sentiments equated with anti-Semitism (Chanes, 2006; Dinnerstein, 2004). Perhaps more disturbing for Jewish Americans, however, was the growth of the minority of the American public who believe that Jews were responsible for Christ's death. In 2009, more than a quarter held that view, up from 19 percent in 1997 (Anti-Defamation League, 2009a; Pew Research Center, 2004).

Even if these trends prove only transitory, they demonstrate how venerable are anti-Jewish stereotypes and how confounding is the phenomenon of anti-Semitism. Jews themselves perceive the continued potential of serious outbreaks of anti-Semitism. In his study of Jewish baby boomers, Waxman found that 90 percent said they had not personally experienced discrimination. Nonetheless 80 percent

---

[7] A 2009 survey revealed, however, that significant numbers of Europeans continued to express anti-Semitic beliefs regarding Jewish power in business and loyalty to Israel, aside from anti-Semitism stemming from Israeli actions in the Middle East (Anti-Defamation League, 2009b).

agreed somewhat or strongly that anti-Semitism "is a serious problem in the United States" (2001:97). An even greater percentage of a national sample of Jewish Americans in 2008 expressed similar concerns about anti-Semitism (American Jewish Committee, 2008). The comprehensive Pew survey of Jewish Americans in 2013 found that about half (54 percent) do not think there is a lot of discrimination against Jews in the United States today. However, a substantial minority (43 percent) said that Jews *do* face much discrimination. This perception is in contrast to that of the general public, only 24 percent of whom perceive much anti-Jewish discrimination.[8] In the survey, 15 percent of Jews said they had experienced in the past year either being called offensive names or being snubbed in a social setting or left out of social activities because they were Jewish.

Perceptions of anti-Semitism, however, differ by generation and by other social variables. Older Jewish Americans, for example, are apt to see it as more prevalent and problematic than younger generations (Cohen, 2010). And Jews in small towns perceive more anti-Semitism than those in large cities. Moreover, Jews are more likely to perceive anti-Semitism in the United States as a whole, rather than in their local communities (Sheskin and Dashefsky, 2011). Conflicting views of contemporary anti-Semitism may also be the product of conflicting definitions of the very meaning of anti-Semitism itself. Consider, for example, the previously mentioned controversy regarding whether expressing displeasure with Israeli policies is tantamount to anti-Semitism.

ASSIMILATION PATTERNS OF JEWISH AMERICANS

Jewish American patterns of assimilation have to a large degree resembled those of other white ethnic groups. Cultural assimilation has taken place increasingly and irrevocably with each succeeding generation. The same may be said of structural assimilation, though less strongly within the realm of primary relations.

CULTURAL ASSIMILATION

Each successive generation of Jewish Americans has displayed cultural characteristics less uniquely Jewish than the previous one. And as with other white ethnic groups, much of this assimilation is related to upward class mobility. As they have moved from the working-class occupations and urban ghettos of the first generation to the middle-class occupations and suburbs of the second and third, Jews have steadily adopted lifestyles and values that characterize middle-class Americans generally. Like Italians of the second generation, the children of Jewish immigrants to some extent found themselves wavering between the confines of the ethnic community and the larger society (Kramer and Leventman, 1961). Those beyond the second generation, however, have clearly chosen the dominant culture.

Perhaps the most significant indicator of cultural assimilation among Jewish Americans is the gradual movement, beginning with the second generation, away from the Orthodox branch of Judaism, encompassing traditional and rigid forms of

---

[8] Jews said that several other minority groups, including gays and lesbians, Muslims, and blacks, face far more discrimination than they do (Pew Research Center, 2013c).

worship and belief, to the Conservative and Reform branches, each reflecting more Americanized practices and doctrines. Despite the predominance of Conservative and, especially, Reform Judaism, Orthodox Judaism continues to survive making up 10 percent of the Jewish population (Pew Research Center, 2013c). It is of note that the division of Jews into three denominational subgroups is, in the main, an American phenomenon. In non-American Jewish communities, people generally are either Orthodox or nonreligious.[9]

Not only has there been a movement toward less rigorous forms of worship, but religious observance in general has diminished with each generation (Gans, 1958, 1967a; Kleiman, 1983; Sklare, 1969; Wertheimer, 1989, 1993). As Jewish Americans increasingly see their Jewish identity in ethnic rather than religious terms, the purely religious aspects of Judaism hold less importance. Only 15 percent, for example, say that being Jewish is mainly a matter of religion (Pew Research Center, 2013c). Thus assimilation into the dominant culture, in a broad sense, continues, while Jews retain their ethnic identity.

STRUCTURAL ASSIMILATION

Jewish Americans have exhibited steady and in some ways substantial progress in the direction of secondary structural assimilation. We have seen that they have begun to participate fully in almost all areas of American society and have also begun to enter elites in most institutional realms, in some cases disproportionately so.

Structural assimilation at the primary level, however, is a more complicated matter. Jewish Americans have traditionally exhibited strong in-group cohesion in the realm of primary relations (Gans, 1958; Lenski, 1963; Ringer, 1967; Sklare and Greenblum, 1967). This stems from attitudes and actions of the larger society and from the in-group preferences of Jews themselves. Sklare (1978) explains the ambiguity among out-groups regarding Jewish-Gentile social relations. Jews have commonly been seen as "cliquish" when they retreat to their own communities but as "pushy" when they make serious efforts to enter Gentile primary groups. Fearing rejection or at best a marginal acceptance, Jews more often than not were reluctant to leave the safety and psychological comfort of the ethnic group. As a result, concentrated Jewish residential areas were unlike black ghettos—which developed primarily because of overt discriminatory actions and policies—in that they evolved through a combination of out-group discrimination and in-group choice. Also, Jews, like other ethnic groups, initially congregated in relatively homogeneous areas where ethnic institutions were close at hand.

Today, discrimination and the attraction of ethnic institutions no longer play major roles in determining residential patterns, but Jewish concentrations are still evident in cities with large Jewish populations. These seem to be a product mostly of

---

[9] In this regard, Konvitz (1978) notes that although the Reform movement began in Germany, it was most fully developed by German Jews in the United States. A fourth branch of American Judaism, Reconstructionism, is subscribed to by only 2 percent of Jews (American Jewish Committee, 2010). It is also important to note that the Orthodox branch of American Judaism is itself split between those who believe that observance of Jewish religious ritual is compatible with the modern world (the "modern Orthodox") and those who separate themselves from the secular society (the "ultra-Orthodox," or *haredim*).

housing markets, age, and socioeconomic factors, however, rather than the desire to live in Jewish areas (Goldscheider, 1986). Moreover, Jewish Americans have generally followed the pattern of other white ethnics in removing to suburbs, where ethnic homogeneity is much less apparent. Waxman (2001) reports that 65 percent of Jewish baby boomers described their neighborhoods as little or not at all Jewish in character. Other aspects of primary structural assimilation are also related to age and to degree of religiosity. For example, when asked about friendship patterns, those who are younger and less religious are less likely to say that most or all of their close friends are Jewish (Pew Research Center, 2013c).

INTERMARRIAGE It is in the realm of intermarriage that Jewish Americans today give the strongest indication of accelerating primary structural assimilation. Like Italians and other white ethnics, Jews are increasingly intermarrying with members of the dominant group as well as other ethnic out-groups.

Until the 1960s, intermarriage among Jewish Americans was very low, indeed, lower than for other white ethnic groups. A radical change in this traditional pattern, however, has become very clear in recent decades (Goldstein, 1980; Kleiman, 1983; Massarik and Chenkin, 1973; Pew Research Center, 2013c; United Jewish Committee, 2003). The 2013 Pew survey found that among those who had married before 1970, only 17 percent had a non-Jewish spouse; by 2013, almost half (44 percent) of all currently married Jewish Americans had a non-Jewish spouse. The tendency toward intermarriage is even stronger among younger Jewish Americans. Nearly six in ten (58 percent) of those who had married since 2000 had a non-Jewish spouse (Pew Research Center, 2013c). Thus more Jews now marry outside the ethnic group than within.

The swell in Jewish intermarriage is attributed to several factors. Jews in recent decades have entered occupational areas and social groups previously closed to them, placing them in social settings where there is increased interaction with non-Jews. Only a third of Jewish baby boomers report that all or most of their close friends are Jewish; and among younger age cohorts, the percentage is even lower (Pew Research Center, 2013c). Moreover, cultural assimilation has weakened ethnic traditions among younger Jews. Attitudes toward marriage outside the ethnic community are therefore not as uncompromisingly negative as they were for older generations. Furthermore, as noted earlier, in the United States generally there is an increasingly tolerant view of ethnic intermarriage.

THE RANGE OF ASSIMILATION Regarding assimilation, the Jewish American population, like all other ethnic groups in American society, is internally variegated. Some individuals remain solidly attached to the ethnic group and retain a strong ethnic identity, whereas others have shed completely their ethnic attachments. Between these two extremes, a range of assimilation is evident among Jewish Americans.

Those who identify themselves as Orthodox Jews, a small percentage, are at one end of the Jewish American population, immersed in Jewish life and religiously observant. This includes, of course, the ultra-Orthodox Jews such as Hasidim who live in relatively insular communities but are only a tiny element of the Jewish American population. At the opposite end are the "peripheral" Jewish Americans (about 30 percent) who are at best nominally Jewish. If they maintain Jewish friends, it is only coincidental. Such people may be in the arts, professions, academia, or other occupational areas in which a freer, more open social atmosphere prevails. In the middle is the largest element of Jewish America, the "moderately affiliated," who engage in some Jewish practices and maintain close Jewish friendships but are not deeply involved in organized Jewish life or ritual. Much of the differing patterns of assimilation are based on generation. Those who are older remain more attached to religious Judaism and their identity as Jews, while those younger increasingly eschew religion, with many identifying themselves as Jewish only on the basis of ethnicity. In fact, among millennials (those born after 1980), almost one-third say they have no religion (Pew Research Center, 2013c). It is important to consider, however, that a trend toward secularism is a general feature of the U.S. population, with those claiming no religious affiliation—about 20 percent—the fastest-growing portion of the U.S. religious landscape (Pew Research Center, 2012c).

Immigrant Jewish populations from Russia and Israel in recent decades have further complicated the range of Jewish assimilation. Israeli immigrants have generally been from among Israel's secular Jewish population, those not committed to religious orthodoxy. As to the Russians, most bring little if any Jewish background with them, and as one researcher has observed, "the challenge of integrating them into a Jewish American community has provided a degree of revitalization for American Jewry" (Sorin, 1997:244). Immigrant Russian Jews, as well as some Israelis, have established institutionally complete enclaves that have further enabled them to confound rapid assimilation (Gold, 1999, 2000; Orleck, 1987).

Looking Ahead: The Future of Ethnicity Among Jewish Americans

Patterns of assimilation and pluralism among Jewish Americans, although in most ways similar to those of other Euro-American groups, have been and will probably continue to be unique in other ways.

Waning Ethnicity: The Strength of Assimilation   Barring the unlikely resurgence of overt and widespread forms of prejudice and discrimination, the forces of assimilation seem no less irresistible for Jews than for other white ethnics in American society. In considering the status of Jewish Americans, Seymour Martin Lipset and Earl Raab concluded two decades ago that although they were not on the verge of disappearance, nonetheless, "current evidence suggests that group identity and cohesiveness are severely eroding for the large majority" (1995:47). In his study of Jewish baby boomers, Waxman, too, concluded that "in terms of social structural significance, as indicated by residential patterns, socioeconomics, language, and mate selection, there is a waning of Judaism and Jewishness for the group as a whole" (2001:117). The more recent Pew survey (2013c) confirms a continuation of these apparent trends.

With upward mobility, geographic dispersion, and increased intermarriage, what is left for most individuals is essentially symbolic ethnicity, having dwindling impact on social life. Studies indicate not only a decline of Jewish identity among younger Jewish Americans (except for the minority who identify as Orthodox Jews) but also a tendency to choose social relations on a nonreligious basis (Pew Research Center, 2013c; Ukeles et al., 2006). Indeed, many Jewish leaders have in recent years expressed concern over the threat to the continued existence of the Jewish American community given the accelerated pace of acceptance of Jews into the mainstream society and the decline of blatant anti-Semitism. Such fears have been reinforced by declining birthrates and increasing intermarriage.

ASSIMILATION AND JEWISH IDENTITY   Despite unmistakable indications of continued cultural and structural assimilation, ethnic identity among Jewish Americans has been sustained in the last several decades by two seminal events of the modern era: the Nazi Holocaust and the founding of Israel as a nation-state. Both events served to make Jews in the United States more conscious of their heritage and instilled a sense of collective commitment to the preservation of the Jewish people. Even among younger Jewish Americans and among those who do not define themselves as Jews by religion, the Holocaust remains a strong element of Jewish identity, though the link with Israel may be waning (Pew Research Center, 2013c; Ukeles et al., 2006).

This sustained Jewish identity, however, must be seen in much the same way as the retention of ethnic identity among Italian Americans and other white ethnic groups: it is mostly symbolic. Jews continue to identify as Jews and to abide by some of the major Judaic rituals, such as attending synagogue during high holy days and having their sons (and increasingly their daughters) bar (bas) mitzvahed. But for most, the expression of their Jewishness does not extend much beyond these few practices. Put simply, a majority of American Jews have drifted away from religious participation (American Jewish Committee, 2004; Goldberg, 1996; Smith, 2005; Wertheimer, 1989, 1993).

Although the decline of religious orthodoxy among most Jewish Americans is very evident, equally evident in recent years is a return to more traditional religious practices by a small but growing minority. Ritterband, in fact, suggests a dichotomy in which the American Jewish community is becoming bimodal: "The more committed Jews have increased their commitment while less committed Jews have more and more opened themselves up to the forces of assimilation" (1995:378). No one has better described the deep divisions that have arisen among Jewish Americans regarding this divergence than Samuel Freedman (2000). The split between those who desire a secularized Judaism and those who call for a reaffirmation of traditional beliefs and practices is, in his estimation, irreconcilable. So profound and abiding are the differences, writes Freedman, that Judaism may very well fragment into divergent faiths sharing only "a common deity and a common ancestry" (2000:354).

Contemporary Jewish ethnicity, even in its symbolic form, is not likely to wither in the near future simply because Jews remain the potential object of differential treatment. Even though their place in American society today is as secure as it has been at any time, the historical roots of anti-Semitism are deep and abiding and therefore continue to serve as a basis of ethnic identity. Despite their positions of prominence in academia, the arts, business, and politics, Jewish Americans may still feel, to some degree, that they are outsiders and thus not entirely part of the centers of secure influence (Cohen, 1985).

The marginality of Jews in American society has been greatly reduced in recent decades, but there nonetheless remains a fundamental cultural difference from their Christian neighbors—religion. This has created a kind of collective schizophrenia for Jewish Americans. As one writer has put it, Jews live "in a state of tension between the two values of integration and survival, and, regardless of where they stand on a survival-integration continuum, they find themselves pulled in both directions" (Liebman, 1973:134). Another has described their position in the American ethnic system as doubly marginal: "marginal to the majority culture, but also marginal among minorities" (Biale, 1998:27). Jewish Americans will therefore continue to be unique among American ethnic groups in their path toward assimilation.

SUMMARY

- Jews collectively represent the great American ethnic success story. Paralleling the immigration and settlement patterns of other groups of the New Immigration, the East European Jews subsequently outpaced all others in upward social mobility.
- The high skill level of Jewish immigrants entering an industrializing economy provided broad opportunities for them, but certain aspects of the Jewish cultural heritage also helped to propel them quickly upward in occupational standing.
- On all measures of socioeconomic status (income, occupation, education), Jewish Americans today rank at or near the top among ethnic groups.
- Anti-Jewish attitudes and actions (anti-Semitism) in the United States have included not only negative stereotyping but also widespread exclusion from certain educational institutions, business areas, and social organizations.
- Anti-Semitism has waned considerably in recent decades. Although some aspects of anti-Jewish prejudice and discrimination remain apparent, these are not organized and popular efforts, as they were in earlier periods.
- Assimilation—both cultural and structural—for Jewish Americans has proceeded with each successive generation. At the secondary level, Jews have begun to participate fully in almost all major institutions of the society; they have also begun to penetrate the leadership echelons of these institutions. At the primary level, Jews continue to exhibit strong in-group cohesion, but sharply increasing rates of intermarriage signify a fundamental change in this pattern.
- As with white ethnics, symbolic ethnicity best describes the character of ethnicity among most younger Jewish Americans.
- Although Jews continue to assimilate into American society, their religious uniqueness remains a source of ethnic identity and out-group recognition.

CRITICAL THINKING

1. Jewish Americans, as this chapter explains, are at or near the top on various measures of socioeconomic status (SES): occupation, education, and family income. But some Asian American groups show similar (or even higher) levels of SES. In fact, in light of Asian Americans' upward mobility and educational success, some observers have referred to them as the "new Jews." How would you explain this similarity?
2. European Jews have immigrated in large numbers in the past not only to the United States but to other societies as well, such as Argentina, Canada, and South Africa. Has the assimilation process and the relative success of Jews in America been replicated in those societies?

3. Although Jewish Americans are among the most assimilated and successful U.S. ethnic groups, many Jews still perceive the threat of anti-Semitism. How might this be explained? What conditions—political, social, economic—might provoke more virulent and widespread anti-Semitism in the United States? Refer to some of the theories of prejudice and discrimination discussed in Chapter 3.

4. Many Jewish Americans are, in terms of religion, non-practicing, and an increasing percentage do not even subscribe to basic Jewish religious beliefs. Yet, most continue to identify themselves in ethnic terms as "Jewish" and very few convert to other religions. What might account for this continued ethnic identity?

Personal/Practical Application

1. In addition to their continuing identity as Jews no matter how observant they may be, Jewish identity also takes precedence over ethnic identity based on national origin (German, Russian, Polish, and so on). Is this creation of religiously based ethnic identity unique in American society or are there other religious groups that might evolve in this way? Consider, for example, Muslim Americans and Mormon Americans. In terms of self-identity, which takes precedence for you, religion or ethnicity?

2. Historically, Jewish Americans have displayed more liberal positions on political and social issues than other religio-ethnic groups. How might this be explained? Examine some public opinion polls on different political and social issues (for example, party identification or abortion) to see if this is still the case.

ARAB AMERICANS

Arabs are a population that is a key element of America's expanding ethnic diversity. Although they actually have a lengthy presence in American society, Arab Americans have become more numerous and more sharply visible only in the past several decades. They serve as an excellent case to demonstrate how the United States has accommodated new and, from a historical standpoint, culturally unprecedented groups during the contemporary period of immigration. They also illustrate well the issues related to adaptation and integration of ethnic groups, as well as to new forms of ethnic division and conflict.

ARABS, MUSLIMS, AND THE MIDDLE EAST

In the United States in recent decades, the terms *Arab, Muslim,* and *Middle East* are often conflated and used interchangeably. Because the media and other institutions do not apply these terms with precision, perceptions of people referred to as Arabs, Arab Americans, Muslims, and Middle Easterners can be confusing and commonly lead to misconceptions. Let's briefly clarify these terms.

THE ARAB WORLD

Arab countries are part of the region of the world referred to as the Middle East or, more precisely, the Middle East and North Africa. Not all countries of the Middle East, however, are Arab countries. Iran and, at times, Turkey are considered part of this world region, and both are among its largest and most populous countries; but neither is part of the Arab world. Although their populations are overwhelmingly Muslim, their languages and cultures are distinctly different from those of Arab countries, like Iraq and Egypt, which share more historical and ethnic similarities. Israel is also part of the Middle East. Although the Israeli population is about one-quarter Arab and much of its Jewish population descends from Arab countries, it is primarily a Jewish state that obviously does not exhibit political, linguistic, or religious similarities to other countries of the region. The Arab League, consisting of

twenty-two countries that share an Arabic culture and language, is usually understood to comprise the so-called Arab world. It is groups from those countries on
which this chapter will focus.

Most of the inhabitants of Arab countries are Muslim—that is, followers of the
religion of Islam. Thus, *Arab*, an ethnic or cultural designation, should not be confused with *Muslim*, a religious designation. Not all Arabs are Muslim. In fact,
throughout history thriving Christian communities have been part of virtually every
Arab country, though their numbers have dwindled in recent years. Significant Jewish communities also characterized countries of the Arab world until the 1960s.

Just as not all Arabs are Muslim, not all Muslims are Arab; in fact, most are not.
Muslims (adherents of Islam) are found in all countries and together constitute the
world's second largest religion. As seen in Figure 12.1, most of the world's 1.6 billion Muslims live not in the Middle East but in South and Southeast Asia, including
Indonesia, Pakistan, India, and Bangladesh. None of these countries is part of the
Arab world. Muslims in Indonesia and Pakistan alone outnumber all those in Arab
nations. Large Muslim populations are also found in Sub-Saharan Africa, in many
countries of Western Europe, and in Russia.

MUSLIMS IN THE UNITED STATES

The number of Muslims in the United States can only be estimated, but all studies are
in agreement that Islam represents a tiny proportion of the religious landscape, in no
case more than 2 percent (Johnson, 2011; Pew Research Center, 2007b). Moreover,
Arab Americans represent only about one-quarter of U.S. Muslims. Almost a third of
American Muslims are African Americans (most of them converts to Islam), and
more than a third are South Asians, mainly Indian and Pakistani Americans. The
most recent immigrants from countries of the Middle East are mostly Muslim, but

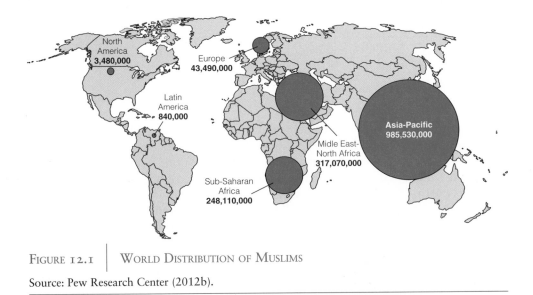

FIGURE 12.1 | WORLD DISTRIBUTION OF MUSLIMS

Source: Pew Research Center (2012b).

even among them, there are significant differences in belief and practice. Some are devoutly religious, whereas others are predominantly secular (Walbridge, 1999).

ARAB IMMIGRANT ORIGINS

The history of Arab immigration to America may be seen as occurring over three broad periods: those who came in the late nineteenth and early twentieth centuries, a second wave in the years immediately following World War II, and the most recent wave, starting in the 1970s.

THE FIRST WAVE

Arabs were a small part of the classic period of immigration in the last decades of the nineteenth and the first two decades of the twentieth centuries, described in Chapter 5 as the New Immigration. Most of these early Arab immigrants were Christian, and thus not different from southern and eastern Europeans, who made up the bulk of that immigrant wave. They were specifically from the Syrian region of the Ottoman Empire and were referred to (and identified themselves) as "Syrians" rather than "Arabs." After World War I, with the breakup of the Ottoman Empire, Greater Syria became what today is a region of the Middle East consisting of Lebanon, Jordan, Syria, Palestine, Israel, and parts of Turkey. A more common modern reference to these initial Arab immigrants is "Lebanese Americans."

It is important to consider that in this early period of Arab immigration, classifications of people from the Middle East were inconsistent and subject to frequent changes, including U.S. Census records. Moreover, religious differences were not noted. As a result it is difficult to delineate with any accuracy the Arab population during these years. Researchers suggest that perhaps 200,000 Arabs lived in the United States before 1940 (Naff, 1983; Orfalea, 2006).

The first wave of Arab immigrants was motivated primarily by the same socioeconomic factors as others of the New Immigration: the promise of economic opportunities. Most followed a path not unlike that of many Italian and other southern and eastern European immigrants of this period. They came with the assumption that immigration to America would be a sojourn—that is, a temporary stay. Following what they hoped would be economic success, they believed they would return to their homelands with the resources to buy land or open a business. Some did, in fact, return, but, like other immigrants, most stayed and established themselves in their new society.

Early Arab immigrants created an easily discernible occupational specialty. While some engaged in the silk trade, which they brought with them from Syria, the majority made their living peddling goods. As itinerant merchants, they radiated out to small towns in virtually every state, selling their items door-to-door. As such, they became a fixture of both rural and urban America. Alixa Naff describes the trade:

> Peddlers were mobile department stores, even when they peddled on foot as most did at first. With well-packed suitcases, frequently one in each hand and one on their back, they carried almost anything a housebound urban housewife or isolated farmwife would need or desire. There were ready-made school clothes, men's work clothes, yard goods, linens, toweling, costume jewelry, and much more. (Naff, 1994:29)

This economic niche eventually led to the establishment of small, family-run businesses as settlement in America gradually lost its sojourning character and became more permanent. One such business was begun in 1926 by a Lebanese immigrant, J. M. Hagger, whose company eventually developed into one of the largest and best-known manufacturers of men's clothing.

Most of these early Arab immigrants blended almost imperceptibly into the U.S. mainstream within a generation, moving into the middle class as entrepreneurs and professionals (Kayal and Kayal, 1975; Naff, 1983, 1994; Walbridge, 1999; Younis, 1995). Two factors seem to have worked in their favor in bringing about comparatively swift and uncomplicated assimilation. For one, except in a few urban areas, their numbers were small compared to other major groups of this immigration period. Second, and of greater importance, they were mostly Christian, practicing Eastern-rite faiths, including Maronite, Melkite (affiliates of the Roman Catholic Church), and Eastern Orthodox. A small number of Muslims, mostly unskilled workers, also entered during the early 1900s, bound for the Ford assembly lines in Detroit. They made up less than 10 percent of the first wave of Arab immigrants (Orfalea, 2006).[1]

THE SECOND WAVE

The most significant increase in the Arab American population in the modern period occurred beginning in the 1970s, but a second, smaller, wave of immigrants from Arab countries arrived in the United States starting in 1948, primarily as a result of the changed political boundaries and subsequent conflicts that arose in the Middle East. To better understand the origins and motivations of Arab immigrants during this and the subsequent period of immigration, we need to briefly review the modern political history of this world region.

World War I led to the fall and dissolution of the Ottoman Empire, which had dominated the Middle East for five centuries.[2] With the defeat of the Ottomans in 1918, Britain and France, victors in World War I, began to carve up this area of the world into spheres of influence, France in much of North Africa, as well as Lebanon and Syria; Britain in Egypt, Palestine, Iraq, and other smaller political entities. These were lands that were now occupied and dependent on British and French colonial regimes.

After World War II, these former colonies or dependencies won independence and began to create nationalistic movements. The result was a hardening of political and religious differences in the region. The creation of the state of Israel in 1948 introduced a new dynamic that would subsequently affect not only countries of the Middle East but the United States as well. A significant Jewish population had lived in Palestine for many decades when it was governed as a British protectorate. A Zionist movement, given additional impetus by the Nazi Holocaust in Europe,

---

[1] Arabs during this time immigrated not only to the United States but also to Canada and several South American countries. As in the United States, they were primarily Christian and followed a similar pattern of adaptation and integration.

[2] Prior to the end of World War I and the breakup of the Ottoman Empire, what is today referred to as the Middle East was generally described as "the Near East."

created a swell of Jewish immigrants after the war, leading to further political pressures to create an independent Jewish nation. In 1948, Israel declared its independence, which led to the first Arab-Israeli war. With the defeat of the Arabs, Palestinians living in areas annexed by the victorious Israelis now found themselves refugees, many of whom came to the United States and other countries.

Arabs in other countries, in addition to those from Palestine, were motivated to leave as a result of the political turbulence that now characterized the region. Most of these refugees or exiles were higher in socioeconomic status than Arab immigrants of the first wave had been on arrival, and they came with higher educational resources. Many were, in fact, students recruited from Middle Eastern countries to study at U.S. universities. The expectation was that they would return to their home countries with a pro-U.S. perspective that would enhance American interests in the region (Haddad, 2011). Many, however, ultimately settled in the United States. The most important characteristic that set this second wave apart from their earlier counterparts, however, was religion: a majority were Muslim, not Christian (Orfalea, 2006).

The Third Wave

As with other non-European groups, the end of restrictive immigration quotas in 1965 led to the entrance of a much more varied immigrant population, of which people from Arab and other Middle Eastern countries benefited. The series of wars and revolutions that continued to occur in the Middle East created political turmoil, which, in turn, prompted a steady stream of immigrants.

Arab immigrants of the past four decades have been more diverse in geographic origin, religion, gender, and social class than those of the first two waves, and for a number of reasons that we will further explore, their absorption into American society has been far more problematic.

GEOGRAPHIC ORIGINS Third wave Arab immigrants have come from more than a dozen countries in the Middle East with the largest numbers from Lebanon, Egypt, Iraq, and the Palestinian territories (Auclair and Batalova, 2013; Suleiman, 1999; U.S. Census Bureau, 2003). Some have been motivated to immigrate to the United States for reasons of economic betterment, but most, even those driven by economic factors, have been impelled in some degree to leave their home countries because of political unrest.

American political involvement in the Middle East has in itself been a catalyst for Arab immigration to the United States. U.S. involvement in the geopolitics of the region has stemmed from two factors: its unwavering support of Israel in the conflicts that ensued after 1948; and its dependence on oil reserves controlled by several Arab countries. U.S. political and military actions in conjunction with those interests have frequently contributed to the region's political instability, thereby creating the conditions for Arab emigration.

RELIGION Beginning with the second wave, religion became a key distinguishing characteristic of Arab immigrants from those of the first wave. This carried over into the third wave, in which the majority have been Muslim. It is also noteworthy that a major source of Middle Eastern immigrants of recent decades has been Iran, following that country's revolution in 1979. Although not an Arab country, Iran is overwhelmingly Muslim and Iranian immigrants, though mostly secular, have therefore been an addition to the U.S. Muslim population (Bozorgmehr, 1997).

Although most who have immigrated in recent decades have been Muslim, many Christian Arabs, like Iraqi Chaldeans and Egyptian Copts, have also been part of the third wave. As seen in Figure 12.2, despite the fact that most new Arab immigrants to the United States are Muslim, a majority of the current Arab American population is still Christian, not Muslim, as is often assumed.

GENDER   The third wave of Arab immigrants has included many more women than had typified earlier Arab immigration to America. Most immigrants of this period have come as family units, rather than as sojourning single men seeking economic opportunities.

SOCIAL CLASS   Arab immigrants of recent decades have generally arrived in America with greater occupational skills and with higher educational qualifications than had typified immigrants of the late nineteenth and early twentieth centuries. Nonetheless, some have come with meager economic resources, particularly those who have fled political upheavals in their origin countries.

DEMOGRAPHIC PATTERNS

Although the U.S. Census describes the Arab American population as 1.8 million, some researchers estimate that Americans with Arab origins may be at least double that number (Arab American Institute, 2012b; Kayyali, 2006a; Telhami, 2002; U.S. Census Bureau, 2010a). Because not all Middle Eastern Americans are Arab in origin, the total Middle Eastern population is also larger than the official count. In any

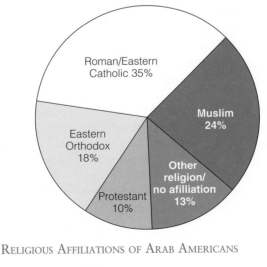

FIGURE 12.2 | RELIGIOUS AFFILIATIONS OF ARAB AMERICANS

Source: Arab American Institute (2002).

case, the Arab American population has grown enormously in the past few decades, having doubled since 1980, mostly as a result of immigration (Asi and Beaulieu, 2013).

As seen in Figure 12.3, the largest specific Arab American groups today are Lebanese, Syrian, and Egyptian. The first two groups are made up mostly of the descendants of those who came in the first wave, with little more than 20 percent foreign-born. For most of the other specific Arab groups—all primarily part of the third wave—more than half are foreign-born (U.S. Census Bureau, 2012a).

Arab Americans are almost entirely an urban population. Arab communities can be found in all regions of the United States, but their major concentrations are in a few large metropolitan areas, especially New York, Detroit, Los Angeles, Chicago, and Washington, D.C.

THE DETROIT ARAB COMMUNITY   The Detroit Arab American community bears closer examination since it represents the largest concentration of Arabs in the United States and one of the largest populations of Arabs outside the Middle East.

Immigration from the Middle East to the Detroit area has a lengthy heritage, with discernible Arab communities established as early as the last decade of the nineteenth century. Although most of the early immigrants were Syrian (today identified as Lebanese) and were Christian, there was a small contingent of Muslims among them. The majority of the Lebanese followed the traditional path of economic adaptation as peddlers who later established small businesses. The Muslims, far fewer in number, were attracted, like European immigrants of that time, by the promise of industrial jobs in the burgeoning auto industry. The first mosque in the United States was built in 1921 in Highland Park, an enclave within the city, not far from the plant of the Ford Motor Company where assembly line production was first implemented.

Today, Detroit is the principal destination of new Arab immigrants to the United States, having absorbed more newcomers in recent years than other American cities with significant Arab populations. Researchers have estimated the Detroit Arab population as around 220,000, but the actual number may be considerably higher

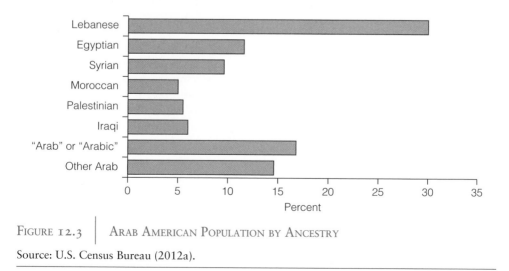

FIGURE 12.3 | ARAB AMERICAN POPULATION BY ANCESTRY

Source: U.S. Census Bureau (2012a).

(Baker and Shryock, 2009; Schopmeyer, 2011). Though referred to collectively as the "Arab American community," in fact it is internally quite diverse in terms of national origin, religion, and social class.

The greatest concentration of Arabs in the metropolitan area is in the suburb of Dearborn, a city of about 100,000. Dearborn's past was shaped primarily by the Ford Motor Company, which established its major production facility there in the late 1920s and where its world headquarters remain. More than a third of Dearborn's population today is Arab American, most of whom are first- and second-generation Muslim immigrants, and at least 60 percent of the students in its schools are from Arab families. It is also the location of the Arab American National Museum, the only one of its kind in the United States. As large and as visible is the Dearborn Arab community, in fact the majority of Detroit's Arab population are scattered in other parts of the metropolitan area, in neighborhoods that cannot be characterized as ethnic enclaves (Baker and Shryock, 2009; Hassoun, 2005).

Today, Arab Detroit is represented by virtually every Middle Eastern country of origin and is almost evenly split between Christians and Muslims (Schopmeyer, 2011). The Lebanese community is the largest and longest resident. Most Lebanese are Christian, but third wave immigrants from Lebanon have been mostly Muslim. Iraqis are a second large group that is religiously mixed and whose members differ in geographical origin as well as time of immigration. Iraqis are divided by religion between Chaldeans—Iraqi Christians—and Muslims.[3] The Chaldeans have a longer history in the city and are generally higher in socioeconomic status than Iraqi Muslims, most of whom have arrived in recent decades. Palestinians, most of whom are Muslim, constitute the third largest component of the Detroit Arab population.

SOCIOECONOMIC STATUS

Like other ethnic populations, Arab Americans are represented in a broad range of social classes. This is seen in Tables 12.1 and 12.2. Some are poor and unskilled, but the majority are solidly part of the middle class. Many are highly educated professionals—doctors, lawyers, engineers, and entrepreneurs—who are relatively well-off.

The higher status of the Lebanese is reflective of their much earlier settlement in the United States and the fact that they have assimilated economically and socially over several generations. Also at the high end of the socioeconomic spectrum are members of third wave groups who arrived with substantial human resources, particularly education and occupational skills. Almost 60 percent of Egyptians in the United States, for example, are foreign-born, but two-thirds of them hold at least a bachelor's degree (more than a quarter hold graduate degrees) and fully half are in professional or managerial occupations.

---

[3] Many Chaldeans do not identify as Arabs, but prefer to see themselves as a distinct ethnic group (Sengstock, 2005; Shryock, Abraham, and Howell, 2011b). Their native language is a modern-day dialect of ancient Aramaic, though most in fact speak Arabic. Their religion—Roman Catholicism—is the most evident cultural feature that distinguishes them from other groups of the Arab region. A sense of difference is driven also by the perception that Chaldeans were a group discriminated against in Iraq (Sengstock, 1983).

TABLE 12.1 | ARAB AMERICAN ECONOMIC ATTAINMENT

	Median Household Income	% Managerial or Professional Occupations	Percent in Poverty	% Unemployed
U.S. total	$50,502	36.0	15.9	10.3
Arab	51,363	43.7	22.8	6.9
Lebanese	65,264	48.5	13.7	5.4
Egyptian	61,571	52.1	18.3	7.8
Syrian	57,196	46.6	13.9	6.6
Moroccan	44,862	35.8	22.6	8.5
Palestinian	53,404	42.1	26.2	8.9
Iraqi	28,004	27.8	43.3	10.3

Source: U.S. Census Bureau (2012a).

TABLE 12.2 | ARAB AMERICAN EDUCATIONAL ATTAINMENT

	High School Graduate (%)	College Graduate (%)	Graduate or Professional Degree (%)
U.S. total	85.9	28.5	10.6
Arab	88.3	45.5	18.1
Lebanese	92.3	46.4	19.5
Egyptian	96.1	66.6	25.8
Syrian	89.2	42.4	19.1
Moroccan	87.7	37.7	15.8
Palestinian	85.5	44.4	17.2
Iraqi	76.0	33.2	9.3

Source: U.S. Census Bureau (2012a).

By contrast, at the other end of the socioeconomic scale are Iraqis, the median household income of whom is well below the national average. Moreover, their poverty rate is extraordinarily high. Three-quarters of Iraqis in the United States are foreign-born, reflecting their relatively recent arrival. Iraqis have fled repeated wars—the eight-year Iran-Iraq war in the 1980s, in which perhaps as many as one million were killed, as well as the U.S.-led Gulf War of 1993, and the U.S. invasion and occupation of Iraq beginning in 2003. Another Arab group low in socioeconomic status are Yemenis, who are mostly young, single males, with fewer occupational skills and educational qualifications than other Arab groups.

Two of the largest groups of the recent immigrant wave have been Palestinians and Lebanese. Many of the former are professionals who found it difficult to practice their skills in the Israeli occupied territories. Most Lebanese of the third wave came with none of the sojourning motivations of the first wave Lebanese, but were pulled to the United States primarily by the disruption of Lebanese society as a result of wars in the region as well as a civil war within Lebanon itself.

While socioeconomic status is varied among the specific groups, when considered in total, Arab Americans exceed the general population on almost all measures. As seen in Tables 12.1 and 12.2, with the exception of Iraqis and Moroccans, median household income of Arab Americans is higher, the percentage of those in managerial and professional occupations is higher, and their rate of unemployment is lower. Only in the rate of poverty do Arab Americans generally exceed the national average; this is due primarily to the immigration in recent years of Yemenis and Iraqis with lower educational attainment and occupational skills. In education, Arab Americans have a higher percentage of high school graduates than the general population, a much higher percentage of college graduates, and a percentage of professional or graduate degrees that is nearly double that of the U.S. population as a whole.

Some of the more recent Arab immigrant groups have played a classic middleman role (discussed in Chapter 2) in their local economies, operating convenience stores, gas stations, and other small enterprises, often in areas underserved by large chain stores. In Detroit, for example, many Lebanese and Syrian Muslims have taken on this entrepreneurial role. Another, more conspicuous, middleman group in Detroit are Chaldeans, who thoroughly dominate the city's supermarkets. Chaldean grocery stores in Detroit date back as far as the early twentieth century, but in recent years they have filled what had become an economic vacuum, as major grocery chains almost completely abandoned the city in favor of suburban locations (Sengstock, 2005).

PREJUDICE AND DISCRIMINATION

Because Muslims predominate among the latest cohort of Arab Americans, they have, more than other immigrant groups and more than their early predecessors, run headlong into the dominant American culture. Islam is a religion that few Americans are familiar with, and they commonly harbor misconceptions about its tenets and rituals. In examining patterns of prejudice and discrimination against Arab Americans, it is important to keep in mind the failure of most Americans to distinguish "Arab" from "Muslim."

STEREOTYPES AND ISLAMOPHOBIA

The activities of Arab terrorists in recent decades in the Middle East and elsewhere have brought to the fore a sinister image of Arab and other Middle Eastern groups, an image that was greatly exacerbated by the attack on the World Trade Center in 2001 and the U.S. war on Iraq beginning in 2003. But the negative image of Arab Americans reaches back much further than simply the last few years. In fact, Arabs have been seen in the West generally as mysterious, violent, and villainous for centuries (Said, 1978). One of the only historical references to the Arab world that people in Western countries are familiar with concerns the Crusades, in which invading Christian knights did battle with Muslims.

In some ways, negative images and anti-Arab prejudice in general resemble anti-Semitism (described in Chapter 11). Like anti-Semitism, an entire litany of historical myths and fabrications has been applied to Muslims, particularly those with origins in the Middle East. Some of the negative images associated with Arabs suggest classical themes familiar to all ethnic minorities, modified to fit present-day circumstances (Stockton, 1994). What has been referred to as **Islamophobia** consists of *the fear of and hostility toward Muslims, based on modern as well as historical images and events.*

The role of the media—especially films and television news—has been critical in presenting and sustaining anti-Arab stereotypes. In the 1980s, Jack Shaheen studied the portrayal of Arabs in television programming and concluded that negative stereotypes were virtually the only TV images of the group. The depiction of Arabs, he found, perpetuated four basic myths: "[T]hey are all fabulously wealthy, they are barbaric and uncultured; they are sex maniacs with a penchant for white slavery; and they revel in acts of terrorism" (Shaheen, 1984:4). A later study (Lind and Danowski, 1998) found little change in these images, except that no longer did the media portray Arabs as fabulously wealthy. Shaheen also studied almost all Hollywood films of the twentieth century that displayed Arab characters and concluded that, like television, the image of the Arab in movies had not changed in over a hundred years: "brute murderers, sleazy rapists, religious fanatics, oil-rich dimwits, and abusers of women" (Shaheen, 2003:172).

Although, as we have seen, not all Arab Americans are Muslim, in the public view, the distinction between the two categories is dim. Adverse images of Muslims have been intensified in recent years in large measure by media portrayals that have been inextricably linked to a government-induced environment of fear, especially following the September 11 attacks and the second Iraq war. In a 2009 national survey, more than 40 percent of Americans openly admitted feeling prejudiced against Muslims (Gallup, 2009) and, whereas a minority believed that U.S. Muslims are sympathetic to the al Qaeda terrorist organization, fewer than half of Americans believed that Muslims are loyal to the United States. Given the tendency for respondents to give survey researchers what they feel are politically correct answers to questions regarding attitudes toward racial and ethnic groups, the actual percentage of those holding prejudicial views of Arab Americans is likely higher.

Other surveys have consistently shown the high level of mistrust and suspicion with which Arabs are generally viewed as part of American society. For example, 59 percent of Americans favor requiring Muslims to carry a special identification, and nearly the same percentage feel that Muslims should be more intensively searched at airports (Saad, 2006). Moreover, a substantial portion of the American public views Islam as more likely than other religions to encourage violence among its believers (Pew Research Center, 2009). It is important to consider that views of Arabs as potential terrorists and as religious fanatics had been commonly held by Americans well before the events of 9/11 (Jones, 2001). Whether Muslim or not, all Arab Americans have borne the effects of these highly negative and apprehensive views and attitudes; no other contemporary American ethnic group has been so intensely targeted.

GENDER STEREOTYPES One of the most controversial and misunderstood aspects of Arab and Muslim culture generally in the United States and other Western societies concerns the role of women. The Western image of Arab Muslim women is a combination of oppression and exoticism. Stereotypes of Arab women as, on the one hand, oppressed and slavishly obedient, and, on the other, as mysterious, alluring, and capable of abetting the evil schemes of their men, have lengthy historical roots, but have been especially pronounced in recent times as Muslims have become more conspicuous to Americans and others in the West. The stereotype of female subordination and oppression is especially palpable. Much of this view stems from media and other depictions of the plight of women in many Arab societies where they have few political rights, are not encouraged to enter the workforce, and remain strictly segregated from men in schools, mosques, and other social settings (Brooks, 1995; Haddad and Esposito, 1998).

By Western standards, the subordinate place of women in Arab societies has given rise to the assumption that these norms are similar throughout the Muslim world, though in fact most of them vary in scope from country to country. More important, this assumption is belied by the place of Arab and Muslim women in American society, where they are integrated into all aspects of social and economic life. In fact, Muslim American women are one of the most highly educated female religious groups in the United States, and Muslim Americans have the highest degree of economic gender parity at the high and low ends of the income spectrum (Gallup, 2009).

Perhaps nothing better symbolizes the contradictory stereotypes of oppression and exoticism than certain items of dress that Muslim women may choose. Many wear a head scarf or veil (*hijab*), which immediately prompts attention and leads to assumptions about women's status. For Americans, this blends naturally into a more generalized perception of Arab and Islamic cultures as primitive and repressive, as opposed to the modern and freedom-enhancing culture of Western societies.[4]

Wearing the hijab has made Arab American women an easy target for harassment and discrimination at school, at work, and in other public places (Bakalian and Bozorgmehr, 2009; Cainkar, 2009; Goodstein, 1997). British scholar Kate Zebiri suggests that the veil "provides a rich and endlessly versatile symbol, perhaps the most powerful symbol of Muslims' otherness and alien values" (2011:181).

The head scarf not only draws attention to Arab American women, but confirms for many their presumed subordinate status. As one researcher has described it, "To many, the veil symbolizes a patriarchal religious culture that universally oppresses Arab women" (Read, 2004:41). Many Americans assume that Muslim women are forced to wear the hijab, but its use is voluntary. In fact, many young Arab American women have chosen to wear it even if their mothers do not. Whereas many view the veil as a symbol of female subjugation, those who defend its wearing argue the very opposite. Islamic scholar Yvonne Haddad and her colleagues explain that women who wear easily identifiable Muslim garb do so as a matter of choice that affords them "freedom, liberation, relief, and even great joy" (2006:9). It has also been

---

[4] Although the subordination of women in most Islamic societies is alien to modern American and Western views of gender roles, it should be remembered that the "liberated" status of women in the United States and other Western societies is a relatively recent historical development. Moreover, Judeo-Christian theologies and practices have a centuries-old legacy of female subordination.

suggested that wearing the hijab enables women to straddle both American and Arab cultures (Haddad, 2007; Williams and Vashi, 2007).

A small minority of Muslim women in the United States and in Europe also wear the *niqab,* a veil that covers most of the face. This commonly leads to even greater suspicion and harassment of the few who wear it. Wearing the niqab has led to much debate, especially in Europe and even in Canada (Kingston, 2012). In France, for example, wearing the niqab in public was declared illegal in 2010, setting off a wave of controversy that has not yet played itself out.

CONTEMPORARY ISLAMOPHOBIA

Until the unfolding of political events in the Middle East during the past four decades, most Americans had had little cause to hold any images of Arabs, other than those presented in early films and literature. The Arab world and the Middle East in general were not part of the history and geography studies of schools at any level and the contributions of Arabs to science, mathematics, and literature were almost entirely neglected. Moreover, Arab American ethnic groups were tiny, with those of the first wave of immigrants blending unobtrusively into the American mainstream. Combined with historically well-founded stereotypes, episodes of terrorism starting in the 1970s, led by Muslim extremists, contributed to the widespread, almost institutionalized, Islamophobia that has characterized American society (and most European societies as well) since then. As the United States launched its so-called "war on international terrorism," a more precisely defined image of the enemy emerged: *Arab* (or *Muslim*) terrorists.

POLITICAL LINKS TO U.S. ISLAMOPHOBIA   Two events of the 1970s seemed to mark the onset of virulent anti-Arab images and the growing definition of Muslims as "the enemy." In 1973, the OPEC nations, most of which were in the Middle East, declared an embargo on the shipment of oil to the United States, in response to American support of Israel in the Arab-Israeli War of that year. This created an economic crisis and sharpened the focus of U.S. foreign policy on the Middle East. A second benchmark incident occurred in 1979 when a group of Iranian students seized the U.S. embassy in Tehran and held sixty-six Americans hostage for fourteen months.

In 1985, anti-Arab hostility and fear once again intensified. In that year, a TWA airliner was hijacked by Lebanese Shiite gunmen, leading to a frenzied response by the media during the seventeen days in which the drama unfolded (Abraham, 1994). A photograph of one of the hijackers holding a gun to the head of the airliner's pilot was flashed in national magazines and newspapers, helping to trigger threats and violence against Arab Americans. A second incident of that year involved the hijacking by Palestine Liberation Front militants of an Italian cruise ship, the *Achille Lauro,* off the coast of Egypt. Since many of the passengers were American tourists, the incident became a U.S. issue, with President Reagan directing the U.S. Navy to prepare for a rescue attempt. Negotiations with the hijackers brought the incident to an end, but not before an American citizen had been shot by the hijackers and his body thrown into the sea. Like the TWA hijacking, this incident was given enormous coverage by the media. With each incident, the Arab/Muslim target of the "war on terror" became clearer for Americans.

Politically induced fear and suspicion of Arabs were stoked again in the early 1990s. The Iraqi invasion and annexation of Kuwait in 1991 was responded to by an American-led coalition that liberated Kuwait in little more than two months. This has become known as the "First Gulf War," distinct from the later war that the United States would wage on Iraq beginning in 2003. Another highly dramatic incident occurred in 1993, when Muslim extremists attempted to bring down the twin towers of the World Trade Center in New York by setting off a bomb in the parking garage of one of the buildings. Although the plot failed, six persons were killed in the attack and more than a thousand injured. Each of these events set off waves of verbal and physical attacks on Arab Americans and their institutions.

Americans' fears of terrorism seemed to reach near-hysterical proportions in the 1990s and Arabs had become the chief source of that public anxiety. Any terrorist activity was almost automatically associated with Muslims or Arabs. In 1995, a truck containing a two and a half ton bomb exploded in front of the Alfred P. Murrah Federal Building in Oklahoma City, housing U.S. government offices as well as a day-care center. In the blast, 168 people were killed and 600 injured. It was the deadliest terrorist attack in U.S. history prior to 9/11. In the immediate response to the bombing, the perpetrators were at once assumed to be Arab or Muslim terrorists. Before any of the facts of the event had been established, television and radio newscasters quickly spread this assumption. No more than hours after the explosion, the bomber and his associate were apprehended: they were white Christian Americans, motivated by antigovernment zealotry. Despite the fact that there were no Arab or Muslim connections to the bombing, in the days immediately following it, dozens of episodes of violence against Arab Americans were reported.

After the events of September 11, 2001, the level of suspicion and fear of Arabs and Muslims generally was raised to even greater heights, and media images were almost entirely negative, focusing on the radical actions of Muslim terrorists (Nacos and Torres-Reyna, 2007). The fact that the vast majority of Arab Americans had no sympathy for the terrorists or the beliefs that motivated them was lost in the ensuing public frenzy.

THE PERCEPTION OF ISLAM AS A "CULTURAL THREAT"   Muslim Americans (and by inference, Arab Americans) have come to be seen by many not only as a security threat but as a cultural threat as well. Those who subscribe to this view see Islam as basically incompatible with Western values and its adherents intent on imposing Islam on other societies. One prominent political scientist referred to the meeting of Islam and the West as a "clash of civilizations" (Huntington, 2004). Such views encourage Arab and Muslim stereotypes of disloyalty, intolerance, and militance and generate the perception of a culture that is hopelessly incapable of ever being woven into the American social fabric.

Prejudice against Muslim Americans, particularly those from Arab countries, has spawned what some have called an Islamophobia network, made up of anti-Muslim bloggers, political pundits, and well-funded organizations that portray America as under attack by Islamic cultural and political forces (Ali et al., 2011; Elliott, 2011; Steinback, 2011). For example, Shariah, a Muslim religious code, is characterized as threatening to U.S. constitutional law, a baseless belief that nonetheless, through organized efforts, has led several states to enact "anti-Shariah" statutes.

In Oklahoma in 2010, for example, voters by a huge majority approved a ballot measure that forbade Shariah law in Oklahoma courts (Jones, 2013). The measure's supporters referred to it as the "Save Our State" amendment and portrayed Shariah as the first step toward the demise of the dominant (American) culture in the state—where Muslims number 20,000 in a population of 3.8 million. In the anti-Muslim fervor, the fact that the U.S. Constitution, as well as Oklahoma law, would have made such a development impossible was ignored. Similar measures have been proposed in other states and communities.

Such organized anti-Islam projects have been successful in perpetuating the negative image of U.S. Muslims and irrational fears of an Islamic cultural threat. A national survey in 2010, for example, revealed that more than one-third of Americans view Islam as "more likely than others to encourage violence." Negative opinions about Islam had actually increased from a similar survey five years earlier. Only 30 percent of Americans said they have a favorable opinion of Islam while more than 38 percent had an unfavorable view (Pew Research Center, 2010c). In another 2010 national poll, nearly one-third of Americans said that Muslims should be barred from running for president, and 28 percent believed that they should not be eligible to sit on the Supreme Court (Altman, 2010). A 2011 survey revealed that almost two-thirds of voters expressed discomfort with the thought of a Muslim being elected U.S. president, almost the same percentage as those who felt uncomfortable with an atheist president (Jones and Cox, 2011).

At the same time that a significant percentage of Americans voice distrust and/or fear of Muslims (again, it is important to keep in mind the common equation among Americans of "Muslim" and "Arab"), a majority admit to having little or no knowledge of the Islamic religion, either its principles or practices. In the 2010 Pew survey cited above, 55 percent said they know "not very much" or "nothing at all" about the religion. This lack of understanding of Islam has provided a fertile breeding ground for anti-Muslim and anti-Arab beliefs and actions. Most Americans would not mistake terrorism by Christian extremists—acts of violence against abortion providers, for example—as reflective of the beliefs and values of Christianity as a whole; such actions and their perpetrators are commonly seen as exceptional. This, however, is not the case in regard to Islam. At the same time that most Americans claim to know little or nothing about Islam, polling data indicate that they see it nonetheless as a religion very different from their own. Moreover, almost half believe that Islam does not teach tolerance of other religions (Panagopoulos, 2006).

As we have seen in previous chapters, once firmly established, ethnic stereotypes are difficult to eradicate and may perpetuate beliefs about ethnic groups for generations. These beliefs, in turn, help to shape ethnic identities that may continue to brand groups even after they have been well-assimilated into the larger society. Consider for example, the persistence of the organized crime-connection stereotype of Italian Americans, discussed in Chapter 10, or the tenacity of anti-Semitism, discussed in Chapter 11.

DISCRIMINATION

Anti-Muslim (and, by implication, anti-Arab) stereotypes and attitudes have become so deeply embedded in contemporary American society that even non-Muslim

Americans acknowledge the fact that Muslims in the United States face far greater discrimination than those of other ethnic minorities (Pew Research Center, 2009a). The forms and intensity of anti-Arab discrimination in recent decades cover a broad spectrum at the individual as well as institutional levels. Actions range from verbal abuse to physical attacks to official government policies.

INDIVIDUAL DISCRIMINATION   At the individual level, acts of discrimination against Arab Americans become particularly widespread following any terrorist incident with Arab or Muslim connections or in the aftermath of Middle Eastern affairs that negatively affect U.S. interests. Immediately following the attacks of 9/11, anti-Arab and anti-Muslim incidents increased many-fold, but seemed to recede as the response to that event moderated (Disha et al., 2011). Nonetheless, Arab Americans during the past decade have encountered repeated acts of discrimination from employers, neighbors, and local police (Bebow, 2003; Cainkar, 2009; Krupa and Bebow, 2003; Schanzer, 2010). Anti-Arab and anti-Muslim stereotypes continue to inspire attacks against mosques, vandalism against Arab-owned businesses and homes, verbal harassment, and even physical assaults (Southern Poverty Law Center, 2013).

Attempts by Islamic groups in the United States to build mosques have commonly been contested by local governments and neighborhood residents. In 2012, the Pew Research Center documented 53 proposed mosques and Islamic centers whose construction had encountered community resistance in recent years. While most of the opposition cited concerns about traffic, noise, and property values, many opponents cited fears of Islam, Shariah law, and terrorism (Pew Research Center, 2012). In Florence, Kentucky, for example, the local Islamic center had proposed to move from a rented storefront to a much larger mosque in an area already zoned for religious use. Opponents set up a "stop the mosque" website and circulated fliers urging neighbors to help "stop the takeover of our country" by Muslims (Pew Research Center, 2012). In a more widely publicized case, residents of Murfreesboro, Tennessee in 2010 sought to halt construction of a new mosque and community center. They alleged in part that Islam is not a religion and that Muslims posed a threat to the neighborhood. In the two years following the beginning of construction, the site was subject repeatedly to vandalism, arson, and bomb threats.

Some hate crimes only threaten violence, but succeed in terrorizing their victims. Consider a case reported by the FBI in 2007 in which an Arab American woman, who managed a large hotel in Philadelphia, was the victim of a threatening note left by one of her employees telling her that "you and your kids will pay," and to "remember 9/11" (FBI, 2007).

All ethnic minority groups suffer from episodes of discrimination at the individual level but it is difficult to accurately gauge their extent since most such incidents go unreported to the FBI, which tracks hate crimes, or to other law enforcement agencies. The Southern Poverty Law Center, which monitors hate crimes against all minority groups in the United States, estimated that anti-Muslim hate crimes in 2011 ranged between three thousand and five thousand. While most of these incidents are carried out by individuals or by small groups, at times they are promoted and financed by well-organized anti-Muslim organizations.

INSTITUTIONAL DISCRIMINATION   In the wake of global events, Arab Americans for the past four decades have been stigmatized as potential terrorists and, therefore, as security risks. As a result, they have been the targets of government actions that in many cases have violated fundamental civil liberties and political rights. Rarely has a group been treated with the degree of suspicion and scrutiny as Arab Americans in recent times. Indeed, some have suggested that anti-Arab and anti-Muslim actions taken to this point are only a small step from internment similar to that suffered by Japanese Americans during World War II (Saito, 2010).

As early as 1972, the federal government began to collect data on immigrants from Arab countries and compile dossiers on Arab American leaders and organizations (Haddad, 2011). With the end of the Cold War in 1989, Islam seemed to replace communism as the "enemy," for which an entire array of official spying and surveillance devices were required. More sweeping government powers were enacted as the United States elevated its "war on terrorism," defined essentially as a war on *Arab* and/or *Muslim* terrorism.

The attacks of 9/11 brought security measures to new heights, including unprecedented restrictions on Arab Americans. The passage of the PATRIOT Act following the attacks removed many of the legal protections for Arabs living in the United States that are presumed to be constitutionally guaranteed. For example, it sanctioned monitoring without notification of telephone calls, emails, and credit card purchases of any individual deemed suspicious. Thousands of legal residents who were of Arab or Muslim origin were detained without charges (and many subsequently deported), simply on the basis of their Arab ethnicity (Saito, 2010). Federal cases against Arabs in the United States increased exponentially, though few proved to have any terrorist connections. Legal scholars have debated the constitutionality of the PATRIOT Act, though it has been upheld by the U.S. Supreme Court and has been reenacted several times by the U.S. Congress (German and Richardson, 2009; Hudson, 2011; Mac Donald, 2003).

Arab Americans have also been the repeated targets of ethnic profiling, in much the same manner as African Americans have been racially profiled. Particularly harsh scrutiny is given Arab Americans when traveling by air. In addition to stringent preflight screening, more severe discriminatory incidents are not uncommon. In 2006, for example, six Muslim religious leaders were taken off a US Airways flight in handcuffs after some passengers and crew members complained of what they deemed "suspicious" behavior. After being detained and questioned by federal agents for several hours, they were subsequently released. An almost identical incident occurred in 2009 when nine Muslim passengers (eight of whom were American-born citizens) were pulled off a domestic airliner and not allowed to rebook their flight even after the FBI had verified that they had done nothing wrong (Robbins, 2009). Such incidents have led to the caustically humorous reference to "flying while Muslim" (Davis, 2009). Some Arab Americans have found their treatment by airport security personnel so embarrassing and humiliating that they have simply discontinued air travel completely, choosing instead to drive great distances. As one Jordanian in Michigan asked, "What father wants their children to see them this way?" (Donnelly, 2003).

TOLERANCE OF ANTI-ARAB PREJUDICE AND DISCRIMINATION   As we have seen in previous chapters, incidents of prejudice and discrimination against virtually all racial and ethnic minorities continue to occur with disturbing regularity in American society. But the social acceptance of those attitudes and actions has been widely condemned and they are seen for the most part as deviant. This is not the case regarding Arab Americans and Muslims. Many politicians, religious leaders, and political commentators routinely express Islamophobic views with impunity (Nimer, 2011). Clearly there is a level of tolerance of anti-Arab and anti-Muslim prejudice and discrimination that no longer typifies the social response to such attitudes and actions aimed at other racial and ethnic groups.

Not only is anti-Arab and anti-Muslim rhetoric tolerated as it is for no other ethnic or religious group, but acts of violence against Arabs and Muslims in general no longer evoke the same degree of media attention and public concern as would ordinarily be expected. When a white supremacist murdered six worshippers in his attack on a Sikh temple near Milwaukee in 2012, it was assumed that the killer mistakenly believed Sikhs to be Muslims. Attacks on Sikhs[5] since 9/11 have, in fact, numbered in the hundreds since Sikh men wear turbans and beards. The Milwaukee killings garnered much national media coverage, in large measure because of the mistaken identity factor. As the journalist Samuel Freedman explained in the aftermath of the event, the mistaken-identity narrative suggested that public reaction might have been different if the attack had been on a mosque and Muslims had been the victims. "It suggests that such a crime would be more explicable, more easily rationalized, less worthy of moral outrage," wrote Freedman. Now that Muslims have been defined as "the enemy," he explained, "violence against them is understood in a mitigated, mediated way" (Freedman, 2012).

EUROPEAN COMPARISONS

Despite consistent and widespread prejudice and discrimination faced by Arab Americans, their plight has not been as turbulent and disruptive as that faced by Arabs and other Middle Eastern groups in Western Europe. After World War II, most Western European countries embarked on an economic rebuilding effort, requiring a labor force that could not be met with native workers alone. Hence, these countries began to recruit workers from former colonies or countries with a labor surplus.[6] Most of those who immigrated from the Middle East, including North Africa, were Muslim, making them highly visible among overwhelmingly Christian populations.

The presence of people radically different culturally has created a situation in which Muslim immigrants generally have been viewed as a social and political problem; this in turn has spurred anti-immigrant, often blatantly racist, views and actions (Benton and Nielsen, 2013; Diffy, 2004; Halász, 2012; Sniderman and Hagendoorn, 2007). Right-wing political parties, whose major policy objectives have been to stanch further immigration of Muslims, have arisen in all Western European countries, and some have even proposed the expulsion of Muslims who have been living

---

[5] Sikhism is a religion founded in the sixteenth century in the Punjab province of India, and its principles and practices are essentially unrelated to Islam. Sikhs have been in the United States since the early nineteenth century and today number around 250,000.

[6] The creation of multiethnic societies in Europe as a result of post–World War II immigration will be discussed in more detail in Chapter 17.

in these countries for two or more generations. The socioeconomic conditions of Muslim ethnic groups are substandard and they remain marginalized in all areas of social and political life.

In worst cases, anti-Islamic movements have given rise to outbreaks of violence. In Germany, physical attacks against Turks, the country's largest immigrant population, have been commonplace for decades. In France, Muslims from North African countries, who compose fully half the immigrant population, have been the primary targets. Racially motivated incidents have occurred even in countries with fewer immigrants and solid traditions of tolerance, like Italy, Sweden, and the Netherlands. In the latter, where most immigrants are from the Middle East or North Africa, anti-Islamic reactions reached a climax in 2004 when a Dutch filmmaker was brutally murdered by a radical Islamist after he had produced a film critical of the Muslim treatment of women (Buruma, 2005; Saunders, 2004).

ASSIMILATION—OR ALWAYS "THE OTHER"?

Will Arab Americans follow a path toward assimilation, or will they continually be seen as "the other," a kind of outlier group in the American ethnic system? In spite of the unique problems faced by Arab Americans in the current social and political atmosphere, there are small but unmistakable indications that this is an ethnic group that, like others before them, is being slowly integrated into the larger society.

GENERATIONAL, RELIGIOUS, AND CLASS DIVIDES

In considering the alternative paths of adaptation and integration on which Arab Americans may advance in the future, we need to reconsider the generational, religious, and social class diversity of this ethnic group, discussed earlier.

ASSIMILATION AND THE EARLY ARAB AMERICANS   Recall that Arab communities had formed in the United States as early as the nineteenth century. Arab Americans whose families derive from that period of immigration have been in the United States for several generations and have followed a passage of assimilation, both culturally and structurally, indistinguishable in most ways from those of southern and eastern European ethnic groups. As explained earlier, as Christians, most who immigrated in the late nineteenth and early twentieth centuries passed almost unnoticed into mainstream America.

The Lebanese Maronite (Roman Catholic) community illustrates an element of the Arab American population that has followed an almost classic passage of assimilation. Second- and third-generation families have moved into professional and managerial occupations and Arabic surnames are no longer common. Language assimilation, too, is very evident, with Arabic no longer understood by many beyond the second generation (Ahdab-Yehia, 1983). Unless it was pointed out, few would know that such luminaries of American society as consumer advocate Ralph Nader, or actress Marlo Thomas, or famed heart surgeon Michael DeBakey, or former football star Doug Flutie, are descendants of that early generation of Arab Americans. Moreover, first wave Arab Americans have been fully assimilated into the American political system, with several having served in the cabinets and executive offices of recent presidential administrations and in the U.S. Congress.

THE RECENT GENERATIONS OF ARAB AMERICANS   Current Arab Americans strike a dramatic contrast with their earlier counterparts. Indeed, those differences cannot be understated. As Christians, most of the first wave Arab immigrants did not face the more daunting difficulties of more recently arrived Arab Americans, the majority of whom are Muslim. Moreover, their presence in American society was not fraught with highly charged political overtones, as is the case today. Arab Americans who have immigrated in the past several decades, along with their children, are highly visible and have arrived in an entirely different historical context from those of the late nineteenth and early twentieth centuries. To expect that the new Arab communities will duplicate first wave Arabs in the absorption process, therefore, is not realistic. They will face considerably more formidable obstacles.

Although it is too early to draw firm conclusions regarding where the new Arabs will fit into the American ethnic system, evidence suggests that they are on an assimilation trajectory, though one that is tortuous and subject to numerous social and political variables.

CULTURAL AND STRUCTURAL ASSIMILATION

CULTURAL ASSIMILATION   On a number of dimensions of social and economic life, first-generation Arab American immigrants and their children appear to be moving toward cultural assimilation. Consider, for example, language. On this, Arab Americans rank high in comparison to other groups with a high percentage of foreign-born. In their study of the Detroit Arab American population, which contains a large number of first-generation immigrants, researchers found that 80 percent speak English well or very well. Moreover, most are bilingual and receive their news and information from both English and Arabic sources (Baker et al., 2004).

Another indication of cultural assimilation among Arab Americans is the large percentage who are becoming American citizens at a rapid rate (Schopmeyer, 2011). In fact, immigrants from the Middle East are much more likely than other immigrant groups to be naturalized U.S. citizens. Among the foreign-born from the Middle East, 59 percent are naturalized citizens, compared to 43 percent among the overall foreign-born population (Terrazas, 2011).

STRUCTURAL ASSIMILATION   Regarding residence and intermarriage, both key indicators of structural assimilation, Arab Americans may be following a path not unlike that of older, more established ethnic groups. One study (Kulczycki and Lobo, 2002), for example, reported a surprisingly high rate of intermarriage among Arab Americans. Over 80 percent of U.S.-born Arabs, the researchers found, had non-Arab spouses. Another study, examining residential patterns among Arab Americans, concluded that they are "less segregated and have greater access to quality neighborhoods than either Hispanics or Black Americans" (Holsinger, 2009:174).

Increasing structural assimilation, as indicated by intermarriage and residential dispersion, may be closely linked to socioeconomic status. The fact that Arab Americans generally rank higher than other ethnic groups in income, occupation, and education provides them with human and capital resources that other groups with large numbers of first- and second-generation immigrants may lack. The relatively high occupational and educational status of many among the newest Arab immigrants will likely enhance the prospects for more rapid structural assimilation, both primary and secondary. Another factor that may increase the pace and scope of structural assimilation is racial status: Arab Americans are, at least officially, white. This factor is discussed in more detail later in this chapter.

Finally, it is significant that less than half of U.S. Muslims say that all or most of their close friends are Muslim. This contrasts sharply with Muslims in other countries, who report that all or most of their friends are Muslim (Pew Research Center, 2011a). Not all U.S. Muslims are Arab, of course, but it is reasonable to assume that Arab American Muslims are not exceptions to this trend.

CONFLICTING AND CONVERGING VALUES   At the individual level, a kind of assimilation can be seen occurring that may at times manifest itself in a unique blending of Arab and American cultures. As a variant of segmented assimilation (discussed in Chapter 4), families may seek to advance economically and socially while preserving ethnic values. This may be particularly challenging for Arab Americans who practice Islam, who often find themselves in a cultural bind, given the requirements and expectations of their religion. For example, devout Muslims are expected to pray five times daily at prescribed times. This may present awkward or conflicting obligations of work or school, where ritual cleansing before prayer, in addition to space, may not be convenient or even permissible. Muslim dietary laws also present challenges to Muslim Americans who must consume halal meat, ritually slaughtered, and are forbidden to consume pork or alcohol. Increasingly, however, Arab American Muslims have been able to accommodate aspects of American life and culture with Islamic traditions and practices.

Consider the odd convergence of football, that most American of sports, with Islam, demonstrated by a Dearborn, Michigan high school. The overwhelming majority of students at Dearborn's Fordson High School are Muslim, not surprising when one considers that, as was noted earlier, almost two-thirds of students attending Dearborn's public schools are Arab American. When in 2011 football practices coincided with Ramadan, the Islamic holy month in which Muslims refrain from eating or drinking during daylight hours, the coach moved the practices to late night, after the fast could be broken and players would have eaten. Thus, they could remain loyal to their faith at the same time they were pursuing typical American goals. The school's principal, himself Arab American, put it this way: "Dearborn is no different from any little big city in the United States. Kids have the same dreams and ambitions as any other kids. They want an education and a lot of them want to play football" (Longman, 2011). Here, then, is an illustration of how seemingly incompatible cultures may intersect and continue to function in a uniquely multicultural fashion.

Opinions of Muslim Americans themselves give indications of a gradual, if uneven, path toward cultural and structural assimilation. An overwhelming majority express satisfaction with their lives in the United States and see their communities as good places to live. In the Detroit Arab American Study (Baker et al., 2004), Arabs and Chaldeans evinced higher levels of confidence in their local school systems, the police, and the U.S. legal system compared to the general population. Two-thirds thought that the quality of life for American Muslims is better than in most Muslim countries and more than 90 percent expressed pride in being

American. Moreover, as seen in Figure 12.4, a majority of U.S. Muslims believe that most Muslims who come to the United States want to adopt American culture; only 20 percent say that Muslims want to be distinct from the larger society.

Muslim Americans seem to be no less committed to basic American values than other ethnic groups. National surveys of Muslim Americans, a large proportion of whom were Arabs, have indicated that they "are decidedly American in their outlook, values and attitudes" (Pew Research Center, 2007b). A majority, for example, subscribe to the American work ethic and most are firm believers in American individualism. While 62 percent of the general public expresses a belief that "people can get ahead if they're willing to work hard," an even greater percentage of Muslim Americans hold that belief (Pew Research Center, 2011a).

Additional evidence of increasing assimilation is the evolving role of Islamic institutions in America. Like other groups that preceded them, Arab Muslims have created religious institutions that reflect American influences. In 2011, there were more than two thousand mosques and Islamic centers in the United States, nearly four times more than their number in 1986 (U.S. Mosque Study, 2011). In addition to its traditional religious functions, however, the mosque in America has taken on meaning and purpose that are somewhat unique in the Islamic world, having begun to reflect American cultural influences. The mosque has become more of a social gathering place for the community and a place in which celebrations (like weddings) take place. Also, rather than a primarily male activity, mosque attendance in the United States is more family oriented and women play a more active role (Gallup, 2009; Haddad, 2011). Similar transformations of traditional religious institutions

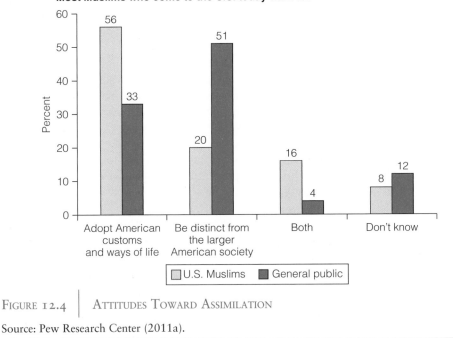

FIGURE 12.4 | ATTITUDES TOWARD ASSIMILATION

Source: Pew Research Center (2011a).

into more Americanized forms typified the experiences of earlier Catholic and Jewish ethnic groups.

ARAB WOMEN IN THE AMERICAN CONTEXT   The growing prominence of Arab American women indicates an increasingly powerful thrust toward assimilation. The place of Arab women in the United States is markedly different from what it is in most Muslim societies, as well as in the large Muslim communities of Western Europe. Their wide-ranging expectations and accomplishments as well as their greater visibility contrast strongly with those of more traditional Islamic settings. A global survey showed that Muslim Americans are more supportive of the role of women in society than are Muslims elsewhere (Pew Research Center, 2011a). Almost all Muslim Americans agree that women should be able to work outside the home, and a majority also do not see any difference between men and women political leaders. These views deviate sharply from those in predominantly Muslim countries. As one Arab American woman executive has put it, "In many of our home countries, socially or politically it would've been harder for Muslim women to take a leadership role. It's actually quite empowering to be Muslim in America" (Knowlton, 2010).

The role and behavior of Arab and Muslim women in America should not be seen as uniformly consistent. On the contrary, there is a wide spectrum of belief, participation, and identity among them. Some choose traditional patterns of behavior, consistent with social roles they may have brought as immigrants from their origin societies. Others find ways of combining traditional ways of belief and identity with more Americanized patterns, and still others may reject entirely the traditions and practices of Islam. These are choices and decisions that continue to be debated in Arab and Muslim communities and by women themselves at the individual level.

DOMINANT GROUP CULTURAL EXCHANGES AND ADAPTATIONS   As explained in earlier chapters, assimilation is not a one-way process, as cultural and structural exchanges and adaptations begin to take place between the dominant group and ethnic minorities. This has become apparent in areas with large and growing Arab American communities.

Companies and advertisers have come to realize this group's market potential and have begun to make outreach efforts. Consider that in the Detroit area, a McDonald's now serves halal Chicken McNuggets, Walgreens features Arabic-language signs in its aisles, and Ikea's women employees are given a company-branded head scarf to wear if they wish (Story, 2007). Public institutions also have begun to accommodate the Arab American community. To see advertisements and public notices posted in both English and Spanish is commonplace in cities like New York, Los Angeles, and Miami, but in Detroit, a third language—Arabic—has joined these two predominant languages in many institutions, like hospitals and banks.

Arab influences are increasingly evident in other spheres of U.S. culture. Middle Eastern fare such as falafel and shwarma, for example, are beginning to join the parade of ethnic foods that have become standard elements of American tastes. And, perhaps most unexpectedly, an Arab American Muslim was crowned Miss USA in 2010.

Although it is exceptional in size, the Detroit Arab American community may be seen as a case of emergent integration into the larger society. Arabs in the Detroit metropolitan area have achieved a high level of political and social incorporation, and cultural exchanges with the larger community are very evident (Bebow, 2003; Ghosh, 2010; Howell and Jamal, 2008; MacFarquhar, 2007; Singer, 2001; Warikoo, 2001). Detroit Arab Americans have been recognized in the local media and in other public institutions as a significant ethnic population that is contributing economically, culturally, and even politically like other ethnic groups that preceded them. This is a development that contrasts sharply with views typically presented in the national media and commonly subscribed to by Americans that convey a problematic image of assimilation for Arab Americans.

Group Identity and Assimilation

A key factor in examining the assimilation process among Arabs is the way in which Americans identify this ethnic group and the way in which Arab Americans identify themselves in terms of race and ethnicity.

Racial and Religious Identity The racial/ethnic classification of Arab Americans remains uncertain and debated both within and outside the Arab American community (Bakalian and Bozorgmehr, 2009; Naber, 2000; Samhan, 1999). For U.S. Census purposes, Arabs are "white," though a number of analysts have suggested that they are being racialized, that is, seen increasingly in racial terms as a separate, non-white, grouping (Cainkar, 2006; Tehranian, 2009).

Arab Americans themselves evince some uncertainty of their place within the American racial/ethnic system. Most self-identify as white, but differences do exist, especially along religious lines (Ajrouch and Jamal, 2007; Terrazas, 2011). Researchers in the Detroit Arab American Study (Baker et al., 2004), for example, found that nearly two-thirds self-identified as "white," while another third identified as "other." One study of Lebanese and Palestinian immigrant children demonstrated this ambivalent racial/ethnic status. Although officially they are white in the currently used racial/ethnic classification scheme, these adolescents distinguish themselves from the white majority. The study concluded that it is not clear if they will "permanently adopt a non-white identity or instead opt for inclusion in the dominant culture as they reach adulthood and have more contacts outside of the community" (Ajrouch, 2004:388).

What is equally uncertain is how the society at large will view Arab Americans as an ethnic or racial category. This may be a confounding factor in the long-term assimilation process. As one researcher has written, "Racial identity is an unfolding, ongoing, contextual, and socially constructed process for Arab Americans" (Cainkar, 2006:268). Even U.S. government agencies responsible for collecting racial and ethnic data have not agreed upon a clear definition of how Arabs and other Middle Easterners fit into the American racial/ethnic classification scheme (Brittingham and de la Cruz, 2005).

Panethnicity In the face of continued marginalization, Arab Americans may develop a stronger collective identity. In earlier periods of immigration, numerous European groups entered American society with a weak sense of ethnic identity that grew stronger not only as a result of census and other classification decisions, but of the experiences of prejudice and discrimination. A panethnic identity among diverse Arab groups in the United States may be developing in the same way. In the Detroit Arab American Study, 70 percent said that the term "Arab American" describes them.[7]

OBSTACLES TO ASSIMILATION

What remains most challenging for a clearer Arab American assimilation path is the level of distrust and suspicion toward this group that continues to be expressed by many Americans (Arab American Institute, 2012; Pew Research Center, 2010c). Arab Americans must continually seek to prove their loyalty and that their values are compatible with those of the larger society. The events of 9/11 will linger in the minds of Americans for many decades, and those events and their aftermath are etched with images of Arabs and Muslims. Future terrorist actions in the United States or abroad will surely intensify those images, making the path to Arab American social incorporation fraught with complications.[8]

The anti-Islamic and anti-Arab rhetoric of recent years and the continued suspicion and lack of understanding of Islamic beliefs and practices continue to raise questions among the most recent generations of the Arab population of whether they will ever be fully accepted as Americans. This uncertain status is faced by *all* Arab Americans, whether Muslim or Christian, as the fear of Islam and the conflation in American minds of "Arab" and "Muslim" have overshadowed religious differences. For Arab immigrants and especially for their children—born and raised in the United States—the question is not whether they will be American, but rather what kind of American will they be, and what kind of American identity will the larger society impose on them?

Arab Americans are not the first ethnic minority to be viewed with widespread suspicion and mistrust as an outsider group. As explained in Chapters 5 and 10, Irish Catholics were the primary targets of American nativism in the nineteenth century, followed by Italian Catholics in the early twentieth. Catholicism, like Islam today, was seen by many Americans as a religion basically incompatible with American culture. Catholic bishops were viewed as agents of the Vatican, intent on undermining American society, and Catholics in general were viewed as violent, seditious, disloyal, and hopelessly incapable of ever integrating into the societal mainstream. As late as the mid-twentieth century, many still clung to the belief that Catholicism was an ideology of conquest, menacing to American democracy (Saunders, 2012).

Irish and Italian Catholics (and East European Jews as well) eventually took their place in the American mainstream and, in the process, fundamentally reconfigured the society's ethnic structure. Moreover, those groups took shape not in their origin societies, but in the context of American society, creating unique social and cultural forms. A similar long-term process may be occurring for Arab Americans, who today represent a challenging case that will once again test the absorptive capabilities of the American ethnic system.

---

[7] It has been suggested that an Arab American identity had begun to develop much earlier as a result of the 1967 Arab-Israeli War during which the United States supported Israel and media portrayals of Arabs were strongly negative (David, 2007).

[8] No other ethnic group in recent times has been so strongly impacted by events outside the United States. What happens in the Middle East affects the assimilation process, just as it affects patterns of immigration and the perpetuation of negative stereotypes.

SUMMARY

- Arab immigration to the United States occurred over three waves, extending from the late nineteenth century to the present. First wave immigrants were from the Syrian region of the Ottoman Empire and were mostly Christians. Second wave immigrants came after World War II mostly in response to political developments in the Middle East; most were Muslim. Third wave Arab immigrants, most of them Muslim, have come to the United States since the early 1970s, from virtually all countries of the Middle East.
- Arab Americans constitute about 1.8 million people, but some researchers suggest a much higher figure.
- In terms of socioeconomic status—income, occupation, and education—Arab Americans collectively rank higher than the general U.S. population.
- A lengthy history of prejudice and discrimination against Arabs and Muslims has characterized most Western societies, and this has been true of the United States as well.
- Arab Americans have experienced a high level of both individual and institutional discrimination, ranging from verbal abuse to physical attacks to official government policies.
- Islamophobia (fear and hostility toward Muslims), based on negative stereotypes, has been evident in the United States in the past several decades, reaching a high point following the terrorist attack of 9/11.
- A high degree of structural and cultural assimilation has occurred for first wave Arab Americans.
- Great obstacles stand in the way of assimilation for more recent generations of Arab Americans, but there are indications nonetheless of movement in that direction.

CRITICAL THINKING

1. Why had Arab Americans been largely invisible as part of the American ethnic system until the occurrence of terrorist actions beginning in the 1970s?
2. Is ethnic profiling of Arab Americans similar to racial profiling of African Americans and Latinos?
3. Why do you suppose there has been so little understanding of the quite varied ethnic composition of the U.S. Muslim population? Why do most Americans associate "Muslim" with "Arab?"
4. Why do right-wing U.S. terrorists (skinheads, neo-Nazis, and the like) not get the same attention by the media and law enforcement agencies as terrorists with Muslim connections?

PERSONAL/PRACTICAL APPLICATION

1. What is your reaction to seeing Muslim women wearing the hijab? Is your reaction different from seeing Orthodox Jewish men wearing kippas (skullcaps) or to Amish women who are easily identified by their bonnets? Make note of the unique garb of other religio-ethnic groups.

2. As we have seen in previous chapters, incidents of prejudice and discrimination against racial and ethnic minorities continue to occur in American society. Nonetheless, it is evident that there has been a noticeable decline of social tolerance of these attitudes and actions. No matter what their real beliefs and feelings may be, people using ethnic slurs, for example, are seen as violating an unwritten social norm. In the case of Arab and Muslim Americans, however, this trend does not seem to apply. Try to take note of how common these references to Arab Americans are made, where and by whom, and how they are reacted to.

THE CHANGING CONTEXT OF AMERICAN RACE AND ETHNIC RELATIONS

CHAPTER 13

Current and Future Issues

What will be the future course of race and ethnic relations in American society? As we will see in Part III, in many other multiethnic societies, problems that derive from ethnic divisions and stratification are far more severe. But understanding that other societies may suffer even greater problems should not cause us to lose sight of the profound and unresolved ethnic-related issues that the United States faces early in the twenty-first century, some relatively new, others long-standing. Currently, three encompassing and interrelated issues seem to clearly stand out regarding American race and ethnic relations:

1. *The changing ethnic configuration.* How is the influx of large numbers of new immigrants and the creation of a more pluralistic ethnic mix affecting various social institutions and what impact will these trends have on future interethnic relations?
2. *Assimilation versus pluralism.* Will—and should—the end product of the society's increasingly diverse ethnic mix be some form of assimilation, in which groups become more culturally alike and socially integrated, or some form of pluralism, in which they maintain or perhaps increase their cultural differences and social boundaries?
3. *The continued economic gap and social boundaries between Euro-American and most Asian American groups on the one hand, and African American, Hispanic American, and Native American groups on the other.* How will these divisions be narrowed, and what role will public policies play in addressing them?

ISSUES OF THE NEWEST IMMIGRATION

As explained in Chapter 5, the large-scale immigration of diverse peoples in the past four decades has had a profound impact on the ethnic flavor of the society. The traditional binary system of American race relations, black versus white, simply no longer applies; the United States is more profoundly multiethnic than at any time in its history. And its future bodes an ever more complex ethnic mix. The newest immigrants and the changes they have sparked have introduced new problems and questions of ethnic relations.

ECONOMIC ISSUES

The economic impact of the Newest Immigration is widely debated. What effects are these immigrants having on the American economy? Does immigration produce net benefits for the labor force, or does it negatively affect native workers? Do immigrants become self-supporting members of the society, or do they become a drain on public services such as schools and hospitals?

IMMIGRATION AND THE LABOR FORCE Whereas some argue that immigrants, particularly the undocumented, constitute an added burden to an already swollen labor pool, others contend that the jobs they typically hold are those that native workers shun—and furthermore, that they create more jobs than they take (Camarota, 2007; Fitz et al., 2013; Kochar, 2006; Meissner, 2010; Peri, 2007). Similarly, some maintain that immigrants overburden the social welfare system, but others hold that they pay in taxes far more than they collect in benefits (Borjas, 1994; Espenshade, 1998; Porter, 2005; Simon, 1991).

The sharpest aspect of the debate concerns the impact of immigration on the jobs and wages of native workers. Some contend that there is a generally negative effect (Borjas, 2004), whereas others have shown that immigration may actually increase the wages of U.S.-born workers (Peri, 2009, 2010; Shierholz, 2010).

There is simply no consensus on either side of this issue. What we do know for certain, however, is that the economic impact of the Newest Immigration is not uniform. Instead, it appears to favor some sectors of the economy and harm others. Most of the immigrants from Mexico (the largest single group) and the Caribbean are unskilled and take their place at the lowest employment levels. They benefit employers in labor-intensive industries, such as clothing manufacturing or other work areas calling for cheap labor; but they depress the job opportunities and wages of native low-status workers with limited education (Borjas, 1998, 2004; Mishel et al., 2009; Smith and Edmonston, 1997; Swain, 2007). Many industries—construction, hotels, restaurants, agriculture, meat- and poultry-processing plants, garment factories—are heavily dependent on such immigrant labor, both legal and undocumented (Schmitt, 2001a).

As we saw in Chapter 5, not all the new immigrants are impoverished and unskilled, however. A large segment of some new groups, particularly Asian Indians, Chinese, Filipinos, and Arabs, are highly trained professionals and managers whose economic impact is far different from that of those who enter with few occupational resources. Many immigrant doctors and nurses, for example, staff U.S. hospitals, which would find it difficult to operate without them. In fact, 15 percent of U.S. health care workers are foreign-born, including fully one-quarter of all doctors (Singer, 2012). The United States has also lagged in producing engineers and other highly trained scientific workers, especially in information technology fields, and many of the newest immigrants are filling these needs (Martin and Midgley, 2010; Schuck, 2007; Singer, 2012).

Those most profoundly affected by immigration are apt to express strongest support or opposition to its rate and extent. On one side are industrialists and business owners, who ordinarily favor an open-door policy with few restrictions on continued immigration. For them, immigrants serve as a needed workforce that helps to control wages. American-born workers at the other end of the class spectrum, however, see immigration from an entirely different perspective. For them, the new groups—regardless of their real impact—are seen as a threat to jobs and wages (Beck, 1996; Portes and Stepick, 1993; Stepick et al., 2003; Swain, 2007).

UNAUTHORIZED IMMIGRATION   In recent years, unauthorized immigration has come under especially sharp public attack, because it is popularly perceived that illegal immigrants not only take jobs from native workers but put an undue burden on public services, like schools and hospitals, at taxpayers' expense (Kotlowitz, 2007). Because such a large proportion of the estimated twelve million unauthorized immigrants originate in Mexico, this is an issue with particular currency in border states, especially Texas, California, and Arizona, which have large Mexican American populations. Punitive government policies have been enacted over the past two decades in these and other states, seeking to limit public services in education, welfare, and health care to immigrants; most have been subsequently ruled unconstitutional. A particularly harsh measure was enacted in Arizona in 2010 giving state and local police wide discretion in efforts to identify and detain illegal immigrants (Archibold, 2010). Its legality was challenged by officials of the U.S. Department of Justice, and its most onerous parts were subsequently struck down. But strong public support for the Arizona statute further inflamed the debate on illegal immigration and prompted similar measures in other states.

As with legal immigration, the economic impact of unauthorized immigrants is hotly contested, with both sides marshalling evidence to support their position. It is evident that many industries dependent on low-skilled workers, especially in the agricultural and service sectors, could not function as they do without a steady infusion of illegal immigrants. According to some estimates, more than 70 percent of farm workers in the United States are undocumented immigrants (Preston, 2007).

Whereas a majority of Americans view illegal immigration as a serious problem, a majority also believe that these immigrants are doing necessary work that Americans reject (Pew Hispanic Center, 2006c). Some have proposed a guest worker system, which, it is felt, would obviate the need for immigrants, particularly from Mexico, to seek undocumented entrance into the United States. To date, this and other legislative reforms aimed at controlling the flow of illegal immigrants have not been enacted. A majority of Americans seem to agree that those immigrants who have entered illegally should be given a way to stay in the country; the much more divisive issue is whether they should be afforded a path to eventual citizenship (Pew Research Center, 2013a; Polling Report, 2013).

Undocumented immigration, like immigration generally, is driven primarily by the promise of jobs, and the deep and prolonged recession of the late 2000s marked a reduction in the flow of those entering illegally from Mexico and other countries (Passel and Cohn, 2008b, 2010). In fact, by 2011, the net migration flow from Mexico to the United States had stopped and may have actually reversed (Passel et al., 2012). Nonetheless, illegal immigration has become a politically charged issue, and during periods of economic downturn, immigrants become inviting scapegoats.

IMMIGRATION AND THE FUTURE LABOR FORCE   Regardless of the ongoing debate regarding both legal and undocumented immigrants and their impact on the economy, what is clear is that future labor needs can be met only with a substantial inflow of immigrants. The American workforce is aging; that is, fewer native-born workers are entering the labor market. Therefore, satisfying future labor force needs will depend heavily on immigration. Between 2005 and 2010, about 40 percent of the growth in the U.S. labor force was due to new immigrants, and that demographic pattern is not likely to change in the foreseeable future (Myers et al., 2013; Passel and Cohn, 2008a; Singer, 2012).

The reliance on immigrants to fill jobs in both high-skilled and low-skilled sectors has already become very apparent. Immigrants make up nearly 17 percent of the total U.S. workforce and can be found in every major occupational sector. Table 13.1 shows the percentage of immigrants in particular occupations, all of which are among those that will be the fastest-growing fields in the next decade. Notice that there is demand in jobs that require both high levels of training (medical scientists or computer software engineers) and that require little training (maids or construction workers).

On a related note, a recent study indicated that immigrants are crucial to sustaining social programs that are financed by taxes on wages, like Medicare and Social Security (Zallman et al., 2013). Without their contributions, these programs would be in greater jeopardy than they currently are. Contrary to the beliefs of those who oppose immigration reform, immigrant workers are currently helping to fortify these programs rather than draining them of resources. Harvard Medical School researchers found that in 2009, immigrants (most of them noncitizens) made almost

TABLE 13.1 | IMMIGRANTS IN THE U.S. WORKFORCE

Occupation	Percent
Medical scientists	44
Computer software engineers	36
Cooks	40
Maids and housekeeping cleaners	53
Construction laborers	42
Electrical engineers	25
Agricultural workers	47
Physicians and surgeons	25

Source: Singer (2012).

15 percent of contributions to the Medicare program but accounted for less than 8 percent of its expenditures, making for a net surplus of almost $14 billion. Even illegal workers in some ways make contributions to Social Security and Medicare, which they will not be able to benefit from in the future (Porter, 2005). A high-ranking official of the Social Security Administration estimates that illegal immigrants in 2010 generated about $12 billion for the Social Security Trust Fund (Tavernise, 2013).

SOCIAL AND CULTURAL ISSUES

Most current immigrants are highly visible, bringing to American society cultural and physical features distant from those of the white majority. This makes them considerably different from previous immigrants who, though always viewed ethnocentrically by those already present, were nonetheless primarily European. The debate, therefore, concerns whether these new cultural and racial strains can—or should—easily blend with the contemporary American ethnic amalgam.

THE DEBATE   Some believe that currently high levels of immigration should be continued or made even higher to allow for additional immigrants. Their view is that immigrants not only create economic activity but are a fresh cultural influence on the society. American society is in fact pluralistic, they hold, and should further develop its multicultural character, creating a more equitable society in the process (Isbister, 1996). Others, however, support highly restrictionist measures, maintaining that, economic issues aside, the new immigrants are causing a radical and unprecedented change in the social and cultural makeup of the society, which can only lead to more racial and ethnic conflict. Some see this change as undesirable in itself, suggesting that the United States remain a primarily European-origin country (Brimelow, 1995; Huntington, 2004). The new groups are viewed as unassimilable and thus a continual drag on the society. These issues have a familiar ring, recalling those that raged during the nineteenth and early twentieth centuries with regard to Irish and southern and eastern European immigrants.

The mixed feelings of Americans regarding the social and cultural impact of immigrants are evident in national surveys. In 1993, 55 percent of Americans felt that the increased diversity of immigrants was a threat to American culture. In 2010, the social divide on immigration remained evident, but most apparent now were the different views of Latinos and whites. Whereas 68 percent of Latinos believed that immigration strengthens the United States, only 43 percent of whites felt that way (Murray, 2010).

LANGUAGE DIVERSITY   Questions and attitudes regarding the social impact of immigrants on American culture have come together most clearly around the issue of language. In Canada, as we will see in Chapter 16, the nucleus of ethnic conflict is the preservation of one group's language. A somewhat similar situation has begun to surface in certain regions of the United States, where large numbers of immigrants, specifically the Spanish-speaking, use their native tongue. Spanish is clearly the leading language in homes where a language other than English is used. Many view the continued use of Spanish by immigrants as offensive and an indication that they do not wish to assimilate. The introduction of Spanish-language signage in public places and the use of bilingual forms by government agencies have created additional resentment.

That these immigrants are able to maintain close links with their origin societies through easy communication and travel and through a constant influx of fellow immigrants from the homeland means that there is no pressing need to quickly relinquish the native language. This makes their situation somewhat different from that of past immigrants. Although the latter also came speaking languages other than English, it was assumed that through the use of English in the school and other institutions, those languages would gradually be abandoned, if not by the first then by the second generation. Today, however, the assumption of language assimilation is being challenged. A greater tolerance of ethnic pluralism has led to efforts to provide public educational and other services in the language of the new groups, especially Spanish. This has created controversy among those who favor or oppose such measures (Baron, 1990; Chavez, 1991; Costantini, 2011; Crawford, 1992; Huntington, 2004; Porter, 1990).

The language issue is particularly acute in cities and states with large immigrant, especially Hispanic, populations where schools must accommodate students whose first language is not English. In New York City, for example, classes are taught not only in Spanish but in Chinese, Haitian Creole, Russian, Korean, Arabic, Vietnamese, Polish, Bengali, and French. More than 40 percent of the parents of children in the New York school system are not native English speakers (Bosman, 2007). In Miami-Dade County, almost 60 percent of public school students speak a language other than English at home (mostly Spanish), almost 20 percent have limited English proficiency, and many schools at all levels are bilingual institutions (Miami-Dade County, 2006; Nazareno, 2000). The issue is not limited to these high-profile immigrant-receiving cities, however. School districts throughout the United States increasingly must deal to some degree with non-English-speaking children.

Some who oppose greater language diversity have sought to have English legally declared the "official" American language. By the mid-1990s, more than twenty states had passed legislation to that effect, and in recent years a number of municipalities have done the same. Those favoring such legislation argue that it will help to reverse (what they see) as a decline in English language use and force immigrants to more quickly adopt the dominant culture. They advocate strict limits on bilingual education, the elimination of voting ballots in languages other than English, and increased language-proficiency standards for prospective citizens. Supporters of the movement to make English the country's official language maintain that without a common language, new immigrants will resist assimilation, and the United States risks becoming linguistically divided, like Canada (Huntington, 2004). In response, critics of the English-only movement charge that efforts to declare English the official language of the United States are a backlash against the new immigrants and essentially a mask for ethnic antagonism. Moreover, they claim that such legislation is useless because it does nothing to help promote the learning of English (Braverman, 1988; Crawford, 1992).

Although English-only proposals seemed to reach a high point in the 1990s, efforts at the local level continue to appear from time to time. For example, in 2009 in Nashville, Tennessee, a proposed amendment to the city's charter called for the restriction of the use of any language other than English in city affairs; it failed to pass by 57 percent to 43 percent. Any English-only proposals at any level of government, of course, are virtually impossible to enforce and, moreover, are not likely to hold up if tested in the courts.

The issue may be moot in any case because a raft of studies indicate that, as in the past, the children of recent immigrants adopt English as their major language and gradually drift away from using their native language outside the home (Alba, 2004; Portes, 2002; Portes and Rumbaut, 2006; Rumbaut et al., 2006; Veltman, 1983). Research among Latinos, as was noted in Chapter 8, shows that first- and second-generation immigrants are usually bilingual, and by the third generation, English has become the primary—and for most, the only—language spoken (Alba, 2005; de la Garza et al., 1992; Latino Coalition, 2003; Pew Hispanic Center, 2002; Rumbaut et al., 2006). In their study of the children of Latino and other immigrants, Portes and Hao (1998) found that knowledge of—and preference for—English is nearly universal among them and they quickly lose fluency in their parents' language. This language shift is shown in Figure 13.1.

Overwhelming evidence indicates that language assimilation, especially among Latinos, is proceeding apace, just as it did for every previous immigrant group in American society. And as we saw in Chapter 8, the vast majority of Latinos believe that citizens and residents of the United States *should* learn English. As Alan Wolfe has described it, "The battle over bilingualism tends to take place over the heads of the immigrants themselves, who in general only want to learn English as fast as they can" (1996:108). Evidence of this is the fact that states offering English-language classes to immigrants cannot keep up with the demand (Santos, 2007). What keeps Spanish alive in the United States, therefore, is not a determination of Latinos to retain their native language but the continuation of large-scale Hispanic immigration.

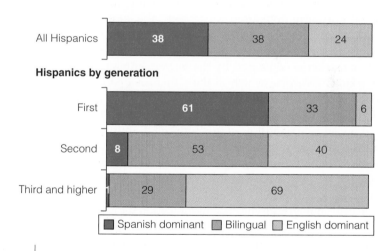

FIGURE 13.1 | PRIMARY LANGUAGE AMONG LATINOS, BY GENERATION (PERCENT)

Notes: N=1,220. Language dominance is a composite measure based on self-described assessments of speaking and reading abilities. "Spanish-dominant" persons are more proficient in Spanish than in English, i.e., they speak and read Spanish "very well" or "pretty well" but rate their ability to speak and read English lower. "Bilingual" refers to persons who are proficient in both English and Spanish. "English-dominant" persons are more proficient in English than in Spanish.

Source: Taylor et al. (2012). Note: As a strictly nonpartism, non-advocacy organization, we do not grant permission for reprints, links, citations or other uses of our data and analysis that in any way imply that Pew Research Center endorses a cause, candidate, issue, party, product, business or religion. To view our Use Policy and citation guidelines, please visit http://www.pewresearch.org/about/use-policy/.

ISSUES OF IMMIGRANT ADAPTATION AND INTEGRATION

As immigration to the United States reaches epic proportions, it is of no little significance that most of the newest immigrants are from non-European societies. Even more significant is the fact that under the U.S. racial/ethnic classification system, most do not fall into the "white" category. How they will be absorbed into the society and the place they will eventually take in the ethnic hierarchy, therefore, are issues of critical importance that have aroused much public controversy.

IMMIGRANT CONTEXTS: OLD AND NEW   For several reasons, the typical patterns of adaptation to American society displayed by past immigrants are not likely to be followed in the same way for the newest immigrants. European immigrants of the nineteenth and early twentieth centuries generally followed a path of eventual assimilation into the mainstream society, leading to upward mobility, over two or three generations. The newest immigrants may follow a more tortuous route and a less certain outcome of social and economic adaptation. This was referred to in Chapter 4 as *segmented assimilation*. Some may follow the traditional assimilation pattern, but others may retain much of their ethnic cultures and social structures for several generations. Still others may find themselves stuck in a low-income status, with stifling job and educational opportunities that confine them to a condition outside the mainstream society. Because most of the newest immigrants are non-white, they automatically face a racial barrier that European immigrant waves were able to overcome relatively quickly.

Further differentiating the newest immigrants from those of past generations are the significantly changed social and economic circumstances newcomers face today. First, the labor market encountered by the newest immigrants is one in which solid industrial working-class jobs, which in the past provided immigrants with opportunities for upward mobility, have disappeared. What has taken its place is a labor market more clearly divided into two segments—high-status (highly skilled professional and technical jobs) and low-status (unskilled, mostly service sector jobs). The latter are poorly paid and essentially dead-end jobs that have been disproportionately occupied by newly entering immigrants. The fear is that a semipermanent immigrant underclass will be created, impeding integration of these groups into the mainstream society for many generations.

Second, recent immigrants enjoy communication and transportation technologies unimagined by past newcomers. To fly from New York to the Dominican Republic, for example, involves a few hours; in the mid-nineteenth century, it took about five weeks to sail from Britain to America in the best of weather conditions (Van Vugt, 1999). Similarly, immigrants today can remain in constant contact with relatives and friends in the origin country by telephone and Internet. In addition to the ease of communication, the proximity of the origin country enables immigrants to retain links with the ethnic culture and to resist rapid cultural and structural assimilation. Immigrants from Mexico, for example, make frequent trips back and forth across the border, and a similar pattern of movement is common among immigrants from the Caribbean.

Third, immigrant communities in the United States are likely to see a continued flow of newcomers from their origin societies for several decades. This is unlike the earlier European immigrant waves, which for the most part ended in the 1920s. Because they were cut off more completely from their societies of origin, European immigrants were faced with a greater incentive to assimilate quickly into the mainstream society. Current immigrants and their children, by contrast, are able to continually interact with coethnics, which will likely make for longer-lasting ethnic communities and cultures (Min, 1999).

The mass media, of course, have a way of homogenizing people quickly and with great force. American popular culture is a global phenomenon, and most immigrants have absorbed a good deal of it before they actually migrate. Gans (1999) refers to this as "anticipatory acculturation." Roberto Suro explains that it is impossible to grow up in Latin America, for example, without being saturated with American images and information from birth. As a result, most immigrants can feel quite comfortable with U.S. culture when they arrive, at the same time feeling no need to discard their "Latino sensibilities" (Suro, 1998:71).

Finally, in recent years major institutions in the society, particularly education and the media, have moved toward an acceptance of ethnic pluralism, or multiculturalism, and away from the Anglo-conformity ideology that typified the dominant response to previous immigrant waves. The economy, too, in various ways has begun to cater to the unique needs and tastes of ethnic groups. In short, American institutions have today made it easier to retain one's ethnicity.

FORCES OF ASSIMILATION   At the same time that the newest immigrants generate unprecedented issues of social and economic adaptation, there are strong indications that they are, in a number of ways, following an assimilation path not radically different from past immigrant groups. As we have seen, language assimilation for both Latinos and Asians is proceeding apace, as are residential integration and interethnic marriage.

Richard Alba has pointed out that assimilation has been incorrectly interpreted by some as a one-way process, in which immigrants surrender their culture and social structure to the dominant society and become carbon copies of the dominant ethnic group (Alba, 1999). In fact, assimilation, as noted in Chapter 4, is a reciprocal process involving a cultural exchange between dominant ethnic groups and immigrant minorities. Immigrants themselves inject change into the receiving society's sociocultural system. As Alba has put it, "[I]mmigrant ethnicity has affected American society as much as American society has affected it" (1999:7). Moreover, a distorted conception of assimilation assumes a homogeneity of American culture that simply does not exist.

"Assimilation," explains Alba, "most often occurs in the form of a series of small shifts that takes place over generations; those undergoing assimilation still carry ethnic markers in a number of ways" (1999:21). Assimilation, in other words, entails a decline of ethnic distinctions, not their absolute disappearance. Thus the remnants of their ethnic cultures and social structures will remain evident, but assimilation will in most regards work for current immigrants as it did for those of past eras, if not as quickly and directly.

That assimilation for America's newest immigrants may be a slower and more tortuous process and that it may not resemble in all ways the assimilation process of the past is clear enough. But recall from Chapter 4 that assimilation is a complex set of variables that includes both cultural and structural components. Moreover, the effects of immigration are often judged on the basis of short-term issues and problems of the immigrants themselves. Those who comprise the immigrant generation should not be confused with their children, the second generation; among the latter, dramatic gains in language assimilation, socioeconomic incorporation, and citizenship are evident (Kasinitz et al., 2008; Myers and Pitkin, 2010). And, above all, despite the increasing acceptance of a multicultural philosophy and the continued ethnic diversification of the society, there remains among immigrants and their children a solid commitment to the most basic American values (Etzioni, 2004).

A REVITALIZED NATIVISM?

Public opinion regarding immigration, as was pointed out earlier, is puzzlingly ambivalent: almost as many believe that immigrants strengthen the society as create economic and cultural problems. Anti-immigrant sentiment today is by no means as virulent as in the earlier part of the twentieth century when it eventually led to a halt to the large-scale immigration of southern and eastern European groups. But the racial and ethnic character of the newest immigrants, when combined with fluctuations in the state of the domestic economy, continually threatens to create a revitalized nativism. Moreover, although most Americans recognize the great historical role played by immigration in shaping their society, they do not necessarily see present-day immigration in the same light. This is revealed consistently in national surveys (Gallup Organization, 2001a, 2002, 2005; Jones, 2001; Morales, 2009; Pew Hispanic Center, 2006c; Pew Research Center, 2013a). Today, immigration and its attendant issues are seen by a majority of Americans as a particularly strong source of social conflict in the United States, stronger even than the division between blacks and whites (Morin, 2009).

SHIFTING PUBLIC VIEWS As noted earlier, tolerance for immigration seems to rise and fall with changing economic conditions. That is, when the economy is robust and unemployment levels are low, public opinion is less opposed to retaining the present level of immigration. But even during economically stable and prosperous times, only a small percentage favor increasing the level of immigration. Most Americans have preferred either to decrease immigration or to maintain it at its current level (Gallup Organization, 2005; Morales, 2009; Murray, 2010; Pew Hispanic Center, 2006c; Polling Report, 2013; Schuck, 2007).

The flow of international events can also affect public attitudes toward immigration. In the wake of the terrorist attacks on the United States in 2001, for example, new thinking about immigration issues surfaced quickly. A national debate ensued regarding whether adequate security measures were in place to sift out undesirable immigrants, how illegal immigration could be better controlled, and whether U.S. policy was too lenient toward immigrants generally.

Most of the highly charged controversy that continues to swirl about immigration focuses on the flow of illegal immigration and what to do to better control the country's borders. A corollary issue concerns the status of undocumented immigrants who are already living in the United States. More specifically, however, the focus of the current immigration debate has fallen mainly on those coming from Mexico and, even more specifically, those who are undocumented.

Future changes that are enacted in immigration laws—whatever their content— are unlikely to bring to an end the debate regarding the number and character of immigrants that should be permitted to enter the United States. Those seeking more stringent limits will continue to argue that immigrants have a negative economic and social influence, whereas those advocating more liberal immigration laws will maintain the opposite. These are hardly new points of debate, however. Historically, the absorption of immigrants has been a persistent theme of American political argument. In one sense, the United States has always been regarded as a "golden door," open to all seeking economic opportunity or political refuge. But the acceptance of new groups has been countered with a tradition of protectionism, which has manifested itself repeatedly in efforts to limit or exclude newcomers. The current public controversy is, therefore, only the latest in a long tradition. As historian Donna Gabaccia has pointed out, "Since the 1960s, celebrations of the United States as a nation of immigrants have encouraged Americans to forget the ferocity of . . . earlier debates about immigration" (2006:3). Those debates, past and present, have been essentially about restriction, that is, how or whether to limit immigration to the United States.

It is important to note that the current public debate in the United States on immigrant issues and the generally skeptical public view of immigration are not without parallel in other societies. An analysis of public opinion regarding immigrants and related issues in Australia, Canada, France, Germany, Great Britain, Japan, and the United States found that despite these countries' different current and historical immigration policies, a majority of the public in each expressed relatively similar views: they want fewer immigrants admitted, especially immigrants of color; they feel that priority should be given to immigrants with special skills rather than to those seeking family unification; and they believe that their country has done more than its share in accepting refugees (Simon and Lynch, 1999).

CULTURAL ASSIMILATION OR PLURALISM? COMPETING GOALS

As we have now seen, the ongoing population shift in American society has provoked fundamental economic, political, and cultural issues. But it has also prompted an ideological issue: given America's increasing ethnic diversity, should the end product of interethnic relations be some form of cultural assimilation, in which diverse groups become more alike, or some form of pluralism, in which they maintain or perhaps increase their differences? The questions of whether disparate ethnic groups in American society should move unwaveringly toward eventual assimilation and, concomitantly, whether ethnic policies should be directed toward assimilationist goals, have arisen many times throughout American history. The political and social conditions of the last four decades, however, have given these issues renewed substance and made them especially salient.

MULTICULTURALISM: THE NEW PLURALISM

In the past, ethnic issues in the United States were generally reduced, by sociologists, policy makers, and laypeople alike, to problems of assimilation (Metzger, 1971). How, it was asked, could diverse peoples achieve a maximum of social integration with a minimum of social travail? Put differently, the direction of ethnic relations was rarely questioned; it was assumed that groups would—and should—move toward some form of assimilation. Today, that assumption is very much in dispute.

In the 1960s, the black civil rights movement, especially its nationalistic phase, gave rise to feelings of pluralism across the entire spectrum of ethnic groups. Heightened collective consciousness among African Americans and black political activism stirred similar movements among Latinos, Asians, American Indians, and even white ethnics. A new ideology of cultural pluralism seemed to materialize in which ethnic differences were not only to be tolerated and respected but perhaps even to be encouraged. In the past five decades, the failure of group boundaries—particularly between whites and blacks—to dissipate, despite higher levels of racial tolerance and the end of blatant forms of discrimination, along with the influx of millions of new immigrants radically different in culture from the European norm, have given added incentive to stronger forms of cultural pluralism. As explained in Chapter 5, this ideology has been referred to in recent years as "multiculturalism." The turn toward a multicultural alternative represents a fundamental shift in the direction of American ethnic attitudes and ideology.

Multicultural ideas and policies are reflected in almost all major societal institutions, including education, business, government, and the mass media. The impact has been especially great in education, where non-European cultures are increasingly acknowledged in school and university curricula and educational institutions generally conform to the multicultural ideology. The essential objective is to give greater recognition to the contributions of non-European groups in American and global historical development and to study those groups' literature and art. These efforts have been met with heated debate among educators and public officials. One side views it as an attempt to demean European culture, and in the process to create sharper group divisions; the other side views it as necessary to counterbalance the traditional emphasis on "white" (that is, European) cultural and historical traditions (Goldberg, 1994; Ravitch, 2002; Schlesinger, 1998; Wood, 2003).

Language, as discussed earlier, has been another flash point of ethnic ideological and policy conflict, pitting proponents of bilingualism in education and public communication against those who view language concessions not only as contrary to traditional American principles and practices but as a dangerous opening to a corporate pluralistic system. It is a vivid illustration of the clash of assimilationist and pluralist perspectives.

A New "Melting Pot"?   In a larger sense, this issue may be hollow. Despite the accentuated cultural diversity of the United States, forces are at work that seem to have homogenizing effects on all groups. Mass communication and transportation increasingly break down cultural singularities and compress them into common forms, spiced with only slight ethnic variations. Billboards that in Des Moines picture white people and are written in English may in Detroit feature blacks or in Miami be written in Spanish, but they advertise the same products for all. Moreover, as the society becomes more ethnically varied, cultural and social influences begin to cross in bewildering combinations. Journalist David Rieff describes, for example, the ethnic fusion occurring in contemporary Los Angeles:

> Just as the Irish, Poles, Jews and Italians, who had rubbed shoulders and more in the wake of the European immigration of 1900, had, by the 1950s and '60s, begun to intermarry en masse, so the process is beginning to take place among the recent arrivals in

L.A. One can find every sort of nonwhite combination in the city now: Hmong and Salvadoran, Ethiopian and Taiwanese, Mexican and Filipino. (1991:20)

In ethnic cauldrons like California, New York, and South Florida, the mix is so intense that ethnicity itself begins to lose significance. As one Asian Indian entrepreneur put it, "California is so cosmopolitan, you don't have to think about ethnicity here. It just doesn't matter—to anyone" (quoted in Jacoby, 2000:25). These centers of ethnic heterogeneity, of course, do not typify the United States; and the likelihood that such demographic conditions and their attendant attitudes will soon follow in the American heartland is at best long-term. Nonetheless, the seed has been planted.

For generations, American society has demonstrated an unflagging ability to absorb aspects of diverse ethnic cultures, thus revealing the reciprocal nature of the assimilation process. Until recently, however, an essentially White Anglo-Saxon Protestant core could be asserted as the "dominant" or "mainstream" American culture. Given the society's increasingly prevalent multicultural ethos, however, it has become more difficult to clearly define what the dominant or mainstream culture actually comprises.[1] Although ethnic variety has been a basic characteristic of American society almost from its founding, that variety has never been more complex and wide-ranging as it is today. It might be said that the dominant culture itself has, in fact, become a loosely defined blend; the mainstream is, most simply, multicultural.

It is in this sociocultural sense that, ironically, the "melting pot"—an inaccurate metaphor of American ethnic relations in previous generations—may be closer than ever to realization today. Steinberg (1989, 2004) suggests that although pluralist principles have become much stronger in recent years, the fact remains that ethnic differences have been diminishing. Thus, though many extol the need to retain ethnic cultures and encounter declining resistance to their retention from the dominant group, societal trends continue to break down and reprocess those cultural differences. What is being created is a multicultural admixture. Moreover, the insistence by ethnic leaders that ethnic cultures must be preserved does not seem to be shared with the same intensity by immigrants themselves. Contrary to conventional thought, immigrants seem most committed to assimilation, not retention of their ethnic cultures (Saad, 1995).

This homogenized culture is brought home strongly when Americans of any ethnic origin travel to the society of their forebears. Most quickly realize how little they have in common with the people of Ireland, if they are Irish Americans; of Poland, if they are Polish Americans; or of African nations, if they are African Americans. It is in such foreign contexts that a unique "Americanness" is most fully revealed. Despite communication and transportation technologies that make it possible to maintain continual contact with their origin societies, the same acculturative process—thoroughly *multi*cultural in content—will likely have a similar effect on the newest immigrants to the United States.

---

[1] As was noted in Chapter 5, in this new multicultural environment even WASP Americans may be evolving into simply one more ethnic group among many. William Greenbaum may have overstated their declining influence only somewhat when he declared that "after having been the society, the Protestants have been relegated to a place within the society, and increasingly they experience a bewildering sense of themselves as a new minority" (1974:412).

Paradoxically, then, the multicultural ideology that now seems predominant in an increasingly diverse American society may actually be giving rise to a new melting pot. But that does not guarantee for all groups an equitable distribution of economic and political resources. This is the third major contemporary issue of race and ethnic relations in the United States, to which we now turn.

THE CONTINUING GAP BETWEEN EURO-AMERICANS AND RACIAL-ETHNIC MINORITIES

Let's briefly review the American ethnic hierarchy that was introduced in Chapter 5. Roughly, the hierarchy is divided into three comprehensive tiers: a top tier of white Protestants, essentially the dominant ethnic group; a second, intermediate tier made up of white ethnics of various national origins (mostly Catholics and Jews) and many Asians, for whom ethnicity remains of fading and for some, only symbolic, significance; and a bottom tier, composed of racial-ethnic minorities, for whom ethnicity continues to render great effect in the distribution of societal rewards and in shaping patterns of social interaction (Figure 13.2). The most important division within this hierarchy is between the bottom tier and the other two.

The third tier of groups continues to maintain a collective place at the bottom of the society's economic and political hierarchies. This does not mean that many individuals in each of these groups have not achieved substantial upward mobility, particularly in the past four decades. Indeed, evidence is unmistakable that the resistance of whites to nonwhite economic, political, and social advancement has greatly diminished, and as a result, the economic and political status of large numbers of racial-ethnic minorities has risen significantly. But in looking at their collective status vis-à-vis Euro-Americans and most segments of the Asian American population, members of these groups remain disproportionately among the poor and the powerless. Moreover, the social boundaries separating the third tier of groups from the first and second, though dissolving slowly, remain in place. Members of third-tier groups continue to encounter higher levels of prejudice and discrimination, and they continue to face barriers to full and unbridled participation in many realms of social and economic life.

COMPENSATORY POLICIES

Efforts to deal with this persistent economic gap and the social distance between majority and minority ethnic groups have, in recent decades, given rise to a number of

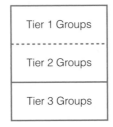

FIGURE 13.2 | THE AMERICAN ETHNIC HIERARCHY

contentious public policies, the most prominent of which are those that fall under the rubric of affirmative action.

Out of the atmosphere of the civil rights movement and the new pluralism of the 1960s and 1970s, the federal government undertook sweeping measures to foster political and economic equity among ethnic groups. Specifically, policies were aimed at improving the social standing of African Americans, against whom official discriminatory measures had been historically applied. Other long-term victims of discrimination—Latinos, Asian Americans, and American Indians—were subsequently brought under the compensatory umbrella and, later, women were added as a target group.

These new public policies, designed as they were to raise group positions, marked a significant departure from the traditionally understood role of government in the area of civil rights. In the past, government's function had been to ensure that everyone was afforded equal opportunities in work and education, regardless of ethnicity. Because this guarantee had been violated for nonwhites, legislation was enacted beginning in the 1960s that was designed to protect members of minority groups from discrimination in schools, workplaces, and other societal institutions. But the question now raised was whether eliminating discriminatory practices alone could counteract the effects of past discrimination. Given the generations of denied opportunities, was it fair to expect African Americans and other racial-ethnic minorities to compete on an equal basis with the white majority? How could these groups ever catch up if they entered the competition burdened by decades of imposed disabilities? President Lyndon Johnson, in proclaiming the government's intention to address this issue, put the matter squarely:

> You do not wipe away the scars of centuries by saying: Now you are free to go where you want, do as you desire, and choose the leaders you please. You do not take a man, who for years has been hobbled by chains, liberate him, bring him to the starting line of the race, saying "you are free to compete with all the others," and still justly believe you have been completely fair. Thus it is not enough to open the gates of opportunity. All our citizens must have the ability to walk through those gates. This is the next and more profound stage of the battle for civil rights. (Johnson, 1965:635)

In response to this dilemma, a series of compensatory measures and programs evolved, generally called **affirmative action**, that were *intended to advance the economic and educational achievement of the minorities that had been most severely and consistently victimized by past discrimination.* That government was now acting to guarantee not simply equality of opportunity but also equality of result evoked a strong public debate, which, almost five decades later, continues to generate much controversy.

As affirmative action policies unfolded in the 1960s, they stipulated that those doing business with the federal government (universities as well as businesses) were required to take steps to increase their minority representation and to establish goals and timetables to meet that objective. The aim was basically to affect minority employment and student admissions. With more widespread and stringent application in the 1970s, affirmative action policies grew increasingly unpopular and created a backlash among majority whites. The arguments on both sides of this issue have not basically changed in the past nearly half century.

AFFIRMATIVE ACTION: THE DEBATE

"AFFIRMATIVE ACTION IS NECESSARY"   Proponents of affirmative action have argued that advancing the position of minorities through the use of goals and preferential hiring practices is necessary if the victims of past discrimination are eventually to attain equity with the majority (Kennedy, 1994; Livingston, 1979; W. Taylor, 1995). Even if direct forms of discrimination no longer prevail, they point out, the indirect and institutional forms continue to perpetuate nonwhite disadvantages in the labor market and in higher education. Simply protecting minority individuals against ethnic discrimination, therefore, is inadequate by itself. Hiring workers or admitting students without regard to ethnicity (a "color-blind" process) will automatically preserve the disproportionate representation of whites because they enter the competition with background advantages accumulated over many generations. Artificial incentives for minorities are needed temporarily, therefore, until the opportunity structure is made more truly equitable. Most important, proponents of affirmative action argue that without such policies, there is a risk of returning to a norm of discrimination. "Affirmative action is not a perfect social tool," writes philosopher Tom Beauchamp, "but it is the best tool yet created as a way of preventing a recurrence of the far worse imperfections of our past policies of segregation and exclusion" (2002:216).

Some maintain that the critical importance of affirmative action lies in providing access for minorities to social networks that aid in securing better jobs and educational opportunities. In this view, what is most important is being enmeshed in key networks through which individuals find jobs, meet influential people, and learn bargaining skills. Individuals may possess formal qualifications, but unless they are able to move into the inner circles of the educational and work worlds, their chances of acquiring top jobs are diminished. African Americans have lacked the connections—social capital—ordinarily developed through ties made in universities, in the corporate world, or through intermarriage, that Euro-Americans have availed themselves of more easily. Affirmative action, it is argued, can therefore cast African Americans and other minorities into networks that they otherwise would have little chance of penetrating (Loury, 1998; Patterson, 1997, 1998).

"AFFIRMATIVE ACTION IS UNFAIR"   Those who oppose affirmative action or who strongly criticize the way it has been carried out argue that, in effect, these programs have become a kind of **reverse discrimination** in which *those previously discriminated against are given preference over others merely on the basis of ethnicity* (Glazer, 1975; Hook, 2002; Yates, 1994). Hence, they maintain, the very objective intended—reducing ethnic discrimination—has been undermined, the victims now being Euro-Americans. Government is seen as strengthening, not weakening, racial and ethnic lines of division. The effect of affirmative action in many cases has been to create quotas favoring minorities, a particularly sensitive issue for certain groups, such as Jews, who themselves were the past victims of quotas limiting their entrance into prestigious colleges and professional schools.

Opponents of affirmative action have pointed out that preferential measures shifting emphasis from equality of opportunity to equality of result create aspects of a corporate pluralistic society, where, as explained in Chapter 4, social benefits are distributed on the basis of group membership, not individual merit. Hence, less qualified minority people may be promoted over better qualified majority people. Moreover, the targeted groups of affirmative action, they argue, are stigmatized because of their special treatment, thus producing negative social and psychological effects on those who are supposedly the beneficiaries of these programs (Sowell, 1990). Some have also suggested that affirmative action programs create a kind of implied inferiority. Whites commonly view blacks as having acquired their positions on the basis of special preference and thus as less than qualified or competent. This not only fuels negative stereotypes by whites, they argue, but also creates self-doubt among blacks and other minorities who are the intended beneficiaries of affirmative action (Heilman, 1996; McWhorter, 2000; Steele, 1991).

In addition to questions of reverse discrimination and effects on employment and educational qualifications, affirmative action programs have been criticized as too sweeping in application and therefore unable to distinguish from among the various targeted minorities those who are truly the past or present victims of discrimination. For instance, although Latinos are covered under the principles of affirmative action, we have seen in Chapter 8 that this broad ethnic category comprises several disparate components, each with different American experiences. Should Cubans, who have not encountered discrimination in work and education, be entitled to the benefits of these programs in the same way as Mexicans in the Southwest, who can invoke a history of discrimination in that region? Or should *any* group whose ancestors came to America as voluntary immigrants be afforded the benefits of affirmative action despite never having been the victims of systemic legal discrimination? Such problems suggest that minority group membership is no longer unambiguous.

Moreover, many have pointed out that the major recipients of the benefits of affirmative action have been middle-class racial and ethnic minorities, not the "truly needy"—that is, those in disadvantaged class positions. Filtering out those truly deserving of compensatory benefits has therefore become more complicated not only between various groups but within them as well.

The issue of affirmative action continues to generate heated debate, with advocates of these programs countering their detractors on various points. As to the lack of qualifications of those employed or admitted through preferential policies, for example, proponents argue that "qualifications" can be variously interpreted, no matter how seemingly objective and valid tests or other sorting mechanisms may seem. Placement tests are often biased in favor of those with a white middle-class background, and in any case they are never wholly adequate in measuring one's potential on the job or in the classroom. Moreover, merit, they point out, has never really been the sole criterion used in filling occupational and educational positions. On the matter of quotas, it is argued that the objective of affirmative action is to facilitate the entrance of minorities into various institutions, not to keep them out, as was the purpose of earlier discriminatory quotas. Furthermore, they point out, preferential treatment is already given certain groups, such as veterans or athletes, in employment or education. For example, legacy preferences, which give advantages to children of alumni in entrance requirements, are used by most top universities and colleges (Golden, 2007; Kahlenberg, 2010). And, whereas opponents see racial minorities as being stigmatized by affirmative action, defenders argue that their absence in jobs and schools would create an even greater stigma (Kennedy, 1994).

The Legal and Political Issues of Affirmative Action

The legal and political questions of affirmative action programs have involved their scope and intent as well as whether they are racially discriminatory and therefore in violation of the Fourteenth Amendment of the Constitution. They have been tested in the courts on numerous occasions since their inception, but judicial decisions have been inconsistent and have not fully clarified these issues.

MAJOR COURT CASES   In defining the scope of affirmative action, an early case of great significance was *Griggs v. Duke Power,* heard by the Supreme Court in 1971. In that case, the Court extended affirmative action to address instances not only of overt, intentional discrimination but structural discrimination as well. If the number of minorities in a company's workforce, for example, did not reflect the proportion of the minority population in the area where the company was located, that fact might be sufficient to demonstrate the effect of such indirect discrimination or, as the term came to be known, "disparate impact." In 1989 in another ruling, however, the Court weakened the business-necessity requirement and shifted the burden of proof for demonstrating such indirect discrimination from the employer to the applicant (Ezorsky, 1991).

As to whether affirmative action programs are themselves discriminatory, the *Bakke* case is perhaps the most momentous, particularly as affirmative action relates to higher education. Allan Bakke was denied admission to the medical school of the University of California at Davis even though his entrance qualifications exceeded those of minority applicants who had been admitted under a special program. As a means of increasing the number of minority students, sixteen of one hundred places in each entering class had been reserved for minority applicants. Bakke, white, maintained that he had been discriminated against on the basis of ethnicity and sued the university for admission. The case eventually reached the U.S. Supreme Court; and in 1978, by a five-to-four decision, the justices ruled in favor of Bakke. Their decision was ambivalent, however, leaving the way open for schools to establish goals for meeting an ethnic balance. In effect, the Court ruled that quotas (which in this case the university was found to have used) were illegal but that the use of race as a criterion of preference was legitimate as long as it was one among many criteria and its purpose was to create a more ethnically balanced student body.

Legal rulings in the 1990s and early 2000s continued to reveal ambivalence on the part of public officials and the courts toward affirmative action policies. Rulings in some cases seemed to uphold the legality of affirmative action, whereas others augured the very opposite trend. Some clarification came in 2003 when the Supreme Court heard two cases regarding affirmative action in undergraduate and law school admissions to the University of Michigan. Not since the *Bakke* case in 1978 had the Court ruled on the use of affirmative action in university admissions, and many believed that the policy's entire legal framework might now be struck down. For undergraduate applicants, the University of Michigan had given an automatic point boost to all minority applicants; the Court rejected this policy. But in the case of the law school it upheld (by a five-to-four vote) the affirmative action policy, using essentially the same rationale as had been used earlier in the *Bakke* decision: the school was justified in using race as one criterion among others as a means of ensuring an ethnically diverse student body. This case seemed to ensure that affirmative action in some form would continue to influence university admissions policies. In a related case in 2013, the Court again ruled that a university could continue to use affirmative action measures to assure a diverse student body, but only if it had exhausted all other means of doing so.

As affirmative action policies have evolved and been applied during the past fifty years, political and legal debate has crystallized around their legitimacy and fairness. As a result, affirmative action has been a key issue separating liberal and conservative sides of the American political spectrum. Liberals have generally supported such measures while conservatives generally have opposed them or have sought to limit their application.

PUBLIC OPINION   Public sentiment regarding affirmative action measures has always been ambivalent. Over the years, a majority of Americans have generally supported affirmative action programs that are designed to help minorities gain better jobs and education, but at the same time they have firmly rejected the use of preferential policies or quotas to further those objectives, viewing them as fundamentally unfair and in violation of principles of fairness (Desilver, 2013; Fineman, 1995; Gallup, 2003; Krysan, 2008; Pew Research Center, 2003, 2007, 2009b; Sniderman and Piazza, 1993). In 2013, public opinion regarding affirmative action indicated a waning support. A national survey showed that while 45 percent of Americans felt these measures were still needed to counteract the effects of discrimination, 45 percent no longer felt they were necessary. Support had been on a downward slope since 1991, when 61 percent favored affirmative action (Montenaro, 2013).

There is also a racial divide regarding affirmative action, with blacks and Latinos more inclined to favor preferential measures in work or college admissions. In the above cited poll, for example, six in ten whites opposed affirmative action, while eight in ten blacks and six in ten Hispanics favored it. But to see the issue as a fundamental split between whites and racial minorities—whites oppose affirmative action, racial minorities support it—is too simple. In one national survey, when asked if college applicants should be admitted solely on the basis of merit, even if the result would be fewer minority students being admitted, 75 percent of non-Hispanic whites answered yes; but 59 percent of Hispanics also answered yes, and even a significant percentage of blacks (44 percent) agreed (Moore, 2003). Racial attitudes, then, are not necessarily the critical—much less the only—factor comprising this issue. People may oppose or feel resentment toward affirmative action policies not because they are antiblack or anti-Hispanic or wish to keep things as they are (surely many do oppose them for these reasons) but because they see them as fundamentally in violation of the value of fairness.

FAIRNESS AND OBJECTIVES   The question of how compensatory measures can be made fairer has been subject to much debate. Some maintain that instead of race or ethnicity, social class should be the major criterion by which people are deemed eligible for compensatory advantages (Kahlenberg, 1996, 2007; Keller, 2013; Michaels, 2006; Schrag, 1995), or that policies should be applied in a "race neutral" fashion (Wilson, 1994). Orlando Patterson (2003) has suggested that in addition to an economic means test, all immigrants need to be excluded from the policy. Others, however, maintain that such measures bring with them the same kinds of inherent problems as those based primarily on race and ethnicity, that is, issues of determining who should be eligible and how they are to be chosen (Hacker, 1994; Kinsley, 1991).

Another confounding aspect of affirmative action concerns the very objective of such policies. Originally, they were intended to provide a temporary advantage to those racial and ethnic minorities who had been the historic victims of systematic discrimination. As they evolved, however, their purpose was transformed into an effort to create schools and workplaces that resembled proportionately the racial and ethnic makeup of the society, state, or community. The question, then, is whether their purpose is to help bring about greater equity among the society's diverse groups or to create more ethnic diversity within the society's major institutions. Efforts at attaining one of these objectives do not necessarily correspond to attainment of the other.

THE FUTURE OF AFFIRMATIVE ACTION

Whether or not affirmative action is sustained through legal actions in the future, it is unlikely that preferential measures in some form will entirely disappear. Transnational corporations, for example, understand that recruiting more minority managers is simply good business in an increasingly ethnically diverse society as well as a global economy (Glater and Hamilton, 1995). Moreover, they recognize that their future labor force will be far more diverse in race and ethnicity than in the past. Universities, too, have recognized the need for an ethnically diverse campus and have taken steps beyond direct affirmative action programs to assure that. But if compensatory policies in some form are likely to remain in place, continuing political and social pressures make it even more likely that they will be recurrently subject to legal and political debate and public scrutiny. Some states, for example, have put the issue to the ballot. By the late 2000s, voters in eight states had already passed a ban on the use of racial, gender, or national origin affirmative action in college admissions, hiring, and contracting.

Affirmative action, in sum, has proven itself one of the most vexing issues of race and ethnicity in America. Yet, it is important to consider the widespread use of such measures in other multiethnic societies (Parikh, 2001; Sowell, 2004; Teles, 2001). Preferential policies designed to bring about more equity among diverse groups have been adopted in countries as divergent as Australia, Britain, India, Nigeria, Malaysia, and, as we will see in Chapter 15, Brazil. Thus we should not think that the problems revolving around this issue are unique to the United States. As countries everywhere are transformed into multiethnic societies, mitigating the inequitable distribution of societal resources among various ethnic groups has become an inescapable predicament.

LOOKING AHEAD

The United States, as we have now seen, has for the past few decades been in the midst of a literal transformation of its ethnic system. Several trends have become evident.

- The racial/ethnic configuration of the United States has become more diverse and complex as new elements, primarily Latino and Asian, make up a sizable and growing proportion of the population.
- The traditional binary, black-white, racial division no longer has relevance as these new groups take their place in the society and in the process complicate racial and ethnic categories.
- Increasing diversity and continual ethnic intermixing make traditional racial and ethnic classification schemes less meaningful and employable.
- The cultural dominance of Euro-Americans continues to prevail, but it is subject to new influences that are moving it in a more multicultural direction.
- Ethnic inequality remains tenacious, specifically the division between Euro-Americans and most Asian Americans, on the one hand, and African Americans, Latinos, and Native Americans, on the other.

Do these social and economic trends portend more contentious and conflictual ethnic relations in the future or might they impel social forces that lead to more equitable and harmonious relations?

A Negative View

Public opinion regarding racial and ethnic issues has vacillated during the past several decades. National surveys have revealed both pessimism as well as more positive evaluations of current and future ethnic relations. A more detailed examination of opinion, however, reveals that long-term ethnic relations are not seen the same way by members of each group. In 2006, when asked, for example, if black-white relations "will always be a problem" in American society or that "a solution will eventually be worked out," a slight majority of blacks (55 percent) answered "always," whereas almost the same percentage of whites (56 percent) answered that a "solution" would be found. These results were not significantly different from what they had been ten years earlier (Jones, 2006).

Equally important, surveys continue to indicate that whites' perception of the condition of social and economic life for racial and ethnic minorities is quite different from the view of the minorities themselves. Blacks and Latinos are generally much less positive than whites about the way ethnic minorities are treated in American society (Associated Press, 2012; Jones, 2006; New America Media, 2007). For example, whereas 57 percent of whites feel that there is at least some discrimination against African Americans, 88 percent of blacks feel that way (Doherty, 2013).

Also of concern is an increasingly apparent division among racial-ethnic minorities themselves. A survey of blacks, Latinos, and Asians revealed the persistence of negative stereotypes and a relatively low level of personal interaction between the groups (New America Media, 2007). These schisms are particularly evident where groups are in competition for jobs and education. In parts of the South, for example, where a growing Latino population has begun to challenge African Americans for blue-collar jobs, strained relations between the two groups have intensified (Swarns, 2006).

Resurgent nativism and new forms of racism have also become evident in recent years. The basis of the new nativism is several-fold, but, as explained earlier, its major source is the changing ethnic composition of American society, driven by continuing immigration from non-European societies. Economic decline, along with increasing public awareness of the country's changing demography, have created fertile ground for opportunistic politicians and commentators to play upon white fears.

The fears of Euro-Americans that they are losing ground to nonwhite ethnic populations have also been inspiration for the appearance among a sizable portion of the population of a new racism, specifically in the form of white resentment. Recall the "group position" theory of prejudice and discrimination described in Chapter 3. When the majority group's position of dominance on the ethnic hierarchy is perceived as threatened, the response is apt to be an increase in anti-minority attitudes and behaviors. Anti-immigrant movements are obvious developments, but fears of displacement by minorities in positions of political and economic power are also likely to be aroused.

In the early twenty-first century, nothing better symbolized an impending change in the U.S. racial/ethnic order than the election of an African American president. In the tumultuous political atmosphere of the time, right-wing opponents of President Obama responded by stoking white resentment, concocting outlandish scenarios of future white disadvantage, and questioning the president's legitimacy and loyalty, often reverting to coarse racist images and slogans. Even his re-election failed to reduce the virulent attacks by political opponents. Moreover, the fact that a self-identified African American occupied the White House did not seem to modify racial attitudes. In fact, a 2012 national survey indicated that a slight majority of Americans expressed prejudice toward blacks, an increase since Obama had first been elected in 2008 (Associated Press, 2012). Further indication of white resentment of progress by racial minorities is found in a 2011 study, which indicated that many whites now see the racial pendulum having swung so far that blacks are now benefiting at their expense, in a kind of zero-sum game (Norton and Sommers, 2011).

Enduring racial attitudes were also brought to the fore by the 2012 killing of Trayvon Martin, an unarmed black youth walking in a predominantly white neighborhood. What that case revealed was the reality of racial profiling and the continued widespread negative stereotyping among whites of African American males.

Finally, the continuing racial/ethnic gap alluded to earlier, specifically between African Americans, Latinos, and Native Americans low on the class hierarchy and the remainder of the population, gives little indication of closing. Moreover, the economic downturn beginning in 2008 took a particularly heavy toll on the jobs, income, and wealth of these already-disadvantaged groups, making their plight even more distressed than in past eras. Indeed, these groups appear to be losing ground economically and, as a result, face the despair of high joblessness and low income for decades (Austin, 2012; Herbert, 2010; McKernan et al., 2013; Weller et al., 2012).

A Positive View

Given the divergent public views of ethnic relations in general and the even greater split in views of how ethnic minorities fare, it is reasonable to assume that ethnic discord at some level will continue to be a persistent feature of American society well into the twenty-first century. Still, there is room for at least some cautious optimism. This view is predicated on five points.

First, we should remember that there have been few times in American history when ethnic issues have not been in the forefront of public consciousness and the focus of public policy. We should therefore not think of ours as a unique era. Moreover, Americans, when asked their view on the future of race and ethnic relations, generally respond with more optimism than pessimism (Taylor, 2008).

Second, although the United States' pluralistic trend and the accommodative responses to it by its major institutions are the source of continued public debate, there is indisputable evidence of a growing ethnic tolerance in the society as a whole. A poll conducted in 1964 and again in 1989 asked a national sample of Americans to rank different American ethnic groups. In both years, the rank order of the groups corresponded closely to the social distance hierarchy that emerged from the earlier Bogardus studies of social distance (described in Chapter 3), indicating much continuity in the social standing of American ethnic groups: the groups closest culturally and racially to the dominant group occupied the highest positions. Perhaps equally significant, however, is that the score for almost all groups rose during the intervening twenty-five years. Especially large increases were noticeable for blacks and Asian American groups. Recall as well the demise of crude racist ideas among whites vis-à-vis racial minorities, noted in Chapter 7.

Another indication of positive change in the trend of American ethnic relations can be detected in the response to the attacks of September 11, 2001, and afterward. In the weeks and months following the catastrophe, anti-Arab sentiment ran very high, as was described in Chapter 12, and Arab Americans were the targets of verbal and in some cases physical abuse. Public opinion polls, moreover, confirmed the readiness of Americans to subject Arab Americans to special security measures, much in the fashion of racial profiling. In the wake of the society's war atmosphere, Arab Americans' fear of being targets for severe discrimination, comparable to the way German Americans were dealt with during World War I and Japanese Americans during World War II, was well grounded. But unlike the tragic actions of those past cases, government officials as well as media, education, and business leaders made direct and forceful appeals to the American public for tolerance, specifically to disassociate the terrorist acts from Arab Americans and Muslim communities in the United States.

These were not overly successful, however, in reducing anti-Arab and anti-Muslim sentiments among a significant part of the American population and preventing the egregious violation of civil rights in numerous cases. More than a decade following the attack, this remains a troubling aspect of American ethnic relations. Nonetheless, there appears to be a heightened awareness of this issue that to date has contained its potentially explosive nature. As we saw in Chapter 12, anti-Muslim attitudes and actions have been more prolific and violent in European countries than in the United States.

Third, despite what at times seems a hopelessly racially and ethnically divided society, the United States may in fact be more socially and politically cohesive than is usually assumed. Closer analysis reveals that on basic values regarding issues such as teaching a common American heritage and values, there is overwhelming approval by all groups, including immigrants (Etzioni, 1998, 2004). Moreover, ethnic leaders often convey the view that race or ethnicity is the sole or major determining factor in shaping people's perspectives on social and political issues. But this view ignores more fundamental class differences that, as we discovered in the preceding few chapters, internally divide popularly conceived ethnic and panethnic categories.

Fourth, intermarriage across racial and ethnic lines, as we have seen, continues, with few exceptions, to rise unabated. More than one-fifth of all American adults say that they have a close relative who is married to someone of a different race and one in seven new marriages is interracial or interethnic (Passel et al., 2010; Pew Research Center, 2006b; Wang, 2012). Moreover, almost all those referred to as "Millennials"—eighteen- to twenty-nine-year-olds—express support for interracial marriage within their families, a level of acceptance significantly greater than in other generations. As well, a majority of Millennials say that at least some of their friends are racially different from themselves (Pew Research Center, 2010).

Thus the social walls separating racial and ethnic groups continue to crumble, and as they do, the division of the society into conveniently demarcated racial and ethnic categories continues to lose meaning. The United States may be entering a postethnic age in which clearly perceived racial and ethnic groups blend into a "beige continuum" (Kington and Nickers, 2001). "The blurring of major ethno-racial boundaries," writes Richard Alba, "is a plausible prospect for the near future" (2009:225). And, rather than being arbitrarily placed into ethnic pigeonholes, individuals may have much greater latitude in choosing their ethnic affiliations (Hollinger, 1995). It is not unreasonable to assume that the hierarchy that currently defines race and ethnicity in American society will be fundamentally reconceptualized in the next few decades.[2]

Finally, as we will see in the next three chapters, the United States is not alone in its ethnic problems. Indeed, few societies of the contemporary world do not face equally daunting problems stemming from an expanding ethnic diversity and its attendant intergroup conflict. And, as will be revealed, the level of American conflict is relatively mild compared to that of many other ethnically divided societies.

SUMMARY

- An issue of major importance in U.S. race and ethnic relations concerns the large-scale immigration that has occurred since the mid-1960s. During this period, more immigrants have entered the society than at any time since the classic period of immigration in the early years of the twentieth century. Most important, the vast majority of the current immigrants are non-European, coming mostly from Latin America and Asia.

---

[2] Some have envisioned a future U.S. racial/ethnic order as simply "nonblack/black." In this scenario, most lighter-skinned and upwardly mobile Latinos and Asians will eventually be absorbed into the "white" category as they assimilate and are redefined as "white," similar to the course followed by southern and eastern Europeans in the early twentieth century (Gans, 2007; Lee and Bean, 2007). Others have suggested that the U.S. racial/ethnic order will come to more closely resemble the Brazilian and Latin American forms (which will be described in Chapter 15), with an intermediate color group defined as neither black nor white (Bonilla-Silva, 2004; Bonilla-Silva and Glover, 2006; Frank et al., 2010). Still others see an as yet undefined "middle" of the U.S. racial/ethnic order—Latinos and Asians—not moving inexorably toward either white or black poles (O'Brien, 2008).

- The influx of a new immigrant wave has prompted much social debate regarding its economic impact and the social integration of the immigrants. Proponents of liberal immigration policies argue that immigrants bring many benefits to the society and have a generally positive economic and social impact. Those who advocate more stringent policies contend that immigrants have a negative economic impact and exert a potentially divisive social influence.

- Immigration has been one of the most virulent issues of national politics for the past decade. Americans have seemed to favor more restrictionist measures, though public opinion has vacillated. Political groups at the national level and in states and communities have contested this issue repeatedly.

- A second major racial/ethnic issue concerns the question of whether American society is—or should be—moving toward greater cultural assimilation or pluralism. Unquestionably, the United States today is a more ethnically diverse society than at any time in its history. Moreover, there appears to be greater tolerance for that diversity, with various societal institutions acknowledging and even sanctioning the expression of ethnic differences. Whether in the long term these differences will dissipate or be sustained is not yet clear.

- The third issue of American race and ethnic relations concerns the continuing socioeconomic gap between Euro-Americans and most Asian Americans, and African Americans, Latinos, and Native Americans. To address this problem, starting in the late 1960s government and other institutions introduced preferential policies aimed at creating greater ethnic equity. During the past five decades, affirmative action has been the subject of much controversy among politicians, scholars, and the general public. An ongoing debate concerns the fairness of these programs and whether they do, in fact, accomplish their ostensible objectives.

- Negative prognoses of the future of race and ethnic relations in the United States stress the continuing economic and social divisions between Euro-Americans and Asian Americans on the one hand and African Americans, Latinos, and Native Americans on the other; the failure of most whites to perceive the actual socioeconomic condition of racial/ethnic minorities; resurgent nativism in response to continuing immigration of mostly non-Europeans; and new forms of racism based on white resentment of advances made by racial and ethnic minorities.

- Positive views of the future of race and ethnic relations stress evidence of greater ethnic tolerance in general, particularly among younger generations; greater social and political cohesion than is usually recognized; rising rates of intermarriage; and the gradual fading of ethnic boundaries.

CRITICAL THINKING

1. If you were to lead a group of foreign students on a cultural tour of the United States, where would you take them to show them the "mainstream" U.S. culture? Or is it no longer possible to define the mainstream in today's multicultural America?

2. Multiculturalism seems to have become the prevalent ethnic ideology in the United States as taught in schools and universities. What has brought about this change from what in the past had been a predominantly assimilationist ideology, specifically Anglo-conformity?

3. Examine the prevalence of immigrants in particular sectors of the U.S. economy. Then explain how immigration is critical to the society's present and future economic well-being.

4. Some sociologists have suggested that the future racial/ethnic system in the United States will be a two-part one, "black and nonblack." This is different from the more recent model, which is essentially "white and nonwhite." Explain the difference between these two models and how such a change in racial/ethnic classification might emerge.

Personal/Practical Application

1. In learning a foreign language, students are implicitly exposed to another culture, different in many ways from their own. If you have studied a foreign language, has it affected your views of cultural differences among ethnic groups and your level of ethnic tolerance?

2. In your view, should the purpose of affirmative action or other compensatory policies (based on race, ethnicity, or social class) be to provide more educational and occupational opportunities to members of disadvantaged groups, or to assure greater diversity in schools and the workforce? How would you defend your argument?

ETHNIC RELATIONS IN COMPARATIVE PERSPECTIVE

Having explored patterns of race and ethnic relations in the United States, we now examine some non-American cases as a basis for comparison. People often assume that what occurs in their own society is unique and that others' societies are simply not like theirs. Conversely, they sometimes assume that what happens in their own society is characteristic of human societies in general. In part, both of these suppositions are valid. To ascertain in what ways U.S. society is similar to and in what ways it is different from others requires that Americans transcend the limited perspective imposed by the social system with which they are most familiar and comfortable. That is, they need to adopt a broader, less parochial approach to human affairs.

The societies examined in this part lend themselves readily to comparison with the United States. It will not be difficult to detect similarities, nor will the differences be obscure. Each is part of a specific geographic region, is a product of a distinct historical tradition, and displays variant bases of ethnic stratification and conflict.

The patterns of race and ethnic relations in each of these societies represent one of the major types described in Chapter 4. South Africa, the focus of Chapter 14, resembles in its recent history the colonial (or extreme inequalitarian pluralistic) type; Brazil, examined in Chapter 15, is an example of the assimilationist type. In Chapter 16, Canada will be seen as a society with many features of a corporate pluralistic, or multicultural, system. In Chapter 17, we explore the rising significance of ethnicity across the globe as nations everywhere, through immigration and political turmoil, become more ethnically diverse. We look first at the impact of this new heterogeneity on western European countries. This will provide us with an opportunity to compare their relatively fresh encounter with ethnic conflict and change to the American experience, in which racial and ethnic issues have been a wellspring of social friction and strife for many generations. Focus then falls on three societies, each of which has experienced extreme ethnic conflict in recent times. Although these particular cases have no U.S. historical counterpart, they nonetheless force us to consider what can happen when fear and distrust of ethnic out-groups become the preeminent characteristics of intergroup relations in a multiethnic society.

SOUTH AFRICA

Society in Transition

Little more than twenty years ago, South Africa, in its system of ethnic relations, was unique in the modern world. As an inequalitarian pluralistic society, it had no peer. Despite persistent intergroup conflict and hostility, almost all multiethnic societies in the post–World War II era adopted policies intended to strengthen ethnic unification. South Africa, however, moved in the opposite direction, compelling greater divisions among its ethnic populations. What made South Africa most singular among contemporary societies was that its racist system was fully legitimized and enforced by the state. South Africa, as we will see, has been a congeries of societal inconsistencies and contradictions. Throughout most of its history, however, one thing was clear and unambiguous—the dominance by whites, the numerical minority, over blacks, the numerical majority.[1] No area of South African life was untouched by the racial policies and ideology of the dominant white group.

Today, South Africa is moving on what appears to be an irreversible course toward a more equalitarian system. Indeed, the scope and depth of change this society has experienced in the past two decades would have been unimaginable just a few years earlier. South Africa, therefore, provides an instructive case of the fluidity of ethnic relations and how they can change dramatically in a relatively short period of time, no matter how rigid and implacable they may seem. Despite the magnitude of change, however, the ultimate shape of the new South Africa as a multiethnic society remains uncertain. As a result, its racial and ethnic system will engage the attention of the world for many years to come. Furthermore, although its system of ethnic separatism has been dismantled officially, enough of it endures informally to make

---

[1] The ethnic terminology used in describing South Africa's population has not been consistent, and as in the United States, it has political connotations. Traditionally, black Africans were called *Bantu*, a name corresponding to their languages. During the apartheid regime, they were officially called *blacks* but referred to themselves as *Africans,* a term used by most scholars as well. Some use the term *black* to describe the entire nonwhite population, including Coloureds and Asians, as well as Africans. Today *black African* or *African/black* is used commonly to refer specifically to Africans, and that usage is adopted in this chapter. As the new South Africa evolves, ethnic terms will likely take on new meanings (see Adam, 1995; Horowitz, 1991; Lever, 1978; Thompson, 2001; and van den Berghe, 1967).

this society still an extraordinary case in its extremes of ethnic division and inequality.

As we look at South Africa, it is important to keep in mind that events of change have moved so swiftly in recent years that any description of its prevailing institutions risks almost immediate obsolescence. South Africa's significance as a case study of race and ethnic relations, however, lies as much in its history as in its current affairs. Furthermore, its system of extreme ethnic separation and stratification is not something of the distant past but has only recently been successfully challenged. It is fresh enough, therefore, to warrant a careful examination and to be instructive to other contemporary multiethnic societies.

THE DEVELOPMENT OF ETHNIC INEQUALITY

South Africa's modern ethnic drama began to unfold in the 1980s; extraordinary social and political changes have taken place continually since then. However, the new South Africa that has emerged can be understood only if these changes are positioned alongside this society's long heritage of intergroup conflict. As in other multiethnic societies, racial/ethnic stratification in South Africa has been the result of a historical sequence of diverse groups coming together, viewing one another ethnocentrically, competing for scarce resources, and ultimately forming a hierarchy in which one imposes its superior power on others.

WHITE SETTLEMENT

Historical evidence indicates that what is today South Africa was inhabited by Bantu-speaking peoples, the society's major African linguistic group, as early as the sixteenth century and probably much earlier (Wilson and Thompson, 1969). Modern South African history, however, involving the confrontation of European and native African peoples, begins with Dutch settlement in 1652. At that time, the Dutch East India Company established a colony in Cape Town as a refreshment station for its ships traveling to and from India. The indigenous people of the area, the Khoikhoi (Hottentots) and the San (Bushmen), were unable to satisfy the labor needs of the colonists who therefore began to import slaves from other parts of Africa and from the Dutch East Indies only a few years after their original settlement. The colonists' negative view of these slaves did not prevent widespread miscegenation; hence, a mixed racial group was created whose descendants are today called Coloureds.

THE TREKKING MOVEMENT As the colony grew, many of the Dutch, German, and French Huguenot colonists chose permanent settlement. At the beginning of the eighteenth century they began to leave Cape Town, establishing farms on land farther into the interior. Thus began the penetration of the northern and eastern frontiers, a movement called "trekking." These settlers became known as Boers, a Dutch word meaning "farmers." The trekking movement lasted 150 years, with the Boers continually pushing farther and farther inland. As they moved into the interior, the major patterns developed that would ultimately affect the shape of South African society—conflict with the indigenous peoples and the emergence of a unique white South African ethnic group.

The Boers came into conflict not only with native African peoples occupying these lands but also with migrating African tribes moving across the continent, westward and southward. These were Bantu-speaking tribes, who were well organized and who presented serious resistance, unlike the Khoikhoi and the San, who had been more easily subdued. The clashes over land and cattle that developed between the Boers and the Bantu, called Kaffir wars, lasted for many decades before the dominance of the Boers was finally established. The disdain for the natives based on their racial features was exacerbated by the bitterness that resulted from these encounters. Additionally, because they greatly outnumbered the Boers, the Bantu were seen as a constant threat whose suppression had to be ensured. Many of the racial attitudes of white South Africans in modern times can be traced to these initial contacts with Africans in the interior.

The Emergence of Boer (Afrikaner) Ethnicity   The second major consequence of the trekking movement was that the Boers increasingly distanced themselves from their European roots. As the settlers pushed inland, Dutch control over them waned. The farther they removed themselves from Cape Town, the more they became "a law unto themselves" (Thompson, 1964:186). As subsistence farmers, they disengaged themselves even further from intercourse with the economic and political institutions of the colony and its civil administration.

With their economic and political isolation, the Boers developed a culture thoroughly distinct from their European heritage. Differences were most apparent in language and religion. A unique Dutch dialect evolved over several generations, producing eventually the Afrikaans language, today only remotely related to the original Dutch. Divergent religious beliefs also contributed to the cultural distance between the Boers and their European ancestors. The Boers resisted the enlightened Protestantism that was emerging in Europe in the late seventeenth and early eighteenth centuries, clinging instead to a primitive Calvinism, whose ideas of predestination and rugged individualism seemed a fitting complement to the pioneering conditions of life in rural South Africa. They increasingly saw themselves in a biblical light as "chosen people," destined to conquer the frontier against the numerically superior Bantu and later the colonial British (Fredrickson, 1981). The chosen people theme continued to run strongly through the ideology of the Boers (today called Afrikaners) and was reflected in their stubborn resistance to changes that threatened to reduce their power.

By the late eighteenth century, the Boers were a unique social entity with a culture and language distinct from any other. They were geographically and intellectually isolated, and their European roots, notes historian Leonard Thompson, "were almost completely severed" (1964:187). Thus was born the Afrikaner ethnic group of South Africa. The Afrikaners' knowledge of and ties to Europe were so minimal that they had become, in effect, "white Africans." Unlike other European settlers on the African continent, the Afrikaners did not see themselves, nor did they even function, as colonials representing and maintaining strong cultural and political ties to a motherland.

This sense of isolation and permanence explains much about the determination of Afrikaners in later years to defend their way of life, and what they came to see as their territory, against the threat of black rule. Not until the late 1980s were the

Afrikaners compelled to alter that view. As we will see, Afrikaner nationalism was tempered and finally gave way to a recognition that only by sharing societal power with those groups it dominated so completely for most of South Africa's history could it survive in the modern world.

BRITISH VERSUS BOERS

Whereas the first century and a half of modern South African history is largely a chronicle of confrontation between white and black—Afrikaner and African—the following 150 years are marked principally by the political rivalry between the Afrikaners and the colonial British.

THE BRITISH ENTRANCE   The British entered the South African scene at the start of the nineteenth century when, in a move designed to protect their colonial interests in the East, they took possession of the Cape colony. The next fifty years saw the development of a competitive and often bitterly hostile relationship between the British and the Afrikaners that, although moderated, would remain an essential element of South Africa's ethnic dynamics. The British were not a large settler group like the Boers, and they brought a strikingly different political and social temperament to South Africa, more liberal and cosmopolitan. Their values clashed with the stern parochialism of the Afrikaners, and a wide gulf opened between them.

Not long after their arrival in South Africa, the British instituted liberalized policies regarding race relations, which exacerbated the already strained relations with the Afrikaners. The British attitude toward the native Africans was no different from the Afrikaners' in its substance—white dominance. But the British saw black Africans in a more paternalistic frame, whereas the Afrikaners saw them as a threat to their ultimate survival as a people. "The abstract difference between British imperial and Afrikaner republican conceptions of how to rule Africans," explains historian George Fredrickson, "might be described as a conflict between the trusteeship ideal and what the Boers called *baaskap*—which in essence meant direct domination in the interest of white settlers without any pretense that the subordinate race was being shielded from exploitation or guided toward civilization" (1981:193). The practice of white supremacy by the Afrikaners was thus consistent with their fundamental belief that force and exploitation were perfectly legitimate means to protect their group interests.

THE SECOND TREKKING MOVEMENT   In 1834, slavery was abolished throughout the British empire. This act was followed during the next two decades by the lifting of other legal restrictions against nonwhites. These radically new racial policies acted as a major incentive for the second and far more substantial migration of Boers into the interior. In 1836, a sweeping movement was launched in which one-quarter of the entire Afrikaans-speaking population would eventually set out on their own into the interior (Robertson and Whitten, 1978). This was the commencement of the Great Trek, comparable in many ways to the westward movement of Americans in the nineteenth century. As the western frontier had served as a kind of social safety valve in the United States, the Boer trek northward and eastward after 1836 provided a means of escape for those who found the British presence a constant source of anguish.

These Voortrekkers, as they were called, were doggedly pursued by British colonial advances, however. After fierce battles with the Zulu tribe in 1838, Boers founded the republic of Natal, which was subsequently annexed by Britain in 1843. Pushing farther northward, the Voortrekkers founded the Orange Free State and the Transvaal, which were eventually recognized as independent states by the British. These states became the heart of the Afrikaner culture, and within them the sense of autonomy and group solidarity was developed to its fullest. In both, the notion of white supremacy was a fundamental principle.

THE BOER WARS The discovery of diamonds around Kimberley in 1867 and of gold in the Transvaal in 1886 once again spurred British advances. Britain had earlier attempted to extend its influence into the Transvaal, resulting in the first Boer War. Although the Boers successfully resisted at that time, the influx of British railroads, mining interests, and workers following the gold discovery led to a more serious challenge to Boer autonomy. The British finally intervened with military force, provoking the second Boer War in 1899, a conflict that would leave permanent scars of bitterness between the two groups. The Boers used guerrilla tactics against the larger and superior British forces; in response, the British interned the Boers in concentration camps where twenty-six thousand, including women and children, died of disease. This is an event of enormous significance in Afrikaner history, symbolizing the martyrdom of the Boer people and their determination to survive against all odds. This second encounter with the British was a key factor in the subsequent development of militant Afrikaner nationalism. As we will see, though they were defeated in war, the Afrikaners eventually asserted political ascendancy and emerged as South Africa's dominant ethnic group.

Unlike the Afrikaners, the British never coalesced into a nation-like unit in South Africa. Their loyalties were fragmented between South Africa and the British Commonwealth (Thompson, 1964). As a result, Afrikaner dominance—though repeatedly diluted by British incursions in education, immigration, and especially the economy throughout the nineteenth century and part of the twentieth—was eventually entrenched and solidified. Although they had been militarily successful in the second Boer War, the British began to lose influence afterward and reached a tenuous political compromise with the numerically superior Afrikaners.

In 1910, the four territories of South Africa—the Cape, Natal, the Orange Free State, and the Transvaal—were merged into a union, at which time the foundation of the society's racial laws was secured. Although the subordination of blacks had always been informally understood and enforced, during this period a succession of government policies institutionalized many segregationist and inequalitarian practices, particularly those regarding participation of Africans in government and the economy.

THE AFRIKANER ASCENDANCY

The modern period of South African history can be said to have begun in 1948 with the political victory of the Afrikaner-led National Party. This ended the almost half century of delicate balance between British and Afrikaner interests and established the Afrikaner political ascendancy. The National Party represented the most extreme element of Afrikaner nationalism, viewing itself as the spearhead of a God-ordained mission to shape South Africa in the interests of Afrikaners and to throw off the yoke of British influence. With the withdrawal of the Union of South Africa from the British Commonwealth in 1961, South Africa's British ties were severed once and for all.

Most important, the Afrikaner-dominated government began to formally erect an *institutionalized system of white supremacy*, called **apartheid**, the dynamics of which we will look at shortly. While other African nations of the post–World War II era were moving toward national unification and a breakdown of tribalism, South Africa was moving in the opposite direction, creating a social order of strict separation between whites and blacks and encouraging the retention of tribal cultures among the African populace. In the modern period of South African history, the Afrikaner-led whites presented to the world a system of formalized ethnic separatism unmatched in rigidity and scope.

ETHNIC STRATIFICATION

In no other modern multiethnic society was ethnic stratification as clear-cut as in South Africa, and in no other was it so inflexible. Until the official demise of apartheid in the early 1990s, the boundaries between ethnic strata were, by law, relatively impermeable. South Africa's racial system was castelike, sustained by endogamy and by the maintenance of separate institutional systems for each ethnic group. Only in the larger economic system did the different ethnic groups come together, and there, as in all other societal areas, the intergroup relations were grossly inequitable. Indeed, along with its rigidity, the other major feature of the South African system of ethnic stratification was the enormous gap between the dominant white group and the several subordinate black groups—a gap that, as we will see, remains very much in evidence today. South Africa, then, was the prototypical inequalitarian pluralistic society, in which only the economic system served as a kind of social glue holding the various groups together.

Undergirding and enforcing the system of ethnic inequality was the South African state—dominated by whites, specifically, the Afrikaners. So complete was white control of the society's political apparatus that blacks were allowed no meaningful participation and were accorded few legal rights. Pierre van den Berghe (1978) referred to the South African sociopolitical system as a *Herrenvolk* democracy: a state that provides most democratic features of political rule to whites while ruling blacks dictatorially. Others described the system as a "race-oligarchy" (Adam, 1971b) or a "pigmentocracy" (Thompson and Prior, 1982). All these appellations denote the fact that skin color was the single overriding criterion of societal power in South Africa.

As was explained in Chapter 2, in almost all multiethnic societies, ethnic status is a significant factor in the distribution of life chances. The allocation of justice, health, occupation, education, living quarters, and so on depends, to some degree or another, on one's ethnic classification. But in no modern society has that classification been so critical in determining people's social place as it was in South Africa under apartheid. To ensure white dominance in all areas of social life required a formal division of the population into specific ethnic categories—whites, Coloureds,

Asians (mostly Indians), and Africans. Though no longer officially recognized, these categories, informally, continue to divide South Africa.

THE WHITES

Whites (or Europeans) constitute less than 10 percent of the population (Table 14.1) but, until 1994, completely dominated government. As the only group with effective political rights, whites for decades filtered all social rewards, siphoning off their disproportionate share before distributing the remainder to the three subordinate groups. Although they no longer dominate the political system, whites continue to maintain key positions of power in the economy and own most of the society's wealth. In an economic sense, they remain the dominant group despite the fact that they have been displaced as the political ruling group.

THE WHITE ETHNIC DIVISION   As previously noted, however, whites are by no means a culturally or even politically unified group. Rather, they comprise two major elements: Afrikaners and the English-speaking, most of the latter of British origin. The two major white subgroups differ not only in language and culture but also, to some degree, in class status. With the victory of the National Party in 1948, Afrikaners for almost four decades dominated all aspects of the political system. The English-speaking whites traditionally maintained the strongest influence in the society's economic institutions—trade, finance, mining, and manufacturing. The English, compared with the Afrikaners, traditionally occupied higher-ranking jobs, were better educated, and were in general wealthier (Adam, 1971b). These discrepancies, however, have diminished in recent years.

The English, whose South African roots do not extend as far back as the Afrikaners', are less cohesive and do not display the same sense of ethnic solidarity. Moreover, though most are of British origin, among the English-speaking whites there are also other, smaller, ethnic groups of European origin, including a Jewish community (Shimoni, 2003).

The Afrikaner-British conflicts of the nineteenth century left a residue of intergroup animosity that continued to manifest itself for much of the twentieth century. On the one hand, Afrikaners resented the economic role of the English, and anti-British sentiment was a staple of Afrikaner nationalism for many generations. On the other hand, the English-speaking whites viewed the Afrikaners as parochial and unenlightened. Other than business and politics, social contacts between the two groups were minimal (Brown, 1966; Lemon, 1987; Thompson and Prior, 1982).

Language was a particularly important factor in separating the two white groups. Afrikaans, the language of the Afrikaners, traditionally was established

TABLE 14.1 | ETHNIC POPULATION OF SOUTH AFRICA

Total Population	Black African	Coloured	Indian/Asian	White
53,000,000	79.8%	9.0%	2.5%	8.7%

Source: Statistics South Africa (2013).

throughout the state school system, and when Afrikaners dominated South Africa politically, it was the language of government and politics. But in business and industry, dominated as they were by the English, the predominant language was English. The language division, however, transcended politics and commerce. For Afrikaners, it was a vital component of Afrikaner nationalism (Lemon, 1987).

THE CONVERGENCE OF WHITE INTERESTS   The class, status, attitudinal, political, and linguistic divisions between the Afrikaners and the English should not be overstated. First, with the ascendancy of the Nationalists in 1948, a significant closing of the gap in wealth and economic influence between the two groups gradually occurred (Van zyl Slabbert, 1975). Class differences, though real, were minimized by the primary importance of one's racial classification. Second, both groups benefited from the castelike system and were united in their desire to preserve white rule. This common interest was strong enough to override the political differences between them. Thus English opposition to apartheid stemmed mainly from the view that it was economically impractical and inefficient, not that it was morally objectionable (Banton, 1967; Lewin, 1963; van den Berghe, 1967). Although the social gap between the English and Afrikaners remained evident, the political differences between them diminished as white dominance was challenged (Horowitz, 1991).

AFRIKANERS AS A NEW MINORITY   Today, with apartheid ended, the political and cultural power of the Afrikaners is in decline. Afrikaners no longer play a major role in government, and the Afrikaans language is under attack in the schools and media, where English is now preferred (Vestergaard, 2001).[2] The civil service, once a stronghold of Afrikaners, now hires mostly blacks (Daley, 1998a). Clearly, this formerly dominant group has been transformed into a minority. And that minority status has fundamentally affected the very self-identity of Afrikaners, many of whom no longer even define themselves in this ethnic fashion. In his fieldwork in South Africa, Mads Vestergaard found that "[w]hen asked, individuals who formerly might have called themselves 'Afrikaner' now increasingly define themselves according to their profession, their geographical location, or simply as 'South African'" (2001:30). As one journalist describes them, many present-day Afrikaners "seem to be searching for a comfortable place in a black-majority society and still have not found it" (Wines, 2007b:A1).

THE COLOUREDS

Out of the sexual and sometimes marital unions of whites and blacks during the early settlement of the Cape was created an intermediate racial group called Coloureds. During the nineteenth century, the Cape continued to evolve along different social and political lines from the frontier settled by the Voortrekkers. As a result, for a long period following the abolition of slavery in 1834, the legal system of the Cape was essentially color-blind, allowing for a great deal of racial intermixing

---

[2] With the fall of Afrikaners as the dominant political group, Afrikaans is increasingly being displaced by English and African languages in all areas of social and political life (Giliomee, 2004; McLarin, 1995; Swarns, 2002).

(Fredrickson, 1981; Mason, 1970). Today, almost all of South Africa's nearly five million Coloureds live in the Western and Northern Cape provinces.

The Coloureds are perhaps the most peculiar element within the South African ethnic system because in all ways except their physical features, they are thoroughly Europeanized. Most speak Afrikaans, are Christian (mostly Protestant), and are in other cultural ways barely distinguishable from the Afrikaners. The sole, and obviously critical, distinction between the two groups is skin color. Many Coloureds, of course, are even physically indistinguishable from the Afrikaners.

The position of the Coloureds has been marginal to both whites and Africans not only physically but also politically and economically. During the apartheid regime, Coloureds maintained a few political privileges that—limited though they were—placed them apart from the Africans. The economic position of the Coloureds has also been somewhat privileged by comparison with that of Africans, though well below the white standard. Most are part of the unskilled proletariat, in both agriculture and industry, though there is a substantial middle class.

The facts that they are thoroughly Europeanized and under apartheid held at least a few token privileges created feelings of marginality to both white and African groups. As Dickie-Clark (1972) has noted, through apartheid they lost their earlier legal equality with whites but gained a somewhat more advantageous position over Africans and Indians. This accounts in some measure for their reluctance to identify or interact more closely with the Africans (Adhikari, 2005; Gevisser, 1994).

THE ASIANS

The bulk of South Africa's Asian population is made up of Indians, mostly descendants of indentured workers brought from India in the late nineteenth century to work in the sugarcane fields of Natal. Following their term of service, many remained and became traders, shopkeepers, and workers at various skill levels (Hart and Padayachee, 2000). Most still live today in Natal, in and around Durban, the province's largest city. Within the South African system they are, like the Coloureds, a marginal group—part of neither the white nor the African populations.

Ethnic Indians in South Africa are a complex group in both culture and class. They are of several linguistic origins and are further divided along religious lines. Within the Indian population is an entire range of classes, from wealthy merchants to unskilled agricultural and industrial workers. All, however, exist in the context of an institutionally separate community. Indians generally occupy a more privileged economic position than either Coloureds or Africans (in most occupational sectors, they earn more than Coloureds and far more than Africans), but most rank well below whites.

South Africa's Indians have functioned as a classic middleman minority. In Chapter 2, such minorities were described as occupying an intermediate economic status between dominant and subordinate ethnic groups. Anthony Lemon has written that "European attitudes to Indians have been characteristically ambivalent; employers found their labour useful, but traders and businessmen feared competition" (1987:245). For a long time the permanence of Indians in South Africa was questioned by the white power structure. Moreover, prohibitions on their movement in South Africa as well as against further immigration limited their size. At the same time that Indians have been viewed suspiciously and with derision by the whites, they have been seen by Africans as exploiters. One scholar, in fact, asserts that in general "Indians are more resented by most Africans than are Whites" (Moodley, 1980:226). By the same token, many Indians expressed fear of black rule following the demise of white rule, and tensions between the two groups have remained evident (McGillivray, 2006; Noble, 1994; Swarns, 2002).

THE AFRICANS

As part of the South African ethnic composition, the African population is by far the largest, almost 80 percent. Although all Africans historically shared a common societal powerlessness, they are not a single culturally unified group. Rather, they derive from several Bantu tribes, the largest among them the Zulu, Xhosa, and Sotho. Even though almost all Africans still speak a native tongue and retain certain other uniquely Bantu traits, the tribal culture for most is residual. Except for those who remain in rural areas, they are assimilated culturally into Western ways. They are at least nominally Christian and in urban areas speak either English or Afrikaans in addition to their Bantu language. Within the urban environment, tribal or subethnic boundaries become indistinct, and the common political and economic oppression they suffered in the past has impelled the formation of an encompassing black African ethnic group.

A crude economic hierarchy is evident among Africans, just as it is among the other three ethnic populations. Occupational differences are particularly acute between those living in urban areas—mostly industrial workers—and those living in rural areas—mainly subsistence farmers or agricultural workers. In addition, a substantial number of migrant laborers from neighboring African nation-states work in the gold and diamond mines.

Most Africans have occupied the lowest-ranking jobs in the labor structure, those requiring little or no skill. With the end of the official policy of racial discrimination, however, this traditional pattern is changing. Job opportunities for Africans have opened in significant numbers, and they have been afforded increasing access to higher education. As a result, the African middle class, growing even before the end of apartheid, is expanding rapidly.

These marked improvements in the status of black Africans, however, should not disguise the fact that, as we will see in a later section, the economic differences between blacks and whites remain vast. As one observer of South Africa described it not long after the end of apartheid, "[P]erhaps no country on earth has cultivated such a wide gulf between the country-club, luxury-car, boutique life style of white privilege and the wretchedness of black denial" (Keller, 1995:A6). That description continues to hold true today.

PREJUDICE AND DISCRIMINATION

In Chapter 4, inequalitarian pluralistic systems were described as having several key traits—separation and duplication of institutions among the society's various ethnic groups, extreme polarization between dominant and subordinate groups, high levels of prejudice and discrimination, and a vastly inequitable distribution of wealth.

The coordinating mechanism of such systems is the absolute authority and power of the dominant group. Under apartheid, the South African system epitomized all of these characteristics.

Starting in 1948 and for the next four decades, political policies of the Afrikaner-dominated National Party were designed to create and sustain inequality among South Africa's four ethnic populations. These policies were grounded in two overriding principles: (1) the subordination of nonwhites in all areas of social, political, and economic life; and (2) the maximum physical and social separation of the four groups, particularly the two major groups, white and African. The latter principle was justified by the premise that peace and order could be ensured "only if people of the various colours mix as little as possible" (Wiechers, 1989:8). So great was the incompatibility of black and white ethnic groups, it was felt, that contact between them could lead only to conflict; hence, minimizing contact would be the logical solution to this perceived problem.

THE DYNAMICS OF APARTHEID

Apartheid provided specific legislation mandating the residential separation of the racial groups, the maintenance of separate social institutions for each, and eventually the formal division of South Africa into separate and independent white and black nations. Indeed, so blatant and uncompromising was the South African system that the very term *apartheid* has, in recent times, come to be applied commonly outside the South African context, used to signify any rigid and durable type of ethnic separation and inequality.

In 1994, the Nationalist regime relinquished power in a peaceful transfer to a democratically elected black-dominated government. This officially ended apartheid. Although it is no longer formally in effect, examining this system of ethnic separation today is instructive in that we may see how extreme inequalitarian pluralism operated in a modern society. Based on an ideology of ethnic incompatibility, this was a complete system of ethnic separatism that shaped all aspects of social life. There have been few comparable cases in the contemporary world.

With the advent of black African political dominance, apartheid was officially dismantled, but its basic principles and underlying framework continue, in large measure, to guide interethnic relations unofficially and informally. Let's look more closely at several of the most important measures that shaped the apartheid system.

RACIAL CLASSIFICATION Where physical distinctions among people are not clear-cut, ethnic boundaries are difficult to define and enforce. To create a classification system that could be used to distinguish the racially dominant and subordinate groups, the Nationalist government soon after it came to power in 1948 installed several formal measures.

The first cornerstone of apartheid was the Population Registration Act of 1950, providing for the official racial categorization of all persons. Questionable cases were decided by race classification boards, made up of whites, who ruled on the identity of such people. Although the registration system was technically applied to everyone in South Africa, it was designed primarily to monitor the movement of Africans and was enforced mainly against them. A "passbook" comprising detailed identification papers was required of all Africans, and failure to produce it on demand was an automatic assumption of criminal status. Over the years, eighteen million arrests were made for passbook violations (Lipton, 1987). The passbook system was a particularly odious element of apartheid for Africans and was a constant reminder of their almost total subordination (Klaaste, 1984). The pass laws were abolished in 1986. Most important, the Population Registration Act was repealed in 1991, removing one of the fundamental legal props of the apartheid system.

PETTY APARTHEID   Physical contact between blacks and whites was minimized through a system of petty apartheid, or what van den Berghe (1967, 1979a) called "micro segregation," consisting of dozens of laws and mandates designed to maintain ethnic separation in almost every conceivable area of social relations. Restaurants, hotels, buses, trains, public toilets, waiting rooms, hospitals, schools, parks, beaches, theaters, and the like were for the most part segregated. Moreover, these separate facilities were invariably unequal in quality and quantity. Crowning the rules of petty apartheid was a policy of strictly enforced endogamy, prohibiting interracial marriage and even sexual intercourse between whites and blacks. In the realms of social life where people come into physical contact with each other, the South African apartheid system was strikingly similar to the Jim Crow system of segregation enforced in the pre-1960s American South. Unlike the Jim Crow system, however, there were no pretensions of "separate but equal."

Most of the measures of petty apartheid were officially repealed beginning in the 1980s. Neighborhoods, hospitals, public transportation, parks, beaches, and other public facilities were desegregated. But these official changes have not been implemented uniformly, nor have social attitudes, cast by decades of socialization to apartheid, been easily reversed. As a result, petty apartheid continues to guide many areas of social life. "To the casual visitor," noted a study group to South Africa in 1986, "apartheid may appear to be on the way out. In its essential elements, it remains very much intact" (Commonwealth Group, 1986:33). Today, that observation still seems valid. Increased interracial encounters do take place in work, school, and other public places, but these do not usually lead to social interaction at any but a formal, superficial level. Social behavior continues to be shaped by group stereotypes and traditional beliefs engendered by years of apartheid (Finchilescu and Tredoux, 2010; Gibson and Claassen, 2010; Lefko-Everett, 2012).

RESIDENTIAL SEGREGATION   A more fundamental and significant level of apartheid concerned the maintenance of separate living areas for whites and nonwhites. Officially mandated residential segregation, then, became another cornerstone of apartheid. Through the Group Areas Act of 1950, each racial group was assigned specifically demarcated living areas. Any area might be proclaimed "reserved" for a particular group, after which residents not belonging to that group had to move elsewhere. As a result, depending on their skin color, people might be arbitrarily removed and sent to their "appropriate" area. Because this measure was designed to assure racial integrity of white areas and to displace nonwhites from them, over the years its effects were felt primarily by the latter.

The designation of specific areas for each ethnic category was closely related to the South African labor system. Africans functioned as the country's unskilled industrial proletariat, and because South African industry was located in traditionally white areas, this spurred the migration of African workers into white neighborhoods. Because the industrial economy could not function without them, some provision had to be made to accommodate Africans without upsetting the ethnic purity of white residential areas. The solution was a migrant laboring system in which Africans were permitted to enter white areas to work but would be housed in separate communities, called townships, adjacent to the white cities. These townships served, in effect, as dormitories for Africans who labored in white areas. Africans who lived in the townships were officially only "temporary" residents or migratory workers, even if they were born in the urban areas and their ancestors had lived there for generations.

The township of Soweto, adjacent to Johannesburg, is perhaps the epitome of this purposefully designed system of residential segregation. Constructed to house Johannesburg's African migrant laborers, Soweto was actually larger than the white population of Johannesburg itself!

Years of restricted movement of Africans created severe social disruption among the urban African populace. In a classic illustration of the self-fulfilling prophecy, social conditions were produced among urban Africans that inevitably led to the decadent behavior commonly attributed to them. Forced to migrate to the cities to find employment, African men lived in squalor and were forbidden to bring their families with them. Demoralization and alienation consequently thrived among them, in turn leading to high rates of crime, violence, and alcoholism in the townships.

Restricting the movement of blacks into white areas was referred to by the Nationalist government as "influx control," and the passbook system was a key device in enforcing this policy. The system, however, failed to stem the urban flow of Africans, and the policy was scrapped in 1986. Nonetheless, residential segregation today remains the norm, in large measure because blacks in general lack the means to acquire housing in what have been traditional white areas (Christopher, 2005; McClinton and Zuberi, 2006).

THE HOMELANDS As the Nationalist government shaped apartheid, it envisioned the culmination of this racially divided system as the complete separation of the ethnic populations in independent nations. This policy of ultimate exclusion, or what van den Berghe (1967, 1979a) called "macro segregation," entailed the division of the society into geographic areas corresponding to race and ethnicity. As a result, certain areas, variously called "homelands," "Bantustans," or "black states," were set aside for Africans, based on the Nationalist version of historic tribal areas. Each homeland roughly corresponded, in this view, to an area originally inhabited by a particular Bantu tribe. The areas set aside for Africans were less than 14 percent of the land area of South Africa, leaving more than 85 percent for whites. All Africans, regardless of where they resided or where they were born, were declared "citizens" of one of the homelands, based on their ethnic or tribal origins. This included those living in officially designated white areas—the majority of the African populace.

When the homelands policy was formulated, the Nationalists expected that eventually all Africans in South Africa would live in one of these areas, each of which would be granted independence. Once this grand design was fully realized, it would be possible for the South African government to declare that there were no Africans in South Africa, only "guest workers" temporarily living in the remaining (white) areas of the country. In short, until all Africans could be "repatriated to their home-lands," those remaining in the white areas were to be considered transient laborers, migrants entitled to few legal rights or privileges. Sebastian Mallaby referred to the homelands as "the most ambitious injustice that apartheid ever devised" (1992:26).

From the outset, the homelands were fictional creations, constructed essentially to provide South Africa with a continued flow of cheap, exploitable labor and to serve as a convenient dumping ground for unemployed, aged, and infirm Africans (Carter, 1980). None of these areas was economically capable of sustaining itself independently of the South African economy. Moreover, Africans living in South Africa's cities had no cultural or even social ties to their official homeland. The Nationalist government acknowledged these facts by encouraging and even subsidizing industries to locate in adjacent border areas or in the homelands themselves so that jobs could be created for Africans living there (Stultz, 1980). None was ever recognized by any international body or nation outside South Africa. With the advent of a black-dominated government in 1994, the homelands as political entities were officially abolished.

APARTHEID AND ECONOMIC INEQUALITY

One of the cruel ironies of apartheid is that it was financed primarily by its victims. All nonwhites—Africans, Coloureds, and Indians—were discriminated against, yet they had to underwrite the oppressive system by accepting artificially low wages and seriously deprived working and living arrangements. Plainly, apartheid enabled whites to live at a level of comfort they would not otherwise have enjoyed.

JOBS AND WAGES   Discriminatory occupational and wage policies guaranteed to whites the higher-ranking jobs and dictated grossly discrepant wages between blacks and whites.[3] As a means of ensuring the most desirable occupations for whites, a system of job reservation was established whereby no black could advance above a white in the same occupational area. In certain areas of work, blacks were totally excluded. To complete the system of labor exploitation, trade union activity and strikes by African workers were forbidden.

The requirements of South African industry forced the circumvention of these measures, however, and dictated official policy changes. Whites were unable to fill all higher-skilled jobs, and the movement of blacks into these positions therefore was necessary. In 1969, for example, only 3 percent of professional and managerial positions were occupied by blacks, but by 1989 blacks accounted for 14 percent of them (SAIRR, 1992). With the end of apartheid, many companies have adopted affirmative action policies and are making efforts to employ more blacks at the managerial level.

THE SOCIOECONOMIC REMNANTS OF APARTHEID   In the post-apartheid period, the economic gap between whites and blacks has declined, but huge discrepancies remain. The unemployment rate is almost six times higher for Africans than for whites, and

---

[3] With the end of apartheid, the protection afforded poor, unskilled whites no longer prevailed, and many of these people became economically desperate (Arnold, 2000).

white household income is more than five and a half times what Africans receive (Figure 14.1). In all occupational areas, the wage gap between whites and blacks remains wide. In fact, the degree of economic inequality in post-apartheid South Africa is one of the highest in the modern world (*Economist,* 2012; Leibbrandt et al., 2010; SAIRR, 2010). Despite the growth of a black middle class, the overwhelming majority of the poor in South Africa are black and the majority of the black population is poor.

The results of the huge discrepancy in resources allocated to blacks and whites during the apartheid era are most apparent in education. Under apartheid, white schools were funded many times higher than black schools. Not surprisingly, the illiteracy rate among blacks is significantly higher than among whites. Although in the post-apartheid era funding no longer favors white schools, black schools continue to experience serious problems including poorly trained, underqualified teachers; lack of resources; and lack of student discipline (Asmal and James, 2001; Dugger, 2009; Fiske and Ladd, 2004; Nattrass and Seekings, 2001; Wines, 2007a). Although schools are increasingly integrated, many whites have abandoned the public educational system in favor of private schools (Daley, 1998a; Swarns, 2000). In brief, racial parity in education remains a distant goal. As seen in Figure 14.2, the percentage of whites with at least a high school diploma is double that of Africans.

At the university level, movement toward greater equality has progressed more quickly. The composition of the student population has changed dramatically in the years since the end of apartheid, with Africans now a majority of students in South African universities (Cloete and Moja, 2005). However, there is a very limited number of students that South African public universities can accommodate (Polgreen, 2012), and despite the increase in African enrollment, there remains a huge discrepancy with whites. In 2011, whereas less than 9 percent of Africans had attended college, 38 percent of whites had done so (Statistics South Africa, 2012a). And, despite

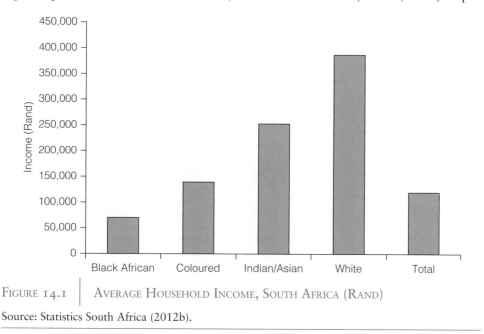

FIGURE 14.1   AVERAGE HOUSEHOLD INCOME, SOUTH AFRICA (RAND)

Source: Statistics South Africa (2012b).

the introduction of affirmative action measures, Africans are proportionately under-represented in universities (Dugger, 2010).

ENFORCEMENT OF APARTHEID: COERCION

As noted in Chapter 4, all societies are characterized by elements of order and conflict. Social control is therefore always a combination of consent and coercion. In a society in which the majority of the population does not accept the legitimacy of the prevailing system, however, coercion becomes the more common of the two. South Africa under apartheid, where four-fifths of the population rejected the system of white domination, illustrated this dramatically.

The South African criminal justice and legal institutions traditionally provided for suppression of any challenges to white rule. Indeed, the use of terror against the African population was routine. Writer Mark Mathabane recalls his own experience: "I remembered the brutal midnight police raids launched into the ghetto to enforce apartheid; the searing images of my father's emasculation as he was repeatedly arrested for the crime of being unemployed; my parents constantly fleeing their own home in the dead of night to escape arrest for living together as husband and wife under the same roof" (1994:38). Police violence against Africans was so pervasive and routine, explains political scientist Gwendolyn Carter, as to be almost unnoticeable except when a well-known person was affected or the circumstances were "so unusual as to attract attention" (1980:12).

The preparedness of the Nationalist government to use force in averting change in the distribution of power was demonstrated repeatedly. At Sharpeville in 1960, the police fired on a crowd of Africans peacefully protesting the pass laws, killing sixty-nine. South African legal expert Albie Sachs pointed out that the Sharpeville massacre, though the best-reported incident of its kind, was "merely one in a series

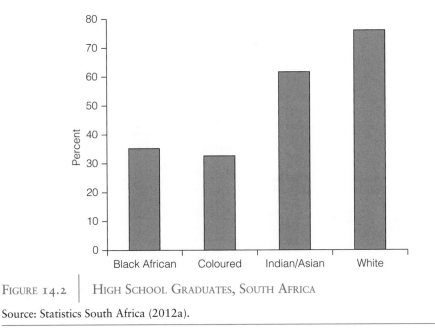

FIGURE 14.2 | HIGH SCHOOL GRADUATES, SOUTH AFRICA

Source: Statistics South Africa (2012a).

of characteristically South African episodes in which police fired on crowds of blacks that were either totally unarmed or armed only with sticks and stones" (1975:227). A second well-reported incident occurred in Soweto in 1976. What began as a student protest against inequalities in the educational system culminated in the slaughter of more than six hundred Africans. Thousands were injured in the disturbances that followed, and schools and universities were closed and industries shut down by strikes. This event represented a spontaneous protest against the entire apartheid system and afterward took on great symbolic importance among the African populace.

A Truth and Reconciliation Commission, created in 1995, revealed the systematic nature of human rights violations that were committed by the Nationalist government in enforcing apartheid. Testimony given by perpetrators as well as victims demonstrated that during the apartheid regime the police considered brutality, torture, and even murder as legitimate tools in stifling political opponents (Daley, 1997; Handley and Herbst, 1997).

SOCIALIZATION TO APARTHEID: IDEOLOGY

South Africa under apartheid is a marvelous illustration of a dominant group's efforts to uphold a system of ethnic supremacy using not only the power of raw force and political manipulation but also material, psychological, and ideological commitment. As in any multiethnic society, most members of the dominant group in South Africa were effectively socialized to accept and justify the prevailing system of ethnic inequality. Socialization to apartheid was especially pronounced among Afrikaners. Traditionally, in every institution—family, school, church, government—the values of Afrikaner nationalism and white dominance were reinforced.

THE RACIST BELIEF SYSTEM South African whites, particularly Afrikaners, believed themselves to be natural overseers of primitive Africans, whose shortcomings were not likely to change significantly. White rule, therefore, was seen as a kind of paternalistic relationship in which more intelligent and civilized peoples asserted their domination over inferior peoples in the interests of both (Mulder, 1972). In this view, notes Leonard Thompson, "It was the duty of the Whites, constituting the civilized, Christian race, to use their control of the state to prevent racial friction and racial bastardization by ensuring that the races would be separated from one another" (1985:190).

This essentially racist ideology underwent alteration during the apartheid regime. Although the crude belief in biological superiority was by no means unvoiced, the white resistance to changes in the system of racial inequality was based mainly on the notion of cultural inferiority and ethnic incompatibility (Adam, 1971a, 1971b; Thompson, 1985; Welsh, 1975). The cultural gap between the races was so vast, it was claimed, that blacks were simply incapable of operating the institutions of a modern industrial society (Schlemmer, 1988). Hence, the continued domination of the society by the white numerical minority was, in this view, essential.

The need to maintain white rule, based on the understood cultural superiority of whites, was thoroughly incorporated into all institutions, especially the school and the mass media. Through the Bantu Education Act of 1953, complete control of black education was given to the central government, thereby ensuring that the history and social science taught would reflect the Nationalist ideology. Blacks, it was thought, would be trained to accept their subordinate position in the society. Prime Minister Hendrik Verwoerd articulated this view:

> The school must equip the Bantu to meet the demands which the economic life of South Africa will impose on him. . . . There is no place for him in the European community above the level of certain forms of labour. Within his own community, however, all doors are open. . . . Until now he has been subject to a school system which drew him away from his own community and misled him by showing him the green pastures of European society in which he is not allowed to graze. . . . What is the use of teaching a Bantu child mathematics when it cannot use it in practice? (quoted in Harrison, 1981:191)

The separation of white and black was deemed necessary to protect whites from cultural pollution by an obviously inferior people. A schoolbook, for example, described the differences between white and black: "The White stands on a much higher plane of civilisation and is more developed. Whites must so live, learn and work that we shall not sink to the cultural level of the non-White" (quoted in Harrison, 1981:204).

In addition to the rationale based on economic and political efficiency, an element of religious mission infused the determination of the Afrikaners, particularly the zealously nationalistic element among them, to maintain white supremacy. As already noted, the Afrikaners traditionally invoked a biblical analogue to portray their place in South Africa, and their self-image of a chosen, superior people is strongly expressed in various aspects of Afrikaner culture and social life. The Dutch Reformed Church served as an integral part of Afrikaner nationalism. For many, the church was the focal point of Afrikaner culture and the religious expression of its nationalism. Unlike most English-language churches, it maintained a close link with the Nationalist state and abided apartheid.

To better comprehend the determination of Afrikaners to preserve their power and, in so doing, their ethnic culture, one needs to keep in mind the unique status of the Afrikaners on the African continent. Unlike traditional European colonial groups in Africa, Afrikaners have no cultural or political ties to a European homeland. Generations of Afrikaners have known nothing but South Africa, and there is no thought of returning to a society of origin, for there no longer is one. The Afrikaners are, in all ways, a group sui generis. The prevailing belief among them was that their way of life would be submerged and ultimately dissolved if the numerically dominant nonwhite populace were afforded equal access to social and political institutions.

THE PRACTICALITY OF RACISM Although these beliefs were the ideological bases of the apartheid system for a large element of the Afrikaner populace, the practical foundations of the system lay mostly in the resolve of the dominant whites, most Afrikaners and English alike, to sustain their supremacy. Whites well understood that a truly democratic political system would inevitably produce an African-dominated government, given the numerical imbalance between the two groups. This, in turn, would spell the end of white economic and social privilege.

Whatever their psychological and ideological commitment to white supremacy may have been, for most whites social pressures ordinarily compelled conformity. As noted in Chapter 3, prejudice and discrimination are most effective and self-sustaining where such thought and behavior are socially expected and rewarded. The economic benefits that flowed to whites from black exploitation were, as we have seen, enormous. Even those who may have been bothered by feelings of moral contradiction would therefore have been reluctant to call for the total dismantling of a system from which they greatly profited.

Even nonresidents of South Africa, used to more fluid social relations, might easily be caught up in the everyday workings of a system of extreme social inequality and separation that may at first have been repugnant. One American sociologist studying in South Africa in the 1970s reflected with astonishment on his own behavior: "One falls in with the culture and those with whom one must cooperate in assumptions and behavior. Even in my short stay I could observe the process occurring in myself: by the time I left I was beginning to unconsciously accept conditions that I found shocking when I first confronted them" (Mechanic, 1978:137).

Not only is it difficult to resist patterns of thought and behavior that reward conformity so generously, but it is especially hard to question a system of exploitation that, so far as it is recognized at all, is perceived from a distance. Because for whites the standard of living has been so comparatively high and the political order essentially democratic, there can arise among them only a limited understanding of black economic and, until recently, political deprivation. Adding to the incomprehension has been the essentially separate societies in which whites and blacks have lived. For the most part, whites have seen blacks only as workers and servants, not as neighbors and schoolmates. Few whites speak any of the various African languages, and few ever venture into black townships. Put simply, blacks and whites in South Africa have in the past lived in different worlds and, for the most part, continue to do so.

BUILDING A NEW SOCIETY

In the last two decades, the ethnic makeup of South Africa has not basically changed, but the relationship among its groups has been altered in a fundamental way. Indeed, if during the apartheid era one were to have suggested the scope and depth of change in the ethnic system that has taken place in South Africa since 1989, that person might have been seen as hopelessly naïve. The unimaginable, however, has occurred. Today, the government is led by blacks, and the most egregious policies of racial separation have been repealed. Official apartheid has ended—an entire generation has now grown up not ever having experienced it—and the society is moving, however tortuously, toward ethnic democracy. What propelled this amazing change in a system so seemingly uncompromising in its commitment to ethnic inequality?

FORCES OF CHANGE

South Africa's panoply of internal contradictions—social, economic, and political—brought the society to a watershed beginning in the late 1980s. A sociopolitical system enforced by coercion no longer seemed viable, and major reforms were now proposed. Much of the impetus behind the reforms was provided by a newly elected president, F. W. de Klerk, who acknowledged the failure of apartheid. As Mallaby describes him, "de Klerk had finally abandoned whites' hopes of extending their privileges beyond the colonial age" (1992:78). In 1992, de Klerk basically declared an official end to the apartheid system. Referring to his National Party, he stated that "[f]or too long we clung to a dream of separate nation states when it was already clear that it could not succeed. For that we are sorry. That is why we are working for a new dispensation" (SAIRR, 1993:25).

The radical transformation of South Africa that began in the late 1980s, however, did not emerge suddenly. Its roots reached back at least a decade when it became evident to the white power structure that changes in the apartheid system would have to be made if the society were to avoid a bloody confrontation. Former prime minister P. W. Botha had publicly declared in 1978 that South Africa must "adapt or die," and reforms in petty apartheid and in the labor system instituted in the 1980s were manifestations of that view. In a desperate effort to preserve white rule, a new constitution was adopted in 1983, creating a tripartite parliamentary body, representing whites, Coloureds, and Asians. But the new constitution was recognized as a sham, giving Coloureds and Indians a few privileges but making no provision at all for the representation of Africans.

INTERNAL CONTRADICTIONS AND CIVIL UNREST   The exclusion of Africans from the constitution was one of the precipitating factors in the civil unrest that occurred in the 1980s (Adam and Moodley, 1986; Dugard, 1992; Mallaby, 1992; Van zyl Slabbert, 1987). Widespread violent protests took place in black townships, the target of which was the system of black local authority. Black officials in the townships were seen as collaborators in the oppressive system and were driven from their communities, forced to resign, and in some cases even murdered. In addition to the violence in the townships, nationwide marches, rent strikes, consumer boycotts, student agitation, and labor protests became commonplace (Meredith, 1987; Murray, 1987). In attempting to control the protests, the government responded with force, using the South African Army as well as the police, who were given broad discretionary powers. Hundreds of blacks were killed in government actions to reassert control in the townships, and in 1985 a state of emergency was declared, providing for arrest and detention without trial and searches without warrant.

INTERNATIONAL PRESSURES   Throughout the Nationalist ascendancy during which apartheid was at its peak, most nations viewed South Africa as a kind of pariah state. Nonetheless, South Africa remained strategically important as a source of vital minerals as well as a major link in the worldwide production and distribution networks of multinational corporations, most of them based in Britain or the United States. Thus, although the apartheid system was condemned, economic relations between South Africa and the Western industrial nations generally were unaffected.

The often brutal suppression of the black protests starting in 1984 led to a stepped-up international campaign that included calls on multinational corporations to divest their interests in South Africa and to force Western nations to impose trade sanctions. Although these measures seemed to only minimally affect the ability of the South African economy to function, they contributed to economic uncertainty and, combined with the civil unrest of the late 1980s, induced the South African business community to call for greater and more rapid social reforms (Blumenfeld, 1987; Harari and Beaty, 1989).

ECONOMIC FORCES   More significant changes in the system of racial inequality were impelled by the demands of economic interdependence. The South African economy required an increasingly skilled workforce that could not be supplied by the white population alone. As noted earlier, this resulted in modifications in the job reservation system, permitting more blacks to fill higher positions. In effect, the system of job reservation, for decades restricting black labor mobility, collapsed (Adam and Moodley, 1986).

Labor needs also compelled relaxation of prohibitions against black trade unions and the right to strike, creating a growing and increasingly significant black labor movement (Thompson and Prior, 1982). African union membership swelled, and strikes by black workers arose frequently starting in the mid-1980s in certain industries, most notably mining, a particularly critical sector of the South African economy.

INTERNAL POLITICAL FORCES   After the Nationalist regime came to power in 1948, opposition black political organizations and movements were declared illegal, forcing them to operate underground or in neighboring African countries. The most important of these were the African National Congress (ANC), the Pan-Africanist Congress (PAC), and the Black Consciousness Movement. In 1990, the de Klerk government restored the legality of these political organizations, providing a new element in the chemistry of change in South Africa.

During apartheid the ANC, the most significant of the black political organizations, had assumed the status of "government in exile" for the African population. Its official leader, Nelson Mandela, was released from prison in 1990 after serving a sentence of twenty-eight years for antiregime political activity. Mandela's release was seen as a threshold. From that point forward, the end of white South African rule and the emergence of a black-majority government were only a matter of time. Shortly thereafter, the South African regime entered into talks with the ANC and other black groups to negotiate the entrance of blacks into the political system and to dismantle the apartheid regime.[4] The historic elections held in 1994, in which all South Africans—black and white—voted for the first time, were won by the ANC, bringing Mandela to power as the country's president. South Africa was now poised to create a nonracial state.

LOOKING AHEAD

As political and social changes continue to unfold, the shape of the society that ultimately will be produced remains unsettled. All of the key issues facing South Africa today, however, in one way or another concern its movement toward greater ethnic equalitarianism.

---

[4] An important stimulus to change was a 1992 referendum, in which white voters presented de Klerk with a mandate to forge ahead with negotiations with black leaders to ultimately construct a new multiracial constitution.

BLACK POLITICAL DOMINANCE   As we have now seen, the legal and structural roots of the apartheid system have been removed, and significant policy changes have muted the most blatant forms of racial discrimination. Most important, blacks have been brought into the political process. Awarding full and equitable voting rights to the black population was the major prerequisite to any peaceful solution of South Africa's ethnic conflict. The great fear of whites was precisely what was enacted: a one-man, one-vote, majority-rule system that, as they realized, spelled the end of white dominance. Astonishingly, this change was brought about peacefully. In light of the generations of racial hostility and the determination of the white regime to retain power, few could have predicted that the resistance to a black-majority political system would evaporate so relatively swiftly and nonviolently.

Today, the ANC's political dominance is almost unchallenged, but a democratic national government now seems solidly embedded in South Africa's constitutional system (Mattes, 2002). Mandela himself retired in 1999, and as the most widely supported and admired figure in South Africa—among whites as well as blacks—he remained a unifying symbol. As one journalist described him, he served as the nation's "moral guide" and, given his history of personal sacrifice, was ideally suited to convince the country of the need for compromise, reconciliation, and change (Duke, 1999b).

Paradoxically, changes effected in the political status of the black population may inevitably provoke more, not less, social unrest. Now that blacks control the political system, expectations are very high that change in their economic status will follow rapidly. This places great pressure on the government to provide jobs, housing, and increased public services, especially education and health care, to an enormous part of the black population that remains impoverished. If expectations exceed the government's ability to meet them, however, blacks may become frustrated and place even greater pressure on leaders to produce economic betterment. As many have observed, it is, ironically, only after conditions have begun to improve that people may be drawn into political action (Hoffer, 1951; Runciman, 1966; Turner and Killian, 1972). Thus failure to more rapidly alleviate black-white economic disparities may foment social and political unrest and place increased stresses and strains on the government.

RACE AND CLASS   Securing political equality is only a first step toward the elimination of ethnically based inequalities in jobs, education, health, and other life chances. As noted earlier, dramatically wide discrepancies remain in the socioeconomic status of whites and blacks. In general, whites continue to live at a level far above that of blacks and control most of South Africa's wealth. In a very real sense, blacks and whites in South Africa live in separate, tightly sealed social worlds. In 2013, in a highly publicized case, a white middle-class couple, along with their two young daughters, left their four-bedroom house in a gated community and moved six miles to a black township. There they lived for one month in a 100-square-foot shack with no electricity, no running water, and no toilets. Their budget was $10 a day, the average income of a black family. Their purpose in making this experiment, they reasoned, was to try to empathize with the daily lives and social conditions of most black South Africans. "Like so many people in South Africa," they explained, "we live in a bubble. We wanted to get outside that bubble" (Polgreen, 2012). Their experiment drew much attention and was met with much controversy. But, at least momentarily, it highlighted the continuing massive socioeconomic gap between whites and most blacks in South Africa.

However, viewing the relationship between race and class in South Africa today as a simple dichotomy—black/poor, white/rich—is a serious distortion. More accurately, the society is one in which there is declining *inter*racial inequality but rising *intra*racial inequality. Specifically, South Africa is divided into three broad classes: an increasingly multiracial class of well-off families, including corporate executives, professionals, and managers; a middle class of mostly urban, industrial, or white-collar workers; and a marginalized class of black unemployed (Besada, 2007b; Seekings and Nattrass, 2005).

Even before the end of apartheid, blacks had been experiencing upward occupational mobility. In the post-apartheid period, despite the still gross economic disparity between white and black populations, there has been a narrowing of the gap, with blacks increasing their share of total income. At the same time, however, inequality *within* the black population has increased. The black middle class, able to take advantage of new opportunities arising with the end of racial discrimination, is rapidly growing, but in the process is leaving behind a depressed black underclass. The gap between well-off and middle-class blacks on one hand and poor blacks on the other thus continues to widen (Besada, 2007b; Mattes, 2002; Seekings and Nattrass, 2005). This is not unlike the internal division within the African American population, discussed in Chapter 7, in which a growing middle class stands in sharp contrast to a marginalized underclass.

WHITE ECONOMIC POWER   Although whites have been displaced in the political realm, they continue to control key elements of the economy. They occupy most of the best jobs in the private sector and hold most of the society's wealth (McGillivray, 2006). Quite simply, as we have already noted, whites exceed Africans and Coloureds by many times on virtually every socioeconomic measure. The ruling black regime desperately needs the capital and managerial and technical skills that are possessed primarily by the white community. Thus it cannot afford to alienate whites, who might threaten to leave and take with them their economic and human resources. In fact, South Africa has suffered a brain drain, particularly among medical professionals. In trying to bring about a more equitable distribution of jobs, health care, housing, and land, the South African government faces the delicate task of raising public revenues—which must come primarily from white-controlled businesses and the wealthier white community—to finance ambitious public projects designed to lift the economic and social conditions of blacks.

PRESSING SOCIAL ISSUES   Failure to sharply reduce socioeconomic inequality may contribute to another serious problem—a dramatically high incidence of crime. The murder rate in South Africa is today the highest in the world—ten times the U.S. rate—and violent crime in general has become rampant (Abramsky, 2005; Bearak, 2011; Besada, 2007b; SAIRR, 2003). Failure to arrest this soaring crime rate may have serious economic ramifications. Those with capital may be less inclined to invest in South Africa, and those with highly needed occupational skills may be induced to emigrate (Shaw and Gastrow, 2001; Sparks, 2004).[5]

Another issue that threatens serious social and economic disruption in the new South Africa is the extraordinarily high rate of HIV/AIDS. One in seven South Africans is infected with the virus, more than in any other country. Life expectancy among all racial groups in the society has actually dropped in recent decades, a reflection of the brutal impact of this disease (Besada, 2007b).

THE CONTINUED SIGNIFICANCE OF RACIAL THINKING   The different, racially defined, social worlds to which whites and blacks in South Africa were traditionally socialized continue, for the most part, to exist in post-apartheid South Africa. "Race," notes Vestergaard, "remains one of the main organizing principles in social life" (2001:30). Racial stereotypes, which formed such a basic part of the dominant ideology—that blacks simply were unable to maintain the institutions of a modern society—have not disappeared. As Vestergaard explains, more than once in his interviews of Afrikaners—after apartheid had been abolished and blacks were in control of the government—did he hear the complaint that the country's ongoing economic problems were due to the fact that blacks "simply don't have what it takes" to run a modern economy. Although they are far more subtle than in the past, expressions of racial thinking continue to color whites' views of social problems such as education and crime.

In addition, in South Africa today, one's race continues to override other social identities. "Races" remain quite real to South Africans, and the four divisions of the population originally created by the apartheid regime continue to identify and define people politically and in everyday social life. "While we have formally, legally discarded race," writes a South African social scientist, "it continues to have an often unacknowledged and unseen power to determine perceptions, experiences and relationships" (Vincent, 2008:1427). As noted earlier, the residue of a thoroughly segregated society with its racist beliefs has clearly affected current patterns of social interaction and attitudes. A strong South African national identity has begun to emerge as the government emphasizes unity and refers to the country as the "Rainbow Nation," highlighting its ethnic diversity. However, racial and ethnic identities—and the beliefs attached to them—remain even stronger (Bornman, 2010). A comprehensive study in 2012 showed that people of all age groups continue to associate most strongly with others who share their racial or ethnic identity (Lefko-Everett, 2012).

Also, similar to U.S. patterns, whites in South Africa claim to accept racial integration and efforts at creating greater racial equality in principle, but seem less willing to accept concrete measures designed to bring about those changes (Durrheim and Dixon, 2010). Recall the attitudes of white Americans toward affirmative action and school integration, for example, discussed in Chapter 7.

It would seem naïve to assume that attitudes and ideas so prevalent among whites during apartheid would be subject to fundamental change in just a few years. As one observer of South Africa has put it, "[G]enerations of whites were brought up to be racists and they are unlikely to be able, even if theoretically willing, to change"

---

[5] Although popular perception would have it that violent crime in South Africa is mostly committed by Africans against whites, in fact most victims are African and poor (Shaw and Gastrow, 2001).

(Arnold, 2000:18). This applies to blacks' views of whites as well, of course. The distrust nurtured during generations of apartheid is not likely to be overcome for many years, perhaps generations.

ETHNIC DIVERSITY   In addition to the overwhelming problems of trying to create a system that will provide for a stable government and thriving economy, South Africa is faced with intraethnic diversity that was camouflaged for decades under the oppression of apartheid. The split of the African population along ethnic lines carries over into the political realm. Also, many Coloureds and Indians feel marginalized by a government that is now led primarily by Africans (Adhikari, 2005; Duke, 1999a; McGillivray, 2006).

Ethnic diversity is heightened also by a steady influx of workers from neighboring countries seeking better opportunities in South Africa. To think of South Africa as a society neatly divided into black and white, then, is a glaring oversimplification. Nothing could better illustrate this than the fact that today South Africa recognizes eleven official languages.

INTO THE FUTURE   Few observers, even the most informed, were able to predict the scope and speed of the changes that occurred in less than a decade. To forecast the specifics of South Africa's future, therefore, would surely be presumptuous. The post-apartheid healing process will remain a long-term affair and will be much dependent on the success of efforts to bridge the socioeconomic gap between whites and blacks. As one observer has put it, South Africa since the end of apartheid has "largely achieved racial peace, but not social harmony" (Besada, 2007b). As change in South Africa continues to unfold, much of the world's attention will be focused on this society well into the twenty-first century.

Despite its lengthy history of extreme ethnic division and conflict, South Africa presents itself today as, ironically, a potential model for other multiethnic societies. If the most oppressive system of ethnic relations in the contemporary world can be fundamentally altered in a mostly peaceful fashion, and if those changes can be sustained, there is strong reason to expect that comparable change can be achieved in other societies with severe ethnic divisions. Contemporary South Africa, then, is a critical laboratory for the study of race and ethnic relations.

SUMMARY

- Among multiethnic societies of the modern period, South Africa has been one of the most extreme cases of inequalitarian pluralism.
- South Africa's ethnic structure is a racial dichotomy, black and white, but there are more specific group subdivisions. The whites, roughly 9 percent of the total population, comprise two major subgroups, Afrikaners and English. Nonwhites comprise three subgroups: Coloureds, a racially mixed group; Asians, mostly ethnic Indians; and black Africans of several different tribal origins. The latter are by far the largest racial-ethnic category, making up almost 80 percent of the population.
- Ethnic stratification in South Africa was traditionally a castelike system in which each ethnic group was institutionally separated by custom and by a legal code.

The system of ethnic separation and inequality, called apartheid, was designed to maintain white supremacy in all areas of social, economic, and political life. Africans were excluded from the political system and were protected by few civil rights. Coloureds and Asians were only slightly better off.

- Starting in the late 1980s, the apartheid system began to be dismantled as black protest and economic uncertainty created strong pressures for reform of the South African regime. The national election of 1994 was the first in which blacks participated as equals, and a black-dominated government, led by Nelson Mandela and his ANC, came to power for the first time. This spelled the end of the apartheid system.
- Though the official system of racial inequality has been abolished and blacks now dominate the political system, the major components of white economic domination remain in place. On all economic measures, the inequality between whites and blacks remains extreme, although the gap is beginning to narrow.
- The key issues of South Africa's immediate future, therefore, concern the manner in which the tremendous economic gap separating whites and blacks will be reduced.
- Racial and ethnic categories, though officially abolished, remain potent dividers. The rigid segregation of apartheid has ended, but racial stereotypes and attitudes and patterns of intergroup relations have been only marginally changed.

COMPARING SOUTH AFRICA AND THE UNITED STATES

Points of Similarity
- In both South Africa and the United States, the dominant group emerged from a process of indigenous subordination, in which a settler group subdued the native population.
- Both societies supported the institution of slavery, though far more extensively and for a longer period in the United States.
- A castelike system of separation was constructed in both societies, with a well-understood racial etiquette and an official racial classification scheme.
- Both societies developed institutionalized systems of racial separation, apartheid in South Africa and Jim Crow in the southern United States.
- Both societies abolished their castelike systems but continue to struggle with efforts to fully integrate socially and economically the formerly subordinate black population.

Points of Difference
- For most of its history, South Africa's dominant ethnic group (whites) was a numerical minority vis-à-vis subordinate Africans, along with Coloureds and Asians. In the United States, by contrast, whites were always a numerical, as well as a sociological, majority.
- The economic and social gap between whites and blacks is today much wider in South Africa than in the United States.
- In multiethnic South Africa, blacks today are the dominant group, at least politically. This is obviously not the case in the United States.
- The United States is an ethnically more varied society than South Africa.

CRITICAL THINKING

1. Was apartheid South Africa more like the U.S. ante-bellum South or like the Jim Crow era? Frame your answer in terms of the different types of inequalitarian pluralism discussed in Chapter 4.
2. Blacks are an overwhelming numerical majority in South Africa, but a numerical minority in the United States. How does that demographic fact explain different long-range outcomes in black-white relations in the two societies?
3. Sociologists and psychologists have observed that South African racial beliefs and attitudes among the different population groups continue to be prevalent, despite the end of apartheid. Consider changes in racial attitudes that have taken place in the United States since the 1960s (Chapter 7) and compare them with post-apartheid South Africa.
4. Contemporary South Africa presents an unusual case of a radically transformed ethnic hierarchy: those who had traditionally reaped the benefits of dominant ethnic status no longer do so. Why was the transformation relatively peaceful, with little conflict among the society's various ethnic groups?

PERSONAL/PRACTICAL APPLICATION

1. As we have seen in this chapter, in the era of South African apartheid, whites retained all the society's social, economic, and political advantages. If you had lived in South Africa during that time and were white, would you have had any incentive to change this system, as grossly unequal as it was? As a follow-up, how do you think white South Africans today feel about their newly acquired minority status? If you were white, would you have any resentment about losing your privileged status or would you accept it as simply a consequence of social change?
2. The experiment described in this chapter of a well-off white family temporarily taking up residence in a poor black township was received in South Africa with mixed views. Some saw it as a sincere and worthy effort, while others saw it as self-serving; one South African referred to it as "poverty pornography." What is your view of this affair? Do you believe that you really could empathize with families living in poverty through a short-lived experiment like this, knowing that you will return to a relatively comfortable lifestyle?

BRAZIL

CHAPTER 15

Racial and Ethnic Democracy?

Shifting our focus from South Africa to Brazil, a more striking contrast in ethnic relations could not be found in the modern world. Instead of an ethnic morass, Brazil has often been praised, by Brazilians and outside observers alike, as a racial and ethnic paradise where people of varied physical features and with diverse cultural heritages live amicably together. Whereas the dominant ideology of South Africa before the end of apartheid was ethnic pluralism in a most extreme form, in Brazil the predominant philosophy has been ethnic assimilation: the more quickly the society's various ethnic and racially defined groups can be merged into one, the better the society will be.

Whether Brazil is in fact a genuine ethnic democracy is, as we will see, by no means universally agreed upon either by scholars or by Brazilians themselves. Another, less laudatory, view of Brazil places it well within the bounds of other multiethnic societies characterized by intergroup conflict and division. In any case, it is clear that the direction of ethnic relations in Brazil is toward assimilation, in which ethnic convergence and the breakdown of distinct physical and cultural groups are the society's long-range goals. More significant, in a number of ways Brazil has advanced further in the realization of these objectives than most multiethnic societies. This fact alone distinguishes it not only from societies with inequalitarian pluralistic systems but also from the United States and other societies like it that have traditionally proclaimed an assimilationist ideology but have often exhibited contradictory patterns of action and policy. If ethnic relations are not as ideally harmonious as Brazilian advocates would have the world believe, nonetheless they are less conflictual than in most other multiethnic societies.

MAJOR FEATURES OF BRAZILIAN SOCIETY

Brazil seems especially appropriate to compare with the United States because the two societies exhibit strongly parallel ethnic developmental patterns. Both were originally colonized by Europeans, who overwhelmed the indigenous population; both imported vast numbers of Africans, who were the mainstays of an institutionalized slave system that lasted until the late nineteenth century; and both were peopled by immigrants from a variety of European societies. There are, of course, important ethnic differences between the two societies as well. Africans taken to Brazil as slaves far exceeded in number those taken to the United States, and as a result, blacks composed the largest element of the Brazilian population until the nineteenth century. Although whites predominate today in both societies, in Brazil the African cultural influence has been more widespread and significant. More important, Brazil's ethnic and racial mix is far more amalgamated than that of the United States.

RACIAL AND ETHNIC FEATURES

THE PORTUGUESE HERITAGE   Brazil's colonial heritage derives from Portugal rather than Spain, unlike its neighbors on the South American continent. Its language and other major cultural elements are therefore Portuguese. As we will see, some hold that the Portuguese cultural legacy largely accounts for Brazil's seemingly more pacific patterns of ethnic relations.

ETHNIC COMPOSITION   In addition to its Portuguese heritage, Brazil is different from most other Latin American societies in its racial and ethnic composition. Except for those countries that are in or border the Caribbean, Brazil is unique in that it comprises a combination of European, indigenous Indian, and African peoples. Specifically, it is the European-African ethnic synthesis that is most distinctive. It is important to consider, however, that although European, African, and Indian racial strains are evident, the Brazilian population today is so thoroughly mixed physically that it is difficult to determine most people's racial derivation with any precision.

REGIONAL VARIATIONS   Brazil is a huge and regionally diverse nation. It covers almost half of the South American continent (about the size of the United States exclusive of Alaska), and its population of nearly two hundred million is almost as great as all the remainder of Latin America, leaving aside Mexico. Given this vastness, social and economic conditions are, predictably, not the same in all regions of the country. Brazil's major ethnic groups are also regionally concentrated, and patterns of ethnic relations therefore vary accordingly.

In the Northeast, most of the population is either black or mulatto (a combination of European and African), and the African cultural heritage is very apparent. This is Brazil's poorest region, an agrarian and largely underdeveloped area that has never fully recovered from the decline of the slave-supported sugar economy of the seventeenth and eighteenth centuries, of which it was the center. The aftereffects of slavery are still felt, and paternalistic black-white relations are more evident than in other regions of Brazil.

The Southeast, dominated by the metropolis of São Paulo, is the most industrialized and urbanized region of Brazil and is the core of its wealth and power. Ethnically it is primarily European, especially Italian, though its industrial preeminence has served as a magnet for out-migrants from other areas, especially the depressed Northeast. This has contributed to an increasing ethnic heterogeneity in recent decades though whites remain an overwhelming majority. The other region of greatest European influence is the extreme South, where German, Polish, Russian, Italian, and Portuguese farming communities are numerous alongside the traditional *estancias*, or cattle ranches.

The Indian influence in Brazil's population is most significant in the extreme North and in the interior, especially the Amazon basin. This is an enormous but sparsely populated region where most of the remaining Indians (around 200,000) live, most in isolated and primitive settings. Indian cultural and physical influences, however, are strongly evident here, and most of the population derives from a combination European-Indian ancestry.

DEVELOPMENT OF THE BRAZILIAN ETHNIC MÉLANGE

Settlement and Slavery

The shaping of Brazil's ethnic mélange resembles that of the United States in two critical respects: (1) European settlers conquered the indigenous population, reducing it eventually to a marginal status; and (2) aside from the native Indians, all its peoples have been immigrants, either voluntary or involuntary.

THE INDIAN POPULATION The indigenous population of Brazil at the time of Portuguese settlement in the early sixteenth century, perhaps 2.5 million, was reduced to virtual insignificance in less than three hundred years (Hemming, 1987). However, unlike the U.S. colonists, who made an unsuccessful attempt to enslave Indians, the Portuguese used the indigenous peoples as slave laborers almost from the outset of colonization. The Indians were devastated in this system because of their susceptibility to European diseases such as smallpox and measles. Another factor contributing to their demise was the intensity of agricultural labor, with which they were unfamiliar.

THE INTRODUCTION OF AFRICANS In the late sixteenth century, sugarcane became Brazil's chief cash crop. As a labor-intensive industry, sugar required large plantations; and as the Indian slave population diminished, the introduction of Africans in their place proved vital. With the development of the world market for sugar, the importation of African slaves—a costly enterprise—became economically feasible (Harris, 1964). Black slaves were many times more valuable than their Indian predecessors, for they seemed better prepared for field labor and were more resistant to European diseases. The African slave population in Brazil grew rapidly, and by the end of the eighteenth century, it numbered around 1.5 million. Estimates of the actual number of slaves imported into Brazil between 1550 and the 1850s, when the slave traffic ended, vary widely from three million to eighteen million (Knight, 1974; Pierson, 1967; Poppino, 1968; Ramos, 1939; van den Berghe, 1978). Because all records of the slave trade were purposely destroyed following the abolition of slavery in 1888, there are no precise data concerning the number of slaves actually imported into Brazil. That far more were brought to Brazil than to the United States, however, is undisputed.

THE BRAZILIAN AND AMERICAN SLAVE SYSTEMS: A COMPARISON

To better understand the roots of Brazil's system of race and ethnic relations, it is necessary to look at its long experience with slavery. It is especially important to compare Portuguese slavery in Brazil with the American slave system because many have argued that contemporary differences in black-white relations in the two societies are traceable in large measure to differences in the historical practice of slavery in each. Indeed, the similarities and differences in the two slave systems and the subsequent effects of each on patterns of race relations together make up one of the most widely debated issues of comparative social history.

BRAZILIAN SLAVERY: A MILDER SYSTEM?   Gilberto Freyre, one of Brazil's most renowned social historians, contended that slavery in Brazil was a milder, more humane form than its North American counterpart. This he attributed to several factors, the most important of which pertain to differences between Anglo-Saxon and Portuguese culture and society. Different practices of slavery, in turn, produced divergent patterns of relations between blacks and whites once slavery was ended in each society.

Freyre held that the Portuguese came to Brazil with a more tolerant attitude toward people of color. Portugal had experienced contact with African peoples in its colonial ventures as early as the fifteenth century and had endured a long period of rule by the Moors even earlier. The Portuguese sense of racial difference, therefore, was not as acute and absolute as that of the Anglo-Saxons, who had had little contact with nonwhites. A greater racial tolerance, according to Freyre's thesis, accounts for the more harmonious, less conflict-ridden nature of race relations in Portuguese colonies and the greater tendency among the Portuguese to intermix with native and slave populations. Freyre considered especially significant the extremely high rate of miscegenation between Portuguese slave masters and African slaves in Brazil. This, he contended, is evidence of a more compassionate and essentially altruistic master-slave relationship (1956, 1963b).

The thesis that the Portuguese slave system in Brazil was basically more humane than its North American counterpart was supported by an American historian, Frank Tannenbaum (1947). He explained that a milder form of slavery was evident in Brazil and in the Spanish colonies of Latin America because of differences between the Portuguese (and Spanish) and the Anglo-Saxons in their religious and legal concepts of slavery. Like Freyre, Tannenbaum maintained that there was a preconceived notion among the Portuguese regarding their treatment of slaves—deriving from principles established by both church and state—that fostered a less harsh slave system. In Brazil, the African did not lack a "soul" but was perceived simply as an unfortunate human being, a slave by accident of fate. In the United States, however, the African was perceived as less than human and thus naturally enslaved. Slavery in Brazil was understood as an economic necessity, hardly a relationship whose rationale derived from religious—or later, biological—principles, as in the United States. In short, in Brazil the human character of slaves was never denied. In the Portuguese view, slavery was a necessary evil. Thus it was not defended so firmly on the notions of racial inferiority and superiority.

The Brazilian and U.S. legal codes dealing with slaves also differed significantly. Portuguese (and Spanish) laws pertaining to slavery protected the slaves' human rights, at least technically. Thus slaves could not be dealt with by their masters with total impunity. Whereas slaves in the United States were chattel, reduced to mere property, in Brazil they were recognized as humans with certain legal rights. They were entitled to own property, marry freely, seek out another master if they were dealt with too harshly, and even buy their freedom, which many apparently did. It is this last-mentioned feature of the slaves' status—their greater opportunities for **manumission** (that is, *to be set free from slavery*) that Tannenbaum contended was an especially critical difference. Compared with the United States, manumission in Brazil was both common and expected. Its much higher incidence, Tannenbaum held, is in itself evidence of a milder form of slavery. More important, however, the accomplishment of manumission brought the freed slave automatic entry into Brazilian society with full and equal rights. In the United States, by contrast, even if slaves were freed, they did not acquire full citizenship but continued to suffer the economic, social, and legal handicaps that attached to their racial status.

The ability of the crown or the church to enforce these protective slave laws, of course, was quite another matter. These were, in effect, paper rights whose effectiveness was not sufficient to materially alter the nature of life for Brazilian slaves. "In Brazil, as everywhere in the colonial world," notes Marvin Harris, "law and reality bore an equally small resemblance to each other" (1964:77). As to the specific role of the church, its position on the issue of slavery was ambivalent and its power to influence Brazilian slave codes, in any case, limited (Marx, 1998).

Neither Freyre nor Tannenbaum contended that slavery in Brazil was not harsh and inherently cruel. The key difference between the Brazilian and American varieties, however, was that "in the Spanish and Portuguese colonies the cruelties and brutalities were against the law, that they were punishable, and that they were perhaps not as frequent as in the British West Indies and the North American colonies" (Tannenbaum, 1947:93). Most important, the openings for freedom were much greater in the Brazilian system, as was the advancement to full citizenship following manumission. Given the different legal and philosophical status of the slave and the freed black in Brazil as compared with the United States, more benign relations were engendered between whites and blacks, and greater efforts at assimilation were made following the formal abolition of slavery.

BRAZILIAN SLAVERY: A HARSHER SYSTEM?   Revisionist thinking on the matter of Brazilian slavery has seriously questioned whether it was in fact a distinctly more humane system than those established in the United States and the Caribbean. Some have even concluded that in many ways it was a system more dehumanizing than the Anglo-Saxon variety. That slaves in Brazil were acknowledged as human and that they possessed certain legal and religious rights did not seem to discourage the common application of brutal and inhumane forms of punishment by masters and the maintenance of severely debased living conditions by any standard. Historian Charles Boxer (1962) has described the brutish and often sadistic cruelty with which slave discipline was enforced by the Portuguese planters. To tie a slave to a cart; flog him; rub his wounds with salt, lemon juice, and urine; and then place him in chains for several days was not considered excessive. Runaways were dealt with, as described by the Brazilian scholar Darcy Ribeiro (2000), by branding with a hot iron, having a tendon cut, living shackled to an iron ball, or being burned alive.

That such punishments did not prevent slaves from continuing to run away on a large scale reveals the desperation to which they were driven. "Every black bore an illusion of flight in his breast," writes Ribeiro, "and was sufficiently bold to attempt flight when there was an opportunity" (2000:77). Other features of Brazilian slavery provide further evidence of its unusually harsh nature. Slaves commonly committed suicide, and insurrections among them were far more frequent than in the United States. Indeed, compared with the comparatively few rebellious actions by American slaves, Brazilian slave insurrections, many of them well organized, appear to have been common. In fact, organized settlements of fugitive slaves, called *quilombos,* sprang up in various locales (Vieira, 1995).

Historical evidence, then, seems to have proved beyond much doubt that "humane" slave owners in Brazil were the exception rather than the rule (Boxer, 1962, 1963; Davis, 1966; Graham, 1970; Marx, 1998). Recent historians have also disputed Freyre's and Tannenbaum's assumption that more frequent miscegenation and manumission as features of the Brazilian slave system substantiate the greater humanity and tolerance of the Portuguese. As for miscegenation, Thomas Skidmore (1993) notes that this was a common occurrence within the American system as well. The early enactment of laws prohibiting interracial marriage in the United States confirms the strong sexual attraction between the two groups, as does the wide range of physical traits of American blacks. Moreover, in Portugal's African colonies like Mozambique, the extent of miscegenation was never as significant as it was in Brazil. Clearly, factors other than the natural attraction of Portuguese men to native and slave women and their greater racial tolerance account for the high rate of interracial sexual encounters in Brazil.

A more compelling explanation lies in the simple fact that colonial Brazil lacked sufficient Portuguese women (Marx, 1998). Unlike North American colonists, who came mostly in family units, Portuguese settlers were mainly unaccompanied males who quite naturally turned to native and slave women to meet their sexual needs. Harris has put the matter succinctly: "In general, when human beings have the power, the opportunity and the need, they will mate with members of the opposite sex regardless of color or the identity of grandfather" (1964:68).

As to the question of more common manumission granted Brazilian slaves, historian Carl Degler (1986) has suggested that this may have been simply a method of relieving slave owners of the responsibilities and costs of caring for elderly or physically unfit slaves. Moreover, slaves were often unable to save enough to buy their freedom. Anthony Marx suggests that, given these conditions within the context of a highly stratified society, manumission in Brazil "did not threaten the social order and freed blacks were often left worse off than slaves" (1998:52). In any event, Harris (1964) notes that although manumission was more common in Brazil, it was not so much more common as to build a case on this fact alone for a presumably milder form of slavery.

Degler (1986) has pointed out that the American and Brazilian slave systems may not have differed so fundamentally even on the matter of slaves' legal and moral status. Though legal decisions were not consistent from time to time and place to place, both church and state, at least technically, recognized slaves as humans in the

United States as well as in Brazil. In neither society, however, was the law, whether church or state, effective in protecting slaves.

The issue of whether the Brazilian system of slavery was more humane than the American may be, in the final analysis, meaningless. How can such an argument be verified in any case? Indeed, the very notion of a "humanized" or "mild" slave system would be interpreted by most as an absurd contradiction. Whatever the circumstances, slavery was dehumanizing and was resisted by the slaves. Moreover, the tremendously high mortality rate among the slave population would seem to render the issue "inconsequential in light of the overriding fact that slavery in Brazil was astonishingly deadly" (Marx, 1998:52).

ABOLITION AND ITS AFTEREFFECTS   Whether the actual conditions of slavery under the Portuguese can explain the seemingly more benign relations between blacks and whites in Brazil than in the United States remains debatable. But the conditions that led to the abolition of slavery in Brazil and the status of blacks following their freedom were quite different from what occurred in the United States, and rendered an effect on subsequent race relations in the two societies.

Slavery endured in Brazil even longer than it did in the United States, ending formally in 1888. Its demise had been assured for several decades, however, and occurred in a gradual, evolutionary manner. This piecemeal abolition resulted in less wrenching social aftereffects than occurred in the United States, where slavery was ended abruptly following a cataclysmic civil war.

With the more common manumission of slaves in Brazil, blacks had already been a part of free Brazilian society for many decades before the formal end of slavery. In 1872, sixteen years before abolition, freed blacks were almost three times as numerous as slaves. By comparison, in the United States at the start of the Civil War, less than half a million out of a total black population of 4.5 million were free (Skidmore, 1972). Hence, blacks and whites in Brazil had experienced a long period of social interaction exclusive of the master-servant relationship. As Tannenbaum explained, in Brazil, "[t]he Negro achieved complete legal equality slowly, through manumission, over centuries, and after he had acquired a moral personality. In the United States he was given his freedom suddenly, and before the white community credited him with moral status" (1947:112). As a result, there was no question of the social standing of free blacks in Brazil and their access to full citizenship. In Brazil, freed slaves were in fact free; in the United States they were only technically free.

Furthermore, blacks in colonial Brazil had played a variety of occupational roles—even as slaves—and their entrance into the competitive labor market was therefore not so precipitous as it was in the United States (Karasch, 1975; Klein, 1969). Harris (1964) argues, in fact, that the continual shortage of skilled and semi-skilled labor in colonial Brazil created the need to fill numerous positions with blacks, which, in turn, contributed more than anything else to the ease and frequency of manumission.

Freed blacks in Brazil, then, were not seen as either a social or an economic threat; therefore, resistance from whites did not strongly materialize. Brazil thus avoided the development of a racial caste system designed to permanently handicap blacks, as was the eventual outcome of abolition in the United States.

Although Brazilian society was not as fundamentally dislocated by the abolition of slavery as the American South had been, the consequences faced by newly freed slaves were in many ways similar. Freed slaves either wandered back to the rural workforce, seeking out their former masters, or migrated to the cities with few urban labor skills. They were further handicapped because they were forced into competition with more skilled European immigrants, especially in the industrializing Southeast.

The release of former slaves into a competitive labor market for which they were ill prepared produced a debilitating effect on black mobility that has remained a basic feature of the Brazilian social structure to the present day. As the noted Brazilian sociologist Florestan Fernandes explained, blacks emerged from slavery materially and psychologically ravaged and "lacked the means to assert themselves as a separate social group or to integrate rapidly into those social groups that were open to them" (1971:28). Further disabling the freed black was the whites' eager relinquishment of any legacy of the slave system and its consequences. The occupational benefits of an industrializing Brazil with its competitive social order was reaped, therefore, primarily by European immigrants.

EUROPEAN IMMIGRATION

Beginning in the late nineteenth century, the ethnic composition of Brazil was altered fundamentally by the large-scale immigration of Europeans. Under the colonial regime, immigration into Brazil had been severely restricted. As a result, the bulk of the population was descended from three major ethnic lines: the Portuguese colonials, the indigenous Indians, and the African slaves. Following Brazilian independence in 1822, immigration remained sporadic until the mid-nineteenth century. An upward trend begun at that time turned into a veritable flood tide of immigrants in the century's last two decades.

With the realization that slavery was doomed, coffee planters increasingly turned to Europeans, especially Italians, as a labor force. The bulk of European immigration beginning in the late nineteenth century, therefore, was directed to the state of São Paulo, where the major coffee-growing estates were located. Growers saw Europeans as better and more reliable workers than native Brazilians, especially freed blacks from the declining Northeast. São Paulo also became the primary destination for European immigrants because it represented the nucleus of burgeoning Brazilian industry. Here were to be found the industrial jobs for which these immigrants were better prepared than were Brazilian blacks.

IMMIGRATION AS A WHITENING STRATEGY   European immigration was encouraged for another reason having nothing to do with the needs of labor. Racially, European immigrants were white and therefore contributed to the "whitening" of the society, an ideal that was now prevalent among the Brazilian political leadership and intelligentsia (Skidmore, 1990). Since abolition, the notion of an increasingly whiter population had been a staple of Brazilian intellectual thought. Abetted by the widespread influence of social Darwinism at the turn of the century, whites felt that Brazil should seek to dilute as quickly and completely as possible the black element of its racial composition. Put simply, the idea was "the whiter Brazil, the better" (Skidmore, 1972). Rather than encouraging a return of blacks to Africa, as was advocated by many in the United States, Brazil chose to import whites. The goal in each case was the same—a dilution of the black population (Marx, 1998).

The mass immigration from Europe gave Brazil's whitening process a giant infusion. So limited had immigration been previously and so substantial had been the importation of African slaves that by the end of the eighteenth century, Africans were a majority of the Brazilian population. European immigration not only increased racial mixing but also reversed this ratio.

In the century between 1850 and 1950, 5 million immigrants came to Brazil. The peak era of immigration, however, occurred between 1870 and 1920, when more than 3 million Europeans entered the society. Almost 2 million entered in the little more than two decades between 1881 and 1903. A second great wave of Europeans arrived during the early decades of the twentieth century (Amaral and Fusco, 2005; de Azevedo, 1950; Skidmore, 1974). Most were absorbed by the state of São Paulo, whose foreign population rose from 3 percent in 1854 to 25 percent by 1886. By 1934 more than half the population were first- and second-generation immigrants (Dzidzienyo and Casal, 1979; Mörner, 1985). Italians were by far the largest immigrant group, but Portuguese, Spanish, and Germans were also significant[1] (Freyre, 1963b; Morse, 1958). Europeans were also attracted to Brazil's extreme southern states, where large numbers of Germans and Poles established cohesive ethnic communities (Luebke, 1987, 1990).

In addition to non-Portuguese Europeans, another distinct ethnic element was added to the Brazilian mosaic beginning in 1908, when Japanese immigrants entered the society in large numbers for the first time. By 1941 almost 190,000 Japanese had immigrated to Brazil, with another 50,000 arriving after World War II. Today the Japanese ethnic community of Brazil numbers more than 1.5 million. Most Japanese originally settled as agricultural workers in the states of São Paulo and Paraná, and they still remain heavily concentrated in those areas. However, they have exited the agricultural sector and today are represented in a variety of businesses and professions (Makabe, 1999).

ETHNIC STRATIFICATION

Brazil's system of ethnic stratification resembles that of the United States in one important regard, the rank order of ethnic categories: whites disproportionately hold the society's power, privilege, and prestige. But several unique features of the Brazilian system distinguish it from that of the United States, as well as other multiethnic societies, making it one of the most far-reaching attempts at racial and ethnic democracy in the modern world.

THE BRAZILIAN MULTIRACIAL SYSTEM

Perhaps the key feature that differentiates the Brazilian system of ethnic relations from that of the United States and other societies with several distinct physical types is the very concept of race and the manner in which people are racially classified.

---

[1] As noted in Chapter 10, until the early 1900s, Brazil and Argentina—not the United States—were the major destinations of Italian emigrants.

DETERMINING "RACE" In the United States, one's racial status is a product of one's descent; that is, individuals, regardless of physical appearance, are assigned to a racial status on the basis of ancestry. This tradition is founded on **hypo-descent**—*a person with any known or recognized ancestry of the subordinate group is automatically classified as part of that group rather than the dominant group* (Harris, 1964). This is the so-called one-drop rule referred to in Chapter 5 that has been enforced customarily and, for most of U.S. history, officially (Hollinger, 2005). Hence, until very recently when the U.S. Census incorporated a multiracial option, the system of racial classification was essentially a dichotomy—white and black. Although the blurring of racial boundaries is occurring, for the most part this system continues to guide perceptions of race in the United States. As a result, one may exhibit generally white physical features yet still be classified as black. For example, in 1983 a Louisiana judge declared a woman who had been raised as a white legally black on the basis of a distant black ancestor. She and her siblings were the great-great-great-great-grandchildren of a black slave and a white planter. The decision was based on a Louisiana law declaring anyone with 1/32 "Negro blood" legally black. The woman was listed as 3/32 black (*Miami Herald*, 1983).

In Brazil, however, this rule of descent is not operative. One is not automatically placed into a racial category on the basis of family of origin. Instead, one's total physical appearance—skin color, hair texture, facial features—is the "obvious" determinant of one's racial classification. But in addition to these strictly physical aspects of racial identity, in Brazil certain social factors are used to classify people racially. The most important of these is social class (Harris et al., 1993). Thus, as one experiences class mobility, one's racial identity may to some extent also change. This changed racial status may be one's own reclassification as well as the perception of others (Carvalho et al., 2004). Popular Brazilian expressions such as "money whitens" connote that as people improve their class status, they are perceived as lighter racially. Ribeiro relates the story of a well-known Afro-Brazilian painter, speaking with a young black man struggling to rise in a diplomatic career. After hearing the young man's complaints of the barriers he faces because of his color, the painter responds, "I understand your case perfectly, my dear boy, I was black once too" (2000:157). To a degree, then, racial status in Brazil is subject to redefinition in different social circumstances.

The imprecise nature of racial classification in Brazil means that racial groups are not castelike as has traditionally been the case in the United States. *Branco* and *prêto*—literally, white and black—merely denote people who are predominantly white or black in appearance, regardless of actual racial origin (Nobles, 2002; Nogueira, 1959; Pierson, 1967). Simply having some traces of black ancestry does not result in the classification of someone as black. Indeed, most people classified as white in Brazil exhibit some evidence of black descent. Many American blacks, then, would be considered white in the Brazilian context.

RACIAL AND ETHNIC CATEGORIES This different perception of race has produced a correspondingly different system of racial categorization. Whereas the United States for most of its history has maintained an essentially biracial system in which people are classified as either white or black,[2] Brazil's multiracial system provides for more than

---

[2] The increasing Asian population in the United States has, of course, forced a third major category into the racial classification scheme used by governments and other institutions. The principal racial division, however, remains black and white.

two categories. Three major racial groupings are recognized in Brazil, but the boundaries between them are neither rigid nor clear-cut. *Branco* (white), *prêto* (black), and *pardo* (mulatto, or brown) are the most encompassing terms, but Brazilians employ literally dozens of more precise terms to categorize people of various mixed racial origins, depending on their physical features. Racial terminology is strongly localized, and from region to region—and even community to community—standards of classification vary.

The most minute physical features are often used to sort out people racially. The complexity and exactness of Brazilian racial typologies can baffle the American familiar with only a two-part (or, increasingly, tripartite) racial scheme. Consider the descriptive categories used by the people of Vila Recôncavo, a town near Salvador in the state of Bahia, studied by anthropologist Harry Hutchinson:

> The *prêto* or *prêto retinto* (black) has black shiny skin, kinky, woolly hair, thick lips and a flat, broad nose.
> The *cabra* (male) and *cabrocha* (female) are generally slightly lighter than the *prêto*, with hair growing somewhat longer, but still kinky and unmanageable, facial features somewhat less Negroid, although often with fairly thick lips and flat nose.
> The *cabo verde* is slightly lighter than the *prêto*, but still very dark. The *cabo verde*, however, has long straight hair, and his facial features are apt to be very fine, with thin lips and a narrow straight nose. He is almost a "black white man."
> The *escuro*, or simply "dark man," is darker than the usual run of mestiços, but the term is generally applied to a person who does not fit into one of the three types mentioned above. The *escuro* is almost a Negro with Caucasoid features.
> The *mulato* is a category always divided into two types, the *mulato escuro* and *mulato claro* (dark and light mulattoes). The *mulato* has hair which grows perhaps to shoulder length, but which has a decided curl and even kink…. The *mulato*'s facial features vary widely; thick lips with a narrow nose, or vice-versa.
> The *sarará* … has a very light skin, and hair which is reddish or blondish but kinky or curled…. His facial features are extremely varied, even more so than the *mulato*'s.
> The *moreno* … is light-skinned but not white. He has dark hair, which is long and either wavy or curly…. His features are much more Caucasoid than Negroid. (1963:28–30)

These are only the primary categories, explained Hutchinson, each of which is further broken down even more precisely. Even the whites are subclassified on the basis of skin, hair, and facial features. Anthropologist Livio Sansone (2003), in his research on the construction of race and ethnicity in Brazil, demonstrated how the dozens of different racial/ethnic terms are fluid and how one's identity can therefore change, depending on the context:

> So, a person that in the United States or Canada is simply black, in Brazil can be *negro* (black) during Carnival and when playing or dancing samba, *escuro* (dark) for his work friends, *moreno* or *negão* (literally, big black man) with his drinking friends, *neguinho* (literally, little black man) for his girlfriend, *preto* for the official statistics, and *pardo* in his birth certificate. (2003:50)

These categories, then, even the major ones, do not denote discrete racial groupings but blend into one another as on a scale (Figure 15.1). As earlier noted, miscegenation has, from the time of the initial Portuguese settlement, been a constant and

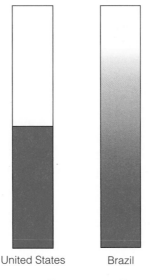

United States   Brazil

FIGURE 15.1 | RACIAL CLASSIFICATION PATTERNS IN BRAZIL AND THE UNITED STATES

widespread practice in Brazil among the society's three major population types—European, African, and Indian. As a result, the racial origins of people are thoroughly interwoven in various combinations. *Black, white,* and *mulatto,* therefore, represent very broad categories, each with a wide range of physical types.

Brazilians, then, are very conscious of color distinctions among people; indeed, they are far more conscious than North Americans or even South Africans, perceiving numerous physical types between black and white.[3] Rather than ancestry, physical distinctions—phenotypes—in different combinations make for almost limitless combinations of what Brazilians call *tipos* (literally, types). Cultural psychologist Jefferson Fish describes the difference between Brazil and the United States in this regard: "The American system tells you about how people's parents are classified, but not what they look like. The Brazilian system tells you what they look like, but not about their parents" (1999:394).

THE "MULATTO ESCAPE HATCH"   Although there is general agreement on the terms *branco* and *prêto,* it is the mulatto, mixed or *pardo,* category, between extreme white and extreme black, that provides the almost endless variety of descriptive combinations throughout Brazil. As shown in Table 15.1, *pardos* officially compose 43 percent of the population, but unofficially the percentage is much higher.[4] Again, it is important to emphasize the indistinct lines between these categories. Because, as we

---

[3] When asked to describe their color in the 1980 census, Brazilians responded with 136 different labels, and in an earlier national household survey, with almost 200 different labels (Andrews, 1991).

[4] In popular discourse, the term *moreno* is more commonly used than the official census term *pardo* (Telles, 2009). The *pardo* category consists not only of mulattoes (those of mixed European and African origins) but also those who are mixed European and Indian, usually referred to as *caboclos,* and those of mixed African and Indian origins, *cafusos,* who are numerically a small element of the Brazilian population.

TABLE 15.1	BRAZILIAN POPULATION BY COLOR (PERCENT)
Branca (white)	47.7%
Parda (mulatto)	43.1
Prêta (black)	7.6
Amarela (Asian)	1.0
Indígina (Indian)	<1.0

Source: Instituto Brasileiro de Geografia e Estatística (2012).

will see shortly, it is socially advantageous in Brazil to be considered white, the population figure of 48 percent white in Table 15.1 is greatly exaggerated in terms of the North American conceptualization of race (Harris et al., 1993). In any case, these census figures indicate a progressive decline of the white percentage, signaling that even when measured by people's self-identification, Brazil is gradually emerging as a racially mixed society.

Might the color continuum of Brazil, as opposed to the more rigid black-white dichotomy of the United States, explain much of the difference in the tone of ethnic relations in the two societies? Degler (1986) contends that the existence of the intermediate—mulatto—category in Brazil, more than anything, accounts for the diminished hostility between blacks and whites. The presence of what he calls "the mulatto escape hatch" makes infeasible the development of segregated institutions, which have so dominated most of the history of black-white relations in the United States. Where there are so many recognized racial combinations and where people's racial status is so arbitrary and personalized, it is impossible to erect castelike structures based on racial definitions. "The presence of the mulatto," explains Degler, "not only spreads people of color through the society, but it literally blurs and thereby softens the line between black and white" (1986:225). Furthermore, where there is such a varied biological mixture of the population, racial ideologies propounding the notion of superior and inferior races are difficult to sustain.

Skidmore (1993), however, casts doubt on the assumption that the presumed multiracial classification system of Brazil is today basically different from the U.S. biracial system. He contends that although there is an increasing awareness of the racial variability—and thus color gradations—of the U.S. population, at the same time there is a growing tendency among Brazilian scholars and others to describe Brazil in bipolar racial terms, that is, simply "white" and "nonwhite." Moreover, Marx (1998) points out that the Brazilian construction of a "triracial" system of categorization, though more fluid than the biracial U.S. system, still provides for the domination of whites over both blacks and browns. Later we will consider how American and Brazilian racial systems may be converging.

THE BRAZILIAN ETHNIC HIERARCHY

At first glance it might seem pointless even to envision an ethnic hierarchy—that is, a ranking structure of ethnic groups—in a society where hard-and-fast lines between ethnic groups are not recognized. Ethnicity in such a social setting would, logically, play no role in the allocation of social rewards. This, in fact, is "officially" how Brazil has been portrayed. Several aspects of the Brazilian system of ethnic relations, however, belie this ideal picture.

THE CLASS-ETHNIC CORRELATION By looking at the general class positions of Brazil's racial/ethnic categories, one can conclude without hesitation that there is indeed an ethnic rank order in this society. Although the boundaries of the three major groupings are not precise, there is little question that whites, mulattoes, and blacks represent a hierarchy of economic, political, and social standing. Whites are clearly at the top, followed in order by mulattoes and blacks. A stranger in Brazil could quickly confirm through simple observation the validity of an old Brazilian rule of thumb, "The darker the skin, the lower the class" (Wagley, 1971). Studies compiled by social scientists have repeatedly corroborated this visual impression (Dzidzienyo and Casal, 1979; Fernandes, 1971; Garcia-Zamor, 1970; Hasenbalg, 1985; Lovell and Wood, 1998; Paixão, 2004; Reeve, 1975, 1977; Silva, 1985; Telles, 2004; Telles and Lim, 1998; Twine, 1998; Wagley, 1971).

In looking at the class-ethnic relationship, one must be careful to keep in mind the considerably different class structure of Brazil compared with those of the United States and other advanced industrial societies. Brazil, like most Latin American societies, traditionally maintained a two-class system: an upper class, made up of landowners as well as wealthy merchants and professionals, and a huge lower class, comprising much of the remaining population. With rapidly advancing industrialization, this traditional class system is gradually breaking down, however, and class patterns are beginning to resemble those of the highly developed countries. As this occurs, a middle class of salaried professionals and white-collar workers is expanding, though it is a much smaller proportion of the population compared to the United States and other postindustrial nations. Brazil remains a society with an extraordinarily wide gap between rich and poor, and nonwhites are disproportionately among the latter.

The ethnic hierarchy in Brazil has customarily paralleled the class system. At the pinnacle of the structure are whites, and "as one moves down the social hierarchy, the number of racially mixed or otherwise nonwhite individuals gradually increases" (Wagley, 1971:121). Although mulattoes and blacks are represented throughout the occupational structure, they are not found in proportion to their share of the population in positions of authority or decision making. Afro-Brazilians have made significant inroads into the political arena since the 1980s, and they are beginning to appear more commonly in high political positions (Davis, 2000). Still, the elite levels of government, the economy, the military, and education are thoroughly dominated by whites. At the opposite end of the hierarchy, whether in rural or urban areas, the bulk of the poor and uneducated are black. Only in sports and entertainment do Afro-Brazilians play a prominent role (Leite Lopes, 2000; Schneider, 1996). In short, although white skin does not guarantee social success, "it always improves one's position over that of a darker person" (Degler, 1986:191).

Numerous studies demonstrate this racial hierarchy, all showing the wide discrepancy between Afro-Brazilians and whites in wages and income, occupations, education, and literacy rate (Beato, 2004; Castro and Guimarães, 1999; Daniel, 2006;

Fernandes, 2005; IBEG, 2006; Lovell and Wood, 1998; Nascimento and Nascimento, 2001; Paixão, 2004; Schneider, 1996; Silva, 1985, 1999; Silva and Hasenbalg, 1999; Skidmore, 1999; Telles, 2009; Wood, 1998). Moreover, the opportunities for non-whites to break the cycle of racial inequality through upward mobility are limited (Daniel, 2006; Telles, 2004).

THE "WHITENING" IDEAL Another clear indication of an ethnic hierarchy in Brazil is the social preference for white over black. Despite the amalgam of skin colors and other physical features in its population, "white" has always been understood as more desirable than "black." The socialization process is quite effective in conveying the notion that to be white is to be socially favored. The previously quoted adages, implying that one gets "whiter" as one moves up in social class, are manifestations of this racial ideology. Brazil's white partiality has found expression at both individual and societal levels.

At the individual level, the white ideal is most evident in the preference for lighter-skinned marital partners. Such a marriage usually earns the darker-skinned partner a higher social status and in any case produces lighter-skinned children, who are assured of better social opportunities. Degler notes that "when a black or mulatto succeeds in marrying lighter, he speaks of 'purging his blood'—a phrase that in itself is revealing of what the process connotes in the minds of Brazilians" (1986:191–92). The white preference is often imparted to children in blunt fashion. Sociologist Oracy Nogueira explains that "from the start it is impressed upon the white child's mind that negroid characteristics make [the] bearer ugly and undesirable for marriage" (1959:171). Whereas marrying into a lighter family and having lighter offspring is the most effective form of whitening for Afro-Brazilians, others include upward economic mobility, associating with white friends and acquaintances, and adopting a white middle-class lifestyle (Andrews, 1991).

At the societal level, the "white is best" attitude has been evident in various governmental policies, particularly those pertaining to immigration. As earlier noted, European immigration was encouraged not only to expand the industrial labor force but also, as Roger Bastide has written, "to submerge the descendants of Africans into a more prolific white population, and, in the last analysis, to change the ethnic composition of the population of the country" (1965:17).

The desire to create a whiter population stems from the response of Brazilian political and intellectual elites in the late nineteenth and early twentieth centuries to racist theories emanating from Europe and America, propounding notions of inferior and superior races. Without having to accept openly the idea of white superiority and black inferiority, they posited that the black (and presumably inferior) element of Brazil's population would be submerged through miscegenation and European immigration. These, they assumed, would progressively "whiten" the population, thus avoiding problems of race. The whitening process is still seen as desirable and is expressed, as we will see, in the tendency to discount racial inequalities and to squelch the development of racial consciousness. Brazilians, writes historian Thomas Skidmore, "are still implicit believers in a whiter Brazil, even though it may no longer be respectable to say so" (1990:28).

PREJUDICE AND DISCRIMINATION

It would be expected that in a society openly and ardently proclaiming the desirability of amalgamation of its various groups, ethnic prejudice and discrimination would be minimal. Brazilians do customarily assert that such prejudice and discrimination in Brazil are insignificant. As with ethnic stratification, however, there are glaring inconsistencies between the official accounts of racial democracy and the realities of individual and group relations.

Visitors to Brazil often comment on the seemingly placid and harmonious nature of interethnic relations. People of the entire range of color and of widely varied cultural origins appear to interact at all social levels with little or no difficulty arising from their racial and ethnic differences. An American journalist stationed in Brazil described his surprise at witnessing one such social encounter, a dance at a neighborhood-based club:

> The crowd was interracial, with both blacks and whites in substantial numbers. At first glance, not much different from the crowd at the average Washington Bullets basketball game. But if you looked closer, you saw that blacks and whites were arriving together, dancing together and leaving together in couples and groups, not in separate cliques. This was different. And if you went down and made your way through the crowd, you saw that up close it was hard even to tell who was black and who was white.... [L]ater ... I pondered what I had found so striking about the evening. I realized that I had been in the midst of an interracial crowd that didn't really feel like an interracial crowd, at least not like one in the United States. (Robinson, 1996:23)

Such impressions extend back even to the period of slavery. Henry Koster, an Englishman traveling in Brazil at the beginning of the nineteenth century, remarked that "it is surprising, though extremely pleasing, to see how little difference is made between a white man, a mulatto and a creole Negro if all are equally poor and if all have been born free" (1966 [1816]:152).

Observant visitors have not been deceived. Relations between ethnic groups in Brazil *are* clearly more benign than those in most multiethnic societies of the modern world. Overt interethnic conflict is rare, and in no institutional area of the society have measures designed to discriminate against people on the basis of ethnicity ever been common or officially sanctioned. But such manifestations of ethnic harmony do not tell the entire story.

PREJUDICE

Closer analysis reveals that prejudice toward nonwhites is a well-established aspect of this society's culture but is often disguised by the customary physical contact between ethnic groups in public places. Moreover, these attitudes are translated into stereotyping of nonwhites and informal rules of social distance.

STEREOTYPING   For several decades, studies have verified the prevalence of negative stereotypes of Brazilian people of color (Andrews, 1991; Bastide, 1965; Bastide and van den Berghe, 1957; Fernandes, 1971; Harris, 1963; Hutchinson, 1963; Pierson, 1967; Sanders, 1981; Saunders, 1972; Telles, 2004). Bastide and van den Berghe (1957) found, for example, that among white middle-class university students in mainly white São Paulo, blacks were commonly viewed as dirty, physically unattractive, superstitious, profligate, immoral, aggressive, lazy, lacking persistence at work, and sexually "perverse." Mulattoes were seen in much the same manner. Almost half of their sample also felt that blacks were intellectually inferior to whites. More than fifty years after Bastide and van den Berghe's observations, another anthropologist found that essentially the same stereotypes—"indolence, intellectual backwardness, and criminal tendencies"—remained "alive and active, though repressed and unstated" (Nascimento, 2007:70).

In Minas Velhas, a small town in the state of Bahia where most of the inhabitants were of mixed European-African origin, Marvin Harris reported that "the superiority of the white man over the Negro is considered to be a scientific fact as well as the incontrovertible lesson of daily experience" (1963:51). Harris noted the abundance of legends, folk stories, and pseudoscientific notions that supported the negative beliefs about blacks. A school textbook used in Minas Velhas, for example, stated that "of all races the white race is the most intelligent, persevering, and the most enterprising," whereas "the Negro race is much more retarded than the others," a view not disputed by any of the village's six teachers (quoted in Harris, 1963:51–52).

Harris explained that although blacks in the town were commonly derided as ignorant and ugly, such views were rarely expressed bitterly or hatefully. Moreover, he found that the black stereotypes here (as well as in other regions) were often contradictory. Blacks could be seen at once as stupid and cunning, honest and dishonest, naïve and shrewd, or lazy yet fit only for hard work. Three white attitudes, however, were constant and basic: (1) the black is inferior to the white; (2) the black does and should play a subservient role to the white; and (3) blacks' physical features are displeasing. "On the whole," noted Harris, "there is an ideal racial ranking gradient in which whites occupy the favorable extreme, Negroes the unfavorable extreme and mulattoes the various intermediate positions" (1964:60).

In a more recent analysis, Telles (2004) reports that negative stereotypes of both blacks and mulattoes remain widely held in Brazil, not simply by whites but by blacks and mulattoes as well. The media are important purveyors of these persistent views. Blacks are seen on Brazilian television in roles that support traditional negative stereotypes, whereas whites, by contrast, are portrayed positively. Telles explains, however, that these views have been modified in the past few decades. Moreover, there is evidence to suggest that some negative stereotypes of blacks in Brazil "may not be as harsh as those in the United States" (2004:153).

Social Distance Studies of social distance also reveal prejudicial attitudes toward nonwhites (Bastide and van den Berghe, 1957; Garcia-Zamor, 1970; Zimmerman, 1963). In their study of white university students in São Paulo, Bastide and van den Berghe (1957) found that almost all accepted the idea of equality of opportunity for everyone, regardless of color. Better than half also accepted the desirability of casual relations between members of different racial groups. But beyond this level, social distance increased markedly. A majority opposed intimacy with blacks beyond simple comradeship, as well as miscegenation with blacks. The most unwavering attitudes involved intermarriage; almost all indicated they would not marry a black or even a light-skinned mulatto.

Intermarriage among members of different racial or ethnic groups, as was explained in Chapters 3 and 4, represents the highest degree of social intimacy and the ultimate stage in the assimilation process. In Chapter 7, we saw that black-white intermarriage in the United States, although increasing, remains uncommon. The same is true in South Africa, of course, given its only recently abolished apartheid system. We would expect a more substantial amount of intermarriage in Brazil, with its historical legacy of miscegenation and its proclamation of racial democracy. Telles (2004) reports that in 1991 nearly a quarter of all Brazilian marriages were interracial. Although this may be accounted for in some part by the ambiguous way in which Brazilians classify people by race, the numbers are simply too large to deny that racial intermarriage is a common occurrence. And, the racially mixed character of most Brazilians, of course, testifies to the practice.

A less sanguine view of racial intermixing in Brazil is that in many ways it is not essentially different from other multiracial societies, like the United States. Although undeniably common during the Portuguese colonial period, researchers have pointed out that in modern times the degree of social tolerance for such unions drops sharply when the contrast in color of the partners is great (Cardoso, 1965; Degler, 1986; Fernandes, 1971; Hutchinson, 1963; Reeve, 1977; Vidal, 1978). Most marriages are between people of similar color, and most intermarriages are between blacks and mulattoes rather than between whites and either blacks or mulattoes. Moreover, most intermarriages occur at the lower end of the class hierarchy (Telles, 2004). Hutchinson (1963), for example, noted that in the small Bahian community he studied, the aristocratic white upper class was completely endogamous. He went on to explain, however, that "no one in any class likes a marriage in which the skin colours are too far apart. Any marriage of dark with light or white will be referred to as *mosca no leite*, or fly in the milk, and a certain repugnance is felt by all" (1963:40). In marriages of black and white, the union is considered fortunate for the former and a case of lowering oneself for the latter (Nogueira, 1959). Moreover, it is the whitening ideal that serves as the incentive for darker-skinned people to marry lighter-skinned partners (Sheriff, 2001). Also, racial intermarriage is more frequent in areas of Brazil where the nonwhite population is great, particularly the Northeast, and where, as a result, whites are most likely to interact with blacks and mulattoes (Telles, 2004).

Brazilians often cite the tradition of miscegenation as proof of their lack of prejudice toward people of different colors. Yet, it has been pointed out that the commonness of this practice may simply reflect the accessibility of black and mulatto women to white men throughout Brazilian history. The mulatto woman is traditionally seen in Brazil as epitomizing female beauty, but this has been interpreted by some as simply a sexual attraction, indicative of an exploitative relationship.

PATTERNS OF DISCRIMINATION

Early comparisons of ethnic relations in the United States and Brazil suggested that in the latter, prejudicial attitudes did not fundamentally affect people's behavior toward members of particular groups (Freyre, 1963a; Harris, 1964). That discrimination against nonwhites in Brazil has been unlike past ethnic discrimination in the United States is clear enough. The forms of blatant derogation, denial, and physical violence historically so well rooted in American society have simply not been present in the same manner or degree in Brazil, and the institutionalized forms of discrimination that characterized the United States before the 1960s are not found at all. Later researchers, however, have pointed out that, although it is rarely virulent and is repeatedly denied, discrimination is very much a part of the Brazilian ethnic system. In addition to the indisputable relationship between skin tone and social class, racial discrimination in various forms has been revealed frequently by social scientists and journalists.

INDIVIDUAL DISCRIMINATION   At the individual level, blacks may be excluded from hotels and clubs and are commonly harassed by the police. Apartment house elevators in São Paulo, Brazil's largest city, frequently are marked "social," for residents and their guests, or "service," for maids and workmen; blacks, no matter their status, are routinely steered into the latter (Schemo, 1995; Telles, 2004). Discrimination is particularly evident in the area of employment (Andrews, 1991; Sanders, 1981). Simple observation as well as careful analysis reveals that those of darker skin are consistently relegated to the lowest-status, poorest-paying jobs in business and government. In São Paulo, jobs requiring contact with the public, such as receptionists, salespeople, and waiters in better restaurants, are often closed to Afro-Brazilians. In screening black job applicants, employment want ads in the recent past would often use the phrase "good appearance" (*boa aparência*), understood by all to mean "white." One observer of Brazilian society claims that such discriminatory hiring practices are in decline, but admits that blatant discrimination in various forms "continues to go unpunished" (Rohter, 2010:73).

INSTITUTIONAL DISCRIMINATION   Less apparent are institutional forms of discrimination as, for example, in education—where the opportunity to break the cycle of racial inequality is greatest. Telles (2004) explains that today some discrimination in schools may be blatant, but mostly it is evident in the form of a self-fulfilling prophecy. Blacks are assumed to be less capable educationally and therefore fewer educational resources are invested in them. In turn, this leads to poorer academic performance, and ultimately to limited job opportunities. At the university level, nonwhites are severely underrepresented as a result of their inferior precollege education. In recent decades, higher education in Brazil has expanded greatly, but at a much faster pace for whites than blacks and mulattoes. Whereas almost no nonwhites had graduated from college in 1960, 2.6 percent had done so by 1999; by comparison, the percentage increase for whites during those years was from 1.4 to 11.0 (Telles, 2004).

Perhaps the most telling evidence of ethnic discrimination in Brazil came forth in 1951, when the Brazilian legislature was prompted to enact an antidiscrimination measure, the Afonso Arinos law. The measure was generally recognized as unenforceable, but its significance lay not so much in its subsequent effect as in the formal recognition for the first time in Brazilian history that discriminatory practices did in fact exist. Ironically, the measure was reportedly prompted by the refusal of a Rio de Janeiro hotel to accommodate a visiting black American entertainer (Andrews, 1991). The law was updated in 1989, with little effect (Nobles, 2005).

CLASS OR RACE?   Although the lower place of blacks and mulattoes in the occupational structure of Brazil is undeniable, many observers—as well as Brazilians themselves—have customarily attributed this situation not to discrimination but to the effects of class disadvantage (Freyre, 1963b; Pierson, 1967). The argument is essentially this: The lower place of blacks and mulattoes is the product of a heritage of slavery. Whites have enjoyed the historic advantage of being at the top of the social hierarchy and therefore continue to disproportionately occupy the society's better jobs and leadership positions. Differences in class, then, generally do reflect differences in color, but this relationship is due to historical accident, not institutional or customary forms of discrimination.

There is little question that class factors weigh heavily in the social placement of Brazilians. But the argument that the lower place of Afro-Brazilians is simply an expression of class has been challenged by many social scientists (Bailey, 2009; Hasenbalg, 1985; Lovell and Wood, 1998; Morse, 1953; Russell-Wood, 1968; Silva, 1985; Skidmore, 1974; Telles, 2004, 2009; Twine, 1998; van den Berghe, 1978). Their studies have shown that one's color is indisputably significant in the distribution of wealth and power, independent of class or region. As Telles has concluded, "[T]he socioeconomic position of nonwhites in Brazilian society is due to both class and race" (2004:220).

STABILITY AND CHANGE

Will the commitment to full assimilation of diverse peoples and the ideology of racial democracy continue to guide the future course of Brazil's ethnic relations? In no society, of course, should we expect to find consistency between ideology and reality in all aspects of ethnic behavior and attitudes. The United States, as we have seen, has maintained for most of its history an equalitarian ideology even though official policies and individual actions have not always reflected that belief system. Brazil's affirmation of equalitarian and assimilationist principles, however, has been so unfailing that as a multiethnic society it promises to be a positive model for others.

BRAZIL AS AN ASSIMILATIONIST SOCIETY

In some ways, Brazil seems to epitomize the assimilationist type of multiethnic society described in Chapter 4.

AMALGAMATION The intermixing of diverse peoples—amalgamation—has advanced further in Brazil than in almost any other nation of the modern world. Several factors have contributed to this dilution of racial lines. First, immigration from Europe substantially increased the white element of the population. As we have seen, before the late nineteenth century whites were a numerical minority in most parts of Brazil, and European immigration played a large part in reversing that ratio. Second, a system of sexual exploitation had been in effect since the first Portuguese arrived in the early 1500s, a system whereby white men enjoyed relatively unlimited access to black and Indian women. Finally, the racially mixed category (mulatto) of Brazil's population has grown larger over the past few decades. This may simply be a result of the increasing tendency for people to identify themselves as neither white nor black, but it may also suggest a steady pattern of increasing intergroup marriage and thus a continuation of racial amalgamation (Sanders, 1982; Telles, 2004).

CULTURAL ASSIMILATION   The "official" view of assimilation in Brazil is that the society is fusing not only into a single racial group but also into a single Brazilian culture, comprising Portuguese, Indian, and African elements. "The Brazilian ideology of interracial and interethnic relations," wrote Nogueira, "is proamalgamation concerning physical traits and pro-assimilation as to cultural characteristics" (1959:172). The assimilationist goals of Brazil have been expressed most articulately by Freyre, who asserted that the society "has not attempted to be or become exclusively European or Christian in its styles" (1963b:156).

African and, to a lesser extent, Indian elements have been blended into the Brazilian culture far more than have African or Native American traits in the United States. African cultural elements are especially evident in religion, food, music, and dance, particularly in those areas of Brazil where the population is heavily of African origin. For example, West African religious rituals have been fused with Catholicism in cults known as *candomblé* and *macumba* in Bahia and Rio de Janeiro (Rohter, 2000).

As for the various European ethnic groups that have populated Brazil since the late nineteenth century, problems of assimilation have seemed less challenging than in other multiethnic societies, although separatist elements have sometimes been evident. For example, the German communities of the southern Brazilian states in the early twentieth century sought to perpetuate their language and to remain unassimilated into what they saw as an "inferior" Luso-Brazilian culture (Luebke, 1983). Moreover, the first two generations of other European groups displayed the resistance to assimilation generally characteristic of early generations of voluntary immigrant groups. By the third generation, however, they had become an integral element of the Brazilian economic and political elites, serving as a prime segment of the expanding Brazilian middle class (Horowitz, 1964; Willems, 1960).

Brazil has clearly achieved a level of physical and cultural assimilation surpassing that of most other multiethnic societies. Although there remain in Brazil strong regional patterns of culture and of ethnic composition, giving the society a pluralistic flavor, the commitment to an assimilationist ideology has been constant and has yielded an increasingly integrated society, physically and culturally. Brazil can proclaim success in creating a common national identity that transcends the racial and ethnic components of its varied population. "One and the same culture takes in all," writes Ribeiro, "and a vigorous self-definition that is more and more Brazilian animates everyone" (2000:169). In this sense, Brazil can be considered an exemplar of an assimilationist society, having transformed its diverse ethnic elements into a culturally homogeneous social entity.

A FAUX RACIAL DEMOCRACY?

Cultural, and to a great extent physical, blending, however, has not translated into the racial democracy that Brazilians so fervently cast their society as. Sociologists now seem in agreement that in several ways Brazil is not unlike other multiethnic and multiracial societies.

CONTINUATION OF THE WHITE IDEAL   As we saw earlier, the amalgamation of Brazilian peoples has proceeded within the context of a white ideal. This essentially racist notion does not negate Brazil's integrative achievement relative to that of other societies, but it certainly reveals its less-than-democratic stimulus. The contradiction between pride in the development of a "racial democracy" and a social preference for whiteness is transparent. Brazilians proclaim that they are moving increasingly toward a racially amalgamated society in which color distinctions are gradually eroding. Yet there is no denying that this is an amalgamation in which the darker hues are—desirably—being diluted by the lighter hues.

The continuation of the white ideal as part of Brazilian culture is no better illustrated than in the phenomenal popularity of Xuxa (pronounced "SHOO-sha"), a Brazilian media superstar of the 1980s and 1990s. As one of the country's most celebrated entertainers, she enjoyed a popularity that at times seemed better described as idolatry. "The star's presence can be so commanding," noted one observer, "that people burst into tears at the sight of her" (Simpson, 1993:2). What made the Xuxa phenomenon so intriguing is that within the racial continuum of Brazil she was so thoroughly atypical: blond, blue-eyed, and fair-skinned. In short, she projected perfectly the white ideal.

LIMITED STRUCTURAL ASSIMILATION   Although cultural assimilation is apparent, structural assimilation of Afro-Brazilians at both the primary and secondary levels has lagged far behind. As explained earlier, blacks and mulattoes have not entered the higher echelons of Brazil's political and economic institutions at anywhere near a level commensurate with their percentage of the population.

Telles (2004) has characterized the contrast between race relations in Brazil and the United States as differences in horizontal and vertical dimensions. On the vertical dimension—people's position on the class hierarchy as determined by income, occupation, and education—African Americans are clearly in advance of Afro-Brazilians. As we observed in Chapter 7, although the collective gap between whites and blacks on most measures of social class have remained wide, expansion of the African American middle class has proceeded apace throughout the past half century. In Brazil, by contrast, blacks remain mired at the bottom of the class hierarchy. Conversely, on the horizontal dimension—integration into various institutions, including intermarriage, residence, and social interaction—the racial boundaries in Brazil are much less rigid than they are in the United States; social relations among black and white Brazilians, therefore, are more fluid.

CONFRONTING PREJUDICE AND DISCRIMINATION   The most glaring inconsistency between the ideology and the reality of ethnic relations in Brazil involves the irrefutable evidence of racial prejudice and discrimination. Efforts at attacking those beliefs and practices have been frustrated, ironically, by the ideology of racial democracy and by the lack of clear lines of racial division.

Traditionally, Brazil's nondiscriminatory ideology had been so preeminent that revelations or accusations of discrimination were customarily met with denials or disbelief. Smith (1963) noted the well-understood principle that any expression of racial discrimination should be attacked as "un-Brazilian." Even the suggestion of a "racial problem" was met with disapproval. In the late 1970s, for example, censors banned a program dealing with racial issues from Brazil's largest television network (Vidal, 1978). Official denials of discrimination have perpetuated the mystique of racial and ethnic democracy and have made any attack on discriminatory practices more difficult (Hanchard, 1994). If officially "there is no racial discrimination," public opinion cannot easily be galvanized to confront it. Only in the past few years have Brazilian officials begun to acknowledge the reality of racial prejudice and discrimination.

A further impediment to the elimination of ethnic discrimination in Brazil has been the fact that it is rarely blatant and aggressive. It might be called "prejudice and discrimination light." For example, although violations of the antidiscrimination law of 1951 were often prominently reported, most were dismissed by the authorities for lack of evidence. Anthropologist Robin Sheriff (2001), in her study of a Rio *favela* (a poor urban area), reports that her informants—poor and nonwhite—were certain that they were commonly discriminated against in applying for jobs for racist reasons, but they were also aware that in most situations they could not prove it. The subtle and mild forms of prejudice and discrimination also may cause blacks themselves to remain relatively unaware of them (Hanchard, 1994).[5]

THE EMPHASIS ON CLASS DISCRIMINATION   That most Afro-Brazilians are at the low end of the class structure makes the issue of discrimination more complex because class and ethnic factors are closely intertwined and difficult to distinguish. Here it is instructive to recall the debate among American sociologists regarding William J. Wilson's thesis of the "declining significance of race" (Chapter 7). As explained earlier, Brazilian whites—and many blacks—prefer to account for prejudice and discrimination against darker-skinned people as a product of their class, not their racial status. Furthermore, the explanation that Afro-Brazilians disproportionately occupy the lower social positions in the society because of class factors is more readily acceptable because this places the blame on "natural" social and economic forces, not those purposely designed to limit the placement of people. Also, failure to improve one's social position can be attributed, as in the popular American ideology, to the individual's shortcomings, further obscuring the effects of structural disadvantages (Hasenbalg and Huntington, 1982/1983). Accusations of class prejudice and discrimination are therefore accepted, and accusations of racism are denied. Blatant episodes of racial discrimination are treated as individual and exceptional cases, not as reflections of a customary pattern (Fontaine, 1981; Hanchard, 1994; Twine, 1998).

THE COLOR CONTINUUM   The difficulty of attacking prejudice and discrimination in Brazil is due not only to the effectiveness of the society's assimilationist and equalitarian ideology but also to the presence of a color continuum, as opposed to the more rigid color lines of the United States. Because racial lines are not inexorably fixed, there is little ground for organizing people on the basis of color. "In assimilationist circumstances," notes Ribeiro, "blackness is diluted on a broad scale of gradations, which breaks solidarity and reduces combativeness by insinuating the idea that the social order is a natural order, even a sacred one" (2000:157). Hence, the development of a strong civil rights movement, aimed at alleviating black deprivation and

---

[5] Telles (2004) found a more moderate level of residential segregation among Brazil's racial groups compared to the United States and other countries with large populations of African and European origin. That blacks, mulattoes, and whites live in similar neighborhoods, he concluded, "may strengthen the perception that race has little or no effect on life chances, at least for individuals of the same social class" (Telles, 1992:195).

discrimination, is undermined. How can there emerge a "black movement" based on racial consciousness and solidarity when there are varying perceptions and definitions of who is black? Blacks in the United States, by contrast, cannot escape racial prejudice and discrimination no matter what the shade of their skin and no matter what their class standing; all are part of the category "blacks." Creating solidarity among them therefore becomes easier despite internal group differences of class and status.[6] Although Brazilian civil rights organizations have emerged from time to time—such as the *Frente Negra Brasileira* (Brazilian Black Front), a movement of the 1930s whose designs were to integrate blacks into mainstream Brazilian society (Fernandes, 1971)—they have not displayed the endurance or the effectiveness of American organizations such as the NAACP or the National Urban League (Hanchard, 1994).

What's more, that people may marry lighter and thus produce lighter children further detracts from building enthusiasm for black political activism. Bastide quotes a black woman who, looking at her lighter-skinned children, declares, "They're white already. What's the use of fighting, forming leagues for the defence of the Negro, and all that?" (1961:12). This low level of racial consciousness and militance is also a result, in large measure, of the feeling of political leaders and intellectuals that to heighten racial awareness, as occurred in the United States during the black civil rights movement, would be "racist" and thus a denial of the idea of Brazil as a racial democracy (Skidmore, 1990). Regardless of its inconsistencies, the idea of racial democracy has been accepted by most Brazilians, and it continues to undermine efforts to galvanize blacks into an effective social movement (Hanchard, 1994; Marx, 1998; Sheriff, 2001; Twine, 1998).

A combination of an effective antiracist ideology, more benign forms of prejudice and discrimination, the complex relationship between color and class, and a multiracial classification scheme has served to make the long-range improvement in the status of Brazilian blacks perhaps even more problematic than raising black status in the United States. To confront racial prejudice and discrimination is, of course, an admission that racial prejudice and discrimination actually exist, which, as Degler has noted, "the national myth explicitly denies" (1986:271). A São Paulo city council member, introducing an antidiscrimination measure, stated that "[e]verybody behaves as if inequality doesn't exist. It's therefore much worse, because unlike in the United States, we've never faced it directly. It's disguised" (Schemo, 1995:A7).

CHANGING PATTERNS OF ETHNIC RELATIONS

During the past decade, noticeable changes have begun to occur in Brazilian race and ethnic relations. Increasingly, racial prejudice and discrimination have been openly acknowledged, even by government officials. As a result, steps have been taken to begin addressing the shortcomings of Afro-Brazilians in economic and social life. These efforts represent a radical shift in policy and in Brazilians' own perspective on race and its societal impact. Most important, the reality of racist thought and

---

[6] As Toplin (1981) has pointed out, however, in the United States, too, lighter-skinned blacks have always fared better than darker-skinned blacks. This pattern began during slave days, when mulattoes were often given house duty and more frequently granted manumission. The continuity of this pattern is evident in the traditional predominance of lighter-skinned blacks as part of the black upper class (Graham, 2000).

practices is no longer routinely denied; the myth of racial democracy has been questioned, stirring a sharp debate within the society. In a sense, Brazilians are now coming to terms with racial inequality in a fashion reminiscent of the United States at the outset of the civil rights movement in the 1960s. There has been a growing awareness of the handicaps faced by Afro-Brazilians and the role that color plays in the distribution of wealth and power (Daniel, 2006). Indeed, the increasing recognition of racial discrimination has led to American-flavored policies and views.

BRAZIL AND THE UNITED STATES: CONVERGENT PATHS?   Are Brazil and the United States becoming more similar on issues of race? Some observers of the two societies maintain that in some ways they do seem to be on convergent paths (Daniel, 2000, 2006; Degler, 1986; Skidmore, 1993, 2003; Toplin, 1981).

First, in the United States, as we saw in Chapter 7, the lower socioeconomic standing of African Americans is increasingly attributed to class factors as the more blatant and legitimized forms of racial discrimination are eradicated. In Brazil, however, the tendency seems to be just the reverse: racial discrimination is becoming more obvious and acknowledged as a factor in black deprivation, and traditional class explanations have been seriously questioned.

Another converging point is that the Brazilian traditional fuzziness in defining racial categories and the denial of race entirely is now under serious question. As a result, "official" racial boundaries appear to be emerging, resembling the American system. This is occurring at the same time that racial definitions and boundaries are becoming less distinct in the United States. G. Reginald Daniel (2000, 2006) points out that whereas the United States has begun to move away from its traditional binary racial classification system to one in which mixed racial, or "multiracial," identities become more commonplace, Brazil has moved in the opposite direction, toward a more traditionally American binary system.

The introduction of affirmative action policies in Brazil represents a third point of convergence with American patterns of race and ethnicity. At a time when these policies have come into serious question, both legally and otherwise, in the United States, they are being advanced for the first time in Brazil (Bailey, 2008; Dzidzienyo, 2005; Fry, 2000; Rohter, 2010; Telles, 2009).

As part of an emergent movement to deal with the reality of racial inequality, the Brazilian government, starting in the early 2000s, adopted affirmative action measures that closely resembled those in operation for decades in the United States. The targets were government employment, university admissions, and businesses competing for public funds (Martins et al., 2004). Affirmative action policies have now either been instituted or are under consideration by all of Brazil's major political and economic institutions.

The most ambitious compensatory efforts have come in higher education, where blacks have been grossly underrepresented. Admission to Brazilian public universities is difficult and highly selective; entrance is determined on the basis of the *vestibular*, the national college entrance exam. Not surprisingly, those who qualify are mostly graduates of private secondary schools, which prepare them well for the exam. The effect of this system is that those admitted to university are likely to be middle class or above—and white. Nationally, no more than 3 percent of students admitted to public universities identify themselves as black or brown, and less than

20 percent come from public schools, which most nonwhite students attend (Downie and Lloyd, 2010; Jeter, 2003; Rohter, 2003). Affirmative action measures have been aimed at addressing these structural inequalities. In 2012, legislation called "the Law of Social Quotas" was enacted that took affirmative action to a new level. It mandated that all public universities were expected to admit half of their entering classes from public schools, thereby raising significantly the number of black university students (Romero, 2012).

THE DILEMMA OF AFFIRMATIVE ACTION IN BRAZIL   The use of affirmative action to address issues of inequality has taken on many of the same controversial trappings as it has in the United States—whether such compensatory policies should reward people on the basis of group membership as opposed to individual merit; whether they can effectively help those most in need; whether they solidify, rather than weaken, racial and ethnic boundaries; and whether they lower standards. Their legal status has also been challenged.

In addition, however, the controversy has displayed a unique Brazilian twist in that it throws into question the very notion of how race is conceived. Classifying people racially using clear-cut lines of division, as in the United States, and subsequently using those classifications to apply affirmative action measures (also like the United States) contradicts the popular notion that racial discrimination does not occur in Brazil, and challenges Brazil's self-proclaimed color blindness. Even more basically, it forces Brazil to rethink its essential perspective on race. Racial and ethnic boundaries have never been clearly defined, as they have in the United States. But to apply and enforce policies of affirmative action makes imperative some system that explicitly defines racial and ethnic groups. To determine who will be the beneficiaries of affirmative action dictates determining who is "black" and who is "white." One of the strongest points of opposition to affirmative action in Brazil is the fear that creating such a system of racial definition will only bring Brazil closer to the U.S. racial model and will create American-style racial divisions, which have never existed in Brazil. As one opponent expressed it, "We had a goal to create a society that didn't look at people in racial terms. Now we are creating a race question" (Davies, 2003).

LOOKING AHEAD

Affirmative action has set off a profound debate within Brazil regarding the society's traditional racial/ethnic system (Bailey, 2008; Dzidzienyo, 2005; Guimarães, 1999; Skidmore, 2003). There are those who favor the adoption of a more American-like approach to racial inequality, using affirmative action to improve the position of blacks. This, as noted, would necessitate the marking of explicit racial boundaries, something never done before. On the other side are those who assert that the system of racial classification in Brazil, along with its ambiguities and inconsistencies, is uniquely Brazilian and should be retained. Racial democracy in this view is, correctly, the society's goal—if not the reality—of race relations. The movement for change in Brazil's racial politics seems intent on finding a way to reconcile these two positions—to promote greater racial equality without completely abandoning the unique Brazilian racial mixture and the ideal of racial democracy (Fry, 2000).

Brazil is embarking on a new course in its racial-ethnic system and will surely be the focus of attention of social scientists, policy makers, and others who continue to wrestle with the complexities of racial and ethnic policies. The very fact that issues of racial inequality are now part of the public discourse represents a fundamental change for Brazil. A black filmmaker, observing the change, puts it this way:

> Whatever happens with quotas, Brazilians are talking about the issue of race and discrimination and there's no going back. My white friends will ask me what I think about racism in Brazil or the quotas, and I always tell them: "The answer is in your question. Three months ago, you would never even have asked." (Jeter, 2003:A1)

Brazil presents a complex and often paradoxical case of ethnic relations. In its commitment to racial amalgamation, it has few peers among current societies. Similarly, its embracing of the theme of ethnic equality has been unwavering since the abolition of slavery. Furthermore, the inconsistencies between ideology and reality cannot repudiate the fact that interethnic relations in Brazil are more benign than in other societies with diverse racial-ethnic groups.

If Brazil is not the racial and ethnic democracy its advocates have proclaimed, it is nonetheless a society in which giant strides have been made toward ethnic assimilation. At the same time, however, Brazil demonstrates the difficulties of mitigating the intergroup conflict that, as noted in Chapter 4, seems endemic to multiethnic societies.

SUMMARY

- Brazil's ethnic composition and the historical development of its ethnic diversity closely parallel the American experience; the main focus of ethnic relations is the confrontation of blacks and whites.
- Brazil was an integral part of the African slave trade, and its system of slavery endured even longer than its American counterpart.
- Other racial and ethnic elements are present, including many European ethnic groups that entered as voluntary immigrants, and indigenous Indians.
- Debate among historians and social scientists concerns differences in the practice and aftereffects of slavery in Brazil and the United States. Some hold that the Portuguese brand of slavery in Brazil was more humane than the Anglo-American version and consequently produced more benign black-white relations following its abolition. Others have challenged that view.
- Ethnic stratification in Brazil features relatively fluid boundaries between groups. Racial classification is not a dichotomy but a continuum in which groups blend into one another. Racial categories are based not on ancestry but on physical appearance and social class.
- Despite blurred lines of division between groups, there is an ethnic hierarchy in which whites maintain a generally advantageous status over blacks and mulattoes, the latter a large intermediate racial grouping.
- Although Brazilians proclaim themselves a society free of prejudice and discrimination based on race and ethnicity, these are present, though in forms milder than those found in the United States. Blacks are the main objects of this largely subtle ethnic antagonism.
- Brazil is an assimilationist society whose long-range objective is ethnic convergence. And it has, in fact, seemed to progress further toward both physical and cultural assimilation than most other multiethnic societies of the world.

- The ideology of racial democracy and the illusion of the nonexistence of ethnic prejudice and discrimination have been challenged in recent years. Brazil has embarked on a national debate regarding the impact of race, and policies have been instituted to address what has now been acknowledged as deeply rooted racial inequality.

COMPARING BRAZIL AND THE UNITED STATES

Points of Similarity
- The historical development of the racial/ethnic system in both societies was parallel: a combination of conquest of the indigenous (Indian) population by Europeans, importation of an African slave population, and immigration primarily from Europe.
- Slavery was institutionalized in both societies for more than two centuries.
- Blacks in both societies (as well as mulattoes in Brazil) are the primary minority ethnic group. In both societies, blacks rank lowest on various social and economic indicators, including income, occupation, education, and political power.

Points of Difference
- Racial divisions are less fixed in Brazil. Numerous racial categories are used to designate people, based on observable physical traits, not on ancestry.
- Throughout its history, Brazil has encouraged amalgamation between whites and blacks, mainly as a way of diluting the black element of its population.
- The African influence, derived from slavery, has remained a more vibrant and expressive part of Brazilian culture.
- Race relations historically have been more benign in Brazil.

CRITICAL THINKING

1. Explain the observation that in Brazil people are not "race" conscious, but "color" conscious. How is this different from the United States?
2. Examine how "race" is conceptualized in Brazil and how it is conceptualized in some Caribbean societies, like Cuba or Puerto Rico. Are there similarities? In turn, are patterns of race relations in those societies comparable to Brazilian patterns? What accounts for the similarities (or differences)?
3. Does the increasing breakdown of racial/ethnic classifications in the United States, as well as the idea of "mixed race" individuals, signify movement in the direction of Brazil's more fluid system of race and ethnic relations?
4. Explain how the adoption of affirmative action measures in Brazil is working to create stronger, more formal, racial boundaries than existed previously.

PERSONAL/PRACTICAL APPLICATION

1. If you were black, would you prefer to live in Brazil or the United States? Explain your preference.
2. How would you use the Brazilian case to demonstrate to someone who conceives of races as real and inflexible divisions of humans that races are in fact not fixed but are created in variable social contexts?

CANADA

CHAPTER 16

Ethnic Model of the Future?

If you were to ask Americans to name the capital of Russia, a society very distant from the United States, most would probably answer correctly. If asked to name the capital of Canada, a society bordering the United States for the entire length of the continent, most would be hard pressed. One reason Americans pay so little attention to and know so little about Canada is that its sociological features are assumed to be replicas of their own. In the American mind, Canada might easily pass for the fifty-first state.

Undeniably, the economic and cultural hegemony of the United States on the North American continent for two centuries has created great cross-border similarities. But the fact that Americans can easily feel at home in Canada tends to disguise some fundamental differences. Common cultural preferences, consumer patterns, political alliances, and, for two-thirds of Canada, language do not make Canadian society simply a northern microcosm of the United States. These differences are nowhere more apparent than in the realm of ethnic relations.

ETHNIC DIMENSIONS OF CANADA

Canada's ethnic structure is today extremely diverse. Moreover, it displays some features of the corporate pluralistic model described in Chapter 4. Essentially this society features three ethnic dimensions:

- the division between two founding groups, British and French, each of whose language and culture have been legally protected since the eighteenth century;
- a large and growing non-English, non-French pastiche of ethnic groups, whose national origins represent virtually every region of the world;
- indigenous groups, referred to as Aboriginal people.

DIMENSION I: THE ENGLISH-FRENCH SCHISM

Most of Canada's French-speaking people (Francophones) live in the province of Quebec, creating for them, in addition to their linguistic and cultural distinction, a territorial base. English-speaking Canadians (Anglophones), by contrast, are dispersed throughout Canada and include groups of various national origins. Those of British origin, however, are recognized as the cofounding group and remain the largest segment of English Canada. The current language breakdown in Canada is shown in Figure 16.1.

In a sense, then, Canada is not one nation but two. As we saw in Chapter 1, when ethnic groups occupy a definable territory—like Francophones in Quebec—they also commonly maintain or aspire to some degree of political autonomy. Hence, they are "nations within nations." In a real sense the Canadian experiment in national dualism is still in process, and the issues separating the English and French segments have by no means been resolved. Indeed, in recent decades the historical discord between these two major cultural groups has periodically intensified, making the union of Canada precarious.

The relations between English and French Canadians constitute the first and most important dimension of ethnicity in Canada, cutting across other lines of ethnic diversity. As such, English-French relations will occupy much of our attention in this chapter.

DIMENSION 2: NON-ENGLISH, NON-FRENCH GROUPS

The various ethnic groups whose origins are neither British nor French together comprise more than one-third of the Canadian population. Canada is a society of virtually dozens of distinct ethnic groups varying in national origin, culture, and racial features. Canada today is one of the world's leading immigrant destinations, and most of the newcomers are *non-European,* or what in Canada are referred to as **visible minorities.** South and East Asian groups are the largest, but the ethnic diversity of Canada's changing population is virtually unmatched by any contemporary

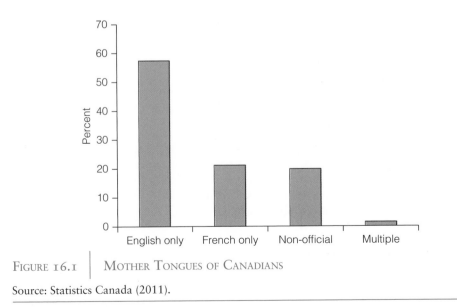

FIGURE 16.1 | MOTHER TONGUES OF CANADIANS

Source: Statistics Canada (2011).

nation. It is the presence of these non-French, non-English ethnic groups that make Canada today a strong example of what was referred to in Chapter 4 as a corporate pluralistic or multicultural society. The important social issues pertaining to these ethnic groups involve the nature and extent of their cultural and structural integration.

DIMENSION 3: ABORIGINAL GROUPS

The third dimension of ethnicity in Canada involves the relations of native (Aboriginal) peoples with the rest of Canadian society. *Canadian Indians,* usually referred to as **First Nations**, and Inuit (Eskimos) are, along with their American counterparts, the original peoples of the continent. Today, they are a relatively tiny proportion of the Canadian population. Their official status, their cultural systems, and the nature of their relations with other groups, however, largely set them apart from the rest of the society. Hence, they must be looked at as a unique component of the Canadian ethnic configuration.

These three dimensions create a complex ethnic picture in which the different elements are interwoven in complicated and, to American eyes, often bewildering combinations. Moreover, this ethnic diversity, particularly in recent decades, has given rise to public policies regarding immigrants and ethnic groups generally that are markedly different from those of the United States. Whereas the United States has, for most of its history, been committed to an assimilationist approach, the Canadian philosophy has traditionally been more pluralistic, and thus more tolerant of the continued expression of cultural differences among diverse groups.

FORMATION OF THE CANADIAN ETHNIC MOSAIC

We can start to make sense of Canada's complex mosaic by looking initially at the historical process by which ethnic diversity in this society has taken shape. We begin with the first and most critical dimension of Canadian ethnic relations, the English-French division.

THE EVOLUTION OF TWO NATIONS: THE ENGLISH-FRENCH DIVISION

The confrontation of English and French groups consumed the affairs of state from the outset of Canada's history and continues to play the preeminent role in ethnic relations.

THE ENGLISH CONQUEST   Both Britain and France established colonies in North America beginning in the early seventeenth century and vied for continental dominance for almost a century and a half. The victory in Quebec in 1759 of the British forces over the French marked the end of France's colonial ventures in North America, but it did not eliminate the presence of a French cultural group. New France (Quebec) was now made politically part of the British colonial empire, but its French inhabitants were granted the right to certain cultural privileges, including the retention of French civil law, the use of the French language, and the practice of Catholicism. The price of this cultural autonomy, however, was that French Canada was placed in the position of a permanent minority within an English milieu. From the moment of British ascendancy in Quebec, French Canadians were consumed with avoiding assimilation into the English-speaking North America that surrounded them. And it is this objective of cultural survival that remains at the heart of today's English-French schism.

During and after the American Revolutionary War, thousands of English-speaking emigrants from the thirteen American colonies who had remained loyal to Britain entered Canada. Many of these United Empire Loyalists, as they were called, settled in Quebec. This gave the province for the first time a substantial English-speaking population, which quickly came into conflict with the French majority over economic and political issues. In an attempt to separate the two groups, Britain established an Upper and a Lower Canada, the former composed mostly of English-speaking people, the latter of French-speaking. These would later become the provinces of Ontario and Quebec. The national duality was thus firmly established, with two peoples, differing essentially in language and culture, juxtaposed geographically and socially.

Although the French in Quebec were not threatened in a numerical sense by the influx of English-speaking people, this period of English-French relations marked the establishment of English dominance of the Quebec economy. With the exit of the French commercial elite following the British conquest, business and financial activities became the domain of the English, and the French Canadians remained mostly on the land as subsistence farmers. This ethnic division of labor became a fixed arrangement that lasted into the modern era, making the French a numerical majority but an economic minority in their own province.

The disparities in power, wealth, and social position between English and French in Lower Canada led to the emergence of a French nationalist political element that sought the colony's independence. An open rebellion ensued in 1837, after which Britain consolidated Upper and Lower Canada into a single union. The underlying objective of this move was the eventual assimilation of the French into the English colonial society. To bring about this end, discriminatory measures were enacted assuring Anglo dominance.

In addition, with the failure of the rebellion, the conservative forces of French Canada were now ascendant. The clergy, in particular, reinforced its influence in the life of the French-speaking people and extended the doctrine that the French-Canadian culture could be sustained only through loyalty to the Catholic faith and the French language. This traditional nationalism, combining a rural-oriented way of life with a staunch Catholicism, held sway among the French-Canadian masses until the mid-twentieth century. The clergy, then, assumed the dominant institutional role in Quebec, promoting détente with the English colonialists as a means of ensuring *la survivance,* the survival of the French culture.[1]

CONFEDERATION   In 1867, the Canadian provinces were linked in a federal system. The effect of confederation was to further isolate Quebec from the rest of Canada and to heighten the minority status of French Canadians. Outside Quebec, Francophones

---

[1] A classic description of this traditional Quebec culture is found in Miner (1939). Later essays are contained in Rioux and Martin (1978).

found themselves defenseless against the Anglophone majority, which chose to ignore the need for French schools and other institutions. The powerlessness of French Canadians was driven home in numerous instances in which minority interests or wishes were swept aside by the dominance of English Canada.

For French Canadians perhaps the most humiliating aspect of the relationship with English Canada was the minority status to which they were relegated in their own province. English control of Quebec's commerce and finance produced an English-speaking business elite, assuring that higher-status positions would remain the reserve of Anglophones. Although the French and English languages were both protected by law, only in Quebec was the principle of bilingualism instituted, and there it was strictly one-sided—Francophones were forced to use English in dealings with the Anglo Quebecers.

The Catholic Church's control of education in the province also handicapped the upward mobility of the French-speaking masses. Expounding the philosophy of French-Canadian cultural survival through retreat to church and land, the clergy emphasized the humanities, classics, and religion in schools and colleges—not the commercial and technical skills that were appropriate to an industrial system. Hence, few French Canadians were prepared for skilled positions in business and science. As French Canada increasingly industrialized, Francophones therefore occupied the least-skilled and lowest-ranking jobs.

THE TWO SOLITUDES Industrialization beginning in the early decades of the twentieth century set in motion social processes that would gradually spell the demise of the pastoral, clerical Quebec culture. Urbanization accompanied the growth of industry. Whereas Quebec had been 60 percent rural in 1900, by 1931 it was 63 percent urban (Legendre, 1980). The focus of this rural-urban movement was Montreal, which tripled in size from 1901 to 1921. Industrialization and urbanization conflicted with the traditional Catholic nationalism, which continued to stress agrarian and ecclesiastical values. The customary culture of Quebec society was clearly no longer compatible with the forces of modernization.

The industrialization of Quebec, however, was not led by French-Canadians but by outsiders—specifically, English-Canadian, American, and British capitalists. Quebecers thus found themselves in a colonial-like situation. The skilled and more desirable jobs naturally went to the Anglophones, and the French speakers, though a numerical majority, were given the unskilled, menial positions. By the 1960s French Canadians were the most poorly paid workers in Quebec, below even the newly arrived immigrants from Europe, and their standard of living was below that of the average Canadian (Royal Commission on Bilingualism and Biculturalism, 1969a).

The English of Quebec, with their own institutional structure—schools, businesses, churches, and neighborhoods—as well as control of the most important commercial and financial institutions, could, by and large, ignore the French around them. They made little effort to learn the French language, for it was unnecessary. On the contrary, because so many jobs were controlled by the Anglophone business class, all the pressure was on the Francophones to learn English. In his classic account of a French-Canadian community of the 1930s, Everett C. Hughes noted that not only was there no need for the English residents of the town to speak

French but to do so in any case would have upset the subordinate-superordinate relationship:

> In fact, the English do not have to learn French to keep their position in industry. The housewife does not have to learn French to keep her housemaid. If they were to speak French in these relationships—except in a joking or patronizing spirit, as is occasionally done—they would be in some measure reversing roles. For they would then be making the greater effort, which generally falls to the subordinate; and they would speak French badly, whereas the subordinate generally speaks English pretty well. (1943:83)

It was a situation described by the novelist Hugh MacLennan (1945) as "two solitudes," French and English living side by side but essentially in different social worlds.

THE QUIET REVOLUTION The decade of the 1960s in Quebec is often referred to as the "Quiet Revolution," for it was then that the transformation of French-Canadian society that had begun in the early part of the century culminated in the emergence of a powerful nationalist ideology espoused by Quebec leaders determined to make Francophones *maîtres chez nous*—"masters in our own house." Though revolutionary only in a figurative sense, the changes brought about during this period basically redefined both the role of government in Quebec and the identity and goals of French-Canadian society.

The Quebec provincial government traditionally had been a force in sustaining the status quo, but it now became the principal vehicle of change. Responsibility for health, welfare, and especially education became its concerns, whereas in the past these had been largely church-sponsored institutions. In the economic realm, the provincial government pursued the objective of improving the position of Francophones and ending Anglophone control of the important segments of the economy (McRoberts, 1988). In short, with the development of a dynamic state led by a new middle class, technologically skilled and ideologically committed to a strong and self-sufficient society, Quebec by 1970 was well on the way toward a social and political transformation.

THE RISE OF THE PARTI QUÉBÉCOIS The new nationalism spawned strong sentiments favoring the separation of Quebec from the rest of Canada and the formation of an independent French-speaking nation. In the late 1960s, elements of the Francophone leadership that had emerged from the Quiet Revolution began to promote the idea that there was no longer any middle ground between partnership within the Canadian confederation and full sovereignty for Quebec; the latter was the only realistic option if Quebecers were to fully control their destiny. To consolidate and advance the achievements of the Quiet Revolution, they felt that Quebec had to fully sever its ties with Canada (Posgate, 1978). Led by René Lévesque, an articulate and charismatic spokesperson, they left the ruling Liberal Party (which, ironically, had initiated the radical changes of the Quiet Revolution) and formed the Parti Québécois (PQ), declaring as its objective the political liberation of the Francophone community.

SOVEREIGNTY-ASSOCIATION Recognizing the need to retain an economic link with the rest of Canada, however, Lévesque created the notion of "sovereignty-association," which proposed that a politically independent Quebec would retain close economic ties to Canada. Nothing short of the removal of Quebec from its historical place in the Canadian union, however, was the avowed aim of the PQ. In its view, the split with English Canada was absolute and irreversible. "First and foremost I am a Québécois," declared Lévesque, "and second—with a rather growing doubt—a Canadian" (Saywell, 1977:4).

By the early 1970s, the growing support among Quebecers for the new, radically nationalist PQ was evident and alarming to English Canada. Even more alarming were the activities of a small extremist separatist group, the *Front de libération du Québec* (FLQ), whose violent tactics and revolutionary ideology provoked a crisis in 1970 prompting the federal prime minister, Pierre Trudeau, to briefly invoke the War Measures Act, giving the police sweeping powers to arrest and detain suspected people.

This was a wrenching experience for Quebec and all of Canada that forced the separatists to moderate their proposals for change. In response, Lévesque stressed the democratic processes by which Quebec's independence was to be attained and emphasized the "economic association" aspect of the envisioned sovereignty-association relationship with Canada. The PQ declared that if it were elected to power, it would put the issue of sovereignty-association before the people of Quebec as a referendum, to accept or reject.

In 1976, only eight years after its formation, the PQ won election as the government of Quebec. Its stunning victory sent shock waves throughout English Canada. For the first time, the contemporary Quebec nationalist movement drew the serious attention of even the United States. What had been feared since 1970 had occurred— the election of a provincial government ideologically committed to the separation of Quebec from Canada, led by the charismatic Lévesque. It now seemed that the division of Canada into two separate nations, in a real rather than symbolic sense, was possible.[2]

THE FIRST DEFEAT OF SOVEREIGNTY-ASSOCIATION   In 1980, the referendum on sovereignty-association that Lévesque had promised was held in Quebec. Lévesque downplayed the independence aspects of the measure and emphasized that approval of sovereignty-association meant only that Quebec would enter into "a new partnership" with Canada. Moreover, the referendum was worded so that approval was only a first step toward political independence. It would give the Quebec government merely a mandate to negotiate a new arrangement with Canada, approval of which in any case would be left to the people through another referendum. Most people, however, interpreted it more simply as a yes or no vote for Quebec sovereignty.

By a margin of three to two, Quebecers rejected the idea of sovereignty-association. Not unexpectedly, virtually all Anglophones voted no, but a slight majority of Francophones did as well. For the moment, then, the intensity of the Quebec nationalist movement had been quelled. The fundamental issue of the relationship between Quebec and the rest of Canada, however, had not been resolved.

---

[2] Often overlooked in accounts of this volatile period is the fact that much of the support for the Parti Québécois was based not on its advocacy of separatism but on its promise to deliver "good government" and on public disaffection with the relationship between Quebec and the federal government (McRoberts, 1988).

SEPARATION: THE RECURRENT ISSUE   Two failed attempts to have Quebec become a signatory to the new Canadian constitution of 1982 led to increased nationalist sentiment among Quebecers in the early 1990s and, when the PQ was elected once again as the Quebec provincial government, the stage was set for a new effort at separating Quebec from Canada. The party leaders had made the sovereignty issue, as before, the core of their electoral program. As in the past, their message was familiar: Quebec would never be afforded the kind of unique place in the Canadian confederation that would allow it to protect and promote its language and culture; thus political independence was the only logical alternative. To temper the uncertainties of going it alone, however, Quebecers were promised that after separation the province would continue to interact closely with the rest of Canada in economic matters, similar to what had been proposed by Levesque in 1980.

In October 1995—fifteen years after the 1980 proposal had been defeated—a referendum on sovereignty was once again held in the province. As in 1980, the measure failed, but this time by the barest of margins—50.6 percent to 49.4 percent. A closer analysis of the results revealed that Francophones in the province had supported the separatist proposal by 60 percent to 40 percent, whereas Anglophones and Allophones (immigrants speaking neither French nor English) had roundly voted against the measure by over 90 percent.

THE CURRENT POLITICAL CLIMATE   The crux of the issue of Quebec separatism is a view of Canada—and thus Quebec's place within it—that is very different between Quebecers and English-speaking Canadians. The latter conceive of Canada as a nation of individual citizens who are equal before the law, regardless of their ancestors' language or ethnic origin, and who live in a federation of provinces with equal constitutional status. Quebecers, in contrast, view Canada as a nation of collectivities defined primarily by language, specifically English and French (Bercuson, 1995). Because they are a cofounding group in Canada and are heavily outnumbered by English-speaking Canadians, most Quebecers feel that they are entitled to special status and should not be subject to the same institutions that are dominated by Anglophones. This is a position that most of the rest of Canada has refused to accept.

In the current political climate in Quebec, the fury and passion of hard-line separatists seems to have been quelled as a younger population, secure in its French language and no longer obsessed with sovereignty, has emerged. Quebec itself in the last two decades has become a more multiethnic province with the infusion of thousands of immigrants, most of whom have little knowledge of—or interest in— the issue of separation. A significant Francophone element, however, continues to harbor separatist sentiments and remains an important political force.

Most Quebecers seem to favor an arrangement in which Quebec would accept its place in Canada if a renewed federalism, recognizing the province's unique status, could be achieved (Séguin, 2005). But the Quebec sovereignty issue has proved to be subject to mercurial changes. It thus remains a perplexing—and perilous— issue for Canada that may resurface with great force at some point in the future. Indeed, the intransigence of an element of Quebec society to accept nothing short of political independence, as well as the low-level animosity that seems so often to define the relations between Quebec and the rest of Canada, promise to eventually push the issue forward again. As one observer has put it, the issue of Quebec separation may simply remain "a recurrent component of the Canadian agenda" (Banting, 1992:161).

LANGUAGE AND IDENTITY: BASIC FEATURES OF THE DUAL NATION

Beyond the politics of Quebec sovereignty, language plays a more basic cultural role in Canada. From the outset of Canadian history, language has served as the most critical feature separating French and English communities; it continues to evoke strong emotions on both sides and thoroughly encompasses all facets of division between Canada's two nations. It would not be overstating the case to assert that the question of language was at the core of all historical conflict between English and French Canada from the time of the British conquest. It is no less so today, and therefore warrants further examination.

LANGUAGE AND THE PRESERVATION OF QUEBEC CULTURE   Language is the very foundation of any people's culture. Hence, it was long felt that assimilation into the dominant English-Canadian society would be inevitable for French-speaking Canadians if measures were not taken to assure the preservation of the French language. Moreover, language for Quebecers, as for many ethnic collectivities in multiethnic societies, is the key symbolic marker setting them off from other groups. Quite simply, it is what defines the French nation in Canada. As Lévesque put it, "Everything else depends on this one essential element and follows from it or leads us infallibly back to it" (1968:14).

In hopes of quelling once-and-for-all Quebec separatist sentiments, the federal government in the late 1960s, led by Prime Minister Trudeau, began to promote the notion of a truly bilingual Canada. Responding to the recommendations of the Royal Commission on Bilingualism and Biculturalism, which had been charged with studying how a more equitable balance could be established between the two founding peoples, the government adopted and avidly supported a policy of official bilingualism. To Quebecers, however, this effort was seen as essentially meaningless: many were already bilingual—of necessity.[3] What was of greatest concern to Quebecers was the preservation of the French language and culture in Quebec, *la survivance*.

QUEBEC LANGUAGE POLICIES   Following the Quiet Revolution, Francophone Quebecers faced two demographic trends that threatened the survival of their culture and language. The first of these trends was a lowered birthrate. Historically, Quebec had maintained the highest fertility rate in Canada. Thus, even though few immigrants from France were attracted to Quebec after the colonial period, the ratio of French to English in Canada had held relatively constant for two centuries. This was

---

[3] Despite the fact that Canada is officially a bilingual country, less than 20 percent of the total population is able to converse in both English and French, and more than half of these bilinguals live in Quebec (Statistics Canada, 2012). Except for Quebec, only New Brunswick among Canada's ten provinces has a large bilingual population.

derisively referred to as the "revenge of the cradle," the implication being that the complete English conquest was foiled by the high French birthrate. Quebec's extraordinarily high birthrate, however, fell dramatically beginning in the 1960s, becoming the lowest in all of Canada.

The second ominous trend was the influx of non-French-speaking immigrants to Quebec, specifically to the city of Montreal. Like Toronto, starting in the 1950s it became an attractive destination for large numbers of European immigrants and, later, immigrants from other world regions as well. Most of these newcomers, however, chose to adopt English as their new language, seeing it as a more practical option in adapting to social and economic life in North America. So, in addition to the declining birthrate of French speakers, immigration now threatened to further dilute the Francophone population.

In response to these demographic threats, aggressive pro-French-language policies were enacted by successive Quebec governments beginning in the 1970s. The basic objective was to bring about the primary use of French in all spheres of Quebec's public life—including, most important, business and education.

The first move in the direction of French-language preeminence in Quebec came in 1974, when the provincial government declared French the official language of the province and made access to any but French-language schools difficult. This created strong opposition from Montreal's English business community as well as the city's large number of immigrants, who sought to maintain their right to choose the language of instruction for their children. Ironically, the measure was unacceptable to the separatists as well, but for the opposite reason: they felt it did not go far enough in establishing the primacy of the French language. In 1977, therefore, Lévesque's PQ government enacted a more sweeping and radical language law, Bill 101, which mandated that French would be the prime, if not sole, language used not only in official matters but also in commerce and industry. Requiring the use of French in the workplace, it was felt, would attack the Anglophone domination of economic institutions directly. No longer could managerial positions be denied Quebecers on the basis of language.

The response of the English-dominated business world of Montreal was indignation. Rather than acquiescing to the rigorous new language regulations, many large corporations chose to move their offices to Toronto and other Canadian cities or at least to limit their Montreal operations. This proved economically costly to the province.

Even more controversial was the effect of the language law on Quebec's educational system. Bill 101 set more stringent restrictions on instruction in English, prompting a further out-migration of Anglophones. But the law's primary target, many believed, was the large immigrant population in Montreal whose mother tongue was neither French nor English (Arnopoulos and Clift, 1980; Levine, 1990). Most of these groups had ordinarily chosen English-language schools for their children, but under the new law, almost all were now required to send their children to French schools.

Over the years, Quebec's language laws have been modified and in many cases legally challenged. Because of its ethnic diversity, Montreal continues to be the focal point of the language debate. But the context of the issue has been fundamentally changed. Despite periodic controversies, the primacy of the French language in

Quebec is no longer disputed. French, today, is the first language of all major institutions, government, business, and education. Moreover, Francophone Quebecers, for the most part, no longer see their language under threat; most, in fact, believe it is important to be bilingual (Patriquin, 2013).

QUÉBÉCOIS ETHNIC IDENTITY   French-speaking Canadians are not simply one more ethnic group within the Canadian mosaic. The French in Canada are a people with a linguistic and cultural autonomy formally recognized from the very founding of the society. Moreover, unlike other Canadian ethnic groups, Francophone Canadians maintain a territorial base. For Canada's French-speaking people who live outside the province of Quebec, the tendencies toward assimilation into the dominant English-speaking society have been evident for many decades (Joy, 1972). More important, the contemporary nationalist movement in Quebec has jettisoned these French-speaking non-Quebecers. It is Quebec alone that has become, in the eyes of the separatists, the only meaningful base for a sovereign French-speaking society in North America.

Manifestations of ethnic identity among Quebecers also illustrate their uniqueness within the Canadian ethnic system. Unlike English Canadians, French Canadians—specifically, those in Quebec—are faced with a dual national identity. Whereas people in the rest of Canada think of themselves simply as "Canadians," those in Quebec are both Canadiens and Québécois (Quebecers).[4] In recent decades, the latter identity has, for most, seemed to take precedence (Centre for Research and Information on Canada [CRIC], 2002). This dual identity demonstrates well the idea of a unique French-Canadian nation in which people's allegiance and consciousness of kind do not focus necessarily on the same national unit perceived by other Canadians (Brunet, 1969). One must keep in mind, of course, that there are different degrees of nationalism among Quebecers. Some are vehemently and radically nationalistic and are unrelenting advocates of separatism, whereas others see themselves more as part of a unified Canada. What characterizes all Francophone Quebecers, however, is an understanding that they are culturally different from other Canadians.

CANADA'S OTHER ETHNIC GROUPS: THE THIRD FORCE

Although the division between English and French Canada overshadows all other ethnic issues, Canada today is far more ethnically diverse than is reflected in this schism. In fact, Canada today comprises a population of more than two hundred different ethnic origins, making it one of the most heterogeneous multiethnic societies in the world. And it is the groups that are neither English nor French in origin that stand out increasingly in the Canadian ethnic picture. They comprise more than one-third of the total Canadian population and thus play an enormous role in all aspects of ethnic relations and policy. They constitute an extremely diverse spectrum, including groups from all parts of Europe, East and Southeast Asia, the Indian subcontinent, the Middle East, the Caribbean, Africa, and Latin America.

---

[4] The inhabitants of the original French colony in Canada called themselves *Canadiens*. Only later was the term *Canadian* adopted by the English. Hence, *Canadien* connotes, in the French-Canadian view, the idea of the original or true Canadians (Brunet, 1969).

It is obvious that the size of these groups as well as their cultural variety have basically altered Canada's ethnic composition. To speak any longer of "English Canada" or even of "French Canada" is a misnomer, as those whose ethnic origins are neither British nor French continue to comprise a greater and greater share of Canada's population. Today, Canada is very much an immigrant society, particularly in its largest cities—Toronto, Montreal, and Vancouver—and the bulk of those immigrants come with neither British nor French roots.

The shaping of Canada's ethnic diversity can be seen historically as a process encompassing two major periods of immigration—the influx of peoples during the late nineteenth and early twentieth centuries, and the post–World War II stream, which continues today. Each period differed in the origins of the major immigrant groups and in the groups' settlement patterns.

THE PERIOD OF WESTERN SETTLEMENT　Although non-British and non-French groups had been present in Canada before the late nineteenth century, they were not sizable enough to make a significant impact on the society's ethnic composition. In 1871, of a total population of 3.7 million, the largest of these groups were the Germans, who numbered 200,000; others were considerably smaller (Hawkins, 1972). Throughout the nineteenth century, the British—English, Scots, Welsh, and Irish—dominated immigration and reinforced the Anglo-Canadian preeminence (exclusive of Quebec, of course) both numerically and culturally.

In the latter years of the nineteenth century, spurred by the promotion of western settlement, large numbers of European immigrants were attracted to the prairie provinces of Saskatchewan, Manitoba, and Alberta. Germans and, to a lesser extent, Dutch were the major immigrant groups until the early 1900s, when people from central and eastern Europe, especially Ukrainians, entered in large numbers. Americans, eager to acquire lands that were no longer available after the closing of the U.S. frontier, were also a significant immigrant group during this time (Troper, 1972).

Although immigrants did settle in Canada's cities, the objectives of nineteenth-century immigration policy were mainly to attract people who would develop the agricultural potential of the vast western domain. This did not mean an open-door policy, however. British peoples, or at least those from northwestern Europe, were deemed most desirable. Ukrainians and others from eastern and central Europe, though technically not among these, were recruited nonetheless because their agricultural background was considered more critical than their national origin. What the west needed above all were farmers; therefore, even nonconforming religious sects like Hutterites, Mennonites, and Doukhobors were promised parcels of land to settle, often as entire communities. Expressly rejected, however, were non-Europeans.

THE CONTEMPORARY PERIOD OF IMMIGRATION　After World War II, the next great phase of immigration began, considerably changing the Canadian ethnic composition. Their societies of origin as well as their Canadian destinations distinguished the new immigrant groups from those of previous eras.

Whereas British, northwestern European, and, selectively, central and eastern European groups had been the major immigrants of the past, the first new arrivals were mainly from southern and eastern European countries, in addition to the continued large-scale entry of the British. The largest among the non-British groups were the Italians, almost a half million of whom had come by 1971 (Iacovetta, 1992).

The contemporary period of immigration also introduced an entirely new dimension to Canada's ethnic mosaic because it included for the first time non-European peoples in significant numbers. Although not officially stated, Canada's immigration policy before 1962 had been, in effect, "white only" (Hawkins, 1989; Richmond, 1976), not unlike that of the United States pre-1965. Discriminatory measures favoring northwestern Europeans, particularly those from the British Isles, were now dropped, profoundly affecting the ethnic makeup of immigration.

Immigrants from Asia in the contemporary period have come from almost every country of that continent, with especially large numbers of Chinese (from the Chinese mainland, Hong Kong, and Taiwan) and East Indians. Other sizable Asian groups are Filipinos, Vietnamese, Pakistanis, Sri Lankans, Bangladeshis, Lebanese, and Iranians. Blacks have come mostly from the Caribbean—especially Jamaica, Trinidad and Tobago, Guyana, and Haiti—and have settled mainly in Toronto and Montreal. Before this period, most of Canada's tiny black population was of American origin. Most Canadian blacks during the colonial period had come with the American loyalists during the Revolutionary War or during the War of 1812, settling mainly in Nova Scotia and New Brunswick (Clairmont and Magill, 1974; Winks, 1971). Also, some American-origin blacks had come to Canada escaping slavery via the Underground Railroad.

Non-European groups (visible minorities) have dominated immigration to Canada during the past fifty years. Whereas blacks and Asians before 1967 had constituted fewer than 4 percent of immigrants, by 1973 they were approximately one-third, and by 1981, one-half. As shown in Figure 16.2, whereas European countries were the major source of immigrants before 1961, today Asian countries are the leading sources.[5] Visible minorities make up nearly 20 percent of the total Canadian population, an increase that is due almost entirely to the immigration of people from non-European countries.

In addition to their diversity, the immigrants of the current era differ from those of previous eras in that their destinations have been almost entirely urban, specifically, Canada's three major metropolises: Toronto, Montreal, and Vancouver. More than a third of all the new immigrants are destined for Toronto alone, making that city the core of Canada's new ethnic character (Lo, 2008; Reitz and Lum, 2006). What had been an overwhelmingly white, Christian, and relatively provincial British-dominated city has been transformed into a cultural and linguistic panorama whose ethnic diversity is virtually unmatched anywhere on the globe. Immigrants today account for almost half of metropolitan Toronto's population, a percentage unsurpassed by any urban area in North America. One of the notable features of the ethnic reconfiguration of large Canadian cities is the relatively conflict-free manner in which it has taken place. This is strikingly different from what has occurred in most European cities that have absorbed large numbers of immigrants in recent decades.

---

[5] Compare Figure 16.2 with Figure 5.3 in Chapter 5. Notice how similar the changing national origins are of immigrants to both Canada and the United States in the past several decades.

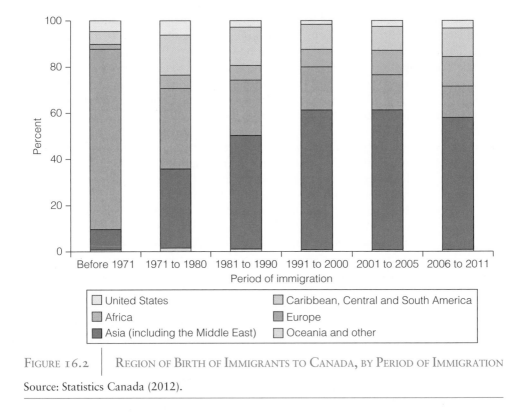

FIGURE 16.2 | REGION OF BIRTH OF IMMIGRANTS TO CANADA, BY PERIOD OF IMMIGRATION

Source: Statistics Canada (2012).

Along with the United States and Australia, Canada is today among the world's primary immigrant-receiving societies. Indeed, its per capita rate of immigration is the highest in the world. Moreover, the contemporary wave of immigration has transformed Canada into a society rivaled in ethnic diversity only by the United States. Today, immigrants make up one in five people in Canada. This proportion is exceeded only by Australia and is significantly higher than the proportion of immigrants that make up the U.S. population (Figure 16.3). As we will see, Canada's groups that are neither British nor French in origin have established themselves as an important political and social element, a so-called Third Force. Their growing numbers and diversity more than anything else have contributed to the current philosophy and government policy of multiculturalism, which is discussed later in this chapter.

ABORIGINAL PEOPLES

The third major element of the Canadian ethnic configuration is the native, or Aboriginal, groups, who together make up about 4 percent of the total population. Among these peoples are Native Indians (or First Nations), Inuit, and Métis, the last of mixed racial origins (Table 16.1).

Canadian Indians are officially designated as "status" or "nonstatus." Status (or "registered") Indians have been classified as Indians under the Canadian Indian Act and are the direct responsibility of the federal government. About one-half live on

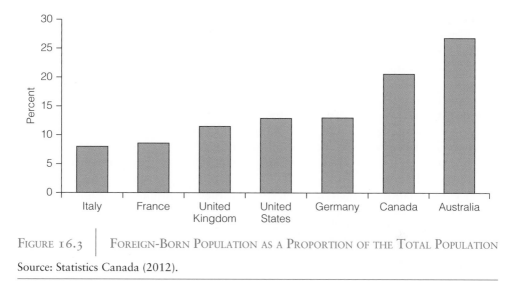

FIGURE 16.3 | FOREIGN-BORN POPULATION AS A PROPORTION OF THE TOTAL POPULATION

Source: Statistics Canada (2012).

reserves (the counterpart of reservations in the United States) established by the government. Like American Indians, Canadian First Nations are ethnically diverse and have long seen themselves as distinct peoples. It is their common relationship with the federal government, however, that links them together (Gibbins, 1997).

The Métis are marginal to both white and Native Indian societies. During the eighteenth and early nineteenth centuries, fur traders, mostly French, often lived with or married Indian women. The offspring of these mixed unions developed a unique culture, partly European and partly Indian. (*Métis* is a French word similar in meaning to the Spanish *mestizo*.) In western Canada, where the intermixture occurred in a geographically and socially isolated environment, the Métis grew in number and developed a distinct ethnic identity, even declaring themselves a nation. Following an abortive rebellion in 1885, however, they became a highly stigmatized group. The Métis were recognized officially as a distinct people until 1940, when the Canadian government reversed this position. Today, most Métis live in the Prairie provinces of Canada, and a majority live in urban areas (Statistics Canada, 2009b, 2012).

TABLE 16.1 | CANADIAN ABORIGINAL POPULATION

North American Indian (First Nations)	851,560	60.8%
Métis	451,795	32.3%
Inuit	59,445	4.2%
Other Aboriginal or more than one Aboriginal identity	37,890	2.7%
Total	1,400,685	100%

Source: Statistics Canada (2013).

Canada's Inuit have remained relatively isolated geographically in the far North and historically have not been subject to colonization to the extent that Native Indians have. Thus, although they do qualify for special political status, technically they are not Indians.

NATIVES AND WHITES: EARLY RELATIONS Except for being markedly less violent, the historical relations between Canada's native peoples and white settlers have been, tragically, not essentially different from those that evolved in the United States. The technological superiority of the whites and their desire for land spawned policies and practices that led to deculturation, dependency, and impoverishment among the Aboriginal populace.

Accommodation characterized the white attitude toward native peoples from the time of initial contact until 1867. The indigenous peoples were vital to the fur trade, and as enterprises like the Hudson's Bay Company expanded their operations, Indians and Inuit were engaged in trapping. As they were drawn into this trade, they became increasingly dependent on the same companies for their subsistence; in the process, their traditional cultures were basically overturned. During this period, however, whites had little direct intent to impose their cultural ways on the native peoples. Their prime concern was profit maximization. As Breton, Reitz, and Valentine have put it, "The traders and whalers wanted productive trappers and hunters, not North American versions of themselves" (1980:73). White domination, then, was real, although not necessarily deliberate.

With confederation and the western expansion, this general attitude of accommodation changed to one of domination. The federal government entered into treaties with the native peoples for the acquisition of occupied lands, and Indians were increasingly relegated to reserves, isolated from the mainstream society. The government subsequently assumed the role of patron, regulating various aspects of Indian life. Moreover, the reserves were increasingly incapable of supporting hunting and fishing economies, forcing Indians into even greater dependence on the government.

Concurrent with government jurisdiction of reserve lands were efforts to assimilate natives into either the English or French culture, resulting in the denigration of indigenous cultures. Aboriginal children were taken from their families and placed into missionary-run residential schools whose intent was to purge them of native ways. Thus, although they were increasingly isolated from white society, they were expected to assimilate into the white culture (Breton et al., 1980). Indeed, until the 1970s, the goal of assimilation, though largely unsuccessful, had been the foundation of government policies toward native peoples for more than a century (Gibbins, 1997).

CHANGING RELATIONS In recent years, the native peoples of Canada have developed a resurgent group consciousness and pride and have become more politically mobilized in movements that seem to parallel those among Indians in the United States, with whom they share generally common origins and cultures. Viewing themselves as a colonized minority in Canada, they have sought the renegotiation of land claims, and in recent years, policies have been instituted that have led to greater native control over their reserves (Asch, 1984; Frideres, 1990; Ponting, 1997; Warry,

2007). In a sense, the political movement among Aboriginal peoples in Canada has displayed many basic similarities to the Quebec nationalist movement. Each has sought greater freedom from the control of the federal government, and each has espoused an assertive and sometimes militant nationalism.

An important event in 1999 moved Canada toward greater autonomy for Aboriginal groups. The federal government entered into an agreement with Inuit leaders that provided for the creation of a new territory called Nunavut ("our land," in the Inuit language), to be carved out of the Northwest Territories. Inuit were given effective control of 135,000 square miles (one-fifth of Canada's landmass) and the equivalent of about $1 billion over a fourteen-year period (Légaré, 1997; Stout, 1997). Although the creation of Nunavut had no bearing on the movement for self-government among Native Indians of southern Canada, it nonetheless marked an important step in giving back control of land to indigenous Canadians.

Two events occurred in the late 1990s that augur additional significant changes in the relations between Aboriginal peoples and the rest of Canada. First, the Canadian government in 1998 offered a historic apology to native peoples for 150 years of paternalistic and sometimes racist treatment. This represented an important symbolic gesture. More substantive, however, was an earlier Supreme Court ruling that validated Indian claims to land in British Columbia. The decision was based on oral histories presented by the Gitxsan tribe, not on written treaties, and therefore may have a profound effect on future treaty negotiations between First Nations and the federal and provincial governments (DePalma, 1998).

ETHNIC STRATIFICATION

Canada's diverse population is arranged in a hierarchy, with some groups exhibiting greater wealth and power than others. But in comparison to the United States and other multiethnic societies, the relative gaps between the various groups on the ethnic hierarchy are not as wide and, in terms of socioeconomic status, it is becoming increasingly difficult to speak of "dominant" and "minority" groups. This increasingly egalitarian ethnic system is a relatively recent development, however. It is in large measure a product of the changed status of Quebec in the Canadian confederation as well as the rapid changes that have occurred in the society's ethnic composition during the past several decades.

The Vertical Mosaic

In a study that has become a classic of Canadian sociology, John Porter (1965) described the system of ethnic stratification in mid-twentieth-century Canada as a "vertical mosaic." A strong relationship, he found, was evident between ethnicity and various measures of social class. The general pattern that emerged from his analysis was a three-part structure made up of British and French charter groups, the former at the top of the income and occupational hierarchies, followed by later-arriving European ethnic groups and, finally, native peoples. The notion of a vertical mosaic suggested that Canadian society was not only ethnically differentiated but ethnically stratified as well. As his work proceeded, Porter explained, "The hierarchical relationship between Canada's many cultural groups became a recurring theme in class and power" (1965:xiii).

The Royal Commission on Bilingualism and Biculturalism in 1969 reaffirmed Porter's basic findings regarding the hierarchy and revealed the concentration of particular ethnic groups in certain occupational areas. British, Jewish, and Asian groups were overrepresented in the managerial, professional, and technical occupations. Ukrainian, Scandinavian, Dutch, German, and Russian groups were overrepresented among farmers (a reflection of these groups' significant presence in the agricultural provinces of Manitoba, Saskatchewan, and Alberta). And Italians, French, Polish, and Hungarians were overrepresented among blue-collar workers (Porter, 1985; Royal Commission on Bilingualism and Biculturalism, 1969b).

THE ENGLISH-FRENCH DIFFERENCE Perhaps the most glaring difference in class position, historically, was between the two charter groups—English and French. Whereas the English were traditionally at the top of the hierarchy, the French were no higher than most later-arriving ethnic groups and actually lower than some. As noted earlier, French Canadians traditionally occupied a lower collective economic position even in Quebec, where they were a numerical majority. Porter found that in Quebec, "by and large the British run the industrial life" (1965:92). As occupational status rose, he noted, so did the proportion of English personnel. The tendency for French Canadians to fill the lower-level working-class positions in the Quebec economy had been vividly demonstrated over two decades earlier by Everett Hughes in his study of a Quebec town. In the town's major industries, Hughes found, the English held "all positions of great authority and perform[ed] all functions requiring advanced technical training" (1943:46). The French, by contrast, predominated in the lower occupational ranks and eventually disappeared in number as the degree of authority rose. "French Canadians as a group," noted Hughes, "do not enjoy that full confidence of industrial directors and executives that would admit them easily to the inner and higher circles of the fraternity—and fraternity it is—of men who run industry" (1943:53). Hughes concluded that the same situation prevailed throughout the province generally. The English-French variance shown by Hughes and Porter was later corroborated by the Royal Commission on Bilingualism and Biculturalism (1969a), which showed that those of French origin were disproportionately in lower-status occupations and earned incomes considerably lower than those of English Canadians.

ETHNICITY AND CLASS IN CANADA TODAY

In the last four decades, the patterns described by Porter have changed dramatically. The requirements of a postindustrial economy have created a more ethnically varied workforce at all occupational levels. Thus, even though collective differences among the various ethnic groups remain evident, ethnic inequality in Canada has been steadily declining (Banting et al., 2007; Brym, 1989; Darroch, 1979; Isajiw et al., 1993; Lautard and Guppy, 1990; Pineo and Porter, 1985). By the 1990s, one researcher had concluded, "The relationship between ethnic origin and class position is in flux, and no ethnic group unequivocally dominates the Canadian class structure" (Nakhaie, 1995:187). Some, in fact, maintain that intergroup differences among white ethnic groups are now virtually nonexistent (Breton, 1989). The large influx of non-British immigrants since World War II has provided further pressures to afford upward mobility and a greater share of power to minority Canadians.

As to the specific differences between Anglophones and Francophones, income inequality has declined markedly, as have occupational differences (Forcese, 1997). Today, there is little meaningful variance in the degree of equality of opportunity enjoyed by the two linguistic groups or in their socioeconomic status in Quebec (Vaillancourt et al., 2007).

Even at the elite level, the dominance of Anglo-Canadians appears to be eroding with the advent of a more ethnically diverse institutional leadership. In the past, non-British groups had been severely underrepresented in the power elites of major Canadian institutions (Clement, 1975; Kelner, 1970; Lautard and Guppy, 1990; Newman, 1975, 1979; Presthus, 1973). Today, elites of government, the economy, education, and labor are increasingly ethnically varied (Ogmundson and Fatels, 1994; Ogmundson and McLaughlin, 1992). And, in Quebec, too, as French has become the primary language of business and commerce, a Francophone economic elite has emerged, making Anglophone dominance largely a phenomenon of the past (Forcese, 1997; Levine, 1990).

This increasingly egalitarian ethnic system, however, does not mean that an ethnic hierarchy in Canada no longer exists. Some studies continue to show differences in income, employment, and poverty between whites and nonwhites, as well as difficulties that recent immigrants are experiencing in the labor market (Block, 2010).

ABORIGINAL GROUPS IN THE CLASS SYSTEM Although their circumstances are much improved from earlier generations, Aboriginal groups continue to occupy the lowest rung on all dimensions of social class and consequently have the poorest life chances of all Canadians. They exhibit the highest unemployment rates in the country, have the lowest levels of income and education, and disproportionately occupy low-level jobs. In the realm of education, for example, one-third of Aboriginal adults have less than a high school diploma, compared to only 13 percent of non-Aboriginals. And, only 9 percent of Aboriginals have a university degree, compared to almost one-quarter of non-Aboriginals (Statistics Canada, 2009c, 2010d).

For First Nations, the standard of living on reserves is characterized by poor and overcrowded housing, low incomes, and low participation in the labor force (Gionet, 2009; Salée, 2006; Siggner and Costa, 2005). Among those who migrate to the cities, a social pattern has developed that is similar to that of American urban Indians—high rates of unemployment, criminality, and alcoholism.

The social consequences of these conditions are predictable. Life expectancy rates are considerably lower, and infant mortality rates, though declining, remain higher among Aboriginal groups than among the general Canadian populace. Standards of housing, education, nutrition, and general health remain below the national averages (Banting et al., 2007; Gionet, 2009; Kendall, 2001; Paperny and Minogue, 2009; Picard, 2010; Statistics Canada, 2009b, 2010a). Moreover, Aboriginal people make up a highly disproportionate number of inmates in Canadian prisons (Brzozowski, 2006).

Despite these dismal facts, the life chances of native peoples in Canada have improved somewhat in recent years, particularly in income and education. Aboriginal people who have moved to urban areas have exhibited especially noticeable rises in socioeconomic status (Siggner and Costa, 2005). But Aboriginal groups on reserves and in remote areas continue to represent a markedly deprived segment of the society and, in many ways, are divorced from mainstream institutions. This is a troublesome and pressing issue of Canadian ethnic relations that is acknowledged by the general public; a 2010 national survey indicated that a majority of Canadians believe that the quality of life for Aboriginals has not improved over the past two decades (Ipsos, 2010a).

PREJUDICE AND DISCRIMINATION

When comparing their own country with their American neighbor in the conduct and character of ethnic relations, Canadians usually see themselves in a favorable light. The prevalent view is that their treatment of and attitudes toward minority ethnic groups have been and remain more tolerant and equitable. This popular Canadian perspective is in some regards accurate. Though incidents of racial and ethnic violence are scattered throughout Canadian history, they have never reached the magnitude, frequency, or intensity of ethnic hostilities in the United States. Moreover, as we will see in the next section, an ethnic ideology more pluralistic in content has traditionally been proclaimed in which the society's various cultural groups have not been forced into a monolithic "Canadian" mold.

Despite its more subdued character, however, ethnic conflict in Canada has been revealed in patterns that are in some ways parallel to those of the United States. Like most other Western societies, Canada does not lack a racist tradition, though it has been more muted in expression and less malignant in consequence. Historically, Canadian racism has been evidenced in its immigration policies and in its treatment, both official and unofficial, of nonwhites, especially Aboriginals.

Immigration and Racism

Historian Howard Palmer has written that "the more one scratches the surface of the period up to 1920, the more difficult it becomes to differentiate between the immigration histories of Canada and the United States" (1976:499). Official government policies regarding who and how many would be admitted and what place they would take in the occupational structure, as well as nativist sentiments and attitudes toward immigration in general, were all basically similar.

One analogous aspect of the early history of immigration in the two societies was the decidedly racist character of the selection process. Like the dominant Anglo group in the United States, "English Canadians," notes Evelyn Kallen, "from the beginning, exercised control of federal immigration policies responsible for determining which ethnic groups would be allowed into Canada, where they would settle, what jobs they could assume, and what ranking and social position would be accorded them within the existing system of ethnic stratification" (2003:97–98). Those deemed most suitable for Canada were, not surprisingly, most like the English Canadians, culturally and physically.

After 1870, when large-scale immigration to Canada began, the preference for British, or at least northwestern European, immigrants was an outspoken national policy. The pretext for this selectivity was the matter of assimilation. Those from Britain and culturally similar societies, it was asserted, were more easily absorbed into the mainstream Canadian society. Others were not culturally or, as some believed, biologically fit. Italians, Jews, and other non-Anglo groups—especially those who immigrated to Canadian cities at the turn of the century—met with a brand of nativism very similar to the U.S. variety (Avery, 1979; Harney and Troper, 1977).

But it was nonwhites, especially Asians, who were least welcome and against whom the most blatantly racist policies and actions were directed. As in the United States, during the late nineteenth and early twentieth centuries, discriminatory policies were enacted to discourage the entry of Chinese and other Asian groups, and in 1923 an exclusionary act totally barred their immigration (Elliott, 1979). Other restrictive measures severely limited entry into Canada of those deemed "unassimilable," understood to mean nonwhites. In 1911, for example, blacks seeking to emigrate from the United States were rejected because they were presumably unable to adapt to Canada's harsh winters (Palmer, 1976).

Discriminatory immigration policies restricting Asians and other nonwhites were not basically changed until the 1960s. Indeed, as recently as 1947, Prime Minister Mackenzie King stated bluntly Canada's intention to encourage population growth through selective immigration. Large-scale Asian immigration, he stated, "would change the fundamental composition of the Canadian population"; therefore, no changes in immigration regulations were to be made (Corbett, 1957:36).

Seeking to enhance its international image and to more realistically meet its need for human resources, Canada abolished its racially discriminatory immigration policy in legislative acts of 1962 and 1967. The selection of immigrants was no longer to be based on nationality or race but on a system of points that objectively evaluated each immigrant's potential economic and social contribution to Canadian society (Hawkins, 1989). As in the United States after it discarded restrictive immigration quotas in 1965, the ethnic origins of immigrants entering Canada changed radically—most were now non-European (Figure 16.2).

ETHNIC ATTITUDES

As we have seen, Canada in the last four decades has become a society far more ethnically diverse than at any time in its history, with visible minorities making up the majority of new immigrants. This has prompted prejudicial attitudes and discriminatory actions, but by American standards these have been less evident and milder in content.

In their national attitudinal study, Berry, Kalin, and Taylor (1977) found that respondents in general reacted very favorably to English and French Canadians but less favorably to non-British and non-French groups. Specifically, northwestern Europeans were judged most favorably, central and southern European groups next, and nonwhite groups least favorably, except for Japanese. This social distance scale was later demonstrated by other studies as well (Mackie, 1980; Pineo, 1987). It is interesting to observe its essential similarity to the ethnic scale that, as noted in Chapter 3, has been consistently reaffirmed in the United States, though the differences among groups are declining in both societies (Reitz and Breton, 1994).

Even with this clear acknowledgment among Canadians of an ethnic hierarchy, Berry and his colleagues found no evidence of extreme ethnic prejudice. Though the rank order of groups was very apparent, the differences among them were not great. Race (that is, physical differences among groups) was found to be an important dimension of group perception among Canadians, but the researchers concluded that "Canadians reject explicit racism" (1977:206). Later studies reaffirmed this conclusion (Berry and Kalin, 2000; Kalin and Berry, 1994). Surveys indicate increasing ethnic tolerance on several measures, including intermarriage and perceptions of visible minorities. Acceptance of marriage between whites and blacks, for example, rose from 55 percent in the mid-1970s to 92 percent in 2007, a rate higher than in the United States (Bibby, 2007; Milan et al., 2010). Today, more than half of all second- and third-generation visible minorities who are married or living-together couples are ethnically mixed.

Despite the lower level of prejudice among Canadians, some suggest that negative views toward nonwhites may simply be more subtle and covert and may reveal themselves when carefully probed (Cannon, 1995; Frideres, 1976; Kallen, 2003). Also, ethnic attitudes and actions of Canadians at times parallel those expressed by Americans. The most obvious example in recent years has been the growing negative attitudes toward Muslims. National polls have shown that Muslims are viewed significantly more unfavorably than other ethnic and religious groups (Geddes, 2013; Jedwab and Al-Yassini, 2012). Anti-Muslim views have taken a far less virulent form, however, and have not blossomed into the Islamophobic outbreaks that have characterized the United States and some western European countries (Geddes, 2009; Jedwab, 2006; Trudeau Foundation, 2006). In fact, a study of twenty-three Western countries found that Canadians had the least bigoted attitudes toward Muslims (Borooah and Mangan, 2007).

DISCRIMINATORY ACTIONS

Although the overt suppression of minority peoples is generally lacking in Canadian history, there are nonetheless some parallels between Canada and the United States in the past treatment of ethnic minorities, particularly nonwhites. In Canada, these racial-ethnic groups have suffered various forms of discrimination including, at one time or another, restrictions in voting, employment, land ownership, housing, and public accommodations (Davis and Krauter, 1971; Kallen, 2003).

Arrant anti-Asian immigration measures were noted earlier. In addition to this official discrimination, violence against Chinese and Japanese workers in British Columbia was common in the late nineteenth and early twentieth centuries. The most blatant discriminatory action against Asians, however, occurred during World War II. As in the United States, those of Japanese origin were forcibly removed from their homes and businesses and placed in internment camps for the duration of the war (Adachi, 1976).

Serious racist actions in the past were also directed at people of East Indian origin. Along with the Chinese and Japanese, East Indians were barred from entry into Canada early in the twentieth century, and those remaining were subjected to discrimination in almost all areas of social life. East Indians in the past were depicted as dirty, sinister, immoral, prone to overcrowding, and generally inferior to whites.

Buchignani suggests that East Indians in Canada before World War II were "an almost ideal type subordinate racial caste" (1980:129). They were denied entry into various occupational fields and could not vote or hold citizenship.

Blacks have never made up a significant element of Canada's ethnic population; today they are about 2.5 percent (Statistics Canada, 2010d). Although their circumstances have in some ways conformed to those of African Americans, negative actions and attitudes toward blacks in Canada have never equaled in scope and intensity U.S. patterns. Studies of urban housing, for example, reveal that blacks in Canadian cities do not experience levels of residential segregation similar to what is so common in the United States (Fong, 1996; Lo, 2008; Walks and Bourne, 2006). As with other ethnic minorities, of course, immigration restrictions and discrimination in schools, housing, and public accommodations were once common features of the Canadian black experience (Walker, 1980). Today, however, the official and obtrusive forms of discrimination of the past have given way to milder and more subtle forms that are not easily verified (Bolaria and Li, 1988; Henry, 1994; Henry et al., 1995; Thompson and Weinfeld, 1995).

A national survey of non-Aboriginal adults in 2003 indicated that, although racial and ethnic discrimination are not absent, the level and frequency of such actions are relatively low. One-third of blacks and 20 percent of South Asians said that they had sometimes or often experienced discrimination in the previous five years, most commonly at work or when applying for a job. The vast majority of respondents, however, reported that they had not experienced discrimination because of their ethnic background (Statistics Canada, 2003).

Whether there are fundamental differences between Canadian and American beliefs and actions toward nonwhite peoples, however, remains a debated issue (Kallen, 2003; Reitz and Breton, 1994). The treatment of Aboriginal people in particular prompts consternation among those who have viewed Canadian ethnic relations as basically different from those in the United States. Also, in recent years, police relations with minorities, particularly blacks, have been a source of some tension in Toronto and Montreal, each with growing black populations.

In sum, Canada, in the past, exhibited actions and attitudes not basically unlike those of other multiethnic societies. In the modern era, however, as the country has become extraordinarily diverse, levels of prejudice and discrimination are comparatively low, and as Berry and Kalin have concluded, "Canadians are generally well-disposed toward the existence of ethnic and racial diversity in their society" (2000:185). Those outside of Canada seem to think the same. A 2010 international survey indicated that eight in ten global citizens described those living in Canada as being "tolerant of people from different racial and cultural backgrounds"—equal to the proportion of Canadians themselves who said this (Ipsos, 2010b).

STABILITY AND CHANGE

Canada addresses ethnic issues differently than does the United States in large measure because it is more pluralistic in its ethnic ideology. The ethnic end product envisioned by Canadians is not a duplicate of that envisioned by Americans. Whereas the United States can be categorized as basically assimilationist in orientation, Canada displays many elements of a corporate pluralistic society.

Melting Pot Versus Mosaic

It has often been noted that one of the key factors differentiating the Canadian and American ethnic systems is how the two regard relations among diverse groups and the path of absorption of those groups into the larger society. Canada has commonly been observed as a society in which there is a greater tolerance of ethnic differences. The popular phrases of comparison are *melting pot,* supposedly characteristic of the United States, and *mosaic,* supposedly characteristic of Canada.

THE MELTING POT   In the United States, as we saw in Chapter 5, the idea of the melting pot—the fusing of many immigrant groups into an American hybrid culture—became popular beginning in the early 1900s. Never, however, was it translated into public policy to any serious extent. In reality, the Anglo-conformity model predominated; that is, the expectation was always that immigrants would conform to the dominant Anglo-Protestant culture. Even in recent years, the assumption that new groups will quickly learn English and generally adopt dominant norms and values has continued to guide social thinking despite increased multicultural rhetoric.

THE MOSAIC   The Canadian ideology, in contrast, has historically favored a more pluralistic outcome of the massing of various ethnic groups. There have been and remain greater awareness and tolerance of ethnic separateness. A simplified view is "unity in diversity." Canada, in this ideal model, is a mosaic, the various pieces fitting together within a common political and economic framework.

In 1971, the mosaic idea was made a basic part of Canada's face to the world when Prime Minister Trudeau announced the policy of "multiculturalism within a bilingual framework." This was in response to the realization that the population was becoming more ethnically diverse as well as to the concerns of non-English, non-French groups that they were being ignored in an officially *bi*cultural Canada. An even stronger commitment to the policy was made in 1988 when Canada became the world's first country to enact a national multiculturalism law. **Multiculturalism—***recognizing and supporting the retention of different ethnic cultures within the larger society*—was firmly established as an official doctrine, and public policies were adopted accordingly (Fleras and Elliott, 2002; Hawkins, 1989; Kymlicka, 2007).

Explaining Canadian Multiculturalism

What accounts for this multicultural ideology, markedly different from the American ethnic ideology, and how strongly is it validated in light of the reality of Canadian ethnic relations?

TWO NATIONS IN ONE   The dual national character of Canada has made the application of assimilationist ideas impractical in any case. Given the historical fact of two founding groups, neither the melting pot nor the Anglo-conformity model could have the same meaning in the Canadian context as in the American. The idea of "Canadianizing" people becomes an empty notion when there are, in fact, two dominant cultural groups and no uniform "Canadian way of life" to serve as a societal reference point.

Although it is also true that one would be hard put to clearly define an "American way of life," there is nonetheless no essential ambiguity with regard to language and other major elements of American culture. In Canada, however, the presence of two founding peoples with distinctly different cultural systems has allowed ethnic groups that entered the society after the British and French to parlay this basic schism into significant freedom to retain their culture and group structure. As Hiller writes, "If one group had dominated, there would have been more accord about the specific nature of the dominant culture; but since the British and French were in conflict themselves, the society had a greater built-in tolerance for the perpetuation of ethnic identities" (1976:107–8).

PARTICULARISM   Some have suggested that one explanation for Canada's greater tolerance of ethnic pluralism lies in the fact that historically Canada retained a British Tory tradition that made for a more particularistic social system, in contrast to the equalitarian and universalistic system of the United States (Clark, 1950; Lipset, 1968, 1990). A traditionally greater emphasis on hierarchy and status restrained pressures to melt down group differences. The necessity for ethnic groups in the United States to "become like others" therefore did not find a strong counterpart in Canada. Moreover, Canada developed as a nation firmly tied to Britain and did not emerge from revolution, like the United States. Thus a unique "Canadian" national identity to which all could relate never materialized. As James Laxer writes, "[T]he consequence ... has been a nation of multiple identities and two main languages" (2003:36).

THE IMPACT OF IMMIGRATION   The impact of immigration on Canadian society has in a way been even more profound than in the United States, creating a greater incentive to adopt a more pluralistic approach to ethnic diversity. Though not as numerous in an absolute sense, immigrants in Canada have been a considerably higher percentage of the population than immigrants in the United States. In the early years of this century, the foreign-born in some Canadian provinces outnumbered the native-born by two to one (McKenna, 1969). As noted earlier, today, the foreign-born continue to make up a much greater percentage of the total population in Canada than in the United States.

MULTICULTURALISM: MYTH VERSUS REALITY   In comparing the ethnic ideologies of Canada and the United States, it is important to recognize that just as the American melting pot has been more myth than reality, so, too, the Canadian ethnic mosaic is not a true reflection of public attitude and policy toward ethnic differences. For both societies, the reality of ethnic relations lies somewhere between these two ideals. As Porter points out, "In practice neither [ideal] has been practicable and neither has been particularly valued by the respective societies despite the rhetoric in prose and poetry that has been devoted to it" (1979:144). Just as Anglo-conformity was the dominant public policy with respect to ethnic groups in the United States after its emergence as a heterogeneous society in the late nineteenth century, so, too, in Canada the first period of heavy immigration at the turn of the century produced an ethnic policy that encouraged assimilation into the dominant Anglo-Canadian group. This basic policy did not change for the next three decades (Burnet, 1976, 1981;

Palmer, 1976). Although immigrants entered a society in which assimilation could not be enforced as strongly as it was in the United States, there was never any question that Anglo-conformity was, except in Quebec, the guiding force of government policy toward the newcomers.

In a comparison of ethnicity in the two societies, Reitz and Breton (1994) suggest that the differences between Canada and the United States regarding assimilation, at least as it pertains to European-origin groups, are not as great as has been commonly assumed. "The fact is," they note, "in both Canada and the United States, the public discourse on immigration reflects both a tolerance for diversity and a bias toward assimilation" (1994:10). Moreover, their analysis suggests there is no evidence to support the notion that ethnic minorities in Canada retain their ethnic identity and culture longer or more strongly than American ethnic minorities. In Canada, as in the United States, ethnicity declines in significance with each passing generation, taking on a largely symbolic function.

Whether succeeding generations of minority ethnic groups in Canada can or will want to resist the forces of assimilation into the dominant Anglo (or in Quebec, French) culture seems doubtful. It is most commonly the case that immigrants and their children are determined to take advantage of the opportunities of their new society rather than preserve the same kind of life they left in their societies of origin. For them, as Freda Hawkins has claimed, "the preservation of cultural heritage is a lesser concern and ... the whole concept of multiculturalism can be confusing" (1989:217). Here, the experience of most ethnic groups in the United States may serve as a valid comparison. Although at times tortuous, a high level of assimilation, both cultural and structural, has been the fate of most groups by the third generation. This seems to be the pattern followed as well by Canadian groups.

Further evidence that multiculturalism in Canada is in large part symbolic lies in the undeniable limits that are imposed on the cultural freedoms that ethnic groups are permitted at the point where cultural practices challenge basic societal values and beliefs. Female circumcision or polygamy, for example, are acceptable and condoned in the origin cultures of many immigrant Canadians, but these practices are clear violations of core Canadian values. In such cases, writes demographer Daniel Stoffman, "there is no room for compromise" (2009:A15). And in this, Canada is little different from its southern neighbor. Stoffman further explains that multiculturalism is not the same as diversity. That Canada, like the United States, is increasingly diverse is obvious; but that should not be confused with official policies that encourage diversity. A variety of ethnic differences arise not because of any government support or encouragement but as a result of simple demographics and societal values that are open to cultural changes, always, however, within limits.

CONFLICTING VIEWS OF MULTICULTURALISM   The policy and ideology of multiculturalism, though now solidly part of the Canadian ethos, remains a debatable issue. Some see it as merely part of a political strategy designed to gain the support of ethnic minorities. There also is concern that it is moving Canada away from its traditional open, individualistic political system toward one in which, as one observer has described it, "the cake has to be sliced very carefully among powerful and competing groups"

(Hawkins, 1989:216). Moreover, in recent years, Quebec, in light of its growing diverse ethnic population, has adopted its own policy of multiculturalism, recognizing the unique needs of these ethnic subcultures while trying to integrate them into Quebec (that is, French-speaking) institutions.

Some view multiculturalism as divisive in effect and stultifying as well for minority ethnic groups (Porter, 1975, 1979). Sustaining and enhancing ethnic pluralism, they maintain, can only hinder the movement of these groups into mainstream institutions, thus perpetuating the system of ethnic stratification. Moreover, as writer Neil Bissoondath (2002) has pointed out, rather than instilling cultural understanding among ethnic groups, the policy of multiculturalism may actually have the opposite effect by reinforcing and perpetuating stereotypes.

MULTICULTURALISM AND PUBLIC OPINION   Despite such concerns and despite the realities of ethnic assimilation, Canadians today overwhelmingly favor the official multicultural policy and view multiculturalism as fundamental to the country's national identity (Adams, 2007; Bloemraad, 2010). "Tolerance towards diversity and the acceptance of pluralism," notes political commentator Richard Gwyn, "have become the defining characteristic of the country and its citizens" (1995:203). National surveys consistently indicate a strong public understanding and acceptance of the principles of cultural pluralism. More than two-thirds disagree with the proposition that "a country in which everyone speaks the same language and has similar ethnic and religious backgrounds is preferable to a country in which people speak different languages and have different ethnic and religious backgrounds" (CRIC, 2004). Similar public attitudes toward ethnic diversity have been consistently evident in national polls (Nanos, 2010; Trudeau Foundation, 2006).

Further proof of the commitment to multiculturalism is found in the comparatively more favorable Canadian attitude toward immigration itself. When measured against Americans (as well as citizens of other immigrant-receiving countries), a greater percentage of Canadians are inclined to see immigration as having a positive effect on the society. National polls repeatedly indicate that a strong majority of Canadians perceive immigration as a fundamental feature of their society and believe that immigrants are vital to economic and social progress (Angus-Reid, 2009; Horowitz, 2010; Nanos, 2010; Trudeau Foundation, 2006).

At the same time that Canadians are solidly committed to cultural pluralism, the expectation that immigrants will assimilate to the society's core values also seems clearly part of the public view. In 1975, 85 percent of Canadians agreed with the statement that "immigrants to Canada have an obligation to learn Canadian ways." Twenty years later the percentage agreeing was 88. Furthermore, this view was no less evident among those who had come to Canada since the 1960s—that is, immigrants themselves (Bibby, 1995). The 2006 poll referred to earlier again confirmed Canadians' acceptance of multiculturalism but also reported the general view that immigrants should abide by mainstream values, particularly as they pertain to gender equality (Trudeau Foundation, 2006). The strong desire for immigrants to integrate into the larger society, however, does not parallel the American belief that immigrants should assimilate as quickly as possible. Nor does it translate into a desire for ethnic groups to remain separate and isolated communities. Rather, the

Canadian approach to immigrant adaptation is something of a middle course, what sociologist Irene Bloemraad calls "multicultural integration" (2010:10).

LOOKING AHEAD

Canada is a North American experiment in ethnic pluralism, exhibiting many of the characteristics of what was described in Chapter 4 as a corporate pluralistic, or multicultural, society. In this, it stands in contrast to its southern neighbor. In considering the future course of ethnic relations in Canada, we might return to the theme of the three ethnic dimensions denoted at this chapter's outset.

The overriding ethnic issue of Canadian society remains the French-English schism. This is the most encompassing dimension of ethnicity in Canada and continues to present the most vexing problems of intergroup relations. Whether Quebec will retain its place in the Canadian union or eventually go the way of independence is a question that in the early twenty-first century remains unresolved. More than any other, this ethnic conflict will dominate internal politics in Canada in the coming years and, in a real sense, will define the future of Canada as a nation-state. Whether a politically and economically centralized society can sustain what are in essence two nations is a question that has been addressed in other corporate pluralistic societies made up of territorially based ethnic groups divided along the lines of language and culture. Judging from those cases, a continuing separatist movement of some scope can be expected as long as Francophone Quebec remains part of Canada. Whatever the nature of its resolution, however, it is not likely to degenerate into warfare or to erupt periodically in violence. The democratic context of Canada makes ethnic conflict amenable to political solution.

A second ethnic issue that will absorb the attention of Canadians in the coming years involves the racial and ethnic characteristics of its new immigrants—mostly non-European and nonwhite. Current studies have indicated that many immigrants have not kept pace with the rest of the workforce in income and jobs. Moreover, problems of discrimination are perceived by some (Banting et al., 2007; Samuel and Basavarajappa, 2006). Will these groups be fully integrated into—or perhaps even eventually dominate—Canadian society?

Although their future status will continue to be of major concern, it is now clear that these groups already have fundamentally altered the ethnic composition and flavor of Canada. As journalist Peter Newman has written, the members of Canada's two founding groups are no longer in the ascendancy, which means that "a new and radically different country has been created" (1995:34). Nothing could better illustrate this than the fact that two of the three most recent occupants of the office of governor general—the highest symbolic political position of the federal government— have been visible minority women, both born outside Canada.

Finally, the place of Aboriginal peoples remains unclear. Whether they will participate with other Canadians as equals in all aspects of citizenship or whether they will exhibit greater autonomy is still an unfolding issue. Also, the continuing socio-economic gap between Aboriginal peoples and the rest of Canadian society remains a troubling reality.

Factors irrespective of ethnicity have in the past impeded the development of a definitive Canadian national identity. Great regional differences as well as the constant specter of the United States, with its enormous economic and cultural influences, have created divergent visions of the country's destiny. But today, it is its approach to issues of ethnic diversity that defines Canada's place among the world's nations and provides its strongest source of national identity. It has accommodated ethnic diversity more successfully than almost any other contemporary multiethnic society, and, should it continue to succeed, will serve as an exemplar for others.[6] Canada will hold the attention of students of ethnic relations in the coming decades, for it may provide an answer to the question of how much diversity can be tolerated by a society that remains a centralized nation-state.

SUMMARY

- Canada is geographically and culturally close to the United States, but its ethnic structure is decidedly different in several respects.
- In its ethnic makeup, Canada is a dual-nation society, French and English, within which additional ethnic groups have taken their place.
- There are three dimensions of ethnicity in Canada: the French-English bifurcation; other ethnic groups, neither French nor English in origin; and Aboriginal peoples.
- The overriding issue of ethnic relations is the relentless conflict between French and English Canadians. The division between the two derives from the French colonial defeat by the British in the eighteenth century. Conflict has revolved around the efforts of French Canadians to retain their language and culture within the context of a surrounding English majority. The rupture intensified in the 1970s with the emergence of a separatist movement in Quebec, where most French-speaking Canadians live, and remains wide today.
- A large and varied cluster of non-British, non-French ethnic groups from virtually all nations of the world immigrated to Canada beginning in the late 1950s, changing the very ethnic character of the society. Today, these groups make up more than one-third of the Canadian population and continue to grow in size as Canada has become one of the world's leading countries of immigration.
- Aboriginal peoples—Indians, Inuit, and Métis—are the third element in Canada's ethnic composition and are a numerically small part of the society. In socioeconomic status, these groups are at the bottom of the Canadian ethnic hierarchy.
- Ethnic stratification in Canada was, in the past, a vertical mosaic in which those of British origin disproportionately occupied the higher occupations and the most important decision-making positions of major institutions. Today French Canadians and, increasingly, those of neither British nor French origin have narrowed those discrepancies considerably and are playing a more prominent role in Canadian economic and political affairs.

---

[6] Michael Ignatieff has explained that "Spaniards avidly follow Canadian politics, because it is full of lessons for meeting the challenge of Basque and Catalan nationalism. Canada matters to Spain and every society in the world seeking to share a political community among peoples of different languages, religion, traditions and cultures" (2005:A19).

- Prejudice and discrimination in Canada today are milder in form and substance than those of the United States, but resembled U.S. forms in the past in immigration policies and in the treatment of nonwhite peoples.
- Canada has maintained a greater tolerance toward ethnic diversity than has the United States and can be categorized as a corporate pluralistic society.
- Canada historically has advanced a mosaic ethnic ideal, embodied today in the official policy of multiculturalism, according to which ethnic minorities are not compelled to fully adopt the cultural ways of the dominant group.

COMPARING CANADA AND THE UNITED STATES

Points of Similarity

- The ethnic development of both societies began with a process of migrant superordination, in which an indigenous population was overcome by a colonizing force. Following conquest, native populations in both societies were reduced to a state of dependence on the dominant group.
- Both societies were populated by successive immigrant waves and current immigration in both is comparable in its societal impact. Both societies today are among the major destinations of global immigration, and most immigrants are non-European.
- Both societies are today among the most ethnically diverse in the world.

Points of Difference

- Canada is a two-nation society, with two distinct cultural groups having charter-group status. In effect, there are two dominant ethnic groups in Canada.
- Canada has adopted a more pluralist approach to the absorption of various ethnic groups into its population, though in fact progressive assimilation of ethnic groups has generally paralleled U.S. patterns.
- Canada did not abide institutionalized slavery and thus did not experience the travails of integrating a former slave population into its mainstream.
- Race and ethnic relations in Canada have historically been more benign and ethnic tolerance higher than in the United States.

CRITICAL THINKING

1. As we have seen in Part II, immigration to the United States during the past four decades has created a demographic shift to a population that continues to become less European in origin. The same has been occurring in Canada, but the contentiousness of immigration and the issues that accompany it have not seemed to arise as strongly as in the United States. What may account for that difference?
2. Can it be argued that Canadian-style multiculturalism, which supports the retention of ethnic cultures, impedes the integration of immigrants into the larger society by confining them to a particular ethnic slot? Reversing the issue, might Canadian-style multiculturalism actually ease the movement of immigrants of different ethnic origin into the larger society?

3. Despite sharing the North American continent, Canada and the United States have followed different paths in their approaches to issues of race and ethnicity. What factors, both historical and contemporary, might account for this divergence?

4. Why would newly emergent multiethnic societies, such as those of Western Europe, be more likely to consider Canada as a potential model for dealing with issues of interethnic relations, rather than the United States?

PERSONAL/PRACTICAL APPLICATION

1. How would you respond to a Canadian who asserts that ethnic prejudice and discrimination are much stronger in the United States? Would you agree or would you argue that different social and political histories and different populations make such comparisons questionable?

2. Canada, as we have seen, is officially a bilingual society. This means that all government services are provided in both English and French and all consumer goods sold nationally must be packaged to accommodate speakers of both languages. What would you think of an officially bilingual United States? Consider that there is already a kind of de facto bilingualism (English and Spanish) in some cities, like Miami and Los Angeles.

GLOBAL ISSUES OF ETHNIC CONFLICT AND CHANGE | CHAPTER 17

During the past half century, societies in all regions of the world have become more ethnically diverse. Those with a history of absorbing a variety of ethnic groups—like the United States and Canada—emerged as even more heterogeneous. Many others that had been relatively homogeneous—such as those in Western Europe—evolved into truly multiethnic societies. This trend has continued unabated.

In addition to increasing diversity, societies that have had a multiethnic structure and tradition for many decades have experienced a rise in ethnic nationalism, sometimes at a ferocious level.

In this chapter we look at these two most evident trends of ethnicity in the contemporary world—the proliferation of ethnic diversity and the resurgence of ethnic nationalism. Both, as we will see, have contributed to a marked rise in ethnic conflict and change.

THE GLOBAL EXPANSION OF ETHNIC DIVERSITY

SHIFTING PATTERNS OF IMMIGRATION

With the movement toward industrialization throughout the West during the one hundred years between the mid-nineteenth and mid-twentieth centuries, global migration occurred mostly within Europe and between Europe and North and South America. Beginning in the 1950s, however, that traditional pattern was replaced by an international migration system encompassing virtually all world regions. Three major immigration streams evolved: (1) from Latin America and the Caribbean countries primarily to the United States, and secondarily to Canada and Western Europe; (2) from south and east Asian countries primarily to the United States, Canada, and Australia, and secondarily to Western Europe; and (3) from North Africa, the Middle East, and southern Europe primarily to northwestern Europe, and secondarily to the United States and Canada. Basically the prevailing pattern of international migration was the movement of people from the developing societies, or Third World, primarily to the developed, industrialized societies (Figure 17.1).

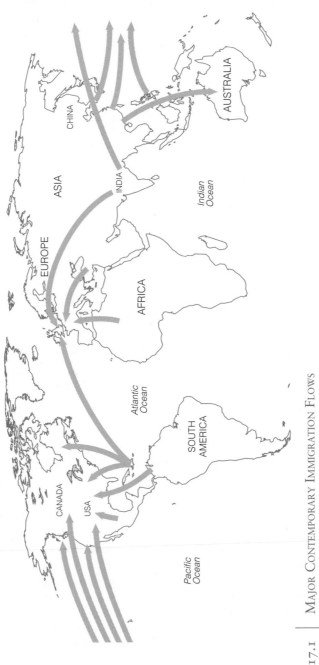

FIGURE 17.1 | MAJOR CONTEMPORARY IMMIGRATION FLOWS

Not only did the direction of world migration change but its scope expanded greatly and its pace accelerated. Whereas international migration had been sharply reduced during the 1930s because of a worldwide economic depression and then by the outbreak of World War II, in the 1950s immigration began anew with great impetus. Nothing comparable had occurred since the great migrations to North America in the nineteenth and early twentieth centuries. Today, an estimated 232 million people are living in a country other than their country of origin, representing more than 3 percent of the world's population (United Nations, 2013).[1] Moreover, international surveys show that another 700 million people would *like* to move to another country permanently if they had the opportunity (Esipova and Ray, 2009). Hence, the flow of international migration is not likely to abate in the near future.

The expansion of immigration and the ensuing ethnic diversity throughout the world in recent decades are attributable primarily to two developments—uneven economic and social development among nations, and inter- and intrasocietal political conflicts.

Uneven Development   International migration has accelerated as a consequence of the widening gap between rich and poor nations. As with previous movements, the primary driving force of contemporary global migration has been the push-pull factors of labor markets: workers migrate to those countries where jobs are more plentiful and lucrative. As noted in Chapters 5 and 16, this factor, more than anything else, accounts for the unprecedented levels of immigration to the United States and Canada during the past fifty years. As we will see, it also accounts for the upsurge in immigration to Western Europe.

The discrepancy between rich and poor nations became evident throughout the 1950s and 1960s when many former European colonies in Africa, Asia, and the Caribbean gained their independence. The economies of these newly independent countries were generally incapable of supporting rapidly growing populations, thereby creating a migration push. Most of the world's population growth since the end of World War II has occurred in the less developed countries. Moreover, gains in economic development made by many of those countries in the 1960s and 1970s slowed or reversed in the 1980s due to economic recession, growing debt, and internal conflicts. These conditions produced further incentives for migration to the wealthier countries.

It has become evident, however, that in recent years immigrant movement has been taking place not only from the developing world to the nations of the developed world, but in the opposite direction as well. With rapidly expanding economies, many countries in Asia and Latin America—traditionally immigrant-sending regions—have become attractive immigrant destinations. This means that people continue to migrate to developed societies, but they are also moving within the developing world (United Nations, 2013). With their economies suffering the effects of deep recession, the United States and Western Europe have begun to experience a slowdown in immigration. By contrast, India, for example, has been attracting entrepreneurs and highly skilled workers, particularly in its booming IT industry. Also,

---

[1] Approximately half of today's international migrants are women, a marked change from the large-scale global migrations of the nineteenth century, when the majority were men (UNDP, 2009).

as they see greater economic opportunities elsewhere, many immigrants to the developed world are increasingly returning to their countries of origin (Migration Information Source, 2011). As part of this global system, few countries today are not either sources or recipients of immigrants.

POLITICAL CONFLICT Wars and internal political strife in many societies have created additional push factors stimulating global migration. From the late 1950s to the present, civil wars, internal uprisings, revolutions, and political violence in various forms created a worldwide flow of refugees unprecedented in modern history. Virtually all parts of the developing world, as well as some countries in the developed world, were affected. Uprisings in the 1950s and 1960s against Soviet occupation in Eastern Europe, revolution in Cuba in 1959, civil wars in the 1970s and 1980s in Central America, political violence in parts of the Caribbean and the South American continent, the war in Vietnam, civil wars in many parts of Africa, the ongoing Arab-Israeli conflict, the Persian Gulf War in 1992, the collapse of the Soviet Empire in the late 1980s, wars in the former Yugoslavia in the 1990s, the U.S. invasion and occupation of Iraq beginning in 2003, and political uprisings throughout the Middle East in recent years all contributed to the swell of refugees alongside the more conventional economic migrants.

IMMIGRATION AND ETHNIC CHANGE IN WESTERN EUROPE

We have looked at the transformative effects of recent immigration to the United States and Canada, but the impact of immigration and the creation of greater ethnic diversity have been dramatically evident in Europe as well. In the past half century, many European nations, relatively homogeneous in the past, have been transformed into authentic multiethnic societies.

IMMIGRATION TRENDS IN EUROPE After World War II, workers from southern European countries—Italy, Portugal, Greece, Spain, Turkey, and Yugoslavia—or from newly independent former colonies such as Algeria and Indonesia migrated to Germany, France, Switzerland, the Netherlands, Belgium, and the Scandinavian countries as these nations experienced their postwar industrial expansion. A similar immigration occurred in Britain with the entry of thousands of people from the newly independent countries of the British Commonwealth—India, Pakistan, Jamaica, and others (Castles, 1992; Castles et al., 1984; Castles and Kosack, 1973; Castles and Miller, 2009; Garson, 1992; Paine, 1974). Labor was in short supply, and immigrants were therefore recruited to work in the economic reconstruction of these countries. This population movement accelerated in the 1960s, leaving previously homogeneous Western European societies with substantial ethnic communities made up of immigrants of vastly different cultures and physical characteristics.

Although by the late 1970s immigration had been severely curtailed by restrictive measures, many of those who had come as temporary workers stayed on, establishing themselves and their families as permanent residents. In the late 1980s another surge in migration to Western Europe occurred, stimulated in large measure by the disintegration of the Soviet Union, but also by the influx of labor migrants and refugees from the developing nations of North Africa, the Middle East, and Asia. By the mid-1990s, immigration to Western European countries once again tapered off, as governments introduced tighter entry procedures and controls (Castles and Miller, 2009; Organisation for Economic Co-operation and Development [OECD], 2001). But the infusion of immigrants had rendered a powerful demographic effect.

Today, Western Europe is home to fifty million immigrants, greater in size than the foreign-born populations of the United States and Canada (Eurostat, 2012). Germany alone has ten million and France more than seven million. Before the economic downturn beginning in 2007, Ireland and Italy, traditional immigrant-sending societies, had become major immigrant destinations (Castles and Miller, 2009; Marrero, 2004; Muenz, 2006; Quinn, 2007). Table 17.1 shows that the percentage of foreign-born in several European countries has already reached a level comparable to that of the United States. The global recession of the late 2000s reduced the flow of immigration to these countries, but they remain home to about one-quarter of the world's migrant population.

PROBLEMS OF SOCIAL INTEGRATION When they began to arrive in the 1960s, immigrant workers were seen by the native populations as temporary residents who would return to their home countries once their labor was no longer needed. This was made especially poignant in Germany, where immigrants were labeled *Gastarbeiter*, literally "guest workers." Although the workers were a vital component of the host society's economic system, little effort was made to integrate them socially or culturally. They were marginal people, and their social conditions were reminiscent of those experienced by successive waves of European immigrants and southern

TABLE 17.1 | FOREIGN-BORN POPULATION OF SELECTED EUROPEAN COUNTRIES

	1960		2011*	
	Number (1000s)	% of Total Population	Number (1000s)	% of Total Population
France	3,507	7.7	7,289	11.2
Germany	2,003	2.8	9,807	12.0
Ireland	73	2.6	557	12.4
Italy	460	0.9	5,350	8.8
Netherlands	447	3.9	1,869	11.2
Norway	62	1.7	568	11.6
Spain	211	0.7	6,556	14.2
Sweden	296	4.0	1,384	14.7
Switzerland	714	13.4	1,940	24.7
United Kingdom	1,662	3.2	7,244	11.6

*Includes citizens of European Union (EU) countries, who are legally permitted to move from one EU country to another.

Source: Eurostat (2012).

black migrants to northern U.S. industrial cities in the late nineteenth and early twentieth centuries. Their housing was substandard; their jobs were the lowest paying and least prestigious; and, except for work, they did not participate in the society's mainstream institutions.

Today, the socioeconomic conditions of these minority ethnic groups are improving, but their absorption into German society has failed to advance very far (Saunders, 2010; Soysal, 2008; Woellert et al., 2009). Few of Germany's 3.5 million Turks have become citizens, even those who are German-born and know nothing of any other society. Until 2000, German citizenship was based on ancestry; that is, only those born into the German community could qualify as a German national. This meant that ethnic Germans, having lived, for example, in Russia for many generations and speaking no German, could obtain German citizenship merely by moving to Germany (which in fact many did in the 1990s). By contrast, very few second- or third-generation Turks, having lived nowhere but Germany and speaking only German, were able to qualify for citizenship. A change in German laws in 2000 making it easier for the foreign-born to obtain citizenship did not result in a significant increase in those naturalizing (Leise, 2007; Oezcan, 2004).

In France, by contrast, an assimilationist model has traditionally shaped the society's response to immigrants, awarding citizenship on the basis of birth in France, not ancestry. Technically, one qualifies as "French" so long as he or she speaks French and adopts French cultural norms and values. Although this would seem to have made the absorption of immigrants a less troublesome process, the results have not been dramatically different from Germany's.

Most immigrant groups in all Western European societies—especially those from Africa, Asia, and the Middle East—continue to experience a high level of economic and social marginalization. Their unemployment rates are high, and they are overrepresented in unskilled, low-paying jobs (OECD, 2004). Furthermore, they live mostly in concentrated areas and participate only minimally in public life.

PREJUDICE AND DISCRIMINATION   Immigrants in Western Europe have encountered high levels of prejudice and discrimination (Campani, 1993; Castles and Miller, 2009; Finn, 2002; Halász, 2010; Peach and Glebe, 1995; Skellington, 1996). Having been relatively homogeneous nations, Western European countries lacked the historical experience of the United States, Canada, and other traditional immigrant societies in absorbing large numbers of ethnic newcomers. The result has been, at best, a begrudging tolerance of immigrants and, at worst, outbreaks of violence. In Germany, physical attacks against immigrants by the early 1990s had become commonplace; in France, North Africans, who compose fully half the immigrant population, are the primary targets. Racially motivated incidents have occurred even in countries with fewer immigrants and solid traditions of tolerance, such as Italy and Sweden.

In some ways the impact of increasing ethnic diversity on Western European societies in recent years has been even more profound than on the United States, and its social, political, and economic consequences have been more wrenching. Questions concerning the social and economic integration of large and distinct ethnic minorities are now in the forefront of political discourse throughout Europe. Should a version of multiculturalism (as in Canada and Australia) continue to be advanced or should stronger efforts be made to assimilate these groups (Vasta, 2007)?

The fact that these new ethnic populations contain large Muslim communities has given added strength to anti-immigrant movements. Most of the Netherlands' immigrants, for example, are from the Middle East or North Africa and are Muslim. The presence of people radically different culturally has created a situation in which Muslim immigrants generally have been viewed as a social and political problem; this, in turn, has spurred anti-immigrant, often racist, views and actions (Duffy, 2004; Erlanger, 2011; Halász, 2010; Sniderman and Hagendoorn, 2007). These reached a climax in 2004 when a Dutch filmmaker was brutally murdered by a radical Islamist after he had produced a film critical of the Muslim treatment of women (Buruma, 2005; Saunders, 2004).

The presence of millions of Muslims in France, Germany, the Netherlands, and other Western European countries has set off virulent social debates regarding the compatibility of Islamic cultures with mainstream cultures and the social integration of these immigrants (Benton and Nielsen, 2013; Eriksen, 2013; Roy, 2005). Whereas most governments have committed to efforts at expanding social and economic opportunities for immigrants, sharp debate has arisen over how—or whether—the newcomers should be integrated. Several governments in 2010, for example, including those of France and Belgium, began to legislate restrictions on one of the most readily visible of Islamic ways, the full-face veil. These measures met with controversy but were mostly supported by the native population (Cody, 2010).[2]

Many believe that the new groups, particularly Muslims, are simply too different culturally and are not prepared to blend into the mainstream. Hence, fears have arisen that European cultures are being threatened, and anti-immigrant sentiment has intensified accordingly. Extreme right-wing parties have called for immigrants to be repatriated to their countries of origin, but few parties at any point on the political spectrum have not openly advocated, at minimum, a halt to further immigration. As was noted in Chapter 12, anti-Muslim prejudice and discrimination in the United States has paralleled its development in Western Europe, particularly since the terrorist attacks of 9/11. But it has been far more turbulent in Western Europe. Moreover, the rise of political parties whose major objectives are to limit or repatriate Muslim immigrants has not had an American counterpart.

THE PERMANENCE OF ETHNIC DIVERSITY   Regardless of efforts to limit immigration, sizable ethnic populations will remain a permanent part of Western European societies. The children and grandchildren of the original immigrants are European-born or reared, and for them a society of origin, or "mother country," is only an abstraction. Moreover, it would be difficult for the economies of Germany, France, and other countries in the region to function without foreign workers. With extremely low birthrates, these societies can fill their future labor needs only through substantial immigration. Western Europe is a central link in the emergent global economy in which the flow of capital and goods and, increasingly, workers is unhindered by national boundaries.

---

[2] Defenders of the ban on wearing the full-face veil have argued that the aim of the measure is not to limit religious freedom but is essential to assure public security and uphold principles of social integration. A leading French advocate of the ban maintains that covering the face makes "identification or participation in economic and social life virtually impossible" (Copé, 2010).

As in North America, the tendency toward greater ethnic and racial diversity, therefore, appears irreversible. These societies are now permanently multiethnic. It is reasonable to expect, therefore, that for many years ethnic issues will be in the forefront of the politics of most Western European countries.

ETHNIC NATIONALISM AND CONFLICT: SOME CONTEMPORARY EXAMPLES

Most people in North America have only a vague understanding of political and social affairs that occur outside the boundaries of their own society. But even those who have kept only casually informed of world events in the last thirty years could not help but become aware of the increasing level of ethnic nationalism, accompanied by political conflict, that has erupted in various parts of the world. Some of these situations have been dealt with through peaceful political means; others have been disastrously violent.

As the world watched the dramatic end of the Cold War in the late 1980s, expectations were high that a new era of global tranquility was on the horizon. With the political map of the world changing frequently and rapidly, however, it became obvious that a new force had been unleashed, one that augured not a more peaceful globe but one enmeshed in ethnic conflict. Indeed, in the early twenty-first century, ethnic nationalism appears to be the world's most ubiquitous, intractable, and devastating sociopolitical force.

THE RESURGENCE OF ETHNIC NATIONALISM

Nationalism, as a form of political and social expression, emerged in the eighteenth century with the appearance of the first modern nation-states in Europe and the United States (Kohn, 1944). Loyalties to family, kin, tribe, or community were extended so as to encompass entire peoples, marked off by common language, religion, and other cultural elements. Benedict Anderson (2006) has referred to these emergent nations as "imagined communities," social constructions of modern history, but amazingly powerful in claiming people's loyalty. In the twentieth century, and especially after the end of World War II, nationalism spread throughout the world among formerly colonized peoples intent on achieving political self-determination. It also advanced among well-established nations where diverse ethnic groups asserted their identity and sought greater social and political autonomy.

In its modern-world expressions, nationalism has essentially been of two types— civic and ethnic.

CIVIC NATIONALISM   **Civic nationalism,** Michael Ignatieff writes, "maintains that the nation should be composed of all those—regardless of race, color, creed, gender, language, or ethnicity—who subscribe to the nation's political creed" (1995:6). In societies that have chosen this form of nationalism, *people possess equal political and social rights and choose to be members of the nation, along with others—regardless of ethnicity— who share broadly similar beliefs and values.* Citizens may be members of various ethnic groups, but what binds them together is not their ethnicity but a common belief in and loyalty to the idea of the nation as a political entity, encompassing everyone who pledges allegiance to it. American patterns of nationalism, as well as those of most other immigrant societies, have historically been expressed along these lines.

Ethnic Nationalism   **Ethnic nationalism** (or as some have referred to it, "ethnonationalism"), by contrast, holds that *people's allegiance is to an ethnic group or nationality into which they have been born or assigned, not to a larger political entity encompassing many ethnic groups or nationalities.* Here, "an individual's deepest attachments are inherited, not chosen" (Ignatieff, 1995:7–8). Basic to ethnic nationalism is the notion that people cannot join the nation simply by living within it or even by accepting its cultural system. One is a member of the nation only through heritage, and those who cannot claim that ethnic heritage do not qualify for membership. Those who are part of the ethnic nation, notes Aviel Roshwald, "are bound to one another by putative ties of blood, not just by juridical categories or ideological affinities" (2006:254). In the past several decades, every continent on the globe has witnessed cases of intense ethnic conflict in one or more countries, provoked by ideologies of ethnic nationalism.

A common form of ethnic nationalism in the contemporary world involves the assertion of national identity by minority ethnic groups and their challenges to majority-group dominance. Within multiethnic societies, a group or groups see themselves as fundamentally different culturally from the dominant group and from the rest of the society. They may view themselves not only as different but as culturally and politically oppressed. Hence, they may seek greater autonomy within or independence from the nation-state of which they are a part. (Recall "secessionist minorities," described in Chapter 2.) Political and cultural movements by Basques and Catalans in Spain; Scots in the United Kingdom; Flemings in Belgium; Tamils in Sri Lanka; Ibo in Nigeria; Kurds in Iraq, Iran, and Turkey; Chechens in Russia; and, as we have seen, Quebecers in Canada all illustrate this form of ethnic nationalism.

By the same token, a dominant ethnic group may try to reduce the economic, political, or cultural influence of another ethnic group that it defines as inimical to its interests or dangerous to the nation. In such cases, the dominant group seeks to entrench its power by further subordinating, expelling, or perhaps even exterminating that rival group. This version of ethnic nationalism gave rise to horrific genocides in the twentieth century, including the Nazi extermination of the European Jewish population and, more recently, the mass killings in Bosnia and Rwanda. Let's look at these latter two cases, each the outcome of extreme ethnic nationalism.

The Rwandan Genocide

Historical Background   Rwanda is a tiny country in central Africa which, like many others on the African continent, is the product of European colonialism and its aftermath. In the early 1900s, both Germany and Belgium sought control over the area that eventually became Rwanda and the neighboring country of Burundi. After World War I, Germany lost control of its African colonies, and Belgium emerged as the dominant colonial power in the region. In the fashion of classic colonialism, Rwanda and Burundi were exploited for their natural resources, specifically coffee and tea production.

When Americans or other Westerners try to make sense of mass killings among rival ethnic groups, the usual explanation is that these are "ancient" rivalries that "naturally" or "inevitably" result in periodic hostilities. That is rarely the case, however, and the Rwandan genocide illustrates this quite well. Its two major ethnic groups, Tutsi and Hutu, both had drifted into the region over several centuries so neither could claim to be the society's original people.[3] Cultural differences between the two were not sharply evident. In fact, they shared a common language and religion. Physical differences—Tutsi taller, Hutu shorter—though originally serving as a rough ethnic marker, had also begun to fade as intermarriage occurred increasingly. In brief, prior to European colonization neither group saw the other in ethnic terms. As sociologist Michael Mann has described them, "Hutus and Tutsis have not been fighting each other since time immemorial" (2005:432).

As the Rwandan state developed, however, an ethnic stratification system evolved, with Tutsis, though only 15 percent of the population, emerging as a ruling elite. As a means of controlling the population, the Belgians during their colonial reign encouraged and solidified this ethnic division in a kind of "divide and rule" strategy. Tutsis were placed in strategic positions in the colonial administration, and Hutus became an ethnic minority despite their numerical dominance (Eltringham, 2004; Human Rights Watch, 2006; Orth, 2006; Semujanga, 2003).

Through a distorted and racialized reading of the history and anthropology of the region, the Belgian colonialists had come to see the Tutsi as a "superior race" to the Hutu. This was a view that, not surprisingly, the Tutsis adopted readily. On the other hand, the Hutu came to be seen as an inferior people who deserved their second-class treatment. As the historian Gérard Prunier writes,

> As a consequence they [Hutu] began to hate *all* Tutsi, even those who were just as poor as they, since *all* Tutsi were members of the "superior race," something which was to translate itself in the post–Second World War vocabulary as "feudal exploiters." (1995:39)

In 1926, the Belgian colonial regime issued ethnic identity cards, thereby establishing an official basis of the ethnic division.

Social and cultural differences between Tutsis and Hutus were, then, essentially political creations of the colonial Belgians who used these putative group characteristics to solidify their control over the population. A racial ideology had been crafted from the myths of the white colonialists, and it would come to shape the relations between the two peoples. The time bomb had been set, as Prunier describes it, and it was only a matter of when it would go off.

Intertribal clashes erupted periodically, many of which resulted in widespread killing and destruction. In 1959, when the Belgians began to leave, Hutus overthrew the Tutsi elite and established a Hutu-led republic. Twenty thousand Tutsis were killed and three hundred thousand driven into exile mostly into the neighboring countries of Burundi, Tanzania, and Uganda. This was a seminal event for both sides: Hutus saw it as a glorious triumph over past injustices, whereas Tutsis saw it as a criminal action. In the several years afterward, Tutsi attempts to regain control

---

[3] A third indigenous tribe in Rwanda, the Twa, was very small (less than 1 percent) and played essentially no political role.

of Rwanda were repelled, but Tutsi leadership in exile never relinquished hope of eventually restoring their power. Meanwhile, waves of interethnic killings in both Rwanda and Burundi continued ceaselessly.

The opportunity for Tutsis to reclaim power in Rwanda came in 1990 when the Tutsi political organization in Uganda, the Rwandan Patriotic Front (RPF), launched an invasion across the border. After a long period of fighting and under pressure from domestic opposition groups, President Juvenal Habyarimana agreed to negotiate a settlement with the RPF and create a power-sharing government that would replace the one-party system from which Tutsis had been excluded. Hutu extremists saw this new political arrangement as a dire threat to their well-being, however, and feared a return of Tutsi dominance. To deal with the perceived threat, they decided to literally exterminate the Tutsi in Rwanda.

THE GENOCIDE   In 1994, the plane carrying the Hutu president of Rwanda was shot down, killing him and others on board. This was a catalytic moment, setting in motion the genocide that would follow for the next six weeks.

What ensued was a virtual orgy of killing. Tutsis were killed only because they were Tutsi, no matter what their political leanings. But murdered as well were thousands of Hutus who were seen as political enemies or sympathizers to opposition parties. Others were simply caught up in the murderous frenzy. The intensity of the killings reached a level of almost unimaginable proportions. Journalist David Rieff recounts the killing as it began and gained momentum:

> Over the radio, the call kept going out for the extermination of every Tutsi man, woman and child in Rwanda, and of the Hutus who opposed this final solution. In the villages, people were hunted down. They died by the tens of thousands in their homes, in their fields and in the churches … in which they had sought refuge. And no matter how many died, the radios kept blaring out the calls for all good Hutus to kill the *imyenzi*, the "cockroaches," who were polluting the Rwandan nation and preventing it from living in peace. (1996:31)

The popular view of the physical differences between Hutu and Tutsi—that the former were generally shorter and stockier and the latter taller and slimmer—led to killing merely on the basis of one's appearance. But because of the frequent intermarriage between the two groups, there were many Hutu-looking Tutsi and Tutsi-looking Hutu. The ethnic identity cards were used by Hutu gangs to determine who was Tutsi and thus to be killed, but merely the appearance of Tutsi ethnic identity could be sufficient cause. "In towns or along the highways," Prunier explains, "Hutu who looked like Tutsi were very often killed, their denials and proffered cards with the 'right' ethnic mention being seen as a typical Tutsi deception" (1995:249).

No one was safe and no one, if identified as Tutsi, was spared. Neighbors killed neighbors; relatives killed relatives; teachers turned over their own students to the militia or even did the killing themselves; hospitals, schools, or churches proved no safe havens for those trying to escape death. Philip Gourevitch describes how militia men, often drunk or fortified with drugs looted from pharmacies, were bused from massacre to massacre. As an added incentive to the killers, "Tutsis' belongings were parceled out in advance—the radio, the couch, the goat, the opportunity to rape a young girl" (1998:115). And the horrors of the actual killings demonstrated the utter depravity of which humans are capable: chopping off body parts and other mutilations; sexual abuse of women before their murder; throwing live babies into pit latrines or smashing their heads against a rock; forcing family members to kill other family members; burning people alive as their relatives were forced to watch—all of these horrific acts were documented by the survivors (Gourevitch, 1998; Prunier, 1995).

Not all Hutu, of course, succumbed to their ideological passions or obediently carried out orders to kill. Numerous cases of heroism on the part of Hutus and attempts to protect potential Tutsi victims were also recorded. But the magnitude and level of brutality of the Rwandan genocide could, in modern history, be compared only to the Nazi Holocaust of the 1940s. Unlike the highly systematic and technologically sophisticated killing machine created by the Nazis, however, much of the actual killing in Rwanda was done with crude weapons, mostly machetes. But its objective was met: the elimination of three-quarters of Rwanda's Tutsi population in less than three months. No one will ever know the number actually killed, though most estimates are in the range of eight hundred thousand to a million.

THE ROLE OF IDEOLOGY   The Rwandan tragedy illustrates the crucial role of ideology in kindling ethnic conflict. Hutu leaders certainly saw themselves benefiting in terms of power and wealth from elimination of the Tutsis and instilled those visions to other Hutus who had hopes of obtaining land and belongings of the victims. Fear, as well, played a strong role in the genocide as the Hutu population had been subjected to a constant barrage of propaganda warning that Tutsis were determined to kill or subjugate all Hutus. Others were coerced into killing, fearing the consequences of refusing to do so (Mironko, 2006). But in the final analysis, Tutsi and Hutu killed each other, as Prunier writes, "more to upbraid a certain vision they ha[d] of themselves, of the others and of their place in the world than because of material interests" (1995:40). Hutu propaganda stressed the inherent evil of the Tutsis. Furthermore, Hutus, in this view, were the "natural" people of Rwanda who had been continually exploited and impoverished by the Tutsi (Mamdani, 2002; Verwimp, 2006). These ideas were continually disseminated through newspapers and, especially, radio starting in 1990.

The strength of the racial ideology that had taken hold of the Hutu majority became horrendously evident. Killing, in this view, was an act of self-defense, and killing women and children was perfectly logical if, as they had been taught to believe, the future was at stake. The power of the racist ideology in this case is further evidenced by the fact that much of the killing was carried out by ordinary peasants in addition to militia and other military units. The latter could not have accomplished the enormity of the slaughter in such a relatively brief time had they not had the cooperation of the civilian population. At the genocide's peak, eight thousand were killed per day, a rate faster than the Nazi Holocaust (Lovgren, 2004).

A TENUOUS PEACE   How does a society recover from such a powerfully destructive event as the Rwandan genocide? More than two decades after the events of 1994, ethnic conflict has been reduced to occasional forays of Hutu partisans across the border. Rwanda appears to be orderly and, by comparison with other central African countries, prospering. A Tutsi-led government is in power, though there is no effective opposition party.

An ambitious program of national unity has been promoted, and efforts have been made to reintegrate into the society many Hutus who had taken part in the killings (Gourevitch, 2009; Lovgren, 2004; McGreal, 2013). For several years, village courts heard testimony from participants and victims with the purpose, as described by Gourevitch, of "establishing a collective accounting of the truth of the crimes in each place where they were committed." By eliciting confessions of the murders, it was hoped that some degree of reconciliation would be brought about. On the surface that program seems to have yielded a tenuous ethnic peace. However, the enormity of the crime created so many victims and perpetrators that, as the current president of Rwanda, Paul Kagame, has admitted, "People's hearts and minds need some time to heal. A very long time indeed. They will probably need a whole generation, and the memories will keep lingering" (quoted in Gourevitch, 2009:42).

The Breakup of Yugoslavia: A Failed Attempt at Corporate Pluralism

Yugoslavia represents the failed attempt of a multiethnic society to construct some form of corporate pluralism. As we saw in Chapter 4, corporate pluralism is a situation in which a number of ethnic groups coexist in a loosely bounded system. These culturally and perhaps linguistically distinct groups are linked together in a federation in which each is recognized officially by a central government that attempts to distribute national political power and economic resources proportionally among them. Each, however, enjoys some degree of political autonomy.

In such cases, each ethnic group also usually occupies a distinct territory or geographical area, where it becomes the dominant group. However, these ethnically concentrated populations usually spill over or are dispersed to other areas of the country, where a majority of the population is composed of members of another ethnic group. As a result, minority ethnic groups develop in those territories. When corporate pluralism breaks down, the federation begins to crumble, and each ethnic territory seeks full political independence. As ethnic divisions intensify, each ethnic group attempts to maximize its territory and to embrace group members who are living as ethnic minorities in other areas. Previously subdued historic rivalries are regenerated, and intergroup hostilities heighten, perhaps leading to violent confrontations. The Yugoslav situation in the 1990s illustrates this process vividly.

YUGOSLAVIA AS A CORPORATE PLURALISTIC SYSTEM For almost thirty-five years following World War II, this ethnically disparate and politically divided country had been held together delicately by Josip Broz Tito (usually referred to simply as Tito), a World War II hero who led the victorious Partisan guerrilla army against the Nazis. After the war, Tito emerged as a unifying authoritarian figure, acting as prime minister and later as president. Though an avowed communist, Tito moved Yugoslavia toward nonalignment and away from the Soviet hegemony in Eastern Europe.

Tito's strong rule and ability to bring the various ethnic factions together had created a multiethnic society that fulfilled the major features of corporate pluralism. Yugoslavia was made up of six republics and two provinces, each different in social and political makeup and history, and each dominated by a particular ethnic group. Political and social institutions were carefully crafted to preserve a sense of ethnic balance and fairness among the various components. None of those measures, however, created more than superficial ethnic harmony. With Tito's death in 1980, the rancor that had been simmering beneath the surface gradually intensified, erupting into full-blown warfare in 1991.

THE DISSOLUTION OF THE YUGOSLAV FEDERATION With the end of the Tito regime, ethnic rivalries previously held in check reemerged, stirred by political leaders in each republic.

The major ethnic split was between the republics of Croatia and Slovenia on the one hand, and Serbia on the other. The former are both Roman Catholic and strongly oriented culturally toward Western Europe; Serbia, in contrast, is Orthodox Christian in religion and more culturally attuned to Eastern Europe. With the collapse of the Yugoslav Communist Party in 1990, Croatia and Slovenia each moved toward independence and seceded from the federation a year later. This left Yugoslavia composed mainly of Serbia. The smaller and less economically viable republics took sides or became pawns in the ensuing confrontations that developed as the three major ethnic states sought to establish boundaries that each interpreted as corresponding to their historical ethnic dominance.

The Serbs and Croats emerged as the major ethnic rivals. The enmity that erupted in the early 1990s was in some ways an extension of animosities that had run deep for many decades. During World War II, fighting in Yugoslavia occurred not only with the German invaders but also among the various ethnic factions themselves, notably between Serbs and Croats, who took opposing sides during the Nazi occupation. More than 1.5 million Yugoslavs died during the war, most as a result of interethnic fighting. Hitler established a puppet fascist regime in Croatia, run by the Ustasha Party, that sought to purge the country of Serbs (and, of course, Jews, though they were a relatively small population). Hundreds of thousands of Serbs were executed by the Ustasha regime (Bennett, 1995). Following the war, these savage killings were bitterly remembered by the Serbs; in their minds all Croats, whether they had been Ustashas or not, were associated with the killings.

Similar actions took place during World War II in Serbia, but against resident Croatians. Serbian guerrilla bands called Chetniks fought against the Nazis (and thus the Ustashas) but also oppressed and slaughtered many Croatians. Because of the atrocities on both sides, Serbs and Croats found themselves in a perpetual state of mutual distrust (Schöpflin, 1993). Indeed, so bloody had been the conflict inside Yugoslavia during World War II and so venomous were the emotions it created on all sides that, as the journalist Misha Glenny accurately forecast, "were it ever to revive, it was always likely to be merciless" (1993:171).

THE ROLE OF PROPAGANDA AND IDEOLOGY Some contend that Yugoslavia's attempt at corporate pluralism was always precarious and, without Tito, was doomed. Others, however, maintain that it did not fall apart of its own volition but rather was destroyed by a few people who incited ethnic emotions that served their political interests, especially Serb president Slobodan Milosevic (Bennett, 1995; Rieff, 1995).

Tito's death was accompanied by an economic crisis, which stirred a new wave of ethnic nationalism in Croatia, Serbia, and Bosnia, the latter another of the Yugoslav republics, about half of whose population consisted of secular Muslims.

Intergroup rivalries were promoted through propaganda campaigns in the media. Serbian television, for example, frequently showed films of the atrocities committed by the Croatian Ustashas during World War II (Human Rights Watch, 1995). In Bosnia, ethnic passions were aroused by accusations against the Muslim population, which was portrayed as a threat to the very survival of Christian Serbs and Croats.

At the outbreak of the conflict, the Serbs were not only the largest of Yugoslavia's ethnic groups but the most dispersed, living in all areas. Serbia alleged that Serbs living outside the Serbian nation—that is, in other parts of Yugoslavia—were victims of oppression, especially in Croatia and Bosnia. Serb leaders and the media engendered fear among Croatian and Bosnian Serbs that, if the territories in which they were living were to become independent states, they would be divested of their rights and property (Bennett, 1995). Hence, the first wave of war, starting in 1991, essentially became one in which these minority populations were "liberated" by Serbs.

ETHNIC CLEANSING IN BOSNIA   The most severe ethnic war within the now-shattered Yugoslavia occurred in Bosnia-Herzegovina (usually referred to simply as Bosnia). When Bosnia declared its independence from Yugoslavia in 1992, Bosnian Serbs refused to accept this arrangement, preferring instead to remain part of Yugoslavia (which by that time was essentially only Serbia). Thus began the efforts of the Bosnian Serbs, aided by Serbia itself, to drive Muslims from the territory. Although they lived in all of Yugoslavia's republics, Muslims were heavily concentrated in Bosnia. Sarajevo, the republic's capital and largest city, was not only heavily Muslim in numbers but culturally as well. The Bosnian Muslims should not be seen as comparable to Muslims in other parts of the world. They are secular, having only a distant link with other more traditional Islamic populations. Most are Slavs (that is, Serbs or Croats) who were converted to Islam during the five centuries of Ottoman rule in Bosnia (Glenny, 1993). Most, moreover, are Sunni Muslims, not in sympathy with fundamentalist groups in the Middle East and North Africa. Like the Serbs and Croats, Bosnian Muslims speak a variant of Serbo-Croatian.

The campaign waged by the Serbs in Bosnia was referred to as "ethnic cleansing." This meant killing Bosnian Muslims and Croatians, or driving them from their homes, in order to prepare their areas for a future "Greater Serbia." In addition to the Serb attack, the Croats also launched an attack on western Bosnia, driving out Serbs and Bosnians.

THE END OF ETHNIC HARMONY   An irony of the Yugoslav conflict is that Bosnia, the area affected most severely in the early 1990s, had for decades been an example of ethnic harmony, a society of "pluralism and tolerance" (Rieff, 1995). Its three major groups—Serbs, Croats, and Muslims—had coexisted peacefully and had created a relatively integrated society. In Sarajevo, Croatian Roman Catholic and Serbian Orthodox churches stood side by side with Muslim mosques, ethnically integrated neighborhoods were commonplace, and intermarriage among the three populations was not at all unusual.

Neighbors and family members, who had been comfortably integrated just weeks before, voiced puzzlement at how quickly their social harmony had dissipated and had been reduced to vicious and bloody hatred. "Growing up before the war here, my boys played with all the neighborhood kids," said a Serbian woman living in Bosnia. "We never knew who was Georgy and who was Muhammed. And I baked cakes for them all. Now neighbors we've known for years don't greet us anymore" (Sudetic, 1994:8A). So intense and bitter were the feelings that surfaced among those of the different ethnic groups that people were driven to exchange their homes with each other to create more ethnically homogeneous areas. Later, as Serbs took control of parts of Bosnia, Croats and Muslims were driven from their homes, which were given to Serb refugees who had fled similar incursions in other areas by Croats or Muslims. Often, as Serbs or Croats advanced on each other, houses of those fleeing were summarily destroyed. Ignatieff describes what he saw as he drove through ravaged villages:

> Toward Novska, you pass Serb house after Serb house, neatly dynamited, beside undisturbed Croat houses and gardens. When you turn toward Lipik, it is the turn of all the Croat houses to be dynamited or firebombed, next to their untouched Serb neighbors. Mile upon mile, the deadly logic of ethnic cleansing unfolds. (1995:35)

Much more dreadful actions were to occur, however, as the fighting intensified. Indeed, quickly the hostilities between Serbs, Croats, and Muslims degenerated into the most ferocious conflict in recent European history. Each side accused the other of atrocities, and in many cases these were confirmed by the United Nations and other nonparticipating observers. The most horrific actions, however, were committed by Serbs in Bosnia as part of their campaign of ethnic cleansing. Mass executions, rape camps, savage mutilations and torture, and wanton destruction of homes, churches, and mosques filled news reports as the Bosnian war intensified starting in 1993. One year later, the United Nations documented 187 mass graves, each containing the bodies of between three thousand and five thousand Muslims murdered by Serbs.

The horrors of the Yugoslav war were eerily reminiscent of the genocide committed by the Nazis fifty years earlier. The massacre in the city of Srebrenica was one of the more thoroughly documented of incidents that seemed so monstrous as to border on the surreal. Srebrenica had served as a United Nations refuge where more than forty thousand people had taken shelter from the war. After it was overrun by the Bosnian Serb army, the aftermath was described as the worst war crime in Europe since World War II. Although no one can be sure how many were slaughtered, American intelligence analysts estimated between five thousand and eight thousand. A few survivors as well as Serbian civilians living in surrounding villages recounted the mass killings. Hundreds were herded into trucks, taken to a field, and shot; others were jammed into a warehouse where they were killed with automatic rifles and shoulder-held grenade launchers. Mass graves of the victims were uncovered later (Engelberg and Weiner, 1995). This incident more than any other prompted the military and diplomatic efforts of the United States and the North Atlantic Treaty Organization (NATO) to stop the war. But it was only one of many such massacres of entire villages.

Muslims were victimized as well by Croats seeking to secure what they claimed was their territory. As with the Serbs, a policy of ethnic cleansing produced gruesome incidents of terror and brutality. One such affair, documented by UN investigators, occurred in late 1993. A so-called death platoon of a few dozen Croatian paramilitaries entered the village of Stupni Do, shouting "Let's kill the Muslims" and "Where are all the pretty girls for us to rape? Bring them out." A journalist describes what ensued:

> With avenging fury, the Croatian death squad crushed the skulls of Muslim children, slit the throats of women and machine-gunned whole families at close range. The next morning Croatian reinforcements finished up the job, torching all 52 houses and dynamiting the community's one small mosque. In less than 48 hours Stupni Do simply ceased to exist. (Nordland, 1993:48–49)

Although Bosnia suffered the most appalling conditions and was the site of the most intense fighting, other regions were affected to a lesser degree in much the same way. Following the Croatian invasion of the Krajina, an area of Croatia populated mostly by Serbs, a hundred thousand Serbs fled what they feared was the Croatian version of ethnic cleansing. Like the Bosnian Muslims, the Krajina Serbs expressed the same disbelief that former Croat friends and neighbors with whom they had lived together and even intermarried for decades could so viciously turn on them without provocation. "I had a very good friend, a Catholic Croat," said a refugee Serb. "I would have given blood for him. But three years ago, he said: 'I like you a lot. But please don't come and see me anymore. My neighbors object to a Serb visiting.'" Another put it: "I'm sorry I didn't burn down my house. I'm sorry I left it to the Croats to burn" (Perlez, 1995a, 1995b).

RESOLUTION AND LEGACY In late 1995, after three and a half years of war, a fragile peace was established—largely through the efforts of the United States—that divided Bosnia into semiautonomous regions along ethnic lines. Under the terms of the agreement, Bosnia was split into two parts—a Muslim-Croat federation made up of 51 percent of the territory, and a Bosnian Serb republic, with 49 percent. NATO forces were deployed to enforce the treaty. Although this temporarily stopped the killing, few seemed to feel that anything short of a more complete partition of Bosnia would secure a long-term resolution to the conflict (Mearsheimer and Van Evera, 1995).

The legacy of the war was that the ethnic mosaic of Yugoslavia's population had been radically—probably irreversibly—altered and interethnic relations reduced to mutual fear and antipathy. In Bosnia, the most severely affected region, at least two hundred thousand people had been killed or had disappeared; almost half of its pre-war population of 4.2 million had been displaced. This, it should be remembered, had been Yugoslavia's most ethnically diverse region. Wholesale population transfers occurred in parts of Croatia and Serbia as well. Most Croats living in Serbia were forced to leave, as were Serbs living in Croatia. Refugees fleeing regions as they repeatedly changed hands reached epic proportions, unseen in Europe since World War II. As many as three million people were driven from their homes. Those brave enough to try to return to their homes in areas now dominated by the opposing ethnic group found them occupied by members of that group, who themselves had been expelled from their homes in other areas.

For those who endured the war, a residue of distrust, resentment, and terror on all sides made the reemergence of interethnic tolerance and accommodation improbable for many decades in the future. Sarajevo, previously the showpiece of ethnic harmony, symbolized this legacy of ethnic hatred. There, Serbs refused to accept

Muslim governance of the city, as the peace agreement provided, and thousands left their homes in protest, some burning them in a final act of defiance. Some even exhumed the remains of dead family members to carry with them, fearing that their graves would be desecrated. In Mostar, another of Bosnia's major cities, two sets of nearly all institutions—schools, hospitals, public transportation—were operating in accordance with the almost total ethnic segregation that had been created (Wood, 2004a).

In effect, Bosnia was left a country partitioned into tightly enclosed ethnic enclaves, reflecting a system of cultural and educational apartheid (Cohen, 2005; Engelhart, 2012a; Hedges, 1996, 1997). The ethnic diversity that had characterized Bosnia before the war was replaced by ethnic homogeneity—Serbs living in Serb areas, Croats in Croat areas, and Muslims in Muslim areas; coexistence had essentially been obliterated.

KOSOVO: EXPANSION OF THE YUGOSLAV CONFLICT   In the late 1990s, the Yugoslav conflict, which had centered mainly in Bosnia, took a new turn. Kosovo, a province of Serbia, became the focus of severe ethnic fighting. For Serbs, Kosovo represented the birthplace of Serbian culture; thus it was of great symbolic importance, even though Serbs living there were only a small numerical minority. It is an area where, in the country's mythology, Serbs held off the Ottoman Turks from invading Europe in the late fourteenth century. The most important monasteries of the Serbian Orthodox Church are also located there.

Two factors made the Kosovo situation different from Bosnia. The Bosnian population before the 1990s had been relatively intermixed ethnically. Most of Kosovo, however, had been made up of ethnic Albanians—about 90 percent of its population of 2.2 million—with the minority Serb population segregated from neighboring Albanians. Very little social interaction occurred between the two populations, and hostile stereotypes thrived among both (Calic, 2000). Second, Bosnia had attained independence before the outbreak of the civil war there, whereas independence was the key issue for Albanians—mostly Muslim—in Kosovo: most desired that the province either become fully independent or become part of a greater Albania.

Under Tito, Kosovo had enjoyed much autonomy even though it remained a Yugoslav province. In 1989, the Milosevic government revoked much of Kosovo's autonomy, making it an integral part of Serbia and imposing repressive rule. Ethnic Albanians lost their jobs to minority Serbs, Albanian-language schools were closed, and the police and Serb army became an occupying force (IOM, 1999; Kinzer, 1992). In the early 1990s, Kosovar Albanian guerrillas—the Kosovo Liberation Army (KLA)—launched a movement for the province's independence from Serbia. Starting in 1998 the Serbs began an offensive intended to crush the insurgents, comparable in many ways to earlier Serb actions in Bosnia referred to as "ethnic cleansing."

In early 1999, the fighting in Kosovo entered a new dimension. NATO countries, which had acted to stop the Bosnian conflict, only after its horrors had become hideously apparent, seemed determined to avoid a similar outcome in Kosovo. Peace talks were initiated and an agreement was offered to the Milosevic regime in Serbia and the rebel forces in Kosovo to end the fighting. The agreement was rejected by the

Serbs, and NATO subsequently began a seventy-eight-day bombing campaign against Serb forces in Kosovo.

The pattern of atrocities that had so grotesquely characterized Serb actions in Bosnia was replicated in Kosovo—mass murders, destruction of entire villages, raping and looting on a massive scale. As a UN war crimes tribunal later explained in its indictment of Yugoslav president Milosevic, the campaign of terror and destruction against Albanian civilians was designed to change the ethnic balance in Kosovo by purging the province of as much of its Albanian population as possible. Within weeks, 90 percent of Kosovo's Albanians were driven from their homes and towns. Most fled to surrounding countries, creating a refugee crisis comparable to that brought about by the Bosnian war. Although by comparison with Bosnia the death and destruction were less expansive, the savagery was no less gruesome and shocking: Serb attackers intent on destroying and looting went from village to village in killing sprees, sparing no one—women, children, infants, the elderly—and torching every house and building in sight (Finn et al., 1999; Kifner, 1999).

At the cessation of fighting in July 1999, war crimes investigators estimated that at least ten thousand people had been slaughtered by Serbian forces during the three-month war, and newly discovered mass graves were being reported daily. The true death toll will probably never be known because many bodies were burned or hidden in an effort by Serbs to destroy evidence or were simply buried by relatives without informing Western officials.

Following removal of the Serbian army and paramilitaries, ethnic Albanians began to return, under the protection of NATO forces that now occupied the province. Not unexpectedly, violent recriminations, including killings and the burning of entire neighborhoods, were carried out by Albanians against Kosovo's Serb population. As the municipal administrator for one ethnically divided town—an American who had served in Bosnia—explained, "There is a degree of hate here that is far greater than anything I found in Bosnia; and also a degree of fear. I am searching for a word that means vitriolic squared" (quoted in Smith, 2000:15). Within a few months, three-quarters of the province's two hundred thousand Serbs had been forced from their homes and villages, most fleeing to Serbia proper. Thus, as in Bosnia, the war had produced an "ethnically cleansed" region. There was little likelihood that Serbs would ever return to Kosovo in significant numbers.[4]

Those who did return were subjected to continual threats and attacks despite the presence of UN peacekeepers. In 2004, five years after Serbian forces were removed, in one episode over two days, dozens of Serbian communities across Kosovo were attacked by ethnic Albanians. More than four hundred homes were ruined and thirty churches were destroyed (Wood, 2004b). In 2008, ethnic antipathies flared again when Kosovo declared its independence from Serbia. Although many countries (including the United States) quickly approved this move,, uncertainty of how Serbia, with its claim of sovereignty, would respond stoked fears of another bloody chapter

---

[4] Another piece in the ethnic puzzle of the disintegrated Yugoslavia—Macedonia—became a focus of global attention in 2001. As in Bosnia and Kosovo, ethnic conflict, in this case between Macedonians and ethnic Albanians, flared into armed conflict as a rebel ethnic Albanian group, intent on uniting the heavily Albanian area of Macedonia with neighboring Albania, initiated the hostilities. The Macedonian government managed to quell the uprising, but only after dozens died in clashes.

in the Yugoslav tragedy. In 2013, Serbia and Kosovo signed a fragile power-sharing agreement, but Serbia continued to refuse to recognize Kosovo's independence.

In sum, the breakup of Yugoslavia and the ethnic animosities and savageries it inflamed, like the Rwandan genocide, demonstrated horrifically the end result of ethnic nationalism carried to its extreme. What remained afterward was a region more intensely divided by ethnic loyalties than before, with little expectation of future reconciliation among the various groups.

NORTHERN IRELAND: DECLINING CONFLICT, ENDURING DIVISION

The ethnic conflict in Northern Ireland demonstrates a variant of ethnic nationalism. Though replete with episodes of frightful violence, it never reached the intensity and ferocity of the Rwandan or Bosnian wars. The total of less than 4,000 killed in Northern Ireland's conflict during its most virulent years seems minuscule in comparison. Moreover, in the past decade, the most combative aspects of the conflict have been brought to a close. Nonetheless, Northern Ireland continues to demonstrate clearly the difficulties of managing intergroup relations where ethnic identities and divisions are intense and historically well defined.

Although the protagonists in Northern Ireland's conflict are Protestants and Catholics, it must be understood at the outset that matters of religious doctrine and dogma are not the basic issues. Rather, this is an ethnic conflict in which religious identities mark off the boundaries of the two major ethnic groups.

To comprehend the basis of Northern Ireland's ethnic conflict, one must take note of the system of ethnic division and stratification that characterize this society. Protestants and Catholics constitute two sharply divided ethnic communities whose differences transcend religion. Each group maintains a relatively separate, cohesive and institutionally complete community (CAIN, 2004; Livingstone et al., 1998; Moxon-Browne, 1983). Although they share some common features, neither community, as Ruane and Todd have described them, "has a developed concept of a general interest which transcends its own particular interests" (1996:78).

Political scientist Richard Rose (1971) referred to Northern Ireland as a "biconfessional society" because almost all people identify themselves as either Protestant or Catholic. Those claiming to be atheists, it is joked, must say whether they are Protestant atheists or Catholic atheists. These identities, however, do not mean the same as they do in the United States or other Western nations. In those societies, "Protestant" and "Catholic" denote denominational affiliations that no longer carry with them significant social consequences. In Northern Ireland, these are ethnic identifications that determine in a fundamental manner one's place in the society and the nature of one's relations with others. Religious affiliation is quite simply one's most important social characteristic, taking precedence over one's social class and even one's gender (Harris, 1972; Mitchell, 2008). In some measure, it is not unlike the significance of racial identity for people in the United States or South Africa. Northern Ireland is a highly polarized society in which ethnicity plays a predominant role in all realms of social life—work, education, politics, residence, leisure activity. Not only do separate institutional structures exist for each group but aspects of ethnic nationalism infuse each group's perspective on Northern Ireland's history, its current political circumstances, and its future social order.

EVOLUTION OF THE CONFLICT The essence of the ethnic conflict in Northern Ireland is this: In 1920, six counties of the Irish province of Ulster, which today make up Northern Ireland, were adamant in retaining their tie to Britain following Ireland's successful movement for independence. Protestants outnumbered Catholics in this area, unlike the remainder of Ireland, and were resolved to remain apart from a sovereign, Catholic-dominated society. As a result, Ireland was partitioned into two states, the Irish Free State (later the Republic of Ireland), with a Catholic majority, and Northern Ireland, with a Protestant majority. Unfortunately, this left a substantial Catholic minority in the six counties.[5] The dispute since that time has concerned this artificially created boundary. Most Protestants seek to maintain the boundary as it is drawn and to remain linked with Britain, whereas most Catholics do not recognize the permanence of the dividing line and visualize an eventually united Ireland.

Today, Protestants are a slight numerical majority of Northern Ireland's 1.8 million, but in a united Ireland they would be outnumbered by about four to one. Hence, since the partition of Ireland, the root of the Protestants' resistance to political change has been the fear that they will be absorbed into a unified Ireland and thus become the minority within a Catholic-majority state. Although these conditions are the immediate sources of the modern conflict, the Protestant-Catholic division has much deeper roots reaching back many centuries. The conflicting ethnic identities and loyalties of Protestants and Catholics in Northern Ireland emerge out of what both groups see as very different and opposing national heritages: British for Protestants, Irish for Catholics. Each, therefore, maintains a distinctly different view of history and national identity.

The beginnings of conflict between Protestants and Catholics in Northern Ireland are traceable to the early seventeenth century, when England embarked on its colonization of Ireland. Over the next two centuries the suppression of Ireland's Catholics, particularly in Ulster, produced antipathies and episodic armed hostilities.

Late in the nineteenth century, sentiment grew in England for the establishment of home rule for Ireland, a measure strongly resisted by the Protestants of Ulster, who saw themselves at the brink of absorption into a Catholic majority. In 1916, Irish nationalists sparked an armed rebellion against British rule in what has become known as the Easter Rising. The rebellion eventually led to the partition of Ireland into North and South, with the latter granted full sovereignty from Britain a year later.

The partition was a compromise measure that fully satisfied neither the North nor the South. The Ulster Protestants wanted no form of autonomy for Ireland, preferring retention of the British union, and the South desired a fully unified Irish nation, including Ulster. Northern Ireland, then, was essentially created by the British, ironically, to avoid bloodshed between Protestants and Catholics (Beckett, 1972). Not until 1925 did all parties—Britain, Ireland, and Northern Ireland—finally accept the arrangement as permanent. Nonetheless, Irish Catholics, in both the North and South, never fully relinquished their belief that Ireland would eventually be unified.

---

[5] The province of Ulster originally comprised nine counties, but with partition it was reduced to six to ensure a more comfortable Protestant majority.

EMERGENCE OF THE "THE TROUBLES"   Increased economic prosperity and educational opportunities following World War II greatly benefited the Catholic minority and led to hopes for an end to the inequities they had long suffered in voting procedures, housing, and employment. In the late 1960s, Catholics adopted peaceful protest tactics similar to those employed by the black civil rights movement in the United States. The protesters even sang the American movement's anthem, "We Shall Overcome." This was not a separatist movement that sought unification with the Republic of Ireland but an effort to secure equal treatment for Catholics in Northern Ireland.

In response, militant Protestants mobilized around a countermovement. The British Army was sent to quell the violence, which had been directed mostly at Catholics, but its use of indiscreet tactics soon turned the Catholic populace against it. The Irish Republican Army (IRA), which had all but disbanded after the successful Irish Rebellion in the 1920s, reemerged as the protector of Catholic neighborhoods and began its campaign against the British Army. Protestant counterparts sprang up, creating the organizational basis for the ethnic violence that would characterize the society for the next three decades, a period referred to by the people of Northern Ireland as "the Troubles."

Throughout the 1970s and 1980s, incidents of sectarian violence, including bombings and assassinations carried out by paramilitary groups on both sides, kept the society in a state of turmoil. Meanwhile, a political middle course, aimed at reason and reconciliation, was stymied. The first real break in the cycle of violence occurred in 1994 when a cease-fire was effected and a so-called peace process begun.

CONFIRMING ETHNIC IDENTITIES   One of the curious aspects of the ethnic division in Northern Ireland concerns the manner in which ethnic identities are established and applied by members of the two communities. In the United States, although there are no "official" ethnic groups, ethnic or racial status is clearly defined in a customary fashion; rarely are there questions about who, for example, is white and who is black. In Northern Ireland, however, there are no apparent physical distinctions between the two ethnic groups, and except for religion, even cultural differences are not readily evident. How, then, do interacting people perceive ethnicity in a society in which it is such a critical social identity?

With obvious physical marks such as skin color not available, cultural cues and symbols, to which the people of Northern Ireland are finely attuned, serve effectively in their place. Names, for example, often suffice as indications of ethnic identity. Certain surnames indicate an English or Scottish (and thus Protestant) or Irish (and thus Catholic) ancestry, and even Christian names are commonly associated with one side or the other. Other cues are accent, dress, demeanor, and—some would insist—even facial features (Fraser, 1977; Lanclos, 2003). Because residence and education are so rigidly segregated, however, the surest indicators of religious affiliation are one's address and school attended. Knowing either of these social facts is ordinarily sufficient to identify a person ethnically (Buckley and Kenny, 1995). Northern Ireland demonstrates dramatically that visible physical differences or even vast cultural differences are not necessary to create and sustain fierce ethnic hostility. Groups need only perceive their differences as unbridgeable.

PREJUDICE AND DISCRIMINATION   Prejudice and discrimination in this society are in many ways similar to patterns of black-white relations in the United States. They are neither officially legitimized nor enforced through the application of coercive power, but they are not difficult to detect.

For Protestants, anti-Catholic stereotypes buttress their prejudices. Many of the stereotypes commonly applied to Catholics are part of what seems to be an almost standard litany employed by dominant groups in Western multiethnic societies— lazy, dirty, shiftless, oversexed, ignorant, quarrelsome, and so on (Bell, 1976, 1996; Fraser, 1977). Fraser writes that of all negative characteristics, the size of Catholic families "seems to provoke Protestant hostility more than anything else" (1977:94). He quotes one young Protestant boy of eleven, already well imbued with anti-Catholic conviction:

> The Catholics have big families and they won't work. Daddy says we're poor because the Government takes his money to give to the Catholics and all their children and their priests to keep them in luxury because they don't work. (1977:96)

For Catholics, anti-Protestant animosity stems primarily from a heritage of discriminatory treatment rather than from any long-standing stereotypes. The objections to and suspicions of Protestants focus on their political power and the society's social and economic issues, which tend to be shaped along religious lines (Barritt and Carter, 1972; Bell, 1996; Pollak, 1993). The animosity of many Catholics toward Protestants can be as deep as that which flows in the opposite direction. O Connor has trenchantly described this mind-set:

> Every Catholic knows someone in their circle who has either reacted with horror when told their child intended to marry a Protestant, tried to prevent such a marriage, or been wounded by such a reaction from a relative. The objectors may say their concern is based on religious conviction, that they fear for their child's religion, their future grandchildren's religion, even their child's safety. But the sheer distaste many show when confronted with the imminent arrival in their family circle of a Protestant cannot be explained as anything other than prejudice. (1993:169–70)

Because the two groups are so segregated in their own communities, it is difficult to ascertain the actual degree of discrimination applied by Protestants against Catholics, or vice versa. Observers of Northern Ireland are generally agreed, however, that ethnic discrimination has been both individual and institutional, particularly in employment and politics, and in both areas Catholics, historically, were the chief victims (Barritt and Carter, 1972; Bell, 1976; Hoare, 1982; Whyte, 1983). During the past few decades, steps have been taken to promote ethnic equality, and as a result Catholics have gradually closed the gap with Protestants on various measures of socioeconomic status (Boyle and Hadden, 1994; Darby, 1995; Edwards, 1995; Osborne and Shuttleworth, 2004; Smith and Chambers, 1991; Smyth and Cebulla, 2008). For example, Catholics have increasingly moved into higher-status professional and managerial jobs once closed to them.[6]

---

[6] Despite the fact that blatant forms of discrimination no longer characterize interethnic relations, Catholics may continue to *perceive* discrimination by Protestants. Because one's religious affiliation is not easily camouflaged, Catholics may not be convinced that "failure to get a job or a home is not governed by that fact" (Jackson and McHardy, 1984:4).

NORTHERN IRELAND AS A PLURAL SOCIETY   Northern Ireland is clearly a society in which the centrifugal forces of pluralism—some imposed, some voluntary—predominate in the relations between the two ethnic groups.

The establishment of separate institutional structures for each ethnic group is a key feature of highly pluralistic societies, and in this regard Northern Ireland is almost a pure case. Protestants and Catholics constitute ethnic communities with a high degree of identification and institutional completeness. Segregation is the general rule in most areas of social life. Each group maintains its own neighborhoods, schools, shops, voluntary associations, and even newspapers. Much social and leisure activity occurs within the churches of both Protestants and Catholics, which further restrains intergroup contact. Social interaction across ethnic lines, therefore, is confined mostly to instrumental contexts like work, and even there it is limited. The separation of the two communities is enforced as well by the norm of endogamy (Gallagher and Dunn, 1991; McGarry and O'Leary, 1995a). A Protestant student at one of the few integrated schools in Belfast confesses that "I'd never met a Catholic before I sat beside one in class" (Irwin, 1994:106). Although group differences may be tolerated in situations where Protestants and Catholics meet, rarely are they disregarded. Social separation is not dictated by law, as it was in South Africa under apartheid or in the pre-1960s American South, but it is effectively enforced by a deep-rooted customary system.

The rigorous ethnic segregation of neighborhoods and schools assures that divergent views are promulgated from generation to generation. The dynamics of Northern Ireland's residential patterns, in fact, seem strikingly similar to those of blacks and whites in American cities (Marger, 1989; Poole, 1982; Shirlow, 2008). Some working-class areas of Belfast are divided literally by a twenty-foot-high reinforced wall (ironically referred to as a "peace line").

Ethnic segregation in education is even more complete than in housing. Almost all Catholic children attend Catholic schools, and Protestants attend state (de facto Protestant) schools (Lloyd et al., 2004). Only at the university level do students of the two groups mix to any significant degree. Noticeably different histories are taught Protestant and Catholic students, one emphasizing the British connection, the other the Irish. Events, personalities, and political institutions are interpreted differently, thus perpetuating the discordant societal views and distinct national identities of the two groups (Murray, 1995).

The high level of residential and educational segregation in Northern Ireland has far-reaching effects on all other aspects of group interrelations, just as it does in the United States. Social contacts are minimized, and the cycle of misunderstanding, suspicion, and at times violence between the two groups is preserved (Murray, 1995).

STABILITY AND CHANGE   Over the years, numerous political schemes had been proposed periodically, aimed at resolving the conflict. But the depth of ethnic division precluded the implementation of any of these scenarios. Thus direct British rule of Northern Ireland—the least unacceptable political option—remained in place (Arthur and Jeffery, 1996). In the 1990s, however, a peace process was begun that seemed, finally, to create a political atmosphere conducive to ending the Troubles.

In 1994, paramilitaries on both sides declared a cease-fire. This was followed by a coming together of representatives of the Irish and British governments and all the political parties in Northern Ireland to begin talks that, it was hoped, would culminate in a permanent political settlement. Additional impetus for the peace process was provided by the United States, particularly the personal efforts of President Bill Clinton. A proposed peace agreement called for the resumption of most government responsibilities by a Northern Ireland government, rather than direct rule by Britain. A set of complex checks and balances provided for a kind of power-sharing system in which one side would be unable to dominate the government, as it had in the past.

A referendum in both Northern Ireland and the Irish Republic was held in 1998, asking whether the political institutions specified by the agreement should be adopted. By a resounding majority, the people of both Northern Ireland and the Republic voted their approval of the measure, though in Northern Ireland a much greater percentage of Catholics than Protestants supported it (Valandro, 2004). This peace accord became known as the "Good Friday Agreement."

For several years after the introduction of the peace accord, efforts at creating a functioning, power-sharing system were repeatedly frustrated by distrust on both sides. In 2007, however, a stable and lasting agreement appeared to have been reached. The most hard-core Protestant leader, Ian Paisley, who for years had vowed never to participate in a government with leaders of Sinn Fein, the IRA-affiliated, mainly Catholic party, committed to a power-sharing arrangement. A government, made up of representatives of Northern Ireland's two major political factions—the Democratic Unionist Party (virtually a Protestant party) and Sinn Fein—was formally agreed to. This marked the end of direct rule by London and the assertion of local power by Northern Ireland leaders. Similar agreements had been attempted in the past, all of which ultimately failed, but observers were optimistic that this may have finally signaled an end to the Troubles.

LOOKING BEYOND Over the years, both Protestants and Catholics had, in a sense, learned to live with their society's deep ethnic division and the often violent episodes that had become part of the social environment. Today, limited eruptions of interethnic violence continue to occur periodically, and the so-called "peace lines" separating Protestant and Catholic neighborhoods still stand. Moreover, there is no guarantee that the peace accord signed in 1998 will continue to withstand challenges by militant groups on both sides. One journalist describes it as "more of a fragile calm than a bona fide peace" (Engelhart, 2012b). Yet, a return to the level of hostility and tension that had so characterized earlier decades seems unlikely.

Public opinion polls have for many years indicated, among both Protestants and Catholics, increasingly conciliatory attitudes and an inclination to accept and even encourage the development of integrated institutions, like schools and neighborhoods (Gallagher and Dunn, 1991; Lloyd et al., 2004; Moxon-Browne, 1983; Ruane and Todd, 1996). Residential mixing, however, remains limited and almost all children still attend religiously segregated schools (Calame and Charlesworth, 2009; Hayes et al., 2007; Knox, 2010). People on both sides may agree in principle that ethnic integration would be desirable, but there seems little desire to create social institutions that would actually reduce the separation between the two communities.

The depth of ethnic division in Northern Ireland, then, is undeniably greater than it is in most other multiethnic societies. Hence, it is difficult to envision the eventual eradication of mutually deep-rooted suspicions. Most important, between the two groups there is no sense of shared community. Institutional segregation remains the norm and gives little indication of diminishing to a significant degree.

Thus, despite the apparent political resolution, so well entrenched are the major differences—real or putative—between this society's two ethnic communities that one cannot realistically see, even in the long run, their full integration. Almost two decades ago, political scientist Brendan O'Leary noted, "Communal tensions, territorial segregation, and endogamous social and sexual relations will remain dominant features of social and political life in Northern Ireland, with or without a definitive constitutional settlement" (1995:713). That observation seems no less compelling today. Even if the power-sharing arrangement proves to be long-lasting, it is unlikely that the distrust between the two communities spawned by the Troubles will be significantly reduced for many years, perhaps even generations (Brown, 2002). The most viable objective, therefore, may be the establishment of a social order in which maximum material and psychological security and optimum political power are provided to people of both groups, divided though they remain.

THE ENDURANCE OF ETHNIC CONFLICT AND CHANGE

Dominance by one group over another, noted Philip Mason, "is as old as the Pharaohs and springs from passions that are common to all men" (1970:337). The question, then, is not whether dominant-subordinate relations will continue to be an integral feature of diverse human societies but whether and to what degree those relations will be based on ethnicity. Given that multiethnicity is an advancing condition of most nations today on all continents, the ethnic basis of inequality appears unwavering. That, in turn, makes it likely that continued ethnic conflict will be a prevailing condition of our time.

Not only within but also between societies, certain trends portend continued ethnic conflict. Economies of global scale and world-encompassing communications networks make national boundaries increasingly permeable, thus creating ever-greater ethnic diversity within most of the world's nation-states. Indeed, as noted in Chapter 1, few societies today cannot to some degree be called multiethnic. These tendencies might lead to greater understanding among diverse peoples—and thus to a breakdown of ethnic barriers—as they come to recognize their problems and motivations as commonly human (Shibutani and Kwan, 1965). But these trends might just as easily serve to rigidify ethnocentric and nationalistic views by demonstrating more dramatically the economic and cultural differences among them. "In deeply divided territories," note McGarry and O'Leary, "increased exposure to the 'other' may make group members more aware of what their group has in common and what separates them from the others. Exposure may cement group solidarity rather than diffuse it" (1995b:855).

FACTORS OF ETHNIC CONFLICT

Figure 17.2 shows the conditions and circumstances that most commonly impel ethnic rivalry and conflict. This is not an exhaustive list but rather an outline of those factors that have consistently appeared as stimulants of conflict in various

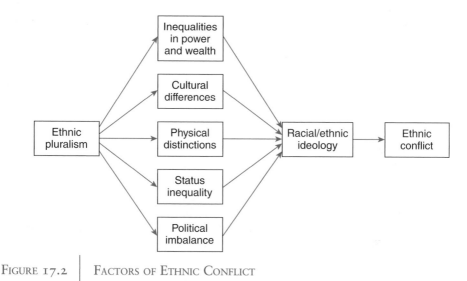

FIGURE 17.2 | FACTORS OF ETHNIC CONFLICT

multiethnic societies. All, to some degree and in some combination, are evident in the cases we have now looked at, as well as in the current and historical relations among ethnic groups in the United States.

In examining these factors, it is important to consider that ethnic conflict will vary in intensity, frequency, and duration. Despite popular belief to the contrary, rare are cases of interminable conflict that reach back centuries; rather, ethnic relations vacillate over long historical periods. Moreover, as explained in Chapter 1, societies are always admixtures of not only conflict and change but also cooperation and stability. Thus these two tendencies are always variable. Intergroup relations in any multiethnic society should not be seen, therefore, as an unending struggle or a constant series of clashes. It should also be remembered that ethnicity is never the only basis of societal conflict, even in highly diverse multiethnic societies. Political, religious, and class divisions are often the source of major conflicts and almost always intersect with ethnic divisions.

INEQUALITIES IN POWER AND WEALTH "Competition for resources," write political scientists David Lake and Donald Rothchild, "typically lies at the heart of ethnic conflict" (1997:100). In a similar vein, Stephen Steinberg suggests that "[i]f there is an iron law of ethnicity, it is that when ethnic groups are found in a hierarchy of power, wealth, and status, then conflict is inescapable" (1989:170). Where there are sharp inequalities in power and wealth among ethnic collectivities, dominant groups may induce stability through various techniques of prejudice and discrimination (discussed in Chapter 3) and the effective transmission of ideology. Eventually, however, a minority group or groups challenge that domination, and conflict then prevails. The United States and each of the other societies we have examined have in recent times experienced a challenge, in some degree, to their prevailing ethnic order.

It is important to understand that inequalities that induce conflict need not be severe or even objectively salient. Only the *perception* of inequality among ethnic groups is necessary. As we saw in Chapter 16, for example, the socioeconomic differences between French-speaking Canadians in Quebec and other Canadians are no longer evident. Despite this shrinking gap over the past several decades, however, the intensity of Quebec nationalism has not been reduced but has actually risen, creating an ongoing threat to Canadian unity.

Also, the grievances of minority ethnic groups may become more impassioned as they actually experience progressive change in their socioeconomic position. As conditions improve, the expectation of and desire for further improvements grow. If those expected changes do not materialize as rapidly as anticipated, minorities may focus on unmet goals rather than on their improved condition relative to that of an earlier period. We noted in Chapter 14 the possibility of such an outcome in South Africa. Susan Olzak (1992) has also postulated that as ethnic stratification breaks down and inequality declines among a society's ethnic groups, more powerful groups may seek to prevent weaker, minority groups from getting their share of the society's resources, thereby instigating greater conflict.

Amy Chua (2004) describes another form of economic inequality that produces ethnic conflict in the modern world. In many contemporary societies there are market-dominant minorities, that is, ethnic minorities who own a widely disproportionate amount of the society's wealth and control much of its commerce. Free markets enable these minorities to accumulate and concentrate great wealth as the impoverished majority looks on. As the majority experience democracy, however, an explosive situation ensues: "[T]he pursuit of free market democracy becomes an engine of potentially catastrophic ethnonationalism, pitting a frustrated 'indigenous' majority, easily aroused by opportunistic vote-seeking politicians, against a resented, wealthy ethnic minority" (2004:6–7).

It is most likely, then, that cooperation and stability will characterize ethnic relations for long periods where groups coexist with only minor differences in power, wealth, and prestige among them. In the contemporary world, of course, such cases are uncommon. Instead, deep economic and political inequalities characterize most multiethnic societies.

CULTURAL DIFFERENCES   In addition to economic and political grievances, ethnic conflict is generated by sustained and fundamental cultural differences. Most basically, where ethnic groups speaking different languages or abiding by sharply divergent norms and values must confront each other in various facets of social life—housing, education, politics, religion—conflict is a given; only its severity is variable.

Conflict arises as well when ethnic groups fear their assimilation into a dominant culture (Lake and Rothchild, 1998). Repeatedly, we discovered such conditions in each of the societies we looked at. Here, too, however, the reality of cultural difference among groups is less critical than the *perception* of difference. As we saw in Northern Ireland, for example, two groups whose cultural variance has faded over many generations (and who even speak a common language) nonetheless have clung to historical perceptions of difference that continue to shape their views and attitudes toward each other. As with economic and political inequality, sociocultural differences, or their perception, are fundamental in contemporary multiethnic societies.

PHYSICAL DISTINCTIONS Where there are physical distinctions among groups, an additional basis of division is in place. But racial differences among groups are not an essential ingredient in the emergence of ethnic divisions and conflict. In Canada, Northern Ireland, Rwanda, and the former Yugoslavia, we saw that ethnic boundaries between racially similar groups can be shaped in a number of ways—and once established, adamantly sustained, often for many generations.

STATUS INEQUALITY Conflict in multiethnic societies is induced by differences in how groups are seen and evaluated. "Strife arises," notes Heribert Adam, "when groups are differentially valued in terms of their intrinsic self-worth" (1992:20). A group's prestige or status is usually tied to its political and economic power, but growing political and economic equality does not always guarantee similar growth in social esteem. As we saw in the United States, African Americans, for example, despite real collective political and economic upward mobility in recent decades, continue in general to be held in lower social esteem by white Americans.

POLITICAL IMBALANCE In multiethnic societies, a strong state is necessary to maintain a balance of power and to negotiate among the competing groups. When that ethnic balance breaks down or groups see their position at risk, the possibility of conflict is heightened (Lake and Rothchild, 1997). The absence of a state capable of ensuring order is likely to lead to an intensification and extension of conflict. This provided the pretext for Britain to rule Northern Ireland directly for several decades, and, as we saw, led to the horrendous ethnic conflict in the former Yugoslavia.

Where the political status of ethnic groups is unbalanced, grievances and fears are easily interpreted in ethnic terms and are commonly exploited by political elites and ethnic activists who attempt to mobilize group members. As ethnicity is fused with political loyalties, intergroup tensions increase and the threat of violence intensifies. Thus, it is not simply ethnic diversity as such that is the cause of conflict, but ethnic politics (Smith, 2004). In Yugoslavia, writes Ignatieff, "[n]ationalist politicians on both sides took the narcissism of minor difference and turned it into a monstrous fable according to which their own side appeared as blameless victims, the other side as genocidal killers" (1995:22). The Rwandan genocide was similarly a product in large measure of political exiles ready to manipulate ethnicity in pursuit of power.

In the early 2000s, the war in Iraq provided another illustration of this outcome. There, following the U.S. invasion, forces were unleashed that created a sectarian-based ethnic conflict. Iraq, though mostly Muslim, is divided into Islamic sects that make for relatively distinct ethnic boundaries. Under the regime of Saddam Hussein, those ethnic divisions had been held in check through the imposition of an oppressive, secular state. With the toppling of the regime, Iraqi Sunnis, who had enjoyed dominant-group status, were now relegated to minority status. Iraqi Shias, the numerical majority, but under Hussein a discriminated-against minority, emerged as the dominant ethnic group (Hashim, 2005).[7]

---

[7] Kurds, a third major Iraqi religio-ethnic group concentrated in the northern region of Iraq, had also suffered under Hussein, and after his toppling assumed a relatively autonomous political position.

As the war progressed, ethnic cleansing occurred with increasing frequency. Paramilitary groups engaged in killing members of opposing sects, Sunni and Shia, and people living in areas of predominance of one or the other sect were driven from their homes into safer areas inhabited by coethnics. By 2007, at least two million people had fled their homes and another two million had fled Iraq, seeking refuge in neighboring countries (Rosen, 2007). All these developments were familiar to those who had experienced the horrors of the Bosnian conflict a decade earlier. A journalist returning to Baghdad after an absence of a year described the changes that had occurred in what had been a mixed neighborhood: "It is now relatively safe for Shiites as most Sunnis have been driven out." In another area, mostly Sunni, dead bodies cannot be picked up because "[t]he police are Shiite and afraid of the area" (Tavernise, 2007). Another journalist (Cave, 2007) described how formerly mixed neighborhoods of Baghdad had been transformed into homogeneous enclaves, where those in the minority sect had been issued death threats and had their houses burned. Even those who had intermarried or had maintained cross-sectarian friendships were threatened.

RACIAL/ETHNIC IDEOLOGIES   Any of the aforementioned factors can be sufficient to set off and to sustain ethnic conflict in a diverse society. What gives them added weight and what mobilizes people to take actions against an out-group is a coherent and well-founded racial or ethnic ideology. As explained in Chapter 1, ideology is a set of beliefs subscribed to by all or most of the society that rationalizes and legitimizes patterns of dominance and subordination. The Rwandan genocide in 1994 and the equally horrifying mass killings in Bosnia were driven by beliefs in the inherent inferiority, evil, and threat of the ethnic other. Similarly, the division in Northern Ireland is sustained by stereotypes and beliefs on both sides that crystallize into obdurate ethnic ideologies.

In a sense, ideology functions as a filter through which various social circumstances are interpreted and subsequently acted upon. It thus becomes a necessary component of ethnic conflict, justifying prejudice and discrimination against ethnic out-groups. As we have seen in previous chapters, racial/ethnic ideologies have played a key role in creating and sustaining a system of ethnic stratification in the United States and in all other multiethnic societies. Usually behaviors that might reflect the beliefs of in-groups about out-groups are held in moderate form and only occasionally manifest themselves in serious acts of violence. But there is never a guarantee that ideologies propounding the inferiority and threat of ethnic out-groups will not degenerate into riots (as have occurred at various times in U.S. history), mass killings, or even genocide, as in Rwanda and the former Yugoslavia. As Daniel Chirot and Clark McCauley have written:

> Humans are predisposed to think of competing groups other than their own in essentializing, that is to say, stereotypical ways, and this obviously leads easily to demonization of entire communities of perceived enemies. Our emotions—anger, shame, fear, resentment—predispose us to violence when we feel threatened, and to mass murder against those who most stand in our way or endanger us. (2006:7)

THE CONTAINMENT OF ETHNIC CONFLICT

Given the pervasiveness of ethnic diversity and inequality, it is perhaps most realistic to think in terms of the management of ethnic tensions and hostilities rather than their elimination. Some ethnic conflicts are resolved with finality, observes political scientist Milton Esman, but "the great majority rise, subside, and recur" (1994:23). It is important to reiterate that conflict is a fundamental condition of human societies and is inherent in all human affairs. It should not be confused with violence, however. Most social conflict is low-intensity and kept in check through the society's political process and legal system. Conflict may, of course, involve the use of violence, but this is ordinarily episodic and does not characterize most intergroup relations. Thus, although ethnic conflict is an essential, ongoing feature of all multiethnic societies, it is usually controlled and limited. Cases in which extreme violence occurs (Rwanda, Yugoslavia, or Iraq, for example) are exceptional, not typical. As sociologist John Darby has put it, "The use of violence marks the failure of normal and functional means of conflict resolution—political exchange, negotiation, compromise" (1995:29). The most achievable objective of multiethnic societies, then, is not to eliminate ethnic conflict entirely but to contain it as much as possible. "The real issue," writes Darby, "is not the existence of conflict, but how it is regulated" (1995:29).

The ability of diverse groups in multiethnic societies to live together in relative harmony while maximizing equality in the distribution of political and economic resources among them depends in large measure on the presence of a central state with sufficient power to contain intergroup conflict to a tolerable degree, but fair enough to win the allegiance of all groups. As Ignatieff has observed, "What keeps ethnic and racial tension within bounds in the world's successful modern multiethnic societies is a state strong enough to make its authority respected" (1995:243). Developments over the past two decades in South Africa, as we have seen, illustrate that efforts in this regard may in some cases yield significant changes. To some extent that may be occurring in Northern Ireland as well. Yugoslavia, in contrast, demonstrates the disaster that may result from the collapse of a strong state holding diverse groups together. The same appears to be the case in contemporary Iraq and may be the fate of Syria following its tragic civil war.[8]

Whatever successes may be achieved in the pursuit of ethnic harmony will probably be overshadowed by discord. It is obvious that with ethnic diversity the prevailing global trend, nations everywhere will for the foreseeable future struggle with issues of political, social, and economic inequalities among racially and culturally diverse groups. To some degree, then, ethnic conflict will be a recurring theme. The spread of pluralistic ideas and movements among ethnic groups within multiethnic societies is an outgrowth of the nationalistic fervor that has swept both the developed and developing worlds, and of international population movements that are occurring on a historically unprecedented scale.

It is possible, of course, that with the advent of a global economy and the emergence of global communications networks ethnic nationalism may prove to be only a

---

[8] In some multiethnic societies, ethnic consciousness may be so strong and divisive that there is little likelihood of reducing or even containing intergroup conflict within a single polity. Historian Jerry Muller (2008) suggests that in such cases, only partition, that is, the breakup into separate and more ethnically homogeneous political entities, may be the most practical and least costly solution.

passing phenomenon, destined to recede as people everywhere are increasingly linked by common needs and cosmopolitan cultural systems. This outcome, however, cannot yet be envisioned. The crucial fact, notes historian Anthony Smith, is that in the twentieth century interethnic conflict became "more intense and endemic ... than at any time in history" (1981:10). Today, there is little indication of the development of a countercurrent. Rather than diminishing, ethnicity as a global phenomenon persists as an exceptionally powerful and tenacious force motivating human behavior. And it will likely follow us throughout the twenty-first century.

SUMMARY

- Shifting patterns of global immigration have emerged in the past sixty years. This movement of people has been mostly from the developing societies to the developed, industrialized societies. Changing labor markets and political strife are major factors that have stimulated global immigration.
- Western European societies have undergone great social changes in the past few decades as a result of immigration and the subsequent creation of ethnic heterogeneity. Large numbers of culturally and racially diverse immigrants have created ethnic minority communities that have not been assimilated into the mainstream.
- Throughout much of the world, ethnic nationalism—distinct from civic nationalism—has been a stimulant of identity and political action.
- Two of the most deadly ethnic conflicts occurred in the 1990s. In Rwanda, a genocide of epic proportions occurred; and in the former Yugoslavia, ethnically divided Serbs, Croats, and Muslims engaged in the most ferocious warfare in Europe since World War II.
- Northern Ireland offers a case in which a historically well-rooted and seemingly endless ethnic conflict may be reaching its end. In recent years it has tortuously moved on a course toward a relatively peaceful and functioning pluralistic system. Nonetheless, ethnic divisions remain strong, and it continues to be a thoroughly segregated and divided society.
- In multiethnic societies where there are significant economic, political, and social inequalities among groups, ethnic conflict in some degree is an inherent feature. Cultural and physical differences among groups also may contribute to ethnic conflict. Racial/ethnic ideologies legitimize and inspire ethnic conflict.
- The conditions of the contemporary world—increasing diversity and ethnic nationalism—make the continuation and even intensification of ethnic conflict very likely.

CRITICAL THINKING

1. Despite globalization, ethnic identity and ethnic nationalism have proliferated in most nations of the world for the past half century. How might this paradox be explained?
2. Immigration, as this chapter has noted, is a global phenomenon, not limited to North America. What may reduce international migrant flows? Or is global migration likely to remain at a high level for the foreseeable future?

3. This chapter mentions the French ban on women wearing the full-face veil in public. Some would consider this an infringement on liberty, but others would argue that it is essential to the common good. Which side in this debate seems most reasonable? Who or what determines the limits that a multiethnic society, no matter how open and liberal, can impose on ethnic cultures?

4. Could a genocide similar to those that occurred in Bosnia and Rwanda in the 1990s occur in a society like the United States? What U.S. political and social conditions make such an occurrence improbable?

Personal/Practical Application

1. This chapter points out that strong ethnic divisions and ethnic conflict are created and sustained not simply by power and coercion but by ideology: beliefs about out-groups that rationalize and legitimize dominance and subordination. No one can escape the influence of their society's ethnic ideology in some degree. Can you account for the source of your beliefs and views regarding ethnic out-groups—parents, peers, schools, religious bodies?

2. The conflicts in Rwanda and the former Yugoslavia described in this chapter entailed some of the most violent and brutal actions witnessed in modern times. These were not the actions, however, only of armies and police, but of ordinary people. Do you believe that Americans could be capable of such horrendous actions against members of ethnic out-groups? Or do you believe that such events are unimaginable in the United States because Americans are "fundamentally different from others"? What circumstances could conceivably lead to extreme ethnic oppression in the United States? Consider the history of oppression against African Americans, American Indians, and Japanese Americans discussed in earlier chapters.

GLOSSARY

accommodation *See* **equalitarian pluralism.**

acculturation *See* **cultural assimilation.**

affirmative action Public policies designed to provide, through preferential measures, wider opportunities in work and education for ethnic minorities and women.

amalgamation *See* **biological assimilation.**

Anglo-conformity The prevalent ethnic ideology in American society, encouraging groups to shed their ethnic uniqueness and take on the ways of the core culture.

Anglo core group The British-origin group that established itself as the American dominant group and whose culture has been understood to be the society's standard.

annihilation Destruction of a minority ethnic group, perhaps culminating in genocide.

anti-Semitism Prejudice and discrimination directed at Jews.

apartheid The former system of extreme racial segregation in South Africa.

ascribed membership Membership in a social category (for example, ethnic group, gender) that is assigned at birth and not subject to fundamental change.

assimilation The process in which a minority ethnic group is integrated into the societal mainstream. *See also* **cultural assimilation; structural assimilation; biological assimilation; psychological assimilation; segmented assimilation.**

assimilationist minority A minority group that seeks integration into the mainstream society.

assimilationist societies Multiethnic societies in which *cultural* assimilation is encouraged for all groups but where *structural* assimilation occurs at different rates, depending on groups' cultural and physical distance from the dominant group.

authoritarian personality A theory suggesting that a certain personality type is inclined to hold prejudicial views and engage in discriminatory behavior.

aversive racism Covert, subtle expressions of prejudice and discrimination against racial and ethnic minorities, often unconscious and unintentional.

bilingualism The use of two languages.

biological assimilation Intermarriage to the point where there is a biological merging of formerly distinct groups.

black power An ideology espoused by militant African Americans during the period of the civil rights movement, which rejected the objective of integration and advocated confrontational tactics.

Bogardus scale A measure of social distance in which respondents are asked whether they would accept a member of an ethnic out-group in varying social contexts, extending from very close encounters to very remote ones.

caste A type of stratification in which status is acquired by heredity and in which mobility is severely constrained by law or custom.

categorical discrimination Discrimination against people on the basis of simply being part of a socially assigned category.

charter group A group that enters a previously unpopulated or underpopulated territory and establishes itself as the dominant group.

Chicano A term sometimes used to describe Mexican Americans.

civic nationalism An ideology in which people are considered members of a nation on the basis of common beliefs and values, regardless of ethnicity.

colonial societies Multiethnic societies in which inequalitarian pluralism, including extreme segregation, is the chief feature of ethnic relations. Also referred to as **segregationist societies**.

compensatory policies Government measures designed to advance the occupational and educational achievement of minorities who had been victimized by past discrimination.

competitive race relations Ethnic relations in which groups, living mostly in segregated areas, are in competition for the society's jobs, education, and other resources.

conflict theories Theories that view societies as held together by the power of dominant classes and ruling elites, able to impose their will on others; stability and order are maintained through coercion, not consensus.

corporate pluralism Ethnic relations in which structural and cultural differences among groups are protected by the state and institutional provisions are made to encourage an ethnically proportionate distribution of power and privilege.

corporate pluralistic societies Multiethnic societies in which groups remain culturally and perhaps physically separated, but in a state of relatively balanced political and economic power. Also referred to as **multicultural** societies.

cultural assimilation The adoption by one ethnic group of another's norms and values.

cultural pluralism The maintenance of many varied ethnic cultures within the framework of a common economic and political system.

cultural racism The belief that the cultures displayed by human groups are not easily changed and therefore can be ranked as superior or inferior.

demographic transition A model that describes population change in a society over the course of several stages, leading eventually to stable or low growth rates.

discrimination Negative actions against a group, aimed at denying its members equal access to societal rewards. *See also* **individual discrimination; institutional discrimination; statistical discrimination; structural discrimination**.

dominant ethnic group That group at the top of the ethnic hierarchy, which receives a disproportionate share of wealth, exercises predominant political authority, and has the greatest influence on shaping the society's cultural system.

dominant-subordinate relations *See* **ethnic stratification**.

dominative racism Actions taken to oppress racial and ethnic minorities and to keep them in a subservient position.

endogamy Marriage within one's social group, regulated on the basis of ethnicity, race, social class, educational level, religion, or other social characteristics.

equalitarian pluralism Ethnic relations in which groups retain their cultural and much of their structural distinctness but participate on an equal basis in a common political and economic system.

ethnic cleansing Expulsion and possibly annihilation of an ethnic group; applied in the 1990s by Serbs against Muslims in Bosnia and against ethnic Albanians in Kosovo.

ethnic group A group within a larger society that displays a unique set of cultural traits and a sense of community among members.

ethnic hierarchy The structural arrangement of diverse groups of a multiethnic society in a rank order, from those at the top, with most of the society's wealth and power, to those at the bottom, with correspondingly little.

ethnic nationalism An ideology holding that people's strongest allegiance is to an ethnic group or nationality into which they have been born or assigned, not to a larger political entity encompassing many ethnic groups or nationalities.

ethnic stereotype An oversimplistic and exaggerated belief about members of an ethnic group, generally acquired secondhand and resistant to change despite contrary evidence.

ethnic stratification A rank order of groups, each made up of people with presumed common cultural or physical characteristics, interacting in patterns of dominance and subordination. Also referred to as **majority-minority** or **dominant-subordinate relations.**

ethnocentrism The belief by members of a group that their culture is superior to others.

expulsion Removal of a group from the society through deportation or internment.

First Nations North American Indian tribes in Canada.

frustration-aggression theory An early theory of prejudice and discrimination explaining these as a means by which people express hostility arising from frustration. Also referred to as **scapegoating.**

genocide An attempt to systematically exterminate an ethnic or racial group.

Gordon's stages of assimilation A theoretical model suggesting that assimilation is not a straightforward movement but rather comprises different degrees of increasingly more profound social and cultural integration.

host society The society of entrance of international migrants. Also referred to as the **receiving society.**

hypo-descent A social policy in which a person with any known or recognized ancestry of the subordinate group is automatically classified as part of that group rather than the dominant group. Also referred to as the **one-drop rule.**

ideology A set of beliefs and values that explain, rationalize, and justify inequalities in wealth, power, and privilege.

indigenous superordination Subordination of a migrant population by an indigenous racial or ethnic group.

individual discrimination *See* **micro-discrimination.**

inequalitarian pluralism Ethnic relations in which groups are separated structurally and coexist in a state of highly unequal access to power and privilege.

institutional discrimination Actions taken against members of particular groups that are the result of the policies and structures of organizations and institutions. May be *direct,* in which discrimination is based on law or custom, or *indirect,* in which it is obscure and often unintended. The latter is referred to as **structural discrimination.**

integration A form of assimilation in which people of diverse ethnic groups participate freely and fully in the institutions of the larger society unconstrained by ethnicity.

intermarriage Marriage of partners of different racial or ethnic origins.

internal colonialism A type of inequalitarian pluralism in which minority ethnic groups are treated in a colonial fashion by the dominant group.

involuntary immigration The forced transfer of people from one society to another to be exploited as captive workers.

Islamophobia Fear of and hostility toward Islam and Muslims, based on false beliefs and negative stereotypes.

Jim Crow The system of black-white segregation maintained in the U.S. southern states from the end of Reconstruction until the 1960s that constituted a castelike form of stratification.

macro-discrimination Discrimination against minority groups not limited to specific cases but firmly incorporated in the society's normative system. *See also* **institutional discrimination.**

majority-minority relations *See* **ethnic stratification.**

manumission The freeing of slaves.

Marxian theory As applied to ethnic relations, the notion that in capitalist societies ethnic antagonism serves the interests of the capitalist class by keeping the working class divided along racial and ethnic lines and thus easier to control.

melting pot An ethnic ideology in the United States suggesting that various ethnic groups contribute to the creation of a hybrid "American."

Merton's paradigm A theoretical model explaining that prejudice and discrimination are variable, depending on a number of situational factors; hence, attitudes and actions toward members of minority groups may fluctuate within different social contexts.

micro-discrimination Actions taken by individuals or small groups to injure or deny access to societal resources (jobs, housing, education) to members of a minority ethnic group.

middleman minorities Certain ethnic groups in multiethnic societies that occupy a middle economic and political status between the dominant group and subordinate groups, often serving as intermediaries between them.

migrant superordination The subordination of an indigenous population by a migrant group.

migration chain A continuous movement of new immigrants to locations that have already been settled by other family members, friends, or coethnics.

militant minority A minority group that seeks to establish itself as the society's dominant group.

minority groups Groups that, on the basis of their physical or cultural traits, are given differential and unequal treatment and receive fewer of the society's rewards.

miscegenation *See* **intermarriage.**

model minority A popular characterization of Asian Americans, recognizing their comparatively successful adaptation to American society.

multiculturalism In the United States, the encouragement of the expression of different ethnic cultures. In Canada and Australia, an official policy recognizing and encouraging the retention of ethnic cultures.

multiethnic society A society composed of numerous cultural, racial, and religious groups.

nativism Anti-immigrant views and actions based on the perceived threat of immigrants to undermine the economic, social, and cultural interests of native-born people.

network factors Links between immigrants in the host society and friends, family, and colleagues in the origin society that contribute to further immigration, creating a migration "chain" or "stream."

New Immigration The second great wave of immigration to the United States, occurring during the late nineteenth and early twentieth centuries, consisting primarily of peoples from southern and eastern European societies.

Newest Immigration The latest and current wave of immigration to the United States, beginning in the 1960s, consisting mostly of peoples from Asian and Latin American societies.

normative theories Theories that explain prejudice and discrimination as conforming responses to social situations in which people find themselves.

Old Immigration The first great wave of immigration to the United States in the early and mid–nineteenth century, consisting mostly of peoples from northwestern Europe, especially Germany and Ireland.

one-drop rule *See* **hypo-descent.**

order theories Theories postulating that a society is a relatively balanced system made up of differently functioning but interrelated parts, held together by a consensus of values among groups and individuals.

panethnic group An inclusive ethnic category made up of several distinct ethnic groups lumped together, usually on the basis of the groups' world region of origin (for example, Asian Americans).

Park's race relations cycle A model of ethnic relations suggesting that groups pass through a sequence of stages—contact, competition, accommodation, and, ultimately, full assimilation.

paternalistic race relations A system such as slavery in which there is maximum social distance and extreme inequality between dominant and subordinate groups.

pluralism The retention of ethnic cultures and boundaries within the larger society. *See also* **cultural pluralism; structural pluralism; equalitarian pluralism; inequalitarian pluralism; corporate pluralism.**

pluralistic minority A minority group that seeks to maintain its cultural ways at the same time it participates in the society's major political and economic institutions.

power-conflict theories Theories that view prejudice and discrimination as emerging from dominant group interests and used to protect and enhance those interests.

prejudice A generalized belief, usually inflexible and unfavorable, applied to members of a particular group.

psychological assimilation A process by which people increasingly identify themselves as part of the larger society rather than an ethnic group.

push-pull factors Economic, social, and political conditions that exert a stimulus for people to emigrate from a society (push) or exert an attraction to immigrate to a receiving society (pull).

race A socially constructed category of humans, classified on the basis of certain arbitrarily selected hereditary characteristics that differentiate them from other human groups.

racial-ethnic group An ethnic group whose members are noticeably different physically from the dominant ethnic group.

racial profiling Actions of law enforcement or other security agencies against people of a particular race or ethnicity, based merely on the presumption that they are statistically more likely to commit crimes. (*See also* **statistical discrimination**.)

racialization The process by which a group comes to be defined as a "race."

racism The belief that humans are subdivided into distinct hereditary groups that are innately different in their social behavior and mental capacities and that can therefore be ranked as superior or inferior.

receiving society *See* **host society**.

Red Power The movement among American Indians during the late 1960s and early 1970s that sought to end the federal policy of termination and to revitalize Indian communities and cultures, often through direct action.

reverse discrimination Actions favoring members of minority groups over members of dominant, or majority, groups, usually in education and work.

secessionist minority A minority group that seeks neither assimilation nor cultural autonomy, but aims for a more complete political independence from the dominant society.

segmented assimilation The notion that immigrant assimilation may take different forms: adoption of the dominant culture, adoption of an oppositional culture, or economic advancement while retaining the ethnic culture.

self-fulfilling prophecy A process in which the false definition of a situation produces behavior that, in turn, makes real the originally falsely defined situation.

social class A category of people with approximately similar incomes and occupations who share similar lifestyles.

social Darwinism A theory, based on the notion of "survival of the fittest," holding that one's economic standing is ultimately the product of one's inherent capabilities.

social distance The degree of intimacy members of one ethnic group are willing to accept with members of other ethnic groups.

social mobility The movement up or down a society's class or ethnic hierarchy by individuals or groups.

social stratification A well-established system of structured inequality in which people and groups receive different amounts of the society's valued resources, based on various social, and sometimes physical, characteristics.

split labor market theory A theory holding that it is workers of the dominant ethnic group who are the chief beneficiaries of prejudice and discrimination, which help to keep ethnic minorities out of desired occupations.

statistical discrimination Discrimination against people based on beliefs about the social category of which they are members; unlike categorical discrimination, it is responsive to counterevidence.

straight-line theory A theory positing that movement toward more complete assimilation proceeds irreversibly with each successive generation.

structural assimilation A process in which social interaction within various institutions and settings increases among different ethnic groups.

structural discrimination Discrimination that results from the normal functioning of a society's institutions, rather than the direct and intended actions or policies of individuals and organizations.

structural pluralism The continued social separation of minority ethnic groups from the dominant group and perhaps from each other.

subordinate groups *See* **minority groups**.

symbolic ethnicity As suggested by Gans, the expression of ethnic identity by third- and fourth-generation people, though ethnicity no longer guides their social behavior.

visible minorities In Canada, ethnic groups that are non-European and non-Aboriginal in origin.

voluntary immigration The movement of people from one society to another by their own choice, most commonly in hope of economic betterment, but in some cases to escape political or religious oppression.

yellow peril An anti-Chinese characterization, referring to the perceived economic and cultural threat of Chinese immigration to the United States in the late nineteenth and early twentieth centuries.

REFERENCES

Abbas, Tahir. 2011. "Islamophobia in the United Kingdom: Historical and Contemporary Political and Media Discourses in the Framing of a 21st-Century Anti-Muslim Racism." Pp. 63–76 in John L. Esposito and Ibrahim Kalin (eds.), *Islamophobia: The Challenge of Pluralism in the 21st Century*. New York: Oxford University Press.

Aboud, Frances. 1988. *Children and Prejudice*. Oxford: Blackwell.

Abraham, Nabeel. 1994. "Anti-Arab Racism and Violence in the United States." Pp. 155–214 in Ernest McCarus (ed.), *The Development of Arab-American Identity*. Ann Arbor: University of Michigan Press.

Abrams, Charles. 1966. "The Housing Problem and the Negro." Pp. 512–24 in Talcott Parsons and Kenneth B. Clark (eds.), *The Negro American*. Boston: Houghton Mifflin.

Abramsky, Sasha. 2005. "One Nation, under Siege." *The American Prospect* (April):48–52.

Abramson, Harold. 1980. "Assimilation and Pluralism." Pp. 150–60 in Stephen Thernstrom (ed.), *Harvard Encyclopedia of American Ethnic Groups*. Cambridge, Mass.: Harvard University Press.

Acosta-Belén, Edna, and Carlos E. Santiago. 2006. *Puerto Ricans in the United States: A Contemporary Portrait*. Boulder, Colo.: Lynne Rienner.

Acuña, Rodolfo. 1972. *Occupied America: The Chicano's Struggle toward Liberation*. San Francisco: Caufield.

Adachi, Ken. 1976. *The Enemy That Never Was*. Toronto: McClelland & Stewart.

Adam, Heribert. 1971a. *Modernizing Racial Domination: South Africa's Political Dynamics*. Berkeley: University of California Press.

———. 1971b. "The South African Power-Elite: A Survey of Ideological Commitment." Pp. 73–102 in Heribert Adam (ed.), *South Africa: Social Perspectives*. London: Oxford University Press.

———. 1995. "The Politics of Ethnic Identity: Comparing South Africa." *Ethnic and Racial Studies* 18:457–75.

Adam, Heribert, and Kogila Moodley. 1986. *South Africa without Apartheid: Dismantling Racial Domination*. Berkeley: University of California Press.

Adams, David Wallace. 1995. *Education for Extinction: American Indians and the Boarding School Experience, 1875–1928*. Lawrence: University Press of Kansas.

Adams, Michael. 2007. *Unlikely Utopia: The Surprising Triumph of Canadian Pluralism*. Toronto: Viking Canada.

Adelman, Robert M. 2005. "The Roles of Race, Class, and Residential Preferences in the Neighborhood Racial Composition of Middle-Class Blacks and Whites." *Social Science Quarterly* 86:209–28.

Adhikari, Mohamed. 2005. *Not White Enough, Not Black Enough: Racial Identity in the South African Coloured Community*. Athens: Ohio University Press.

Adorno, T. W., et al. 1950. *The Authoritarian Personality*. New York: Harper & Row.

Agbayani-Siewert, Pauline, and Linda Revilla. 1995. "Filipino Americans." Pp. 134–68 in Pyong Gap Min (ed.), *Asian*

Americans: Contemporary Trends and Issues. Thousand Oaks, Calif.: Sage.

Ahdab-Yehia, May. 1983. "The Lebanese Maronites: Patterns of Continuity and Change." Pp. 148–62 in Sameer Y. Abraham and Nabeel Abraham (eds.), *Arabs in the New World: Studies on Arab-American Communities.* Detroit: Wayne State University, Center for Urban Studies.

Ajrouch, Kristine J. 2004. "Gender, Race, and Symbolic Boundaries: Contested Spaces of Identity among Arab American Adolescents." *Sociological Perspectives* 47:371–91.

Ajrouch, Kristine J., and Amaney Jamal. 2007. "Assimilating to a White Identity: The Case of Arab Americans." *International Migration Review* 41:860–79.

Alba, Francisco. 2013. "Mexico: The New Migration Narrative." Washington, D.C.: Migration Policy Institute.

Alba, Richard. 1976. "Social Assimilation among American Catholic National-Origin Groups." *American Sociological Review* 41:1030–46.

———. 1981. "The Twilight of Ethnicity among American Catholics of European Ancestry." *Annals of the American Academy of Political and Social Science* 454:86–97.

———. 1985. *Italian Americans: Into the Twilight of Ethnicity.* Englewood Cliffs, N.J.: Prentice Hall.

———. 1990. *Ethnic Identity: The Transformation of White America.* New Haven, Conn.: Yale University Press.

———. 1992. "Ethnicity." Pp. 575–84 in Edgar F. and Mary L. Borgatta (eds.), *Encyclopedia of Sociology.* Vol. 2. New York: Macmillan.

———. 1994. "Identity and Ethnicity among Italians and Other Americans of European Ancestry." Pp. 21–44 in Lydio Tomasi, Piero Gastaldo, and Thomas Row (eds.), *The Columbus People: Perspectives in Italian Immigration to the Americas and Australia.* Staten Island, N.Y.: Center for Migration Studies.

———. 1995. "Assimilation's Quiet Tide." *The Public Interest* (Spring):3–18.

———. 1999. "Immigration and the American Realities of Assimilation and Multiculturalism." *Sociological Forum* 14:3–25.

———. 2004. "Language Assimilation Today: Bilingualism Persists More Than in the Past, but English Still Dominates." Albany, N.Y.: Lewis Mumford Center for Comparative Urban and Regional Research.

———. 2005. "Bilingualism Persists, but English Still Dominates." Migration Information Source, Migration Policy Institute (February 1). http://www.migrationinformation.org/

———. 2009. *Blurring the Color Line: The New Chance for a More Integrated America.* Cambridge, Mass.: Harvard University Press.

Alba, Richard, and Dalia Abdel-Hady. 2005. "Galileo's Children: Italian Americans' Difficult Entry into the Intellectual Elite." *Sociological Quarterly* 46:3–18.

Alba, Richard D., and Reid M. Golden. 1986. "Patterns of Ethnic Marriage in the United States." *Social Forces* 65:202–23.

Alba, Richard D., and John R. Logan. 1993. "Minority Proximity to Whites in Suburbs: An Individual-Level Analysis of Segregation." *American Journal of Sociology* 98:1388–1427.

Alba, Richard D., John R. Logan, and Brian J. Stults. 2000. "How Segregated Are Middle-Class African Americans?" *Social Problems* 47:545–58.

Alba, Richard D., John R. Logan, Brian J. Stults, Gilbert Marzan, and Wenquan Zhang. 1999. "Immigrant Groups in the Suburbs: A Reexamination of Suburbanization and Spatial Assimilation." *American Sociological Review* 64:446–60.

Alba, Richard D., and Gwen Moore. 1982. "Ethnicity in the American Elite." *American Sociological Review* 47:373–83.

Alba, Richard D., and Victor Nee. 1999. "Rethinking Assimilation Theory for a New Era of Immigration." Pp. 137–60 in Charles Hirschman, Philip Kasinitz, and Josh DeWind (eds.), *The Handbook of International Migration: The American Experience.* New York: Russell Sage Foundation.

———. 2003. *Remaking the American Mainstream: Assimilation and Constemporary Immigration.* Cambridge, Mass.: Harvard University Press.

Alberts, Heike C. 2006. "The Multiple Transformations of Miami." Pp. 135–52 in Heather A. Smith and Owen J. Furuseth (eds.), *Latinos in the New South: Transformations of Place.* Aldershot, England: Ashgate.

Albini, Joseph L. 1971. *The American Mafia: Genesis of a Legend.* New York: Appleton-Century-Crofts.

Ali, Wajahat, Eli Clifton, Matthew Duss, Lee Fang, Scott Keyes, and Fais Shakir. 2011. "Fear, Inc.: The Roots of the Islamophobia Network in America." Center for American Progress. http://www.americanprogress.org/

Alland, Alexander, Jr. 2002. *Race in Mind: Race, IQ, and Other Racisms.* New York: Palgrave Macmillan.

Allen, Robert L. 1970. *Black Awakening in Capitalist America.* Garden City, N.Y.: Doubleday.

Allport, Gordon W. 1958. *The Nature of Prejudice.* Garden City, N.Y.: Doubleday.

Altemeyer, Bob. 1996. *The Authoritarian Specter.* Cambridge, Mass.: Harvard University Press.

Altman, Alex. 2010. "TIME Poll: Majority Oppose Mosque, Many Distrust Muslims." *Time* (August 19). http://www.time.com/time/printout/0,8816,2011799,00.html

Alvarez, Rodolfo. 1973. "The Psycho-Historical and Socioeconomic Development of the Chicano Community in the United States." *Social Science Quarterly* 53:920–42.

Amaral, Ernesto Friedrich, and Wilson Fusco. 2005. "Shaping Brazil: The Role of International Migration." Washington, D.C.: Migration Policy Institute.

Ameri, Anan, and Holly Arida (eds.). 2012. *Daily Life of Arab Americans in the 21st Century.* Santa Barbara: Greenwood.

American Jewish Committee. 2003. *American Jewish Year Book 2003.* Vol. 103. New York: American Jewish Committee.

———. 2004. "2004 Annual Survey of American Jewish Opinion" (August 28–September 1). New York: American Jewish Committee.

———. 2008. "2008 Annual Survey of American Jewish Opinion." http://www.ajc.org

———. 2010. "2010 Annual Survey of American Jewish Opinion." http://www.ajc.org

Anderson, Benedict. 2006. *Imagined Communities: Reflections on the Origin and Spread of Nationalism.* Rev. ed. London: Verso.

Anderson, Charles H. 1970. *White Protestant Americans: From National Origins to Religious Group.* Englewood Cliffs, N.J.: Prentice Hall.

Anderson, Elijah. 1999. *Code of the Street Decency, Violence, and the Moral Life of the Inner City.* New York: Norton.

Anderson, Jon Lee. 2001. "Home Fires." *The New Yorker* (February 12):40–47.

Andrews, George Reid. 1991. *Blacks & Whites in São Paulo, Brazil 1888–1988.* Madison: University of Wisconsin Press.

Angus Reid Global Monitor. 2009. "Immigration Seen Negatively in Britain and U.S." (August 20). http://www.angusreidglobal.com/

Anti-Defamation League. 1993. *Highlights from an Anti-Defamation League Survey on Racial Attitudes in America.* New York: Anti-Defamation League of B'nai B'rith.

———. 1998. *Survey on Anti-Semitism and Prejudice in America.* New York: Anti-Defamation League of B'nai B'rith.

———. 2009a. *American Attitudes toward Jews in America.* New York: Anti-Defamation League of B'nai B'rith.

———. 2009b. *Attitudes toward Jews in Seven European Countries.* New York: Anti-Defamation League of B'nai B'rith.

Antonovsky, Aaron. 1960. "The Social Meaning of Discrimination." *Phylon* 21:81–95.

Aoyagi, Kiyotaka, and Ronald P. Dore. 1964. "The Buraku Minority in Urban Japan." Pp. 95–107 in *Transactions of the Fifth World Congress of Sociology.* Vol. 3. Louvain: International Sociological Association.

Aponte, Robert, and Marcelo Siles. 1994. "Latinos in the Heartland: The Browning of the Midwest." *JSRI Research Report No. 5.* Julian Samora Research Institute. East Lansing: Michigan State University.

Arab American Institute. 2012a. "The American Divide: How We View Arabs and Muslims." Arab American Institute.

———. 2012b. "Demographics." http://b.3cdn. net/aai/44b17815d8b386bf16_v0m6iv4b5 .pdf

Arango, Joaquín. 2013. *Exceptional in Europe? Spain's Experience with Immigration and Integration.* Washington, D.C.: Migration Policy Institute.

Arias, Elizabeth. 2001. "Change in Nuptiality Patterns among Cuban Americans: Evidence of Cultural and Structural Assimilation." *International Migration Review* 35:525–56.

Arnold, Guy. 2000. *The New South Africa.* New York: St. Martin's.

Arnopoulos, Sheila McLeod, and Dominique Clift. 1980. *The English Fact in Quebec.* Montreal: McGill-Queen's University Press.

Arthur, Paul. 1980. *Government and Politics of Northern Ireland.* Burnt Mill, England: Longman House.

Arthur, Paul, and Keith Jeffery. 1996. *Northern Ireland Since 1968.* 2nd ed. Oxford: Blackwell.

Asch, Michael. 1984. *Home and Native Land: Aboriginal Rights and the Canadian Constitution.* Toronto: Methuen.

Asi, Maryam, and Daniel Beaulieu. 2013. *Arab Households in the United States: 2006–2010.* Washington, D.C.: U.S. Census Bureau.

Asmal, Kader, and Wilmot James. 2001. "Education and Democracy in South Africa Today." *Daedalus* 130(Winter):185–204.

Associated Press. 2012. "The Associated Press Racial Attitudes Survey" (October 29). http://surveys. ap.org/

Auclair, Gregory, and Jeanne Batalova. 2013. "Middle Eastern and North African Immigrants in the United States." *Migration Information Source* (September 26). Washington, D.C.: Migration Policy Institute.

Aunger, Edmund A. 1981. *In Search of Political Stability: A Comparative Study of New Brunswick and Northern Ireland.* Montreal: McGill-Queen's University Press.

Austin, Algernon. 2009. "High Unemployment: A Fact of Life for American Indians." Economic Policy Institute. http://www.epi.org/

———. 2012. "Reversal of Fortune: Economic Gains of 1990s Overturned for African Americans from 2000-07." Washington, D.C.: Economic Policy Institute.

Avery, Donald. 1979. *Dangerous Foreigners: European Immigrant Workers and Labour Radicalism in Canada 1896–1932.* Toronto: McClelland & Stewart.

Bahr, Howard M., Bruce A. Chadwick, and Joseph H. Stauss. 1979. *American Ethnicity.* Lexington, Mass.: Heath.

Bailey, Stanley R. 2008. "Unmixing for Race Making in Brazil." *American Journal of Sociology* 114:577–614.

———. 2009. *Legacies of Race: Identities, Attitudes, and Politics in Brazil.* Stanford: Stanford University Press.

Baily, Samuel L. 1999. *Immigrants in the Lands of Promise: Italians in Buenos Aires and New York City, 1870–1914.* Ithaca, N.Y.: Cornell University Press.

Bakalian, Anny, and Mehdi Bozorgmehr. 2009. *Backlash 9/11: Middle Eastern and Muslim Americans Respond.* Berkeley: University of California Press.

Baker, Wayne, and Andrew Shryock. 2009. "Citizenship and Crisis." Pp. 3–19 in Detroit Arab American Study Team, *Citizenship and Crisis: Arab Detroit After 9/11.* New York: Russell Sage Foundation.

Baker, Wayne, Sally Howell, Amaney Jamal, Ann Chih Lin, Andrew Shryock, Ron Stockton, and Mark Tessler. 2004. "Preliminary Findings from the Detroit Arab American Study." Ann Arbor: University of Michigan. http://www.ur.umich.edu/

Balgamis, A. Deniz, and Kemal H. Karpat (eds.). 2008. *Turkish Migration to the United States: From Ottoman Times to the Present.* Madison: Center for Turkish Studies, University of Wisconsin Press.

Baltzell, E. Digby. 1958. *Philadelphia Gentlemen: The Making of a National Upper Class.* New York: Free Press.

———. 1964. *The Protestant Establishment: Aristocracy and Caste in America.* New York: Vintage.

———. 1991. *The Protestant Establishment Revisited.* New Brunswick, N.J.: Transaction.

———. 1994. *Judgment and Sensibility: Religion and Stratification.* New Brunswick, N.J.: Transaction.

Bamshad, Michael J., and Steve E. Olson. 2003. "Does Race Exist?" *Scientific American* 289 (December):78–85.

Banfield, Edward C. 1968. *The Unheavenly City.* Boston: Little, Brown.

———. 1990 [1974]. *The Unheavenly City Revisited.* Prospect Heights, Ill.: Waveland Press.

Banting, Keith G. 1992. "If Quebec Separates: Restructuring Northern North America." Pp. 159–78 in R. Kent Weaver (ed.), *The Collapse of Canada?* Washington, D.C.: Brookings Institution.

Banting, Keith, Thomas J. Courchene, and F. Leslie Seidle. 2007. "Conclusion: Diversity, Belonging and Shared Citizenship." Pp. 647–87 in Keith Banting, Thomas J. Courchene, and F. Leslie Seidle (eds.), *Belonging? Diversity, Recognition and Shared Citizenship in Canada.* Montreal: Institute for Research and Public Policy.

Banton, Michael. 1967. *Race Relations.* London: Tavistock.

———. 1970. "The Concept of Racism." Pp. 17–34 in Sami Zubaida (ed.), *Race and Racialism.* London: Tavistock.

———. 1983. *Racial and Ethnic Competition.* Cambridge, England: Cambridge University Press.

Barlett, Donald L., and James B. Steele. 2002. "Wheel of Misfortune." *Time* (December 16):42–58.

Baron, Dennis. 1990. *The English-Only Question: An Official Language for Americans?* New Haven, Conn.: Yale University Press.

Baron, Harold. 1969. "The Web of Urban Racism." Pp. 134–76 in Louis L. Knowles and Kenneth Prewitt (eds.), *Institutional Racism in America.* Englewood Cliffs, N.J.: Prentice Hall.

Barone, Michael. 1985. "Italian Americans and Politics." Pp. 378–84 in Lydio F. Tomasi (ed.), *Italian Americans: New Perspectives in Italian Immigration and Ethnicity.* Staten Island, N.Y.: Center for Migration Studies.

Barrera, Mario. 1979. *Race and Class in the Southwest: A Theory of Racial Inequality.* South Bend, Ind.: University of Notre Dame Press.

Barritt, Denis. 1982. *Northern Ireland: A Problem to Every Solution.* London: Quaker Peace and Service.

Barritt, Denis P., and Charles F. Carter. 1972. *The Northern Ireland Problem: A Study in Group Relations.* 2nd ed. London: Oxford University Press.

Barta, Russell. 1979. "The Representation of Poles, Italians, Latins and Blacks in the Executive Suites of Chicago's Largest Corporations." Pp. 418–30 in U.S. Commission on Civil Rights, *Civil Rights Issues of Euro-Ethnic Americans in the United States: Opportunities and Challenges.* Washington, D.C.: U.S. Government Printing Office.

Barth, Ernest A. T., and Donald L. Noel. 1972. "Conceptual Frameworks for the Analysis of Race Relations." *Social Forces* 50:333–48.

Barth, Fredrik. 1969. *Ethnic Groups and Boundaries.* Boston: Little, Brown.

Bastide, Roger. 1961. "Dusky Venus, Black Apollo." *Race* 3(November):10–18.

———. 1965. "The Development of Race Relations in Brazil." Pp. 9–29 in Guy Hunter (ed.), *Industrialization and Race Relations.* London: Oxford University Press.

Bastide, Roger, and Pierre van den Berghe. 1957. "Stereotypes, Norms and Interracial Behavior in São Paulo, Brazil." *American Sociological Review* 22:689–94.

Bayer, Patrick, Robert McMillan, and Kim S. Rueben. 2004. "What Drives Racial Segregation? New Evidence Using Census Microdata." *Journal of Urban Economics* 56:514–35.

Bean, Frank D., and Gillian Stevens. 2003. *America's Newcomers and the Dynamics of Diversity*. New York: Russell Sage Foundation.

Bean, Frank D., and Marta Tienda. 1987. *The Hispanic Population of the United States*. New York: Russell Sage Foundation.

Bearak, Barry. 2011. "Death by a Thousand Blows." *New York Times Magazine*. (June 5):32.

Beato, Lucila Bandeira. 2004. "Inequality and Human Rights of African Descendants in Brazil." *Journal of Black Studies* 34:766–86.

Beauchamp, Tom. 2002. "In Favor of Affirmative Action." Pp. 209–23 in Steven M. Cahn (ed.), *The Affirmative Action Debate*. 2nd ed. New York: Routledge.

Bebow, John. 2003. "Metro Arabs Pay Stiff Price for Feds' Focus on Terrorism." *Detroit News* (November 2). http://www.detnews.com

Beck, Roy. 1994. "The Ordeal of Immigration in Wausau." *Atlantic Monthly* 273(April):84–97.

———. 1996. *The Case against Immigration: The Moral, Economic, Social, and Environmental Reasons for Reducing U.S. Immigration Back to Traditional Levels*. New York: Norton.

Beckett, J. C. 1972. "Northern Ireland." Pp. 11–24 in Institute for the Study of Conflict (ed.), *The Ulster Debate*. London: Bodley Head.

Bell, Daniel. 1962. *The End of Ideology*. Rev. ed. New York: Free Press.

———. 1975. "Ethnicity and Social Change." Pp. 141–174 in Nathan Glazer and Daniel P. Moynihan (eds.), *Ethnicity: Theory and Experience*. Cambridge, Mass.: Harvard University Press.

Bell, Geoffrey. 1976. *The Protestants of Ulster*. London: Pluto Press.

Bell, J. Bowyer. 1996. *Back to the Future: The Protestants and a United Ireland*. Dublin: Poolbeg.

Belluck, Pam. 1995. "Healthy Korean Economy Draws Immigrants Home." *New York Times* (August 22):A1, B4.

———. 2002. "Mixed Welcome as Somalis Settle in a Maine City." *New York Times* (October 15):A16.

Benedict, Ruth. 1959 [1940]. *Race: Science and Politics*. New York: Viking.

Bennett, Christopher. 1995. *Yugoslavia's Bloody Collapse: Causes, Course and Consequences*. New York: New York University Press.

Benton, Meghan, and Anne Mark Nielsen. 2013. "Integrating Europe's Muslim Minorities: Public Anxieties, Policy Responses." *Migration Information Source* (May 10). Washington, D.C.: Migration Policy Institute.

Bercuson, David J. 1995. "Why Quebec and Canada Must Part." *Current History* 94:123–26.

Berg, Charles Ramìrez. 2002. *Latino Images in Film: Stereotypes, Subversion, Resistance*. Austin: University of Texas Press.

Berger, Peter L. 1963. *Invitation to Sociology: A Humanistic Perspective*. Garden City, N.Y.: Doubleday.

Berke, Richard L. 1998. "Voters Found Real Chance at Peace Irresistible." *New York Times* (May 24):A1, A4.

Berkhofer, Robert F., Jr. 1979. *The White Man's Indian*. New York: Vintage.

Bernard, Jesse. 1951. "The Conceptualization of Intergroup Relations with Special Reference to Conflict." *Social Forces* 19:243–51.

Bernstein, Nina. 2005. "Record Immigration Is Changing the Face of New York's Neighborhoods." *New York Times* (January 24):A16.

Bernstein, Richard. 1988. "Asian Newcomers Hurt by Precursors' Success." *New York Times* (July 10):12.

Berreman, Gerald D. 1966. "Structure and Function of Caste Systems." Pp. 277–307 in George De Vos and Hiroshi Wagatsuma (eds.), *Japan's Invisible Race: Caste in Culture and Personality*. Berkeley and Los Angeles: University of California Press.

———. 1972. "Race, Caste, and Other Invidious Distinctions in Social Stratification." *Race* 13:385–414.

Berry, Brewton, and Henry L. Tischler. 1978. *Race and Ethnic Relations*. 4th ed. Boston: Houghton Mifflin.

Berry, John W., and Rudolf Kalin. 2000. "Racism: Evidence from National Surveys." Pp. 172–85 in Leo Driedger and Shiva S. Halli (eds.), *Race and Racism: Canada's Challenge*. Montreal: McGill-Queen's University Press.

Berry, John W., Rudolf Kalin, and D. M. Taylor. 1977. *Multiculturalism and Ethnic Attitudes in Canada*. Ottawa: Minister of Supply and Services.

Bertrand, Marianne, and Sendhil Mullainathan. 2004. "Are Emily and Greg More Employable than Lakisha and Jamal? A Field Experiment on Labor Market Discrimination." *American Economic Review* 94:991–1013.

Besada, Henry. 2007a. "Enduring Political Divides in South Africa." *CIGI Technical Paper*, No. 3. Waterloo, Ontario: Centre for International Governance Innovation.

———. 2007b. "Fragile Stability: Post-Apartheid South Africa." *Working Paper* No. 27. Waterloo, Ontario: Centre for International Governance Innovation.

Betancur, John J., Teresa Cordova, and Maria de los Angeles Torres. 1993. "Economic Restructuring and the Process of Incorporation of Latinos into the Chicago Economy." Pp. 109–32 in Rebecca Morales and Frank Bonilla (eds.), *Latinos in a Changing U.S. Economy*. Newbury Park, Calif.: Sage.

Béteille, André. 1969. *Castes Old and New: Essays in Social Structure and Social Stratification*. Bombay: Asian Publishing Home.

Biale, David. 1998. "The Melting Pot and Beyond: Jews and the Politics of Identity." Pp. 17–33 in David Biale, Michael Galchinsky, and Susannah Heschel (eds.), *Insider/Outside: American Jews and Multiculturalism*. Berkeley: University of California Press.

Bibby, Reginald W. 1995. *The Bibby Report: Social Trends Canadian Style*. Toronto: Stoddart.

———. 2007. "Racial Intermarriage: Canada & the U.S." Project Canada Press Release #9, University of Lethbridge, Alberta.

Bilefsky, Dan. 2006. "Latvia Fears New 'Occupation' by Russians but Needs the Labor." *New York Times* (November 16):A17.

———. 2007. "After 8 Years in Limbo, Frustrated Kosovo Awaits Its Future." *New York Times* (December 11):A18.

———. 2013. "In Kosovo, Ethnic Barriers Linger as a New Accord Is Taking Effect." *New York Times* (June 12):A5.

Bissoondath, Neil. 2002. *Selling Illusions: The Cult of Multiculturalism in Canada*. Rev. ed. Toronto: Penguin.

Blackburn, Daniel G. 2000. "Why Race Is Not a Biological Concept." Pp. 3–26 in Berel Lang (ed.), *Race and Racism in Theory and Practice*. Boston: Rowman & Littlefield.

Blackwell, James E. 1976. "The Power Basis of Ethnic Conflict in American Society." Pp. 179–96 in Lewis A. Coser and Otto N. Larsen (eds.), *The Use of Controversy in Sociology*. New York: Free Press.

Blalock, Hubert M., Jr. 1967. *Toward a Theory of Minority-Group Relations*. New York: Wiley.

Blau, Joseph L., and Salo W. Baron (eds.). 1963. *The Jews of the United States 1790–1840: A Documentary History*. New York: Columbia University Press.

Blau, Peter M. 1977. "A Macrosociological Theory of Social Structure." *American Journal of Sociology* 83:26–54.

Blauner, Robert. 1969. "Internal Colonialism and Ghetto Revolt." *Social Problems* 16:393–408.

———. 1972. *Racial Oppression in America*. New York: Harper & Row.

———. 1992. "Talking Past Each Other: Black and White Languages of Race." *The American Prospect* 10(Summer):55–64.

———. 2001. *Still the Big News: Racial Oppression in America*. Philadelphia: Temple University Press.

Block, Sheila. 2010. "Ontario's Growing Gap: The Role of Race and Gender." Ottawa: Canadian Centre for Policy Alternatives.

Bloemraad, Irene. 2006. *Becoming a Citizen: Incorporating Immigrants and Refugees in the United States and Canada*. Berekely: University of California Press.

———. 2012. *Understanding "Canadian Exceptionalism" in Immigration and Pluralism Policy*. Washington, D.C.: Migration Policy Institute.

Bloomgarden, Laurence. 1957. "Who Should Be Our Doctors?" *Commentary* (January):506–15.

Blumer, Herbert. 1958. "Race Prejudice as a Sense of Group Position." *Pacific Sociological Review* 1:3–6.

———. 1965. "Industrialization and Race Relations." Pp. 200–53 in Guy Hunter (ed.), *Industrialization and Race Relations*. London: Oxford University Press.

Bobo, Lawrence D. 1999. "Prejudice as Group Position: Microfoundations of a Sociological Approach to Racism and Race Relations." *Journal of Social Issues* 55:445–72.

———. 2004. "Inequalities That Endure? Racial Ideology, American Politics, and the Peculiar Role of the Social Sciences." Pp. 13–42 in Maria Krysan and Amanda E. Lewis (eds.), *The Changing Terrain of Race and Ethnicity*. New York: Russell Sage Foundation.

Bobo, Lawrence D., and Ryan A. Smith. 1998. "From Jim Crow Racism to Laissez Faire Racism: An Essay on the Transformation of Racial Attitudes in America." Pp. 182–220 in Wendy F. Katkin, Ned Landsman, and Andrea Tyree (eds.), *Beyond Pluralism: Essays on the Conception of Groups and Group Identities in America*. Urbana: University of Illinois Press.

Bogardus, Emory S. 1925a. "Measuring Social Distances." *Journal of Applied Sociology* 9:299–308.

———. 1925b. "Social Distance and Its Origins." *Journal of Applied Sociology* 9:216–26.

———. 1930. "A Race Relations Cycle." *American Journal of Sociology* 35:612–17.

———. 1959. *Social Distance*. Yellow Springs, Ohio: Antioch Press.

———. 1968. "Comparing Racial Distance in Ethiopia, South Africa, and the United States." *Sociology and Social Research* 52:149–56.

Boller, Paul F., Jr. 2004. *Presidential Campaigns: From George Washington to George W. Bush*. New York: Oxford University Press.

Bonacich, Edna. 1972. "A Theory of Ethnic Antagonism: The Split Labor Market." *American Sociological Review* 37:547–59.

———. 1973. "A Theory of Middleman Minorities." *American Sociological Review* 38:583–94.

———. 1976. "Advanced Capitalism and Black-White Race Relations in the United States: A Split Labor Market Interpretation." *American Sociological Review* 41:34–51.

Bonacich, Edna, and John Modell. 1980. *The Economic Basis of Ethnic Solidarity: Small Business in the Japanese American Community*. Berkeley: University of California Press.

Bonilla-Silva, Eduardo. 2003. *Racism without Racists: Color-Blind Racism and the Persistence of Racial Inequality in the United States*. Lanham, Md.: Rowman & Littlefield.

———. 2004. "From Bi-Racial to Tri-Racial: Towards a New System of Racial Stratification in the USA." *Ethnic and Racial Studies* 27:931–50.

Bonilla-Silva, Eduardo, and David Dietrich. 2011. "The Sweet Enchantment of Color-Blind Racism in Obamerica." *Annals of the American Academy of Political and Social Science* 634(1):190–206.

Bonilla-Silva, Eduardo, and Karen S. Glover. 2006. "'We Are All Americans': The Latin Americanization of Race Relations in the United States." Pp. 149–83 in Maria Krysan and Amanda E. Lewis (eds.), *The Changing Terrain of Race and Ethnicity*. New York: Russell Sage Foundation.

Borjas, George J. 1994. "Tired, Poor on Welfare." Pp. 76–80 in Nicolaus Mills (ed.), *Arguing Immigration*. New York: Touchstone.

———. 1998. "Do Blacks Gain or Lose from Immigration?" Pp. 51–74 in Daniel S. Hamermesh and Frank D. Bean (eds.), *Help or Hindrance? The Economic Implications of Immigration for African Americans*. New York: Russell Sage Foundation.

———. 2004. "Increasing the Supply of Labor Through Immigration." Washington, D.C.: Center for Immigration Studies.

Bornman, Elirea. 2010. "Emerging Patterns of Social Identification in Postapartheid South Africa." *Journal of Social Issues* 66:237–54.

Borooah, Vani, and John Mangan. 2007. "Love Thy Neighbour: How Much Bigotry Is There in Western Countries?" Paper presented at Public Choice World Meeting, Amsterdam.

Bosman, Julie. 2007. "School Translators Can Help Parents Lost in the System." *New York Times* (August 13):B1, B2.

Boswell, Thomas D., and Terry-Ann Jones. 2006. Caribbean Hispanics: Puerto Ricans, Cubans, and Dominicans." Pp. 123–50 in Ines M. Miyares and Christopher A. Airriess (eds.), *Contemporary Ethnic Geographies in America*. Lanham, Md.: Rowman & Littlefield.

Bowles, Samuel, and Herbert Gintis. 1976. *Schooling in Capitalist America: Educational Reform and the Contradictions of Economic Life*. New York: Basic Books.

Bowser, Benjamin P., and Raymond G. Hunt (eds.). 1981. *Impacts of Racism on White Americans*. Beverly Hills, Calif.: Sage.

Boxer, Charles R. 1962. *The Golden Age of Brazil*. Berkeley: University of California Press.

———. 1963. *Race Relations in the Portuguese Colonial Empire, 1415–1825*. Oxford: Clarendon Press.

Boyle, Kevin, and Tom Hadden. 1994. *Northern Ireland: The Choice*. New York: Penguin.

Bozorgmehr, Mehdi. 1997. "Internal Ethnicity: Iranians in Los Angeles." *Sociological Perspectives* 40:387–408.

Bozorgmehr, Mehdi, Claudia Der-Martirasian, and Georges Sabagh. 1996. "Middle Easterners: A New Kind of Immigrant." Pp. 345–78 in Roger Waldinger and Mehdi Bozorgmehr (eds.), *Ethnic Los Angeles*. New York: Russell Sage Foundation.

Branch-Brioso, Karen, Tim Henderson, and Alfonso Chardy. 2000. "Despite Hispanic Majority, White Non-Hispanics Hold Sway." *Miami Herald* (September 3):A1.

Braverman, Harold, and Louis Kaplan. 1967. "A Study of Religious Discrimination by Social Clubs." Pp. 211–21 in Milton L. Barron (ed.), *Minorities in a Changing World*. New York: Knopf.

Breton, Raymond. 1964. "Institutional Completeness of Ethnic Communities and the Personal Relations of Immigrants." *American Journal of Sociology* 70:193–205.

———. 1989. "Canadian Ethnicity in the Year 2000." Pp. 149–52 in James Frideres (ed.), *Multiculturalism and Intergroup Relations*. New York: Greenwood.

Breton, Raymond, Jeffrey G. Reitz, and Victor F. Valentine. 1980. *Cultural Boundaries and the Cohesion of Canada*. Montreal: Institute for Research on Public Policy.

Brewer, Marilynn. 1979. "Ingroup Bias in the Minimal Intergroup Situation: Cognitive-Motivational Analysis." *Psychological Bulletin* 86:307–32.

Briggs, John W. 1978. *An Italian Passage*. New Haven, Conn.: Yale University Press.

Brimelow, Peter. 1995. *Alien Nation*. New York: HarperCollins.

Brink, William, and Louis Harris. 1963. *The Negro Revolution in America*. New York: Simon & Schuster.

———. 1967. *Black and White: A Study of Racial Attitudes Today*. New York: Simon & Schuster.

Brittingham, Angela, and G. Patricia de la Cruz. 2005. *We the People of Arab Ancestry in the United States*. Washington, D.C.: U.S. Census Bureau.

Brockman, Joshua. 2001. "Gathered Vestiges of a Covered-Wagon Diaspora." *New York Times* (February 4):37, 41.

Brown, Anna, and Mark Hugo Lopez. 2013. "Mapping the Latino Population, by State, County, and City." Washington, D.C.: Pew Research Center.

Brown, Cecil. 1981. "Blues for Blacks in Hollywood." *Mother Jones* (January):20–28, 59.

Brown, Dee. 1972. *Bury My Heart at Wounded Knee*. New York: Bantam.

Brown, Douglas. 1966. *Against the World: A Study of White South African Attitudes*. London: Collins.

Brown, Michael K., Martin Carnoy, Elliott Currie, Troy Duster, David B. Oppenheimer, Marjorie M. Shultz, and David Wellman. 2003. *Whitewashing Race: The Myth of a*

Color-Blind Society. Berkeley: University of California Press.

Brown, Paul. 2002. "Peace but No Love as Northern Ireland Divide Grows." *Guardian Weekly* (January 10–16):8.

Brown, Rupert. 1995. *Prejudice: Its Social Psychology.* Oxford: Blackwell.

Brubaker, Rogers. 2001. "The Return of Assimilation? Changing Perspectives on Immigration and Its Sequels in France, Germany, and the United States." *Ethnic and Racial Studies* 24:531–48.

———. 2004. *Ethnicity without Groups.* Cambridge, Mass.: Harvard University Press.

Brunet, Michel. 1969. "Canadians and Canadiens." Pp. 284–93 in Ramsay Cook (ed.), *French-Canadian Nationalism.* Toronto: Macmillan of Canada.

Brym, Robert J. 1989. *From Culture to Power: The Sociology of English Canada.* Toronto: Oxford University Press.

Brzozowski, Jodi-Anne, Andrea Taylor-Butts, and Sara Johnson. 2006. "Victimization and Offending among the Aboriginal Population in Canada." *Juristat: Canadian Centre for Justice Statistics* 26(3):1–31.

Buchanan, Susy. 2007. "Indian Blood." *Southern Poverty Law Center Intelligence Report.* http://www.splcenter.org/

Buchanan, William, and Hadley Cantril. 1953. *How Nations See Each Other.* Urbana: University of Illinois Press.

Buchignani, Norman L. 1980. "Accommodation, Adaptation, and Policy: Dimensions of the South Asian Experience in Canada." Pp. 121–50 in K. Victor Ujimoto and Gordon Hirabayashi (eds.), *Visible Minorities and Multiculturalism: Asians in Canada.* Toronto: Butterworths.

Buckley, Anthony D., and Mary Catherine Kenny. 1995. *Negotiating Identity: Rhetoric, Metaphor, and Social Drama in Northern Ireland.* Washington, D.C.: Smithsonian Institution Press.

Burck, Charles G. 1976. "A Group Profile of the Fortune 500 Chief Executive." *Fortune* (May):173–77, 308–12.

Bureau of Indian Affairs (BIA). 2009. "Indian Entities Recognized and Eligible to Receive Services from the United States Bureau of Indian Affairs." *Federal Register*, Vol. 74, No. 153 (August 11).

Burkey, Richard M. 1978. *Ethnic and Racial Groups: The Dynamics of Dominance.* Menlo Park, Calif.: Cummings.

Burkholz, Herbert. 1980. "The Latinization of Miami." *New York Times Magazine* (September 21):44–47, 84–88, 98–100.

Burma, John. 1954. *Spanish-Speaking Groups in the United States.* Durham, N.C.: Duke University Press.

Burnet, Jean. 1976. "Ethnicity: Canadian Experience and Policy." *Sociological Focus* 9:199–207.

Burns, John F. 1988. "Ottawa Will Pay Compensation to Uprooted Japanese-Canadians." *New York Times* (September 23): A10.

———. 2013. "New Violence in Belfast May Be About More Than the Flag." *New York Times* (January 19):A4.

Buruma, Ian. 2005. "Final Cut." *The New Yorker* (January 3):26–32.

Butterfield, Fox. 1986. "Why Asians Are Going to the Head of the Class." *New York Times* (August 3):12, 18–23.

Button, James W. 1978. *Black Violence.* Princeton, N.J.: Princeton University Press.

CAIN Web Service. 2004. "Religion in Northern Ireland." http://cain.ulst.ac.uk/ni/religion.htm

Cainkar, Louise. 2006. "The Social Construction of Difference and the Arab American Experience." *Journal of American Ethnic History* 25:244–78.

———. 2009. *Homeland Insecurity: The Arab American and Muslim American Experience After 9/11.* New York: Russell Sage Foundation.

Calame, Jon, and Esther Charlesworth. 2009. *Divided Cities: Belfast, Beirut, Jerusalem, Mostar, and Nicosia.* Philadelphia: University of Pennsylvania Press.

Camarillo, Albert. 1979. *Chicanos in a Changing Society.* Cambridge, Mass.: Harvard University Press.

Camarillo, Albert M., and Frank Bonilla. 2001. "Hispanics in a Multicultural Society: A New American Dilemma?" Pp. 103–34 in Neil J. Smelser, William Julius Wilson, and Faith Mitchell (eds.), *America Becoming: Racial Trends and Their Consequences.* Vol. 1. Washington, D.C.: National Academy Press.

Camarota, Steven A. 2007. "Immigrant Employment Gains and Native Losses, 2000–2004." Pp. 139–56 in Carol M. Swain (ed.), *Debating Immigration.* New York: Cambridge University Press.

Campani, Giovanna. 1993. "Immigration and Racism in Southern Europe: The Italian Case." *Ethnic and Racial Studies* 16:507–35.

Campbell, Angus. 1971. *White Attitudes Toward Black People.* Ann Arbor: Institute for Social Research, University of Michigan.

Campbell, Lori Ann, and Robert L. Kaufman. 2006. "Racial Differences in Household Wealth: Beyond Black and White." *Research in Social Stratification and Mobility* 24:131–52.

Campos-Flores, Arian, and Howard Fineman. 2005. "A Latin Power Surge." *Newsweek* (May 30):25–31.

Cancio, A. Silva, T. David Evans, and David Maune. 1996. "Reconsidering the Declining Significance of Race: Racial Differences in

Early Career Wages." *American Sociological Review* 61:541–46.

Cannon, Margaret. 1995. *The Invisible Empire: Racism in Canada.* Toronto: Random House of Canada.

Caplan, Nathan, Marcella H. Choy, and John K. Whitmore. 1991. *Children of the Boat People: A Study of Educational Success.* Ann Arbor: University of Michigan Press.

———. 1992. "Indochinese Refugee Families and Academic Achievement." *Scientific American* 266 (February):36–42.

Caplan, Nathan, John K. Whitmore, and Marcella H. Choy. 1989. *The Boat People and Achievement in America.* Ann Arbor: University of Michigan Press.

Cárdenas, Vanessa, and Sophia Kerby. 2012. *"The State of Latinos in the United States."* Washington, D.C.: Center for American Progress.

Cardoso, Fernando Henrique. 1965. "Color Prejudice in Brazil." *Présence Africaine* 25:120–28.

Cardwell, Cary. 1998. "The Miami Herald's Big News." *Hispanic Business* (October):80–81.

Carilli, Theresa. 1998. "Still Crazy After All These Years: Italian Americans in Mainstream U.S. Films." Pp. 111–24 in Yahya R. Kamalipour and Theresa Carilli (eds.), *Cultural Diversity and the U.S. Media.* Albany: State University of New York Press.

Cariño, Benjamin V. 1987. "The Philippines and Southeast Asia: Historical Roots and Contemporary Linkages." Pp. 305–25 in James T. Fawcett and Benjamin V. Cariño (eds.), *Pacific Bridges: The New Immigration from Asia and the Pacific Islands.* Staten Island, N.Y.: Center for Migration Studies.

Carmichael, Stokely, and Charles V. Hamilton. 1967. *Black Power: The Politics of Liberation in America.* New York: Vintage.

Carnevale, Anthony P., and Jeff Strohl. 2013. "Separate and Unequal: How Higher Education Reinforces the Intergenerational Reproduction of White Racial Privilege." Washington, D.C.: Georgetown University Center on Education and the Workforce.

Carroll, Joseph. 2001. "Public Overestimates U.S. Black and Hispanic Populations." Gallup News Service (June 4).

———. 2006. "Whites, Minorities Differ in Views of Economic Opportunities in U.S." *The Gallup Poll Briefing* (July):22–24.

———. 2007a. "Most Americans Approve of Interracial Marriages." Princeton, N.J.: Gallup News Service (August 16).

———. 2007b. "Whites, Blacks, Hispanics Assess Race Relations in the U.S." Princeton, N.J.: Gallup News Service (August 6).

———. 2007c. "Hispanics Support Requiring English Proficiency for Immigrants." Princeton, N.J.: Gallup News Service (July 5).

Carstensen, Fred, William Lott, Stan McMillen, Bobur Alimov, Na Li Dawson, and Tapas Ray. 2000. "The Economic Impact of the Mashentucket Pequot Tribal Nation Operations on Connecticut." Storrs: Connecticut Center for Economic Analysis, University of Connecticut.

Carter, Gwendolyn M. 1980. *Which Way Is South Africa Going?* Bloomington: Indiana University Press.

Carvalho, José Alberto Magno de, Charles H. Wood, and Flávia Cristina Drumond Andrade. 2004. "Estimating the Stability of Census-Based Racial/Ethnic Classifications: The Case of Brazil." Paper presented at meetings of the American Sociological Association, San Francisco.

Castles, Stephen. 1992. "Migrants and Minorities in Post-Keynesian Capitalism: The German Case." Pp. 36–54 in Malcolm Cross (ed.), *Ethnic Minorities and Industrial Change in Europe and North America.* Cambridge, England: Cambridge University Press.

Castles, Stephen, Heather Booth, and Tina Wallace. 1984. *Here for Good: Western Europe's New Ethnic Minorities.* London: Pluto Press.

Castles, Stephen, and Godula Kosack. 1973. *Immigrant Workers and Class Structure in Western Europe.* London: Oxford University Press.

Castles, Stephen, and Mark J. Miller. 2009. *The Age of Migration: International Population Movements in the Modern World.* 4th ed. Basingstoke, Aldershot, England: Palgrave Macmillan.

Castro, Max J. 1992. "The Politics of Language in Miami." Pp. 109–32 in Guillermo J. Grenier and Alex Stepick III (eds.), *Miami Now! Immigration, Ethnicity, and Social Change.* Gainesville: University Press of Florida.

Castro, Nadya Araújo, and Antonio Sérgio Alfredo Guimarães. 1999. "Racial Inequalities in the Labor Market and the Workplace." Pp. 83–108 in Rebecca Reichmann (ed.), *Race in Contemporary Brazil: From Indifference to Inequality.* University Park: Pennsylvania State University Press.

Caudill, William, and George De Vos. 1956. "Achievement, Culture and Personality: The Case of the Japanese Americans." *American Anthropologist* 58:1103–26.

Cave, Damien. 2007. "In New Tactic, Militants Burn Houses in Iraq." *New York Times* (March 12):A1, A6.

Cawthorne, Alexandra. 2010. "Weathering the Storm: Black Men in the Recession." Washington, D.C.: Center for American Progress.

Celis, William, III. 1992. "Hispanic Rate for Dropouts Remains High." *New York Times* (October 14):A1, B8.

Centre for Research and Information on Canada (CRIC). 2002. *Portraits of Canada 2001.* Montreal: Centre for Research and Information on Canada.

———. 2004. "Canadians Reject Ban on Religious Symbols or Clothes in Schools; Majority Sees Racial or Religious Background of Party Leaders as Irrelevant." (July 1). http://www .library.carleton.ca/find/data/centre-research-and-information-canada-cric

Chanes, Jerome A. 1994. "Interpreting the Data: Antisemitism and Jewish Security in the United States." *Patterns of Prejudice* 28:87–101.

———. 2004. *Antisemitism: A Reference Handbook.* Santa Barbara: ABC-CLIO.

———. 2006. "Anti-Semitism." Pp. 64–91 in David Singer and Lawrence Grossman (eds.), *American Jewish Year Book 2006.* New York: American Jewish Committee.

Chang, Mariko. 2010. *Lifting as We Climb: Women of Color, Wealth, and America's Future.* Oakland, Calif.: Insight Center for Community Economic Development.

Chao, Melody Manchi, Chi-yue Chiu, and Jamee S. Lee. 2010. "Asians as the Model Minority: Implications for US Government's Policies." *Asian Journal of Social Psychology* 13:44–52.

Charles, Camille Z. 2003. "Dynamics of Racial Residential Segregation." *Annual Review of Sociology* 29:167–207.

Chavez, Linda. 1991. *Out of the Barrio: Toward a New Politics of Hispanic Assimilation.* New York: Basic Books.

———. 1998. "Our Hispanic Predicament." *Commentary* (June):47–50.

Chen, David W. 1999. "Asian Middle Class Alters a Rural Enclave." *New York Times* (December 27):A1.

Cherribi, Sam. 2011. "Islamophobia in the Netherlands, Austria, and Germany." Pp. 47–62 in John L. Esposito and Ibrahim Kalin (eds.), *Islamophobia: The Challenge of Pluralism in the 21st Century.* New York: Oxford University Press.

Chesley, Roger, and Brenda J. Gilchrist. 1993. "Simmering Distrust." *Detroit Free Press* (February 22):1A, 5A.

Chinchilla, Norma Stoltz, and Nora Hamilton. 2007. "Central America: Guatemala, Honduras, Nicaragua." Pp. 328–39 in Mary C. Waters and Reed Ueda (eds.), *The New Americans: A Guide to Immigration Since 1965.* Cambridge, Mass.: Harvard University Press.

Chinn, Jeff, and Robert Kaiser. 1996. *Russians as the New Minority: Ethnicity and Nationalism in the Soviet Successor States.* Boulder, Colo.: Westview Press.

Chirot, Daniel, and Clark McCauley. 2006. *Why Not Kill Them All? The Logic and Prevention of Mass Political Murder.* Princeton, N.J.: Princeton University Press.

Chiswick, Barry R. 1984. "The Labor Market Status of American Jews: Patterns and Determinants." Pp. 131–53 in *American Jewish Year Book 1985.* New York: American Jewish Committee.

———. 1993. "The Skills and Economic Status of American Jewry: Trends over the Last Half-Century." *Journal of Labor Economics* 11:229–42.

Chowkwanyun, Merlin, and Jordan Segall. 2012. "The Rise of the Majority-Asian Suburb." *The Atlantic Cities.* http://www.theatlanticcities .com/politics/2012/08/rise-majority-asian-s uburb/3044/

Choy, Catherine Ceniza. 2007. "Philippines." Pp. 556–69 in Mary C. Waters and Reed Ueda (eds.), *The New Americans: A Guide to Immigration Since 1965.* Cambridge, Mass.: Harvard University Press.

Christopher, A. J. 2005. "The Slow Pace of Desegregation in South African Cities, 1996–2001." *Urban Studies* 42:2305–20.

Christopher, Robert C. 1989. *Crashing the Gates: The De-WASPing of America's Power Elite.* New York: Simon & Schuster.

Chua, Amy. 2004. *World on Fire: How Exporting Free Market Democracy Breeds Ethnic Hatred and Global Instability.* New York: Anchor.

Clairmont, Donald H., and Dennis W. Magill. 1974. *Africville: The Life and Death of a Canadian Black Community.* Toronto: McClelland & Stewart.

Clark, Kenneth B. 1965. *Dark Ghetto.* New York: Harper & Row.

Clark, S. D. 1950. "The Canadian Community." Pp. 375–89 in George Brown (ed.), *Canada.* Berkeley: University of California Press.

Clearfield, Esha, and Jeanne Batalova. 2007. "Foreign-Born Health-Care Workers in the United States." Washington, D.C.: Migration Policy Institute.

Clement, Wallace. 1975. *The Canadian Corporate Elite: An Analysis of Economic Power.* Toronto: McClelland & Stewart.

Clemetson, Lynette. 2002. "20 Years Later, Michigan Killing Still Galvanizes Asian Americans." *New York Times* (June 18).

Cloete, Nico, and Teboho Moja. 2005. "Transformation Tensions in Higher Education: Equity, Efficiency, and Development." *Social Research* 72:693–723.

Cody, Edward. 2010. "Anti-Islamic Sentiments Surface in Wake of Restrictions on Veils." *Washington Post* (May 15).

Cohen, Jeffrey E. 2010. "Perceptions of Anti-Semitism among American Jews, 2000–05: A Survey Analysis." *Political Psychology* 3:85–107.

Cohen, Lenard J. 2005. "The Balkans Ten Years After: From Dayton to the Edge of Democracy." *Current History* 104:365–73.

Cohen, Patricia. 2010. "Discussing That World That Prompts Either a Fist Pump or a Scowl." *New York Times* (January 23):C1.

Cohen, Steven M. 1985. *The 1984 National Survey of American Jews: Political and Social Outlooks.* New York: American Jewish Committee.

———. 1987. *Ties and Tensions: The 1986 Survey of American Jewish Attitudes toward Israel and Israelis.* New York: American Jewish Committee.

———. 1995. "Jewish Continuity over Judaic Content: The Moderately Affiliated American Jew." Pp. 395–416 in Robert M. Selter and Norman J. Cohen (eds.), *The Americanization of the Jews.* New York: New York University Press.

Cohen, Steven M., and Arnold M. Eisen. 2000. *The Jew Within: Self, Family, and Community in America.* Bloomington: Indiana University Press.

Coleman, James S., et al. 1966. *Equality of Educational Opportunity.* Washington, D.C.: U.S. Government Printing Office.

Collins, Lauren. 2011. "England, Their England." *The New Yorker* (July 4):28–34.

Collins, Randall. 2001. "Ethnic Change in Macro-Historical Perspective." Pp. 13–46 in Elijah Anderson and Douglas S. Massey (eds.), *Problem of the Century: Racial Stratification in the United States.* New York: Russell Sage Foundation.

Commonwealth Group of Eminent Persons. 1986. *Mission to South Africa: The Commonwealth Report.* Harmondsworth, England: Penguin.

Conley, Dalton. 1999. *Being Black, Living in the Red: Race, Wealth, and Social Policy in America.* Berkeley: University of California Press.

Connor, Walker. 1972. "Nation-Building or Nation-Destroying?" *World Politics* 24:319–55.

Constable, Pamela, and Luz Lazo. 2013. "Hispanics and Asian Americans Celebrate New Electoral Clout with Inaugural Galas, Wish Lists." *Washington Post* (January 19).

Cook, Fred J. 1971. "The Black Mafia Moves into the Numbers Racket." *New York Times Magazine* (April 4):26–27, 107–12.

Copé, Jean François. 2010. "Tearing away the Veil." *New York Times* (May 5):A31.

Corbett, David C. 1957. *Canada's Immigration Policies: A Critique.* Toronto: University of Toronto Press.

Cormack, Robert, and Robert Osborne. 1994. "The Evolution of a Catholic Middle Class." Pp. 65–85 in Adrian Guelke (ed.), *New Perspectives on the Northern Ireland Conflict.* Aldershot, England: Avebury.

Cornelius, Wayne. 2002. "Ambivalent Reception: Mass Public Responses to the 'New' Latino Immigration to the United States." Pp. 165–89 in Marcelo M. Suárez-Orozco and Mariela M. Páez (eds.), *Latinos: Remaking America.* Berkeley: University of California Press.

Cornell, Stephen. 1988. *The Return of the Native: American Indian Political Resurgence.* New York: Oxford University Press.

———. 1990. "Land, Labour and Group Formation: Blacks and Indians in the United States." *Ethnic and Racial Studies* 13:368–88.

———. 2008. "The Political Economy of American Indian Gaming." *Annual Review of Law and Social Science* 4:63–82.

Corral, Oscar, and Andres Viglucci. 2005. "Mariel: From Turmoil to Triumph." *Miami Herald* (April 3). http://www.miamiherald.com/

Corso, Regina. 2010. "Oprah Regains Her Position as America's Favorite Television Personality." *Harris Poll* #10, January 25. http://www.harrisinteractive.com

Cortés, Carlos E. 1994. "The Hollywood Curriculum on Italian Americans: Evolution of an Icon of Ethnicity." Pp. 89–108 in Lydio F. Tomasi, Piero Gastaldo, and Thomas Row (eds.), *The Columbus People: Perspectives in Italian Immigration to the Americas and Australia.* New York: Center for Migration Studies.

Cose, Ellis. 1993. *The Rage of a Privileged Class.* New York: Harper Perennial.

———. 2011. *The End of Anger: A New Generation's Take on Race and Rage.* New York: HarperCollins.

Coser, Lewis A. 1956. *The Functions of Social Conflict.* New York: Free Press.

Costantini, Cristina. 2011. "Spanish in Miami: Dicienda 'Hola' or Saying 'Hello.'" *Huffington Post* (November 29). http://www.huffingtonpost.com/2011/11/29/Spanish-in-Miami-dicienda_n_1118713.html

Cottrell, Catherine A., and Steven L. Neuberg. 2005. "Different Emotional Reactions to Different Groups: A Sociofunctional Threat-based Approach to 'Prejudice.'" *Journal of Personality and Social Psychology* 88:770–89.

Covello, Leonard. 1967. *The Social Background of the Italo-American School Child.* Leiden, the Netherlands: E. J. Brill.

Coward, John M. 1999. *The Newspaper Indian: Native American Identity in the Press, 1820–90.* Urbana: University of Illinois Press.

Cox, Oliver C. 1948. *Caste, Class and Race.* New York: Monthly Review Press.

Crawford, James. 1992. *Hold Your Tongue: Bilingualism and the Politics of "English Only."* Reading, Mass.: Addison-Wesley.

Crawford, Michael H. 2001. *The Origins of Native Americans: Evidence from Anthropological Genetics.* Cambridge: Cambridge University Press.

Crispino, James A. 1980. *The Assimilation of Ethnic Groups: The Italian Case.* Staten Island, N.Y.: Center for Migration Studies.

Cumberland, Charles C. 1968. *Mexico: The Struggle for Modernity.* New York: Oxford University Press.

Dahrendorf, Ralf. 1959. *Class and Class Conflict in Industrial Society.* Stanford, Calif.: Stanford University Press.

———. 1968. *Essays in the Theory of Society.* Stanford, Calif.: Stanford University Press.

Daley, Suzanne. 1997. "Torturer's Testimony Gives South Africa a New Lesson in the Banality of Evil." *New York Times* (November 9):1, 12.

———. 1998a. "At Afrikaner Schools, a Backlash Against Blacks." *New York Times* (April 13): A4.

———. 1998b. "South Africa's 'White Tribe' Sees Its Dark Past as Clouding a Black Future." *New York Times* (February 22):1, 16.

D'Amico, Ronald, and Nan L. Maxwell. 1995. "The Continuing Significance of Race in Minority Male Joblessness." *Social Forces* 73:969–91.

Daniel, G. Reginald. 2000. "Multiracial Identity in Brazil and the United States." Pp. 153–79 in Paul Spickard and W. Jeffrey Burroughs (eds.), *We Are a People: Narrative and Multiplicity in Constructing Ethnic Identity.* Philadelphia: Temple University Press.

———. 2006. *Race and Multiraciality in Brazil and the United States: Converging Paths?* University Park: Pennsylvania State University Press.

Daniels, Roger. 1977. *The Politics of Prejudice: The Anti-Japanese Movement in California and the Struggle for Japanese Exclusion.* 2nd ed. Berkeley: University of California Press.

———. 1981. *Concentration Camps: North America.* Malabar, Fla.: Kriege.

Dao, James. 1998. "Indians Using New Wealth for More Effective Lobbying." *New York Times* (February 9):A16.

Darby, John. 1995. *Northern Ireland: Managing Difference.* London: Minority Rights Group International.

Darroch, A. Gordon. 1979. "Another Look at Ethnicity, Stratification and Social Mobility in Canada." *Canadian Journal of Sociology* 4: 1–26.

Davey, Monica. 2009. "In Twist, Tribe Fights for College Nickname." *New York Times* (December 9):A16, A25.

David, Gary. 2007. "The Creation of 'Arab American': Political Activism and Ethnic (Dis)Unity." *Critical Sociology* 33:833–62.

Davidson, James D., Ralph E. Pyle, and David V. Reyes. 1995. "Persistence and Change in the Protestant Establishment, 1930–1992." *Social Forces* 74:157–75.

Davies, Rodrigo. 2003. "Brazil Takes Affirmative Action in HE." *Education Guardian* (August 4).

http://education.guardian.co.uk/higher/worldwide/story/0,9959,1012157,00.html

Davis, Darién J. 2000. *Afro-Brazilians: Time for Recognition.* London: Minority Rights Group International.

Davis, David Brion. 1966. *The Problem of Slavery in Western Culture.* Ithaca, N.Y.: Cornell University Press.

Davis, James. 2001. *Who Is Black? One Nation's Definition.* 10th anniversary ed. University Park: Pennsylvania State University Press.

Davis, Morris, and Joseph F. Krauter. 1971. *The Other Canadians: Profiles of Six Minorities.* Toronto: Methuen.

Davis, Yvonne R. 2009. "'Flying While Arab' Continues to Soar." *Huffington Post* (March 16). http://www.huffingtonpost.com/yvonne-r-davis/flying-while-arab-continu_b_174367.html

Dawidowicz, Lucy S. 1986. *The War Against the Jews, 1933–1945.* 10th anniversary ed. New York: Free Press.

Dawsey, Darrell. 1992. "Fatal Shooting Escalates Tension Between Blacks, Arab Americans." *Detroit News* (September 19):1C.

Dean, John P., and Alex Rosen. 1955. *A Manual of Intergroup Relations.* Chicago: University of Chicago Press.

de Azevedo, Fernando. 1950. *Brazilian Culture.* Trans. William Rex Crawford. New York: Macmillan.

Degler, Carl. 1986 [1971]. *Neither Black nor White: Slavery and Race Relations in Brazil and the United States.* Madison: University of Wisconsin Press.

de la Garza, Rodolfo O., Louis DeSipio, F. Chris Garcia, John Garcia, and Angelo Falcon. 1992. *Latino Voices: Mexican, Puerto Rican, and Cuban Perspectives on American Politics.* Boulder, Colo.: Westview Press.

de la Merced, Michael J. 2006. "Florida's Seminole Tribe Buys Hard Rock Cafes and Casinos." *New York Times* (December 8):C3.

Della Pergola, Sergio, Uzi Rebhun, and Mark Tolts. 2000. "Prospecting the Jewish Future: Population Projections, 2000–2080." Pp. 103–46 in *American Jewish Year Book 2000.* New York: American Jewish Committee.

DePalma, Anthony. 1998. "Canadian Indians Celebrate Vindication of Their History." *New York Times* (February 9):A1, A8.

de Roche, Constance, and John de Roche. 1991. "Black and White: Racial Construction in Television Police Dramas." *Canadian Ethnic Studies* 23(3):69–91.

Desilver, Drew. 2013. "As Supreme Court Defers Affirmative Action Ruling, Deep Divides Persist." Pew Research Center (June 24). http://www.pewresearch.org/fact-tank/2013/06/24/as-supreme-court-defers-affirmative-action-deep-divides-persist/

DeSipio, Louis. 2006. "Latino Civic and Political Participation." Pp. 447–79 in Marta Tienda and Faith Mitchell (eds.), *Hispanics and the Future of America*. Washington, D.C.: National Academies Press.

Destexhe, Alain. 1995. *Rwanda and Genocide in the Twentieth Century*. New York: New York University Press.

Detroit News. 1990. "Pointes' Past Bias Still Keeps Some Minorities Away." (February 28):1A, 6A.

Deutsch, Karl W. 1966. *Nationalism and Social Communication*. Cambridge, Mass.: MIT Press.

Devine, P., and A. J. Elliot. 1995. "Are Stereotypes Really Fading? The Princeton Trilogy Revisited." *Personality and Social Psychology Bulletin* 21:1139–50.

Devine, P. G., and S. J. Sherman. 1992. "Intuitive vs. Rational Judgment and the Role of Stereotyping in the Human Condition: Kirk or Spock?" *Psychological Inquiry* 3:153–59.

De Vos, George, and Lola Romanucci-Ross (eds.). 1975. *Ethnic Identity: Cultural Continuities and Change*. Palo Alto, Calif.: Mayfield.

De Vos, George, and Hiroshi Wagatsuma. 1966. *Japan's Invisible Race: Caste in Culture and Personality*. Berkeley: University of California Press.

Dhingra, Pawan. 2012. *Life Behind the Lobby: Indian American Motel Owners and the American Dream*. Stanford, Calif.: Stanford University Press.

Diaz, Guarione M. (ed.). 1980. *Evaluation and Identification of Policy Issues in the Cuban Community*. Miami: Cuban National Planning Council.

Dickie-Clark, H. F. 1972. "The Coloured Minority of Durban." Pp. 25–38 in Noel P. Gist and Anthony G. Dworkin (eds.), *The Blending of Races: Marginality and Identity in World Perspective*. New York: Wiley.

Dillon, Sam. 2009. "'No Child' Law Is Not Closing a Racial Gap." *New York Times* (April 29): A1, A16.

Ding, Erin Chan. 2008. "An Awakened Culture." *Detroit Free Press* (May 4):1A, 6A.

Dinnerstein, Leonard. 2004. "Is There a New Anti-Semitism in the United States?" *Society* 41(January/February):53–58.

Dinnerstein, Leonard, Roger L. Nichols, and David M. Reimers. 2003. *Natives and Strangers: A Multicultural History of Americans*. 4th ed. New York: Oxford University Press.

Dinnerstein, Leonard, and David M. Reimers. 1999. *Ethnic Americans: A History of Immigration*. 4th ed. New York: Columbia University Press.

Disha, Ilir, James C. Cavendish, and Ryan D. King. 2011. "Historical Events and Spaces of Hate: Hate Crimes Against Arabs and Muslims in Post-9/11 America." *Social Problems* 58:21–46.

DiversityInc. 2013. "Where's the Diversity in Fortune 500 CEOs?" http://www.diversityinc.com/facts/wheres-the-diversity-in-fortune-500-ceos/

Dobbin, Christine. 1996. *Asian Entrepreneurial Minorities: Conjoint Communities in the Making of the World Economy, 1570–1940*. Richmond, Surrey: Curzon.

Dobyns, H. F. 1966. "Estimating Aboriginal American Population: An Appraisal of Techniques with a New Hemispheric Estimate." *Current Anthropology* 7:395–416.

Dockterman, Daniel. 2011. "A Demographic Portrait of Puerto Ricans, 2009." Washington, D.C.: Pew Hispanic Center.

Doherty, Carroll. 2013. "For African Americans, Discrimination Is Not Dead." Pew Research Center (June 28). http://www.pewresearch.org/fact-tank/2013/06/28/for-african-americans-discrimination-is-not-dead/

Dollard, John. 1937. *Caste and Class in a Southern Town*. New Haven, Conn.: Yale University Press.

Dollard, John, et al. 1939. *Frustration and Aggression*. New Haven, Conn.: Yale University Press.

Domhoff, G. William. 2010. *Who Rules America? Challenges to Corporate and Class Dominance*. New York: McGraw-Hill.

Domhoff, G. William, and Richard L. Zweigenhaft. 2012. "Diversity and the New CEOs." *The Society Pages: Social Science That Matters*. http://thesocietypages.org/papers/new-ceos/

Donnelly, Francis X. 2003. "Metro Arabs, Muslims Suffer Harassment, Hatred." *Detroit News* (November 4). http://www.detnews.com/

Dorman, James H. 1980. "Ethnic Groups and 'Ethnicity': Some Theoretical Considerations." *Journal of Ethnic Studies* 7:23–36.

Dougherty, Tim. 2003. "Se Habla English." *Hispanic Business* (July/August):66–68, 72.

Dovidio, John F. 2001. "On the Nature of Contemporary Prejudice: The Third Wave." *Journal of Social Issues* 57:829–49.

Dovidio, John F., and S. L. Gaertner, 1998. "On the Nature of Contemporary Prejudice: The Causes, Consequences, and Challenges of Aversive Racism." Pp. 3–32 in J. L. Eberhardt and S. T. Fiske (eds.), *Confronting Racism: The Problem and the Response*. Thousand Oaks, Calif.: Sage.

Dovidio, John, Agata Gluszek, Melissa-Sue John, Ruth Ditlmann, and Paul Lagunes. 2010. "Understanding Bias Toward Latinos: Discrimination, Dimensions of Difference, and Experience of Exclusion." *Journal of Social Issues* 66:59–78.

Downie, Andrew, and Marion Lloyd. 2010. "At Brazil's Universities Affirmative Action Faces Crucial Tests." *Chronicle of Higher Education* (August 1). http://chronicle.com/article/At-Brazils-Universities/123720

D'Souza, Dinesh. 1995. *The End of Racism: Principles for a Multiracial Society*. New York: Free Press.

Duany, Jorge. 1998. "Reconstructing Racial Identity: Ethnicity, Color, and Class Among Dominicans in the United States and Puerto Rico." *Latin American Perspectives* 25:147–72.

Duckitt, J. 1989. "Authoritarianism and Group Identification: A New View of an Old Construct." *Political Psychology* 10:63–84.

Duffy, Andrew. 2004. "The Dutch Transformation." *Toronto Star* (October 1):A6.

Dugard, John. 1992. "The Law of Apartheid." Pp. 3–31 in John Dugard (ed.), *The Last Years of Apartheid: Civil Liberties in South Africa*. New York: Ford Foundation.

Dugger, Celia W. 2009. "Eager Students Fall Prey to Apartheid's Legacy." *New York Times* (September 20):A1.

———. 2010. "Campus That Apartheid Ruled Faces a Policy Rift." *New York Times* (November 22):A1.

Duncan, Brian, V. Joseph Hotz, and Stephen J. Trejo. 2006. "Hispanics in the U.S. Labor Market." Pp. 228–90 in Marta Tienda and Faith Mitchell (eds.), *Hispanics and the Future of America*. Washington, D.C.: National Academies Press.

Dunn, L. C. 1956. "Race and Biology." Pp. 31–67 in Leo Kuper (ed.), *Race, Science and Society*. New York: Columbia University Press.

Durand, Jorge, Edward Telles, and Jennifer Flashman. 2006. "The Demographic Foundations of the Latino Population." Pp. 66–99 in Marta Tienda and Faith Mitchell (eds.), *Hispanics and the Future of America*. Washington, D.C.: National Academies Press.

Durrheim, Kevin, and John Dixon. 2010. "Racial Contact and Change in South Africa." *Journal of Social Issues* 66:273–88.

Duthu, N. Bruce. 2008. "Broken Justice in Indian Country." *New York Times* (August 11):A21.

Dye, Thomas R. 1995. *Who's Running America? The Clinton Years*. 6th ed. Englewood Cliffs, N.J.: Prentice Hall.

Dzidzienyo, Anani. 2005. "The Changing World of Brazilian Race Relations?" Pp. 137–155 in Anani Dzidzienyo and Suzanne Oboler (eds.), *Neither Enemies nor Friends: Latinos, Blacks, Afro-Latinos*. New York: Palgrave Macmillan.

Dzidzienyo, Anani, and Lourdes Casal. 1979. *The Position of Blacks in Brazilian and Cuban Society*. London: Minority Rights Group.

Eagly, Alice H., and Amanda B. Diekman. 2005. "What Is the Problem? Prejudice as an Attitude-in-Context." Pp. 19–35 in John F. Dovidio, Peter Glick, and Laurie A. Rudman (eds.), *On the Nature of Prejudice: Fifty Years After Allport*. Malden, Mass.: Blackwell.

Eberstadt, Fernando. 2004. "A Frenchman or a Jew?" *New York Times Magazine* (February 29):48–51, 61.

Eckstein, Susan Eva. 2009. *The Immigrant Divide: How Cuban Americans Changed the US and Their Homeland*. New York: Routledge.

Economist, The. 2012. "Race in South Africa: Still an Issue." (February 4).

Edmonston, Barry, Sharon M. Lee, and Jeffrey S. Passel. 2002. "Recent Trends in Intermarriage and Immigration and Their Effects on the Future Racial Composition of the U.S. Population." Pp. 227–55 in Joel Perlmann and Mary C. Waters (eds.), *The New Race Question: How the Census Counts Multiracial Individuals*. New York: Russell Sage Foundation.

Edsall, Thomas Byrne, with Mary D. Edsall. 1992. *Chain Reaction: The Impact of Race, Rights, and Taxes on American Politics*. New York: Norton.

Egan, Timothy. 2007. "Little Asia on the Hill." *New York Times* (January 7):4A–2A, 27, 35.

Ehrlich, Howard J. 1973. *The Social Psychology of Prejudice*. New York: Wiley.

Elkins, Stanley M. 1976. *Slavery: A Problem in American Institutional and Intellectual Life*. 3rd ed. Chicago: University of Chicago Press.

Elliott, Andrea. 2011. "Behind an Anti-Shariah Push." *New York Times* (July 31):A1.

Elliott, Jean Leonard. 1979. "Canadian Immigration: A Historical Assessment." Pp. 160–72 in Jean Leonard Elliott (ed.), *Two Nations, Many Cultures*. Scarborough, Ontario: Prentice Hall of Canada.

Eltringham, Nigel. 2004. *Accounting for Horror: Post-Genocide Debates in Rwanda*. London: Pluto Press.

Employment and Immigration Canada. 1988. *Profiles of Canadian Immigration*. Ottawa: Minister of Supply and Services.

Engelberg, Stephen, and Tim Weiner. 1995. "Srebrenica: The Days of Slaughter." *New York Times* (October 29):1, 6–7.

Engelhardt, Tom. 1975. "Ambush at Kamikaze Pass." Pp. 522–31 in Norman R. Yetman and C. Hoy Steele (eds.), *Majority and Minority: The Dynamics of Racial and Ethnic Relations*. 2nd ed. Boston: Allyn & Bacon.

Engelhart, Katie. 2012a. "Divided They Stall." *Maclean's* (January 9):46–49.

———. 2012b. "Troubled Times." *Maclean's* (October 8):28–30.

Entman, Robert M. 1997. "African Americans According to TV News." Pp. 29–36 in Everette E. Dennis and Edward C. Pease (eds.), *The Media in Black and White*. New Brunswick, N.J.: Transaction.

Eriksen, Thomas Hylland. 2013. *Immigration and National Identity in Norway*. Washington, D.C.: Migration Policy Institute.

Erlanger, Steven. 2011. "For the Dutch, Persistent Questions of Identity." *New York Times* (August 21).7.

Eschbach, Karl. 1995. "The Enduring and Vanishing American Indian: American Indian Population Growth and Intermarriage in 1990." *Ethnic and Racial Studies* 18:89–108.

Esipova, Neli, and Julie Ray. 2009. "700 Million Worldwide Desire to Migrate Permanently." *Gallup* (November 2). http://www.gallup.com/

Esman, Milton J. 1994. *Ethnic Politics.* Ithaca, N.Y.: Cornell University Press.

Espenshade, Thomas J. 1998. "U.S. Immigration and the New Welfare State." Pp. 231–50 in David Jacobson (ed.), *The Immigration Reader: America in a Multidisciplinary Perspective.* Malden, Mass.: Blackwell.

Espiritu, Yen Le. 1992. *Asian American Panethnicity: Bridging Institutions and Identities.* Philadelphia: Temple University Press.

Etzioni, Amitai. 1998. "Some Diversity." *Society* (July/August):59–61.

———. 2004. "Assimilation to the American Creed." Pp. 211–220 in Tamar Jacoby (ed.), *Reinventing the Melting Pot: The New Immigrants and What It Means to Be American.* New York: Basic Books.

Eurostat. 2012. *Migrants in Europe: A Statistical Portrait of the First and Second Generation.* Luxembourg: European Union.

Ezorsky, Gertrude. 1991. *Racism and Justice: The Case for Affirmative Action.* Ithaca, N.Y.: Cornell University Press.

Fabricant, Florence. 1993. "Riding Salsa's Coast-to-Coast Wave of Popularity." *New York Times* (June 2):B1, B7.

Fagan, Richard R., Richard A. Brody, and Thomas J. O'Leary. 1968. *Cubans in Exile.* Stanford, Calif.: Stanford University Press.

Fallows, Marjorie R. 1979. *Irish Americans: Identity and Assimilation.* Englewood Cliffs, N.J.: Prentice Hall.

Farhi, Paul. 2006. "U.S., Media Settle with Wen Ho Lee." *Washington Post* (June 3):A01.

Farley, Reynolds. 1985. "The Residential Segregation of Blacks from Whites: Trends, Causes, and Consequences." In U.S. Commission on Civil Rights, *Issues in Housing Discrimination: A Consultation/Hearing of the United States Commission on Civil Rights.* Vol. 1: Papers Presented. Washington, D.C.: U.S. Government Printing Office.

———. 1993. "The Common Destiny of Blacks and Whites: Observations about the Social and Economic Status of the Races." Pp. 197–233 in Herbert Hill and James E. Jones Jr. (eds.), *Race in America: The Struggle for Equality.* Madison: University of Wisconsin Press.

———. 2011. "The Waning of American Apartheid?" *Contexts* 10(August):36–43.

Farley, Reynolds, Suzanne Bianchi, and Diane Colasanto. 1979. "Barriers to the Racial Integration of Neighborhoods: The Detroit Case." *Annals of the American Academy of Political and Social Science* 441:97–113.

Farley, Reynolds, and Albert Hermalin. 1972. "The 1960s: Decade of Progress for Blacks?" *Demography* 9:353–70.

Farley, Reynolds, Howard Schuman, Suzanne Bianchi, Diane Colasanto, and Shirley Hatchett. 1978. "Chocolate City, Vanilla Suburbs: Will the Trend Toward Racially Separate Communities Continue?" *Social Science Research* 7:319–44.

Farley, Reynolds, Charlotte Steeh, Maria Krysan, Tara Jackson, and Keith Reeves. 1994. "Stereotypes and Segregation: Neighborhoods in the Detroit Area." *American Journal of Sociology* 100:750–80.

Fauman, S. Joseph. 1958. "Occupational Selection among Detroit Jews." Pp. 119–36 in Marshall Sklare (ed.), *The Jews: Social Patterns of an American Group.* Glencoe, Ill.: Free Press.

FBI (Federal Bureau of Investigation). 2007. "Decidedly Uncivil, Part 2: Muslim Mother Target of Hate Crime." http://www.fbi.gov/news/stories/2007/may/hate050207

Feagin, Joe R., and Clairece Booher Feagin. 1978. *Discrimination American Style: Institutional Racism and Sexism.* Englewood Cliffs, N.J.: Prentice Hall.

Feagin, Joe R., and Melvin P. Sikes. 1994. *Living with Racism: The Black Middle-Class Experience.* Boston: Beacon.

Fein, Helen. 1993. *Genocide: A Sociological Perspective.* London: Sage.

Fenwick, Rudy. 1982. "Ethnic-Culture and Economic Structure: Determinants of French-English Earnings Inequality in Quebec." *Social Forces* 61:1–23.

Fernandes, Danielle Cireno. 2005. "Race, Socioeconomic Development and the Educational Stratification Process in Brazil." *Research in Social Stratification and Mobility* 22:365–422.

Fernandes, Florestan. 1971. *The Negro in Brazilian Society.* New York: Atheneum.

Fernández Kelly, M. Patricia, and Richard Schauffler. 1996. "Divided Fates: Immigrant Children and the New Assimilation." Pp. 30–53 in Alejandro Portes (ed.), *The New Second Generation.* New York: Russell Sage Foundation.

File, Thom. 2013. *"The Diversifying Electorate—Voting Rates by Race and Hispanic Origin in 2012 (and Other Recent Elections)."* Washington, D.C.: U.S. Census Bureau.

Finchilescu, Gilian, and Colin Tredoux. 2010. "The Changing Landscape of Intergroup Relations in South Africa." *Journal of Social Issues* 66:223–36.

Fineman, Howard. 1995. "Race and Rage." *Newsweek* (April 2):23–33.

Finn, Peter. 2002. "A 'Keep Out' Sign for Foreigners." *Washington Post National Weekly Edition* (April 8–14):9.

Finn, Peter, David Finkel, R. Jeffrey Smith, and Michael Dobbs. 1999. "Across Kosovo, Death at Every Turn." *Washington Post National Weekly Edition* (June 21):15–18.

Fischer, Claude S., Michael Hout, Martin Sanchez Jankowski, Samuel R. Lucas, Ann Swidler, and Kim Voss. 1996. *Inequality by Design: Cracking the Bell Curve Myth*. Princeton, N.J.: Princeton University Press.

Fish, Jefferson M. 1999. "Mixed Blood." Pp. 391–97 in Robert M. Levine and John J. Crocitti (eds.), *The Brazil Reader: History, Culture, Politics*. Durham, N.C.: Duke University Press.

Fishman, Joshua A. 1987. "What Is Happening to Spanish on the U.S. Mainland?" *Ethnic Affairs* 1 (Fall):12–23.

Fiske, Edward B., and Helen F. Ladd. 2004. *Elusive Equity: Education Reform in Post-Apartheid South Africa*. Washington, D.C.: Brookings Institution Press.

Fiske, Susan T., Jun Xu, Amy C. Cuddy, and Peter Glick. 1999. "(Dis)respecting Versus (Dis)liking: Status and Interdependence Predict Ambivalent Stereotypes of Competence and Warmth." *Journal of Social Issues* 55:473–89.

Fitz, Marshall, Philip E. Wolgin, and Patrick Oakford. 2013. "Immigrants Are Makers, Not Takers." Washington, D.C.: Center for American Progress. http://www .americanprogress.org/issues/immigration/ news/2013/02/08/52377

Fitzpatrick, Joseph P. 1987. *Puerto Rican Americans: The Meaning of Migration to the Mainland*. 2nd ed. Englewood Cliffs, N.J.: Prentice Hall.

———. 1995. "Puerto Rican New Yorkers, 1990." *Migration World* 23(1):16–19.

Fixico, Donald L. 2000. *The Urban Indian Experience in America*. Albuquerque: University of New Mexico Press.

Fleras, Augie, and Jean Leonard Elliott. 2002. *Engaging Diversity: Multiculturalism in Canada*. 2nd ed. Toronto: Nelson Thomson Learning.

Fletcher, Michael A. 1997. "Coming to Terms with the Racial Melting Pot." *Washington Post National Weekly Edition* (May 5):31.

Foderaro, Lisa W. 2010. "Unlikely Group Charges Bias at University." *New York Times* (September 15):A1, A22.

Foerster, Robert S. 1919. *The Italian Emigration of Our Times*. Cambridge, Mass.: Harvard University Press.

Fogel, Robert William, and Stanley L. Engerman. 1974. *Time on the Cross*. Boston: Little, Brown.

Foner, Nancy, and Richard Alba. 2006. "The Second Generation from the Last Great Wave of Immigration: Setting the Record Straight." *Migration Information Source*, Migration Policy Institute (October 1). http://www .migrationinformation.org/

Fong, Eric. 1996. "A Comparative Perspective on Racial Residential Segregation: American and Canadian Experiences." *Sociological Quarterly* 37:199–226.

Fong, Timothy P. 1994. *The First Suburban Chinatown: The Remaking of Monterey Park, California*. Philadelphia: Temple University Press.

Fontaine, Pierre-Michel. 1981. "Transnational Relations and Racial Mobilization: Emerging Black Movements in Brazil." Pp. 141–62 in John F. Stack, Jr. (ed.), *Ethnic Identities in a Transnational World*. Westport, Conn.: Greenwood.

Forcese, Dennis. 1997. *The Canadian Class Structure*. 4th ed. Toronto: McGraw-Hill Ryerson.

Fordham, Signithia, and John U. Ogbu. 1986. "Black Students' School Success: Coping with the Burden of 'Acting White.'" *Urban Review* 18:176–206.

Forero, Juan. 2001. "Prosperous Colombians Flee, Many to U.S., to Escape War." *New York Times* (April 10):A1, A6.

Forman, Tyrone A. 2004. "Color-Blind Racism and Racial Indifference: The Role of Racial Apathy in Facilitating Enduring Inequalities." Pp. 43–66 in Maria Krysan and Amanda E. Lewis (eds.), *The Changing Terrain of Race and Ethnicity*. New York: Russell Sage Foundation.

Foroutan, Naika. 2013. *Identity and (Muslim) Integration in Germany*. Washington, D.C.: Migration Policy Institute.

Fortune. 1936. "Jews in America." 13 (February):79–85, 128–44.

Frady, Marshall. 1968. *Wallace*. New York: World.

Frank, Reanne, Ilana Redstone Akresh, and Bo Lu. 2010. "Latino Immigrants and the U.S. Racial Order: How and Where Do They Fit In?" *American Sociological Review* 75:378–401.

Franklin, John Hope. 1980. *From Slavery to Freedom*. 5th ed. New York: Knopf.

Franklin, Raymond S., and Solomon Resnik. 1973. *The Political Economy of Racism*. New York: Holt, Rinehart & Winston.

Fraser, Morris. 1977. *Children in Conflict: Growing Up in Northern Ireland*. New York: Basic Books.

Fraser, Steven (ed.). 1995. *The Bell Curve Wars: Race, Intelligence, and the Future of America*. New York: Basic Books.

Frazier, E. Franklin. 1939. *The Negro Family in the United States*. Chicago: University of Chicago Press.

———. 1949. *The Negro in the United States.* New York: Macmillan.

Fredrickson, George M. 1971. *The Black Image in the White Mind.* New York: Harper & Row.

———. 1981. *White Supremacy: A Comparative Study in American and South African History.* New York: Oxford University Press.

———. 2002. *Racism: A Short History.* Princeton, N.J.: Princeton University Press.

Freedman, Maurice. 1955. "The Chinese in Southeast Asia." Pp. 388–411 in Andrew W. Lind (ed.), *Race Relations in World Perspective.* Honolulu: University of Hawaii Press.

Freedman, Samuel G. 2000. *Jew vs. Jew: The Struggle for the Soul of American Jewry.* New York: Touchstone.

———. 2012. "If the Sikh Temple Had Been a Mosque." *New York Times* (August 11):A12.

Frey, William H. 2010a. "Race and Ethnicity." Pp. 50–63 in *The State of Metropolitan America.* Washington, D.C.: Brookings Institution Press.

———. 2010b. "New Racial Segregation Measures for Large Metropolitan Areas: Analysis of the 1990–2010 Decennial Censuses." University of Michigan, Institute for Social Research, Population Studies Center. http://www.psc.isr .umich.edu/

Freyre, Gilberto. 1956. *The Masters and the Slaves.* New York: Knopf.

———. 1963a. "Ethnic Democracy: The Brazilian Example." *Americas* 15 (December):1–6.

———. 1963b. *New World in the Tropics: The Culture of Modern Brazil.* New York: Vintage.

Frideres, James S. 1976. "Racism in Canada: Alive and Well." *Western Canadian Journal of Anthropology* 6:124–45.

———. 1990. "Policies on Indian People in Canada." Pp. 98–119 in Peter S. Li (ed.), *Race and Ethnic Relations in Canada.* Toronto: Oxford University Press.

Fried, Morton H. 1965. "A Four-Letter Word That Hurts." *Saturday Review* (October 2): 21–23, 35.

Frisbie, W. Parker, and Lisa Neidert. 1977. "Inequality and the Relative Size of Minority Populations: A Comparative Analysis." *American Journal of Sociology* 82:1007–30.

Fry, Peter. 2000. "Politics, Nationality, and the Meanings of 'Race' in Brazil." *Daedalus* 129 (Spring):83–118.

Fry, Richard. 2009. "College Enrollment Hits All-Time High, Fueled by Community College Surge." Pew Research Center (October 29). http://pewsocialtrends.org/pubs/747/college-enrollment-hits-all-time-high-fueled-by-community-college-surge/

Fry, Richard, and Paul Taylor. 2013. "Hispanic High School Graduates Pass Whites in Rate of College Enrollment." Pew Hispanic Center. http://www.pewhispanic.org/2013/05/09 /hispanic-high-school-graduates-pass-whites-in-rate-of-college-enrollment/

Fugita, Stephen S., and David J. O'Brien. 1991. *Japanese American Ethnicity: The Persistence of Community.* Seattle: University of Washington Press.

Furnivall, J. S. 1948. *Colonial Policy and Practice.* Cambridge, England: Cambridge University Press.

Gabaccia, Donna R. 2006. "Today's Immigration Policy Debates: Do We Need a Little History?" Washington, D.C.: Migration Policy Institute (November 1). http://www.migrationinformation .org.

Gallagher, A. M., and S. Dunn. 1991. "Community Relations in Northern Ireland: Attitudes to Contact and Integration." Pp. 7–22 in Peter Stringer and Gillian Robinson (eds.), *Social Attitudes in Northern Ireland.* Belfast: Blackstaff Press.

Gallo, Patrick J. 1974. *Ethnic Alienation: The Italian-Americans.* Rutherford, N.J.: Fairleigh Dickinson University Press.

Gallup. 2009a. *Muslim Americans: A National Portrait.* Gallup Center for Muslim Studies. http://www.gallup.com/strategicconsulting/ 153572/Report-Muslim-Americans-National-Portrait.aspx

———. 2009b. *Religious Perceptions in America: With an In-Depth Analysis of U.S. Attitudes Toward Muslims and Islam.* Gallup Muslim West Facts Project.

———. 2010a. "Race Relations." (June 25). http://www.gallup.com/poll/1687/Race-Relations.aspx

———. 2010b. "In U.S., Religious Prejudice Stronger Against Muslims." Washington, D.C.: Gallup Center for Muslim Studies. http://www .gallup.com/poll/125312/Religious-Prejudice-Stronger-Against-Muslims.aspx?v

———. 2011. *Muslim Americans: Faith, Freedom, and the Future.* Abu Dhabi: Gallup Center.

Gallup, George, Jr. 1982. *The Gallup Poll: Public Opinion 1981.* Wilmington, Del.: Scholarly Resources.

———. 1988. *The Gallup Poll: Public Opinion 1987.* Wilmington, Del.: Scholarly Resources.

Gallup Organization. 2001a. "Americans Ambivalent About Immigrants." Gallup News Service, *Poll Analyses* (May 3).

———. 2001b. *Black-White Relations in the United States: 2001 Update.* Gallup Poll Social Audit (July 10).

———. 2001c. "Racial or Ethnic Labels Make Little Difference to Blacks, Hispanics." Gallup News Service, *Poll Analyses* (September 11).

———. 2003. "Gallup Brain: The Darkest Hours of Racial Unrest." (June 3). http://www.gallup .com/poll/8539/gallup-brain-darkest-hours-racial-unrest.aspx

———. 2005. "Focus On: Immigration." (February 1). http://www.gallup.com/

Gambino, Richard. 1975. *Blood of My Blood: The Dilemma of the Italian-Americans.* Garden City, N.Y.: Anchor.

———. 1977. *Vendetta*. Garden City, N.Y.: Doubleday.

———. 1998. "America's Most Tolerated Intolerance: Bigotry Against Italian Americans." Pp. 155–73 in Gregory R. Campbell (ed.), *Many Americas: Critical Perspectives on Race, Racism, and Ethnicity*. Dubuque, Iowa: Kendall/Hunt.

Gamson, William A. 1975. *The Strategy of Social Protest*. Homewood, Ill.: Dorsey.

Gans, Herbert J. 1958. "The Origin and Growth of a Jewish Community in the Suburbs: A Study of the Jews of Park Forest." Pp. 205–58 in Marshall Sklare (ed.), *The Jews: Social Patterns of an American Group*. New York: Free Press.

———. 1967a. *The Levittowners*. New York: Pantheon.

———. 1967b. "Some Comments on the History of Italian Migration and on the Nature of Historical Research." *International Migration Review* 1:5–9.

———. 1979. "Symbolic Ethnicity: The Future of Ethnic Groups and Cultures in America." *Ethnic and Racial Studies* 2:1–20.

———. 1982. *The Urban Villagers: Group and Class in the Life of Italian Americans*. Updated and expanded ed. New York: Free Press.

———. 1992. "Comment: Ethnic Invention and Acculturation, A Bumpy-Line Approach." *Journal of American Ethnic History* 12:43–52.

———. 1999. "Filling in Some Holes: Six Areas of Needed Immigration Research." *American Behavioral Scientist* 42:1302–13.

———. 2007. "The Possibility of a New Racial Hierarchy in the Twenty-First-Century United States." Pp. 266–74 in David B. Grusky and Szonja Szelényi (eds.), *The Inequality Reader: Contemporary and Foundational Readings in Race, Class, and Gender*. Boulder, Colo.: Westview.

Garcia, F. Chris, and Rudolph P. de la Garza. 1977. *The Chicano Political Experience: Three Perspectives*. North Scituate, Mass.: Duxbury.

Garcia, Guy. 1999. "Another Latin Boom, But Different." *New York Times* (June 27):AR25, AR27.

Garcia, Juan Ramon. 1980. *Operation Wetback*. Westport, Conn.: Greenwood.

Garcia-Passalacqua, Juan Manuel. 1994. "The Puerto Ricans: Migrants or Commuters?" Pp. 103–13 in Carlos Antonio Torre, Hugo Rodríguez Vecchini, and William Burgos (eds.), *The Commuter Nation: Perspectives on Puerto Rican Migration*. Rìo Piedras, Puerto Rico: Editorial de la Universidad de Puerto Rico.

Garcia-Zamor, Jean-Claude. 1970. "Social Mobility of Negroes in Brazil." *Journal of Inter-American Studies* 12:242–54.

Gardner, Howard. 1983. *Frames of Mind: The Theory of Multiple Intelligences*. New York: Basic Books.

Gardner, Robert W., Bryant Robey, and Peter C. Smith. 1985. "Asian Americans: Growth, Change, and Diversity." *Population Bulletin* 40 (October):1–43.

Garson, Jean-Pierre. 1992. "Migration and Interdependence: The Migration System Between France and Africa." Pp. 80–93 in Mary M. Kritz, Lin Lean Lim, and Hania Zlotnik (eds.), *International Migration Systems: A Global Approach*. Oxford: Clarendon Press.

Geddes, John. 2009. "What Canadians Think of Sikhs, Jews, Christians, Muslims . . . ." *Maclean's* (May 4):20–24.

———. 2013. "Land of Intolerance." *Maclean's* (October 14):22–23.

Geertz, Clifford. 1994. "Primordial and Civic Ties." Pp. 29–34 in John Hutchinson and Anthony D. Smith (eds.), *Nationalism*. Oxford: Oxford University Press.

Gerdes, Wylie. 1993. "Chaldean Merchants Decry Slaying, Look for Answers." *Detroit Free Press* (December 2):1B, 2B.

German, Michael, and Michelle Richardson. 2011. *Reclaiming Patriotism: A Call to Reconsider the Patriot Act*. New York: American Civil Liberties Union.

Gerth, Hans. 1940. "The Nazi Party: Leadership and Composition." *American Journal of Sociology* 45:517–41.

Geschwender, James A. 1978. *Racial Stratification in America*. Dubuque, Iowa: W. C. Brown.

Gevisser, Mark. 1994. "Who Is a South African?" *New York Times* (April 26):A19.

Ghosh, Bobby. 2010. "Arab-Americans: Detroit's Unlikely Saviors." *Time* (November 13). http://www.time.com/time/magazine/article/0,9171,2028057-3,00.html-racial-unrest-aspx

Gibbins, Roger. 1997. "Historical Overview and Background: Part I." Pp. 19–34 in J. Rick Ponting (ed.), *First Nations in Canada: Perspectives on Opportunity, Empowerment, and Self-Determination*. Toronto: McGraw-Hill Ryerson.

Gibson, James L., and Christopher Claassen. 2010. "Racial Reconciliation in South Africa: Interracial Contact and Changes over Time." *Journal of Social Issues* 66:255–72.

Gilbert, G. M. 1951. "Stereotype Persistence and Change Among College Students." *Journal of Abnormal and Social Psychology* 46:245–54.

Gilens, Martin. 1999. *Why Americans Hate Welfare: Race, Media, and the Politics of Anti-poverty Policy*. Chicago: University of Chicago Press.

Giliomee, Hermann. 2004. "The Rise and Possible Demise of Afrikaans as Public Language." *Nationalism and Ethnic Politics* 10:25–58.

Ginsberg, Eli. 1978. "Jews in the American Economy: The Dynamics of Opportunity." Pp. 109–19 in Gladys Rosen (ed.), *Jewish Life in America: Historical Perspectives*.

NewYork: Institute of Human Relations, American Jewish Committee.

Gionet, Linda. 2009. *First Nations People: Selected Findings of the 2006 Census*. Ottawa: Statistics Canada.

Giuliano, Laura, David I. Levine, and Jonathan Leonard. 2008. "Manager Race and the Race of New Hires." University of Miami, Department of Economics, Working Paper #0722.

Glaberson, William. 2001. "Who Is a Seminole, and Who Gets to Decide?" *New York Times* (January 29):A1, A14.

Glater, Jonathan D., and Martha M. Hamilton. 1995. "Affirmative Action's Defenders." *Washington Post* (March 27–April 2):20.

Glazer, Nathan. 1957. *American Judaism*. Chicago: University of Chicago Press.

———. 1958. "The American Jews and the Attainment of Middle-Class Rank: Some Trends and Explanations." Pp. 138–46 in Marshall Sklare (ed.), *The Jews: Social Patterns of an American Group*. Glencoe, Ill.: Free Press.

———. 1971. "Blacks and Ethnic Groups: The Difference and the Political Difference It Makes." *Social Problems* 18:444–61.

———. 1975. *Affirmative Discrimination: Ethnic Identity and Public Policy*. New York: Basic Books.

———. 1997. *We Are All Multiculturalists Now*. Cambridge, Mass.: Harvard University Press.

Glazer, Nathan, and Daniel P. Moynihan. 1970. *Beyond the Melting Pot*. 2nd ed. Cambridge, Mass.: MIT Press.

———. (eds.). 1975. *Ethnicity: Theory and Experience*. Cambridge, Mass.: Harvard University Press.

Gleason, Philip. 1964. "Immigration and American Catholic Intellectual Life." *Review of Politics* 26:147–73.

Glenn, Norvall D. 1963. "Occupational Benefits to Whites from the Subordination of Negroes." *American Sociological Review* 28:443–48.

———. 1966. "White Gains from Negro Subordination." *Social Problems* 14:159–78.

Glenny, Misha. 1993. *The Fall of Yugoslavia: The Third Balkan War*. New York: Penguin.

Glock, Charles Y., and Rodney Stark. 1966. *Christian Beliefs and Anti-Semitism*. New York: Harper & Row.

Golab, Caroline. 1977. *Immigrant Destinations*. Philadelphia: Temple University Press.

Gold, Steven J. 1999. "From 'The Jazz Singer' to 'What a Country!' A Comparison of Jewish Migration to the United States, 1880–1930 and 1965–1998." *Journal of American Ethnic History* 18:115–41.

———. 2000. "Israeli Americans." Pp. 409–20 in Peter Kivisto and Georganne Rundblad (eds.), *Multiculturalism in the United States: Current Issues, Contemporary Voices*. Thousand Oaks, Calif.: Pine Forge.

———. 2010. *The Store in the Hood: A Century of Ethnic Business and Conflict*. Lanham, MD: Rowan & Littlefield.

Goldberg, J. J. 1996. *Jewish Power: Inside the American Jewish Establishment*. Reading, Mass.: Addison-Wesley.

Golden, Daniel. 2007. *The Price of Admission: How America's Ruling Class Buys Its Way into Elite Colleges and Who Gets Left Outside the Gates*. New York: Three Rivers Press.

Goldscheider, Calvin. 1986. *Jewish Continuity and Change: Emerging Patterns in America*. Bloomington: Indiana University Press.

———. 2002. *Israel's Changing Society: Population, Ethnicity, and Development*. 2nd ed. Boulder, Colo.: Westview Press.

Goldstein, Sidney. 1980. "Jews in the United States: Perspectives from Demography." Pp. 3–59 in *American Jewish Yearbook 1981*. New York: American Jewish Committee.

Golub, Jennifer L. 1990. *What Do We Know About Black Anti-Semitism?* New York: American Jewish Committee.

Goodman, Mary Ellen. 1964. *Race Awareness in Young Children*. Rev. ed. New York: Collier.

Goodnough, Abby, and David Gonzalez. 2006. "Among Cuban-Americans, the Hyphen Remains." *New York Times* (August 3):A1, A8.

Goodstein, Laurie. 1997. "Women in Islamic Headdress Find Faith and Prejudice, Too." *New York Times* (November 3):A1, A14.

Gordon, Larry. 2013. "UC Takes Fewer State Seniors." *Los Angeles Times* (April 19):AA1, AA4.

Gordon, Milton M. 1964. *Assimilation in American Life: The Role of Race, Religion, and National Origins*. New York: Oxford University Press.

———. 1975. "Toward a General Theory of Racial and Ethnic Group Relations." Pp. 84–110 in Nathan Glazer and Daniel P. Moynihan (eds.), *Ethnicity: Theory and Experience*. Cambridge, Mass.: Harvard University Press.

———. 1981. "Models of Pluralism: The New American Dilemma." *Annals of the American Academy of Political and Social Science* 454:178–88.

Gossett, Thomas F. 1963. *Race: The History of an Idea in America*. Dallas, Tex.: Southern Methodist University Press.

Gould, Stephen Jay. 1983. *The Mismeasure of Man*. New York: Norton.

———. 1984. "Human Equality Is a Contingent Fact of History." *Natural History* (November):26–32.

Gourevitch, Philip. 1998. *We Wish to Inform You That Tomorrow We Will Be Killed with Our Families*. New York: Farrar, Straus and Giroux.

———. 2009. "The Life After." *The New Yorker* (May 4):36–49.

Gozdziak, Elzbieta, and Susan F. Martin. 2005. *Beyond the Gateway: Immigrants in a Changing America*. Lanham, Md.: Lexington.

Graham, Richard. 1970. "Brazilian Slavery Reexamined: A Review Article." *Journal of Social History* 3:431–53.

Grasser, Edward. 2013. "Navajo Farm Exports: $2–$3 Million Per Year." *Progressive Economy* (September 4). GlobalWorks Foundation. http://progressive-economy.org /2013/09/04/navajo-farm-exports-2-3-million-per-year/

Graves, Joseph L., Jr. 2004. *The Race Myth: Why We Pretend Race Exists in America*. New York: Dutton.

Graves, Sherryl Browne. 1999. "Television and Prejudice Reduction: When Does Television as a Vicarious Experience Make a Difference?" *Journal of Social Issues* 55:707–25.

Grebler, Leo, Joan W. Moore, and Ralph C. Guzman. 1970. *The Mexican-American People: The Nation's Second Largest Minority*. New York: Free Press.

Greeley, Andrew M. 1971. *Why Can't They Be Like Us?* New York: Dutton.

———. 1972. *That Most Distressful Nation: The Taming of the American Irish*. Chicago: Quadrangle.

———. 1974. *Ethnicity in the United States: A Preliminary Reconnaissance*. New York: Wiley.

———. 1977. *The American Catholic: A Social Portrait*. New York: Harper & Row.

———. 1981. *The Irish Americans: The Rise to Money and Power*. New York: Harper & Row.

Greeley, Andrew M., and Paul B. Sheatsley. 1974. "Attitudes toward Racial Integration." Pp. 241–50 in Lee Rainwater (ed.), *Social Problems and Public Policy*. Chicago: Aldine.

Greenbaum, William. 1974. "America in Search of a New Ideal: An Essay on the Rise of Pluralism." *Harvard Educational Review* 44:411–40.

Greenberg, Bradley S., and Jeffrey E. Brand. 1998. "U.S. Minorities and the News." Pp. 3–22 in Yahya R. Kamalipour and Theresa Carilli (eds.), *Cultural Diversity and the U.S. Media*. Albany: State University of New York Press.

Grenier, Guillermo J., Lisandro Pérez, Sung Chang Chun, and Hugh Gladwin. 2007. "There Are Cubans, There Are Cubans, and There Are Cubans: Ideological Diversity Among Cuban Americans in Miami." Pp. 93–111 in Martha Montero-Sieburth and Edwin Meléndez (eds.), *Latinos in a Changing Society*. Westport, Conn.: Praeger.

Grieco, Elizabeth. 2004. "The Foreign Born from the Dominican Republic in the United States." *Migration Information Source* (October 1). Migration Policy Institute. http:// www .migrationinformation.org/

Grogan, Maura, with Rebecca Morse and April Youpee-Roll. 2011. *Native American Lands and Natural Resource Development*. Revenue Watch Institute. http://www.revenuewatch .org/sites/default/files/RWI_Native_American_ Lands_2011.pdf

Gross, Jane. 1989. "Diversity Hinders Asians' Influence." *New York Times* (June 25):18.

Guarnizo, Luis Eduardo, and Marilyn Espitia. 2007. "Colombia." Pp. 371–85 in Mary C. Waters and Reed Ueda (eds.), *The New Americans: A Guide to Immigration Since 1965*. Cambridge, Mass.: Harvard University Press.

Guimarães, Antonio Sèrgio Alfredo. 1999. "Measures to Combat Discrimination and Racial Inequality in Brazil." Pp. 139–53 in Rebecca Reichmann (ed.), *Race in Contemporary Brazil: From Indifference to Inequality*. University Park: Pennsylvania State University Press.

Gutiérrez, Ramon. 2004. "Internal Colonialism: An American Theory of Race." *Du Bois Review* 1:281–95.

Gutman, Herbert G. 1976. *The Black Family in Slavery and Freedom, 1750–1925*. New York: Pantheon.

Gwyn, Richard. 1995. *Nationalism without Walls: The Unbearable Lightness of Being Canadian*. Toronto: McClelland & Stewart.

Hacker, Andrew. 1994. "Education: Ethnicity and Achievement." Pp. 214–29 in Nicolaus Mills (ed.), *Debating Affirmative Action: Race, Gender, Ethnicity, and the Politics of Inclusion*. New York: Delta.

———. 1995. *Two Nations: Black and White, Separate, Hostile, Unequal*. Expanded and updated ed. New York: Ballantine.

Haddad, Yvonne. 1983. "Arab Muslims and Islamic Institutions in America: Adaptation and Reform." Pp. 65–81 in Sameer Y. Abraham and Nabeel Abraham (eds.), *Arabs in the New World: Studies in Arab-American Communities*. Detroit: Wayne State University, Center for Urban Studies.

Haddad, Yvonne Yazbeck. 2007. "The Post-9/11 Hijab as Icon." *Sociology of Religion* 68:253–267.

———. 2011. *Becoming American? The Forging of Arab and Muslim Identity in Pluralist America*. Waco, TX: Baylor University Press.

Haddad, Yvonne Yazbeck, Jane I. Smith, and Kathleen M. Moore. 2006. *Muslim Women in America: The Challenge of Islamic Identity Today*. New York: Oxford University Press.

Haddad, Yvonne Yazbeck, and Adair T. Lummis. 1987. *Islamic Values in the United States: A Comparative Study*. New York: Oxford University Press.

Hagan, William T. 1961. *American Indians*. Chicago: University of Chicago Press.

———. 1992. "Full Blood, Mixed Blood, Generic, and Ersatz: The Problem of Indian Identity." Pp. 278–88 in Roger L. Nichols (ed.), *The American Indian: Past and Present*. 4th ed. New York: McGraw-Hill.

Halász, Karalin. 2010. "Europe." Pp. 150–177 in Preti Taneja (ed.), *State of the World's Minorities and Indigenous Peoples 2010*. London: Minority Rights Group International.

Halpern, Ben. 1958. "America Is Different." Pp. 23–39 in Marshall Sklare (ed.), *The Jews: Social Patterns of an American Group*. Glencoe, Ill.: Free Press.

Halter, Marilyn. 2007. "Africa: West." Pp. 283–94 in Mary C. Waters and Reed Ueda (eds.), *The New Americans: A Guide to Immigration Since 1965*. Cambridge, Mass.: Harvard University Press.

Hanchard, Michael George. 1994. *Orpheus and Power: The Movimento Negro of Rio de Janeiro and São Paulo, Brazil, 1945–1988*. Princeton, N.J.: Princeton University Press.

Handley, Antoinette, and Jeffrey Herbst. 1997. "South Africa: The Perils of Normalcy." *Current History* 96 (May):222–26.

Handlin, Oscar. 1957. *Race and Nationality in American Life*. Boston: Little, Brown.

———. 1962. *The Newcomers: Negroes and Puerto Ricans in a Changing Metropolis*. Garden City, N.Y.: Doubleday.

Hannerz, Ulf. 1969. *Soulside: Inquiries into Ghetto Culture and Community*. New York: Columbia University Press.

Harari, Oren, and David Beaty. 1989. *Lessons from South Africa: A New Perspective on Public Policy and Productivity*. New York: Harper & Row.

Harding, John. 1968. "Stereotypes." Pp. 259–61 in David L. Sills (ed.), *International Encyclopedia of the Social Sciences*. Vol. 15. New York: Macmillan.

Harlow, Caroline Wolf. 2005. *Hate Crime Reported by Victims and Police. U.S. Department of Justice, Bureau of Justice Statistics, Special Report*. Washington, D.C.: U.S. Government Printing Office.

Harmon, Amy. 2007. "In DNA Era, New Worries About Prejudice." *New York Times* (November 11):1, 24.

Harney, Robert F., and Harold Troper. 1977. "Introduction." *Canadian Ethnic Studies* 9:1–5.

Harris, David R. 1999. "'Property Values Drop When Blacks Move In, Because …': Racial and Socioeconomic Determinants of Neighborhood Desirability." *American Sociological Review* 64:461–79.

Harris, Marvin. 1963. "Race Relations in Minas Velhas, a Community in the Mountain Region of Central Brazil." Pp. 47–81 in Charles Wagley (ed.), *Race and Class in Rural Brazil*. 2nd ed. New York: Columbia University Press.

———. 1964. *Patterns of Race in the Americas*. New York: Norton.

Harris, Marvin, Josildeth Gomes Consorte, Joseph Lang, and Bryan Byrne. 1993. "Who Are the Whites? Imposed Census Categories and the Racial Demography of Brazil." *Social Forces* 72:451–62.

Harris, Rosemary. 1972. *Prejudice and Tolerance in Ulster: A Study of Neighbours and Strangers in a Border Community*. Manchester, England: Manchester University Press.

Harrison, David. 1981. *The White Tribe of Africa: South Africa in Perspective*. London: British Broadcasting Corporation.

Hart, Keith, and Vishnu Padayachee. 2000. "Indian Business in South Africa After Apartheid: New and Old Trajectories." *Comparative Studies in Society and History* 42:683–712.

Hartley, E. L. 1946. *Problems in Prejudice*. New York: Kings Crown.

Harvard Project on American Indian Economic Development. 2008. *The State of the Native Nations: Conditions Under U.S. Policies of Self-Determination*. New York: Oxford University Press.

Hasenbalg, Carlos A. 1985. "Race and Socioeconomic Inequalities in Brazil." Pp. 25–11 in Pierre-Michel Fontaine (ed.), *Race, Class, and Power in Brazil*. Los Angeles: Center for Afro-American Studies, University of California.

Hasenbalg, Carlos, and Suellen Huntington. 1982/1983. "Brazilian Racial Democracy: Reality or Myth?" *Humboldt Journal of Social Relations* 10:129–42.

Hashim, Ahmed S. 2005. "Iraq: From Insurgency to Civil War?" *Current History* 104:10–18.

Haskins, Ron. 2009. "Moynihan Was Right: Now What?" *The Annals of the American Academy of Political and Social Science* 621(1):281–14.

Hassoun, Rosina J. 2005. *Arab Americans in Michigan*. East Lansing: Michigan State University Press.

Hattam, Victoria. 2005. "Ethnicity and the Boundaries of Race: Rereading Directive 15." *Daedalus* 134 (January):61–69.

Hawkins, Freda. 1972. *Canada and Immigration: Public Policy and Public Concern*. Montreal: McGill-Queen's University Press.

———. 1989. *Critical Years in Immigration: Canada and Australia Compared*. Kingston and Montreal: McGill-Queen's University Press.

Hay, Michelle A. 2009. *"I've Been Black in Two Countries": Black Cuban Views on Race in the US*. El Paso: LFB Scholarly Publishing.

Hayes, B., I. McAllister, and L. Dowds. 2007. "Integrated Education, Intergroup Relations, and Political Identities in Northern Ireland." *Social Problems* 54:454–82.

Heath, Brad, Oralander Brand-Williams, and Shawn D. Lewis. 2002. "Wealth Doesn't Stop Race Divide." *Detroit News* (November 3): 1A, 12A–13A.

Hechter, Michael. 1975. *Internal Colonialism: The Celtic Fringe in British National Development, 1536–1966.* Berkeley: University of California Press.

Hedges, Chris. 1996. "Serbs Left in Bosnia See No Rest for the Dead." *New York Times* (January 18):A6.

———. 1997. "Bosnia Journal: Ethnic Diversity Distorts History, Art, Language." *New York Times* (November 25):1, 14.

Heer, David. 1996. *Immigration in America's Future: Social Science Findings and the Policy Debate.* Boulder, Colo.: Westview Press.

Heilman, Madeline E. 1996. "Affirmative Action's Contradictory Consequences." *Journal of Social Issues* 52(4):105–9.

Heilman, Samuel. 1999. "Separated but Not Divorced." *Society* 36 (May/June):8–14.

Heilman, Samuel C., and Steven M. Cohen. 1989. *Cosmopolitans & Parochials: Modern Orthodox Jews in America.* Chicago: University of Chicago Press.

Hein, Jeremy. 1995. *From Vietnam, Laos, and Cambodia: A Refugee Experience in the United States.* New York: Twayne Publishers.

———. 2006. *Ethnic Origins: The Adaptation of Cambodian and Hmong Refugees in Four American Cities.* New York: Russell Sage Foundation.

Heisler, Martin O. 1991. "Hyphenating Belgium: Changing State and Regime to Cope with Cultural Division." Pp. 177–95 in Joseph V. Montville (ed.), *Conflict and Peacemaking in Multiethnic Societies.* New York: Lexington Books.

Hemming, John. 1987. *Amazon Frontier: The Defeat of the Brazilian Indians.* Cambridge, Mass.: Harvard University Press.

Hendrix, Steve. 2011. "The Woman Without a Face: Controversy Dogs Muslim Veil." *Washington Post* (April 16): D2.

Henry, Frances. 1978. *The Dynamics of Racism in Toronto: Research Report.* Toronto: York University.

———. 1994. *The Caribbean Diaspora in Toronto: Learning to Live with Racism.* Toronto: University of Toronto Press.

Henry, Frances, Carol Tator, Winston Mattis, and Tim Rees. 1995. *The Colour of Democracy: Racism in Canadian Society.* Toronto: Harcourt Brace Canada.

Herberg, Will. 1960. *Protestant-Catholic-Jew.* Garden City, N.Y.: Doubleday.

Herbert, Bob. 2010. "Blacks in Retreat." *New York Times* (January 19):A27.

Hernández, Ramona. 2007. "Living on the Margins of Society: Dominicans in the United States." Pp. 34–57 in Martha Montero-Sieburth and Edwin Meléndez (eds.), *Latinos in a Changing Society.* Westport, Conn.: Praeger.

Herrnstein, Richard J., and Charles Murray. 1994. *The Bell Curve: Intelligence and Class Structure in American Life.* New York: Free Press.

Hershberg, Theodore, et al. 1979. "A Tale of Three Cities: Blacks and Immigrants in Philadelphia: 1850–1880, 1930 and 1970." *Annals of the American Academy of Political and Social Science* 441:55–81.

Herskovitz, Melville. 1941. *The Myth of the Negro Past.* New York: Harper & Brothers.

Higham, John. 1963. *Strangers in the Land: Patterns of American Nativism, 1860–1925.* 2nd ed. New York: Atheneum.

Hilberg, Raul. 2003. *The Destruction of the European Jews.* 3rd ed. New Haven, Conn.: Yale University Press.

Hiller, Harry H. 1976. *Canadian Society: A Sociological Analysis.* Scarborough, Ontario: Prentice Hall of Canada.

Hipsman, Faye, and Doris Meissner. 2013. "Immigration in the United States: New Economic, Social, Political Landscapes with Legislative Reform on the Horizon." *Migration Information Source* (April). Washington, D.C.: Migration Policy Institute.

Hirschman, Charles. 1975. *Ethnic and Social Stratification in Peninsular Malaysia.* Washington, D.C.: American Sociological Association.

Hirschman, Charles, Philip Kasinitz, and Josh DeWind. 1999. "Immigrant Adaptation, Assimilation, and Incorporation." Pp. 127–36 in Charles Hirschman, Philip Kasinitz, and Josh DeWind (eds.), *The Handbook of International Migration: The American Experience.* New York: Russell Sage Foundation.

Hirschman, Charles, and Douglas S. Massey. 2008. "Places and Peoples: The New American Mosaic." Pp. 1–21 in Douglas S. Massey (ed.), *New Faces in New Places: The Changing Geography of American Immigration.* New York: Russell Sage.

Hitt, Jack. 2005. "The Newest Indians." *New York Times Magazine* (August 21):36–41.

Hoare, Anthony G. 1982. "Problem Region and Regional Problem." Pp. 195–223 in Frederick W. Boal and J. Neville H. Douglas (eds.), *Integration and Division: Geographical Perspectives on the Northern Ireland Problem.* London: Academic Press.

Hochschild, Jennifer L. 1995. *Facing Up to the American Dream: Race, Class, and the Soul of the Nation.* Princeton, N.J.: Princeton University Press.

Hochschild, Jennifer, Vesla Weaver, and Traci Burch. 2012. *Creating a New Racial Order: How Immigration, Multiracialism, Genomics, and the Young Can Remake Race in America.* Princeton: Princeton University Press.

Hoefer, Michael, Nancy Rytina, and Bryan C. Baker. 2012. "Estimates of the Unauthorized Immigrant Population Residing in the United States: January 2011." Washington, D.C.: U.S. Department of Homeland Security, Office of Immigration Statistics.

Hoffer, Eric. 1951. *The True Believer.* New York: Harper & Row.

Hollinger, David A. 1995. *Postethnic America: Beyond Multiculturalism.* New York: Basic Books.

———. 2005. "The One Drop Rule and the One Hate Rule." *Daedalus* 134 (Winter):18–28.

Holsinger, Jennifer L. 2009. *Residential Patterns of Arab Americans: Race, Ethnicity, and Spatial Assimilation.* El Paso, TX: LFB Scholarly Publishing.

Holzer, Harry J. 2001. "Racial Difference in Labor Market Outcomes Among Men." Pp. 98–123 in Neil J. Smelser, William Julius Wilson, and Faith Mitchell (eds.), *America Becoming: Racial Trends and Their Consequences.* Vol. 2. Washington, D.C.: National Academy Press.

———. 2009. "The Labor Market and Young Black Men: Updating Moynihan's Perspective." *Annals of the American Academy of Political and Social Science* 621(1):47–69.

Hook, Sidney. 2002. "Reverse Discrimination." Pp. 224–30 in Steven M. Cahn (ed.), *The Affirmative Action Debate*, 2nd ed. New York: Routledge.

Hordge-Freeman, Elizbeth. 2013. "What's Love Got to Do with It? Racial Features, Stigma and Socialization in Afro-Brazilian Families." *Ethnic and Racial Studies* 36:1507–1523.

Horowitz, Donald L. 1975. "Ethnic Identity." Pp. 111–40 in Nathan Glazer and Daniel P. Moynihan (eds.), *Ethnicity: Theory and Experience.* Cambridge, Mass.: Harvard University Press.

———. 1991. *A Democratic South Africa? Constitutional Engineering in a Divided Society.* Berkeley: University of California Press.

Horowitz, Irving Louis. 1964. *Revolution in Brazil.* New York: Dutton.

Horowitz, Juliana Menasce. 2010. "Widespread Anti-Immigrant Sentiment in Italy." Pew Research Center (January 12). http://www .pewresearch.org/

Horton, John. 1995. *The Politics of Diversity: Immigration, Resistance, and Change in Monterey Park, California.* Philadelphia: Temple University Press.

Hostetler, John A. 1993. *Amish Society.* 4th ed. Baltimore: Johns Hopkins University Press.

Howard, John R. 1970. *Awakening Minorities.* New Brunswick, N.J.: Transaction.

Howe, Irving. 1976. *World of Our Fathers.* New York: Simon & Schuster.

Howell, Sally, and Andrew Shryock. 2003. "Cracking Down on Diaspora: Arab Detroit and America's 'War on Terror'." *Anthropological Quarterly* 76:443–62.

Howell, Sally, and Amaney Jamal. 2008. "Detroit Exceptionalism and the Limits of Political Incorporation." Pp. 45–79 in Katherine Pratt Ewing (ed.), *Being and Belonging: Muslims in the United States Since 9/11.* New York: Russell Sage Foundation.

———. 2009. "Belief and Belonging." Pp. 103–34 in Detroit Arab American Study Team, *Citizenship and Crisis: Arab Detroit After 9/11.* New York: Russell Sage Foundation.

Huber, Joan, and William H. Form. 1973. *Income and Ideology.* New York: Free Press.

Hudson, David L., Jr. 2011. "Debate on Patriot Act and First Amendment Continues." First Amendment Center. http://www .firstamendmentcenter.org/debate-on-patriot- act-and-first-amendment-continues

Hughes, David R., and Evelyn Kallen. 1974. *The Anatomy of Racism: Canadian Dimensions.* Montreal: Harvest House.

Hughes, Everett C. 1943. *French Canada in Transition.* Chicago: University of Chicago Press.

Hughes, Everett C., and Helen M. Hughes. 1952. *Where Peoples Meet: Racial and Ethnic Frontiers.* Glencoe, Ill.: Free Press.

Human Resources and Skill Development Canada. 2013. *Well-being in Canada.* http://www4. hrsdc.gc.ca/.3ndic.1t.4r@-eng.jsp?iid=38

Human Rights Watch. 1995. *Slaughter Among Neighbors: The Political Origins of Communal Violence.* New Haven, Conn.: Yale University Press.

———. 1999. *Leave None to Tell the Story: Genocide in Rwanda.* New York: Human Rights Watch.

———. 2006. *The Rwandan Genocide: How It Was Prepared.* New York: Human Rights Watch.

Hunt, Chester, and Lewis Walker. 1974. *Ethnic Dynamics.* Homewood, Ill.: Dorsey.

Huntington, Samuel P. 2004. *Who Are We? The Challenges to America's National Identity.* New York: Simon & Schuster.

Hurh, Won Moo, and Kwang Chung Kim. 1989. "The 'Success' Image of Asian Americans: Its Validity and Its Practical and Theoretical Implications." *Ethnic and Racial Studies* 12:512–38.

Hutchinson, Edward P. 1956. *Immigrants and Their Children, 1850–1950.* New York: Wiley.

Hutchinson, Harry. 1963. "Race Relations in a Rural Community of the Bahian Reconcavo." Pp. 16–46 in Charles Wagley (ed.), *Race and Class in Rural Brazil.* 2nd ed. New York: Columbia University Press.

Iacovetta, Franca. 1992. *Such Hardworking People: Italian Immigrants in Postwar Toronto*. Montreal and Kingston: McGill-Queen's University Press.

Ianni, Francis A. J. 1974. *Black Mafia: Ethnic Succession in Organized Crime*. New York: Simon & Schuster.

Iceland, John. 2009. *Where We Live Now: Immigration and Race in the United States*. Berkeley: University of California Press.

Iceland, John, and Daniel H. Weinberg. 2002. *Racial and Ethnic Residential Segregation in the United States: 1980–2000*. Census 2000 Special Reports. Washington, D.C.: U.S. Government Printing Office.

Iceland, John, and Rima Wilkes. 2006. "Does Socioeconomic Status Matter? Race, Class, and Residential Segregation." *Social Problems* 53:248–73.

Ignatieff, Michael. 1995. *Blood and Belonging: Journeys into the New Nationalism*. New York: Farrar, Straus, & Giroux.

———. 2005. "Balancing Foreign and Domestic." *Toronto Star* (June 2):A19.

Ignatiev, Noel. 1995. *How the Irish Became White*. New York: Routledge.

Illa, Hernan Iglesias. 2010. *Miami*. Buenos Aires: Planeta/Seix Barral.

Indian Health Service. 2010. "Indian Health Disparities." http://info.ihs.gov/

Institute for Social Research. 1964. *Discrimination without Prejudice. A Study of Promotion Practices in Industry*. Ann Arbor: Institute for Social Research, University of Michigan.

Instituto Brasileiro de Geografia e Estatística. 2005. *Censoá Demográfico 2000*. http://www.ibge .gov.br/

———. 2006. "PME Color and Race." (November 17, 2006). http://www.ibge.gov.br/

———. 2012. *Censo 2010*. http://www.ibge.gov.br/

International Organization for Migration (IOM). 1999. *Migration in Central and Eastern Europe: 1999 Review*. Geneva: International Organization for Migration.

Iorizzo, Luciano J., and Salvatore Mondello. 1980. *The Italian Americans*. Boston: Twayne.

Ipsos. 2010a. "Twenty Years After Oka Crisis, Majority (57%) of Canadians Don't Believe Quality of Life for Aboriginals in Canada is Getting Better." (July 14). http://www.ipsosna .com/news-polls/pressrelease.aspx?id=4866

———. 2010b. "What the World Thinks of Canada: Canada and the World in 2010— Immigration and Diversity." (June 22). http://ipsos-na.com/

Irwin, Colin. 1994. "The Myths of Segregation." Pp. 104–18 in Adrian Guelke (ed.), *New Perspectives on the Northern Ireland Conflict*. Aldershot, England: Avebury.

Isaacs, Harold R. 1989. *Idols of the Tribe: Group Identity and Political Change*. Cambridge, Mass.: Harvard University Press.

Isajiw, Wsevolod W., Aysan Sev'er, and Leo Driedger. 1993. "Ethnic Identity and Social Mobility: A Test of the 'Drawback Model.'" *Canadian Journal of Sociology* 18:179–98.

Isbister, John. 1996. *The Immigration Debate: Remaking America*. West Hartford, Conn.: Kumarian Press.

Jackman, Mary R. 1994. *The Velvet Glove: Paternalism and Conflict in Gender, Class, and Race Relations*. Berkeley: University of California Press.

———. 2005. "Rejection or Inclusion of Outgroups?" Pp. 89–105 in John F. Dovidio, Peter Glick, and Laurie A. Rudman (eds.), *On the Nature of Prejudice: Fifty Years After Allport*. Malden, Mass.: Blackwell.

Jackson, Harold, and Anne McHardy. 1984. *The Two Irelands*. Rev. ed. London: Minority Rights Group.

Jacobson, Matthew Frye. 1998. *Whiteness of a Different Color: European Immigrants and the Alchemy of Race*. Cambridge, Mass.: Harvard University Press.

Jacoby, Russell, and Naomi Glauberman (eds.). 1995. *The Bell Curve Debate: History, Documents, Opinions*. New York: Times Books.

Jacoby, Tamar. 1998. *Someone Else's House: America's Unfinished Struggle for Integration*. New York: Free Press.

———. 2000. "In Asian America." *Commentary* (July/August):21–28.

Jaffe, A. J. 1992. *The First Immigrants from Asia: A Population History of the North American Indians*. New York: Plenum.

Jaffe, A. J., Ruth M. Cullen, and Thomas D. Boswell. 1980. *The Changing Demography of Spanish Americans*. New York: Academic Press.

Jamal, Amaney, and Nadine Naber (eds.). 2008. *Race and Arab Americans Before and After 9/11: From Invisible Citizens to Visible Subjects*. Syracuse: Syracuse University Press.

Jedwab, Jack. 2006. "Christians, Muslims and Jews in Canada: Perceptions, Partnerships and Perspectives." Montreal: Association for Canadian Studies.

Jedwab, Jack, and Ayman Al-Yassini. 2012. "Canadians Regard the Internet as the Place Where Racism Is Most Prevalent" (March 21). Association for Canadian Studies/Canadian Race Relations Foundation. http://www .acs-aec.ca/en/

Jencks, Christopher. 1972. *Inequality: A Reassessment of the Effect of Family and Schooling in America*. New York: Basic Books.

Jencks, Christopher, and Meredith Phillips. 1998. "The Black-White Test Score Gap: An Introduction." Pp. 1–55 in Christopher Jencks and Meredith Phillips (eds.), *The Black-White Test Score Gap*. Washington, D.C.: Brookings Institution Press.

Jensen, Arthur. 1969. "How Much Can We Boost IQ and Scholastic Achievement?" *Harvard Educational Review* 39:1–123.

Jeter, Jon. 2003. "Affirmative Action Debate Forces Brazil to Take Look in the Mirror." *Washington Post* (June 16):A1.

Jiménez, Tomás R. 2011. "Immigrants in the United States: How Well Are They Integrating into Society?" Washington, D.C.: Migration Policy Institute.

Johnson, Colleen Leahy. 1985. *Growing Up and Growing Old in Italian-American Families.* New Brunswick, N.J.: Rutgers University Press.

Johnson, Kirk. 2004. "Hispanic Voters Declare Their Independence." *New York Times* (November 9):A1, A16.

Johnson, Lyndon B. 1965. *Public Papers of the Presidents. Lyndon B. Johnson II.* Washington, D.C.: U.S. Government Printing Office.

Johnson, Tom. 2011. "Muslims in the United States." Washington, D.C.: Council on Foreign Relations. http://www.cfr.org/united-states/muslims-united-states/p25927

Jones, James M. 1996. *Prejudice and Racism.* 2nd ed. New York: McGraw-Hill.

Jones, Jeffrey M. 2001. "Americans Have Mixed Opinions About Immigration." *The Gallup Poll Monthly* (July):28–32.

———. 2003. "Majority of Americans Say Anti-Semitism a Problem in U.S." *The Gallup Poll Tuesday Briefing* (June 11):18–19.

———. 2004. "Blacks More Pessimistic Than Whites About Economic Opportunities." Washington, D.C.: The Gallup Poll.

———. 2006. "White, Blacks, Hispanics Disagree About Way Minority Groups Treated. *The Gallup Poll Briefing* (July):30–32.

———. 2008. "Majority of Americans Say Racism Against Blacks Widespread." *Gallup Poll* (August 4). http://www.gallup.com/

———. 2010. "Americans Felt Uneasy Toward Arabs Even Before September 11." *Gallup Poll* (September 28).

———. 2011. "Record-High 86% Approve of Black-White Marriages." *Gallup Poll* (September 12). http://www.gallup.com/

———. 2013a. "Americans Rate Racial and Ethnic Relations in U.S. Positively." *Gallup Poll* (July 17). http://www.gallup.com/

———. 2013b. "U.S. Blacks, Hispanics Have No Preferences on Group Labels." *Gallup Poll* (July 26). http://www.gallup.com/

Jones, Maldwyn Allen. 1992. *American Immigration.* 2nd ed. Chicago: University of Chicago Press.

Jones, Mike. 2013. "Sharia Law Is Another Imagined Threat." *Tulsa World* (September 1). http://www.tulsaworld.com/article.aspx/Mike_Jones_Sharia_law_is_another_imagined_threat/20130901_213_G1_Ntihtn379936?subj=7

Jones, Robert P., and Daniel Cox. 2011. *The 2011 American Values Survey: The Mormon Question, Economic Inequality, and the 2012 Presidential Campaign.* Washington, D.C.: Public Religion Research Institute.

Jordan, Winthrop. 1969. *White over Black.* Baltimore: Penguin.

Josephy, Alvin M., Jr. 1969. *The Indian Heritage of America.* New York: Knopf.

Josephy, Alvin M., Jr., Joane Nagel, and Troy Johnson. 1999. *Red Power: The American Indians' Fight for Freedom.* Lincoln: University of Nebraska Press.

Joy, Richard J. 1972. *Languages in Conflict.* Toronto: McClelland & Stewart.

Jubulis, Mark A. 2001. *Nationalism and Democratic Transition: The Politics of Citizenship and Language in Post-Soviet Latvia.* Lanham, Md.: University Press of America.

Kahlenberg, Richard. 1996. *The Remedy: Class, Race, and Affirmative Action.* New York: Basic Books.

———. 2007. *Rescuing Brown v. Board of Education: Profiles of Twelve School Districts Pursuing Socioeconomic School Integration.* New York: Century Foundation.

———. 2010. "Elite Colleges, or Colleges for the Elite?" *New York Times* (September 29):A39.

Kalin, Rudolf, and J. W. Berry. 1994. "Ethnic and Multicultural Attitudes." Pp. 293–321 in J. W. Berry and J. A. Laponce (eds.), *Ethnicity and Culture in Canada: The Research Landscape.* Toronto: University of Toronto Press.

Kallen, Evelyn. 2003. *Ethnicity and Human Rights in Canada: A Human Rights Perspective on Ethnicity, Racism, and Systemic Inequality.* 3rd ed. Don Mills, Ontario: Oxford University Press.

Kane, Thomas J. 2004. "College-Going and Inequality." Pp. 319–53 in Kathryn M. Neckerman (ed.), *Social Inequality.* New York: Russell Sage Foundation.

Kantrowitz, Nathan. 1973. *Ethnic and Racial Segregation in the New York Metropolis.* New York: Praeger.

———. 1979. "Racial and Ethnic Residential Segregation in Boston 1830–1970." *Annals of the American Academy of Political and Social Science* 441:41–54.

Karasch, Mary. 1975. "From Porterage to Proprietorship: African Occupations in Rio de Janeiro, 1808–1850." Pp. 369–93 in Stanley L. Engerman and Eugene D. Genovese (eds.), *Race and Slavery in the Western Hemisphere: Quantitative Studies.* Princeton, N.J.: Princeton University Press.

Karlins, M., T. L. Coffman, and G. Walters. 1969. "On the Fading of Social Stereotypes: Studies in Three Generations of College Students." *Journal of Personality and Social Psychology* 13:1–16.

Kasinitz, Philip, John Mollenkopf, and Mary C. Waters. 2006. "Becoming American/Becoming New Yorkers: The Second Generation in a Majority Minority City." Migration Policy

Institute (October). http://www
.migrationinformation.org/

Kasinitz, Philip, John H. Mollenkopf, Mary C. Waters, and Jennifer Holdaway. 2008. *Inheriting the City: The Children of Immigrants Come of Age*. New York: Russell Sage Foundation.

Katz, Daniel, and Kenneth Braly. 1933. "Racial Stereotypes of One Hundred College Students." *Journal of Abnormal and Social Psychology* 28:280–90.

Katznelson, Ira. 2005. *When Affirmative Action Was White: An Untold History of Racial Inequality in Twentieth Century America*. New York: Norton.

Kaus, Mickey. 1995. *The End of Equality*. 2nd ed. New York: Basic Books.

Kayal, Philip, and Joseph Kayal. 1975. *The Syrian-Lebanese in America: A Study in Religion and Assimilation*. Boston: Twayne.

Kayyali, Randa A. 2006a. *The Arab Americans*. Westport, Conn.: Greenwood Press.

———. 2006b. "The People Perceived as a Threat to Security: Arab Americans Since September 11" (July 1). Washington, D.C.: Migration Policy Institute.

Kehoe, Alice B. 1999. "American Indians." Pp. 48–74 in Elliott Robert Barkan (ed.), *A Nation of Peoples: A Sourcebook on America's Multicultural Heritage*. Westport, Conn.: Greenwood Press.

Keil, Hartmut. 1991. "Socialist Immigrants from Germany and the Transfer of Socialist Ideology and Workers' Culture." Pp. 315–38 in Rudolph J. Vecoli and Suzanne M. Sinke (eds.), *A Century of European Migrations, 1830–1930*. Urbana: University of Illinois Press.

Keister, Lisa A. 2003. "Religion and Wealth: The Role of Religious Affiliation and Participation in Early Adult Asset Accumulation." *Social Forces* 82:175–207.

———. 2007. "Upward Wealth Mobility: Exploring the Roman Catholic Advantage." *Social Forces* 85:1195–1226.

Keller, Bill. 1995. "After Apartheid, Change Lags Behind Expectations." *New York Times* (April 27):A1, A6.

———. 2013. "Affirmative Reaction." *New York Times* (June 9):A27.

Keller, Suzanne. 1953. "The Social Origins and Career Lines of Three Generations of American Business Leaders." Ph.D. dissertation, Columbia University.

Kelly, Mary E., and Joane Nagel. 2002. "Ethnic Re-identification: Lithuanian Americans and Native Americans." *Journal of Ethnic and Migration Studies* 28:275–89.

Kelner, Merrijoy. 1970. "Ethnic Penetration into Toronto's Elite Structure." *Canadian Review of Sociology and Anthropology* 7:128–37.

Kendall, Joan. 2001. "Circles of Disadvantage: Aboriginal Poverty and Underdevelopment in Canada." *The American Review of Canadian Studies* 31:42–59.

Kennedy, Randall. 1994. "Persuasion and Distrust: The Affirmative Action Debate." Pp. 48–67 in Nicolaus Mills (ed.), *Debating Affirmative Action: Race, Gender, Ethnicity, and the Politics of Inclusion*. New York: Delta.

Kenny, Kevin. 2000. *The American Irish: A History*. Harlow, England: Longman.

Kent, Mary Mederios. 2007. "Immigration and America's Black Population." *Population Bulletin 62 (December)*. Washington, D.C.: Population Reference Bureau.

Kephart, William M., and William W. Zellner. 1994. *Extraordinary Groups: An Examination of Unconventional Life-Styles*. 2nd ed. New York: St. Martin's.

Keshena, Rita. 1980. "The Role of American Indians in Motion Pictures." Pp. 106–11 in Gretchen M. Bataille and Charles L. P. Silet (eds.), *The Pretend Indians: Images of Native Americans in the Movies*. Ames: Iowa State University Press.

Keung, Nicholas. 2006. "Multiculturalism Accepted, to a Point: Poll." *Toronto Star* (November 15).

Khan, M. A. Muqtedar. 2003. "Constructing the American Muslim Community." Pp. 179–98 in Yvonne Yazbeck Haddad, Jane I. Smith, and John L. Esposito (eds.), *Religion and Immigration: Christian, Jewish, and Muslim Experiences in the United States*. Walnut Creek, Calif.: AltaMira Press.

Kibria, Nazli. 2002. *Becoming Asian American: Second-Generation Chinese and Korean American Identities*. Baltimore: Johns Hopkins University Press.

Kiester, Edwin, Jr. 1968. *The Case of the Missing Executive*. New York: Institute of Human Relations, American Jewish Committee.

Kifner, John. 1999. "Emptying a City of All but Bodies." *New York Times* (May 29):1, 6.

Kikumura, Akemi, and Harry H. L. Kitano. 1973. "Interracial Marriage: A Picture of the Japanese-Americans." *Journal of Social Issues* 29:67–81.

Killian, Lewis M. 1953. "The Adjustment of Southern White Migrants to Northern Urban Norms." *Social Forces* 32:66–69.

———. 1975. *The Impossible Revolution, Phase II: Black Power and the American Dream*. New York: Random House.

———. 1985. *White Southerners*. Rev. ed. Amherst: University of Massachusetts Press.

Kim, Illsoo. 1981. *New Urban Immigrants: The Korean Community in New York*. Princeton, N.J.: Princeton University Press.

———. 1987. "The Koreans: Small Business in an Urban Frontier." Pp. 219–42 in Nancy Foner (ed.),

New Immigrants in New York. New York: Columbia University Press.

King, James C. 1981. *The Biology of Race*. Berkeley: University of California Press.

King, Martin Luther, Jr. 1964. *Why We Can't Wait*. New York: Harper & Row.

King, Ryan D., and Melissa F. Weiner. 2007. "Group Position, Collective Threat, and American Anti-Semitism." *Social Problems* 54:47–77.

Kingston, Anne. 2012. "Who Are We to Judge?" *Maclean's* (January 23):50–53.

Kington, Raymond S., and Herbert W. Nickers. 2001. "Racial and Ethnic Differences in Health: Recent Trends, Current Patterns, Future Directions." Pp. 253–310 in Neil J. Smelser, William Julius Wilson, and Faith Mitchell (eds.), *America Becoming: Racial Trends and Their Consequences*. Vol. 2. Washington, D.C.: National Academy Press.

Kinsley, Michael. 1991. "Class, Not Race." *The New Republic* (August 19–26):4.

Kinzer, Stephen. 1992. "Ethnic Conflict Is Threatening in Yet Another Region of Yugoslavia: Kosovo." *New York Times* (November 9):8.

Kirscht, John P., and Ronald C. Dillehay. 1967. *Dimensions of Authoritarianism: A Review of Research and Theory*. Lexington: University of Kentucky Press.

Kirschten, Dick. 1992. "Building Blocs." *National Journal* 24(September 26):2173–77.

Kitano, Harry. 1976. *Japanese Americans: The Evolution of a Subculture*. 2nd ed. Englewood Cliffs, N.J.: Prentice Hall.

Kitano, Harry H. L., and Roger Daniels. 1995. *Asian Americans: Emerging Minorities*. 2nd ed. Englewood Cliffs, N.J.: Prentice Hall.

Klaaste, Aggrey. 1984. "Exiles in Their Native Land." *New York Times Magazine* (June 24):34–35, 51–53, 68–69.

Klausner, Samuel Z. 1988. *Succeeding in Corporate America: The Experience of Jewish MBAs*. New York: American Jewish Committee.

Kleiman, Dena. 1983. "Less Than 40% of Jews in Survey Observe Sabbath." *New York Times* (February 6):45.

Klein, Herbert S. 1969. "The Colored Freedman in Brazilian Slave Society." *Journal of Social History* 3:30–52.

Kleinknecht, William. 1996. *The New Ethnic Mobs: The Changing Face of Organized Crime in America*. New York: Free Press.

Klineberg, Otto. 1968. "Prejudice: The Concept." Pp. 439–47 in David L. Sills (ed.), *International Encyclopedia of the Social Sciences*. Vol. 12. New York: Macmillan.

Knight, Franklin. 1974. *The African Dimension in Latin American Societies*. New York: Macmillan.

Knowlton, Brian. 2010. "Muslim Women Gain Higher Profile in U.S." *New York Times* (December 27).

Knowlton, Clark. 1972. "The New Mexican Land War." Pp. 258–70 in Edward Simmen (ed.), *Pain and Promise: The Chicano Today*. New York: New American Library.

Knox, Colin. 2010. "Peace Building in Northern Ireland: A Role for Civil Society." *Social Policy and Society* 10:13–28.

Kochhar, Rakesh. 2006. "Growth in the Foreign-Born Workforce and Employment of the Native Born." (August 10). Washington, D.C.: Pew Hispanic Center.

Kochhar, Rakesh, Ana Gonzalez-Barrera, and Daniel Dockterman. 2009. "Through Boom and Bust: Minorities, Immigrants and Homeownership." Washington, D.C.: Pew Hispanic Center.

Koenig, Barbara A., Sandra Soo-Jin Lee, and Sarah S. Richardson (eds.). 2008. *Revisiting Race in a Genomic Age*. New Brunswick, N.J.: Rutgers University Press.

Kohn, Hans. 1944. *The Idea of Nationalism: A Study in Its Origins and Background*. New York: Macmillan.

Kolenda, Pauline. 1985. *Caste in Contemporary India: Beyond Organic Solidarity*. Prospect Heights, Ill.: Waveland Press.

Konvitz, Milton B. 1978. "The Quest for Equality and the Jewish Experience." Pp. 28–60 in Gladys Rosen (ed.), *Jewish Life in America: Historical Perspectives*. New York: Institute of Human Relations, American Jewish Committee.

Korman, Abraham K. 1988. *The Outsiders: Jews and Corporate America*. Lexington, Mass.: Lexington Books.

Kosmin, Barry A., et al. 1991. *Highlights of the CJF 1990 National Jewish Population Survey*. New York: Council of Jewish Federations.

Kosmin, Barry A., and Ariela Keysar. 2006. *Religion in a Free Market: Religious and Non-Religious Americans*. Ithaca, N.Y.: Paramount Market Publishing.

Koster, Henry. 1966 [1816]. *Travels in Brazil*. Carbondale: Southern Illinois University Press.

Kotlowitz, Alex. 2007. "All Immigration Politics is Local." *New York Times Magazine* (August 5):31–37, 52, 57.

Kottak, Conrad Phillip, and Kathryn A. Kozaitis. 1999. *On Being Different: Diversity and Multiculturalism in the North American Mainstream*. Boston: McGraw-Hill College.

Kovel, Joel. 1970. *White Racism: A Psychohistory*. New York: Pantheon.

Kramer, Judith R., and Seymour Leventman. 1961. *Children of the Gilded Ghetto*. New Haven, Conn.: Yale University Press.

Kristof, Nicholas D. 1995. "Japanese Outcasts Better Off Than in Past but Still Outcast." *New York Times* (November 30):A1, A8.

———. 1998. "New Freedoms Feed Ethnic Frictions." *New York Times* (June 25):A6.

Kristol, Irving. 1970. "The Negro Today Is Like the Immigrant Yesterday." Pp. 139–57 in Nathan Glazer (ed.), *Cities in Trouble*. Chicago: Quadrangle.

Kroeber, A. L. 1939. *Cultural and Natural Areas of Native North America*. Berkeley: University of California Press.

Krupa, Gregg, and John Bebow. 2003. "Immigration Crackdown Snares Arabs." *Detroit News* (November 3). http://www.detnews.com/

Krysan, Maria. 2008. "Data Update to Racial Attitudes in America." An update and Web site to complement *Racial Attitudes in America: Trends and Interpretations*. Rev. ed., Howard Schuman, Charlotte Steeh, Lawrence Bobo, and Maria Krysan. 1997. Harvard University Press. http://igpa.uillinois.edu/programs/racial-attitudes

Krysan, Maria, and Reynolds Farley. 2002. "The Residential Preferences of Blacks: Do They Explain Persistent Segregation?" *Social Forces* 80:937–80.

Krysan, Maria, and Nakesha Faison. 2011. *Racial Attitudes in America: Trend and Interpretations: 2011 Update*. Institute of Government and Public Affairs, University of Illinois. http:igpa.uillinois.edu

Kulczycki, Andrzej, and Lobo, Arun Peter. 2002. "Patterns, Determinants, and Implications of Intermarriage among Arab Americans." *Journal of Marriage and Family* 64:202–10.

Kuper, Leo. 1968. "Segregation." Pp. 144–50 in David L. Sills (ed.), *International Encyclopedia of the Social Sciences*. Vol. 14. New York: Macmillan.

———. 1969. "Ethnic and Racial Pluralism: Some Aspects of Polarization and Depluralization." Pp. 159–87 in Leo Kuper and M. G. Smith (eds.), *Pluralism in Africa*. Berkeley: University of California Press.

———. 1981. *Genocide*. New York: Penguin.

Kuper, Leo, and M. G. Smith (eds.). 1969. *Pluralism in Africa*. Berkeley: University of California Press.

Kurlansky, Mark. 1999. *The Basque History of the World*. New York: Vintage.

Kurokawa, Minako (ed.). 1970. *Minority Responses*. New York: Random House.

Kutner, Bernard, Carol Wilkens, and Penny Rechtman Yarrow. 1952. "Verbal Attitudes and Overt Behavior." *Journal of Abnormal and Social Psychology* 47:649–52.

Kymlicka, Will. 2007. *Multicultural Odysseys: Navigating the New International Politics of Diversity*. Oxford: Oxford University Press.

Lacy, Karyn R. 2007. *Blue-Chip Black: Race, Class, and Status in the New Black Middle Class*. Berkeley: University of California Press.

La Ferla, Ruth. 2001. "Latino Style Is Cool. Oh, All Right: It's Hot." *New York Times* (April 16): B1, B3.

———. 2007. "We, Myself and I." *New York Times* (April 5):E1, E7.

LaGumina, Salvatore J. 1988. *From Steerage to Suburb: Long Island Italians*. Staten Island, N.Y.: Center for Migration Studies.

———. 1996. "Anti-Italian Discrimination—Update." Pp. 293–304 in Mary Jo Bona and Anthony Julian Tamburri (eds.), *Through the Looking Glass: Italian and Italian/American Images in the Media*. Staten Island, N.Y.: American Italian Historical Association.

Lai, James S., Wendy K. Tam-cho, Thomas P. Kim, and Okiyoshi Takeda. 2003. "Campaigns, Elections, and Elected Officials." Pp. 317–30 in Don T. Nakanishi and James S. Lai (eds.), *Asian American Politics: Law, Participation and Policy*. Lanham, Md.: Rowman & Littlefield.

Lake, David A., and Donald Rothchild. 1997. "Containing Fear: The Origins and Management of Ethnic Conflict." Pp. 97–131 in Michael E. Brown, Owen R. Coté Jr., Sean M. Lynn-Jones, and Steven E. Miller (eds.), *Nationalism and Ethnic Conflict*. Cambridge, Mass.: MIT Press.

———. 1998. "Spreading Fear: The Genesis of Transnational Ethnic Conflict." Pp. 3–32 in David A. Lake and Donald Rothchild (eds.), *The International Spread of Ethnic Conflict: Fear, Diffusion, and Escalation*. Princeton, N.J.: Princeton University Press.

Lanclos, Donna M. 2003. *At Play in Belfast*. New Brunswick, N.J.: Rutgers University Press.

Landry, Bart. 1987. *The New Black Middle Class*. Berkeley: University of California Press.

Langberg, Mark, and Reynolds Farley. 1985. "Residential Segregation of Asian Americans in 1980." *Sociology and Social Research* 70 (October):71–75.

Langer, Gary. 1989. "Polling on Prejudice: Questionable Questions." *Public Opinion* 12 (May/June):18–19, 57.

LaPiere, Richard T. 1934. "Attitudes vs. Actions." *Social Forces* 13:230–37.

Lapointe, Joe. 2004. "The Little Rink on the Prairie." *New York Times* (November 5):C1, C12.

———. 2006. "Bonding over a Mascot." *New York Times* (December 29):C1, C12.

LaRuffa, Anthony L. 1982. "Media Portrayals of Italian-Americans." *Ethnic Groups* 4:191–206.

———. 1988. *Monte Carmelo: An Italian-American Community in the Bronx*. New York: Gordon and Breach.

Lautard, Hugh, and Neil Guppy. 1990. "The Vertical Mosaic Revisited: Occupational Differentials Among Canadian Ethnic Groups." Pp. 189–208 in Peter S. Li (ed.), *Race and Ethnic Relations in Canada*. Toronto: Oxford University Press.

Laxer, James. 2003. *The Border: Canada, the U.S., and Dispatches from the 49th Parallel*. Toronto: Doubleday Canada.

Learsi, Rufus. 1954. *The Jews in America: A History*. Cleveland, Ohio: World.

Lee, Everett S. 1966. "A Theory of Migration." *Demography* 3:47–57.

Lee, Jennifer, and Frank D. Bean. 2007. "Remaking the Color Line: Immigration and America's New Racial/Ethnic Divide." *Social Forces* 86:561–87.

———. 2010. *The Diversity Paradox: Immigration and the Color Line in Twenty-First Century America*. New York: Russell Sage Foundation.

Lee, Sharon M. 2001. *Using the New Racial Categories in the 2000 Census*. Washington, D.C.: Population Reference Bureau.

Lee, Sharon M., and Barry Edmonston. 2005. *New Marriages, New Families: U.S. Racial and Hispanic Intermarriage*. Population Bulletin 60, no. 2. Washington, D.C.: Population Reference Bureau.

Lee, Sharon M., and Marilyn Fernandez. 1998. "Trends in Asian American Racial/Ethnic Intermarriage: A Comparison of 1980 and 1990 Census Data." *Sociological Perspectives* 41:323–42.

Lee, Sharon M., and Keiko Yamanaka. 1990. "Patterns of Asian American Intermarriage and Marital Assimilation." *Journal of Comparative Family Studies* 21:287–305.

Lefko-Everett, Kate. 2012. *Ticking Time Bomb or Demographic Divide? Youth and Reconciliation in South Africa. SA Reconciliation Barometer Survey: 2012 Report*. South Africa: Institute for Justice and Reconciliation.

Légaré, André. 1997. "The Government of Nunavut (1999): A Prospective Analysis." Pp. 404–31 in J. Rick Ponting (ed.), *First Nations in Canada: Perspectives on Opportunity, Empowerment, and Self-Determination*. Toronto: McGraw-Hill Ryerson.

Legendre, Camille. 1980. *French Canada in Crisis: A New Society in the Making?* London: Minority Rights Group.

Lehrer, Evelyn L. 1999. "Religion as a Determinant of Educational Attainment: An Economic Perspective." *Social Science Research* 28:358–79.

Leibbrandt, Murray, Ingrid Woolard, Arden Finn, and Jonathan Argent. 2010. "Trends in South African Income Distribution and Poverty Since the Fall of Apartheid." OECD Social, Employment and Migration Working Papers No. 101. http://www.oecd.org/eco /economicsdepartmentworkingpapers.htm

Leise, Eric. 2007. "Germany Strives to Integrate Immigrants with New Policies." Migration Policy Institute (July 9). http://www .migrationinformation.org/

Leite Lopes, José Sergio. 2000. "Class, Ethnicity, and Color in the Making of Brazilian Football." *Daedalus* 129 (Spring):239–70.

Leiter, Robert. 2000. "American Jewry Is (a) Fading or (b) Reviving. It's All in How You Read the Signs." *New York Times* (September 23):B9.

Lemon, Anthony. 1987. *Apartheid in Transition*. Aldershot, England: Gower.

Lenski, Gerhard. 1963. *The Religious Factor*. Rev. ed. Garden City, N.Y.: Doubleday.

———. 1966. *Power and Privilege: A Theory of Social Stratification*. New York: McGraw-Hill.

Leonhardt, David. 2013. "Hispanics, the New Italians." *New York Times* (April 21):SR5.

Lever, Henry. 1968. "Ethnic Preferences of White Residents in Johannesburg." *Sociology and Social Research* 52:157–73.

———. 1978. *South African Society*. Johannesburg: Jonathan Ball.

Lévesque, René. 1968. *An Option for Quebec*. Toronto: McClelland & Stewart.

Levin, Jack, and William Levin. 1982. *The Functions of Discrimination and Prejudice*. 2nd ed. New York: Harper & Row.

Levine, Bruce. 1992. *The Spirit of 1840: German Immigrants, Labor Conflict, and the Coming of the Civil War*. Urbana: University of Illinois Press.

Levine, Edward M. 1966. *The Irish and Irish Politicians*. Notre Dame, Ind.: University of Notre Dame Press.

Levine, Gene N., and Colbert Rhodes. 1981. *The Japanese American Community: A Three-Generation Study*. New York: Praeger.

Levine, Marc V. 1990. *The Reconquest of Montreal: Language Policy and Social Change in a Bilingual City*. Philadelphia: Temple University Press.

Levine, Naomi, and Martin Hochbaum (eds.). 1974. *Poor Jews: An American Awakening*. New Brunswick, N.J.: Transaction.

Levitan, Sar A., William B. Johnston, and Robert Taggart. 1975. *Still a Dream: The Changing Status of Blacks Since 1960*. Cambridge, Mass.: Harvard University Press.

Levitt, Peggy. 2007. "Dominican Republic." Pp. 399–411 in Mary C. Waters and Reed Ueda (eds.), *The New Americans: A Guide to Immigration Since 1965*. Cambridge, Mass.: Harvard University Press.

Lewin, Julius. 1963. *Politics and Law in South Africa*. London: Merlin.

Lewis, Oscar. 1965. *La Vida*. New York: Random House.

Lichter, S. Robert, and Linda Lichter. 1982. *Italian-American Characters in Television Entertainment*. West Hempstead, N.Y.: Commission for Social Justice.

Lichter, S. Robert, et al. 1987. "Prime-Time Prejudice: TV's Images of Blacks and Hispanics." *Public Opinion* 10 (July/August):13–16.

Lieberson, Stanley. 1961. "A Societal Theory of Race Relations." *American Sociological Review* 26:902–10.

———. 1980. *A Piece of the Pie: Blacks and White Immigrants Since 1880.* Berkeley: University of California Press.

Lieberson, Stanley, and Mary C. Waters. 1988. *From Many Strands: Ethnic and Racial Groups in Contemporary America.* New York: Russell Sage Foundation.

Liebman, Charles S. 1973. "American Jewry: Identity and Affiliation." Pp. 127–52 in David Sidorsky (ed.), *The Future of the Jewish Community in America.* New York: Basic Books.

Light, Ivan, and Edna Bonacich. 1988. *Immigrant Entrepreneurs: Koreans in Los Angeles 1965–1982.* Berkeley: University of California Press.

Lijphart, Arend. 1977. *Democracy in Plural Societies: A Comparative Exploration.* New Haven, Conn.: Yale University Press.

Lind, Rebecca Ann, and James A. Danowski. 1998. "The Representation of Arabs in U.S. Electronic Media." Pp. 157–68 in Yahya R. Kamalipour and Theresa Carilli (eds.), *Cultural Diversity and the U.S. Media.* Albany: State University of New York Press.

Lippmann, Walter. 1922. *Public Opinion.* New York: Macmillan.

Lipset, Seymour Martin. 1960. *Political Man.* New York: Anchor.

———. 1968. *Revolution and Counterrevolution.* New York: Basic Books.

———. 1987. "Blacks and Jews: How Much Bias?" *Public Opinion* 10 (July/August):4–5, 57–58.

———. 1990. *Continental Divide: The Values and Institutions of the United States and Canada.* New York: Routledge.

Lipset, Seymour Martin, and Earl Raab. 1995. *Jews and the New American Scene.* Cambridge, Mass.: Harvard University Press.

Lipton, Merle. 1987. "Reform: Destruction or Modernization of Apartheid?" Pp. 34–55 in Jesmond Blumenfeld (ed.), *South Africa in Crisis.* London: Croom Helm.

Livingston, John C. 1979. *Fair Game? Inequality and Affirmative Action.* San Francisco: Freeman.

Livingstone, David N., Margaret C. Keane, and Frederick W. Boal. 1998. "Space for Religion: A Belfast Case Study." *Political Geography* 17:145–70.

Lloyd, Katrina, Paula Devine, Ann-Marie Gray, and Deirdre Heenan (eds.). 2004. *Social Attitudes in Northern Ireland: The Ninth Report.* London: Pluto Press.

Lo, Lucia. 2008. "DiverCity Toronto: Canada's Premier Gateway City." Pp. 97–127 in Marie Price and Lisa Benton-Short (eds.), *Migrants to the Metropolis: The Rise of Immigrant Gateway Cities.* Syracuse, N.Y.: Syracuse University Press.

Logan, John R. 2001. *From Many Shores: Asians in Census 2000.* Albany, N.Y.: Lewis Mumford Center for Comparative Urban and Regional Research.

———. 2007. "Who Are the Other African Americans? Contemporary African and Caribbean Immigrants in the United States." Pp. 49–67 in Yoku Shaw-Taylor and Steven A. Tuch (eds.), *The Other African Americans: Contemporary African and Caribbean Immigrants in the United States.* Lanham, MD: Rowman & Littlefield.

———. 2011. "Separate and Unequal: The Neighborhood Gap for Blacks, Hispanics and Asians in Metropolitan America." US2010 Project. http://www.s4.brown.edu/us2010/

Logan, John R., and Glenn Deane. 2003. *Black Diversity in Metropolitan America.* Albany, N.Y.: Lewis Mumford Center for Comparative Urban and Regional Research.

Logan, John, Brian Stults, and Reynolds Farley. 2004. "Segregation of Minorities in the Metropolis: Two Decades of Change." *Demography* 41:1–22.

Logan, John R., and Wenquan Zhang. 2009. "Cubans and Dominicans: Is There a Latino Experience in the United States?" Pp. 191–207 in Margarita Cervantes-Rodrigues, Ramón Grosfoguel, and Eric Mielants (eds.), *Caribbean Migration to Western Europe and the United States.* Philadelphia: Temple University Press.

Logan, John R., and Brian J. Stults. 2011. "The Persistence of Segregation in the Metropolis: New Findings from the 2010 Census." Census Brief prepared for Project US2010. http://www.s4.brown.edu/us2010/

Logan John R., and Richard N. Turner. 2013. "Hispanics in the United States: Not Only Mexicans." US2010 Project. http://www.s4brown.edu/us2010

Longman, Jeré. 2011. "All-Nighters Keep Football Team Competitive During Ramadan." *New York Times* (August 11):B13.

Lopata, Helena Znaniecki. 1976. *Polish Americans: Status Competition in an Ethnic Community.* Englewood Cliffs, N.J.: Prentice Hall.

Lopez, Adalberto. 1980. "The Puerto Rican Diaspora: A Survey." Pp. 313–43 in Adalberto Lopez (ed.), *The Puerto Ricans: Their History, Culture, and Society.* Cambridge, Mass.: Schenkman.

Lopez, David E. 1996. "Language: Diversity and Assimilation." Pp. 139–63 in Roger Waldinger and Mehdi Bozorgmehr (eds.), *Ethnic Los Angeles.* New York: Russell Sage Foundation.

Lopez, Mark Hugo, and Ana Gonzalez-Barrera. 2013. "Inside the 2012 Latino Electorate." Pew Hispanic Center. http://www.pewhispanic.org/2013/06/03/inside-the-2012-latino-electorate/

Lopez, Mark Hugo, and Gretchen Livingston. 2009. "Hispanics and the Criminal Justice System." Pew Hispanic Center (April 7). http://pewhispanic.org/

Lopez, Mark Hugo, and Susan Minushkin. 2008. "2008 National Survey of Latinos: Hispanics See Their Situation in U.S. Deteriorating: Oppose Key Immigration Enforcement Measures." Pew Hispanic Center (September 18). http://pewhispanic.org/

Lopez, Mark Hugo, Rich Morin, and Paul Taylor. 2010. "Illegal Immigration Backlash Worries, Divides Latinos." Washington, D.C.: Pew Hispanic Center.

Lopez, Mark Hugo, and Gabriel Velasco. 2011. "A Demographic Portrait of Puerto Ricans, 2009." Washington, D.C.: Pew Hispanic Center.

Lopreato, Joseph. 1970. Italian Americans. New York: Random House.

Los Angeles Times. 1997. "Los Angeles Times Poll Alert," Study No. 396 (June 29):1–12.

Louie, Miriam Ching. 2000. "Immigrant Asian Women in Bay Area Garment Sweatshops: 'After Sewing, Laundry, Cleaning and Cooking, I Have No Breath Left to Sing.'" Pp. 226–42 in Timothy P. Fong and Larry H. Shinagata (eds.), Asian Americans: Experiences and Perspectives. Upper Saddle River, N.J.: Prentice Hall.

Loury, Glenn. 1998. "How to Mend Affirmative Action." The Public Interest (Spring):33–43.

Lovato, Roberto. 2007. "Why the Latino Community's Political Clout Is Rising." New America Media (September 11). http://www.alternet.org/

Lovejoy, Paul E. 1996. "The Volume of the Atlantic Slave Trade: A Synthesis." Pp. 37–64 in Patrick Manning (ed.), Slave Trades, 1500–1800: Globalization of Forced Labour. Aldershot, England: Variorum.

Lovell, Peggy A., and Charles H. Wood. 1998. "Skin Color, Racial Identity, and Life Chances in Brazil." Latin American Perspectives 25 (May):90–109.

Lovgren, Stefan. 2004. "Rwanda, Ten Years Later: Justice Is Elusive, Despite Peace." National Geographic News (April 6). http://www.nationalgeographic.com/

Luebke, Frederick C. 1983. "A Prelude to Conflict: The German Ethnic Group in Brazilian Society, 1890–1917." Ethnic and Racial Studies 6:1–17.

———. 1987. Germans in Brazil: A Comparative History of Cultural Conflict during World War I. Baton Rouge: Louisiana State University Press.

———. (ed.). 1990. Germans in the New World: Essays in the History of Immigration. Urbana: University of Illinois Press.

Lurie, Nancy Oestreich. 1991. "The American Indian: Historical Background." Pp. 132–46 in Norman R. Yetman (ed.), Majority and Minority: The Dynamics of Race and Ethnicity in American Life. 5th ed. Boston: Allyn & Bacon.

Lyman, Stanford. 1968a. "Contrasts in the Community Organizations of Chinese and Japanese in North America." Canadian Review of Sociology and Anthropology 5:51–67.

———. 1968b. "The Race Relations Cycle of Robert Park." Pacific Sociological Review 11:16–22.

———. 1970. "Strangers in the Cities: The Chinese on the Urban Frontier." Pp. 61–100 in Charles Wollenberg (ed.), Ethnic Conflict in California History. Los Angeles: Tinnon-Brown.

———. 1974. Chinese Americans. New York: Random House.

Lynn, Richard, and Tatu Vanhanen. 2002. IQ and the Wealth of Nations. Westport, Conn.: Praeger.

MacDonald, J. Fred. 1983. Blacks and White TV: Afro-Americans in Television Since 1948. Chicago: Nelson-Hall.

Mac Donald, Heather. 2003. "Straight Talk on Homeland Security." City Journal (Summer). http://www.city-journal.org/

MacFarquhar, Neil. 2007. "In Arab Capital of U.S., Ethnic Divide Remains." New York Times (January 23):A12.

Macías, Ysidro. 1972. "The Chicano Movement." Pp. 137–43 in Edward Simmen (ed.), Pain and Promise: The Chicano Today. New York: New American Library.

Mack, Raymond W. 1963. Race, Class, and Power. New York: American.

Mackie, Marlene. 1973. "Arriving at 'Truth' by Definition: The Case of Stereotype Inaccuracy." Social Problems 20:431–47.

———. 1980. "Ethnic Stereotypes and Prejudice: Alberta Indians, Hutterites and Ukrainians." Pp. 233–46 in Jay E. Goldstein and Rita M. Bienvenue (eds.), Ethnicity and Ethnic Relations in Canada. Toronto: Butterworths.

MacLennan, Hugh. 1945. Two Solitudes. New York: Duell, Sloan & Pearce.

Madon, Stephanie, Max Guyll, Kathy Abonfadel, Eulices Montiel, Alison Smith, Polly Palumbo, and Lee Jussim. 2001. "Ethnic and National Stereotypes: The Princeton Trilogy Revisited and Revised." Personality and Social Psychology Bulletin 27:996–1010.

Makabe, Tomoko. 1981. "The Theory of the Split Labor Market: A Comparison of the Japanese Experience in Brazil and Canada." Social Forces 59:786–809.

———. 1999. "Ethnic Hegemony: The Japanese Brazilians in Agriculture, 1908–1968." *Ethnic and Racial Studies* 22:702–23.

Mallaby, Sebastian. 1992. *After Apartheid: The Future of South Africa*. New York: Times Books.

Mamdani, Mahmood. 2001. *When Victims Become Killers: Colonialism, Nativism, and the Genocide in Rwanda*. Princeton, N.J.: Princeton University Press.

Mangiafico, Luciano. 1988. *Contemporary American Immigrants: Patterns of Filipino, Korean, and Chinese Settlement in the United States*. New York: Praeger.

Mann, Michael. 2005. *The Dark Side of Democracy: Explaining Ethnic Cleansing*. Cambridge, England: Cambridge University Press.

Manning, Jennifer E. 2013. *Membership of the 113th Congress: A Profile*. Washington, D.C.: Congressional Research Service.

Manning, Patrick. 1992. "The Slave Trade: The Formal Demography of a Global System." Pp. 117–41 in Joseph E. Inikori and Stanley L. Engerman (eds.), *The Atlantic Slave Trade: Effects on Economies, Societies, and Peoples in Africa, the Americas, and Europe*. Durham, N.C.: Duke University Press.

Marden, Charles F., and Gladys Meyer. 1978. *Minorities in American Society*. 5th ed. New York: Van Nostrand.

Marger, Martin N. 1974. *The Force of Ethnicity: A Study of Urban Elites*. Journal of University Studies, Ethnic Monograph Series. Detroit, Mich.: Wayne State University.

———. 1979. "A Reexamination of Gordon's Ethclass." *Sociological Focus* 11:21–32.

———. 1987. *Elites and Masses: An Introduction to Political Sociology*. 2nd ed. Belmont, Calif.: Wadsworth.

———. 1989. "Factors of Structural Pluralism in Multiethnic Societies: A Comparative Case Study." *International Journal of Group Tensions* 19:52–68.

Margolis, Maxine L. 1998. *An Invisible Minority: Brazilians in New York City*. Boston: Allyn & Bacon.

Marks, Jonathan. 1995. *Human Biodiversity: Genes, Race, and History*. New York: Aldine de Gruyter.

Marrero, Pilar. 2004. "Spain, Here We Come." *AlterNet* (December 10). http://www.alternet.org/

Marrow, Helen B. 2007. "South America: Ecuador, Peru, Brazil, Argentina, Venezuela." Pp. 593–611 in Mary C. Waters and Reed Ueda (eds.), *The New Americans: A Guide to Immigration Since 1965*. Cambridge, Mass.: Harvard University Press.

Marston, Wilfred G., and Thomas L. Van Valey. 1979. "The Role of Residential Segregation in the Assimilation Process." *Annals of the American Academy of Political and Social Science* 441:13–25.

Martin, Patrick. 2011. "Dissecting Syria." *Globe and Mail* (June 15):A10–A11.

Martin, Philip, and Elizabeth Midgley. 2010. "Immigration in America 2010." *Population Bulletin Update (June)*. Washington, D.C.: Population Reference Bureau.

Martin, Philip, and Jonas Widgren. 2002. *International Migration: Facing the Challenge*. Population Bulletin 57, no. 1. Washington, D.C.: Population Reference Bureau.

Martinelli, Phyllis Cancilla. 1989. *Ethnicity in the Sunbelt: Italian American Migrants in Scottsdale*, Arizona. New York: AMS Press.

Martinez, Thomas M. 1972. "Advertising and Racism: The Case of the Mexican American." Pp. 94–105 in Edward Simmen (ed.), *Pain and Promise: The Chicano Today*. New York: New American Library.

Martins, Sérgio da Silva, Carolos Alberto Medeiros, and Elisa Larkin Nascimento. 2004. "Paving Paradise: The Road from 'Racial Democracy' to Affirmative Action in Brazil." *Journal of Black Studies* 34:787–816.

Martire, Greg, and Ruth Clark. 1982. "Anti-Semitism in America." *Public Opinion* (April/May):56–59.

Marx, Anthony W. 1998. *Making Race and Nation: A Comparison of South Africa, the United States, and Brazil*. Cambridge, England: Cambridge University Press.

Maske, Mark. 2013. "Redskins Name Change Would Have to Pass Muster with NFL, Sponsors." *Washington Post* (February 24). http://articles.Washingtonpost.com/2013-02-24/sports/37276036_1_nfl-owners-team-names-washington-redskins

Mason, Philip. 1970. *Patterns of Dominance*. New York: Oxford University Press.

Massarik, Fred, and Alvin Chenkin. 1973. "United States National Jewish Population Study: A First Report." Pp. 264–306 in *American Jewish Year Book 1973*. New York: American Jewish Committee.

Massey, Douglas S. 1979. "Effects of Socioeconomic Factors on the Residential Segregation of Blacks and Spanish Americans in United States Urbanized Areas." *American Sociological Review* 44:1015–22.

———. 1981a. "Dimensions of the New Immigration to the United States and the Prospects for Assimilation." *Annual Review of Sociology* 7:57–85.

———. 1981b. "Hispanic Residential Segregation: A Comparison of Mexicans, Cubans, and Puerto Ricans." *Sociology and Social Research* 65:311–22.

———. 1999. "Why Does Immigration Occur? A Theoretical Synthesis." Pp. 34–52 in Charles Hirschman, Philip Kasinitz, and Josh DeWind (eds.), *The Handbook of International*

Migration: The American Experience. New York: Russell Sage Foundation.

———. 2001. "Residential Segregation and Neighborhood Conditions in U.S. Metropolitan Areas." Pp. 391–434 in Neil J. Smelser, William Julius Wilson, and Faith Mitchell (eds.), *America Becoming: Racial Trends and Their Consequences.* Vol. 1. Washington, D.C.: National Academy Press.

———. 2004. "Segregation and Stratification: A Biosocial Perspective." *Du Bois Review* 1:7–25.

———. 2008. "Assimilation in a New Geography." Pp. 343–53 in Douglas S. Massey (ed.), *New Faces in New Places: The Changing Geography of American Immigration.* New York: Russell Sage.

Massey, Douglas S., Joaquin Arango, Graeme Hugo, Ali Kouaouci, Adela Pellegrino, and J. Edward Taylor. 1993. "Theories of International Migration: A Review and Appraisal." *Population and Development Review* 19:431–66.

Massey, Douglas S., and Brooks Bitterman. 1985. "Explaining the Paradox of Puerto Rican Segregation." *Social Forces* 64:306–31.

Massey, Douglas S., Camille Z. Charles, Garvey F. Lundy, and Mary J. Fischer. 2003. *The Source of the River: The Social Origins of Freshmen at America's Selective Colleges and Universities.* Princeton, N.J.: Princeton University Press.

Massey, Douglas S., and Nancy A. Denton. 1987. "Trends in the Residential Segregation of Blacks, Hispanics, and Asians: 1970–1980." *American Sociological Review* 52:802–25.

———. 1989a. "Hypersegregation in U.S. Metropolitan Areas: Black and Hispanic Segregation Along Five Dimensions." *Demography* 26:373–91.

———. 1989b. "Residential Segregation of Mexicans, Puerto Ricans, and Cubans in Selected U.S. Metropolitan Areas." *Sociology and Social Research* 73:73–83.

———. 1993. *American Apartheid: Segregation and the Making of the Underclass.* Cambridge, Mass.: Harvard University Press.

Mathabane, Mark. 1994. "'Like the Second Coming.'" *Newsweek* (May 9):38.

Mather, Mark. 2007. "Education and Occupation Separates Two Kinds of Immigrants in the United States." Population Reference Bureau (September). http://www.prb.org/

———. 2010. "U.S. Children in Single-Mother Families." *PRB Data Brief.* Washington, D.C.: Population Reference Bureau.

Mather, Mark, and Linda A. Jacobsen. 2010. "Hard Times for Latino Men in U.S." Population Reference Bureau (February). http://www.prb.org/

Mattes, Robert. 2002. "South Africa: Democracy without the People?" *Journal of Democracy* 13 (January):22–36.

McAneny, Leslie, and Lydia Saad. 1994. "America's Public Schools: Still Separate? Still Unequal?" *Gallup Poll Monthly* (May):23–27.

McCaffrey, Lawrence J. 1997. *The Irish Catholic Diaspora in America.* Washington, D.C.: Catholic University of America Press.

McClatchy, Valentine Stuart. 1978 [1919–25]. *Four Anti-Japanese Pamphlets.* New York: Arno Press.

McClinton, Fareed, and Tukufu Zuberi. 2006. "Racial Residential Segregation in South Africa and the United States." Paper presented at 2006 Annual Meeting of the Population Association of America.

McFalls, Joseph A., Jr. 2003. *Population: A Lively Introduction.* 4th ed. Washington, D.C.: Population Reference Bureau.

McGarry, John, and Brendan O'Leary. 1995a. *Explaining Northern Ireland: Broken Images.* Oxford: Blackwell.

———. 1995b. "Five Fallacies: Northern Ireland and the Liabilities of Liberalism." *Ethnic and Racial Studies* 18:837–61.

McGillivray, Liz. 2006. "South Africa Ten Years After Apartheid: Making Progress, but Still Struggling with Racial Integration." ORC Worldwide. http://www.orcinc.com/readroom/southafrica.html

McGreal, Chris. 2013. "Rwanda Genocide 20 Years On: 'We Live with Those Who Killed Our Families. We Are Told They're Sorry. But Are They?'" *The Observer* (May 18). http://www.theguardian.com/world/2013/may/12/rwanda-genocide-20-years-on

McKee, James B. 1993. *Sociology and the Race Problem: The Failure of a Perspective.* Urbana: University of Illinois Press.

McKenna, Barrie. 2006. "From Hard Knocks to the Hard Rock." *The Globe and Mail* (December 8):B1, B10.

McKenna, Marian C. 1969. "The Melting Pot: Comparative Observations in the United States and Canada." *Sociology and Social Research* 53:433–47.

McKernan, Signe-Mary, Caroline Ratcliffe, Eugene Steuerle, and Sisi Zhang. 2013. "Less Than Equal: Racial Disparities in Wealth Accumulation." Washington, D.C.: Urban Institute.

McLarin, Kimberly J. 1995. "To Preserve Afrikaners' Language, Mixed-Race South Africans Join Fray." *New York Times* (June 28):A4.

McLemore, S. Dale. 1973. "The Origins of Mexican-American Subordination in Texas." *Social Science Quarterly* 53:656–79.

McRae, Kenneth D. 1974. *Consociational Democracy: Political Accommodation in Segmented Societies.* Toronto: McClelland & Stewart.

———. 1983. *Conflict and Compromise in Multilingual Societies: Switzerland*. Waterloo, Ontario: Wilfrid Laurier University Press.

———. 1986. *Conflict and Compromise in Multilingual Societies. Vol. 2: Belgium*. Waterloo, Ontario: Wilfrid Laurier University Press.

McRoberts, Kenneth. 1988. *Quebec: Social Change and Political Crisis*. 3rd ed. Toronto: McClelland & Stewart.

McWhorter, John H. 2000. *Losing the Race: Self-Sabotage in Black America*. New York: Free Press.

McWilliams, Carey. 1948. *A Mask for Privilege: Anti-Semitism in America*. Boston: Little, Brown.

———. 1968 [1948]. *North from Mexico: The Spanish-Speaking People of the United States*. New York: Greenwood.

Meagher, Timothy J. 2001. *Inventing Irish America: Generation, Class, and Ethnic Identity in a New England City, 1880–1928*. Notre Dame, Ind.: University of Notre Dame Press.

Mearsheimer, John J., and Stephen Van Evera. 1995. "When Peace Means War." *The New Republic* (December 18):16–21.

Mechanic, David. 1978. "Apartheid Medicine." Pp. 127–38 in Ian Robertson and Phillip Whitten (eds.), *Race and Politics in South Africa*. New Brunswick, N.J.: Transaction.

Meissner, Doris. 2010. "5 Myths About Immigration." *Washington Post* (May 2). http://www.washingtonpost.com/wpdyn/content/article/2010/04/30/AR2010043001106.html

Meredith, Martin. 1987. "The Black Opposition." Pp. 77–89 in Jesmond Blumenfeld (ed.), *South Africa in Crisis*. London: Croom Helm.

Merton, Robert K. 1949. "Discrimination and the American Creed." Pp. 99–126 in R. H. MacIver (ed.), *Discrimination and National Welfare*. New York: Harper & Row.

———. 1968. *Social Theory and Social Structure*. New York: Free Press.

Metzger, L. Paul. 1971. "American Sociology and Black Assimilation: Conflicting Perspectives." *American Journal of Sociology* 76:627–47.

Miami-Dade County. 2006. *FY2005–06 Children and Families Budget and Resource Allocation Report*. http://www.miamidade.gov/budget/FY2005-06/PDF/ch_condition.pdf

Miami Herald. 1983. "Slave's Descendant Ruled Legally Black in Louisiana." (May 19):1A, 12A.

Michaels, Walter Benn. 2006. *The Trouble with Diversity: How We Learned to Love Identity and Ignore Inequality*. New York: Holt.

Migration Policy Institute. 2006. *Immigration and America's Future: A New Chapter. Report of the Independent Task Force on Immigration and America's Future*. Washington, D.C.: Migration Policy Institute.

Migration Information Source. 2011. "Highly Skilled Migrants Seek New Destinations as Global Growth Shifts to Emerging Economies." (December 1). Washington, D.C.: Migration Policy Institute.

Milan, Anne, Hélène Maheux, and Tina Chui. 2010. "A Portrait of Couples in Mixed Unions." Ottawa: Statistics Canada. http://www.statcan.gc.ca/pub/11-008-x/2010001/article/11143-eng.htm

Milbrath, Lester. 1982. *Political Participation*. 2nd ed. Washington, D.C.: University Press of America.

Miller, James Nathan. 1984. "Ronald Reagan and the Techniques of Deception." *Atlantic Monthly* (February):62–68.

Miller, Kerby A. 1985. *Emigrants and Exiles: Ireland and the Irish Exodus to North America*. New York: Oxford University Press.

Mills, C. Wright. 1956. *The Power Elite*. New York: Oxford University Press.

Min, Pyong Gap. 1988. *Ethnic Business Enterprise: Korean Small Business in Atlanta*. New York: Center for Migration Studies.

———. 1991. "Cultural and Economic Boundaries of Korean Ethnicity: A Comparative Analysis." Paper presented at the annual meeting of the American Sociological Association, Cincinnati, Ohio.

———. 1995. "Korean Americans." Pp. 199–231 in Pyong Gap Min (ed.), *Asian Americans: Contemporary Trends and Issues*. Thousand Oaks, Calif.: Sage.

———. 1999. "A Comparison of Post-1965 and Turn-of-the-Century Immigrants in Intergenerational Mobility and Cultural Transmission." *Journal of American Ethnic History* 18 (Spring):65–94.

———. 2006. "Major Issues Related to Asian American Experiences." Pp. 80–107 in Pyong Gap Min (ed.), *Asian Americans: Contemporary Trends and Issues*. 2nd ed. Thousand Oaks, Calif.: Pine Forge.

———. 2007. "Korea." Pp. 491–503 in Mary C. Waters and Reed Ueda (eds.), *The New Americans: A Guide to Immigration Since 1965*. Cambridge, Mass.: Harvard University Press.

Miner, Horace. 1939. *St. Denis: A French Canadian Parish*. Chicago: University of Chicago Press.

Minority Rights Group International. 2010. *State of the World's Minorities and Indigenous Peoples 2010*. London: Minority Rights Group International.

Mironko, Charles K. 2006. "Ibitero: Means and Motive in the Rwandan Genocide." Pp. 163–89 in Susan E. Cook (ed.), *Genocide in Cambodia and Rwanda: New Perspectives*. New Brunswick, N.J.: Transaction.

Mishel, Lawrence, Jared Bernstein, and Heidi Shierholz. 2009. *The State of Working*

America 2008/2009. Washington, D.C.: Economic Policy Institute.

Mitchell, Claire. 2008. "Religious Change and Persistence." Pp. 135–155 in *Northern Ireland After the Troubles: A Society in Transition*. Manchester: Manchester University Press.

Mitrano, John R., and James G. Mitrano. 1996. "Mammas, Papas, Traditions, and Secrets: The Marketing of Italian Americans." Pp. 71–84 in Mary Jo Bona and Anthony Julian Tamburri (eds.), *Through the Looking Glass: Italian and Italian/American Images in the Media*. Staten Island, N.Y.: American Italian Historical Association.

Mittelbach, Frank G., and Joan W. Moore. 1968. "Ethnic Endogamy: The Case of the Mexican Americans." *American Journal of Sociology* 74:50–62.

Miyares, Ines M. 2006. "Central Americans: Legal Status and Settledness." Pp. 175–90 in Ines M. Miyares and Christoper A. Airriess (eds.), *Contemporary Ethnic Geographies in America*. Lanham, Md.: Rowman & Littlefield.

Model, Suzanne. 2008. *West Indian Immigrants: A Black Success Story?* New York: Russell Sage.

Molina, Fernando. 2010. "The Historical Dynamics of Ethnic Conflicts: Confrontational Nationalisms, Democracy and the Basques in Contemporary Spain." *Nations and Nationalism* 16:240–60.

Moll, Luis C., and Richard Ruiz. 2002. "The Schooling of Latino Children." Pp. 362–72 in Marcelo M. Suárez-Orozco and Mariela M. Páez (eds.), *Latinos: Remaking America*. Berkeley: University of California Press.

Monger, Randall. 2010. "U.S. Legal Permanent Residents: 2009." *Annual Flow Report*, Office of Immigration Statistics, U.S. Department of Homeland Security. Washington, D.C.: U.S. Government Printing Office.

Montagu, Ashley. 1963. *Race, Science, and Humanity*. Princeton, N.J.: Van Nostrand.

———. 1972. *Statement on Race*. 3rd ed. New York: Oxford University Press.

———. (ed.). 1975. *Race and IQ*. New York: Oxford University Press.

Montero, Darrel. 1980. *Japanese Americans: Changing Patterns of Ethnic Affiliation over Three Generations*. Boulder, Colo.: Westview Press.

Montonaro, Domenico. 2013. "NBC News-Wall Street Journal Poll: Affirmative Action Support at Historic Low."(June 11). http://firstread .nbcnews.com/

Moodley, Kogila A. 1980. "Structural Inequality and Minority Anxiety: Responses of Middle Groups in South Africa." Pp. 217–35 in Robert M. Price and Carl G. Rotberg (eds.), *The Apartheid Regime: Political Power and Racial Domination*. Berkeley: Institute of International Studies, University of California.

Moore, David W. 2003. "Public: Only Merit Should Count in College Admissions." *The Gallup Poll Tuesday Briefing: Government and Public Policy* (June 24):43–44.

———. 2006. "Immigration." *The Gallup Poll Briefing* (April):29–30.

Moore, Joan W. 1970. "Colonialism: The Case of the Mexican-Americans." *Social Problems* 17:463–72.

Moore, Joan W., with Harry Pachon. 1976. *Mexican Americans*. 2nd ed. Englewood Cliffs, N.J.: Prentice Hall.

Moore, Michael. 1987. "The Man Who Killed Vincent Chin." *Detroit Magazine, Detroit Free Press* (August 30):12–20.

Moquin, Wayne, and Charles Van Doren (eds.). 1974. *A Documentary History of the Italian Americans*. New York: Praeger.

Morales, Armando. 1972. *Ando Sangrando*. Fair Lawn, N.J.: Burdick.

Morales, Lymari. 2009. "Americans Return to Tougher Immigration Stance." *Gallup Poll* (August 5). http://www.gallup.com/

Morgan, Curtis. 1995. "Across the Racial Divide." *Washington Post National Weekly Edition* (October 16–22):6–10.

———. 1998a. "Language Barriers Frustrate Residents." *Miami Herald* (June 22). http://www.herald.com

———. 1998b. "Mixed Messages on Campus Diversity." *Washington Post National Weekly Edition* (October 26):34.

Morin, Rich. 2009. "What Divides America?: Immigration and Income—Not Race—Are Seen as Primary Sources of Social Conflict." Pew Research Center (September 24). http://pewresearch.org/

———. 2013. "Sign of Things to Come? Integration Without Blacks in New York City Neighborhoods." Washington, D.C.: Pew Research Center (August 7).

Morin, Richard, and Michael H. Cottman. 2001. "The Invisible Slap." *Washington Post National Weekly Edition* (July 2–8):6–7.

Mörner, Magnus. 1985. *Adventurers and Proletarians: The Story of Migrants in Latin America*. Pittsburgh: University of Pittsburgh Press.

Morning, Ann. 2005. "Race." *Contexts* 4 (Fall):44–45.

Morse, Richard M. 1953. "The Negro in São Paulo, Brazil." *Journal of Negro History* 38:290–306.

———. 1958. *From Community to Metropolis: A Biography of São Paulo, Brazil*. Gainesville: University of Florida Press.

Motel, Seth, and Eileen Patten. 2012. "Characteristics of the 60 Largest Metropolitan Areas by Hispanic Population." Washington, D.C.: Pew Hispanic Center.

Moxon-Browne, Edward. 1983. *Nation, Class and Creed in Northern Ireland*. Aldershot, England: Gower.

Moynihan, Daniel P. 1965. *The Negro Family: The Case for National Action.* Washington, D.C.: U.S. Government Printing Office.

Myers, Dowell, Stephen Levy, and John Pitkin. 2013. "The Contributions of Immigrants and Their Children to the American Workforce and Jobs of the Future." Washington, D.C.: Center for American Progress.

Muenz, Rainer. 2006. "Europe: Population and Migration in 2005." Migration Policy Institute (June). http://www.migrationinformation.org/

Mulder, C. P. 1972. "The Rationale of Separate Development." Pp. 48–63 in N. J. Rhoodie (ed.), *South African Dialogue.* Philadelphia: Westminister.

Mullen, Brian, and Tirza Leader. 2005. "Linguistic Factors: Antilocutions, Ethnonyms, Ethnophaulisms, and Other Varieties of Hate Speech." Pp. 192–207 in John F. Dovidio, Peter Glick, and Laurie A. Budman (eds.), *On the Nature of Prejudice: Fifty Years After Allport.* Malden, Mass.: Blackwell.

Muller, Jerry Z. 2008. "Us and Them: The Enduring Power of Ethnic Nationalism." *Foreign Affairs* 87 (March/April):18–35.

Muller, Thomas. 1993. *Immigrants and the American City.* New York: New York University Press.

Mumford Center for Comparative Urban and Regional Research. 2001a. *Census 2000: Sortable List of Dissimilarity Scores.* Albany: Mumford Center, State University of New York at Albany. http://www.albany.edu/

———. 2001b. *Ethnic Diversity Grows, Neighborhood Integration Lags Behind.* Albany, N.Y.: Lewis Mumford Center.

———. 2001c. *The New Ethnic Enclaves in America's Suburbs.* Albany: Mumford Center, State University of New York at Albany. http://www.albany.edu/

Murguía, Edward. 1975. *Assimilation, Colonialism and the Mexican American People.* Austin: Center for Mexican American Studies, University of Texas.

Murguía, Edward, and W. Parker Frisbie. 1977. "Trends in Mexican American Intermarriage: Recent Findings in Perspective." *Social Science Quarterly* 58:374–89.

Murray, Charles. 2008. "Poverty and Marriage, Income Inequality and Brains. *Pathways* (Winter):21–24.

Murray, Dominic. 1995. "Culture, Religion and Violence in Northern Ireland." Pp. 215–29 in Seamus Dunn (ed.), *Facets of the Conflict in Northern Ireland.* Basingstoke, England: Macmillan.

Murray, Mark. 2010. "On Immigration, Racial Divide Runs Deep." NBC/MSNBC/Telemundo. http://www.msnbc.msn/com/

Murray, Martin. 1987. *South Africa: Time of Agony, Time of Destiny.* London: Verso.

Mydans, Seth. 1992a. "New Unease for Japanese-Americans." *New York Times* (March 4):7.

———. 1992b. "A Target of Rioters, Koreatown Is Bitter, Armed and Determined." *New York Times* (May 3):1, 16.

———. 1998. "Indonesia Turns Its Chinese into Scapegoats." *New York Times* (February 2): A6.

Myers, Dowell, and John Pitkin. 2010. "Assimilation Today: New Evidence Shows the Latest Immigrants to America Are Following in Our History's Footsteps." Washington, D.C.: Center for American Progress.

Myrdal, Gunnar. 1944. *An American Dilemma: The Negro Problem and Modern Democracy.* New York: Harper & Row.

Naber, Nadine. 2000. "Ambiguous Insiders: An Investigation of Arab American Invisibility." *Ethnic and Racial Studies* 23:37–61.

———. 2012. *Arab America: Gender, Cultural Politics, and Activism.* New York: New York University Press.

Nackerud, Larry, Alyson Spring, Christopher Larrison, and Alicia Issac. 1999. "The End of the Cuban Contradiction in U.S. Refugee Policy." *International Migration Review* 33:176–92.

Nacos, Brigitte L., and Oscar Torres-Reyna. 2007. *Fueling Our Fears: Stereotyping, Media Coverage, and Public Opinion of Muslim Americans.* Lanham, Md.: Rowman & Littlefield.

Naff, Alixa. 1983. "Arabs in America: A Historical Overview." Pp. 9–29 in Sameer Y. Abraham and Nabeel Abraham (eds.), *Arabs in the New World: Studies on Arab-American Communities.* Detroit: Wayne State University Press.

———. 1994. "The Early Arab Immigrant Experience." Pp. 23–36 in Ernest McCarus (ed.), *The Development of Arab-American Identity.* Ann Arbor: University of Michigan Press.

Nagel, Joane. 1995. "American Indian Ethnic Renewal: Politics and the Resurgence of Identity." *American Sociological Review* 60:947–65.

Nakanishi, Don T. 2001. "Political Trends and Electoral Issues of the Asian Pacific American Population." Pp. 170–99 in Neil J. Smelser, William Julius Wilson, and Faith Mitchell (eds.), *America Becoming: Racial Trends and Their Consequences.* Vol. I. Washington, D.C.: National Academy Press.

Nakhaie, M. Reza. 1995. "Ownership and Management Position of Canadian Ethnic Groups in 1973 and 1989." *Canadian Journal of Sociology* 20:167–92.

Nanos, Nik. 2010a. "Canadians Strongly Support Immigration, but Don't Want Current Levels

Increased." *Options Politiques* (July–August): 10–14.

———. 2010b. "Canadians Strongly Support Immigration, but Don't Want Current Levels Increased." *Policy Options* (August). Montreal: Institute for Research on Public Policy. http://archive.irpp.org/po/archive /jul10/nanos.pdf

Nascimento, Abdias do, and Elisa Larkin Nascimento. 2001. "Dance of Deception: A Reading of Race Relations in Brazil." Pp. 105–56 in Charles V. Hamilton, Lynn Huntley, Neville Alexander, Antonio Sérgio Alfredo Guimarães, and Wilmot James (eds.), *Race and Inequality in Brazil, South Africa, and the United States.* Boulder, Colo.: Lynne Rienner.

Nascimento, Elisa Larkin. 2007. *The Sorcery of Color: Identity, Race, and Gender in Brazil.* Philadelphia: Temple University Press.

Nash, Manning. 1962. "Race and the Ideology of Race." *Current Anthropology* 3:285–88.

National Advisory Commission on Civil Disorders. 1968. *Report of the National Advisory Commission on Civil Disorders.* New York: Bantam.

National Association of Latino Elected and Appointed Officials. 2011. *2011 Directory of Latino Elected Officials.* http://www.naleo .org/directory.html

National Conference. 1995. *Taking America's Pulse: A Summary Report of the National Conference Survey on Inter-Group Relations.* New York: National Conference.

National Conference of Christians and Jews. 1978. *A Study of Attitudes toward Racial and Religious Minorities and Toward Women* (conducted by Louis Harris and Associates, Inc.). New York: National Conference of Christians and Jews.

National Indian Gaming Commission. 2012. "Growth in Gaming Revenues." http://www .Nigc.gov/Gaming_Revenue_Reports.aspx

National Urban League. 2001. *The State of Black America 2001.* New York: National Urban League.

Nattrass, Nicoli, and Jeremy Seekings. 2001. "'Two Nations'? Race and Economic Inequality in South Africa Today." *Daedalus* 130 (Winter):45–70.

Navarro, Mireya. 1999. "Miami's Generations of Exiles Side by Side, Yet Worlds Apart." *New York Times* (February 11):A1, A25.

Nazareno, Analisa. 2000. "How Immigrants Reshaped Schools." *Miami Herald* (January 2). http://www.herald.com

Neal, Derek. 2006. "Why Has Black-White Skill Convergence Stopped?" Pp. 511–76 in Eric Hanushek and Finis Welch (eds.), *Handbook of the Economics of Education.* Vol. 1. Amsterdam: Elsevier.

Nelli, Humbert S. 1970. *The Italians in Chicago, 1880–1930.* New York: Oxford University Press.

Neusner, Jacob. 2003. "Immigration and Religion in America: The Experience of Judaism." Pp. 105–16 in Yvonne Yazbeck Haddad, Jane I. Smith, and John L. Esposito (eds.), *Religion and Immigration: Christian, Jewish, and Muslim Experiences in the United States.* Walnut Creek, Calif.: AltaMira Press.

New America Media. 2007. *Deep Divisions, Shared Destiny: A Poll of African Americans, Hispanics and Asian Americans on Race Relations Sponsored by New America Media and Nine Founding Ethnic Media Partners.* San Francisco: New America Media.

Newcomer, Mabel. 1955. *The Big Business Executive.* New York: Columbia University Press.

Newman, Peter C. 1975. *The Canadian Establishment.* Vol. 1. Toronto: McClelland & Stewart.

———. 1979. *The Canadian Establishment.* Vol. 2. Toronto: McClelland & Stewart.

———. 1995. "A Country of Many Cultures and Flavors." *Maclean's* (July 24):34.

Newman, William M. 1973. *American Pluralism: A Study of Minority Groups and Social Theory.* New York: Harper & Row.

Newport, Frank. 2013a. "Gulf Grows in Black-White Views of U.S. Justice System Bias." *Gallup Poll* (July 22). http://www.gallup.com/

———. 2013b. "In U.S., 87% Approve of Black-White Marriage, vs. 4% in 1958." *Gallup Poll* (July 25). http://www.gallup.com/

Newport, Frank, David W. Moore, and Lydia Saad. 1999. "Long-Term Gallup Poll Trends: A Portrait of American Public Opinion Through the Century." *Poll Releases* (December 20):The Gallup Organization.

New York Times. 1987. "Black Woman Who Lost Job Is Asked Back by Pharmacist." (August 15):A7.

NIAF (National Italian American Foundation). 2001. *National Survey: American Teen-Agers and Stereotyping.* Washington, D.C.: National Italian American Foundation.

———. 2004. *Italian American Congressional Delegation Summary of Italian Ancestry (Year 2003–2005).* Washington, D.C.: National Italian American Foundation.

———. 2012. "The Italian American Congressional Delegation." Washington, D.C.: National Italian American Foundation. http:// www.niaf.org/publicpolicy/italian-american-congressional-delegation.asp

Niebuhr, Gustav. 1997. "A Shift to Rigorous Tradition Gains Influence in Judaism." *New York Times* (May 1):A1, A14.

Nieves, Evelyn. 2007a. "Indian Reservation Reeling in Wave of Youth Suicides and Attempts." *New York Times* (June 9):A9.

———. 2007b. "Putting to a Vote the Question 'Who Is Cherokee?'" *New York Times* (March 3):A9.

Nimer, Mohamed. 2011. "Islamophobia and Anti-Americanism: Measurements, Dynamics, and Consequences." Pp. 77–92 in John L. Esposito and Ibrahim Kalin (eds.), *Islamophobia: The Challenge of Pluralism in the 21st Century*. New York: Oxford University Press.

Nisbett, Richard E. 2007. "All Brains Are the Same Color." *New York Times* (December 9):4.

———. 2009. *Intelligence and How to Get It: Why Schools and Culture Count*. New York: Norton.

Noble, Kenneth B. 1994. "Fearing Domination by Blacks, Indians of South Africa Switch Loyalties." *New York Times* (April 22):A8.

———. 1995. "Attacks Against Asian-Americans Are Rising." *New York Times* (December 13): A14.

Nobles, Melissa. 2002. "Lessons from Brazil: The Ideational and Political Dimensions of Multi-raciality." Pp. 300–17 in Joel Perlmann and Mary C. Waters (eds.), *The New Race Question: How the Census Counts Multiracial Individuals*. New York: Russell Sage Foundation.

Noel, Donald L. 1968. "A Theory of the Origin of Ethnic Stratification." *Social Problems* 16:157–72.

Nogueira, Oracy. 1959. "Skin Color and Social Class." In Vera Rubin (ed.), *Plantation Systems of the New World*. Pan American Union Social Science Monograph No. 7. Washington, D.C.: Pan American Union.

Nordheimer, Jon. 1987. "Black Cubans: Apart in Two Worlds." *New York Times* (December 2):26.

Nordland, Rod. 1993. "'Let's Kill the Muslims!'" *Newsweek* (November 8):48–51.

North, Robert D. 1965. "The Intelligence of the American Negro." Pp. 334–40 in Arnold M. Rose and Caroline B. Rose (eds.), *Minority Problems*. New York: Harper & Row.

Nostrand, Richard L. 1973. "Mexican American and Chicano: Emerging Terms for a People Coming of Age." *Pacific Historical Review* 62:389–406.

Norton, Michael I. and Samuel R. Sommers. 2011. "Whites See Racism as a Zero-Sum Game That They Are Now Losing." *Perspectives on Psychological Science* 6:215–18.

Novak, Michael. 1972. *The Rise of the Unmeltable Ethnics*. New York: Macmillan.

O'Brien, Eileen. 2008. *The Racial Middle: Latinos and Asian Americans Living Beyond the Racial Divide*. New York: New York University Press.

Oberschall, Anthony. 2007. *Conflict and Peace Building in Divided Societies: Responses to Ethnic Violence*. New York: Routledge.

O Connor, Fionnuala. 1993. *In Search of a State: Catholics in Northern Ireland*. Belfast: Blackstaff Press.

Oezcan, Veysel. 2004. "Germany: Immigration in Transition." Migration Policy Institute (July). http://www.migrationinformation.org/

Ogmundson, R. 1990. "Perspectives on the Class and Ethnic Origins of Canadian Elites: A Methodological Critique of the Porter/Clement/Olsen Tradition." *Canadian Journal of Sociology* 15:165–77.

Ogmundson, R., and L. Fatels. 1994. "Are the Brits in Decline? A Note on Trends in the Ethnic Origins of the Labour and Church Elites." *Canadian Ethnic Studies* 26:108–12.

Ogmundson, R., and J. McLaughlin. 1992. "Trends in the Ethnic Origins of Canadian Elites: The Decline of the BRITS?" *Canadian Review of Sociology and Anthropology* 29:227–42.

Ojito, Mirta. 1997. "Dominicans, Scrabbling for Hope." *New York Times* (December 16):C27.

———. 2000. "Best of Friends, Worlds Apart." *New York Times* (June 5):A1.

Okamoto, Dina G. 2003. "Toward a Theory of Panethnicity: Explaining Asian American Collective Action." *American Sociological Review* 68:811–42.

O'Leary, Brendan. 1995. "Introduction: Reflections on a Cold Peace." *Ethnic and Racial Studies* 18:695–714.

Oliver, Melvin L., and Thomas M. Shapiro. 1995. *Black Wealth/White Wealth: A New Perspective on Racial Inequality*. New York: Routledge.

———. 2008. "Sub-prime as a Black Catastrophe." *American Prospect* (October): A9–A11.

Olsen, Marvin E. 1970. "Power Perspectives on Stratification and Race Relations." Pp. 296–305 in Marvin E. Olsen (ed.), *Power in Societies*. New York: Macmillan.

Olzak, Susan. 1992. *The Dynamics of Ethnic Competition and Conflict*. Stanford, Calif.: Stanford University Press.

Omi, Michael, and Howard Winant. 1994. *Racial Formation in the United States: From the 1960s to the 1990s*. New York: Routledge.

Ong, Paul, Kye Young Park, and Yasmin Tong. 1992. "The Korean-Black Conflict and the State." Pp. 264–94 in Paul Ong, Edna Bonacich, and Lucie Cheng (eds.), *The New Asian American Immigration*. Philadelphia: Temple University Press.

Onishi, Norimitsu. 1995. "Japanese in America Looking Beyond Past to Shape Future." *New York Times* (December 25):1, 9.

———. 2009. "Japan's Outcasts Still Wait for Acceptance." *New York Times* (January 16): A1, A6.

Orfalea, Gregory. 2006. *The Arab Americans: A History*. Northampton, Mass.: Olive Branch Press.

Orfield, Gary. 1985. "Ghettoization and Its Alternatives." Pp. 161–96 in Paul E. Peterson (ed.), *The New Urban Reality*. Washington, D.C.: Brookings Institution Press.

———. 2002. "Commentary." Pp. 389–97 in Marcelo M. Suárez-Orozco and Mariela M. Páez (eds.), *Latinos: Remaking America*. Berkeley: University of California Press.

Organization for Economic Co-operation and Development (OECD). 2004. *Trends in International Migration: Annual Report*. Paris: OECD.

Orleck, Annalise. 1987. "The Soviet Jews: Life in Brighton Beach, Brooklyn." Pp. 273–304 in Nancy Foner (ed.), *New Immigrants in New York*. New York: Columbia University Press.

Orth, Richard. 2006. "Rwanda's Hutu Extremist Insurgency: An Eyewitness Perspective." Pp. 215–56 in Susan E. Cook (ed.), *Genocide in Cambodia and Rwanda: New Perspectives*. New Brunswick, N.J.: Transaction.

Osborne, Bob, and Ian Shuttleworth (eds.). 2004. *Fair Employment in Northern Ireland: A Generation On*. Belfast: Blackstaff Press.

Owen, Carolyn A., Howard C. Eisner, and Thomas R. McFaul. 1981. "A Half-Century of Social Distance Research: National Replication of the Bogardus Studies." *Sociology and Social Research* 66:80–99.

Pace, David. 2000. "Casino Boom Bypasses Indians." Associated Press (August 31). http://wire.ap.org

Pachon, Harry P., and Joan W. Moore. 1981. "Mexican Americans." *Annals of the American Academy of Political and Social Science* 454:111–24.

Padilla, Elena. 1958. *Up from Puerto Rico*. New York: Columbia University Press.

Pager, Devah, and Diana Karafin. 2009. "Bayesian Bigot? Statistical Discrimination, Stereotypes, and Employer Decision Making." *The Annals of the American Academy of Political and Social Science* 621:70–93.

Paine, Suzanne. 1974. *Exporting Workers: The Turkish Case*. London: Cambridge University Press.

Paixão, Marcelo. 2004. "Waiting for the Sun: An Account of the (Precarious) Social Situation of the African Descendant Population in Contemporary Brazil." *Journal of Black Studies* 34:743–65.

Palmer, Howard. 1976. "Mosaic vs. Melting Pot? Immigration and Ethnicity in Canada and the United States." *International Journal* 31:488–528.

Panagopoulos, Costas. 2006. "Arab and Muslim Americans and Islam in the Aftermath of 9/11." *Public Opinion Quarterly* 70:608–24.

Paperny, Anna Mehler, and Sara Minogue. 2009. "Life on the Mean Streets of Iqaluit." *Globe and Mail* (August 15):A8.

Parikh, Sunita. 2001. "Affirmative Action, Caste and Party Politics in Contemporary India." Pp. 297–323 in John D. Skrentny (ed.), *Color Lines: Affirmative Action, Immigration, and Civil Rights Options for America*. Chicago: University of Chicago Press.

Parisi, Domenico, and Daniel T. Lichter. 2007. "Hispanic Segregation in America's New Rural Boomtowns." Population Reference Bureau (September). http://www.prb.org/Publications/Articles/2007/HispanicSegregation.aspx

Park, Robert E. 1924. "The Concept of Social Distance." *Journal of Applied Sociology* 8:339–44.

———. 1928. "Human Migration and the Marginal Man." *American Journal of Sociology* 33:881–93.

———. 1950. *Race and Culture*. Glencoe, Ill.: Free Press.

Parker, Christopher S., and Matt A. Barrato. 2013. *Change They Can't Believe In: The Tea Party and Reactionary Politics in America*. Princeton: Princeton University Press.

Parrado, Emilio A., and William Kandel. 2008. "New Hispanic Migrant Destinations: A Tale of Two Industries." Pp. 99–123 in Douglas S. Massey (ed.), *New Faces in New Places: The Changing Geography of American Immigration*. New York: Russell Sage Foundation.

Passel, Jeffrey S., and D'Vera Cohn. 2008a. *Immigration to Play Lead Role in Future U.S. Growth*. Washington, D.C.: Pew Research Center.

———. 2008b. "Trends in Unauthorized Immigration: Undocumented Inflow Now Trails Legal Inflow." Washington, D.C.: Pew Hispanic Center.

———. 2008c. "U.S. Population Projections: 2005–2050." Washington, D.C.: Pew Research Center.

———. 2009. "A Portrait of Unauthorized Immigrants in the United States." Washington, D.C.: Pew Hispanic Center.

———. 2010. "U.S. Unauthorized Immigration Flows Are Down Sharply Since Mid-Decade." Washington, D.C.: Pew Research Center.

Passel, Jeffrey, D'Vera Cohn, and Ana Gonzalez-Barrera. 2012. "Net Migration from Mexico Falls to Zero—and Perhaps Less." (May 3). Washington, D.C.: Pew Hispanic Center.

Passel, Jeffrey S., Wendy Wang, and Paul Taylor. 2010. "Marrying Out: One-in-Seven New U.S. Marriages Is Interracial or Interethnic." Pew Research Center (June 4). http://pewsocialtrends.org/assets/pdf/755-marrying-out.pdf

Patriquin, Martin. 2013. "Tongue-Tied No Longer." *Maclean's* (June 10):20–26.

Pattillo, Mary. 2005. "Black Middle-Class Neighborhoods." *Annual Review of Sociology* 31:305–29.

Pattillo-McCoy, Mary. 1999. *Black Picket Fences: Privilege and Peril among the Black Middle Class.* Chicago: University of Chicago Press.

Patterson, Orlando. 1995a. "The Culture of Caution." *The New Republic* (November 27):22–26.

———. 1995b. "The Paradox of Integration." *The New Republic* (November 6):24–27.

———. 1997. *The Ordeal of Integration: Progress and Resentment in America's "Racial" Crisis.* Washington, D.C.: Civitas/Counterpoint.

———. 1998. "Affirmative Action: Opening Up Workplace Networks to Afro-Americans." *Brookings Review* (Spring):17–23.

———. 2003. "Affirmative Action: The Sequel." *New York Times* (June 22):A27.

———. 2006. "The Last Race Problem." *New York Times* (December 30):A23.

Peach, Ceri, and Günther Glebe. 1995. "Muslim Minorities in Western Europe." *Ethnic and Racial Studies* 18:26–45.

Pearlin, Leonard I. 1954. "Shifting Group Attachments and Attitudes Toward Negroes." *Social Forces* 33:47–50.

Pedraza-Bailey, Silva. 1985. *Political and Economic Migrants in America: Cubans and Mexicans.* Austin: University of Texas Press.

Pendall, Rolf, Lesley Freiman, Dowell Myers, and Selma Hepp. 2012. *Demographic Challenges and Opportunities for U.S. Housing Markets.* Washington, D.C.: Bipartisan Policy Center.

Perez, Anthony Daniel, and Charles Hirschman. 2009. "The Changing Racial and Ethnic Composition of the U.S. Population: Emerging American Identities." *Population Development Review* 35(1):1–51.

Pérez, Lisandro. 1980. "Cubans." Pp. 256–61 in Stephen Thernstrom (ed.), *Harvard Encyclopedia of American Ethnic Groups.* Cambridge, Mass.: Harvard University Press.

———. 2007. "Cuba." Pp. 387–98 in Mary C. Waters and Reed Ueda (eds.), *The New Americans: A Guide to Immigration Since 1965.* Cambridge, Mass.: Harvard University Press.

Pérez-Stable, Marifeli, and Miren Uriarte. 1993. "Cubans and the Changing Economy of Miami." Pp. 133–59 in Rebecca Morales and Frank Bonilla (eds.), *Latinos in a Changing U.S. Economy.* Newbury Park, Calif.: Sage.

Peri, Giovanni. 2007. *How Immigrants Affect California Employment and Wages.* San Francisco: Public Policy Institute of California.

———. 2009. "The Effect of Immigrants on U.S. Employment and Productivity." *FRBSF Economic Letter* (August 30). San Francisco: Federal Reserve Bank of San Francisco.

———. 2010. "The Effect of Immigration on Productivity: Evidence from US States." NBER Working Paper No. 15507 (November). http://www.nber.org/papers/w15507

Perlez, Jane. 1995a. "Serbs Become Latest Victims in Changing Fortunes of War." *New York Times* (August 7):A1, A4.

———. 1995b. "Thousands of Serbian Civilians Are Caught in Soldiers' Gunfire." *New York Times* (August 9):A1, A6.

Pessar, Patricia R. 1995. *A Visa for a Dream: Dominicans in the United States.* Boston: Allyn & Bacon.

Petersen, William. 1980. "Concepts of Ethnicity." Pp. 234–42 in Stephen Thernstrom (ed.), *Harvard Encyclopedia of American Ethnic Groups.* Cambridge, Mass.: Harvard University Press.

———. 1997. *Ethnicity Counts.* New Brunswick, N.J.: Transaction.

Peterson, Iver. 2004. "Would-Be Tribes Entice Investors." *New York Times* (March 29):A1, A20.

Petit, Arthur G. 1980. *Images of the Mexican American in Fiction and Film.* College Station: Texas A&M Press.

Pettigrew, Thomas F. 1960. "Social Distance Attitudes of South African Students." *Social Forces* 38:246–53.

———. 1975. "Black and White Attitudes Toward Race and Housing." Pp. 92–126 in Thomas Pettigrew (ed.), *Racial Discrimination in the United States.* New York: Harper & Row.

———. 1979. "Racial Change and Social Policy." *Annals of the American Academy of Political and Social Science* 441:114–31.

———. 1980. "Prejudice." Pp. 820–29 in Stephen Thernstrom (ed.), *Harvard Encyclopedia of American Ethnic Groups.* Cambridge, Mass.: Harvard University Press.

———. 1981. "Race and Class in the 1980s: An Interactive View." *Daedalus* 110:233–55.

Peukert, Detlev J. K. 1987. *Inside Nazi Germany: Conformity, Opposition and Racism in Everyday Life.* London: Penguin.

Pew Forum on Religion & Public Life. 2006. "Prospects for Inter-Religious Understanding." (March 22). Washington, D.C.: Pew Forum on Religion & Public Life.

———. 2009. "Religiously Mixed Couples: Cupid's Arrow Often Hits People of Different Faiths" (February 10). http://www.pewforum.org/Religiously-Mixed-Couples-Cupids-Arrow-Often-Hits-People-of-Different-Faiths.aspx

———. 2012a. "Asian Americans: A Mosaic of Faiths." Washington, D.C.: The Pew Forum on Religion and Public Life.

———. 2012b. "Faith on the Hill: The Religious Composition of the 113th Congress." (November 16). http://www.pewforum.org

Pew Hispanic Center. 2002. *The 2002 National Survey of Latinos.* Washington, D.C.: Pew Hispanic Center.

———. 2004a. *Survey Brief: Assimilation and Language.* Washington, D.C.: Pew Hispanic Center.

———. 2004b. *The 2004 National Survey of Latinos: Politics and Civic Participation.* Washington, D.C.: Pew Hispanic Center.

———. 2006a. "Cubans in the United States." (August 25). Washington, D.C.: Pew Hispanic Center.

———. 2006b. "Hispanic Attitudes Toward Learning English." (June 7). Washington, D.C.: Pew Hispanic Center.

———. 2006c. "The State of American Public Opinion on Immigration in Spring 2006: A Review of Major Surveys." Washington, D.C.: Pew Hispanic Center.

———. 2007. "The Latino Electorate: A Widening Gap Between Voters and the Large Hispanic Population in the U.S." (July 24). Washington, D.C.: Pew Hispanic Center.

———. 2009. "Between Two Worlds: How Young Latinos Come of Age in America." (December 11). http://pewhispanic.org/reports/report.php?ReportID=117

———. 2010. "Hispanics and Arizona's New Immigration Law." (April 29). http://pewresearch.org/pubs/1579/arizona-immigration-law-fact-sheet-hispanic-population-opinion-discrimination

———. 2013. "A Nation of Immigrants: A Portrait of the 40 Million, Including 11 Million Unauthorized." Washington, D.C.: Pew Hispanic Center.

Pew Research Center for the People and the Press. 2003. "Conflicted Views of Affirmative Action." Washington, D.C.: Pew Research Center for the People and the Press.

———. 2004. "Belief That Jews Were Responsible for Christ's Death Increases." *Survey Reports* (April 2). http://people-press.org/reports

———. 2006a. "America's Immigration Quandary." (March 30). Washington, D.C.: Pew Research Center for the People and the Press.

———. 2006b. "Guess Who's Coming to Dinner." *Social and Demographic Trends* (March 14). http://pewsocialtrends.org/pubs/304/guess-whos-coming-to-dinner

———. 2007a. "Blacks See Growing Values Gap Between Poor and Middle Class." Washington, D.C.: Pew Research Center for the People and the Press.

———. 2007b. "Muslim Americans: Middle Class and Mostly Mainstream." Washington, D.C.: Pew Research Center.

———. 2007c. "Trends in Political Values and Core Attitudes: 1987–2007." Washington, D.C.: Pew Research Center for the People and the Press.

———. 2009a. "Muslims Widely Seen As Facing Discrimination." (September 9). Pew Research Center. http://www.pewforum.org/2009/09/09/muslims-widely-seen-as-facing-discrimination/

———. 2009b. "Public Backs Affirmative Action, but Not Minority Preferences." (June 2). Pew Research Center. http://www.pewresearch.org/2009/06/02/public-backs-affirmative-action-but-not-minority-preferences/

———. 2010a. "Almost All Millennials Accept Interracial Dating and Marriage." (February 1). http://pewresearch.org/pubs/1480/millennials-accept-iinterracial-dating-marriage-'friends-different-race-generation

———. 2010b. "Blacks Upbeat About Black Progress, Prospects." (January 12). http://pewsocialtrends.org/pubs/749/blacks-upbeat-about-black-progress-obama-election

———. 2010c. "Public Remains Conflicted over Islam." (August 24). Washington, D.C.: Pew Research Center for the People & the Press.

———. 2011a. "Muslim Americans: No Signs of Growth in Alienation or Support for Extremism." (August 30). Washington, D.C.: Pew Research Center for the People & the Press.

———. 2011b. "Wealth Gaps Rise to Record Highs Between Whites, Blacks and Hispanics." Washington, D.C.: Pew Research Center.

———. 2012a. "Controversies over Mosques and Islamic Centers Across the U.S." Washington, D.C.: Pew Research Center's Forum on Religion & Public Life.

———. 2012b. "Muslims." Pew Research Religion & Public Life Project. http://www.pewforum.org/2012/12/18/global-religious-landscape-muslim/

———. 2012c. "'Nones' On the Rise." (October 9). Washington, D.C.: Pew Research Center's Religion & Public Life Project.

———. 2013a. "Most Say Immigration Policy Needs Big Changes." Pew Research Center. http://www.people-press.org/2013/05/09/most-say-immigration-policy-needs-big-changes/

———. 2013b. *The Rise of Asian Americans* (Updated Edition). Washington, D.C.: Pew Research Center for the People & the Press. http://www.pewsocialtrends.org/asianamericans

———. 2013c. *A Portrait of Jewish Americans: Findings from a Pew Research Center Survey of U.S. Jews.* Washington, D.C.: Pew Research Center's Religion & Public Life Project.

———. 2013d. "King's Dream Remains an Elusive Goal: Many Americans See Racial Disparities." Washington, D.C.: Pew Research Center. http://www.pewsocialtrends.org/files/2013/08/final_full_report_racial_disparities.pdf

Philliber, William W., and Clyde B. McCoy (eds.). 1981. *The Invisible Minority: Urban*

Appalachians. Lexington: University Press of Kentucky.

Picard, André. 2010. "Native Women at Heart of Diabetes Epidemic." *Globe and Mail* (January 19):L4.

Picca, Leslie Houts, and Joe R. Feagin. 2007. *Two-Faced Racism: Whites in the Backstage and Frontstage*. New York: Routledge.

Pierson, Donald. 1967 [1942]. *Negroes in Brazil: A Study of Race Contact at Bahia*. Carbondale: Southern Illinois University Press.

Pineo, Peter C. 1987. "The Social Standing of Ethnic and Racial Groupings." Pp. 256–72 in Leo Driedger (ed.), *Ethnic Canada: Identities and Inequalities*. Toronto: Copp Clark Pitman.

Pineo, Peter C., and John Porter. 1985. "Ethnic Origin and Occupational Attainment." Pp. 357–92 in Monica Boyd et al. (eds.), *Ascription and Achievement: Studies in Mobility and Status Attainment in Canada*. Ottawa: Carleton University Press.

Pinkney, Alphonso. 1963. "Prejudice Toward Mexican and Negro Americans: A Comparison." *Phylon* 24:353–59.

———. 2000. *Black Americans*. 5th ed. Englewood Cliffs, N.J.: Prentice Hall.

Pitt-Rivers, Julian. 1987. "Race, Color, and Class in Central America and the Andes." Pp. 298–305 in Celia S. Heller (ed.), *Structured Social Inequality: A Reader in Comparative Social Stratification*. 2nd ed. New York: Macmillan.

Piven, Frances Fox, and Richard A. Cloward. 1971. *Regulating the Poor: The Functions of Public Welfare*. New York: Random House.

Polgreen, Lydia. 2012. "Fatal Stampede in South Africa Points Up Crisis at Universities." *New York Times* (January 11):A1, A9.

———. 2013. "Trading Privilege for Privation, Family Hits A Nation's Nerve." *New York Times* (Septebmer 16):A1.

Poll, Solomon. 1969. *The Hasidic Community of Williamsburg*. New York: Schocken.

Pollack, Andrew. 1992. "Asian Immigrants New Leaders in Silicon Valley." *New York Times* (January 14):A1, C5.

Pollak, Andy. 1993. *A Citizen's Inquiry: The Opsahl Report on Northern Ireland*. Dublin: Lilliput Press for Initiative '92.

Polling Report. 2013. "Immigration." PollingReport.com (June 23). http://www .pollingreport.com

Ponting, J. Rick. 1997. "Historical Overview and Background: Part II." Pp. 35–67 in J. Rick Ponting (ed.), *First Nations in Canada: Perspectives on Opportunity, Empowerment, and Self-Determination*. Toronto: McGraw-Hill Ryerson.

Poole, Michael A. 1982. "Religious Residential Segregation in Urban Northern Ireland." Pp. 182–308 in Frederick W. Boal and J. Neville H. Douglas (eds.), *Integration and Division:*

Geographical Perspectives on the Northern Ireland Problem. London: Academic Press.

Poppino, Rollie E. 1968. *Brazil: The Land and People*. New York: Oxford University Press.

Porter, Eduardo. 2005. "Illegal Immigrants Are Bolstering Social Security with Billions." *New York Times* (April 5):A1, C6.

Porter, John. 1965. *The Vertical Mosaic: An Analysis of Social Class and Power in Canada*. Toronto: University of Toronto Press.

———. 1975. "Ethnic Pluralism in Canadian Perspective." Pp. 267–304 in Nathan Glazer and Daniel P. Moynihan (eds.), *Ethnicity: Theory and Experience*. Cambridge, Mass.: Harvard University Press.

———. 1979. *The Measure of Canadian Society*. Toronto: Gage.

———. 1985. "Canada: The Societal Context of Occupational Allocation." Pp. 29–65 in Monica Boyd et al. (eds.), *Ascription and Achievement: Studies in Mobility and Status Attainment in Canada*. Ottawa: Carleton University Press.

Porter, Judith D. R. 1971. *Black Child, White Child*. Cambridge, Mass.: Harvard University Press.

Porter, Rosalie Pedalino. 1990. *Forked Tongue: The Politics of Bilingual Education*. New York: Basic Books.

Portes, Alejandro. 1969. "Dilemmas of a Golden Exile: Integration of Cuban Refugee Families in Milwaukee." *American Sociological Review* 34:505–18.

———. 2002. "English-Only Triumphs, but the Costs Are High." *Contexts* 1(Spring):10–15.

———. 2007. "The Fence to Nowhere." *American Prospect* (October):26–29.

Portes, Alejandro, and Robert L. Bach. 1985. *Latin Journey: Cuban and Mexican Immigrants in the United States*. Berkeley: University of California Press.

Portes, Alejandro, J. M. Clark, and R. L. Bach, 1977. "The New Wave: A Statistical Profile of Recent Cuban Exiles to the United States." *Estudios Cubanos/Cuban Studies* 7:1–32.

Portes, Alejandro, Juan M. Clark, and Robert D. Manning. 1985. "After Mariel: A Survey of the Resettlement Experiences of 1980 Cuban Refugees in Miami." *Estudios Cubanos/ Cuban Studies* 15:37–59.

Portes, Alejandro, and Lingxin Hao. 1998. "E Pluribus Unum: Bilingualism and Loss of Language in the Second Generation." *Sociology of Education* 71:269–94.

Portes, Alejandro, and Rafael Mozo. 1985. "The Political Adaptation Process of Cubans and Other Ethnic Minorities in the United States: A Preliminary Analysis." *International Migration Review* 19:35–63.

Portes, Alejandro, and Rubén Rumbaut. 2006. *Immigrant America: A Portrait*. 3rd ed. Berkeley: University of California Press.

Portes, Alejandro, and Richard Schauffler. 1996. "Language and the Second Generation: Bilingualism Yesterday and Today." Pp. 8–29 in Alejandro Portes (ed.), *The New Second Generation*. New York: Russell Sage Foundation.

Portes, Alejandro, and Alex Stepick. 1993. *City on the Edge: The Transformation of Miami.* Berkeley: University of California Press.

Portes, Alejandro, and Cynthia Truelove. 1987. "Making Sense of Diversity: Recent Research on Hispanic Minorities in the United States." *Annual Review of Sociology* 13:359–85.

Portes, Alejandro, and Min Zhou. 1993. "The New Second Generation: Segmented Assimilation and Its Variants." *Annals of the American Academy of Political and Social Science* 530 (November):74–96.

———. 1994. "Should Immigrants Assimilate?" *The Public Interest* (Summer):18–33.

Posgate, Dale. 1978. "The Quiet Revolution." Pp. 50–57 in Norman Penner et al. (eds.), *Keeping Canada Together.* Toronto: Amethyst.

Poston, Dudley L., Jr., Michael Xinxiang Mao, and Mei-Yu Yu. 1994. "The Global Distribution of the Overseas Chinese Around 1990." *Population and Development Review* 20:631–45.

Pressley, Sue Anne. 1999. "The Star of David Sets in the South's Small Towns." *Washington Post National Weekly Edition* (May 31):29.

Presthus, Robert. 1973. *Elite Accommodation in Canadian Politics.* Toronto: Macmillan.

Preston, Julia. 2007. "Government Set for a Crackdown on Illegal Hiring." *New York Times* (August 8):A1, A16.

Price, John A. 1973. "The Stereotyping of North American Indians in Motion Pictures." *Ethnohistory* 20 (Spring):153–71.

Price, Marie. 2006. "Andean South Americans and Cultural Networks." Pp. 191–212 in Ines M. Miyares and Christopher A. Airriess (eds.), *Contemporary Ethnic Geographies in America.* Lanham, Md.: Rowman & Littlefield.

Pritchett, Wendell. 2002. *Brownsville, Brooklyn: Blacks, Jews, and the Changing Face of the Ghetto.* Chicago: University of Chicago Press.

Prunier, Gérard. 1995. *The Rwanda Crisis: History of a Genocide.* New York: Columbia University Press.

Public Opinion. 1987. "Opinion Roundup." (July/August):25–39.

Pulley, Brett. 1999. "Tribes Weighing Tradition vs. Casino Growth." *New York Times* (March 16):A1, A20.

Qian, Zhenchao, Sampson Lee Blair, and Stacey D. Ruf. 2001. "Asian American Interracial and Interethnic Marriages: Differences by Education and Nativity." *International Migration Review* 35:557–86.

Qian, Zhenchao, and Daniel T. Lichter. 2007. "Social Boundaries and Marital Assimilation: Interpreting Trends in Racial and Ethnic Intermarriage." *American Sociological Review* 72:68–94.

Quinley, Harold E., and Charles Y. Glock. 1979. *Anti-Semitism in America.* New York: Free Press.

Quinn, Eamon. 2007. "Ireland Learns to Adapt to a Population Growth Spurt." *New York Times* (August 19):3.

Raab, Earl. 1989. *What Do We Really Know About Anti-Semitism and What Do We Want to Know? Working Papers on Contemporary Anti-Semitism.* New York: American Jewish Committee.

Raab, Earl, and Seymour Martin Lipset. 1971. "The Prejudiced Society." Pp. 31–45 in Gary T. Marx (ed.), *Racial Conflict: Tension and Change in American Society.* Boston: Little, Brown.

Ramakrishnan, Karthick, and Taeku Lee. 2012. "The Policy Priorities and Issue Preferences of Asian Americans and Pacific Islanders." National Asian American Survey.

Ramcharan, Subhas. 1982. *Racism: Nonwhites in Canada.* Toronto: Butterworths.

Ramirez, J. Martin, and Bobbie Sullivan. 1987. "The Basque Conflict." Pp. 120–39 in Jerry Boucher, Dan Landis, and Karen Arnold Clark (eds.), *Ethnic Conflict: International Perspectives.* Newbury Park, Calif.: Sage.

Ramos, Arthur. 1939. *The Negro in Brazil.* Trans. Richard Pattee. Washington, D.C.: Associated Publishers.

Ramsey, Patricia G. 1987. "Young Children's Thinking about Ethnic Differences." Pp. 56–72 in Jean S. Phinney and Mary Jane Rotheram (eds.), *Children's Ethnic Socialization: Pluralism and Development.* Newbury Park, Calif.: Sage.

Rattansi, Ali. 2007. *Racism: A Very Short Introduction.* Oxford: Oxford University Press.

Ravitch, Diane. 2002. "A Considered Opinion: Diversity, Tragedy, and the Schools." *Brookings Review* 20(1):2–3.

Rawlings, Lynette, Laura Harris, and Margery Austin Turner. 2004. *Race and Residence: Prospects for Stable Neighborhood Integration.* Washington, D.C.: Urban Institute.

Read, Jen'nan Ghazal. 2004. *Culture, Class and Work Among Arab-American Women.* New York: LFB Scholarly Publishing LLC.

Redfield, Robert. 1958. "Race as a Social Phenomenon." Pp. 66–71 in Edgar T. Thompson and Everett C. Hughes (eds.), *Race: Individual and Collective Behavior.* Glencoe, Ill.: Free Press.

Reeve, R. Penn. 1975. "Black Economic Mobility in a Brazilian Town." *Plural Societies* 6:45–50.

———. 1977. "Race and Social Mobility in a Brazilian Industrial Town." *Luso Brazilian Review* 14:236–53.

Reich, Michael. 1978. "The Economics of Racism." Pp. 381–88 in Richard C. Edwards, Michael Reich, and Thomas E. Weisskopf (eds.), *The Capitalist System*. 2nd ed. Englewood Cliffs, N.J.: Prentice Hall.

Reimer, Dana G. 2006. "Korean Culture and Entrepreneurship." Pp. 233–50 in Ines M. Miyares and Christopher A. Airriess (eds.), *Contemporary Ethnic Geographies in America*. Lanham, Md.: Rowman & Littlefield.

Reimers, Cordelia. 2006. "Economic Well-Being." Pp. 291–361 in Marta Tienda and Faith Mitchell (eds.), *Hispanics and the Future of America*. Washington, D.C.: National Academies Press.

Reitz, Jeffrey G., and Raymond Breton. 1994. *The Illusion of Difference: Realities of Ethnicity in Canada and the United States*. Toronto: C. D. Howe Institute.

Reitz, Jeffrey G., and Janet M. Lum. 2006. "Immigration and Diversity in a Changing Canadian City: Social Bases of Intergroup Relations in Toronto." Pp. 15–50 in Eric Fong (ed.), *Inside the Mosaic*. Toronto: University of Toronto Press.

Rex, John. 1970. *Race Relations in Sociological Theory*. New York: Schocken.

Ribeiro, Darcy. 2000. *The Brazilian People: The Formation and Meaning of Brazil*. Trans. Gregory Rabassa. Gainesville: University Press of Florida.

Richer, Stephen, and Pierre E. Laporte. 1979. "Culture, Cognition, and English-French Competition." Pp. 75–85 in Jean Leonard Elliott (ed.), *Two Nations, Many Cultures*. Scarborough, Ontario: Prentice Hall of Canada.

Richmond, Anthony H. 1976. "Immigration, Population, and the Canadian Future." *Sociological Focus* 9:125–36.

Rieff, David. 1991. "The New Face of L.A." *Los Angeles Times Magazine* (September 15): 14–20, 46–48.

———. 1993. *The Exile: Cuba in the Heart of Miami*. New York: Touchstone.

———. 1995. *Slaughterhouse: Bosnia and the Failure of the West*. New York: Touchstone.

———. 1996. "An Age of Genocide." *The New Republic* (January 29):27–36.

Riis, Jacob A. 1957 [1890]. *How the Other Half Lives*. New York: Hill & Wang.

Ringer, Benjamin B. 1967. *The Edge of Friendliness*. New York: Basic Books.

Rioux, Marcel, and Yves Martin (eds.). 1978. *French-Canadian Society*. Toronto: Macmillan of Canada.

Rippley, La Vern J. 1984. *The German Americans*. Lanham, Md.: University Press of America.

Ritterband, Paul. 1995. "Modern Times and Jewish Assimilation." Pp. 377–94 in Robert M. Seltzer and Norman J. Cohen (eds.), *The Americanization of the Jews*. New York: New York University Press.

Rivlin, Gary. 2007. "Moving Off the Reservation." *New York Times* (September 22):B1, B4.

Robbins, Liz. 2009. "9 Muslims Are Pulled from Plane and Denied Re-entry; Airline Belatedly Apologizes." *New York Times* (January 3): A9.

Roberts, Sam. 2005. "More Africans Enter U.S. Than in Days of Slavery." *New York Times* (February 21):A1, A18.

Robinson, Eugene. 1996. "Over the Brazilian Rainbow." *Washington Post National Weekly Edition* (December 18–24):23.

Roche, John Patrick. 1982. "Suburban Ethnicity: Ethnic Attitudes and Behavior among Italian Americans in Two Suburban Communities." *Social Science Quarterly* 63:145–53.

Rodríguez, Clara. 1989. *Puerto Ricans: Born in the U.S.A.* Boston: Unwin & Hyman.

———. 1996. "Racial Themes in the Literature: Puerto Ricans and Other Latinos." Pp. 105–25 in Gabriel Haslip-Viera and Sherrie L. Baver (eds.), *Latinos in New York: Communities in Transition*. Notre Dame, Ind.: University of Notre Dame Press.

Rodriguez, Gregory. 1998. "Minority Leader." *The New Republic* (October 19):21–23.

Rodriguez, Richard. 1983. *Hunger of Memory: The Education of Richard Rodriguez*. New York: Bantam.

Rodríguez-Vecchini, Hugo. 1994. "Foreword: Back and Forward." Pp. 29–102 in Carlos Antonio Torre, Hugo Rodríguez-Vecchini, and William Burgos (eds.), *The Commuter Nation: Perspectives on Puerto Rican Migration*. Río Piedras, Puerto Rico: Editorial de la Universidad de Puerto Rico.

Roediger, David R. 2005. *Working Toward Whiteness: How America's Immigrants Became White*. New York: Basic Books.

Rogoff, Leonard. 2001. *Homelands: Southern Jewish Identity in Durham and Chapel Hill, North Carolina*. Tuscaloosa: University of Alabama Press.

Rohter, Larry. 2000. "Catholics Battle Brazilian Faith in 'Black Rome.'" *New York Times* (January 10):A13.

———. 2003. "Racial Quotas in Brazil Touch Off Fierce Debate." *New York Times* (April 5):A5.

———. 2010. *Brazil on the Rise: The Story of a Country Transformed*. New York: Palgrave Macmillan.

Rolle, Andrew. 1980. *The Italian Americans: Troubled Roots*. New York: Free Press.

Romanucci-Ross, Lola, and George De Vos (eds.). 1995. *Ethnic Identity: Creation, Conflict and Accommodation*. 3rd ed. Walnut Creek, Calif.: Alta Mira.

Romero, Simon. 2012. "Brazil Enacts Affirmative Action Law for Universities." *New York Times* (August 31):A4.

Roper Center. 2001. "Mending the Fabric." *Public Perspective* (May/June):22–31.

Rose, Hilary, and Steven Rose. 1978. "The IQ Myth." *Race and Class* 20:63–74.

Rose, Peter I. 1968. *The Subject Is Race: Traditional Ideologies and the Teaching of Race Relations*. New York: Oxford University Press.

———. 1977. *Strangers in Their Midst*. Merrick, N.Y.: Richwood.

———. 1985. "Asian Americans: From Pariahs to Paragons." Pp. 181–212 in Nathan Glazer (ed.), *Clamor at the Gates: The New American Immigration*. San Francisco: ICS Press.

Rose, Richard. 1971. *Governing Without Consensus: An Irish Perspective*. Boston: Beacon.

———. 1976. *Northern Ireland: Time of Choice*. Washington, D.C.: American Enterprise Institute for Public Policy Research.

Rosen, Bernard C. 1959. "Race, Ethnicity, and the Achievement Syndrome." *American Sociological Review* 24:47–60.

Rosen, Nir. 2007. "The Flight from Iraq." *New York Times Magazine* (May 13):32–41, 56, 74–78.

Roshwald, Aviel. 2006. *The Endurance of Nationalism: Ancient Roots and Modern Dilemmas*. Cambridge, England: Cambridge University Press.

Ross, Susan Dente. 2003. "Images of Irish Americans: Invisible, Inebriated, or Irascible." Pp. 131–40 in Paul Martin Lester and Susan Dente Ross (eds.), *Images That Injure: Pictorial Stereotypes in the Media*. Westport, Conn.: Praeger.

Roy, Olivier. 2005. "Europe's Response to Radical Islam." *Current History* 104:360–64.

Royal Commission on Bilingualism and Biculturalism. 1969a. Report. Vol. 3, *The World of Work*. Ottawa: Queen's Printer.

———. 1969b. Report. Vol. 4, *The Cultural Contribution of the Other Ethnic Groups*. Ottawa: Queen's Printer.

Royko, Mike. 1971. *Boss: Richard J. Daley of Chicago*. New York: Dutton.

Rudman, Laurie A. 2005. "Rejection of Women? Beyond Prejudice as Antipathy." Pp. 106–20 in John F. Dovidio, Peter Glick, and Laurie A. Rudman (eds.), *On the Nature of Prejudice: Fifty Years After Allport*. Malden, Mass.: Blackwell.

Rudolph, Susanne Hoeber, and Lloyd I. Rudolph. 1993. "Modern Hate." *The New Republic* (March 22):24–29.

Rudwick, Elliott M. 1964. *Race Riot at East St. Louis*. Carbondale: Southern Illinois University Press.

Rudwick, Elliott, and August Meier. 1969. "Negro Retaliatory Violence in the Twentieth Century." Pp. 406–17 in August Meier and Elliott Rudwick (eds.), *The Making of Black America: Essays in Negro Life and History*. Vol. 2. New York: Atheneum.

Rumbaut, Rubén G. 1996. "Origins and Destinies: Immigration, Race, and Ethnicity in Contemporary America." Pp. 21–42 in Silvia Pedraza and Rubén Rumbaut (eds.), *Origins and Destinies: Immigration, Race, and Ethnicity in America*. Belmont, Calif.: Wadsworth.

———. 1999. "Assimilation and Its Discontents: Ironies and Paradoxes." Pp. 172–95 in Charles Hirschman, Philip Kasinitz, and Josh DeWind (eds.), *The Handbook of International Migration: The American Experience*. New York: Russell Sage Foundation.

———. 2011. "Pigments of Our Imagination: The Racialization of the Hispanic-Latino Category." *Migration Information Source* (April 27). Washington, D.C.: Migration Policy Institute.

Rumbaut, Rubén G., Douglas S. Massey, and Frank D. Bean. 2006. "Linguistic Life Expectancies: Immigrant Language Retention in Southern California." *Population and Development Review* 32:447–60.

Runciman, W. G. 1966. *Relative Deprivation and Social Justice*. Berkeley: University of California Press.

Russell-Wood, A. J. R. 1968. "Race and Class in Brazil, 1937–1967, a Re-Assessment: A Review." *Race* 10:185–91.

Ryan, William. 1975. *Blaming the Victim*. Rev. ed. New York: Vintage.

Ryu, Mikyung. 2009. *Minorities in Higher Education 2009 Supplement*. Washington, D.C.: American Council on Education.

Saad, Lydia. 1995. "Immigrants See U.S. as Land of Opportunity." *Gallup Poll Monthly* (July):19–33.

———. 2006. "Anti-Muslim Sentiments Fairly Commonplace." *Gallup Tuesday Briefing* (August 10):39–41.

———. 2007a. *"Black-White Educational Opportunities Widely Seen as Equal."* Washington, D.C.: The Gallup Poll (July 2).

———. 2007b. *"A Downturn in Black Perceptions of Racial Harmony."* Washington, D.C.: The Gallup Poll (July 6).

Sachs, Albie. 1975. "The Instruments of Domination in South Africa." Pp. 223–49 in Leonard Thompson and Jeffrey Butler (eds.), *Change in Contemporary South Africa*. Berkeley: University of California Press.

———. 1992. "Watch Out—There's a Constitution About." *South Africa International* 22 (April):184–89.

Said, Edward W. 1978. *Orientalism*. New York: Pantheon.

———. 1981. *Covering Islam*. New York: Pantheon.

Saito, Natsu Taylor. 2010. "Internments Then and Now: Constitutional Accountability in Post-9/11 America." *Duke Forum for Law & Social Change* 2:71, 72–102.

Saito, Yoshitaka, and George Farkas. 2004. "The *Burakumin*: An Updated Review." *International Journal of Contemporary Sociology* 41:232–50.

Sakamoto, Arthur, Jeng Liu, and Jessie M. Tzeng. 1998. "The Declining Significance of Race among Chinese and Japanese American Men." Pp. 225–46 in Kevin T. Leicht (ed.), *Research in Social Stratification and Mobility*. Vol. 16. Stamford, Conn.: JAI Press.

Sakamoto, Arthur, and Yu Xie. 2006. "The Socioeconomic Attainment of Asian Americans." Pp. 54–77 in Pyong Gap Min, (ed.), *Asian Americans: Contemporary Trends and Issues*. 2nd ed. Thousand Oaks, Calif.: Pine Forge.

Salée, Daniel. 2006. "Quality of Life of Aboriginal People in Canada: An Analysis of Current Research." *IRPP Choices* 12(6).

Samhan, Helen Hatab. 1999. "Not Quite White: Race Classification and the Arab American Experience." Pp. 209–23 in Michael N. Suleiman (ed.), *Arabs in America: Building a New Future*. Philadelphia: Temple University Press.

Sampson, Robert J. 2009. "Racial Stratification and the Durable Tangle of Neighborhood Inequality." *The Annals of the American Academy of Political and Social Science* 621:260–80.

Samuda, Ronald J. 1975. *Psychological Testing of American Minorities*. New York: Harper & Row.

Samuel, John, and Kogalur Basavarajappa. 2006. "The Visible Minority Population in Canada: A Review of Numbers, Growth and Labour Force Issues." *Canadian Studies in Population* 33:241–69.

Sandberg, Neil. 1974. *Ethnic Identity and Assimilation: The Polish American Community*. New York: Praeger.

———. 1986. *Jewish Life in Los Angeles*. Lanham, Md.: University Press of America.

Sandefur, Gary D., Molly Martin, Jennifer Eggerling-Boeck, Susan E. Mannon, and Ann M. Meier. 2001. "An Overview of Racial and Ethnic Demographic Trends." Pp. 40–102 in Neil J. Smelser, William Julius Wilson, and Faith Mitchell (eds.), *America Becoming: Racial Trends and Their Consequences*. Vol. 1. Washington, D.C.: National Academy Press.

Sanders, Thomas G. 1981. "Racial Discrimination and Black Consciousness in Brazil." *American Universities Field Staff Reports*, No. 42. Hanover, N.H.: American Universities Field Staff.

———. 1982. "Brazilian Population in 1982: Growth, Migration, Race, Religion." *UFSI Reports*, No. 42.

Sansone, Livio. 2003. *Blackness Without Ethnicity: Constructing Race in Brazil*. New York: Palgrave Macmillan.

Santos, Fernanda. 2007. "Demand for English Lessons Outstrips Supply." *New York Times* (February 27):A1, C14.

Sarlin, Stuart H., and Eugene D. Tate. 1976. "'All in the Family': Is Archie Funny?" *Journal of Communication* 26:61–67.

Sassen, Saskia. 1998. *Globalization and Its Discontents*. New York: New Press.

Saulny, Susan, and Robbie Brown. 2009. "Case Recalls Tightrope Blacks Walk with Police." *New York Times* (July 24):A1, A3.

Saunders, Doug. 2004. "Dutch Lash Out at Muslims After Slaying." *Globe and Mail* (November 11):A1, A15.

———. 2010. "Germany's Multiculturalism Dilemma a Cautionary Tale for Canada." *Globe and Mail* (October 18):A21.

———. 2012. "Catholics Then, Muslims Now." *New York Times* (September 18):A25.

Saunders, John. 1972. "Class, Color, and Prejudice: A Brazilian Counterpoint." Pp. 141–65 in Ernest Q. Campbell (ed.), *Racial Tensions and National Identity*. Nashville, Tenn.: Vanderbilt University Press.

Saywell, John. 1977. *The Rise of the Parti Québécois 1967–76*. Toronto: University of Toronto Press.

Schanzer, David, Charles Kurzman, and Ebrahim Moosa. 2010. "Anti-Terror Lessons of Muslim-Americans." Duke University and the University of North Carolina at Chapel Hill. http://www.sanford.duke.edu/news/Schanzer_Kurzman_Moosa_Anti-Terror_Lessons.pdf

Schemo, Diana Jean. 1995. "Elevators Don't Lie: Intolerance in Brazil." *New York Times* (August 30):A7.

Schermerhorn, R. A. 1949. *These Our People: Minorities in American Culture*. Boston: Heath.

———. 1970. *Comparative Ethnic Relations: A Framework for Theory and Research*. New York: Random House.

Schlemmer, Lawrence. 1988. "South Africa's National Party Government." Pp. 7–54 in Peter L. Berger and Bobby Godsell (eds.), *A Future South Africa: Visions, Strategies and Realities*. Cape Town: Human & Rousseau Tafelberg.

———. 2001. *Race Relations and Racism in Everyday Life*. South African Institute of Race Relations. http://www.sairr.org.za/

Schlesinger, Arthur M., Jr. 1998. *The Disuniting of America: Reflections on a Multicultural Society*. 2nd ed. New York: Norton.

Schmidhauser, John R. 1960. *The Supreme Court: Its Politics, Personalities, and Procedures.* New York: Holt, Rinehart & Winston.

Schmidt, Peter. 2007. "What Color Is an A?" *The Chronicle of Higher Education* (June 1): A24–A28.

Schmitt, Eric. 2001a. "Americans (a) Love (b) Hate Immigrants." *New York Times* (January 14): 1, 3.

———. 2001b. "Census Figures Show Hispanics Pulling Even with Blacks." *New York Times* (March 8):A1, A4.

Schneider, Ronald M. 1996. *Brazil: Culture and Politics in a New Industrial Powerhouse.* Boulder, Colo.: Westview Press.

Schooler, Carmi. 1976. "Serfdom's Legacy: An Ethnic Continuum." *American Journal of Sociology* 81:1265–86.

Schöpflin, George. 1993. "The Rise and Fall of Yugoslavia." Pp. 172–203 in John McGarry and Brendan O'Leary (eds.), *The Politics of Ethnic Conflict Regulation.* London: Routledge.

Schopmeyer, Kim. 2011. "Arab Detroit After 9/11: A Changing Demographic Portrait." Pp. 29–63 in Nabeel Abraham, Sally Howell, and Andrew Shryock (eds.), *Arab Detroit 9/11: Life in the Terror Decade.* Detroit: Wayne State University Press.

Schrag, Peter. 1995. "So You Want to Be Color Blind: Alternative Principles for Affirmative Action." *The American Prospect* (Summer):38–43.

Schuck, Peter H. 2007. "The Disconnect Between Public Attitudes and Policy Outcomes in Immigration." Pp. 17–31 in Carol M. Swain (ed.), *Debating Immigration.* New York: Cambridge University Press.

Schuman, Howard. 1982. "Free Will and Determinism in Public Beliefs About Race." Pp. 345–50 in Norman R. Yetman and C. Hoy Steele (eds.), *Majority and Minority: The Dynamics of Race and Ethnicity in American Life.* 3rd ed. Boston: Allyn & Bacon.

Schuman, Howard, Charlotte Steeh, Lawrence Bobo, and Maria Krysan. 1997. *Racial Attitudes in America: Trends and Interpretations.* Rev. ed. Cambridge, Mass.: Harvard University Press.

Schwartz, Mildred A. 1967. *Trends in White Attitudes Toward Negroes.* Chicago: National Opinion Research Center.

Schwartz, Nelson D., and Michael Cooper. 2013. "Racial Diversity Efforts Ebb for Elite Careers, Analysis Finds." *New York Times* (May 28): A1.

Schwarz, Adam. 1994. *A Nation in Waiting: Indonesia in the 1990s.* St. Leonard's, N.S.W.: Allen & Unwin.

Sears, David O. 2000. "Egalitarian Values and Contemporary Racial Politics." Pp. 75–117 in David O. Sears, Jim Sidanius, and Lawrence Bobo (eds.), *Racialized Politics: The Debate About Racism in America.* Chicago: University of Chicago Press.

Seekings, Jeremy, and Nicoli Nattrass. 2005. *Class, Race and Inequality in South Africa.* New Haven, Conn.: Yale University Press.

Séguin, Rhéal. 2005. "54% in Quebec Back Sovereignty." *The Globe and Mail* (April 27): A1, A8.

Selznick, Gertrude J., and Stephen Steinberg. 1969. *The Tenacity of Prejudice: Anti-Semitism in Contemporary America.* New York: Harper & Row.

Semujanga, Josias. 2003. *Origins of Rwandan Genocide.* Amherst, N.Y.: Humanity Books.

Senese, Guy B. 1991. *Self-Determination and the Social Education of Native Americans.* New York: Praeger.

Sengstock, Mary C. 1983. "Detroit's Iraqi-Chaldeans: A Conflicting Conception of Identity." Pp. 136–146 in Sameer Y. Abraham and Nabeel Abraham (eds.), *Arabs in the New World: Studies on Arab-American Communities.* Detroit: Wayne State University, Center for Urban Studies.

———. 2005. *Chaldeans in Michigan.* East Lansing: Michigan State University Press.

Sengupta, Somini, and Vivian S. Toy. 1998. "United Ethnically, and by an Assault." *New York Times* (October 7):A21.

Shaheen, Jack G. 1984. *The TV Arab.* Bowling Green, Ohio: Bowling Green State University Popular Press.

———. 2003. "Reel Bad Arabs: How Hollywood Vilifies a People." *Annals of the American Academy of Political and Social Science* 588 (July):171–93.

Shapiro, Thomas M. 2004. *The Hidden Cost of Being African American: How Wealth Perpetuates Inequality.* New York: Oxford University Press.

Shapiro, Thomas M., Tatjana Meschede, and Laura Sullivan. 2010. "The Racial Wealth Gap Increases Fourfold." Research and Policy Brief (May), Institute on Assets and Social Policy, Brandeis University.

Sharkey, Patrick. 2009. *Neighborhoods and the Black-White Mobility Gap.* Economic Mobility Project. (July). http://www.economicmobility.org

Sharot, Stephen. 1973. "The Three-Generations Thesis and the American Jews." *British Journal of Sociology* 24:151–64.

Shaw, Mark, and Peter Gastrow. 2001. "Stealing the Show? Crime and Its Impact in Post-Apartheid South Africa." *Daedalus* 130 (Winter):235–58.

Sheatsley, Paul B. 1966. "White Attitudes toward the Negro." Pp. 303–24 in Talcott Parsons and Kenneth B. Clark (eds.), *The Negro American.* Boston: Houghton Mifflin.

Sheriff, Robin E. 2001. *Dreaming Equality: Color, Race and Racism in Urban Brazil*. New Brunswick, N.J.: Rutgers University Press.

Sherkat, Darren E. 2004. "Religious Intermarriage in the United States: Trends, Patterns, and Predictions." *Social Science Research* 33:606–25.

Sheskin, Ira M., and Arnold Dashefsky. 2007. "Jewish Population in the United States, 2007." Pp. 133–220 in David Singer and Lawrence Grossman (eds.), *American Jewish Year Book 2007*. New York: American Jewish Committee.

———. 2011. "Jewish Population in the United States, 2011." *Current Jewish Population Reports*, no. 4. Mandell L. Berman Institute-North American Jewish Data Bank. Storrs, CT: University of Connecticut.

Shibutani, Tamotsu, and Kian M. Kwan. 1965. *Ethnic Stratification: A Comparative Approach*. New York: Macmillan.

Shierholz, Heidi. 2010. *"Immigration and Wages: Methodological Advancements Confirm Modest Gains for Native Workers."* Washington, D.C.: Economic Policy Institute.

Shimoni, Gideon. 2003. *Community and Conscience: The Jews in Apartheid South Africa*. Lebanon, N.H.: Brandeis University Press.

Shinagawa, Larry Hajime, and Gin Yong Pang. 1996. "Asian American Panethnicity and Intermarriage." *Amerasia Journal* 22:127–52.

Shirlow, Peter. 2008. "Belfast: A Segregated City." Pp. 73–87 in Colin Coulter and Michael Murray (eds.), *Northern Ireland After the Troubles: A Society in Transition*. Manchester: Manchester University Press.

Shryock, Andrew, Nabeel Abraham, and Sally Howell. 2011a. "The New Order and Its Forgotten Histories." Pp. 381–93 in Nabeel Abraham, Sally Howell, and Andrew Shryock (eds.), *Arab Detroit 9/11: Life in the Terror Decade*. Detroit: Wayne State University Press.

———. 2011b. "The Terror Decade in Arab Detroit: An Introduction." Pp. 1–25 in Nabeel Abraham, Sally Howell, and Andrew Shryock (eds.), *Arab Detroit 9/11: Life in the Terror Decade*. Detroit: Wayne State University Press.

Sibley, Mulford Q. 1963. *The Quiet Battle*. Chicago: Quadrangle.

Sigelman, Lee, Timothy Bledsoe, Susan Welch, and Michael W. Combs. 1996. "Making Contact? Black-White Social Interaction in an Urban Setting." *American Journal of Sociology* 101:1306–32.

Siggner, Andrew J., and Rosalinda Costa. 2005. *"Aboriginal Conditions in Census Metropolitan Areas, 1981–2001."* Ottawa: Statistics Canada.

Silberman, Charles E. 1985. *A Certain People: American Jews and Their Lives Today*. New York: Summit.

Silva, Nelson do Valle. 1985. "Updating the Cost of Not Being White in Brazil." Pp. 42–55 in Pierre-Michel Fontaine (ed.), *Race, Class, and Power in Brazil*. Los Angeles: Center for Afro-American Studies, University of California, Los Angeles.

———. 1999. "Racial Differences in Income: Brazil, 1988." Pp. 67–82 in Rebecca Reichmann (ed.), *Race in Contemporary Brazil: From Indifference to Inequality*. University Park: Pennsylvania State University Press.

Silva, Nelson do Valle, and Carlos A. Hasenbalg. 1999. "Race and Educational Opportunity in Brazil." Pp. 53–65 in Rebecca Reichmann (ed.), *Race in Contemporary Brazil: From Indifference to Inequality*. Park: University Pennsylvania State University Press.

Simmons, Ozzie G. 1971. "The Mutual Images and Expectations of Anglo-Americans and Mexican-Americans." Pp. 62–71 in Nathaniel N. Wagner and Marsha J. Haug (eds.), *Chicanos: Social and Psychological Perspectives*. St. Louis: Mosby.

Simon, Julian L. 1991. "The Case for Greatly Increased Immigration." *The Public Interest* 102 (Winter):89–103.

Simon, Rita James. 1978. *Continuity and Change: A Study of Two Ethnic Communities in Israel*. Cambridge, England: Cambridge University Press.

Simon, Rita J., and James P. Lynch. 1999. "A Comparative Assessment of Public Opinion toward Immigrants and Immigration Policies." *International Migration Review* 33:455–67.

Simpson, Amelia. 1993. *Xuxa: The Mega-Marketing of Gender, Race, and Modernity*. Philadelphia: Temple University Press.

Simpson, George Eaton. 1968. "Assimilation." Pp. 428–44 in David L. Sills (ed.), *International Encyclopedia of the Social Sciences*. Vol. 1. New York: Macmillan.

Simpson, George Eaton, and J. Milton Yinger. 1985. *Racial and Cultural Minorities: An Analysis of Prejudice and Discrimination*. 5th ed. New York: Harper & Row.

Singer, Audrey. 2010. "Immigration." Pp. 64–75 in *The State of Metropolitan America*. Washington, D.C.: Brookings Institution.

———. 2012. *"Immigrant Workers in the U.S. Labor Force."* Washington, D.C.: Brookings Institution.

Singer, Mark. 2001. "Home Is Here: America's Largest Arab Community in the Aftermath of September 11th." *The New Yorker* (October 15):62–70.

Skellington, Richard. 1996. *"Race" in Britain Today*. 2nd ed. London: Sage.

Skidmore, Thomas. 1972. "Toward a Comparative Analysis of Race Relations Since Abolition in Brazil and the United States." *Journal of Latin American Studies* 4 (May):1–28.

———. 1974. *Black into White: Race and Nationality in Brazilian Thought.* New York: Oxford University Press.

———. 1990. "Racial Ideas and Social Policy in Brazil, 1870–1940." Pp. 7–36 in Richard Graham (ed.), *The Idea of Race in Latin America, 1870–1940.* Austin: University of Texas Press.

———. 1993. "Bi-racial U.S.A. vs. Multi-racial Brazil: Is the Contrast Still Valid?" *Journal of Latin American Studies* 25:373–86.

———. 1999. *Brazil: Five Centuries of Change.* New York: Oxford University Press.

———. 2003. "Racial Mixture and Affirmative Action: The Case of Brazil and the United States." *American Historical Review* 108:1391–96.

Sklare, Marshall. 1969. "The Ethnic Church and the Desire for Survival." Pp. 101–17 in Peter I. Rose (ed.), *The Ghetto and Beyond.* New York: Random House.

———. 1971. *America's Jews.* New York: Random House.

———. 1978. "Jewish Acculturation and American Jewish Identity." Pp. 167–88 in Gladys Rosen (ed.). *Jewish Life in America: Historical Perspectives.* New York: American Jewish Committee.

Sklare, Marshall, and Joseph Greenblum. 1967. *Jewish Identity in the Suburban Frontier: A Study of Group Survival in the Open Society.* New York: Basic Books.

Skocpol, Theda, and Vanessa Williamson. 2012. *The Tea Party and the Remaking of Republican Conservatism.* New York: Oxford University Press.

Skrentny, John David. 1996. *The Ironies of Affirmative Action: Politics, Culture, and Justice in America.* Chicago: University of Chicago Press.

———. 2002. *The Minority Rights Revolution.* Cambridge, Mass.: Harvard University Press.

Slawson, John, and Lawrence Bloomgarden. 1965. *The Unequal Treatment of Equals: The Social Club . . . Discrimination in Retreat.* New York: Institute of Human Relations, American Jewish Committee.

Smedley, Audrey. 2007. *Race in North America: Origin and Evolution of a Worldview.* 3rd ed. Boulder, Colo.: Westview Press.

Smith, Anthony D. 1981. *The Ethnic Revival.* Cambridge, England: Cambridge University Press.

Smith, Dan. 2004. "Trends and Causes of Armed Conflict." Pp. 2–16 in *Berghof Handbook for Conflict Transformation.* Berlin: Berghof Research Center for Constructive Conflict Management.

Smith, David J., and Gerald Chambers. 1991. *Inequality in Northern Ireland.* Oxford: Clarendon Press.

Smith, Elmer Lewis. 1958. *The Amish People.* New York: Exposition Press.

Smith, Heather A., and Owen J. Furuseth. (eds.) 2006. *Latinos in the New South: Transformations of Place.* Aldershot, England: Ashgate.

Smith, James P. 2001. "Race and Ethnicity in the Labor Market: Trends over the Short and Long Term." Pp. 52–97 in Neil J. Smelser, William Julius Wilson, and Faith Mitchell (eds.), *America Becoming: Racial Trends and Their Consequences.* Vol. 2. Washington, D.C.: National Academy Press.

Smith, James P., and Barry Edmonston (eds.). 1997. *The New Americans: Economic, Demographic, and Fiscal Effects of Immigration.* Washington, D.C.: National Academy Press.

Smith, M. G. 1965. *The Plural Society in the British West Indies.* Berkeley: University of California Press.

———. 1969. "Institutional and Political Conditions of Pluralism." Pp. 27–65 in Leo Kuper and M. G. Smith (eds.), *Pluralism in Africa.* Berkeley: University of California Press.

Smith, R. Jeffrey. 2000. "Peacekeeping Without Peace in Kosovo." *Washington Post National Weekly Edition* (June 19):15.

Smith, T. Lynn. 1963. *Brazil: People and Institutions.* Baton Rouge: Louisiana State University Press.

———. 1974. *Brazilian Society.* Albuquerque: University of New Mexico Press.

Smith, Tom W. 1980. *A Compendium of Trends on General Social Survey Questions* (NORC Report No. 129). Chicago: National Opinion Research Center.

———. 1991. *What Do Americans Think About Jews?* New York: American Jewish Committee.

———. 2000. *Taking America's Pulse II: NCCJ's 2000 Survey of Intergroup Relations in the United States.* New York: National Conference for Community and Justice.

———. 2005. *Jewish Distinctiveness in America: A Statistical Portrait.* New York: American Jewish Committee.

Smith, Tom W., and Seokho Kim. 2004. *The Vanishing Protestant Majority.* GSS Social Change Report No. 49. Chicago: NORC/ University of Chicago.

Smooha, Sammy. 1978. *Israel: Pluralism and Conflict.* London: Routledge & Kegan Paul.

Smyth, Jim, and Andreas Cebulla. 2008. "The Glacier Moves? Economic Change and Class Structure." Pp. 175–91 in Colin Coulter and Michael Murray (eds.), *Northern Ireland After the Troubles: A Society in Transition.* Manchester, England: Manchester University Press.

Sniderman, Paul M., and Edward G. Carmines. 1997. *Reaching Beyond Race*. Cambridge, Mass.: Harvard University Press.

Sniderman, Paul M., and Louk Hagendoorn. 2007. *When Ways of Life Collide*. Princeton, N.J.: Princeton University Press.

Sniderman, Paul M., and Thomas Piazza. 1993. *The Scar of Race*. Cambridge, Mass.: Belknap Press.

Snipp, C. Matthew. 1986. "The Changing Political and Economic Status of the American Indians: From Captive Nations to Internal Colonies." *American Journal of Economics and Sociology* 45:145–57.

———. 2002. "American Indians: Clues to the Future of Other Racial Groups." Pp. 189–214 in Joel Perlmann and Mary C. Waters (eds.), *The New Race Question: How the Census Counts Multiracial Individuals*. New York: Russell Sage Foundation.

Solis-Garza, Luis A. 1972. "Cesar Chavez: The Chicano Messiah?" Pp. 297–305 in Edward Simmen (ed.), *Pain and Promise: The Chicano Today*. New York: New American Library.

Sorensen, Theodore C. 1965. *Kennedy*. New York: Harper & Row.

Sorin, Gerald. 1997. *Tradition Transformed: The Jewish Experience in America*. Baltimore: Johns Hopkins University Press.

South African Institute of Race Relations (SAIRR). 2005. "The Poor Get Poorer." Johannesburg: South African Institute of Race Relations. http://www.sairr.org.za/

———. 2010. *The Long Shadow of Apartheid: Race in South Africa Since 1994*. Johannesburg: South African Institute of Race Relations.

Southern Poverty Law Center, 2013. "FBI: Bias Crimes Against Muslims Remain at High Levels." *Intelligence Report 149*. Montgomery, Ala.: Southern Poverty Law Center.

Sowell, Thomas. 1981. *Ethnic America*. New York: Basic Books.

———. 1990. *Preferential Policies: An International Perspective*. New York: Morrow.

———. 2004. *Affirmative Action Around the World: An Empirical Study*. New Haven, Conn.: Yale University Press.

Soysal, Levent. 2008. "The Migration Story of Turks in Germany: From the Beginning to the End." Pp. 199–225 in Resat Kasaba (ed.), *Turkey in the Modern World*, Vol. 4. Cambridge: Cambridge University Press.

Sparks, Allister. 2004. *Beyond the Miracle: Inside the New South Africa*. Chicago: University of Chicago Press.

Spear, Allan H. 1967. *Black Chicago*. Chicago: University of Chicago Press.

Spencer, Kyle. 2012. "For Asians, School Tests Are Vital Steppingstones." *New York Times* (October 27):A18.

Spicer, Edward H. 1980. "American Indians." Pp. 58–122 in Stephen Thernstrom (ed.), *Harvard Encyclopedia of American Ethnic Groups*. Cambridge, Mass.: Harvard University Press.

———. 1982. *The American Indians*. Cambridge, Mass.: Harvard University Press.

Stack, Carol. 1996. *Call to Home: African Americans Reclaim the Rural South*. New York: Basic Books.

Stafford, Walter W. 2001. "The National Urban League Survey: Black America's Under-35 Generation." In *The State of Black America 2001*. New York: National Urban League. http://nul.iamempowered.com/

Stampp, Kenneth M. 1956. *The Peculiar Institution: Slavery in the Ante-Bellum South*. New York: Knopf.

Stanley, Alessandra. 2011. "Ciao, Jersey; Hello, Italy: No Culture Shock Here." *New York Times* (August 6):C1, C7.

Stanley, David T., Dean E. Mann, and Jameson W. Doig. 1967. *Men Who Govern*. Washington, D.C.: Brookings Institution Press.

Staples, Brent. 2012. "Young, Black, Male, and Stalked by Bias." *New York Times* (April 15): SR10.

Statistics Canada. 2003. *Ethnic Diversity Survey: Portrait of a Multicultural Society*. Ottawa: Statistics Canada.

———. 2007a. *Mother Tongue, Census 2006*. Ottawa: Statistics Canada.

———. 2007b. *Immigration and Citizenship, Census. 2006*. Ottawa: Statistics Canada.

———. 2007c. *Population by Knowledge of Official Languages, 2006 Census*. Ottawa: Statistics Canada.

———. 2008. *Canada's Ethnocultural Mosaic, 2006 Census*. Ottawa: Statistics Canada.

———. 2009a. *Aboriginal Peoples, Census, 2006*. Ottawa: Statistics Canada.

———. 2009b. *2006 Census: Aboriginal Peoples in Canada in 2006: Inuit, Métis and First Nations, 2006 Census*. Ottawa: Statistics Canada.

———. 2009c. *2006 Census: Educational Portrait of Canada, 2006*. Ottawa: Statistics Canada.

———. 2010a. *Aboriginal Statistics at a Glance*. Ottawa: Statistics Canada.

———. 2010b. *Canada's Ethnocultural Mosaic, 2006 Census: National Picture*. Ottawa: Statistics Canada.

———. 2010c. *Immigration and Citizenship, 2006 Census*. Ottawa: Statistics Canada.

———. 2010d. *2006 Profile of Aboriginal Children, Youth and Adults: Census Highlights*. Ottawa: Statistics Canada.

———. 2012. *Census 2011*. http://www12.statcan.gc.ca

———. 2013. *Aboriginal Peoples in Canada: First Nations People, Metís and Inuit. National Household Survey, 2011*. Ottawa: Minister of Industry.

Statistics South Africa. 2007. *Community Survey 2007*. Pretoria: Statistics South Africa.

———. 2012a. *Census 2011*. Pretoria: Statistics South Africa.

———. 2012b. *Income and Expenditure of Households 2010/2011*. Pretoria: Statistics South Africa.

———. 2013. *Mid-Year Population Estimates 2013*. Pretoria: Statistics South Africa.

Steeh, Charlotte, and Howard Schuman. 1992. "Young White Adults: Did Racial Attitudes Change in the 1980s?" *American Journal of Sociology* 98:340–67.

Steele, Shelby. 1991. *The Content of Our Character: A New Vision of Race in America*. New York: Harper Perennial.

Steger, Wilbur. 1973. "Economic and Social Costs of Residential Segregation." Pp. 83–113 in Marion Clawson (ed.), *Modernizing Urban Land Policy*. Baltimore: Johns Hopkins University Press.

Steinback, Robert. 2011. "The Anti-Muslim Inner Circle." *Intelligence Report* (Summer). Southern Poverty Law Center.

Steinberg, Stephen. 1989. *The Ethnic Myth: Race, Ethnicity and Class in America*. Updated and expanded ed. New York: Atheneum.

———. 1995. *Turning Back: The Retreat from Racial Justice in American Thought and Policy*. Boston: Beacon.

———. 2004. "The Melting Pot and the Color Line." Pp. 235–47 in Tamar Jacoby (ed.), *Reinventing the Melting Pot: The New Immigrants and What It Means to Be American*. New York: Basic Books.

Steiner, Stan. 1970. *La Raza: The Mexican Americans*. New York: Harper & Row.

Steinfels, Peter. 1992. "Debating Intermarriage, and Jewish Survival." *New York Times* (October 18):1, 16.

Stember, Charles H. 1966. *Jews in the Mind of America*. New York: Basic Books.

Stepick, Alex, and Carol Dutton Stepick. 2002. "Power and Identity: Miami Cubans." Pp. 75–92 in Marcelo M. Suárez-Orozco and Mariela M. Páez (eds.), *Latinos: Remaking America*. Berkeley: University of California Press.

Stepick, Alex, Guillermo Grenier, Max Castro, and Marvin Dunn. 2003. *This Land Is Our Land: Immigrants and Power in Miami*. Berkeley: University of California Press.

Sterngold, James. 1998a. "Blacks Have Had Good Times and Bad on Prime-Time TV." *New York Times* (December 29): 1, 12.

———. 1998b. "Prime-Time TV's Growing Racial Divide Frustrates Industry's Blacks." *New York Times* (December 29):12.

Stevens, Gillian, Mary E. M. McKillip, and Hiromi Ishizawa. 2006. "Intermarriage in the Second Generation: Choosing Between Newcomers and Natives." Migration Policy Institute (October 1). http://www.migrationinformation.org/

Stevens, Gillian, and Michael K. Tyler. 2002. "Ethnic and Racial Intermarriage in the United States: Old and New Regimes." Pp. 221–42 in Nancy A. Denton and Stewart E. Tolnay (eds.), *American Diversity: A Demographic Challenge for the Twenty-First Century*. Albany: State University of New York Press.

Stockton, Ronald. 1994. "Ethnic Archetypes and the Arab Image. Pp. 119–53 in Ernest McCarus (ed.), *The Development of Arab-American Identity*. Ann Arbor: University of Michigan Press.

Stoddard, Ellwyn R. 1973. *Mexican Americans*. New York: Random House.

Stodghill, Ron. 2007. "Room at the Top?" *New York Times* (November 1):C1, C11.

Stoffman, Daniel. 2009. "An Ideology, Not a Fact." *Globe and Mail* (August 22):A15.

Stoll, Michael. 2006. "Job Sprawl, Spatial Mismatch and Black Employment Disadvantage." *Journal of Policy Analysis and Management* 25:827–54.

Stone, John. 1973. *Colonist or Uitlander: A Study of the British Immigrant in South Africa*. Oxford: Clarendon Press.

Stonequist, Everett V. 1937. *The Marginal Man*. New York: Charles Scribner's Sons.

Stoney, Sierra, and Jeanne Batalova. 2013. "Central American Immigrants in the United States." *Migration Information Source* (March 18). Washington, D.C.: Migration Policy Institute.

Story, Louise. 2007. "Overture to an Untapped Market." *New York Times* (April 28):B1, B4.

Stout, Cameron W. 1997. "Canada's Newest Territory in 1999." *Canadian Social Trends* (Spring):13–18.

Strodtbeck, Fred L. 1958. "Family Interaction, Values, and Achievement." Pp. 147–65 in Marshall Sklare (ed.), *The Jews: Social Patterns of an American Group*. Glencoe, Ill.: Free Press.

Stultz, Newell M. 1980. "Some Implications of African 'Homelands' in South Africa." Pp. 194–216 in Robert M. Price and Carl G. Rotberg (eds.), *The Apartheid Regime: Political Power and Racial Domination*. Berkeley: Institute of International Studies, University of California.

Stutz, Howard. 2012. "Seminoles Have Become a Force in Indian Gaming." *Las Vegas Review-Journal* (February 12).

Suárez-Orozco, Marcelo M., and Mariela M. Páez. 2002. "The Research Agenda." Pp. 1–37 in Marcelo M. Suárez-Orozco and Mariela M. Páez (eds.), *Latinos: Remaking America*. Berkeley: University of California Press.

Sudetic, Chuck. 1994. "Serbs of Sarajevo Stay Loyal to Bosnia." *New York Times* (August 26):8A.

Suleiman, Michael W. 1999. "Introduction: The Arab Immigrant Experience." Pp. 1–24 in Michael Suleiman (ed.), *Arabs in America: Building a New Future.* Philadelphia: Temple University Press.

Sunahara, M. Ann. 1980. "Federal Policy and the Japanese Canadians: The Decision to Evacuate, 1942." Pp. 93–120 in K. Victor Ujimoto and Gordon Hirabayashi (eds.), *Visible Minorities and Multiculturalism: Asians in Canada.* Toronto: Butterworths.

Suro, Roberto. 1989. "Employers Are Looking Abroad for the Skilled and the Energetic." *New York Times* (July 16):IV-4.

———. 1998. *Strangers Among Us: How Latino Immigration Is Transforming America.* New York: Knopf.

Swain, Carol M. 2001. "Affirmative Action: Legislative History, Judicial Interpretations, Public Consensus." Pp. 318–47 in Neil J. Smelser, William Julius Wilson, and Faith Mitchell (eds.), *America Becoming: Racial Trends and Their Consequences.* Vol. 1. Washington, D.C.: National Academy Press.

———. 2007. "The Congressional Black Caucus and the Impact of Immigration on African American Unemployment." Pp. 175–88 in Carol M. Swain (ed.), *Debating Immigration.* New York: Cambridge University Press.

Swarns, Rachel L. 2000. "Hesitantly, Students in South Africa Reach Across the Racial Divide." *New York Times* (November 12):10.

———. 2002a. "In a New South Africa, an Old Tune Lingers." *New York Times* (October 7): A1, A9.

———. 2002b. "A Hit Song Puts Ethnic Tension at Center Stage." *New York Times* (June 13).

———. 2004. "'African-American' Becomes a Term for Debate." *New York Times* (August 29):1, 14.

———. 2006. "Bridging a Racial Rift That Isn't Black and White." *New York Times* (October 3):A1, A25.

———. 2008. "Quiet Political Shifts as More Blacks Are Elected." *New York Times* (October 14):A1, A18.

Szymanski, Albert. 1976. "Social Discrimination and White Gain." *American Sociological Review* 41:403–14.

Taeuber, Karl E. 1975. "Racial Segregation: The Persisting Dilemma." *Annals of the American Academy of Political and Social Science* 422:87–96.

———. 1979. "Housing, Schools, and Incremental Segregative Effects." *Annals of the American Academy of Political and Social Science* 441:157–67.

Taeuber, Karl E., and Alma F. Taeuber. 1965. *Negroes in Cities.* Chicago: Aldine.

Tajitsu Nash, Phil, and Frank Wu. 1997. "Asian Americans under Glass: Where the Furor over the President's Fundraising Has Gone Awry—and Racist." *The Nation* (March 31):15–16.

Tannenbaum, Frank. 1947. *Slave and Citizen: The Negro in the Americas.* New York: Knopf.

Tasker, Fredric. 1980. "Anti-Bilingualism Measure Approved in Dade County." *Miami Herald* (November 5):1.

Tavernise, Sabrina. 2007. "It Has Unraveled So Quickly." *New York Times* (January 28):1, 18.

———. 2013. "For Medicare, Immigrants Offer Surplus, Study Finds." *New York Times* (May 30):A12, A15.

Taylor, D. Garth. 1979. "Housing, Neighborhoods, and Race Relations: Recent Survey Evidence." *Annals of the American Academy of Political and Social Science* 441:26–40.

Taylor, D. Garth, Paul B. Sheatsley, and Andrew M. Greeley. 1978. "Attitudes Toward Racial Integration." *Scientific American* 238 (June):42–49.

Taylor, Paul. 2008. "Race, Ethnicity and Campaign '08." Pew Research Center (January 17). http://www.pewresearch.org/2008/01/17/race-ethnicity-and-campaign-08/

———. 2012. "The Growing Electoral Clout of Blacks Is Driven by Turnout, Not Demographics." Pew Research Center. http://www.pewsocialtrends.org/files/2013/01/2012_Black_Voter_Project_revised_1-9.pdf

Taylor, Paul, and D'Vera Cohn. 2012. "A Milestone En Route to a Majority Minority Nation." Pew Research Social & Demographic Trends (November 7). http://www.pewsocialtrends.org/2012/11/07/a-milestone-en-route-to-a-majority-minority-nation/

Taylor, Paul, Ana Gonzalez-Barrera, Jeffrey Passel, and Mark Hugo Lopez. 2012. "An Awakened Giant: The Hispanic Electorate Is Likely to Double by 2030." Washington, D.C.: Pew Hispanic Center.

Taylor, Paul, Mark Hugo Lopez, Jessica Hamar Martínez, and Gabriel Velasco. 2012. "When Labels Don't Fit: Hispanics and Their Views of Identity." Washington, D.C.: Pew Hispanic Center.

Taylor, Ronald L. 1979. "Black Ethnicity and the Persistence of Ethnogenesis." *American Journal of Sociology* 84:1401–23.

Taylor, William L. 1995. "Affirmative Action: The Questions to Be Asked." *Poverty and Race* (May/June):2–3.

Tehranian, John. 2009. *Whitewashed: America's Invisible Middle Eastern Minority.* New York: New York University Press.

Teles, Steven M. 2001. "Positive Actions or Affirmative Action? The Persistence of Britain's Antidiscrimination Regime." Pp. 241–70 in John D. Skrentny (ed.), *Color Lines: Affirmative Action, Immigration, and Civil Rights Options for America*. Chicago: University of Chicago Press.

Telhami, Shibley. 2002. "Arab and Muslim America: A Snapshot." *Brookings Review* 20(1):14–15.

Telles, Edward E. 1992. "Residential Segregation by Skin Color in Brazil." *American Sociological Review* 57:186–97.

———. 2004. *Race in Another America: The Significance of Skin Color in Brazil*. Princeton, N.J.: Princeton University Press.

———. 2009. "The Social Consequences of Skin Color in Brazil." Pp. 9–24 in Evelyn Nakano Glenn (ed.), *Shades of Difference: Why Skin Color Matters*. Stanford, Calif.: Stanford University Press.

Telles, Edward E., and Nelson Lim. 1998. "Does It Matter Who Answers the Race Question? Racial Classification and Income Inequality in Brazil." *Demography* 35:465–74.

tenBroek, Jacobus, Edward N. Barnhart, and Floyd W. Matson. 1968. *Prejudice, War and the Constitution*. Berkeley: University of California Press.

Terrazas, Aaron. 2011. "Middle Eastern and North African Immigrants in the United States." Washington, D.C.: Migration Policy Institute.

Thernstrom, Stephan, and Abigail Thernstrom. 1997. *America in Black and White: One Nation, Indivisible*. New York: Simon & Schuster.

Thistlethwaite, Frank. 1991. "Migration from Europe Overseas in the Nineteenth and Twentieth Centuries." Pp. 17–57 in Rudolph J. Vecoli and Suzanne M. Sinke (eds.), *A Century of European Migrations, 1830–1930*. Urbana: University of Illinois Press.

Thomas, Piri. 1967. *Down These Mean Streets*. New York: Knopf.

Thomas, Robert K. 1966. "Colonialism: Classic and Internal." *New University Thought* 4 (Winter):37–44.

Thomas, W. I., and Florian Znaniecki. 1918. *The Polish Peasant in Europe and America*. Vol. 1. New York: Knopf.

Thompson, John Herd, and Morton Weinfeld. 1995. "Entry and Exit: Canadian Immigration Policy in Context." *Annals of the American Academy of Political and Social Science* 538:185–98.

Thompson, Leonard M. 1964. "The South African Dilemma." Pp. 178–218 in Louis Hartz (ed.), *The Founding of New Societies*. New York: Harcourt, Brace & World.

———. 1985. *The Political Mythology of Apartheid*. New Haven, Conn.: Yale University Press.

———. 2001. *A History of South Africa*. 3rd ed. New Haven, Conn.: Yale University Press.

Thompson, Leonard M., and Andrew Prior. 1982. *South African Politics*. New Haven, Conn.: Yale University Press.

Thomson, Dale. 1995. "Language, Identity, and the Nationalist Impulse: Quebec." *Annals of the American Academy of Political and Social Science* 538 (March):69–82.

Thornton, Russell. 1987. *American Indian Holocaust and Survival: A Population History Since 1492*. Norman: University of Oklahoma Press.

———. 1995. "North American Indians and the Demography of Contact." Pp. 213–30 in Vera Lawrence Hyatt and Rex Nettleford (eds.), *Race, Discourse, and the Origin of the Americas: A New World View*. Washington, D.C.: Smithsonian Institution Press.

———. 2001. "Trends among American Indians in the United States." Pp. 135–69 in Neil J. Smelser, William Julius Wilson, and Faith Mitchell (eds.), *America Becoming: Racial Trends and Their Consequences*. Vol. 1. Washington, D.C.: National Academy Press.

Thurow, Lester. 1969. *Poverty and Discrimination*. Washington, D.C.: Brookings Institution Press.

Time. 1960. "Grosse Pointe's Gross Points." 75 (April 25):25.

———. 1987. "The New Whiz Kids." 131 (August 31):42–51.

Tinker, John N. 1973. "Intermarriage and Ethnic Boundaries: The Japanese American Case." *Journal of Social Issues* 29:49–66.

Tobar, Hector. 2001. "A Battle over Who Is Indian." *Los Angeles Times* (January 4):A1, A7.

Tobias, Phillip V. 1995. "Race." Pp. 711–15 in Adam Kuper and Jessica Kuper (eds.), *The Social Science Encyclopedia*. 2nd ed. London: Routledge.

Toplin, Robert Brent. 1981. *Freedom and Prejudice: The Legacy of Slavery in the United States and Brazil*. Westport, Conn.: Greenwood.

Torrecilha, Ramon S., Lionel Cantú, and Quon Nguyen. 1999. "Puerto Ricans in the United States." Pp. 230–54 in Anthony Gary Dworkin and Rosalind J. Dworkin (eds.), *The Minority Report: An Introduction to Racial, Ethnic, and Gender Relations*. 3rd ed. Orlando: Harcourt Brace.

Train, Arthur. 1974 [1912]. "Imported Crime." Pp. 184–88 in Wayne Moquin and Charles Van Doren (eds.), *A Documentary History of the Italian Americans*. New York: Praeger.

Troper, Harold M. 1972. *Only Farmers Need Apply*. Toronto: Griffen House.

Trudeau Foundation. 2006. Backgrounder-Environics Research Group Poll for the Trudeau Foundation. http://www .foundationtrudeau.ca

Trueheart, Charles, and Dennis McAuliffe, Jr. 1995. "A Way of Life at Risk." *Washington Post National Weekly Edition* (September 18–24):18–19.

Tuch, Steven A., and Michael Hughes. 2011. "Whites' Racial Policy Attitudes in the Twenty-First Century: The Continuing Significance of Racial Resentment." *Annals of the American Academy of Political and Social Science* 634(1):134–52.

Tucker, Clyde, and Brian Kojetin. 1996. "Testing Racial and Ethnic Origin Questions in the CPS Supplement." *Monthly Labor Review* (September):3–7.

Tumin, Melvin M. 1964. "Ethnic Group." P. 243 in Julius Gould and William L. Kolb (eds.), *Dictionary of the Social Sciences*. New York: Free Press.

Turner, Jonathan H., and Edna Bonacich. 1980. "Toward a Composite Theory of Middleman Minorities." *Ethnicity* 7:144–58.

Turner, Ralph H., and Lewis M. Killian. 1972. *Collective Behavior*. 2nd ed. Englewood Cliffs, N.J.: Prentice Hall.

Tuttle, William M., Jr. 1970. *Race Riot: Chicago in the Red Summer of 1919*. New York: Atheneum.

Twine, France Winddance. 1998. *Racism in a Racial Democracy: The Maintenance of White Supremacy in Brazil*. New Brunswick, N.J.: Rutgers University Press.

Tyler, Gus. 1975. "Introduction: A People on the Move." Pp. 1–18 in Gus Tyler (ed.), *Mexican-Americans Tomorrow*. Albuquerque: University of New Mexico Press.

Ubelaker, D. H. 1988. "North American Indian Population Size, A.D. 1500 to 1985." *American Journal of Physical Anthropology* 77:289–94.

Ukeles, Jacob B., Ron Miller, and Pearl Beck. 2006. "Young Jewish Adults in the United States Today: Harbingers of the American Jewish Community of Tomorrow?" Washington, D.C.: American Jewish Committee.

Ukeles, Jacob B., Steven M. Cohen, and Ron Miller. 2013. *Jewish Community Study of New York: 2011: Special Report on Jewish Poverty*. New York: UJA Federation of New York.

United Jewish Committee. 2003. *National Jewish Population Survey 2000–01 (NJPS)*. New York: United Jewish Committee.

UNDP (United Nations Development Programme). 2009. *Human Development Report 2009*. New York: United Nations Development Programme.

United Nations. 2006. *International Migration, 2006*. http://www.un.org/esa/population/publications/2006Migration_Chart/2006IttMig_chart.htm

———. 2013. "232 Million International Migrants Living Abroad Worldwide—New UN Global Migration Statistics Reveal." September 11.

New York: United Nations Department of Economic and Social Affairs.

U.S. Census Bureau. 1900. *Census of Population*. Washington, D.C.: U.S. Government Printing Office.

———. 1920. *Census of Population*. Washington, D.C.: U.S. Government Printing Office.

———. 1979. *The Social and Economic Status of the Black Population in the United States: An Historical Overview, 1790–1978*. Current Population Reports. Series P-23, No. 80. Washington, D.C.: U.S. Government Printing Office.

———. 1997. *Statistical Abstract of the United States, 1997*. Washington, D.C.: U.S. Government Printing Office.

———. 2000. *Projections of the Resident Population by Race*, Hispanic Origin, and Nativity. http://www.census.gov/population/nation/summary/np-t5-g.txt

———. 2003. *The Arab Population: 2000*. Washington, D.C.: U.S. Government Printing Office.

———. 2004a. *Income, Poverty, and Health Insurance Coverage in the United States: 2003*. Washington, D.C.: U.S. Government Printing Office.

———. 2004b. *Statistical Abstract of the United States, 2004*. Washington, D.C.: U.S. Government Printing Office.

———. 2006. *Hispanic-Owned Firms: 2002*. Washington, D.C.: U.S. Government Printing Office.

———. 2007a. *The American Community—American Indians and Alaska Natives: 2004*. Washington, D.C.: U.S. Government Printing Office.

———. 2007b. *The American Community—Asians: 2004*. Washington, D.C.: U.S. Government Printing Office.

———. 2007c. *The American Community—Hispanics: 2004*. Washington, D.C.: U.S. Government Printing Office.

———. 2007d. *Selected Social Characteristics in the United States: 2006*. 2006 American Community Survey. http://factfinder.census.gov/

———. 2008a. *National Population Projections*. http://www.census.gov/population/www/projections/summarytables.html

———. 2008b. *Net Worth and the Assets of Households: 2002*. Current Population Reports. P70–115. Washington, D.C.: U.S. Government Printing Office.

———. 2010a. *American Community Survey*. http://factfinder.census.gov/servlet/DatasetMainPageServlet?_program=ACS&_submenuId=datasets_2&_lang=en

———. 2010b. *Income, Poverty, and Health Insurance Coverage in the United States: 2009*. Washington, D.C.: U.S. Government Printing Office.

———. 2010c. *Race and Hispanic Origin of the Foreign-Born Population in the United States: 2007.* Washington, D.C.: U.S. Government Printing Office.

———. 2010d. *Statistical Abstract of the United States 2010.* Washington, D.C.: U.S. Government Printing Office.

———. 2010e. *Survey of Business Owners-Asian-Owned Firms: 2007.* http://www.census.gov/econ/sbo/get07sof.html?2

———. 2010f. *Voting and Registration in the Election of November 2008.* Washington, D.C.: U.S. Government Printing Office.

———. 2011. *The Hispanic Population: 2010.* Washington, D.C.: U.S. Government Printing Office.

———. 2012a. *American Community Survey.* 2012. http://factfinder2.census.gov

———. 2012b. *The American Indian and Alaska Native Population: 2010.* Washington, D.C.: U.S. Government Printing Office.

———. 2012c. *The Asian Population: 2010.* Washington, D.C.: U.S. Government Printing Office.

———. 2012d. *Income, Poverty, and Health Insurance Coverage in the United States: 2011.* Washington, D.C.: U.S. Government Printing Office.

———. 2012e. *Statistical Abstract of the United States: 2012.* Washington, D.C.: U.S. Government Printing Office.

———. 2012f. *The Two or More Races Population.* Washington, D.C.: U.S. Government Printing Office.

U.S. Commission on Civil Rights. 1970. *Mexican Americans and the Administration of Justice in the Southwest.* Washington, D.C.: U.S. Government Printing Office.

———. 1976. *Puerto Ricans in the Continental United States: An Uncertain Future.* Washington, D.C.: U.S. Government Printing Office.

———. 1977. *Window Dressing on the Set: Women and Minorities in Television.* Washington, D.C.: U.S. Government Printing Office.

———. 1979. *Window Dressing on the Set: An Update.* Washington, D.C.: U.S. Government Printing Office.

———. 1986. *Recent Activities Against Citizens and Residents of Asian Descent.* Washington, D.C.: U.S. Government Printing Office.

———. 1988. *The Economic Status of Americans of Asian Descent: An Exploratory Investigation.* Washington, D.C.: Clearinghouse Publication.

———. 1992. *Civil Rights Issues Facing Asian Americans in the 1990s.* Washington, D.C.: U.S. Government Printing Office.

———. 2003. *A Quiet Crisis: Federal Funding and Unmet Needs in Indian Country.* Washington, D.C.: U.S. Government Printing Office.

———. 2004. *Broken Promises: Evaluating the Native American Health Care System.* Washington, D.C.: U.S. Government Printing Office.

U.S. Department of Education. 1991. *Indian Nations at Risk: An Educational Strategy for Action.* Washington, D.C.: U.S. Department of Education.

———. 2003. *The Condition of Education 2003.* Washington, D.C.: U.S. Government Printing Office.

———. 2010. *Status and Trends in the Education of Racial and Ethnic Groups.* Washington, D.C.: National Center for Education Statistics.

U.S. Department of Homeland Security. 2003. *Yearbook of Immigration Statistics, 2002.* Washington, D.C.: U.S. Government Printing Office.

———. 2004. *Yearbook of Immigration Statistics, 2003.* Washington, D.C.: U.S. Government Printing Office.

———. 2010. *Yearbook of Immigration Statistics: 2009.* Washington, D.C.: U.S. Government Printing Office.

———. 2012. *Yearbook of Immigration Statistics.* Washington, D.C.: U.S. Government Printing Office.

U.S. Department of Housing and Urban Development. 2013. *"Housing Discrimination Against Racial and Ethnic Minorities 2012."* Washington, D.C.: U.S. Government Printing Office.

U.S. Immigration and Naturalization Service. 1984. *Statistical Yearbook of the Immigration and Naturalization Service 1984.* Washington, D.C.: U.S. Government Printing Office.

———. 1996. *Statistical Yearbook of the Immigration and Naturalization Service 1996.* Washington, D.C.: U.S. Government Printing Office.

Utter, Jack. 1993. *American Indians: Answers to Today's Questions.* Lake Ann, Mich.: National Woodlands.

Vaillancourt, Francois, Dominique Lemay, and Luc Vaillancourt. 2007. *Laggards No More: The Changed Socioeconomic Status of Francophones in Quebec.* Toronto: C.D. Howe Institute.

Valandro, Franz. 2004. *The Peace Process in Northern Ireland.* Frankfurt am Main: Peter Lang.

Valdez, Avelardo. 1983. "Recent Increases in Intermarriage by Mexican-American Males." *Social Science Quarterly* 64:136–44.

Valentino, Benjamin A. 2004. *Final Solutions: Mass Killing and Genocide in the Twentieth Century.* Ithaca, N.Y.: Cornell University Press.

Van Ansdale, Debra, and Joe R. Feagin. 2001. *The First R: How Children Learn Race and Racism.* Lanham, Md.: Rowman & Littlefield.

van den Berghe, Pierre L. 1967. *South Africa: A Study in Conflict*. Berkeley: University of California Press.

———. 1970. *Race and Ethnicity: Essays in Comparative Sociology*. New York: Basic Books.

———. 1976. "Ethnic Pluralism in Industrial Societies: A Special Case?" *Ethnicity* 3:242–55.

———. 1978. *Race and Racism: A Comparative Perspective*. 2nd ed. New York: Wiley.

———. 1979a. "The Impossibility of a Liberal Solution in South Africa." Pp. 56–67 in Pierre van den Berghe (ed.), *The Liberal Dilemma in South Africa*. New York: St. Martin's.

———. 1979b. "Nigeria and Peru: Two Contrasting Cases in Ethnic Pluralism." *International Journal of Comparative Sociology* 20:162–74.

———. 1981. *The Ethnic Phenomenon*. New York: Elsevier.

———. 1997. "Rehabilitating Stereotypes." *Ethnic and Racial Studies* 20:1–16.

Vander Zanden, James W. 1983. *American Minority Relations*. New York: Knopf.

Van Valey, Thomas L., Wade Clark Roof, and Jerome Wilcox. 1977. "Trends in Residential Segregation: 1960–1970." *American Journal of Sociology* 82:826–44.

Van Vugt, William E. 1999. *Britain to America: Mid-Nineteenth-Century Immigrants to the United States*. Urbana: University of Illinois Press.

van zyl Slabbert, F. 1975. "Afrikaner Nationalism, White Politics, and Political Change in South Africa." Pp. 3–18 in Leonard Thompson and Jeffrey Butler (eds.), *Change in Contemporary South Africa*. Berkeley: University of California Press.

———. 1987. "Incremental Change or Revolution?" Pp. 399–409 in Jeffrey Butler, Richard Elphick, and David Welsh (eds.), *Democratic Liberalism in South Africa: Its History and Prospect*. Middletown, Conn.: Wesleyan University Press.

Varadarajan, Tunku. 1999. "A Patel Motel Cartel?" *New York Times Magazine* (July 4): 36–39.

Vasta, Ellie. 2007. "From Ethnic Minorities to Ethnic Majority Policy: Multiculturalism and the Shift to Assimilationism in the Netherlands." *Ethnic and Racial Studies* 30:713–40.

Vecoli, Rudolph J. 1964. "Contadini in Chicago: A Critique of the Uprooted." *Journal of American History* 51:404–17.

———. 1996. "Are Italian Americans Just White Folks?" Pp. 3–17 in Mary Jo Bona and Anthony Julian Tamburri (eds.), *Through the Looking Glass: Italian and Italian/American Images in the Media*. Staten Island, N.Y.: American Italian Historical Association.

Vedantam, Shankar. 2010. "Shades of Prejudice." *New York Times* (January 19):A27.

Veltman, Calvin. 1983. *Language Shift in the United States*. Berlin: Mouton.

Verba, Sidney, and Norman H. Nie. 1972. *Participation in America*. New York: Harper & Row.

Verba, Sidney, Kay Lehman Scholzman, and Henry Brady. 1995. "Race, Ethnicity, and Political Participation." Pp. 354–78 in Paul E. Peterson (ed.), *Classifying by Race*. Princeton, N.J.: Princeton University Press.

Verwimp, Philip. 2006. "Peasant Ideology and Genocide in Rwanda under Habyarimana." Pp. 1–40 in Susan E. Cook (ed.), *Genocide in Cambodia and Rwanda: New Perspectives*. New Brunswick, N.J.: Transaction.

Vestergaard, Mads. 2001. "Who's Got the Map? The Negotiation of Afrikaner Identities in Post-Apartheid South Africa." *Daedalus* 130 (Winter):19–44.

Vidal, David. 1978. "Many Blacks Shut Out of Brazil's Racial 'Paradise.'" *New York Times* (June 5):1, 10.

Vidmar, Neil, and Milton Rokeach. 1974. "Archie Bunker's Bigotry: A Study in Selective Perception and Exposure." *Journal of Communication* 24:36–47.

Vieira, Rosângela Maria. 1995. "Black Resistance in Brazil: A Matter of Necessity." Pp. 227–40 in Benjamin Bowser (ed.), *Racism and Anti-Racism in the World Perspective*. Thousand Oaks, Calif.: Sage.

Viglucci, Andres. 2000. "Hispanic Wave Forever Alters Small Town in North Carolina." *Miami Herald* (January 2). http://www.herald.com

Viglucci, Andres, and Diana Marrero. 2000. "Poll Reveals Widening Split over Elian." *Miami Herald* (April 9). http://www.herald.com

Vincent, Louise. 2008. "The Limitations of 'Inter-Racial Contact': Stories from Young South Africa." *Ethnic and Racial Studies* 31:1426–51.

Vitello, Paul. 2007. "Call Him Harry: Village Mayor Symbolizes Indian-Americans' Political Rise." *New York Times* (August 1):A20.

Wagley, Charles. 1971. *An Introduction to Brazil*. Rev. ed. New York: Columbia University Press.

Wagley, Charles, and Marvin Harris. 1958. *Minorities in the New World: Six Case Studies*. New York: Columbia University Press.

Wagner, Nathaniel N., and Marsha J. Haug. 1971. *Chicanos: Social and Psychological Perspectives*. St. Louis: Mosby.

Wakin, Daniel J. 2007. "Increasingly in the West, the Players Are from the East." *New York Times* (April 4):B1, B8.

Walbridge, Linda S. 1999. "Middle Easterners and North Africans." Pp. 391–410 in Elliott R. Barkan (ed.), *A Nation of Peoples: A*

Sourcebook on America's Multicultural Heritage. Westport, Conn.: Greenwood Press.

Waldinger, Roger. 2007. *Between Here and There: How Attached Are Latino Immigrants to Their Native Country?* Washington, D.C.: Pew Hispanic Center.

Walker, James W. St. G. 1980. *A History of Blacks in Canada: A Study Guide for Teachers and Students.* Ottawa: Minister of Supply and Services.

Walks, R. Alan, and Larry S. Bourne. 2006. "Ghettos in Canada's Cities? Racial Segregation, Ethnic Enclaves and Poverty Concentration in Canadian Urban Areas." *Canadian Geographer* 50:273–98.

Wang, L. Ling-chi. 2003. "Race, Class, Citizenship, and Extraterritoriality: Asian Americans and the 1996 Campaign Finance Scandal." Pp. 281–95 in Don T. Nakanishi and James S. Lai (eds.), *Asian American Politics: Law, Participation and Policy.* Lanham, Md.: Rowman & Littlefield.

Wang, Wendy. 2012. *"The Rise of Intermarriage: Rates, Characteristics Vary by Race and Gender."* Washington, D.C.: Pew Research Center.

Ward, David. 1971. *Cities and Immigrants: A Geography of Change in Nineteenth Century America.* New York: Oxford University Press.

Ward, Lewis B. 1965. "The Ethnics of Executive Selection." *Harvard Business Review* 43 (March/April):6–39.

Warikoo, Niraj. 2001. "Arab Americans Fight Redistricting." *Detroit Free Press* (July 6):A1, A14.

———. 2011. "Metro Arabs and Muslims Thrive as Influence Rises." *Detroit Free Press* (September 8):10A–11A.

Warner, W. Lloyd, and James Abegglen. 1963. *Big Business Leaders in America.* New York: Atheneum.

Warner, W. Lloyd, and Leo Srole. 1945. *The Social Systems of American Ethnic Groups.* New Haven, Conn.: Yale University Press.

Warry, Wayne. 2007. *Ending Denial: Understanding Aboriginal Issues.* Peterborough, ON: Broadview.

Washburn, Wilcomb E. 1975. *The Indian in America.* New York: Harper & Row.

Washington Post. 2008. "Off-Reservation Casinos Pit Indians Against Gambling Opponents." (November 26).

Washington Post-ABC News. 2012. *Washington Post-ABC News Poll* (April 8). http://www.washingtonpost.com/

Waters, Mary C. 1990. *Ethnic Options: Choosing Identities in America.* Berkeley: University of California Press.

———. 1999. *Black Identities: West Indian Immigrant Dreams and American Realities.* New York: Russell Sage Foundation.

Waters, Mary C., and Karl Eschbach. 1995. "Immigration and Ethnic and Racial Inequality in the United States." *Annual Review of Sociology* 21:419–46.

Watson, James L. 1977. *Between Two Cultures: Migrants and Minorities in Britain.* Oxford: Basil Blackwell.

Wax, Murray L. 1971. *Indian Americans: Unity and Diversity.* Englewood Cliffs, N.J.: Prentice Hall.

Waxman, Chaim I. 2001. *Jewish Baby Boomers: A Communal Perspective.* Albany: State University of New York Press.

Webster, Peggy Lovell, and Jeffrey W. Dwyer. 1988. "The Cost of Being Nonwhite in Brazil." *Sociology and Social Research* 72:136–38.

Weiss, Lowell. 1994. "Timing Is Everything." *Atlantic Monthly* (January):32–37.

Weissbach, Lee Shai. 2005. *Jewish Life in Small-Town America: A History.* New Haven, Conn.: Yale University Press.

Weller, Christian E., and Amanda Logan. 2009. "Leveling the Playing Field: How to Ensure Minorities Share Equitably in the Economic Recovery and Beyond." Washington, D.C.: Center for American Progress.

Weller, Christian E., Julie Ajinkya, and Jane Farrell. 2012. *The State of Communities of Color in the U.S. Economy.* Washington, D.C.: Center for American Progress.

Welsh, David. 1975. "The Politics of White Supremacy." Pp. 51–78 in Leonard Thompson and Jeffrey Butler (eds.), *Change in Contemporary South Africa.* Berkeley: University of California Press.

Wertheimer, Jack. 1989. "Recent Trends in American Judaism." Pp. 63–162 in *American Jewish Year Book 1989.* New York: American Jewish Committee.

———. 1993. *A People Divided: Judaism in Contemporary America.* New York: Basic Books.

Westerman, Marty. 1989. "Death of the Frito Bandito." *American Demographics* 11 (March):28–32.

Westie, Frank P. 1964. "Race and Ethnic Relations." Pp. 576–618 in Robert E. L. Faris (ed.), *Handbook of Modern Sociology.* Chicago: Rand McNally.

Westoff, Charles F. 2007. "Immigration and Future Population Change in America." Pp. 165–72 in Carol M. Swain (ed.), *Debating Immigration.* New York: Cambridge University Press.

Weston, Mary Ann. 1996. *Native Americans in the News: Images of Indians in the Twentieth Century Press.* Westport, Conn.: Greenwood Press.

Wexler, Alan. 1995. *Atlas of Westward Expansion.* New York: Facts on File.

Whalen, Carmen Teresa. 2005. "Colonialism, Citizenship, and the Making of the Puerto

Rican Diaspora: An Introduction." Pp. 1–42 in Carmen Teresa Whalen and Víctor Vásquez-Hernández (eds.), *The Puerto Rican Diaspora: Historical Perspectives*. Philadelphia: Temple University Press.

Whatley, Monica, and Jeanne Batalova. 2013. "Indian Immigrants in the United States." *Migration Information Source* (August 21). Washington, D.C.: Migration Policy Institute.

Whitaker, Mark. 1995. "Whites v. Blacks." *Newsweek* (October 16):28–35.

White, George. 1993. "Anger Flares on Both Sides of Counter." *Los Angeles Times* (June 8):H7.

White, Michael J. 1987. *American Neighborhoods and Residential Differentiation*. New York: Russell Sage Foundation.

Whitefield, Mimi. 2000. "Hispanic Firms: Global View Spurs Growth of Hispanic Companies." *Miami Herald* (October 7). http://www.herald.com

Whitfield, Stephen J. 1995. "Movies in America as Paradigms of Accommodation." Pp. 79–94 in Robert M. Seltzer and Norman J. Cohen (eds.), *The Americanization of the Jews*. New York: New York University Press.

Whyte, John. 1983. "How Much Discrimination Was There Under the Unionist Regime, 1921–68?" Pp. 1–36 in Tom Gallagher and James O'Connell (eds.), *Contemporary Irish Studies*. Manchester, England: Manchester University Press.

———. 1990. *Interpreting Northern Ireland*. Oxford: Clarendon Press.

Wiechers, Marinus. 1989. *South African Political Terms*. Cape Town: Tafelberg.

Wilder, Esther I., and William H. Walters. 1998. "Ethnic and Religious Components of the Jewish Income Advantage, 1969 and 1989." *Sociological Inquiry* 68:426–36.

Wilkerson, Isabel. 1993. "Black Mediator Serves as a Bridge to Koreans." *New York Times* (June 2):A7.

Wilkes, Rima, and John Iceland. 2004. "Hyper-segregation in the Twenty-First Century." *Demography* 41:23–36.

Wilkie, Mary E. 1977. "Colonials, Marginals and Immigrants: Contributions to a Theory of Ethnic Stratification." *Comparative Studies in Society and History* 19:67–95.

Wilkinson, Charles. 2005. *Blood Struggle: The Rise of Modern Indian Nations*. New York: Norton.

Willems, Emilio. 1960. "Brazil." Pp. 119–46 in International Sociological Association (ed.), *The Positive Contribution by Immigrants*. Paris: UNESCO.

Williams, Rhys H., and Gira Vashi. 2007. "Hijab and American Muslim Women: Creating the Space for Autonomous Selves. *Sociology of Religion* 68:269–87.

Williams, Robin M., Jr. 1964. *Strangers Next Door: Ethnic Relations in American Communities*. Englewood Cliffs, N.J.: Prentice Hall.

———. 1977. *Mutual Accommodation: Ethnic Conflict and Cooperation*. Minneapolis: University of Minnesota Press.

———. 1979. "Structure and Process in Ethnic Relations: Increased Knowledge and Unanswered Questions." Paper presented at the annual meeting of the American Sociological Association, Boston.

———. 1994. "The Sociology of Ethnic Conflicts: Comparative International Perspectives." *Annual Review of Sociology* 20:49–79.

———. 2003. *The Wars Within: Peoples and States in Conflict*. Ithaca, N.Y.: Cornell University Press.

Willie, Charles V. 1978. "The Inclining Significance of Race." *Society* 15 (July/August):10–15.

Wilson, Clint C., II, and Félix Gutiérrez. 1995. *Race, Multiculturalism, and the Media: From Mass to Class Communication*. 2nd ed. Thousand Oaks, Calif.: Sage.

Wilson, Kenneth L., and Alejandro Portes. 1980. "Immigrant Enclaves: An Analysis of the Labor Market Experiences of Cubans in Miami." *American Journal of Sociology* 86:295–315.

Wilson, Monica, and Leonard Thompson. 1969. *The Oxford History of South Africa*. New York: Oxford University Press.

Wilson, William J. 1973. *Power, Racism, and Privilege: Race Relations in Theoretical and Sociohistorical Perspectives*. New York: Free Press.

———. 1980. *The Declining Significance of Race*. 2nd ed. Chicago: University of Chicago Press.

———. 1987. *The Truly Disadvantaged: The Inner City, the Underclass, and Public Policy*. Chicago: University of Chicago Press.

———. 1994. "Race-Neutral Programs and the Democratic Coalition." Pp. 159–73 in Nicolaus Mills (ed.), *Debating Affirmative Action: Race, Gender, Ethnicity, and the Politics of Inclusion*. New York: Delta.

———. 1996. *When Work Disappears: The World of the New Urban Poor*. New York: Knopf.

———. 1999. "Jobless Poverty: A New Form of Social Dislocation in the Inner-City Ghetto." Pp. 133–50 in Phyllis Moen, Donna Dempster-McClain, and Henry A. Walker (eds.), *A Nation Divided: Diversity, Inequality, and Community in American Society*. Ithaca, N.Y.: Cornell University Press.

———. 2009. *More than Just Race: Being Black and Poor in the Inner City*. New York: Norton.

Wines, Michael. 2007a. "One Test and 600,000 Destinies in South Africa." *New York Times* (December 30):4.

———. 2007b. "Song Wakens Injured Pride of Afrikaners." *New York Times* (February 27): A1, A4.

Winks, Robin. 1971. *The Blacks in Canada*. New Haven, Conn.: Yale University Press.

Wirth, Louis. 1945. "The Problem of Minority Groups." Pp. 347–72 in Ralph Linton (ed.),

The Science of Man in the World Crisis. New York: Columbia University Press.
———. 1956 [1928]. *The Ghetto.* Chicago: University of Chicago Press.

Wittke, Carl. 1956. *The Irish in America.* Baton Rouge: Louisiana State University Press.
———. 1967. *We Who Built America: The Saga of the Immigrant.* Cleveland: Case Western Reserve University Press.

Woellert, Franziska, Steffen Kröhnert, Lilli Sippel, and Reiner Klingholz. 2009. *Ungenutzte Potenziale: Zur Lage der Integration in Deutschland.* Berlin: Berlin-Institut.

Wolfe, Alan. 1996. *Marginalized in the Middle.* Chicago: University of Chicago Press.
———. 1998. *One Nation After All: What Middle-Class Americans Really Think About: God, Country, Family, Racism, Welfare, Immigration, Homosexuality, Work, the Right, the Left, and Each Other.* New York: Viking.

Wolfinger, Raymond E. 1966. "Some Consequences of Ethnic Politics." Pp. 42–54 in M. Kent Jennings and L. Harmon Zeigler (eds.), *The Electoral Process.* Englewood Cliffs, N.J.: Prentice Hall.

Wong, Francisco Raimundo. 1974. "The Political Behavior of Cuban Migrants." Ph.D. dissertation, University of Michigan.

Wong, Morrison G. 1986. "Post-1965 Asian Immigrants: Where Do They Come From, Where Are They Now, and Where Are They Going?" *Annals of the American Academy of Political and Social Science* 487:150–68.

Wood, Charles H., and Jose Alberto Magno de Carvalho. 1988. *The Demography of Inequality in Brazil.* Cambridge, England: Cambridge University Press.

Wood, Floris W. (ed.). 1990. *An American Profile—Opinions and Behavior, 1972–1989.* Detroit, Mich.: Gale Research.

Wood, Nicholas. 2004a. "An Effort to Unify a Bosnian City Multiplies Frictions." *New York Times* (March 15):A4.
———. 2004b. "Kosovo Smolders After Mob Violence." *New York Times* (March 24):A8.

Wood, Peter. 2003. *Diversity: The Invention of a Concept.* San Francisco: Encounter Books.

Woodrum, Eric. 1981. "An Assessment of Japanese American Assimilation, Pluralism, and Subordination." *American Journal of Sociology* 87:157–69.

Woodward, C. Vann. 1974. *The Strange Career of Jim Crow.* 3rd ed. New York: Oxford University Press.

Worsnop, Richard. 1992. "Native Americans." *CQ Researcher* 2(May 8):385–408.

Wright, Kai. 2009. "The Assault on the Black Middle Class." *The American Prospect* (July/August):A7–A10.

Wright, Lawrence. 1994. "One Drop of Blood." *The New Yorker* (July 25):46–50, 52–55.

Wright, Ronald. 1992. *Stolen Continents: The Americas Through Indian Eyes Since 1492.* Boston: Houghton Mifflin.

Wright, Stephen C., and Donald M. Taylor. 2003. "The Social Psychology of Cultural Diversity: Social Stereotyping, Prejudice, and Discrimination." Pp. 432–57 in Michael A. Hogg and Joel Cooper (eds.), *The Sage Handbook of Social Psychology.* London: Sage.

Wu, Frank. 2002. *Yellow: Race in America Beyond Black and White.* New York: Basic Books.
———. 2003. "Profiling Principle: The Prosecution of Wen Ho Lee and the Defense of Asian Americans." Pp. 297–316 in Don T. Nakanishi and James S. Lai (eds.), *Asian American Politics: Law, Participation and Policy.* Lanham, Md.: Rowman & Littlefield.

Wyman, Hastings. 2000. "Speaking of Religion . . . ." *Washington Post National Weekly Edition* (August 28):22.

Wynter, Leon E. 2002. *American Skin: Pop Culture, Big Business, and the End of White America.* New York: Crown.

Yaffee, James. 1968. *The American Jews.* New York: Random House.

Yankelovich Partners. 2001. *American Attitudes Toward Chinese Americans and Asian Americans.* New York: Committee of 100.

Yans-McLaughlin, Virginia. 1982. *Family and Community: Italian Immigrants in Buffalo, 1880–1930.* Urbana: University of Illinois Press.

Yates, Steven. 1994. *Civil Wrongs: What Went Wrong with Affirmative Action.* San Francisco: ICS Press.

Yinger, J. Milton. 1981. "Toward a Theory of Assimilation and Dissimilation." *Ethnic and Racial Studies* 4:249–64.

Yinger, John. 1996. *Closed Door, Opportunities Lost: The Continuing Costs of Housing Discrimination.* New York: Russell Sage Foundation.

Young, Donald. 1932. *American Minority Peoples.* New York: Harper & Row.

Younis, Adele L. 1995. *The Coming of the Arabic-Speaking People to the United States.* Staten Island, N.Y.: Center for Migration Studies.

Yzerbyt, Vincent, and Olivier Corneille. 2005. "Cognitive Process: Reality Constraints and Integrity Concerns in Social Perception." Pp. 175–91 in John F. Dovidio, Peter Glick, and Laurie A. Budman (eds.), *On the Nature of Prejudice: Fifty Years After Allport.* Malden, Mass.: Blackwell.

Zack, Naomi. 2001. "Different Forms of Mixed Race: Microdiversity and Destabilization." Pp. 49–58 in Curtis Stokes, Theresa Meléndez, and Genice Rhodes-Reed (eds.), *Race in 21st Century America.* East Lansing: Michigan State University Press.

Zallman, Leah, Steffie Woolhandler, David Himmelstein, David Bor, and Danny

McCormick. 2013. "Immigrants Contributed an Estimated $115.2 Billion More to the Medicare Trust Fund than They Took out in 2002–09." *Health Affairs* 32:1153–60.

Zebiri, Kate. 2011. "Orientalist Themes in Contemporary British Islamophobia." Pp. 173–90 in John L. Esposito and Ibrahim Kalin (eds.), *Islamophobia: The Challenge of Pluralism in the 21st Century*. New York: Oxford University Press.

Zenner, Walter P. 1991. *Minorities in the Middle: A Cross-Cultural Analysis*. Albany: State University of New York Press.

Zhou, Min. 2004. "Assimilation the Asian Way." Pp. 139–54 in Tamar Jacoby (ed.), *Reinventing the Melting Pot: The New Immigrants and What It Means to Be American*. New York: Basic Books.

———. 2009. "Intragroup Diversity: Asian American Population Dynamics and Challenges of the Twenty-first Century." Pp. 25–44 in Huping Ling (ed.), *Asian America: Forming New Communities: Expanding Boundaries*. New Brunswick, N.J.: Rutgers University Press.

Zhou, Min, and Carl L. Bankston III. 1998. *Growing Up American: How Vietnamese Children Adapt to Life in the United States*. New York: Russell Sage Foundation.

Zhou, Min, and Susan S. Kim. 2007. "After-School Institutions in Chinese and Korean Immigrant Communities: A Model for Others?" Migration Policy Institute (May 3). http://www.migrationinformation.org/

Zimmerman, Ben. 1963. "Race Relations in the Arid Sertao." Pp. 82–115 in Charles Wagley (ed.), *Race and Class in Rural Brazil*. 2nd ed. New York: Columbia University Press.

Zuñiga, Victor, and Rubén Hernández-León (eds.). 2005. *New Destinations of Mexican Immigration in the United States: Community Formation, Local Responses and Inter-Group Relations*. New York: Russell Sage Foundation.

Zureik, Elia T. 1979. *The Palestinians in Israel: A Study in Internal Colonialism*. London: Routledge & Kegan Paul.

Zweigenhaft, Richard L. 1984. *Who Gets to the Top? Executive Suite Discrimination in the Eighties*. New York: American Jewish Committee.

———. 1987. "Minorities and Women of the Corporation: Will They Attain Seats of Power?" Pp. 37–62 in G. W. Domhoff and T. R. Dye (eds.), *Power Elites and Organizations*. Newbury Park, Calif.: Sage.

Zweigenhaft, Richard L., and G. William Domhoff. 1982. *Jews in the Protestant Establishment*. New York: Praeger.

———. 2006. *Diversity in the Power Elite: How It Happened, Why It Matters*. Lanham, Md.: Rowman & Littlefield.

INDEX